D0161512

Perspectives in Amino Acid and Protein Geochemistry

Perspectives in Amino Acid and Protein Geochemistry

EDITED BY
Glenn A. Goodfriend
Matthew J. Collins
Marilyn L. Fogel
Stephen A. Macko
John F. Wehmiller

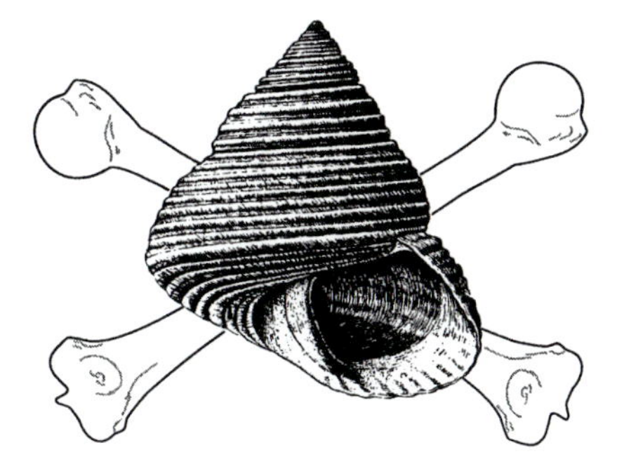

2000

OXFORD
UNIVERSITY PRESS

Oxford New York
Athens Auckland Bangkok Bogotá Buenos Aires Calcutta
Cape Town Chennai Dar es Salaam Delhi Florence Hong Kong Istanbul
Karachi Kuala Lumpur Madrid Melbourne Mexico City Mumbai
Nairobi Paris São Paulo Shanghai Singapore Taipei Tokyo Toronto Warsaw

and associated companies in
Berlin Ibadan

Published by Oxford University Press, Inc.
198 Madison Avenue, New York, New York 10016

Library of Congress Cataloging-in-Publication Data
Perspectives in amino acid and protein geochemistry / edited by
Glenn A. Goodfriend ... [et al.].
p. cm.
Includes bibliographical references and index.
ISBN 0-19-513507-5
1. Biomolecules, Fossil. 2. Amino acids. 3. Proteins. 4. Biogeochemistry.
I. Goodfriend, Glenn A., 1951–
QP517.F66 P47 2000
577′.14—dc21 00-022978

9 8 7 6 5 4 3 2
Printed in the United States of America
on acid-free paper

Preface

Twenty years have passed since the last compilation of studies on amino acid and protein geochemistry—*The Biogeochemistry of Amino Acids*, edited by P. E. Hare, T. C. Hoering, and K. King, Jr (Wiley; 1980). The present volume re-examines the state of this field in 2000 and its evolution over the last 20 years.

The Biogeochemistry of Amino Acids grew out of a conference held in 1978, on the occasion of the retirement of P. H. Abelson from the Carnegie Institution of Washington. The present volume has grown out of a conference held on April 5–7, 1998, at the American Geophysical Union headquarters in Washington, D.C., upon the retirement of P. E. Hare, following a career of 35 years at the Geophysical Laboratory of the Carnegie Institution of Washington. Ed Hare was appointed to the Geophysical Laboratory staff in 1963 by the then director, P. H. Abelson. While papers presented at this conference form the core of the present volume, other contributions are also included in order to round out the coverage of the field of amino acid and protein geochemistry and its development over the last 20 years.

Because this volume is associated with this conference, its production was made possible by the financial and logistic support for the 1998 conference provided by Dr Maxine Singer (President of the Carnegie Institution of Washington), Dr Charles Prewitt (then Director of the Geophysical Laboratory and a member of the conference organizing committee), Sue Schmidt (Geophysical Laboratory), and Karen Unger (then of the American Geophysical Union). Support for publication of this volume was provided by the Geophysical Laboratory of the Carnegie Institution of Washington (Wesley Huntress, Director), the Department of Geology of the University of Delaware, Dr P. H. Abelson, and Dr J. F. Wehmiller. Production of this volume was greatly assisted by the editors of Oxford University Press: the acquiring editor, Joyce Berry, and production editor Lisa Stallings.

The editors thank the following individuals, who contributed to this publication by serving as reviewers for one or more of the chapters: Daniel Belknap, Giorgio Belluomini, Jay Brandes, Henry Fricke, Julie Brigham-Grette, Michael Engel, Tim Filley, Richard Hoover, David Hopkins, Anne Katzenberg, Richard Keil, William Manley, William McCoy, Gifford Miller, Richard Mitterer, June Mirecki, Eric Oches, Tom O'Donnell, Peggy Ostrom, Laurent Richard, Lisa Robbins, Steve Roof, Denis-Didier Rousseau, Jeff Seewald, Peter Shewry, Anne Sigleo, Noreen Tuross, Derek Walton, Harold White, Linda York, and Sue Ziegler.

Glenn A. Goodfriend
Matthew J. Collins
Marilyn L. Fogel
Stephen A. Macko
John F. Wehmiller

Contents

Contributors

Jan P. Amend, Department of Earth and Planetary Sciences, Washington University, St. Louis, MO 63130, U.S.A.

S. Andrews, Department of Geology, University of South Florida, Tampa, FL 33620, U.S.A. (present address: Schreuder, Inc., 110 West Country Club Dr., Tampa, FL 33612, U.S.A.)

Jeffrey L. Bada, University of California at San Diego, Scripps Institution of Oceanography, La Jolla, CA 92093, U.S.A.

Bonnie A. B. Blackwell, Department of Chemistry, Williams College, Williamstown, MA 01267, U.S.A.

D. Q. Bowen, Department of Earth Sciences, Cardiff University, Cardiff, Wales CF1 3YE, U.K.

Jay A. Brandes, Marine Sciences Institute, The University of Texas at Austin, Port Aransas, TX 78373, U.S.A.

Richard R. Burky, Radiocarbon Laboratory, Department of Anthropology, Institute of Geophysics and Planetary Physics, University of California, Riverside, CA 92521, U.S.A.

Laureano Canoira, Laboratory of Biomolecular Stratigraphy, Madrid School of Mines, Ríos Rosas 21, E-28003 Madrid, Spain

George D. Cody, Geophysical Laboratory, Carnegie Institution of Washington, 5251 Broad Branch Road, NW, Washington, DC 20015-1305, U.S.A.

F. Javier Coello, Laboratory of Biomolecular Stratigraphy, Madrid School of Mines, Ríos Rosas 21, E-28003 Madrid, Spain

Matthew J. Collins, Fossil Fuels and Environmental Geochemistry, NRG, Drummond Building, University of Newcastle, Newcastle upon Tyne NE1 7RU, U.K.

J. R. Cronin, Department of Chemistry and Biochemistry, Arizona State University, Tempe, AZ 85287-1604, U.S.A.

Julie Czywczynski, Center for Geochronological Research, Institute of Arctic and Alpine Research, University of Colorado, Boulder, CO 80309-0450, U.S.A.

Marilyn L. Fogel, Geophysical Laboratory, Carnegie Institution of Washington, 5251 Broad Branch Road, NW, Washington, DC 20015, U.S.A.

Irving Friedman, U.S. Geological Survey, Denver Federal Center, Denver, CO 80225, U.S.A.

Pilar García-Alonso, Laboratory of Biomolecular Stratigraphy, Madrid School of Mines, Ríos Rosas 21, E-28003 Madrid, Spain

Daniel P. Glavin, University of California at San Diego, Scripps Institution of Oceanography, La Jolla, CA 92093, U.S.A.

Dagmar Gnieser, Department of Geosciences, University of Massachusetts, Amherst, MA 01003, U.S.A. (present address: Via Plose 38, 39042 Bressanone-Milan, Italy)

Lucio Godinez-Orta, Centro Interdisciplinario de Ciencias Marinas, Playa el Conchalito, Apdo. 476, La Paz, B.C.S. 23000, Mexico

Glenn A. Goodfriend, Department of Earth and Environmental Sciences, George Washington University, 2029 G Street, NW, Washington, DC 20052, U.S.A.

Charles V. Griffin, Private Consultant, Blackwood, SA 5051, Australia

Jochen Halfar, Department of Geological and Environmental Sciences, Stanford University, Stanford, CA 94305, U.S.A. (present address: Institut für Geologie und Paläontologie, Universität Stuttgart, Herdweg 51, 70174 Stuttgart, Germany)

P. E. Hare, Geophysical Laboratory, Carnegie Institution of Washington, 5251 Broad Branch Road, NW, Washington, DC 20015-1305, U.S.A., and Departments of Anthropology and Earth Sciences, University of California, Riverside, CA 92521, U.S.A.

Charles P. Hart, Center for Geochronological Research, Institute of Arctic and Alpine Research, and Department of Geological Sciences, University of Colorado at Boulder, Boulder, CO 80309-0450, U.S.A.

Robert M. Hazen, Geophysical Laboratory, Carnegie Institution of Washington, 5251 Broad Branch Road, NW, Washington, DC 20015-1305, U.S.A.

John I. Hedges, School of Oceanography, Box 357940, University of Washington, Seattle, WA 98195-7940, U.S.A.

Beverly J. Johnson, Department of Geology, Bates College, 214 Carnegie Science Hall, Lewiston, ME 04240, U.S.A.

Darrell S. Kaufman, Departments of Geology and Environmental Sciences, Northern Arizona University, Flagstaff, AZ 86011-4099, U.S.A.

Richard G. Keil, School of Oceanography, Box 357940, University of Washington, Seattle, WA 98195-7940, U.S.A.

Ronald W. L. Kimber, CSIRO Land and Water, Adelaide, SA 5064, Australia

Donna L. Kirner, Radiocarbon Laboratory, Department of Anthropology, Institute of Geophysics and Planetary Physics, University of California, Riverside, CA 92521, U.S.A.

W. John Kress, Department of Botany, National Museum of Natural History, Smithsonian Institution, Washington, DC 20560, U.S.A.

Keith A. Kvenvolden, U.S. Geological Survey, 345 Middlefield Road, MS 999, Menlo Park, CA 94025, U.S.A.

William M. Last, Department of Geological Sciences, University of Manitoba, Winnipeg, MB, Canada R3T 2N2

Juan F. Llamas, Laboratory of Biomolecular Stratigraphy, Madrid School of Mines, Ríos Rosas 21, E-28003 Madrid, Spain

Stephen A. Macko, Department of Environmental Sciences, University of Virginia, Charlottesville, VA 22903, U.S.A.

William F. Manley, Center for Geochronological Research, Institute of Arctic and Alpine Research, University of Colorado, Boulder, CO 80309-0450, U.S.A.

William D. McCoy, Department of Geosciences, University of Massachusetts, Amherst, MA 01003, U.S.A.

Andrew R. Millard, Department of Archaeology, University of Durham, South Road, Durham DH1 3LE, U.K.

Gifford H. Miller, Center for Geochronological Research, Institute of Arctic and Alpine Research, and Department of Geological Sciences, University of Colorado at Boulder, Boulder, CO 80309-0450, U.S.A.

Colin V. Murray-Wallace, School of Geosciences, University of Wolongong, NSW 2522, Australia

Wayne L. Newell, U.S. Geological Survey, Reston, VA 20192, U.S.A.

Eric A. Oches, Department of Geology, University of South Florida, 4202 E. Fowler Ave., Tampa, FL 33620, U.S.A.

Michael L. O'Neal, Department of Education, Loyola College in Maryland, Baltimore, MD 21210, U.S.A.

José E. Ortiz, Laboratory of Biomolecular Stratigraphy, Madrid School of Mines, Ríos Rosas 21, E-28003 Madrid, Spain

P. H. Ostrom, Department of Geological Sciences, Michigan State University, East Lansing, MI 48824, U.S.A.

Paul M. Peterson, Department of Botany, National Museum of Natural History, Smithsonian Institution, Washington, DC 20560, U.S.A.

S. Pizzarello, Department of Chemistry and Biochemistry, Arizona State University, Tempe, AZ 85287-1604, U.S.A.

Michael S. Riley, School of Earth Science, University of Birmingham, Edgbaston, Birmingham B15 2TT, U.K.

E. Brendan Roark, Center for Geochronological Research, Institute of Arctic and Alpine Research, and Department of Geological Sciences, University of Colorado at Boulder, Boulder, CO 80309-0450, U.S.A. (present address: Department of Geography, University of California, Berkeley, CA 94720, U.S.A.)

L. L. Robbins, Department of Geology, University of South Florida, Tampa, FL 33620, U.S.A. (present address: U.S. Geological Survey, 600 4th Street S., St. Petersburg, FL 33701, U.S.A.)

George Russell, Department of Botany, National Museum of Natural History, Smithsonian Institution, Washington, DC 20560, U.S.A.

N. W. Rutter, Department of Geology, University of Alberta, Edmonton, AB, Canada T6G 2E3.

Everett L. Shock, Department of Earth and Planetary Sciences, Washington University, St. Louis, MO 63130, U.S.A.

John R. Southon, Center for Accelerator Mass Spectrometry, University of California, Lawrence Livermore National Laboratory, Livermore, CA 94551, U.S.A.

Hilmar A. Stecher III, Department of Geology, University of Delaware, Newark, DE 19716, U.S.A. (present address: Shannon Point Marine Center, Western Washington University, 1900 Shannon Point Road, Anacortes, WA 98221, U.S.A.)

R. E. Taylor, Radiocarbon Laboratory, Department of Anthropology, Institute of Geophysics and Planetary Physics, University of California, Riverside, Riverside, CA 92521, U.S.A.

Mark A. Teece, Geophysical Laboratory, Carnegie Institution of Washington, 5251 Broad Branch Road, NW, Washington, DC 20015, U.S.A., and Smithsonian Center for Materials Research and Education, Silver Hill Road, Smithsonian Institution, Suitland, MD 20746, U.S.A. (present address: Department of Chemistry, SUNY—College of Environmental Science and Forestry, 1 Forestry Dr., Syracuse, NY 13210, U.S.A.)

Trinidad Torres, Laboratory of Biomolecular Stratigraphy, Madrid School of Mines, Ríos Rosas 21, E-28003 Madrid, Spain

Elizabeth Tsamakis, School of Oceanography, Box 357940, University of Washington, Seattle, WA 98195-7940, U.S.A.

Noreen Tuross, Smithsonian Center for Materials Research and Education, Silver Hill Road, Smithsonian Institution, Suitland, MD 20746, U.S.A.

Emma R. Waite, Fossil Fuels and Environmental Geochemistry, NRG, Drummond Building, University of Newcastle, Newcastle upon Tyne NE1 7RU, U.K.

John F. Wehmiller, Department of Geology, University of Delaware, Newark, DE 19716, U.S.A.

Hatten S. Yoder, Jr, Geophysical Laboratory, Carnegie Institution of Washington, 5251 Broad Branch Road, NW, Washington, DC 20015-1305, U.S.A.

Linda L. York, Department of Geology, University of Delaware, Newark, DE 19716, U.S.A.

Perspectives in Amino Acid and Protein Geochemistry

Introduction

Glenn A. Goodfriend

The origin of what is now recognized as the field of amino acid geochemistry can probably be traced to P. H. Abelson's 1954 paper "Organic constituents of fossils", in which he investigated whether the mineral matrix of shells, teeth, and bones might serve to preserve ancient amino acids and proteins in fossil material. Employing paper chromatographic techniques, Abelson demonstrated the preservation of amino acids in a variety of Miocene mollusk shells, as well as a wide age range of bones and teeth. Subsequent studies showed that the amino acid content decreases with increasing age and that protein-bound amino acids progressively hydrolyze to free amino acids (Abelson, 1955). The subsequent development of a liquid chromatographic system for amino acid analysis (Spackman et al., 1958) and its later modification to a high-performance ion-exchange liquid chromatographic system in the early 1970s (summarized in Hare et al., 1985) led to rapid advances in the study of amino acid geochemistry. Among these was the discovery of the epimerization of isoleucine to alloisoleucine (Hare and Mitterer, 1966), which initiated the field of amino acid racemization dating (Hare and Abelson, 1968).

Proteins are fundamental to the functioning of life and are composed of amino acids. Consequently, an understanding of amino acid synthesis is critical to an understanding of the origin of life. Early work focused on the formation of amino acids in a reducing atmosphere of methane, ammonia, water, and hydrogen subjected to electrical discharges (Miller, 1953). Despite the high yield of nitrogenous compounds in these experiments, the systems appear too reactive, polymerizing the very compounds needed to sustain biochemical pathways. Recently, interest has shifted towards the possible origin of life under conditions of high temperature and pressure—conditions associated with hydrothermal vents or deep in the earth. In this volume, Amend and Shock (Chapter 3) consider how the mixing of hot vent fluids with sea water renders the synthesis of amino acids thermodynamically favorable. Brandes et al. (Chapter 4) demonstrate that it is possible to mimic, abiotically, a key step in the biochemical synthesis of amino acids, namely the synthesis of amino acids from α-keto acids under a range of conditions. The highly pH-dependent nature of the reaction offers scope for a primitive form of directed synthesis. A number of studies have also been carried out on the possible extraterrestrial origin of amino acids, based on analysis of meteorites. In the Hare et al. (1980) volume *Biochemistry of Amino Acids* that resulted from the 1978 conference *Advances in the Biogeochemistry of Amino Acids*, Cronin et al. (1980) and Harada and Hare (1980) presented the results of amino acid analyses of meteorites. In the present volume, Kvenwolden et al. (Chapter 1) present a review of the history of such studies on the Murchison meteorite, and Cronin and Pizzarello (Chapter 2) summarize the most recent work in this area, which benefits from advances in analytical techniques such as combined gas chromatography/isotope-ratio mass spectrometry (GC–IRMS). Whether these meteoritic amino acids represent extraterrestrial material or terrestrial contaminants remains controversial, however, despite these significant methodological advances of recent years.

The influential study of DeNiro and Epstein (1978) on the use of stable carbon isotopes for the reconstruction of diets and paleodiets was published the same year as the 1978 conference. Soon thereafter, similar work on nitrogen isotopes was published (DeNiro and Epstein, 1981). These studies led to a whole new field—the application of stable isotopes to paleodiet and paleoclimate reconstruction—that is not represented in *Biogeochemistry of Amino Acids*. In the present volume, Millard (Chapter 5) considers factors controlling changes in the nitrogen isotopic composition of humans during weaning, which is applicable to the reconstruction of the age at weaning from analysis of fossil human remains. Recent work in organic geochemistry has led to methods that isolate specific compounds for isotopic analysis rather than involving bulk organic isotope analyses. This advance is significant because the various amino acids differ significantly in their stable carbon-isotope composition (Macko et al., 1983). Burky et al. (Chapter 6) consider the problem of the isolation of pure, uncontaminated organic compounds (specifically, aminomalinate) from bone, for ^{14}C analysis. Using a chiral column for gas chromatography, D- and L-enantiomers of amino acids may be separated for isotopic analysis; this provides a powerful test of possible contamination (Engel et al., 1994). This methodology has also been applied to amino acids in meteorites to evaluate the question of indigeneity (Cronin and Pizzarello, Chapter 2).

A number of papers in *Biogeochemistry of Amino Acids* focused on the diagenesis of proteins and amino acids in fossil bone, shell, and other materials. Since then, research has tended to focus more on applications of amino acid geochemistry. However, one field that has flourished since that time concerns the fate of proteins and amino acids in the oceans and ocean sediments (e.g., Lee, 1988; Hedges and Keil, 1995). In the present volume, Keil et al. (Chapter 7) present a survey of the amino acid compositions and concentrations in a range of marine sediment types and relate these to their sources and the depositional setting. Short-term diagenetic changes (tens to thousands of years) in amino acids are examined in two of the studies presented here. Teece et al. (Chapter 8) examine the amino acid composition of dried herbarium specimens in museums, as reconstructed from a time-series of samples; preservation is shown to be very good under these conditions. Blackwell et al. (Chapter 9), studying bones in saline lake environments in Australia, demonstrate very rapid degradation of amino acids, often accompanied by mineralization of the bones. In their study on Pleistocene and modern mollusk shells, Robbins et al. (Chapter 10) compare the relative abundances of hydrophilic and hydrophobic proteins and the amino acids making up these protein fractions. In Chapter 11, Collins and Riley present a model of the diagenesis of proteins in fossils that provides a basis for the modeling of amino acid racemization kinetics.

At the time of publication of *Biogeochemistry of Amino Acids*, the kinetics of amino acid racemization were generally modeled as a reversible first-order reaction (Bada and Schroeder, 1972). However, it had been recognized that racemization rates in older samples were slower than predicted from such models, and alternative kinetic models had been proposed (e.g., Wehmiller and Belknap, 1978). Several factors complicate the application of kinetic models of amino acid racemization, as outlined below.

- Although racemization of free amino acids in solution does follow reversible first-order kinetics (Kriausakul and Mitterer, 1978), the complex mixtures of proteins that occur in biogenic minerals such as shells and bones means that each amino acid moiety will have its own kinetic patterns, the sum of which (as measured by sample hydrolysates) is not expected to follow first-order kinetics (Goodfriend, 1991). Thus, most workers have approached this problem of kinetics with a purely empirical, curve-fitting technique, e.g., some have recently proposed parabolic kinetic models (Mitterer and Kriausakul, 1989), or power-function transformations of D/L values (Goodfriend et al., 1995). In this volume, a variant of this latter model, in which the D/L values are first transformed to their first-order kinetic (FOK) equivalents (i.e., values that are linear with respect to time under the FOK model) and then this value is transformed to a power function, is shown to give a better fit to racemization data from marine bivalves (Manley et al., Chapter 16) and ostracodes (Kaufman, Chapter 12). The complexity of racemization in biominerals has discouraged the development of kinetic models that mimic the actual processes involved. However, Collins and Riley (Chapter 11) boldly present a new kinetic model of racemization based on a simplification of these processes and including diagenetic changes.
- Aspartic acid (Asp) and glutamic acid (Glu) present particular problems for kinetic studies, because during the hydrolysis of proteins for amino acid analysis, asparagine (Asn) is converted to aspartic acid and glutamine (Gln) to glutamic acid. The result is that the analyzed material of Asp or Glu is the sum of Asp + Asn and Glu + Gln, respectively. This has been considered to contribute to the observed non-linear kinetics of Asp (Brinton and Bada, 1995; Goodfriend and Hare, 1995). In this volume, Waite and Collins (Chapter 14) explore the consequences of this for precision and refinement of human age estimates based on *in vivo* racemization of Asp in teeth—a method developed by Helfman and Bada (1975).
- Pre-radiocarbon calibrations for racemization rates (e.g., based on Th-U dates of corals) are often available only for certain sites. In order to estimate ages at other sites, temperature differences between these sites and the calibration site must be taken into account. This has commonly been done by looking at differences in mean annual air temperatures between sites, based on weather records. However, long-term *in situ* temperature measurements by Wehmiller et al. (Chapter 17) have shown that there is considerable variability among sites, and that precise racemization-rate differences between sites can be calculated only on the basis of such *in situ* temperature measurements.
- It has been recognized that diagenesis plays a key role in the observed kinetics of racemization (Kriausakul and Mitterer, 1980). Racemization rates of different amino acid moieties differ depending upon their position within a peptide or protein or whether or not they are free (not attached to other amino acids by peptide bonds). Over time, hydrolysis of proteins results in changes in the positions of the amino acids and thus results in changes in racemization rates of the overall mix of amino acids. In bone, these patterns of diagenesis tend to be unpredictable, leading to problems in the use of racemization for dating (e.g., Elster et al., 1991). Teeth seem to offer a more consistent pattern than bones under some circumstances (Torres et al., Chapter 24) but not in others: in saline lake environments, amino acid composition may be

greatly altered and most amino acids may be lost within a matter of years (Blackwell et al., Chapter 9). In contrast, studies on diagenesis and racemization in ostrich eggshells have shown these patterns to be very predictable, leading to excellent precision in the amino acid racemization dates of such materials (Brooks et al., 1990). In the present volume, Miller et al. (Chapter 13) show that eggshells of the Australian emu and the extinct *Genyornis* are similarly well behaved, with similar kinetic patterns in both modern samples and fossils of various ages.

- Some materials may be subject to contamination by later addition of amino acids, and this will clearly affect their apparent racemization kinetics and possible usefulness for dating. Kimber and Griffin (Chapter 15) examine racemization in soils and find that rates may differ greatly according to the age and state of degradation of the soil. In bones, the appearance of anomalously large amounts of some D-amino acids that comprise bacterial cell walls in relatively young samples suggests that bacterial decomposition of bone organic matter may result in substantial additions of these D-amino acids (Blackwell et al., Chapter 9).

The use of amino acid racemization dating was well established at the time of the 1978 conference, although the method had first been applied less than 10 years earlier. Amino acid racemization dating has been most widely used for correlation of marine interglacial deposits, mainly on the basis of analyses of mollusk shells (e.g., Wehmiller, 1982; Miller and Mangerud, 1985; Wehmiller and Miller, 2000). We are pleased to report that now, over 20 years later, the method is still alive and well, as evidenced by a wide variety of new applications. The method has been applied to a variety of new materials. Studies on bird eggshells were noted above (Miller et al., Chapter 13, and references above). Ostracodes had first been studied some years ago (McCoy, 1988), but recent methodological advances involving reverse-phase high-performance liquid chromatographic analysis of minute samples has made the study of these tiny carbonate fossils more practical (Kaufman, Chapter 12). Results are reported here for racemization in calcareous algae (Goodfriend et al., Chapter 22). Organic endoskeletons of colonial anemones (Goodfriend, 1997) and speleothems (Lauritzen et al., 1994) have also been dated using amino acid racemization. New sites continue to be studied using racemization, and are fitted into regional stratigraphic frameworks (northern Delaware Bay: O'Neal et al., Chapter 21; the Mediterranean coast of Spain: Torres et al., Chapter 19). Summaries of regional marine chronostratigraphic frameworks based on racemization analysis are presented here for the U.K. (Bowen, Chapter 18) and Australia (Murray-Wallace, Chapter 20). Racemization has also been used for correlation of Quaternary terrestrial sequences. Oches et al. (Chapter 23) present an overview of the chronostratigraphy of the loess deposits of Europe based on racemization in land snail shells, while Torres et al. (Chapter 24) present a chronology for European cave bears, based on racemization in teeth.

Time-scales of the application of racemization for dating of marine and terrestrial have also changed in recent years. Racemization dating has been used to estimate ages over the last several hundred years—samples too young to date precisely using the radiocarbon method (Goodfriend, 1992). Goodfriend et al. (Chapter 22) apply racemization to littoral zone cores to determine the ages of recent shells and to identify the locations of stratigraphic breaks in the cores and evidence of reworking.

Thus, the last 20 years since publication of *Biogeochemistry of Amino Acids* has seen a large number of developments in the field of amino acid and protein geochemistry, as well as a continuation and refinement of methodologies that had already been established. Nevertheless, amino acid and protein geochemistry remains one of the less well understood aspects of organic geochemistry (cf. Engel and Macko, 1993). It is hoped that this survey of this field will serve as a stimulus to further research, by summarizing areas of knowledge and pointing out gaps in our understanding.

REFERENCES

Abelson P.H. (1954) Organic constituents of fossils. *Yrbk Carnegie Instn Wash.* **53**, 97–101.

Abelson P.H. (1955) Organic constituents of fossils. *Yrbk Carnegie Instn Wash.* **54**, 107–109.

Bada J.L. and Schoeder R.A. (1972) Racemization of isoleucine in calcareous marine sediments: kinetics and mechanism. *Earth Planet. Sci. Lett.* **15**, 1–11.

Brinton K.L.F. and Bada J.L. (1995) Comment on "Aspartic acid racemization and protein diagenesis in corals over the last 350 years" by G.A. Goodfriend, P.E. Hare, and E.R.M. Druffel. *Geochim. Cosmochim. Acta* **59**, 415–416.

Brooks A.S., Hare P.E., Kokis J.E., Miller G.H., Ernst R.D., and Wendorf F. (1990) Dating Pleistocene archeological sites by protein diagenesis in ostrich eggshell. *Science* **248**, 60–64.

Cronin J.R., Gandy W.E., and Pizzarello S. (1980) Amino acids of the Murchison meteorite. In *Biogeochemistry of Amino Acids* (eds P.E. Hare et al.), pp. 153–168. Wiley.

DeNiro M.J. and Epstein S. (1978) Influence of diet on the distribution of carbon isotopes in animals. *Geochim. Cosmochim. Acta* **42**, 495–506.

DeNiro M.J. and Epstein S. (1981) Influence of diet on the distribution of nitrogen isotopes in animals. *Geochim. Cosmochim. Acta* **45**, 341–351.

Elster H., Gil-Av E., and Weiner S. (1991) Amino acid racemization of fossil bone. *J. Archaeol. Sci.* **18**, 605–617.

Engel M.H. and Macko S.A. (eds). (1993) *Organic Geochemistry: Principles and Applications.* Plenum Press.

Engel M.H., Goodfriend G.A., Qian Y. and Macko S. A. (1994) Indigineity of organic matter in fossils: a test using stable isotope analysis of amino acid enantiomers in Quaternary mollusk shells. *Proc. Natl. Acad. Sci., U.S.A.* **91**, 10475–10478.

Goodfriend G.A. (1991) Patterns of racemization and epimerization of amino acids in land snail shells over the course of the Holocene. *Geochim. Cosmochim. Acta* **55**, 293–302.

Goodfriend G.A. (1992) Rapid racemization of aspartic acid in mollusk shells and potential for dating over recent centuries. *Nature* **357**, 399–401.

Goodfriend G.A. (1997) Aspartic acid racemization and amino acid composition of the organic endoskeleton of the deep-water colonial anemone *Gerardia*: determination of longevity from kinetic experiments. *Geochim. Cosmochim. Acta* **61**, 1931–1939.

Goodfriend G.A. and Hare P.E. (1995) Reply to the comment by K. L. F. Brinton and J. L. Bada on "Aspartic acid racemization and protein diagenesis in corals over the last 350 years". *Geochim. Cosmochim. Acta* **59**, 417–418.

Goodfriend G.A., Kashgarian M. and Harasewych M.G. (1995) Aspartic acid racemization and the life history of deep-water slit shells. *Geochim. Cosmochim. Acta* **59**, 1125–1129.

Harada K. and Hare P.E. (1980) Analysis of amino acids from the Allende meteorite. In *Biogeochemistry of Amino Acids* (eds P.E. Hare et al.), pp. 169–181. Wiley.

Hare P. E. and Abelson P.H. (1968) Racemization of amino acids in fossil shells. *Yrbk Carnegie Instn Wash.* **66**, 526–528.

Hare P.E. and Mitterer R.M. (1966) Nonprotein amino acids in fossil shells. *Yrbk Carnegie Instn Wash.* **65**, 362–364.

Hare P.E., Hoering T.C. and King K., Jr (eds). (1980) *Biogeochemistry of Amino Acids.* Wiley.

Hare P.E., St. John P.A. and Engel M.H. (1985) Ion-exchange separation of amino acids. In *Chemistry and Biochemistry of Amino Acids* (ed. G.C. Barrett), pp. 415–425. Chapman and Hall.

Hedges J.I. and Keil R.G. (1995) Sedimentary organic matter preservation: an assessment and speculative synthesis. *Mar. Chem.* **49**, 81–115.

Helfman P.M. and Bada J.L. (1975) Aspartic acid racemization in tooth enamel from living humans. *Proc. Natl. Acad. Sci., U.S.A.* **72**, 2891–2894.

Kriausakul N. and Mitterer R.M. (1978) Isoleucine epimerization in peptides and proteins: kinetic factors and application to fossil proteins. *Science* **201**, 1011–1014.

Kriausakul N. and Mitterer R.M. (1980) Some factors affecting the racemization of isoleucine in peptides and proteins. In *Biogeochemistry of Amino Acids* (eds P.E. Hare et al.), pp. 283–296. Wiley.

Lauritzen S.-E., Haugen J.E., Lovlie R. and Gilie-Nielsen H. (1994) Geochronological potential of isoleucine epimerization in calcite speleothems. *Quat. Res.* **41**, 52–58.

Lee C. (1988) Amino acid chemistry and amine biogeochemistry in particulate material and sediments. In *Nitrogen Cycling in Coastal Marine Environments* (eds T.H. Blackburn and J. Sorensen), pp. 126–141. Wiley.

McCoy W.D. (1988) Amino acid racemization in fossil non-marine ostracod shells: a potential tool for the study of Quaternary stratigraphy, chronology, and palaeotemperature. In *Ostracoda in the Earth Sciences* (eds P. DeDeckker, J.-P. Colin, and J.-P. Peypouquet), pp. 219–229. Elsevier.

Macko S.A., Estep M.L.F., Hare P.E. and Hoering T.C. (1983) Stable nitrogen and carbon isotopic composition of individual amino acids isolated from cultured microorganisms. *Yrbk Carnegie Instn Wash.* **82**, 404–410.

Miller S.L. (1953) A production of amino acids under possible primitive Earth conditions. *Science* **117**, 528–529.

Miller G.H. and Mangerud J. (1985) Aminostratigraphy of European marine interglacial deposits. *Quat. Sci. Rev.* **4**, 215–278.

Mitterer R.M. and Kriausakul N. (1989) Calculation of amino acid racemization ages based on apparent parabolic kinetics. *Quat. Sci. Rev.* **8**, 353–357.

Spackman D.H., Stein W.H. and Moore S. (1958) Automatic recording apparatus for use in the chromatography of amino acids. *Anal. Chem.* **30**, 1190–1206.

Wehmiller J.F. (1982) A review of amino acid racemization studies in Quaternary mollusks: stratigraphic and chronologic applications in coastal and interglacial sties, Pacific and Atlantic coasts, United States, United Kingdom, Baffin Island, and tropical islands. *Quat. Sci. Rev.* **1**, 83–120.

Wehmiller J.F. and Belknap D.F. (1978) Alternative kinetic models for the interpretation of amino acid enantiomeric ratios in Pleistocene mollusks: examples from California, Washington, and Florida. *Quat. Res.* **9**, 330–348.

Wehmiller J.F. and Miller G.H. (2000) Aminostratigraphic dating methods in Quaternary geology, In *Quaternary Geochronology, Methods and Applications.* (eds J.S. Noller, J.M. Sowers, and W.R. Lettis). American Geophysical Union: American Geophysical Union Reference Shelf, vol. 4, pp. 187–222.

I. EXTRATERRESTRIAL AMINO ACIDS AND ORIGIN OF LIFE

1. Extraterrestrial amino acids in the Murchison meteorite: re-evaluation after thirty years

Keith A. Kvenvolden, Daniel P. Glavin, and Jeffrey L. Bada

The search for amino acids in meteorites, utilizing modern analytical techniques, began in the early 1960s (Calvin and Vaughn, 1960; Briggs, 1961; Briggs and Mamikunian, 1963) with identifications first obtained by Degens and Bajor (1962) and Kaplan et al. (1963). These early observations were quickly followed by others (Anders et al., 1964; Vallentyne, 1965). A problem immediately arose, however, in that the qualitative distributions of the observed amino acids were disturbingly similar to those of human fingerprints, leading to P. B. Hamilton's memorable quotation "What appears to be the pitter-patter of heavenly feet is probably instead the print of an earthly thumb." (Hayes, 1967). In spite of the problem of fingerprint contamination, Hayes (1967) stated that the collective evidence indicated that meteorites might contain indigenous glycine, phenylalanine, and β-alanine, the last of which is a compound not found in fingerprints. This conclusion summarized the state of knowledge in 1969.

However, in 1969 the fall of the Murchison meteorite changed forever the course of organic cosmochemistry, because, for the first time, a terrestrially uncontaminated meteorite was available for study by state-of-the-art analytical techniques. The Murchison meteorite fell at about 11:00 A.M. (local time) on September 28, 1969, near Murchison, Victoria, Australia (figure 1-1). The parent object broke up during flight and scattered many fragments over an area of about 13 km^2, with many pieces breaking further upon impact. Most fractured surfaces of the individual pieces had fusion crusts. Notable and distinctive features of many of the individual stones were networks of deep cracks extending into their interiors. Several stones were picked up soon after the fall, and many more were collected afterward, particularly during February and March, 1970.

In an initial study of the Murchison meteorite, a C2 carbonaceous chondrite, Kvenvolden et al. (1970) selected the fragments that had the fewest cracks, the least obvious exterior contamination, and which were the most massive. The ablation crust of one massive fragment that showed signs of minimal exterior contamination was chiseled away. The inside pieces were then pulverized for 15 sec in a disc mill. From a portion of this pulverized sample came the discovery of 6 protein and 12 non-protein amino acids (Kvenvolden et al., 1970, 1971). A portion of this pulverized sample, as well as a pulverized sample of the outside surface (not analyzed at that time), were stored in the dark at room temperature in glass vials sealed with Teflon-lined screw caps. These samples, stored for about 27 years, are the basis of the re-analysis of the Murchison meteorite reported here.

METHODS

During the past three decades, analytical procedures for the study of amino acids in meteorites have improved, providing increased sensitivity and expanded capabilities for identification of individual amino acid isomers. For example, by 1983, 52 amino acids had been positively identified as being present in samples of the Murchison meteorite (Cronin and Pizzarello, 1983).

Methods in 1970

Because the methods used in the discovery of 18 amino acids in the Murchison meteorite have been published previously (Kvenvolden 1970, 1971) only a brief outline is given here. A 10 g portion of the pulverized piece of the meteorite was refluxed with distilled water. The recovered extract and rinses were combined, evaporated to dryness, and hydrolyzed with 6 M HCl at 100°C. This hydrolysate was evaporated to dryness and desalted by ion-exchange chromatography on Dowex 50 (H^+) followed by Dowex 2 (OH^-). Portions of the recovered product were analyzed by conventional ion-exchange chromatography (with ninhydrin as the detector reagent), and by high-resolution capillary gas chromatography, and by gas chromatography combined with mass spectrometry of *N*-trifluoroacetyl-D-2-butyl ester derivatives of the amino acids, separated on two different capillary columns, one coated with UCON 75-H-90,000 and the other with XE-60.

Methods in 1997

The extraction procedure used to isolate amino acids from the powdered Murchison meteorite

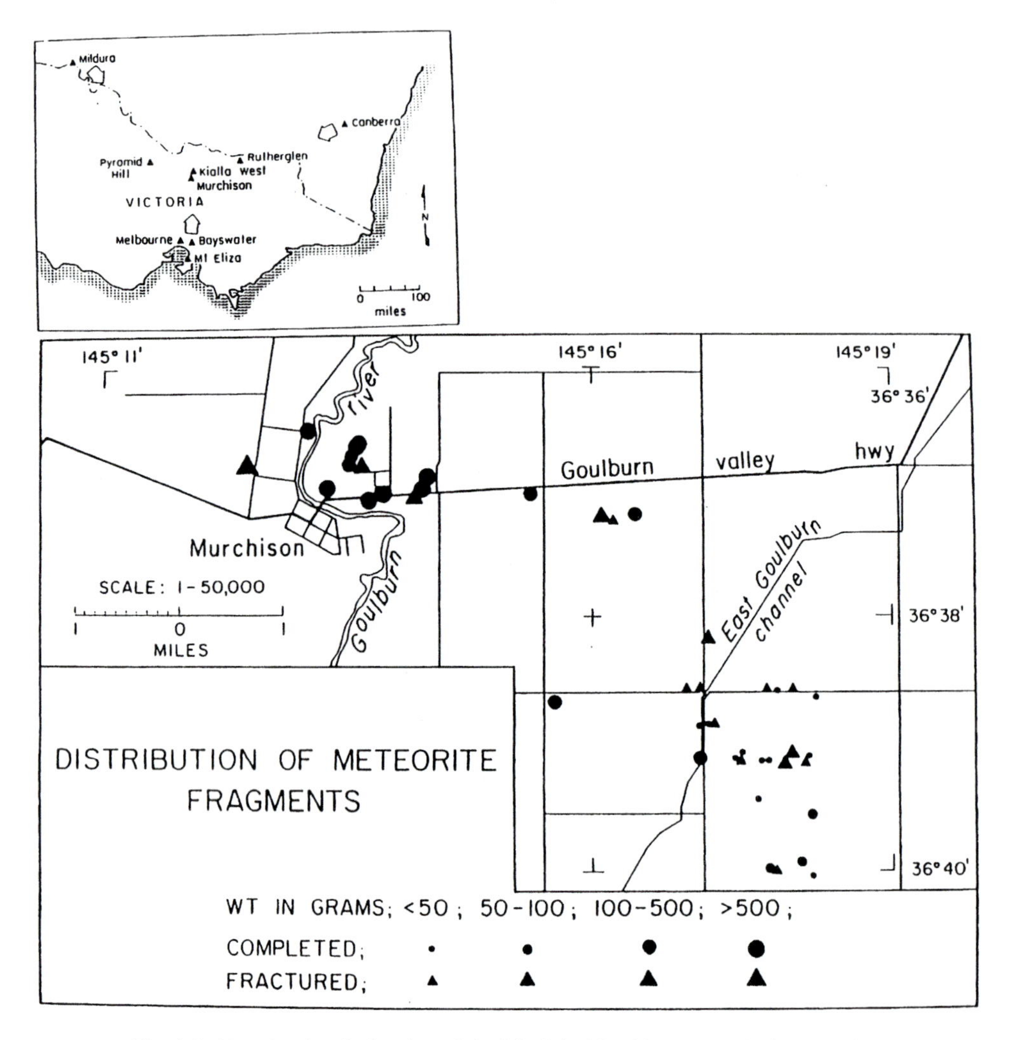

Fig. 1-1. Map showing the location of the fall of the Murchison meteorite in Australia.

sample was similar to the 1970 methodology, although, because of greatly improved sensitivity of the analytical method (see Bada et al., 1998), much smaller amounts of sample were required. Prior to use, all glassware and tools were heated overnight in an annealing oven. Approximately 100 mg of the pulverized Murchison meteorite sample was transferred into a clean test tube, sealed in 1 ml of double-distilled water and heated at 100°C for 24 hr. The water supernatant was then transferred to a glass test tube, dried under vacuum, and acid-hydrolyzed in 6 N HCl at 100°C for 24 hr in a sealed test tube. The HCl was then removed under vacuum and the hydrolysate desalted using Bio-Rad AG50W-X8 cation-exchange resin prior to analysis. The final desalted residue was resuspended in 100 μl double-distilled water, transferred to a sterile capped Eppendorf vial and stored in a freezer (−2°C). The following day, 10 μl of the thawed sample was dried down under vacuum in 10 μl of 0.4 M sodium borate buffer (pH = 9.5) to remove ammonia, resuspended in 20 μl double-distilled water and then derivatized by OPA/NAC (*o*-phthalaldehyde/*N*-acetyl-L-cysteine, both obtained from Sigma Aldrich). The fluorescent amino acid derivatives were then separated with high-performance liquid chromatography (HPLC) using a C-18 reverse-phase Phenomenex column and detected with a highly sensitive Shimadzu RF-535 UV-fluorescence detector (detection limit $\sim 10^{-15}$ mol). Individual amino acids were then identified by comparison of the retention times with those of known standards. The reaction of α-dialkyl amino acids such as α-aminoisobutyric acid (AIB) and isovaline with the OPA/NAC reagent requires longer to reach completion. Therefore we carried out derivatizations for 1 and 15 min to provide an additional means of confirming the presence of these amino acids (Zhao and Bada, 1989, 1995).

RESULTS

The results obtained during the period from 1970 to 1975 (Kvenvolden et al., 1970, 1971; Pollock et al., 1975) are compared with the results obtained in 1997 on portions of the same inside-pulverized sample of the Murchison meteorite, and also on portions of the outside-pulverized sample of the same specimen, not previously analyzed.

Results (1970–1975)

Quantitation (relative to the weight of the meteorite sample) of five abundant protein amino acids was determined by the conventional amino acid analyzer (ninhydrin detection) and the identification confirmed by gas chromatography/mass spectrometry. The presence of aspartic acid was detected by gas chromatography/mass spectrometry, but its concentration was too low to determine with the ninhydrin detector of the amino acid analyzer; however, a qualitative estimation can be made from the gas chromatogram (figure 1-2). The concentrations of these amino acids are shown in table 1-1.

Twelve non-protein amino acids were identified (Kvenvolden et al., 1971). Although the concentrations of these amino acids were not originally reported, an estimation of amounts can be made from the gas chromatogram (figure 1-2), using average gas-chromatographic response factors. The 12 non-protein amino acids and their estimated concentrations are listed in table 1.

By 1971, therefore, it was known that the Murchison meteorite contained all of the amino acid structural isomers with two and three carbon atoms. By 1986, Cronin and Pizzarello (1983), Cronin et al. (1985), and Cronin and Pizzarello (1986) had shown that all amino acid structural isomers with two to five carbon atoms and all α-amino acids with six and seven carbon atoms are present in the Murchison meteorite.

In addition to the distribution of structural isomers of amino acids, the distribution of optical isomers (enantiomers) of several chiral amino

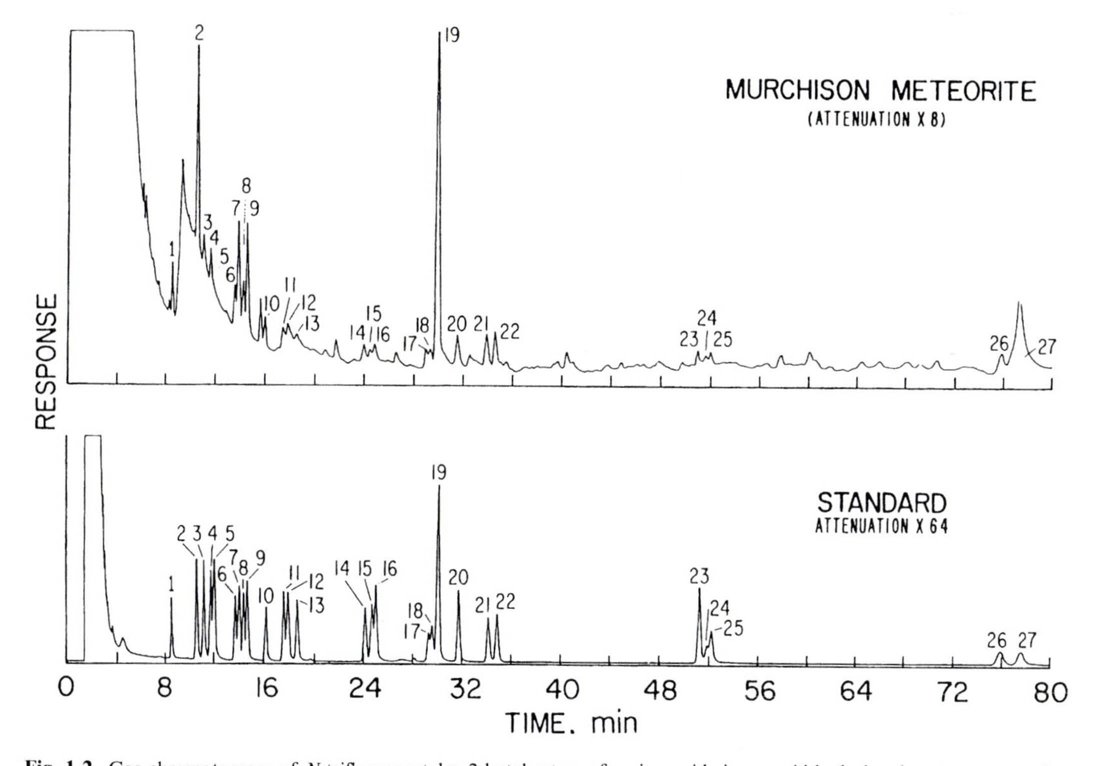

Fig. 1-2. Gas chromatogram of *N*-trifluoroacetyl-D-2-butyl esters of amino acids in an acid-hydrolyzed water extract of an inside sample of the Murchison meteorite compared with a standard. Gas chromatography was done on a high-resolution capillary column coated with UCON 75 H 90,000. Identification is as follows: 1, (DL)-isovaline; 2, α-aminoisobutyric acid (AIB); 3, D-valine; 4, L-valine; 5, (DL)-N-methylalanine; 6, D-α-amino-*n*-butyric acid; 7, D-alanine; 8, L-α-amino-*n*-butyric acid; 9, L-alanine; 10, *N*-methylglycine; 11, *N*-ethylglycine; 12, D-norvaline; 13, L-norvaline; 14, D-β-aminoisobutyric acid; 15, L-β-aminoisobutyric acid; 16, (DL)-β-amino-*n*-butyric acid; 17, D-pipecolic acid; 18, L-pipecolic acid; 19, glycine; 20, β-alanine; 21, D-proline; 22, L-proline; 23, γ-amino-*n*-butyric acid; 24, D-aspartic acid; 25, L-aspartic acid; 26, D-glutamic acid; and 27, L-glutamic (masked by an unidentified peak in the sample). Identifications are the same for the standard. From Kvenvolden et al. (1971).

TABLE 1-1 Amino acid concentrations and D/L ratios of chiral amino acids; results obtained in 1970–1975.

Amino acid	Concentration (μg/g)	D/L ratio[a]
Protein amino acid		
glycine	6	not chiral
alanine	3	1
glutamic acid	3	0.8
valine	2	0.8 ± 0.1
proline	1	0.7 ± 0.1
aspartic acid	~0.4	ND
Non-protein amino acid		
α-amino-*n*-butyric acid	1	~1
β-aminoisobutyric acid	0.5	~1
norvaline	0.2	~1
pipecolic acid	0.4	~1
isovaline	2	1
N-methylalanine	<0.1	ND
β-amino-*n*-butyric aid	0.3	ND
N-methylglycine	0.4	not chiral
β-alanine	0.5	not chiral
N-ethylglycine	0.3	not chiral
α-aminoisobutyric acid	7	not chiral
γ-amino-*n*-butyric acid	0.4	not chiral

[a] ND, Not determined.

acids were determined (Kvenvolden et al., 1970). The D/L ratios of the enantiomers were calculated from gas chromatograms (figure 1-2) and are listed in table 1-1.

In addition to four chiral protein amino acids, seven chiral non-protein amino acids were identified, and, of these, four could be separated into their respective D- and L-isomers using the same methods as for the protein amino acids (table 1-1). Each of the four non-protein amino acids appeared to consist of approximately equal amounts of the D- and L-isomers (Kvenvolden et al., 1971).

Accurate measurement of the enantiomeric ratios of all of these amino acids was difficult to achieve. The fact that the *N*-trifluoracetyl-D-2-butyl ester derivatives of the amino acids were prepared with 2-butanol that was 96% enriched in the D-isomer complicated the measurement of the D/L ratios of amino acids. In the final analyses, the amino acids appeared to have a slight excess of L-amino acids. It was concluded, however, that the slight excess might indicate minor terrestrial contamination, and that all of the chiral amino acids were probably present in the meteorite as racemic mixtures in which each amino acid is composed of approximately equal amounts of the D- and L-isomers (Kvenvolden et al., 1971).

Besides the eight chiral amino acids for which the stereochemistry was determined, three others, *N*-methylalanine, β-amino-*n*-butyric acid, and isovaline, each having an asymmetric carbon atom, were identified in the Murchison meteorite; however, these three compounds could not be resolved into their enantiomeric components by the methods used for the other chiral amino acids. Therefore, modified procedures (Pollock, 1972, 1974) were applied specifically to isovaline (2-amino-2-methylbutanoic acid). This amino acid was singled out because, unlike all of the other chiral amino acids found up to that time in this meteorite, it has no proton (hydrogen atom) on its asymmetric α-carbon atom. Thus the compound cannot undergo racemization by the normal process involving proton abstraction and addition. Detailed analyses by Pollock et al. (1975) produced results indicating that isovaline in the Murchison meteorite is present as a racemic mixture. This result implied that all chiral amino acids were originally synthesized as racemic mixtures in the meteorite and were not involved in the process of racemization of an original mixture containing an excess of either the D- or the L-isomer of each amino acid.

Results (1997)

The HPLC-based analyses of the Murchison meteorite outside and inside samples are shown in figure 1-3. The most abundant protein amino acids identified include glycine, alanine, glutamic acid and aspartic acid. Three of the original 12 non-protein amino acids, β-alanine, α-aminoisobutyric acid, and isovaline, found in the Murchison meteorite in 1970 were also identified by HPLC. Many other peaks are present in the HPLC chromatograms, but we have concentrated on identifying only the major peaks. The concentrations (μg/g) of the various amino acids were determined from the average of three separate analyses (table 1-2). Both protein and non-protein amino acid abundances are consistently higher in the outside sample relative to the inside sample.

The enantiomeric abundances of four chiral amino acids were also determined by calculating D/L ratios from the HPLC chromatograms. In each case, a racemic standard was run in parallel with the sample and used to make enantiomeric ratio corrections. A comparison of the enantiomeric ratios for both outside and inside Murchison metorite samples is shown in table 1-2. The enantiomeric ratios of the protein amino acids aspartic acid and glutamic acid and alanine in the outside sample are much lower (30–60%) than in the inside. The only chiral non-protein amino acid measured was isovaline, and this amino acid has a D/L ratio of 1.0 ± 0.1 in both outside and inside samples.

DISCUSSION

Re-analysis of the same sample of Murchison meteorite, after 27 years in storage, using more sensitive techniques reveals results similar to those

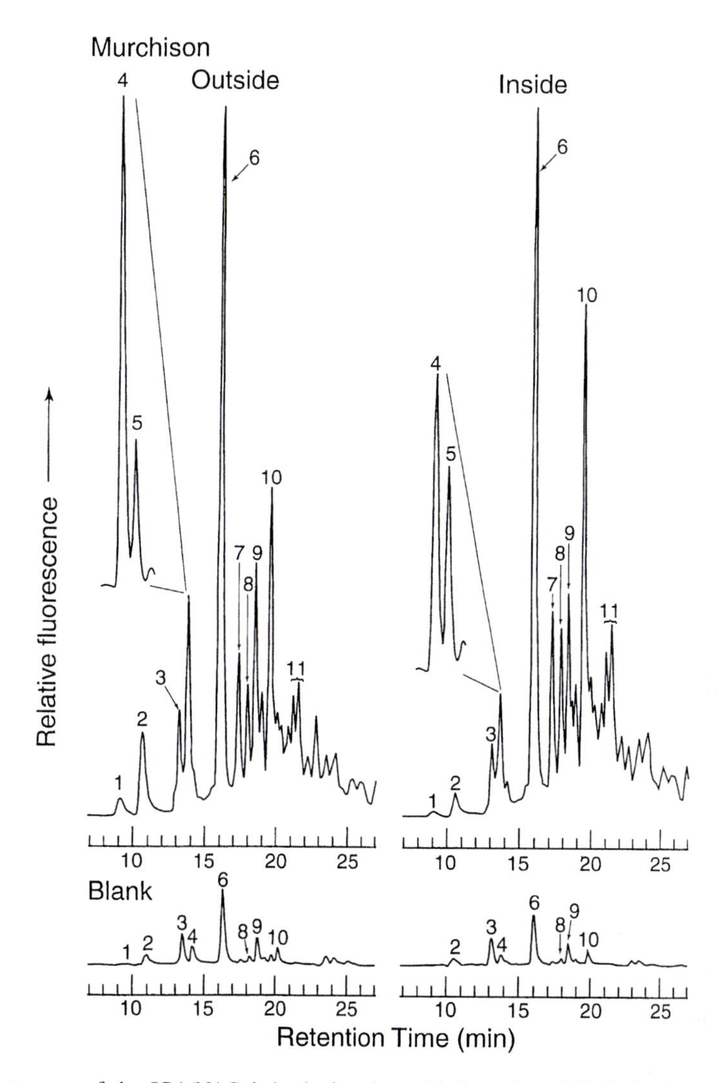

Fig. 1-3. HPLC chromatograms of the OPA/NAC-derivatized amino acids from the acid-hydrolyzed water extract of an outside (left) and inside (right) sample of the Murchison meteorite. A blank carried through the same extraction procedure is shown at the bottom. The enantiomers of the various amino acids, except glutamic acid, were separated using the gradient described in Bada et al. (1998). For glutamic acid, the following conditions were used for separation: (A) 50 mM sodium acetate (pH = 8), (B) methanol: gradient, 0–16 min: 0–10% B, 16–22 min: 10–40% B; flow rate, 1 ml/min. Peaks were identified by comparison of the retention times with those of an amino acid standard run at the same time. The peak identifications are as follows: 1, D-aspartic acid; 2, L-aspartic acid; 3, (D + L) serine; 4, L-glutamic acid; 5, D-glutamic acid; 6, glycine; 7, β-alanine; 8, D-alanine; 9, L-alanine; 10, AIB; and 11, (D + L) isovaline.

shown in table 1-2, in which the concentrations of eight amino acids and the D/L ratios of two amino acids are compared. Although the concentrations of the eight amino acids are not the same, they are comparable, especially in view of the different methodologies and measurement techniques that were used in 1970–1975 and in 1997. Also, the distribution of concentrations, especially of the protein amino acids, is similar, glycine being the most abundant, alanine and glutamic acid being of intermediate abundance, and aspartic acid being the least abundant. D/L ratios for alanine, glutamic acid, and isovaline are also comparable in the inside samples measured in 1970 and 1997. Thus the observations made in 1997 support the original reports (Kvenvolden et al., 1970, 1971; Pollock et al., 1975).

The presence of higher concentrations of protein amino acids in the outside sample relative to the inside sample of the meteorite (table 1-2) is consistent with the view that the outside sample has been more affected by terrestrial contamination than has

TABLE 1-2 Amino acid concentrations and D/L ratios of chiral amino acids; comparison of results from 1970–1975 and 1997

Concentration (μg/g)	1970–1975[a]	1997 Inside	1997 Outside
glycine	6	4.0 ± 1.0	6.0 ± 1.0
alanine	3	2.0 ± 0.5	4.0 ± 0.5
glutamic acid	3	2.0 ± 1.0	5.0 ± 0.5
aspartic acid	~0.4	0.3 ± 0.1	3.0 ± 0.5
β-alanine	0.5	1.5 ± 0.1	3.0 ± 1.0
α-aminoisobutyric acid	7	3.5 ± 0.5	5.0 ± 1.0
γ-amino-*n*-butyric acid	0.4	0.8 ± 0.1	2.0 ± 1.0
isovaline	2	2.0 ± 1.0	6.0 ± 2.0
D/L ratios			
alanine	1	0.9 ± 0.1	0.6 ± 0.1
glutamic acid	0.8	0.7 ± 0.1	0.3 ± 0.1
aspartic acid	ND	0.40 ± 0.05	0.30 ± 0.05
isovaline	1	1.0 ± 0.1	1.0 ± 0.1

[a] ND, Not determined.

the inside sample. Terrestrial contamination would consist mainly of L-isomers of protein amino acids. The inside sample may be slightly contaminated, in view of the observation of relatively high D/L ratios of alanine and glutamic acid compared to the lower D/L ratios of protein amino acids in the outside sample.

Kvenvolden et al. (1970, 1971) proposed terrestrial contamination superimposed on racemic mixtures of indigenous amino acids as an explanation for the observations on the inside sample. The 1997 work on inside and outside samples of the meteorite supports, but does not prove, that the *in situ* amino acids are racemic. However, Pizzarello and Cronin (1998) have clearly shown that alanine in the Murchison meteorite is racemic, thus supporting the observation first made in 1970. The low D/L ratios of aspartic acid in both inside and outside samples (0.4 and 0.3, respectively) are not consistent with the higher D/L ratios (0.7–1) measured for the other chiral amino acids. These low ratios may be the result of extensive contamination by this one amino acid, but this explanation is tenuous at best.

The concentrations of non-protein amino acids in the inside and outside samples (table 1-2) also present a problem. Higher concentations of α-aminoisobutyric acid and isovaline in the outside sample cannot be explained by terrestrial contamination. These two amino acids are extremely rare on Earth, and thus cannot account for the increased concentration of these amino acids in the outside sample. Thus, these higher concentrations of non-protein amino acids in the outside sample relative to the inside sample are inconsistent with the argument for terrestrial contamination based on the occurrence of the protein amino acids. A consistent argument would require each of the non-protein amino acids to be in approximately equal abundance in the outside and inside samples. In spite of this inconsistency, we believe that terrestrial contamination offers the best explanation for the distribution and D/L ratios of protein amino acids in the meteorite samples that we analyzed.

The D/L ratios of four amino acids found in the outside sample of the Murchison meteorite (table 1-2) are the same as the ratios previously reported by Engel and Nagy (1982). They reported the following D/L ratios: alanine, 0.6; glutamic acid, 0.3; aspartic acid, 0.3; and isovaline, 1. The fact that the values are the same is certainly coincidental, and the comparison is made here merely to emphasize that distribution of amino acids found in the Murchison meteorite is certainly sample-dependent.

Engel and Nagy (1982) have interpreted their results as follows: "the partially racemic amino acids in this Murchison meteorite stone are probably due to extraterrestrial stereoselective synthesis or decomposition reactions, although the possibility of unusual terrestrial contamination cannot be excluded". Bada et al. (1983), on reviewing the previous work, concluded that the results can be most easily understood as terrestrial contamination superimposed on the indigenous amino acids. Thus a controversy exists which has not yet been resolved. The results of the 1997 work reported here (comparing protein amino acids in the inside and outside samples of the meteorite) bolster the argument for terrestrial contamination but do not prove the case.

Nevertheless, the idea that stereoselective processes have affected the amino acids in meteorites, as first suggested by Engel and Nagy (1982), has gained support through studies by Cronin and Pizzarello (1997). They showed that α-methyl-α-amino alkanoic acids in the Murchison meteorite have up to 9.1% L-enantiomeric excess, with isovaline having a D/L ratio of approximately 0.85.

Previous work, including that reported here (table 1-2), led to the view that isovaline is racemic, with a D/L ratio of 1. Cronin and Pizzarello (1997) suggest possible extraterrestrial, stereoselective syntheses to account for their observations, but these processes have only affected the α-methyl-α-amino alkanoic acids, which, lacking a proton on the chiral carbon, are resistant to racemization. On the other hand, these workers have shown that the α-(H)-amino acids such as α-amino-*n*-butyric acid, norvaline, and also alanine (Pizzarello and Cronin, 1998), are present as racemates with D/L ratios of 1. This later observation gives credence to the original interpretations that the *in situ* protein amino acids in the Murchison meteorite are probably present as racemic mixtures (Kvenvolden et al. 1970, 1971). Thus, it appears that the amino acids found in this meteorite signal a complex history including both stereoselective extraterrestrial processes and (often) terrestrial contamination.

CONCLUSIONS

Re-analysis of portions of an inside sample of the Murchison meteorite after 27 years shows comparable concentrations of four protein amino acids and similar relative distributions of their concentrations. Also, D/L ratios of two protein amino acids and one non-protein amino acid are essentially the same. These results are not unexpected given the fact that portions of the same inside sample were used for analysis; however, different methods were applied.

A comparison of results from an inside and an outside sample of the meteorite shows that both the protein and non-protein amino acid abundances are consistently higher in the outside sample, and that the D/L ratios of protein amino acids are lower in the outside sample. Although the arguments are not yet totally consistent, we believe that the measured distributions and D/L ratios of the protein amino acids in the meteorite samples can best be explained by terrestrial contamination superimposed on indigenous protein amino acids present as racemic mixtures. The inside sample has been less affected by terrestrial contamination.

Acknowledgments: The 1997 analyses were supported by the National Aeronautics and Space Administration Specialized Center of Research and Training (NSCORT) in Exobiology at the University of California at San Diego.

REFERENCES

Anders E. et al. (1964) Contaminated meteorite. *Science* **146**, 1157–1161.

Bada J. L. et al. (1983) On the reported optical activity of amino acids in the Murchison meteorite. *Nature* **301**, 494–496.

Bada J. L., Glavin D. P., and McDonald G. D. (1998) A search for endogenous amino acids in Martian meteorite ALH84001. *Science* 279, 362–365.

Briggs M. H. (1961) Organic constituents of meteorites. *Nature* **191**, 1137–1140.

Briggs M. H. and Mamikunian G. (1963) Organic constituents of the carbonaceous chondrites. *Space Sci. Rev.* **1**, 647–682.

Calvin M. and Vaughn S. K. (1960) Extraterrestrial life: some organic constituents of meteorites and their significance for possible extraterrestrial biological evolution. In *Space Research* (ed. H. K. Bijl), vol. I, pp. 1171–1191. North-Holland Publishing Co., Amsterdam.

Cronin J. R. and Pizzarello S. (1983) Amino acids in meteorites. *Adv. Space Res.* **3**, 5–18.

Cronin J. R. and Pizzarello S. (1986) Amino acids of the Murchison meteorite. III. Seven carbon acyclic primary α–amino alkanoic acids. *Geochim. Cosmochim. Acta* **50**, 2419–2427.

Cronin J. R. and Pizzarello S. (1997) Enantiomeric excesses in meteoritic amino acids. *Science* **275**, 951–955.

Cronin J. R., Pizzarello S., and Yuen G. U. (1985) Amino acids of the Murchison meteorite. II. Five carbon acyclic primary β-, γ-, and δ-amino alkanoic acids. *Geochim. Cosmochim. Acta* **49**, 2259–2265.

Degens E. T. and Bajor M. (1962) Amino acids and sugars in the Brudeheim and Murray meteorites. *Naturwiss.* **49**, 605–606.

Engel M. H. and Nagy B. (1982) Distribution and enantiomeric composition of amino acids in the Murchison meteorite. *Nature* **296**, 837–840.

Hayes J. M. (1967) Organic constituents of meteorites—a review. *Geochim. Cosmochim. Acta* **31**, 1395–1440.

Kaplan I. R., Degens E. T., and Reuter J. H. (1963) Organic compounds in stony meteorites. *Geochim. Cosmochim. Acta* **17**, 805–834.

Kvenvolden K. A. et al. (1970) Evidence for extraterrestrial amino acids and hydrocarbons in the Murchison meteorite. *Nature* **228**, 923–926.

Kvenvolden K. A., Lawless J. G., and Ponnomperuma C. (1971) Nonprotein amino acids in the Murchison meteorite. *Proc. Natl Acad. Sci. USA* **68**, 486–490.

Pizzarello S. and Cronin J. R. (1998) Alanine enantiomers in the Murchison meteorite. *Nature* **394**, 236.

Pollock G. E. (1972) Resolution by gas–liquid chromatography of diastereomers of five nonprotein amino acids known to occur in the Murchison meteorite. *Anal. Chem.* **44**, 2368–2372.

Pollock G. E. (1974) Correction. Resolution by gas–liquid chromatography of diastereomers of five nonprotein amino acids known to occur in the Murchison meteorite. *Anal. Chem.* **46**, 614.

Pollock G. E., Cheng C.-N., Cronin S. E., and Kvenvolden K. A. (1975) Stereoisomers of isovaline in the Murchison meteorite. *Geochim. Cosmochim. Acta* **39**, 1571–1573.

Vallentyne J. R. (1965) Two aspects of the geochemistry of amino acids. In *The Origin of Prebiological Systems and of their Molecular Matrices* (ed. S. W. Fox), pp. 105–125. Academic Press.

Zhao M. and Bada J. L. (1989) Extraterrestrial amino acids in Cretaceous/Tertiary boundary sediments at Stevns Klint, Denmark. *Nature* **339**, 463–465.

Zhao M. and Bada J. L. (1995) Determination of α-dialkyl amino acids and their enantiomers in geological samples by high-performance liquid chromatography after derivatization with a chiral adduct of *o*-phthaldialdehyde. *J. Chromatogr.* **A690**, 55–63.

2. Chirality of meteoritic organic matter: a brief review

J. R. Cronin and S. Pizzarello

It is a question of long-standing interest as to whether meteoritic organic compounds are indicative of extraterrestrial life (Berzelius, 1834), terrestrial contamination (Hamilton, 1965), abiotic chemistry (Kvenvolden et al., 1970), or some combination of these. In attempting to answer this question, that is, in attempting to understand the origin and implications of meteoritic organic compounds, direct measurements of their chirality or the associated phenomenon of optical activity have frequently been carried out. The basis for this approach to the question is the exquisite stereoselectivity of terrestrial biochemistry, that is, given the two structural possibilities (enantiomers) presented by chiral compounds, organisms use only one. On the other hand, abiotic processes, whether synthetic or degradative, fail to distinguish between enantiomers. An exception arises when pre-existing asymmetry somehow bears on a process; however, such circumstances are rare for natural abiotic processes.

EARLY POLARIMETRIC ANALYSES

Early attempts to measure optical rotation in organic compounds from meteorites by polarimetric analysis gave uniformly negative results. Organic solvent extracts of the Cold Bokkeveld meteorite (Mueller, 1953) and the Mokoia and Haripura meteorites (Briggs and Mamikunian, 1963) failed to show optical rotation, as did both organic solvent and aqueous extracts from eight carbonaceous chondrites of various petrologic types (Kaplan et al., 1963).

Subsequently, conflicting results were obtained from a series of polarimetric analyses carried out at the time of the controversial claim that "organized elements" observed in carbonaceous chondrites represent the remains of indigenous cells. Optical activity was reported in saponified organic extracts of the Orgueil meteorite (Nagy, 1965, 1966; Nagy et al., 1964); however, it was not observed by others in similar, purified extracts and the original observation was attributed to impurities and analytical artifacts (Hayatsu, 1965, 1966; Meinschein et al., 1966). A comprehensive review of this work concluded that the presence of optically active compounds in Orgueil had not been proved (Hayes, 1967).

EARLY ENANTIOMERIC ANALYSES

As Kvenvolden et al. point out in Chapter 1 of this volume, a new era in the organic chemistry of carbonaceous chondrites was entered with the fall of the Murchison meteorite in 1969. Substantial amounts of this fresh meteorite were available for analysis, the pitfalls posed by terrestrial contamination had become apparent to analysts, and gas chromatographic (GC) methods that allowed high-sensitivity analysis of individual enantiomers were available for the first time. The first enantiomeric analyses of organic compounds from the Murchison and Murray meteorites (type CM) were carried out by the National Aeronautics and Space Administration (NASA)–Ames group and focused on amino acids. The methods employed and the results obtained are described in detail in Chapter 1 of this volume and are only briefly summarized here.

It is important to remember that contamination problems had seriously compromised analyses of meteoritic amino acids carried out earlier in the 1960s and that many in the community doubted that amino acids were in fact indigenous to meteorites. The finding of several biologically unusual amino acids in Murchison (Kvenvolden et al., 1970, 1971) strongly suggested that amino acids were of meteoritic origin; an important motivation for carrying out chiral analyses was to support the case for indigenous amino acids by demonstrating their presence as racemates or near-racemates. The results obtained by the NASA–Ames group are summarized in table 2-1.

Shimoyama et al. (1979) carried out chiral analyses of amino acids from the Yamato-74662 carbonaceous chondrite with essentially the same objective and found the enantiomers of alanine, aspartic acid, and glutamic acid to be present in approximately equal abundance. Shimoyama later obtained a similar result for alanine and α-amino butyric acid from Yamato-791198 (Shimoyama et al., 1985).

TABLE 2-1 Early enantiomeric analyses of amino acids from the Murchison and Murray meteorites

Amino acid	Murchison	L-Enantiomeric excess (%)	Ref.[a]	Murray[b]	Ref.[a]
alanine	~50% D	~0	1	D≅L	4
glutamic acid	~45% D	~10	1	D≅L	4
valine	~40–47% D	~6–20	1	app. D≅L	4
proline	~40–43% D	~14–20	1	app. D≅L	4
α-aminobutyric acid	D≅L	~0	2	D≅L	4
pipecolic acid	D≅L	~0	2	D≅L	4
β-aminoisobutyric acid	D≅L	~0	2	app. D≅L	4
norvaline	D≅L	~0	2		
isovaline	racemic	0	3		

[a] 1, Kvenvolden et al. (1970); 2, Kvenvolden et al. (1971); 3, Pollock et al. (1975); 4, Lawless et al. (1971).
[b] app., Apparently.

Determining exact enantiomer ratios was not a priority in the early analyses; indeed, this was not possible in most cases because of incomplete resolution. An exception was the attempt by Pollock et al. (1975) to obtain an accurate determination of the enantiomer ratio of isovaline (discussed later). Lawless (1973) had previously pointed out that the enantiomer ratio of isovaline, which lacks an α-hydrogen atom and is therefore resistant to racemization, held special interest owing to its "memory" of any original enantiomeric asymmetry.

Analyses of other organic compounds from Murchison gave similar results. The *R*- and *S*-isomers of methyl succinic acid were found to be present in Murchison in "nearly equal concentration" when analyzed gas chromatographically as the di-(+)-2-butyl esters (Lawless et al., 1974). Later, Shimoyama and Shigematsu (1994) resolved six additional Murchison C_5–C_7 dicarboxylic acids as the di-(*S*)-2-butyl esters and found them to have enantiomer ratios of "nearly one." Peltzer and Bada (1978) carried out gas chromatographic analyses of several α-hydroxy acids from Murchison as the diastereomeric *N*-TFA-L-prolyl methyl esters and found four to have D/L ratios ranging from 0.82 to 0.93. These values were considered to be "essentially racemic," and were presented as evidence for the indigenousness.

These early enantiomeric analyses, along with the finding of many compounds that are rare or unknown in the biosphere, provide a strong argument for the indigenousness of the Murchison amino-, hydroxy-, and dicarboxylic acids. Although the results usually did not show the compounds analyzed to be exactly racemic, in view of the accuracy and/or precision of the methods and the possibility of terrestrial contamination contributing small amounts of the natural enantiomer, the conclusion was that they are racemic. This interpretation of the data was consistent with the expectation for products of chemical evolution and with the earlier failure to demonstrate convincingly optical rotation in meteorite extracts.

Although this conclusion was widely accepted, Engel and Nagy (1982) were unconvinced by the explanations for deviations in the data from the perfectly racemic condition and carried out additional GC, gas chromatography–mass spectrometry (GC–MS), and high-performance liquid chromatography (HPLC) amino acid analyses on Murchison samples they believed to be free of terrestrial contamination. Although they found isovaline to be racemic and α-amino-*n*-butyric acid apparently so, five protein amino acids (alanine, glutamic acid, proline, aspartic acid, and leucine) present in an aqueous extract, a hydrolyzed aqueous extract, and an acid extract of the meteorite showed substantial L-excesses. Their results are summarized in table 2-2.

The magnitudes of the L-excesses reported for the protein amino acids were in startling contrast to the earlier results (Kvenvolden et al., 1970; Lawless et al., 1971). Nevertheless, they concluded that such excesses are characteristic of these amino acids as native to the meteorite and are the result of an extraterrestrial stereoselective synthesis and/or decomposition. The possibility of terrestrial contamination was not excluded but was deemed unlikely because of the failure to detect several protein amino acids, e.g., tyrosine, methionine, etc., in their preparations.

This conclusion was criticized (Bada et al., 1983), largely on the grounds that the sampling procedure had failed to avoid surficial microbial contamination. It was pointed out that (1) calculations based on the mass of surface material removed indicated that the analyzed sample did not entirely exclude material from the exterior zone of contamination; (2) the amino acids reported to have low D/L ratios are plausible constituents of a microbial free amino acid pool; and (3) the lower D/L ratio observed after acid hydrolysis of the extract and acid extraction of the meteorite residue is consistent with the

TABLE 2-2 Enantiomeric analyses of Murchison amino acids by Engel and Nagy (1982)

Amino acid	D/L	L-Enantiomeric excess (%)	Method
glutamic acid	0.30 ± 0.02	54	GC–CIMS single-ion plots
aspartic acid	0.30 ± 0.04	54	GC–CIMS single-ion plots
proline	0.30 ± 0.02	54	GC–CIMS single-ion plots
leucine	0.166 ± 0.021	72	Integration of GC trace
alanine	0.60 ± 0.03	25	GC–CIMS single-ion plots
isovaline	0.99 ± 0.04	~0	HPLC
α-aminobutyric acid	~1	~0	HPLC

hydrolysis of protein from contaminant microorganisms.

GAS CHROMATOGRAPHY ISOTOPE RATIO MASS SPECTROMETRY

A few years later, Engel et al. (1990) bolstered the case for a meaningful L-alanine excess in Murchison by measuring carbon isotope ratios ($\delta^{13}C$) for the individual alanine enantiomers by using a gas chromatography–combustion–isotope-ratio mass spectrometry (GC–C–IRMS) in conjunction with a chiral GC phase. In this work, they reported an alanine L-enantiomer excess of 8% and $\delta^{13}C$ values for the L- and D-enantiomers of +27 and +30‰, respectively. On the basis of these data (table 2-3), they argued for an indigenous L-excess on the grounds that terrestrial contamination to this extent would have lowered the L-enantiomer $\delta^{13}C$ value more than was observed (an 8% excess of terrestrial L-alanine with a value of −28‰ lowers the overall $\delta^{13}C$ to +21‰).

Recently, Engel and Macko (1997) extended this approach to ^{15}N and, in this case, argued for indigenous L-excesses in both alanine and glutamic acid of 33 and 54%, respectively, based on the similar $\delta^{15}N$ values obtained for their D- and L-enantiomers (table 2-3).

Pizzarello and Cronin (1998) pointed to a potential problem with the use of GC–C–IRMS data in enantiomeric analyses of amino acid mixtures of the complexity of those extracted from CM chondrites.

TABLE 2-3 Enantiomeric analyses of Murchison amino acids by GC–C–IRMS

Amino acid	Enantiomeric excess (%)	$\delta^{13}C$ (‰)[a]	$\delta^{15}N$ (‰)[a]	Ref.[b]
L-alanine	8	+27	ND	1
D-alanine	–	+30	ND	1
L-alanine	33	ND	+57	2
D-alanine	–	ND	+60	2
L-glutamic acid	54	ND	+58	2
D-glutamic acid	–	ND	+60	2

[a] ND, Not determined.
[b] 1, Engel et al. (1990); 2, Engel and Macko (1997).

The ability to infer the absence of contamination from the enantiomer isotope ratios depends on the molecular purity of the chromatographic peaks, i.e., on each peak being composed of only the enantiomer of a single amino acid of either meteoritic or both meteoritic and contaminant origin. If another meteoritic amino acid coelutes with one of the enantiomers, it will contribute to both the overall isotopic ratio as well as the apparent enantiomer ratio, making the inference of purity impossible. Since meteoritic amino acids are isotopically heavier than their terrestrial counterparts, coelution can, in effect, mask the contribution of the lighter terrestrial contaminant. Such coelution problems are commonplace in the analysis of meteoritic amino acids. Although Engel and Macko (1997) state that "GC–MS and GC–C–IRMS analyses indicate no obvious coelutions," several other amino acids are known to have retention times very close to those of D- and L-alanine on Chirasil-L-Val, the phase used for their GC–C–IRMS work. Since three of these elute between L-alanine and sarcosine and baseline resolution of these latter amino acids was not achieved in their analysis, the possible interference of these meteoritic amino acids remains a concern.

ENANTIOMERIC ANALYSES OF NON-PROTEIN AMINO ACIDS

It is clear that the potential for terrestrial contamination and the complexity of the meteoritic amino acid mixture make chiral analyses of amino acids that are common to both meteorites and the biosphere problematic. In order to avoid the first, and minimize the second, of these problems, Cronin and Pizzarello (1997) focused their attention on those amino acids that are unknown or, at least, of very limited occurrence in the biosphere and they introduced a preliminary fractionation prior to GC–MS analysis. They found small L-enantiomer excesses (2–9%) in a subset of the amino acids, the α-methyl-α-amino acids, from Murchison and somewhat smaller enantiomeric excesses (1–6%) in the same amino acids obtained from the Murray (CM) carbonaceous chondrite (Pizzarello and Cronin,

TABLE 2-4 Recent enantiomeric analyses of amino acids from the Murchison and Murray meteorites by Cronin and Pizzarello

Amino acid	Murchison % L-ee (corr.)	Murray % L-ee (corr.)
2-amino-2,3-dimethylpentanoic acid		
2S,3S/2R,3R	7.0	1.0
2S,3R/2R,3S	9.1	2.2
2-amino-2-methylpentanoic acid	2.8	1.4
2-amino-2-methylbutanoic acid (isovaline)	8.4	6.0
2-amino-2,3-dimethylbutanoic acid	2.8	1.0
2-amino-2-methylhexanoic acid	4.4	1.8
2-amino pentanoic acid (norvaline)	0.4[a]	0.8[a]
2-amino butanoic acid (butyrine)	0.4[a]	−0.4[a]
2-amino propanoic acid (alanine)	1.2	0.4
2-amino-3-methylbutanoic acid (valine)	2.2	−0.4[a]

[a] Not significant.

1998, 2000). Interestingly, enantiomeric excesses of the magnitude seen in the α-methyl-α-amino acids were not observed in most α-unsubstituted amino acids, including alanine (Pizzarello and Cronin, 1998).

The data from both meteorites are combined in table 2-4. Amino acids were extracted from meteorite powders with hot water, isolated by adsorption on a cation-exchange resin, and fractionated by reverse-phase chromatography prior to analysis as their *N*-TFA/PFP-isopropyl esters by GC–MS on Chirasil-L-Val. The enantiomeric excesses are mean values calculated from integrated single-ion plots and, when justified by the absence of extraneous fragment ions in enantiomer difference spectra, on total-ion plots of data obtained from multiple chromatographic runs. The data were corrected in each case for any deviations from zero in the mean values shown by the corresponding racemic amino acid standards.

The probability of terrestrial contamination differs among the amino acids and should vary with their abundance in the biosphere. Table 2-5 summarizes the results of literature searches for reported natural occurrences of the amino acids discussed here. In four cases, the amino acids have never been reported and so the contribution of L-enantiomers by contamination is considered to be highly improbable. Furthermore, the observation that the ubiquitous protein amino acids alanine and valine are racemic (Murray) or show only small enantiomeric excesses (Murchison) indicates that contamination in general is minimal, making an external contribution to the biologically unknown amino acids even more unlikely.

The unrecognized coelution of another meteoritic amino acid with one of the amino acid enantiomers of interest was deemed to be a more likely cause of error in this work. To minimize this possibility, the amino acids were fractionated by reverse-phase chromatography prior to GC–MS analysis. This limited the possibilities of gas chromatographic coelution to amino acids of similar polarity, i.e., to α-amino isomers, amino position isomers (e.g., β-, γ-, δ-isomers), cyclic or unsaturated amino acid analogs, and/or homologues of similar carbon number. Particular attention was given to these in the mass spectral analysis of each enantiomeric peak. The coelution problem and data analysis are discussed at length elsewhere (Pizzarello and Cronin, 2000).

Pizzarello and Cronin (2000) provide two possible explanations for the appearance of significant enantiomeric excesses only in the α-methyl-α-amino

TABLE 2-5 Terrestrial occurrence of meteoritic amino acids

Amino acid	Occurrence	Amino acid	Occurrence
2-amino-2,3-dimethyl pentanoic acid diastereomers	unknown	2-amino pentanoic acid (norvaline)	rare
2-amino-2-methylpentanoic acid	unknown	2-amino butanoic acid (butyrine)	common
2-amino-2-methylbutanoic acid	rare	2-amino propanoic acid (alanine)	ubiquitous
2-amino-2,3-dimethylbutanoic acid	unknown	2-amino-3-methylbutanoic acid (valine)	ubiquitous
2-amino-2-methylhexanoic acid	unknown		

acids: (1) their racemization was prevented by the α-methyl substituent (Pollock et al., 1975); consequently, they retained the original enantiomeric excesses even though exposed to conditions under which the α-H-α-amino acids racemized, for example, during aqueous processing of the meteorite parent body (Bunch and Chang, 1980); (2) the α-H and α-methyl-α-amino acids are products of different formation processes, for example, presolar formation of the α-methyl amino acids and parent-body formation of those with α-H atoms (see below). This possibility is consistent with observations that the two classes of amino acids vary in relative amount among different Murchison specimens (Cronin and Pizzarello, 1983) and differ in their ease of extraction from meteorite powders (S. Pizzarello, unpublished data), observations suggesting that they may be associated with different phases within the meteorite matrix.

These results are in general agreement with the earlier work of Kvenvolden et al. (1970, 1971); however, the finding of a significant L-enantiomer excess in isovaline contrasts with the results of Pollock et al. (1975), who found this amino acid to be racemic in Murchison. That result was obtained by GC analysis alone on a complex fraction containing numerous other components and cannot be considered definitive, because of the possibility of unrecognized coelution. It should be noted that Engel and Nagy (1982) also found isovaline to be racemic in an aqueous extract of Murchison analyzed using chiral HPLC. The finding of little or no enantiomeric excess of alanine, α-amino butyric acid, of valine from the Murray meteorite, however, agrees with earlier results (Lawless et al., 1971). The α-methyl amino acids from Murray were not enantiomerically analyzed previously.

The larger (8–33%) L-alanine excesses in the Murchison meteorite, reported by others (Engel and Nagy, 1982; Engel et al., 1990; Engel and Macko, 1997) were not observed. Engel and coworkers carried out their enantiomer separations without prior fractionation of the meteoritic amino acids (Engel et al., 1990; Engel and Macko, 1997). As noted above, several amino acids in the whole meteorite extract have GC retention times similar to those of the alanine enantiomers on Chirasil-L-Val, and coelution of one or more of these may have compromised both their enantiomer ratios and their isotopic values. Fractionation of the meteorite extracts by reverse-phase chromatography removes these amino acids from the alanine-containing fraction (Pizzarello and Cronin, 1998).

ORIGIN OF ENANTIOMERIC EXCESSES

Engel et al. (1990) invoked the Vester–Ulbricht hypothesis (reviewed by Bonner, 1991) as a possible explanation for the L-alanine excesses they observed, i.e., that the β-decay of cosmic ray-produced ^{14}C gave rise to circularly polarized photons (Bremsstrahlung), which in turn brought about selective destruction of the D-enantiomer.

More recently, Cronin and Pizzarello (1997) and Engel and Macko (1997) suggested that enantiomeric excesses in meteoritic amino acids might have originated in the presolar cloud as a result of its irradiation by ultraviolet circularly polarized light (UVCPL) from a neutron star, a concept proposed by others (Rubenstein et al., 1983; Bonner and Rubenstein, 1987) and recently discussed in the context of interstellar grains and cometary organic matter (Greenberg, 1996).

Substantial deuterium enrichments in the meteoritic amino acids suggest that either the amino acids or their precursors are of interstellar cloud origin (Epstein et al., 1987). The presence of graphite, SiC, and Si_3N_4 (supernova grains) in carbonaceous chondrites (Amari and Zinner, 1990) provides evidence for the supernova precursor of a neutron star; consequently, exposure of meteoritic organic matter, at least at an early stage in its formation, to electromagnetic radiation from a neutron star does not seem unlikely.

Roberts (1984) raised a theoretical objection to the production of CPL by neutron stars soon after Rubenstein et al. (1983) set out their proposal, and Mason (1997) has criticized the Rubenstein–Bonner hypothesis as a mechanism for the production of enantiomeric excesses on the grounds that the Kuhn–Condon zero sum rule (cancellation of the CD spectrum over the full electromagnetic spectrum) prevents photochemical discrimination by the broad-band emission of a neutron star. In response to the latter, Bonner (1991) pointed out that even though there may be no net absorption difference, the particular absorption bands excited are not the same for enantiomers and therefore their tendencies to undergo photolysis will not necessarily be the same.

Recently, Bailey et al. (1998) have suggested a modified form of the Rubenstein–Bonner hypothesis that avoids the objections noted above. They have observed substantial circular polarization in the infrared from the Orion molecular cloud and interpreted this as originating not from a neutron star, but from scattering of unpolarized star light by magnetically aligned spheroidal interstellar grains (Mie scattering). They calculate that UV light would be similarly circularly polarized, although difficult to observe. They point out that, in the case of many stars, there is a major drop in UV intensity below 220 nm such that the selective destruction of amino acid enantiomers of a particular configuration would be possible as a result of the excitation of only their longest wavelength band, i.e., the objection based on the Kuhn–Condon rule would not apply. They suggest the origin of UVCPL by Mie scattering as an alternative to the neutron-star hypothesis, but one with the same implications for the chirality of interstellar organic matter.

In laboratory experiments, UVCPL photolysis (212.8 nm) has been shown to give rise to enantiomeric excesses in racemic leucine of 1.98% (right-handed) and 2.50% (left-handed) with 59 and 75% overall decomposition, respectively (Flores et al., 1977). Resolution of racemic mixtures by UV photolysis is a race between overall decomposition and the selective decomposition of a particular enantiomer; thus, the enantiomeric purity achieved depends on the extent of photolysis and the magnitude of the anisotropy factor, g, where $g = [\varepsilon_r - \varepsilon_l]/\varepsilon$, ε_r and ε_l are the extinction coefficients of an enantiomer for right- and left-CPL, and ε is the extinction coefficient of the racemate. In principle, the limiting enantiomeric excesses that should be obtained for leucine ($g = 0.0244$; Flores et al., 1977) range from 1.7% at 80% decomposition to 3.2% at 95% decomposition (Balavoine et al., 1974), values in fair agreement with experiment and comparable to most of the meteoritic values. The order-of-magnitude larger L-excesses reported for alanine (Engel and Macko, 1997) and the isovaline and 2-amino-2,3-dimethyl pentanoic acid excesses (Cronin and Pizzarello, 1997; Pizzarello and Cronin, 2000) appear to be outside the range achievable by UVCPL photolysis given the g-values of α-amino alkanoic acids (Katzin and Gulyas, 1968).

IMPLICATIONS FOR THE FORMATION OF METEORITIC AMINO ACIDS

In the 20 years that have passed since meteoritic amino acids were discussed (Cronin et al., 1980) in *Biogeochemistry of Amino Acids*, the predecessor of this volume, both molecular (Peltzer and Bada, 1978; Cronin et al., 1993) and isotopic (deuterium) analyses (Epstein et al., 1987; Cronin and Chang, 1993) have contributed to a better understanding of their origin. Amino acid formation has come to be viewed as a two-step process in which their molecular precursors (aldehydes, ketones, HCN, and ammonia) were formed in the presolar cloud and, after incorporation via planetesimals into an icy, volatile-rich primitive parent body, underwent Strecker reactions during its aqueous processing (Cronin et al., 1995).

The presence of enantiomeric excesses in the Murchison amino acids is difficult to reconcile with this hypothesis if UVCPL provided the necessary asymmetric influence as suggested above. If their asymmetry was achieved by preferential photolysis, it must have been manifest after (or during) formation of the amino acids; however, if this occurred in the early solar system it seems likely that the amino acids would, for the most part, have been shielded from UV light by the outer layers of the parent body. On the other hand, if exposure to UVCPL occurred in the presolar cloud, only the precursor ketones of the α-methyl amino acids would have been affected. In this case, there would have been only a secondary effect on the chiral α-carbons of the amino acids formed later in the parent body, and then only in the case of amino acids derived from chiral aldehydes/ketones. Consequently, if the asymmetric influence was, in fact, UVCPL, formation of the non-racemic α-methyl-α-amino acids seems likely to have been an entirely presolar process.

IMPLICATIONS FOR THE ORIGIN OF LIFE

As Pasteur realized almost 150 years ago (Geison, 1995), understanding the origin of homochirality may be key to understanding the origin of life. Since then, numerous origin theories, both biotic and abiotic, have been proposed (reviewed by Bonner, 1991). According to the former, life initially was based on achiral molecules and/or racemates, and the use of specific enantiomers came about through evolution; whereas the latter theories, however, propose that a tendency towards homochirality was inherent in chemical evolution. The finding of L-enantiomeric excesses in the amino acids of meteorites supports the latter concept. Since it is likely that the accretion of dust, meteorites and cometary material provided the early Earth with a significant fraction of its organic inventory (Chyba and Sagan, 1992), amino acids delivered in this way may have provided the initial enantiomeric excesses necessary for amplification by further chemical evolution.

The α-methyl amino acids, which have not been viewed as important for the origin of life (they are generally unimportant in terrestrial biochemistry), are abundant in carbonaceous chondrites and may be well suited to such a role. Polymerization accompanied by the formation of regular secondary structure, for example, α-helices and β-sheets, has been shown to be an effective way of amplifying modest initial enantiomeric excesses (Bonner et al., 1981; Brack and Spach, 1981) and α-methyl-α-amino acids are known to have strong helix-inducing and stabilizing effects (Altman et al., 1988; Formaggio et al., 1995). Furthermore, any hard-won gains in enantiomeric excess achieved in the prebiotic world for α-methyl amino acids would be stable against racemization, whereas this would not be the case for amino acids lacking an α-substituent (Bada and Miller, 1987). Although a transition to α-H amino acids obviously must have occurred at some later stage, perhaps for biosynthetic reasons, the α-substituted amino acids conceivably could have played a key role in the chemical evolution of homochirality and perhaps were significant molecules in the biochemistry of a pre-RNA world.

SUMMARY

The data obtained to date are strongly indicative of enantiomeric excesses in meteoritic amino acids and, if correct, replete with meaning with respect to both prebiotic chemistry on the early Earth and the origin of life. They are not conclusive, however. What needs to be done to establish, with certainty, that amino acids, or at least some amino acids, are not racemic in carbonaceous chondrites? Two approaches have been fruitfully employed to date: (1) GC–MS analysis of nonterrestrial amino acids with careful attention to possible coelution errors, but without isotopic measurements; and (2) GC–C–IRMS analysis of protein amino acids common to meteorites, without accompanying data supporting the purity of the enantiomers on which the isotopic measurements were made. Combining these approaches will achieve the desired end of establishing whether or not the amino acids indigenous to carbonaceous chondrites are to some extent non-racemic. The case remains to be made that a set of meteoritic amino acids have statistically significant enantiomeric excesses, that the enantiomers have identical isotopic ratios, and that both measurements were made on enantiomers that were free of coeluting compounds.

Acknowledgments: The authors gratefully acknowledge research support from the NASA Exobiology Program (NAGW-1899; NAG5-4131).

REFERENCES

Altman E., Altman K. H., Nebel K., and Mutter M. (1988) Conformational studies on host–guest peptides containing chiral α-methyl-α-amino acids. *Intl. J. Pept. Protein Res* **32**, 344–351.

Amari S. and Zinner E. (1990) Supernova grains from meteorites. *Nature* **345**, 238–240.

Bada J. L. and Miller S. L. (1987) Racemization and the origin of optically active compounds in living organisms. *BioSystems* **20**, 21–26.

Bada J. L., Cronin J. R., Ho M.-S., Kvenvolden K. A., Lawless J. G., Miller S. L., Oró J., and Steinberg S. (1983) On the reported optical activity of amino acids in the Murchison meteorite. *Nature* **301**, 494–497.

Bailey J., Chrysostomou A., Hough J. H., Gledhill T. M., McCall A., Clark S., Ménard F., and Tamura M. (1998) Circular polarization in star formation regions: Implications for biomolecular homochirality. *Science* **281**, 672–674.

Balavoine G., Moradpour A., and Kagan H. B. (1974) Preparation of chiral compounds with high optical purity by irradiation with circularly polarized light, a model reaction for the prebiotic generation of optical activity. *J. Amer. Chem. Soc.* **96**, 5152–5158.

Berzelius J. J. (1834) Ueber meteorsteine. *Ann. Phys. Chem.* **33**, 113–148.

Bonner W. A. (1991) The origin and amplification of biomolecular chirality. *Orig. Life Evol. Biosphere* **21**, 59–111.

Bonner W. A. and Rubenstein E. (1987) Supernovae, neutron stars and biomolecular chirality. *BioSystems* **20**, 99–111.

Bonner W. A., Blair N. E., and Dirbas F. M. (1981) Experiments on the abiotic amplification of optical activity. *Orig. Life* **11**, 119–134.

Brack A. and Spach G. (1981) Enantiomer enrichment in early peptides. *Orig. Life* **11**, 135–142.

Briggs M. H. and Mamikunian G. (1963) Organic constituents of the carbonaceous chondrites. *Space Sci. Rev.* **1**, 647–682.

Bunch T. E. and Chang S. (1980) Carbonaceous chondrites. II. Carbonaceous chondrite phyllosilicates and light element geochemistry as indicators of parent body processes and surface conditions. *Geochim. Cosmochim. Acta* **44**, 1543–1577.

Chyba C. F. and Sagan C. (1992) Endogenous production, exogenous delivery, and impact-shock synthesis of organic molecules: an inventory for the origins of life. *Nature* **355**, 125–132.

Cronin J. R. and Chang S. (1993) Organic matter in meteorites: molecular and isotopic analyses of the Murchison meteorite. In *The Chemistry of Life's Origins* (eds J. M. Greenberg, C. X. Mendoza-Gomez, and V. Pirronello), pp. 209–258. Kluwer.

Cronin J. R. and Pizzarello S. (1983) Amino acids in meteorites. *Adv. Space Res.* **3**, 5–18.

Cronin J. R. and Pizzarello S. (1997) Enantiomeric excesses in meteoritic amino acids. *Science* **275**, 951–955.

Cronin J. R., Gandy W. E., and Pizzarello S. (1980) Amino acids of the Murchison meteorite. In *Biogeochemistry of the Amino Acids* (eds P. E. Hare, T. C. Hoering, and K. King), pp. 153–168. Wiley.

Cronin J. R., Pizzarello, S., Epstein, S., and Krishnamurthy R. V. (1993) Molecular and isotopic analyses of the hydroxy acids, dicarboxylic acids, and hydroxydicarboxylic acids of the Murchison meteorite. *Geochim. Cosmochim. Acta* **57**, 4745–4752.

Cronin J. R., Cooper G. W., and Pizzarello S. (1995) Characteristics and formation of amino acids and hydroxy acids of the Murchison meteorite. *Adv. Space Res.* **15**, 91–97.

Engel M. H. and Macko S. A. (1997) Isotopic evidence for extraterrestrial non-racemic amino acids in the Murchison meteorite. *Nature* **389**, 265–268.

Engel M. H. and Nagy B. (1982) Distribution and enantiomeric composition of amino acids in the Murchison meteorite. *Nature* **296**, 837–840.

Engel M. H., Macko S. A., and Silfer J. A. (1990) Carbon isotope composition of individual amino acids in the Murchison meteorite. *Nature* **348**, 47–49.

Epstein S., Krishnamurthy R. V., Cronin J. R., Pizzarello S., and Yuen G. U. (1987) Unusual stable isotope ratios in amino acid and carboxylic acid extracts from the Murchison meteorite. *Nature* **326**, 477–479.

Flores J. J., Bonner W. A., and Massey G. A. (1977) Asymmetric photolysis of (*R,S*)-leucine with circularly polarized light. *J. Amer. Chem. Soc.* **99**, 3622–3625.

Formaggio F., Crisma M., Bonora G. M., Pantano M., Valle G., Toniolo C., Aubry A., Bayeul D., and Kamphuis J. (1995) (*R*)-Isovaline homopeptides adopt the left-handed 3_{10}-helical structure *Peptide Res.* **8**, 6–14.

Geison G. L. (1995) *The Private Science of Louis Pasteur*. Princeton University Press.

Greenberg J. M. (1996). Chirality in interstellar dust and in comets: life from dead stars. In *Physical Origin of Homochirality in Life* (ed. D. B. Cline), pp. 185–210. American Institute of Physics.

Hamilton P. B. (1965) Amino acids on hands. *Nature* **205**, 284–285.

Hayatsu R. (1965) Optical activity in the Orgueil meteorite. *Science* **149**, 443–447.

Hayatsu R. (1966) Artifacts in polarimetry and optical activity in meteorites. *Science* **153**, 859–861.

Hayes J. M. (1967) Organic constituents of meteorites—a review. *Geochim. Cosmochim. Acta* **31**, 1395–1440.

Kaplan I. R., Degens E. T., and Reuter J. H. (1963) Organic compounds in stony meteorites. *Geochim. Cosmochim. Acta* **27**, 805–834.

Katzin L. I. and Gulyas E. (1968) Absorption, rotatory dispersion, and circular dichroism studies on some hydroxy and amino acids. *J. Amer. Chem. Soc.* **90**, 247–251.

Kvenvolden K., Lawless J., Pering K., Peterson E., Flores J., Ponnamperuma C., Kaplan, I. R., and Moore C. (1970) Evidence for extraterrestrial amino acids and hydrocarbons in the Murchison meteorite. *Nature* **228**, 923–926.

Kvenvolden K. A., Lawless, J. G., and Ponnamperuma C. (1971) Nonprotein amino acids in the Murchison meteorite. *Proc. Natl Acad. Sci. U.S.A.* **68**, 486–490.

Lawless J. G. (1973) Amino acids in the Murchison meteorite. *Geochim. Cosmochim. Acta* **37**, 2207–2212.

Lawless J. G., Kvenvolden K. A., Peterson E., Ponnamperuma C., and Moore C. (1971) Amino acids indigenous to the Murray meteorite. *Science* **173**, 626–627.

Lawless J. G., Zeitman B., Pereira W. E., Summons R. E., and Duffield A. M. (1974) Dicarboxylic acids in the Murchison meteorite. *Nature* **251**, 40–41.

Mason S. F. (1997) Extraterrestrial handedness. *Nature* **389**, 804.

Meinschein W. G., Frondel C., Laur P., and Mislow K. (1966) Meteorites: optical activity in organic matter. *Science* **154**, 377–380.

Mueller G. (1953) The properties and theory of genesis of the carbonaceous complex within the Cold Bokevelt meteorite. *Geochim. Cosmochim. Acta* **4**, 1–10.

Nagy B. (1965) Optical activity in the Orgueil meteorite. *Science* **150**, 1846.

Nagy, B. (1966) The optical rotation of lipids extracted from soils, sediments, and the Orgueil carbonaceous meteorite. *Proc. Natl Acad. Sci. U.S.A.* **56**, 389–398.

Nagy B., Murphy M. T. J., Modzeleski V. E., Rouser G., Claus G., Hennessy D. J., Colombo U., and Gazzarini F. (1964) Optical activity in saponified organic matter isolated from the interior of the Orgueil meteorite. *Nature* **202**, 228–233.

Peltzer E. T. and Bada J. L. (1978) α-Hydroxycarboxylic acids in the Murchison meteorite. *Nature* **272**, 443–444.

Pizzarello S. and Cronin J. R. (1998) Alanine enantiomers in the Murchison meteorite. *Nature* **394**, 236.

Pizzarello S. and Cronin J. R. (2000) Non-racemic amino acids in the Murray and Murchison meteorites. *Geochim. Cosmochim. Acta* **64**, 329–338.

Pollock G. E., Chang C.-N., Cronin S. E., and Kvenvolden K. A. (1975) Stereoisomers of isovaline in the Murchison meteorite. *Geochim. Cosmochim. Acta* **39**, 1571–1573.

Roberts J. A. (1984) Supernovae and life. *Nature* **308**, 318.

Rubenstein E., Bonner W. A., Noyes H. P., and Brown G. S. (1983) Supernovae and life. *Nature* **306**, 118.

Shimoyama A. and Shigematsu R. (1994) Dicarboxylic acids in the Murchison and Yamato-791198 carbonaceous chondrite. *Chem. Lett.* **1994**, 523–526.

Shimoyama A., Ponnamperuma C., and Yanai K. (1979) Amino acids in the Yamato carbonaceous chondrite from Antarctica. *Nature* **282**, 394–396.

Shimoyama A., Harada K., and Yanai K. (1985) Amino acids from the Yamato-791198 carbonaceous chondrite from Antarctica. *Chem. Lett.* **1985**, 1183–1186.

3. Thermodynamics of amino acid synthesis in hydrothermal systems on early Earth

Jan P. Amend and Everett L. Shock

Since their discovery more than 20 years ago (Corliss et al., 1979), submarine hydrothermal systems have received relentless attention, first as possible sites (Ingmanson and Dowler, 1977; Corliss et al., 1981; Baross and Hoffman, 1985) and now as probable sites of abiotic organic synthesis (Shock, 1990, 1996; Huber and Wächtershäuser, 1997, 1998; Russell and Hall, 1997; Amend and Shock, 1998; Shock and Schulte, 1998; Shock et al., 1998). However, laboratory experiments demonstrating organic synthesis predate this discovery by about 150 years; in 1828, Friedrich Wöhler synthesized urea from a heated solution of ammonia and cyanic acid (Wöhler, 1828). Nevertheless, a concerted effort to investigate experimentally the formation and stability of aqueous organic compounds from inorganic precursors did not take effect until approximately 1950 (for a review, see Shock, 1992). Of these laboratory efforts, Miller's amino acid synthesis experiments (Miller, 1953, 1957; Miller and Urey, 1959) may be the most familiar. Miller's famous spark-discharge experiments, in which a flask of boiling water represented the primitive ocean and another flask filled with reduced gases represented the atmosphere, are almost solely responsible for bringing to the forefront the connection between the origin of life and the abiotic formation of amino acids under early Earth conditions. Although unintentional because Miller's focus was lightning-driven chemistry in a reduced atmosphere, his experiments were the first to show that a variety of amino acids could be synthesized from inorganic precursors in a hot aqueous solution, owing to the fact that the flask representing the "ocean" was heated to 100°C. The atmospheric constraints imposed by Miller have not withstood the test of time, but the emphasis on amino acid synthesis as a vital step in understanding the origin of life has persisted.

Starting with the serendipitously successful hydrothermal organic synthesis experiments of French (1964) and continuing to the present, the justification and relevance of aqueous organic synthesis experiments at elevated temperatures (and occasionally pressures) can be attributed to the proposal that hydrothermal systems were plausible environments for the origin of life. Their role in the origin and evolution of life is consistent with current views (1) that hydrothermal systems have probably existed throughout Earth's history—their activity and abundance were quite possibly more pronounced in the Hadean than subsequently, (2) that disequilibrium in near- or subsurface fluid mixing zones provides the chemical energy required to drive otherwise endergonic (table 3-1) synthesis and polymerization processes, (3) that deep marine systems would have been sheltered from early heavy bombardment by meteorites and from the partial vaporization of the oceans, and (4) that the deepest-branching organisms in the Archaea and Bacteria domains of the phylogenetic tree, and therefore the last common ancestor as well, were obligately thermophilic, autotrophic rather than heterotrophic, and relied on chemosynthesis rather than photosynthesis. In other words, it appears that submarine hydrothermal systems provided (and may still provide) many, if not all, of the fundamental chemical conditions required for the emergence of the first self-replicating cells on Earth.

Although we are far from understanding all of the processes that led from an abiotic world to the one we inhabit today, significant steps have been taken to identify environments and geochemical conditions that may facilitate stepwise reactions from carbon dioxide, ammonia, hydrogen, and other inorganic species to complex biochemical systems. In this Chapter, we focus on amino acids and calculate the overall Gibbs free energies (ΔG_r) of their net synthesis reactions in hydrothermal systems on early Earth. The natural sites envisioned for these anabolic aqueous processes are subsurface zones of mixing between a warm, acid, slightly oxidized ocean and hot, slightly alkaline, significantly reduced hydrothermal fluids. To permit these calculations, geochemical constraints were provided by current views on the composition of the Hadean atmosphere and ocean (Walker, 1985; Kasting and Ackerman, 1986; Kasting, 1993a,b,c; Russell and Hall, 1997), together with a model hydrothermal fluid resulting from sea water equilibration with olivine-rich gabbro (the most abundant type of rock in the oceanic crust) (McCollom and Shock, 1998). Uncertainty and speculation surround such calculations because they depend on conditions that prevailed before the time of the oldest portions of the existing rock record. As a consequence, we have

TABLE 3-1 Glossary of terms

Archaea: one of two phylogenetic domains of prokaryotes, the other being the Bacteria.
Autotroph: an organism able to utilize CO_2 or CO as the sole sources of carbon.
Bacteria: one of two phylogenetic domains of prokaryotes, the other being the Archaea.
Chemolithoautotroph: an organism that obtains its metabolic energy from the oxidation of inorganic compounds and can grow on CO_2 or CO as its only carbon sources.
Chemosynthesis: the use of chemical energy in autotrophs to drive the synthesis of biomass (in contrast to *photosynthesis*, in which light energy is used).
Domain: the highest level of phylogenetic classification, consisting of the Archaea, Bacteria, and Eukarya.
Endergonic: refers to an energy-consuming reaction, i.e., one in which the value of ΔG_r is positive.
Exergonic: refers to an energy-yielding reaction, i.e., one in which the value of the overall Gibbs free energy (ΔG_r) (see text) is negative.
Hadean Eon: the segment of Earth history that predates the oldest (~3.8 Ga) rock sequence known on Earth.
Heterotroph: an organism that utilizes reduced organic compounds as its source of carbon.
Thermophile: an organism with an optimum growth temperature of at least 45°C.

no direct constraints on the composition of Hadean sea water or hydrothermal fluids, and cannot infer uniquely their host environments, ambient temperatures, redox conditions, or the mineralogy of the oceanic crust. Therefore, it may be beneficial first to explore analogous calculations for present-day marine hydrothermal systems using well-constrained data.

AMINO ACID SYNTHESIS IN PRESENT-DAY HYDROTHERMAL SYSTEMS

Disequilibrium in and around submarine hydrothermal systems provides sufficient chemical energy for the synthesis, from inorganic precursors, of numerous aqueous organic compounds (Shock and Schulte, 1998), including many of the common amino acids (Amend and Shock, 1998). Presently, mixing of cold, oxidized sea water with hot, reduced hydrothermal fluids establishes disequilibrium in the subsurface (McCollom and Shock, 1997) that supports the metabolism of autotrophic organisms and consequently drives these formation reactions. It has been shown (Amend and Shock, 1998) that the autotrophic synthesis of all 20 common amino acids (table 3-2) is energetically favored in a mixed hydrothermal solution at 100°C relative to synthesis in surface sea water. In fact, the net synthesis reactions from aqueous CO_2, H_2, NH_4^+, and H_2S of 11 amino acids are exergonic at and near 100°C in current submarine hydrothermal systems. Although the standard state Gibbs free energies (ΔG_r^0) of these reactions (r) are strongly temperature dependent, it is the significant reducing potential of hydrothermal fluids that dominates the overall reaction energetics.

To quantify these energetics, Amend and Shock (1998) calculated values of the overall Gibbs free energies (ΔG_r) of net amino acid synthesis reactions from the equation

$$\Delta G_r = \Delta G_r^0 + RT \ln Q_r, \tag{3-1}$$

where R and T represent the gas constant and temperature in Kelvin, respectively, and Q_r denotes the activity product. Values of Q_r required to evaluate ΔG_r with eqn 3-1 can be determined from

$$Q_r = \Pi\, a_i^{\upsilon_{i,r}}, \tag{3-2}$$

where a_i stands for the activity of the ith species, and $\upsilon_{i,r}$ represents the stoichiometric reaction coefficient of the ith species in reaction r, which is negative for reactants and positive for products.

The standard state[1] term (ΔG_r^0) can be computed at any temperature and pressure by combining the *apparent* standard Gibbs free energies of formation[2] (ΔG_i^0) of species i in reaction r at these conditions in accord with the expression

$$\Delta G_r^0 = \Sigma \upsilon_{i,r} \Delta G_i^0. \tag{3-3}$$

Values of ΔG_i^0 at the temperature and pressure of interest for the species in the reactions given in table 3-2 and in other reactions throughout this Chapter are readily calculated using the SUPCRT92 software package (Johnson et al., 1992), which uses thermodynamic data from Helgeson et al. (1978) for minerals, and from Shock and Helgeson (1988), Shock et al. (1989, 1992, 1997), and Amend and Helgeson (1997a,b) for gases and aqueous species. For all 20 reactions in table 3-2, values of ΔG_r^0 at 250 bar and temperatures from 0–200°C are negative, except for a few reactions at the upper temperature limit (fig. 3-1). It can also be seen in figure 3-1 that the values of ΔG_r^0 exhibit strong temperature dependencies, increasing with increasing temperature.

Evaluating $RT \ln Q_r$ in eqn 3-1 for the reactions in table 3-2 requires the activities of the reactants and products to be known at the environmental conditions of interest. As an example, we describe the approach used for the net synthesis of aqueous leucine [$(CH_3)_2CHCH_2CHNH_2COOH$)]. Using activities of the aqueous species in sea water and vent fluids, we calculated Q_r in a present-day mixed hydrothermal solution (Amend and Shock, 1998) with the expression

TABLE 3-2 Net amino acid synthesis reactions[a]

Amino acid (abbrev.)	Chemical formula	Net amino acid synthesis reaction
alanine (Ala)	$C_3H_7NO_2$	$3CO_2(aq) + NH_4^+ + 6H_2(aq) \rightarrow Ala(aq) + H^+ + 4H_2O(liq)$
arginine$^+$ (Arg$^+$)	$C_6H_{15}N_4O_2^+$	$6CO_2(aq) + 4NH_4^+ + 11H_2(aq) \rightarrow Arg^+ + 3H^+ + 10H_2O(liq)$
asparagine (Asn)	$C_4H_8N_2O_3$	$4CO_2(aq) + 2NH_4^+ + 6H_2(aq) \rightarrow Asn(aq) + 2H^+ + 5H_2O(liq)$
aspartate$^-$ (Asp$^-$)	$C_4H_6NO_4^-$	$4CO_2(aq) + NH_4^+ + 6H_2(aq) \rightarrow Asp^- + 2H^+ + 4H_2O(liq)$
cysteine (Cys)	$C_3H_7NO_2S$	$3CO_2(aq) + NH_4^+ + H_2S(aq) + 5H_2(aq) \rightarrow Cys(aq) + H^+ + 4H_2O(liq)$
glutamate$^-$ (Glu$^-$)	$C_5H_8NO_4^-$	$5CO_2(aq) + NH_4^+ + 9H_2(aq) \rightarrow Glu^- + 2H^+ + 6H_2O(liq)$
glutamine (Gln)	$C_5H_{10}N_2O_3$	$5CO_2(aq) + 2NH_4^+ + 9H_2(aq) \rightarrow Gln(aq) + 2H^+ + 7H_2O(liq)$
glycine (Gly)	$C_2H_5NO_2$	$2CO_2(aq) + NH_4^+ + 3H_2(aq) \rightarrow Gly(aq) + H^+ + 2H_2O(liq)$
histidine (His)	$C_6H_9N_3O_2$	$6CO_2(aq) + 3NH_4^+ + 10H_2(aq) \rightarrow His(aq) + 3H^+ + 10H_2O(liq)$
isoleucine (Ile)	$C_6H_{13}NO_2$	$6CO_2(aq) + NH_4^+ + 15H_2(aq) \rightarrow Ile(aq) + H^+ + 10H_2O(liq)$
leucine (Leu)	$C_6H_{13}NO_2$	$6CO_2(aq) + NH_4^+ + 15H_2(aq) \rightarrow Leu(aq) + H^+ + 10H_2O(liq)$
lysine$^+$ (Lys$^+$)	$C_6H_{15}N_2O_2^+$	$6CO_2(aq) + 2NH_4^+ + 14H_2(aq) \rightarrow Lys^+ + H^+ + 10H_2O(liq)$
methionine (Met)	$C_5H_{11}NO_2S$	$5CO_2(aq) + NH_4^+ + H_2S(aq) + 11H_2(aq) \rightarrow Met(aq) + H^+ + 8H_2O(liq)$
phenylalanine (Phe)	$C_9H_{11}NO_2$	$9CO_2(aq) + NH_4^+ + 20H_2(aq) \rightarrow Phe(aq) + H^+ + 16H_2O(liq)$
proline (Pro)	$C_5H_9NO_2$	$5CO_2(aq) + NH_4^+ + 11H_2(aq) \rightarrow Pro(aq) + H^+ + 8H_2O(liq)$
serine (Ser)	$C_3H_7NO_3$	$3CO_2(aq) + NH_4^+ + 5H_2(aq) \rightarrow Ser(aq) + H^+ + 3H_2O(liq)$
threonine (Thr)	$C_4H_9NO_3$	$4CO_2(aq) + NH_4^+ + 8H_2(aq) \rightarrow Thr(aq) + H^+ + 5H_2O(liq)$
tryptophan (Trp)	$C_{11}H_{12}N_2O_2$	$11CO_2(aq) + 2NH_4^+ + 23H_2(aq) \rightarrow Trp(aq) + 2H^+ + 20H_2O(liq)$
tyrosine (Tyr)	$C_9H_{11}NO_3$	$9CO_2(aq) + NH_4^+ + 19H_2(aq) \rightarrow Tyr(aq) + H^+ + 15H_2O(liq)$
valine (Val)	$C_5H_{11}NO_2$	$5CO_2(aq) + NH_4^+ + 12H_2(aq) \rightarrow Val(aq) + H^+ + 8H_2O(liq)$

[a] At the temperatures and pHs of the mixed hydrothermal solutions considered here, 16 of the 20 amino acids have a net charge of 0, 2 are present as anions (aspartate$^-$ and glutamate$^-$), and 2 are present as cations (arginine$^+$ and lysine$^+$).

$$Q_r = \frac{a_{Leucine} a_{H^+} (a_{H_2O})^{10}}{(a_{CO_2})^6 a_{NH_4^+} (a_{H_2})^{15}}. \quad (3\text{-}4)$$

Values of the activities of aqueous H^+, CO_2, NH_4^+ H_2, H_2S, and leucine as a function of temperature corresponding to different degrees of mixing are depicted as curves in figure 3-2. Combining ΔG_r^0 and Q_r in eqn 3-1 yields values of ΔG_r such as those plotted versus temperature in figure 3-3 for the leucine synthesis reaction in a hydrothermal system. It can be seen in this figure that at low temperatures, where the mixed hydrothermal solution is dominated by sea water, the reaction is highly endergonic ($\Delta G_r > 0$) At temperatures between approximately 40° and 160°C, however, this reaction is exergonic and can release energy in excess of 100 kJ per mole leucine produced. At elevated temperatures, the high value of a_{H_2} in the vent fluid provides the thermodynamic drive to reduce CO_2 to amino acids. The slight increase in ΔG_r between ~40 and 200°C can be attributed predominantly to the increase in ΔG_r^0 with increasing temperature.

Because sea water and vent fluids have been extensively sampled and analyzed, the activities of the aqueous species involved in the reactions given in table 3-2 are well constrained. Such tight constraints are not possible for early Earth conditions. Nevertheless, owing to extensive modeling (Levine et al., 1982; Canuto et al., 1983; Kasting et al., 1983; Stevenson, 1983; Walker, 1983, 1985; Holland, 1984; Kasting and Ackerman, 1986; Zahnle et al., 1988; Grotzinger and Kasting, 1993; Kasting, 1993a,b,c; Russell and Hall, 1997; Sagan and Chyba, 1997; Brandes et al., 1998; McCollom and Shock, 1998), limits can now be placed on the compositions of the early Earth atmosphere, ocean, and even hydrothermal fluids. Below, we incorporate these modeling efforts into an analysis of amino acid synthesis energetics during the Hadean Eon. We also discuss the consequences of certain proposed atmospheric compositions for these energetics.

GEOCHEMICAL CONDITIONS DURING THE HADEAN EON

The timing of the first appearance of life on Earth has been discussed extensively (Schopf and Packer, 1987; Schidlowski, 1988; Pace, 1991; Schopf, 1993; de Duve, 1995; Mojzsis et al., 1996; Russell and Hall, 1997; Rosing, 1999). Nevertheless, it remains unknown. The oldest, well-preserved microfossils date back to ~3.5 billion years (Ga) ago (Schopf, 1993). However, carbon-isotope ratios in the oldest known sediment sequences in West Greenland (Schidlowski, 1988; Mojzsis et al., 1996) indicate biological activity on Earth as early as 3.8 Ga ago. If so, the emergence of the first self-replicating cell

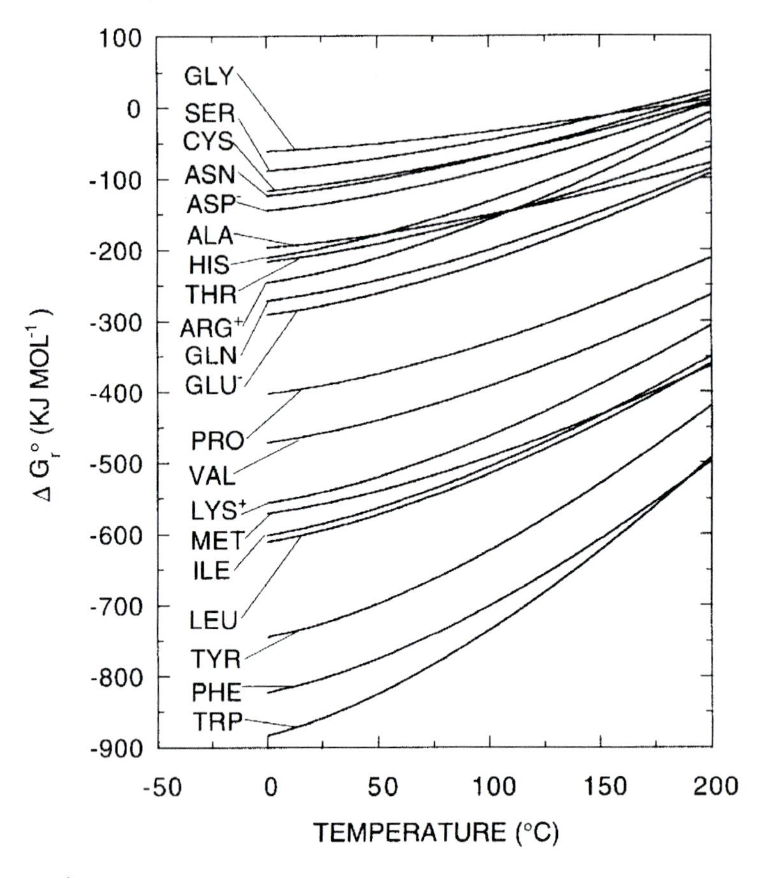

Fig. 3-1. Values of ΔG_r^0 for the net amino acid synthesis reactions (table 3-2) in aqueous solution as functions of temperature at 250 bar. This pressure was chosen to approximate conditions in hydrothermal systems. The key to the amino acid abbreviations used here is given in table 3-2.

occurred during the Hadean Eon (~4.6–3.8 Ga). This would have been contemporaneous with the heaviest bombardment of Earth by comets and asteroids (Chyba, 1989; Sleep et al., 1989). In part because of the protection afforded against such impacts, but also because of favorable energetics for organic synthesis reactions (Ingmanson and Dowler, 1977; Shock, 1990, 1992; Schulte and Shock, 1995; Amend and Shock, 1998; Shock and Schulte, 1998), it has been argued that life may have emerged in submarine hydrothermal systems, where entrained sea water and vent fluids mix (Corliss et al., 1981; Baross and Hoffman, 1985; Shock, 1996; Huber and Wächtershäuser, 1997, 1998; Russell and Hall, 1997).

Ocean chemistry

The constraints on the composition of the Hadean ocean are intimately coupled to the chemistry of the atmosphere. Arguments to the contrary notwithstanding (Ohmoto et al., 1993; Ohmoto, 1997) it is generally agreed that the Earth's Hadean atmosphere contained little free O_2 and was dominated by CO_2 and N_2, with significant traces of H_2, CO, H_2O, NH_3, and reduced sulfur species (Walker, 1985; Kasting, 1993a,b,c). The current popular paradigm states that the partial pressures of CO_2 and N_2 in the atmosphere may have been 10 and 1 bar, respectively (Walker, 1985; Kasting, 1993b). On the basis of radiative–convective climate modeling and estimates of very high CO_2 levels in the early atmosphere, Kasting and Ackerman (1986) suggested that the surface would have had a temperature of ~85–110°C.

The elevated concentration of CO_2 in the atmosphere would also have directly affected the ocean chemistry. For example, in accord with the reaction

$$Ca^{2+} + 2HCO_3^- \leftrightarrow \text{calcite} + CO_2(g) + H_2O(liq) \quad (3\text{-}5)$$

Kasting (1993c) determined the activity of HCO_3^- in an ocean saturated with calcite ($CaCO_3$) and in equilibrium with a 10 bar CO_2 atmosphere. In this calculation, the activity of Ca^{2+} was assumed to be equivalent to the present value in sea water. Here,

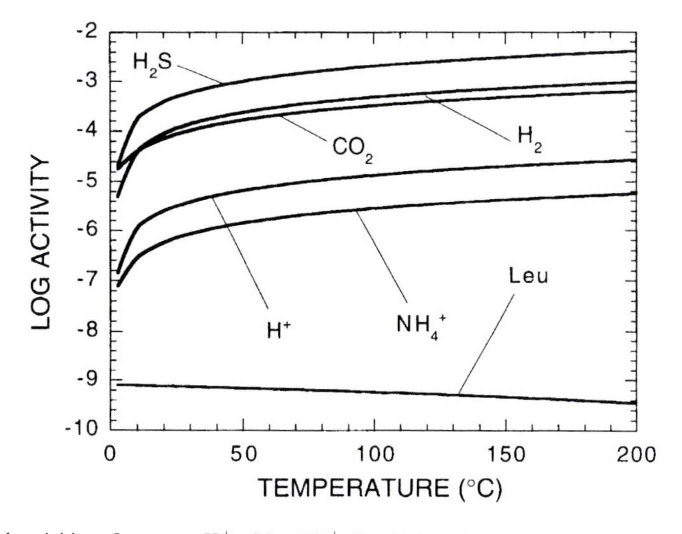

Fig. 3-2. Calculated activities of aqueous H^+, CO_2, NH_4^+, H_2, H_2S, and leucine as a function of temperature in a mixed hydrothermal solution (see text).

we re-evaluate $a_{HCO_3^-}$ by computing the equilibrium constant, K, for reaction (5) at 85°C (Kasting used 25°C).[3] It can be seen in figure 3-4 that using thermodynamic data at 85°C instead of 25°C lowers $a_{HCO_3^-}$ at a fixed $a_{Ca^{2+}}$ by almost an order of magnitude. At $a_{Ca^{2+}} = \sim 1.8 \times 10^{-3}$ (broken line in figure 3-4)[4], the value of $a_{HCO_3^-}$ is calculated to be 1.9×10^{-2} (for reference, Kasting (1993c) obtained a 0.13 mol/l bicarbonate concentration).

The pH of the ocean can be determined simultaneously from the carbonate equilibrium written as

$$CO_2(g) + H_2O(liq) \leftrightarrow HCO_3^- + H^+. \quad (3\text{-}6)$$

At 85°C, log K for this reaction is −8.3, which, when combined with a fugacity of CO_2 (f_{CO_2}) equal to 10, $a_{HCO_3^-}$ as calculated above, and water activity of unity, yields a pH of 5.6, consistent with Kasting (1993c). It should be pointed out, however, that contrary to assertions made by Kasting (1993c), CO_2, not HCO_3^-, would have been the dominant inorganic carbon species under the geochemical conditions for the Hadean ocean described

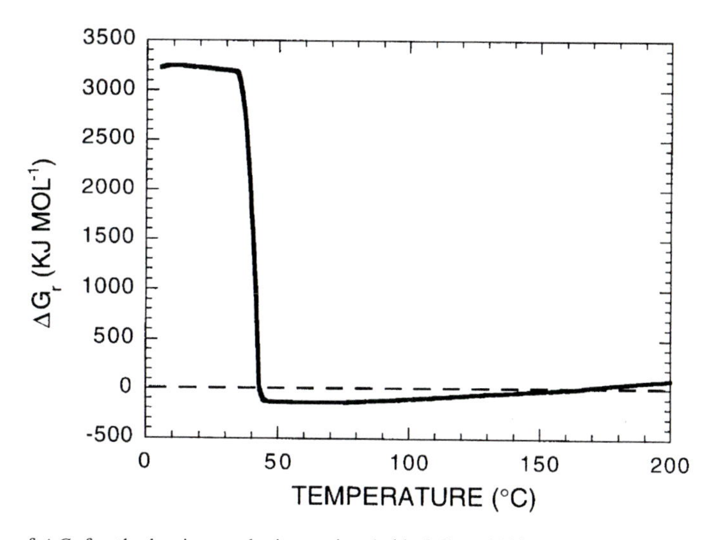

Fig. 3-3. Values of ΔG_r for the leucine synthesis reaction (table 3-2) at 250 bar as a function of temperature in present-day hydrothermal solutions. The dashed line represents equilibrium ($\Delta G_r = 0$).

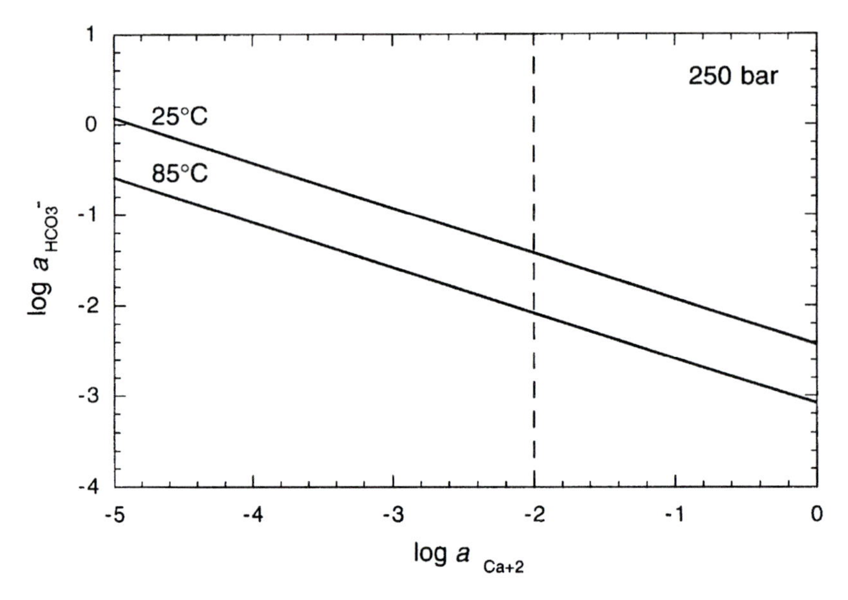

Fig. 3-4. Log $a_{HCO_3^-}$ vs. log $a_{Ca^{2+}}$ at 250 bar and 25 and 85°C. The broken line represents the present-day value of $a_{Ca^{2+}}$ in sea water.

here. In fact, at 85°C and in equilibrium with $f_{CO_2} = 10$ we calculate a value of a_{CO_2} in the ocean equal to 9.1×10^{-2}.

To calculate the activity of H_2 in the Hadean atmosphere, Kasting (1993c) uses an approach developed by Walker (1977) to "balance the hydrogen budget". In this model, the source of hydrogen is volcanic outgassing and the sink is escape to space. By incorporating mixing models in the stratosphere and diffusion rates across the homopause, as well as assuming that the rate of outgassing on the early Earth was the same as today, Kasting evaluated a constant concentration of H_2 throughout the atmosphere, consistent with a fugacity (f_{H_2}) of $\sim 1.8 \times 10^{-3}$. This would have resulted in an extremely reducing ocean with an activity of H_2 equal to $\sim 1.1 \times 10^{-6}$ and, consequently, a phenomenal potential for organic synthesis (see below).

An ocean buffered at $a_{H_2} = 1.1 \times 10^{-6}$ would yield a very different speciation of sulfur than the sulfate-dominated oxidized sea water of today. It can be calculated that at equilibrium, under such reducing conditions, H_2S would be the dominant aqueous sulfur species, and the activity of SO_4^{2-} at 85°C would be approximately five orders of magnitude less than that of H_2S. The consequences on the synthesis energetics of cysteine and methionine (the two sulfurous amino acids) are discussed below. Although a sulfide ocean rather than one dominated by sulfate may well have existed during the Hadean, one of the models described below for calculating energetics includes constraints on fluid compositions consistent with sulfate-dominated sea water. In these calculations, we set the value of a_{H_2} (7.2×10^{-10}) to be consistent with $a_{SO_4^{2-}}/a_{H_2S}$ equal to 100.

Although the dominant nitrogen species in the Hadean atmosphere was undoubtedly N_2, Brandes et al. (1998) use experimental results to argue that mineral-catalyzed reduction of N_2, NO_2^-, and NO_3^- at high temperatures in crustal and marine hydrothermal systems may have served as a major source of NH_3 and/or NH_4^+ for the ocean and atmosphere. They further estimate that the ocean could have been a reservoir of 5×10^7 mol reduced nitrogen. Assuming an average ocean depth of 3.5 km across the entire surface of the Earth, we calculate that the corresponding activity of NH_4^+ may have been as high as 2.8×10^{-4}.

In the calculations of reaction energetics for amino acid synthesis described below, we consider the consequences of three different ocean compositions (table 3-3). All three have the same values of pH and activities of aqueous CO_2 and NH_4^+ consistent with equilibrium arguments formulated above. However, the values of a_{H_2} and a_{H_2S} differ significantly. The value of a_{H_2} in Ocean-1 is in equilibrium with the f_{H_2} proposed by Kasting (1993a); a_{H_2} in Ocean-2 is set arbitrarily at a value consistent with $a_{CO_2}/a_{CH_4} = 100$ (see below); and the value of a_{H_2} in Ocean-3 is set by the sulfate/sulfide ratio described above. Values of a_{H_2S} in the three oceans are equilibrium values corresponding to their respective redox states, assuming total sulfur in the Hadean ocean was the same as it is today. Ocean-1 and Ocean-2 are too reduced for any appreciable sulfate activity; only the oxidation

TABLE 3-3 Activities of aqueous inorganic species in three proposed Hadean ocean compositions and in a model submarine hydrothermal fluid

	Ocean-1	Ocean-2	Ocean-3	400°C Hydrothermal fluid
a_{CO_2}	9.1×10^{-2}	9.1×10^{-2}	9.1×10^{-2}	1.2×10^{-5}
pH	5.6	5.6	5.6	6.9
a_{H_2}	1.1×10^{-6}	5.3×10^{-8}	7.2×10^{-10}	0.16
$a_{NH_4^+}$	2.8×10^{-4}	2.8×10^{-4}	2.8×10^{-4}	3.2×10^{-3}[a]
a_{H_2S}	2.3×10^{-2}	2.3×10^{-2}	3.3×10^{-5}	7.3×10^{-3}

[a] This value refers to the activity of NH_3(aq), which is the stable species at the pH and temperature of this fluid.

state in Ocean-3 is high enough to lower the equilibrium value of a_{H_2S}.

Disequilibrium in the atmosphere and ocean

Atmospheric models for the Hadean Earth are far from thermodynamic equilibrium, and are maintained in their disequilibrium states by photochemical reactions (Levine et al., 1982; Canuto et al., 1983; Kasting et al., 1983; Sagan and Chyba, 1997). The fugacity of H_2 is so high in these atmospheres that CO_2 should be reduced to methane (Kasting, 1993c). However, any methane that may be formed is rapidly oxidized to CO_2 (or CO) in reactions involving hydroxyl radicals generated by photolysis of H_2O vapor. The input of solar energy maintains the far-from-equilibrium composition of the atmosphere, despite the thermodynamic drive to form methane.

Likewise, the values of a_{CO_2} and a_{H_2} given in table 3-3 for Ocean-1 are far from equilibrium with respect to methane. Equilibrium ($\Delta G_r = 0$) at 85°C for Ocean-1 dictates that the value of a_{CH_4}/a_{CO_2} would be ~1600. This is determined in accord with the reaction

$$CO_2(aq) + 4H_2(aq) \leftrightarrow CH_4(aq) + 2H_2O(liq) \quad (3\text{-}7)$$

for which the law of mass action can be expressed as

$$K = \frac{a_{CH_4}(a_{H_2O})^2}{a_{CO_2}(a_{H_2})^4}. \quad (3\text{-}8)$$

At 85°C and 250 bar, K for reaction (7) is equal to 1.2×10^{27} In other words, if there was a way of synthesizing methane from aqueous CO_2 and H_2 in equilibrium with the atmosphere proposed by Kasting, the progress of reaction (7) would lower the overall free energy of the system to the point at which the activity of CH_4(aq) was 1600 times that of CO_2(aq).

Below the atmosphere/ocean interface, the effectiveness of photochemical reactions diminishes enormously. Therefore, the mechanism that destroys methane in the atmosphere would be largely absent from the oceans. Nevertheless, reduction of CO_2 to methane is kinetically inhibited on the Earth at temperatures such as those proposed for the Hadean ocean, despite the enormous thermodynamic drive for reaction to occur (3-7) (Shock, 1992). As an example, partial pressures of methane in sedimentary basins are up to four orders of magnitude below their equilibrium value given the concentrations of CO_2(aq) and prevailing oxidation states (Shock, 1988, 1989). On the other hand, the concentrations of many petroleum hydrocarbons and organic acids dissolved in basin brines are consistent with metastable equilibrium involving those same concentrations of CO_2(aq) and partial pressures of H_2 (Helgeson et al., 1993; Shock, 1994). Thus, abundant evidence from nature shows that a metastable equilibrium between CO_2 and many organic compounds is attained, even if a stable equilibrium with CH_4 is kinetically inhibited.

Inhibition of methane formation opens the door to the synthesis of metastable organic molecules with an oxidation state intermediate between CO_2 and CH_4. Shock and Schulte (1998) have shown that disequilibrium states of this kind could result in the formation of numerous aqueous organic compounds, dominated by organic acids; consequently, the concentrations of aqueous inorganic carbon species, such as CO_2 or HCO_3^- would be below the detection limit. In fact, a metastable equilibrium in Ocean-1 at 85°C would be dominated by a concentrated mixture of long-chain organic acids.

Kasting and others have argued persuasively for the existence of photochemical destruction of reduced gases in the Hadean atmosphere, without considering the consequences of the resulting redox disequilibrium in the coexisting Hadean ocean. We have adopted constraints consistent with Kasting's model of the atmosphere for our Ocean-1, but note with caution that this composition relies on the assumption that a metastable equilibrium involving aqueous CO_2, HCO_3^- and organic compounds is somehow avoided. We are unaware of geologic settings in which it can be shown that this is the case in the temperature range adopted by Kasting and Ackerman (1986). Alternatively, we suggest that rampant, spontaneous, metastable production of organic compounds could have been avoided in the Hadean ocean if the a_{H_2} was considerably less than 10^{-6}. Therefore, we adopt, in the Ocean-2 model, a value for a_{H_2} in Hadean sea water equal to 5.3×10^{-8} which is consistent with a CO_2-dominant ocean ($a_{CO_2}/a_{CH_4} = 100$) As described above, even this ocean would contain little sulfate at equilibrium, so we constructed the Ocean-3 model, which would have an a_{H_2} low enough to stabilize sulfate. The equilibrium ratio of a_{CO_2}/a_{CH_2} in Ocean-3 would be $\sim 3 \times 10^9$.

Chemistry of hydrothermal fluids

Mixing of sea water and hydrothermal fluids in and around submarine hydrothermal systems on the Hadean Earth established chemical disequilibria. Constraints on the composition of the hydrothermal fluid used in this study (table 3-3) are based primarily on models of water–rock interactions (McCollom and Shock, 1998) in which the consequences of sea water reacting with an olivine-rich gabbro were evaluated at temperatures from 300 to 900°C and pressures from 250 to 2000 bar. Olivine-rich gabbro is the most common rock type in present-day lower oceanic crust; it is more mafic than basalt and may represent the type of rock composition with which submarine hydrothermal fluids equilibrated in the Hadean.

In the absence of more reliable data on the composition of the Hadean lower oceanic crust, we have adopted the resultant composition of a hydrothermal fluid generated at 400°C and 750 bar in equilibrium with olivine gabbro, which has a pH of 6.9 and a value of a_{H_2} equal to 0.16 (McCollom and Shock, 1998). Similarly high activities of H_2 were calculated for 350°C peridotite-hosted submarine hydrothermal fluids (Wetzel and Shock, 2000). These H_2 activities seem realistic in view of the high concentrations of dissolved H_2 measured in several active hydrothermal systems along the mid-Atlantic ridge (J. L. Charlou, personal communication, 1999). The stable equilibrium ratio of a_{CH_4} to a_{CO_4} can be calculated for reaction 3-7 using the value of a_{H_2} (0.16) generated in the model, assuming equilibrium among H_2O, O_2, and H_2, and the value of K for reaction 3-7 (1.32×10^7) at 400°C and 750 bar. From this ratio (9.55×10^3), and the assumption that CO_2 in sea water is the only source of carbon in the hydrothermal fluid, we compute that a_{CO_2} in the fluid at 400°C and 750 bar is 1.2×10^{-5}. Our mixing model then assumes that this hydrothermal fluid is isothermally and isochemically transported to an environment at shallower depth, where it can interact with sea water at 250 bar.

If equilibrium is reached, the activity of NH_3 in a rock-buffered hydrothermal fluid at 400°C can be equal to 3.2×10^{-3} (Shock, 1992). This value is qualitatively consistent with the conclusions reached by Brandes et al. (1998), i.e., that the activity of NH_3 on the abiotic Earth was greater in hydrothermal environments than anywhere else, including the oceans. Unlike the ammonia in Hadean sea water, which would have occurred predominantly as NH_4^+ at 85°C and a pH of 5.6, NH_3 would have predominated in hydrothermal fluids at 400°C and a pH of 6.9.

The placing of constraints on the activity of H_2S in hydrothermal fluids presents a considerable challenge. In the absence of reliable evidence to the contrary, we assume that the value of a_{H_2S} was approximately the same during the Hadean as it is today. On the basis of reported compositions of vent fluids at 21° N on the East Pacific Rise (Von Damm et al., 1985; Von Damm, 1990) and speciation calculations carried out using the computer program EQ3 (Wolery, 1992), with the same thermodynamic data used in the rest of this study, we determined a value for a_{H_2S} equal to 7.3×10^{-3}. At 400°C and pH 6.9, H_2S is the dominant species, with lesser amounts of HS^-.

Subsurface mixing of sea water and hydrothermal fluid

In modern submarine vent environments, mixing of sea water and hydrothermal fluids, combined with sluggish kinetics for oxidation/reduction reactions, provides geochemical energy sources for chemolithoautotrophic microorganisms (McCollom and Shock, 1997). In analogous systems on the Hadean Earth, such as those considered here, the chemical disequilibrium states due to mixing provide potential energy for organic synthesis reactions (Shock and Schulte, 1998). The temperature and composition of the resultant solution, and consequently the state of disequilibrium, depend on the mixing ratio of the two aqueous fluids. For example, a 100°C mixed hydrothermal solution can be generated if 20 kg 85°C sea water is mixed with 1 kg 400°C hydrothermal fluid. Similarly, a 200°C solution results from mixing ~1.7 kg 85°C sea water per kg hydrothermal fluid.

The calculated activities of aqueous CO_2, H_2, H^+, NH_4^+, and H_2S for mixed hydrothermal solutions at 100 and 200°C are given in table 3-4 for the three different proposed ocean compositions from table 3-3. With the exception of the value for a_{H_2S} in the mixture generated with Ocean-3, the compositions of the three mixed hydrothermal solutions are nearly identical; this is so even though the values of a_{H_2} in the three ocean models differ by more than three orders of magnitude. The high activity of H_2 in the 400°C hydrothermal fluid simply overwhelms those of the oceans. This comes as no surprise, because lower H_2 concentrations in present-day hydrothermal fluids dominate the redox state of mixtures with sea water at temperatures down to ~40°C (McCollom and Shock, 1997). In the three ocean models considered in the present study, sea water dilutes the H_2 and does not contribute to its "budget" in the mixed solutions. The only demonstrable differences in calculations with Ocean-1 or -2 rather than Ocean-3 are the values of ΔG_r for the synthesis of cysteine and methionine. The values of ΔG_r for these two sulfurous amino acids would be more negative by 12.5 and 5.5 kJ/mol at 100 and 200°C, respectively, if we used Ocean-1 or -2.

TABLE 3-4 Activities of aqueous inorganic species in mixed hydrothermal solutions at 100 and 200°C

	100°C			200°C		
	Ocean-1 + hydrothermal fluid	Ocean-2 + hydrothermal fluid	Ocean-3 + hydrothermal fluid	Ocean-1 + hydrothermal fluid	Ocean-2 + hydrothermal fluid	Ocean-3 + hydrothermal fluid
a_{CO_2}	8.6×10^{-2}	8.6×10^{-2}	8.6×10^{-2}	5.8×10^{-2}	5.8×10^{-2}	5.8×10^{-2}
pH	5.6	5.6	5.6	5.7	5.7	5.8
a_{H_2}	7.8×10^{-3}	7.8×10^{-3}	7.8×10^{-3}	6.0×10^{-2}	6.0×10^{-2}	6.0×10^{-2}
$a_{NH_4^+}$	4.2×10^{-4}	4.2×10^{-4}	4.2×10^{-4}	1.3×10^{-3}	1.3×10^{-3}	1.3×10^{-3}
a_{H_2S}	2.3×10^{-2}	2.3×10^{-2}	3.8×10^{-4}	1.7×10^{-2}	1.7×10^{-2}	2.7×10^{-3}

To ensure that the dominant species of each inorganic compound were used in the amino acid synthesis reactions, we carried out speciation calculations as functions of temperature and pH. It can be seen in figure 3-5 that at the pH values of the mixed solution over the entire temperature range considered, CO_2(aq) dominates over HCO_3^- and H_2S(aq) dominates over HS^-. The line representing equal activities of NH_4^+ and NH_3, however, crosses the pH values of the mixed solutions, as shown in this figure. At the values of pH for temperatures ≤ 210°C, NH_4^+ dominates over NH_3, but the reverse is true above this temperature. However, since the temperatures of the mixed hydrothermal solutions considered in this study are 85–200°C, only values of $a_{NH_4^+}$ were used in the amino acid synthesis calculations.

CALCULATION OF REACTION ENERGETICS FOR AMINO ACID SYNTHESIS

The overwhelming effect of the hydrothermal fluid on the composition of cooler solutions generated by mixing with sea water, as summarized in table 3-4, means that variable Hadean ocean compositions

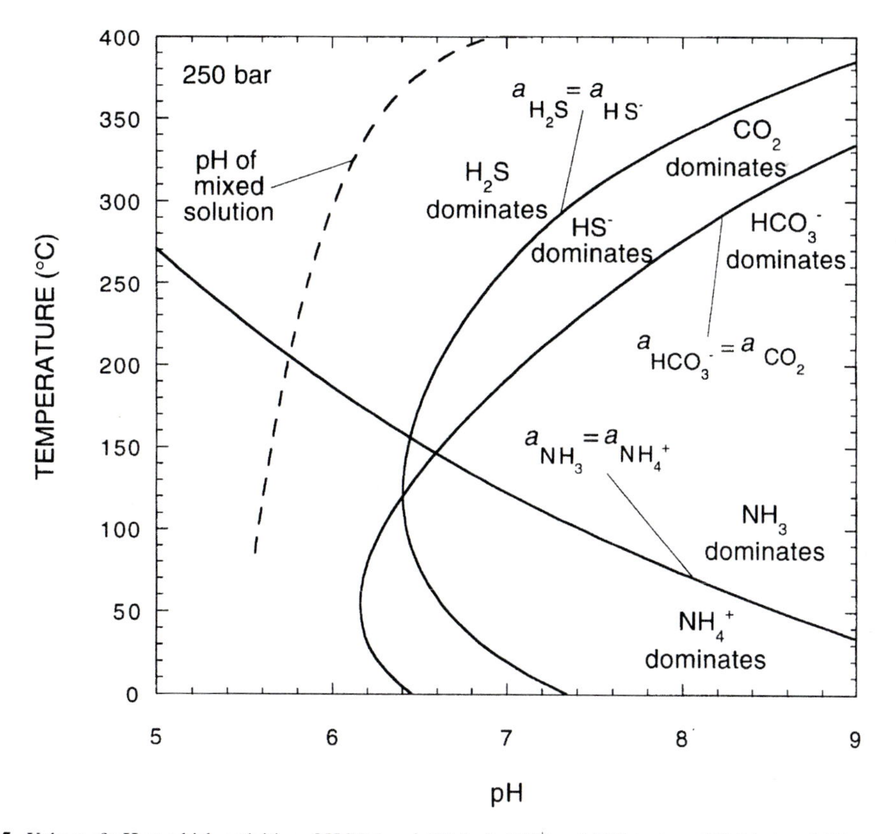

Fig. 3-5. Values of pH at which activities of HCO_3^- and CO_2(aq), NH_4^+ and NH_3(aq), and H_2S(aq) and HS^- would be equal (solid curves), and pH values for the mixed hydrothermal solution (broken curve) as functions of temperature at 250 bar.

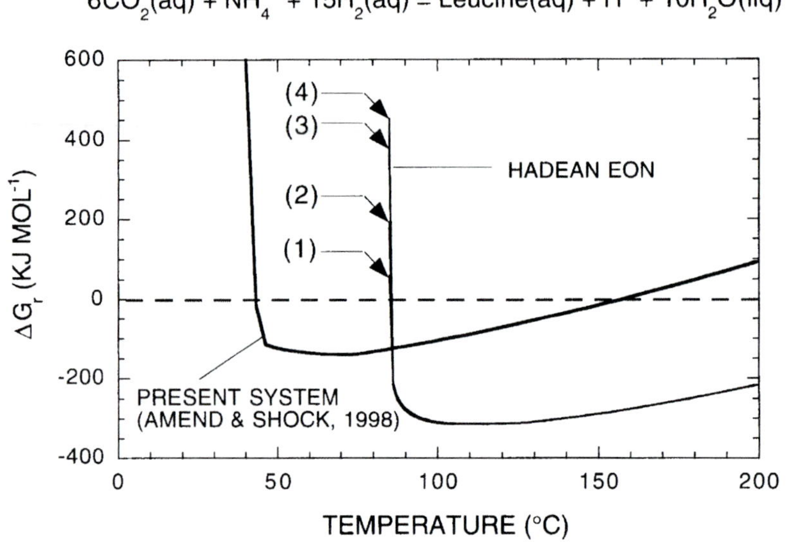

Fig. 3-6. Values of ΔG_r for the leucine synthesis reaction (table 3-2) at 250 bar as a function of temperature in the present-day system and in mixed hydrothermal solutions generated from 400°C fluid and four (labeled 1–4) models of Hadean ocean compositions (see text). The numbered arrows point to the values of ΔG_r computed for leucine synthesis in the corresponding Hadean sea water without mixing. The broken line represents equilibrium. A leucine activity of 10^{-6} was used to carry out these calculations.

have little if any effect on the potential for amino acid synthesis in Hadean hydrothermal systems. This is demonstrated by the example of leucine synthesis shown in figure 3-6. Values of ΔG_r for the leucine synthesis reaction (table 3-2) as a function of temperature in mixed solutions generated from the compositions of the 400°C hydrothermal fluid and the three ocean models given in table 3-3 are shown as curves in this figure; for reference, the curve for the present from figure 3-3 is reproduced. In addition, a fourth Hadean ocean composition is also considered, in which the activity of NH_4^+ is arbitrarily decreased by ten orders of magnitude. It can clearly be seen in figure 3-6 that the four curves, representing four significantly different ocean compositions, are almost indistinguishable from one another. Even a change by ten orders of magnitude in the activity of one of the reactants (NH_4^+) has no discernible effect on the energetics. Leucine synthesis is endergonic at the constant activities of the oceanic models at 85°C, but even the input of minute amounts of hydrothermal fluid (represented in this figure by an increase in temperature) renders the value of ΔG_r negative. The tremendous reducing potential of the hydrothermal fluid not only drives the synthesis of leucine, but, by analogy, drives that of the other amino acids as well. It can also be noted from figure 3-6 that, after its steep initial drop, ΔG_r decreases more gradually with increasing temperature, before minimizing at ~110°C. Analogous curves for the other amino acids (not shown) show similar trends, although the absolute values will differ. At present, the input from submarine hydrothermal fluids also drives leucine synthesis, but at higher mixing ratios of hydrothermal fluid to sea water, owing to the highly oxidized composition of present-day sea water (Amend and Shock, 1998).

Values of ΔG_r as functions of amino acid activities were computed for all 20 net synthesis reactions in mixed hydrothermal solutions. For these calculations, the mixed solution with 'Ocean-3' was used. In figures 3-7 and 3-8, values of ΔG_r at 250 bar in 100 and 200°C mixed hydrothermal solutions, respectively, are plotted. It can be seen in figure 3-7 that, with the exception of Arg^+, Cys, Ser, and Asn at high activities, all values of ΔG_r are negative over the range of activities investigated. For example, even at activities of 10^{-3} for each amino acid,[5] 17 of the 20 synthesis reactions are exergonic, releasing as much as ~380 kJ mol^{-1} in the case of Phe. At 200°C and 250 bar (figure 3-8), values of ΔG_r are slightly less negative for each amino acid synthesis reaction than the values at 100°C, but most of the reactions are still exergonic at the conditions investigated. At activities of 10^{-3} for each amino acid, 12 of the 20 synthesis reactions remain exergonic, yielding, for example, ~295 kJ/mol of Phe produced. The formation from inorganic precursors of these amino acids, even at elevated activities, lowers the overall energy state of the system at the conditions likely to have prevailed

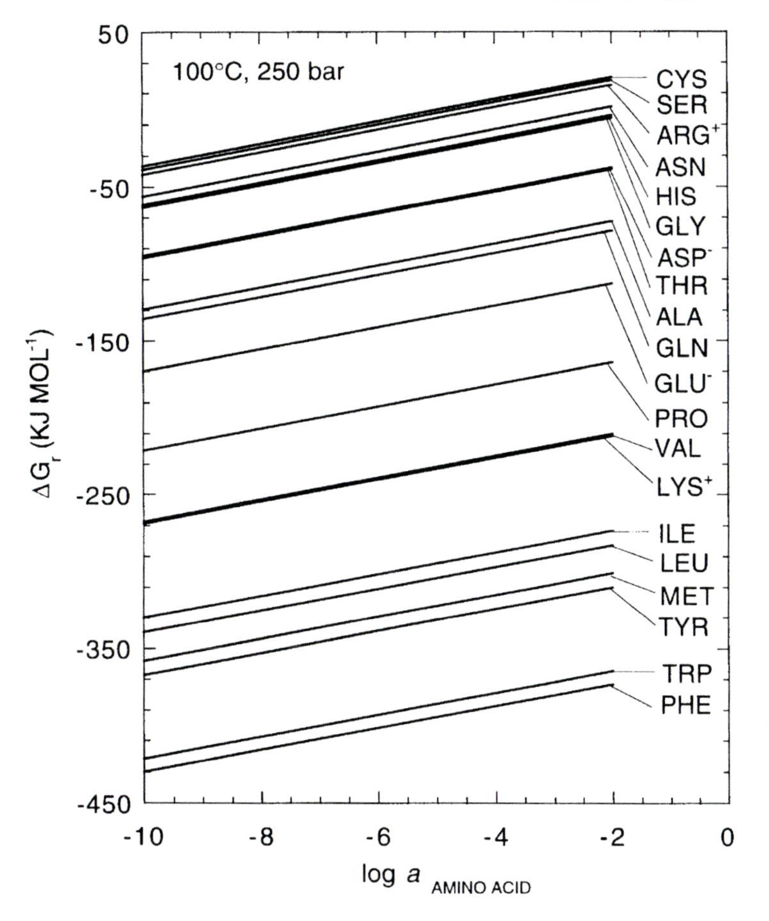

Fig. 3-7. Values of ΔG_r for the amino acid synthesis reactions given in table 3-2 at 100°C and 250 bar as a function of amino acid activity using the mixed hydrothermal solution with 'Ocean-3.'

in hydrothermal environments on the early Earth. In fact, the conditions for organic synthesis appear to have been much more favorable at this time than they are even today in environments in which hydrothermal input directly supports chemolithoautotrophic microorganisms (Amend and Shock, 1998).

AMINO ACID SYNTHESIS FROM CO

Huber and Wächtershäuser (1997, 1998) suggest that the reduction of aqueous CO to organic compounds at hydrothermal sites may have represented the first steps in a chemolithoautotrophic origin of life. It has also been proposed (Kasting, 1993c) that CO may have been a significant trace gas in the Hadean atmosphere. If so, aqueous CO in the oceans in equilibrium with the atmosphere may have served as an alternate carbon source to CO_2 in organic synthesis reactions on the early Earth. Here, we calculate values of ΔG_r in mixed hydrothermal solutions for the 20 amino acid synthesis reactions, listed in table 3-5, using CO as the carbon source. The approach used is analogous to the one described above for the reactions from CO_2. Values for ΔG_r^0 for the 20 reactions in table 3-5 at 250 bar as a function of temperature are depicted as curves in figure 3-9. It can be seen in this figure that the values of ΔG_r^0 are negative for all 20 reactions within the temperature range considered. Like the values of ΔG_r^0 represented by the curves in figure 3-1, those shown in figure 3-9 also increase with increasing temperature.

To calculate values of ΔG_r for the amino acid synthesis reactions given in table 3-5, a value of $a_{CO} = 4.9 \times 10^{-7}$ in Hadean sea water is used. This value is calculated from the equilibrium constant (4.9×10^{-4}) at 85°C and 250 bar for the reaction

$$CO(g) \leftrightarrow CO(aq) \tag{3-9}$$

and the value of $f_{CO} = 10^{-3}$ based on the suggestion by Kasting (1993b) that the concentration of CO in the Hadean atmosphere was approximately four orders or magnitude less than that of CO_2.

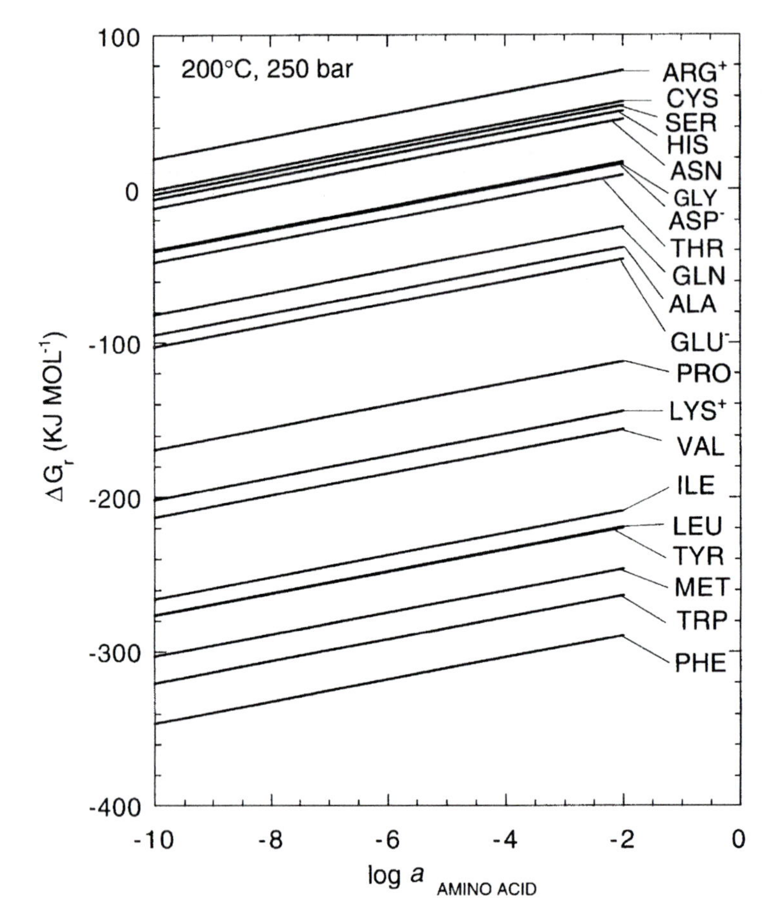

Fig. 3-8. Values of ΔG_r for the amino acid synthesis reactions given in table 3-2 at 200°C and 250 bar as a function of amino acid activity using the mixed hydrothermal solution with 'Ocean-3.'

In the absence of data to the contrary, we set the activity of CO in the highly reduced 400°C hydrothermal fluid to zero.[6] Values of ΔG_r for the leucine synthesis reaction (table 3-5) as a function of temperature in the four model mixed solutions (see text above) are plotted in figure 3-10. The curves from figure 3-6 representing ΔG_r for leucine synthesis from CO_2 are included for comparison. As with the four curves representing leucine synthesis from CO_2, those corresponding to synthesis from CO are also indistinguishable from one another. In can be seen in figure 3-10 that, at the temperatures and geochemical conditions considered, the synthesis of leucine from CO relative to CO_2 is less exergonic. By analogy, this is also true for the other amino acids.

To calculate values for ΔG_r as functions of amino acid activities for the 20 synthesis reactions (table 3-5), the composition of the mixed solution obtained with 'Ocean-3' was used, together with the CO activity described above. Values of ΔG_r at 250 bar in 100°C and 200°C mixed hydrothermal solutions are plotted in figures 3-11 and 3-12, respectively. Though not as exergonic as the amino acid synthesis reactions from CO_2 (see figures 3-7 and 3-8), most of those from CO also have negative values of ΔG_r. The exceptions are among the synthesis reactions of amino acids with relatively high nominal oxidation states (e.g., arginine, cysteine, serine, histidine, asparagine, aspartate, and glycine).

These calculations are based on a relatively conservative estimate for f_{CO} in the Hadean atmosphere. If f_{CO} were 1–2 orders of magnitude higher, the corresponding equilibrium activity in the sea water would translate into values of ΔG_r for amino acid synthesis from CO that rival those for synthesis from CO_2. Because of the intrinsic reactivity of CO relative to CO_2, net amino acid synthesis from CO may represent the more likely abiotic synthesis pathway of the two considered here.

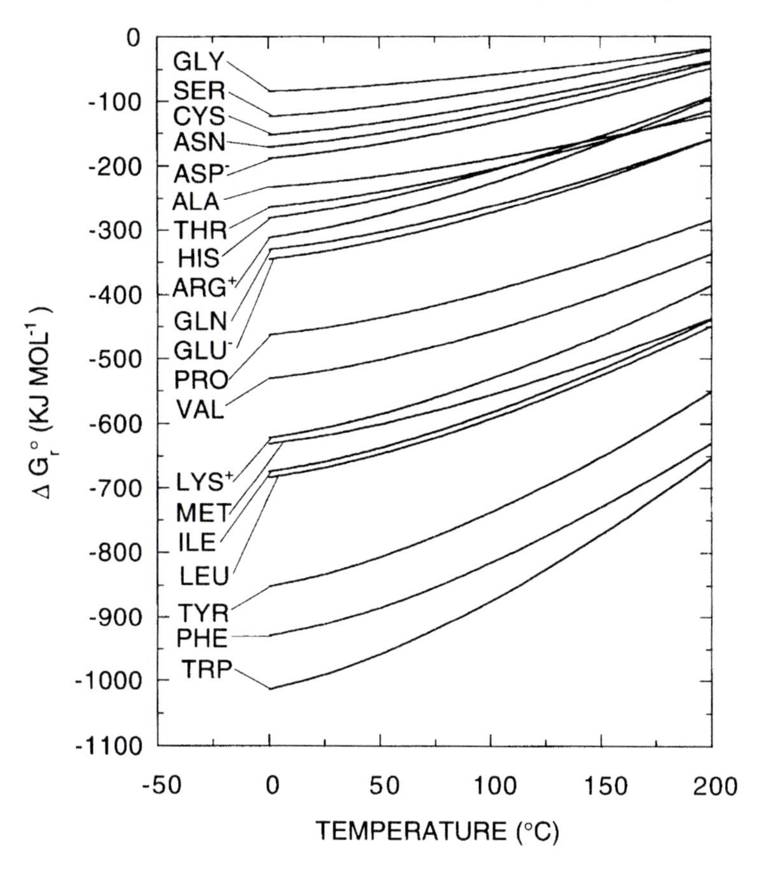

Fig. 3-9. Values of ΔG_r^0 for the net amino acid synthesis reactions with CO as the carbon source (table 3-5) in aqueous solution as a function of temperature at 250 bar. The key to the amino acid abbreviations used here is given in table 3-2.

TABLE 3-5 Net amino acid synthesis reactions using CO as the carbon source

$3CO(aq) + NH_4^+ + 3H_2(aq) \rightarrow Ala(aq) + H^+ + H_2O(liq)$
$6CO(aq) + 4NH_4^+ + 5H_2(aq) \rightarrow Arg^+ + 3H^+ + 4H_2O(liq)$
$4CO(aq) + 2NH_4^+ + 2H_2(aq) \rightarrow Asn(aq) + 2H^+ + H_2O(liq)$
$4CO(aq) + NH_4^+ + 2H_2(aq) \rightarrow Asp^- + 2H^+$
$3CO(aq) + NH_4^+ + H_2S(aq) + 2H_2(aq) \rightarrow Cys(aq) + H^+ + H_2O(liq)$
$5CO(aq) + NH_4^+ + 4H_2(aq) \rightarrow Glu^- + 2H^+ + H_2O(liq)$
$5CO(aq) + 2NH_4^+ + 4H_2(aq) \rightarrow Gln(aq) + 2H^+ + 2H_2O(liq)$
$2CO(aq) + NH_4^+ + H_2(aq) \rightarrow Gly(aq) + H^+$
$6CO(aq) + 3NH_4^+ + 4H_2(aq) \rightarrow His(aq) + 3H^+ + 4H_2O(liq)$
$6CO(aq) + NH_4^+ + 9H_2(aq) \rightarrow Ile(aq) + H^+ + 4H_2O(liq)$
$6CO(aq) + NH_4^+ + 9H_2(aq) \rightarrow Leu(aq) + H^+ + 4H_2O(liq)$
$6CO(aq) + 2NH_4^+ + 8H_2(aq) \rightarrow Lys^+ + H^+ + 4H_2O(liq)$
$5CO(aq) + NH_4^+ + H_2S(aq) + 6H_2(aq) \rightarrow Met(aq) + H^+ + 3H_2O(liq)$
$9CO(aq) + NH_4^+ + 11H_2(aq) \rightarrow Phe(aq) + H^+ + 7H_2O(liq)$
$5CO(aq) + NH_4^+ + 6H_2(aq) \rightarrow Pro(aq) + H^+ + 3H_2O(liq)$
$3CO(aq) + NH_4^+ + 2H_2(aq) \rightarrow Ser(aq) + H^+$
$4CO(aq) + NH_4^+ + 4H_2(aq) \rightarrow Thr(aq) + H^+ + H_2O(liq)$
$11CO(aq) + 2NH_4^+ + 12H_2(aq) \rightarrow Trp(aq) + 2H^+ + 9H_2O(liq)$
$9CO(aq) + NH_4^+ + 10H_2(aq) \rightarrow Tyr(aq) + H^+ + 6H_2O(liq)$
$5CO(aq) + NH_4^+ + 7H_2(aq) \rightarrow Val(aq) + H^+ + 3H_2O(liq)$

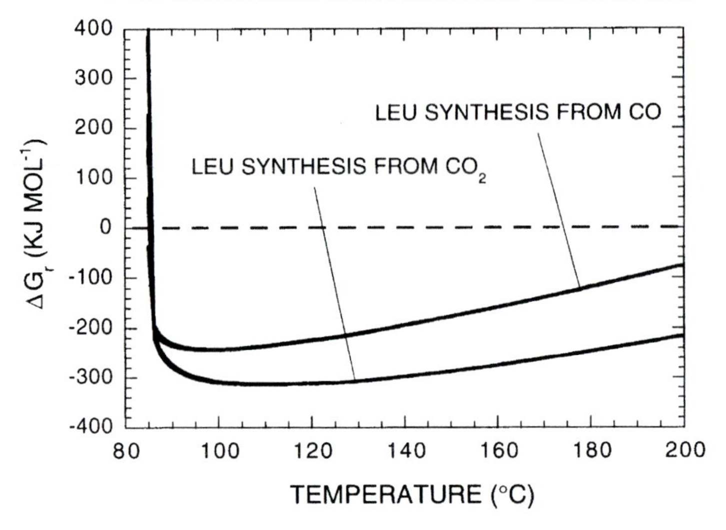

Fig. 3-10. Values of ΔG_r for the leucine synthesis reaction from CO and CO_2 given in Tables 3-5 and 3-2, respectively, at 250 bar as functions of temperature in mixed hydrothermal solutions generated from 400°C fluid and four models of Hadean ocean composition (see text). The broken line represents equilibrium. A leucine activity of 10^{-6} was used to carry out these calculations.

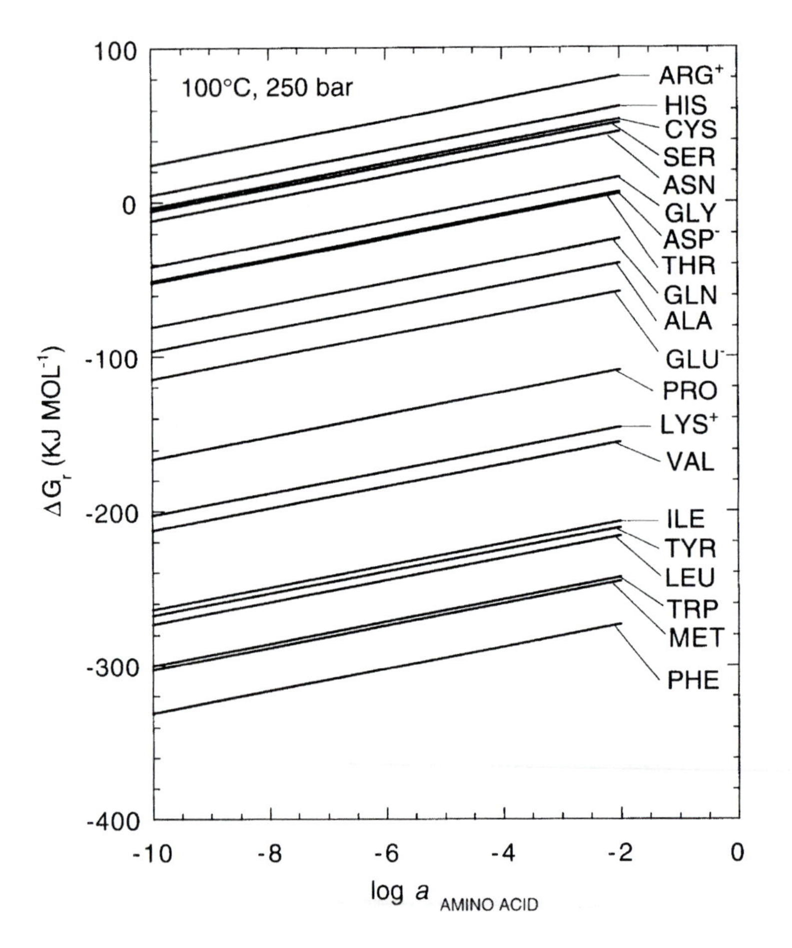

Fig. 3-11. Values of for the amino acid synthesis reactions given in table 3-5 at 100°C and 250 bar as a function of amino acid activity using the mixed hydrothermal solution with 'Ocean-3.'

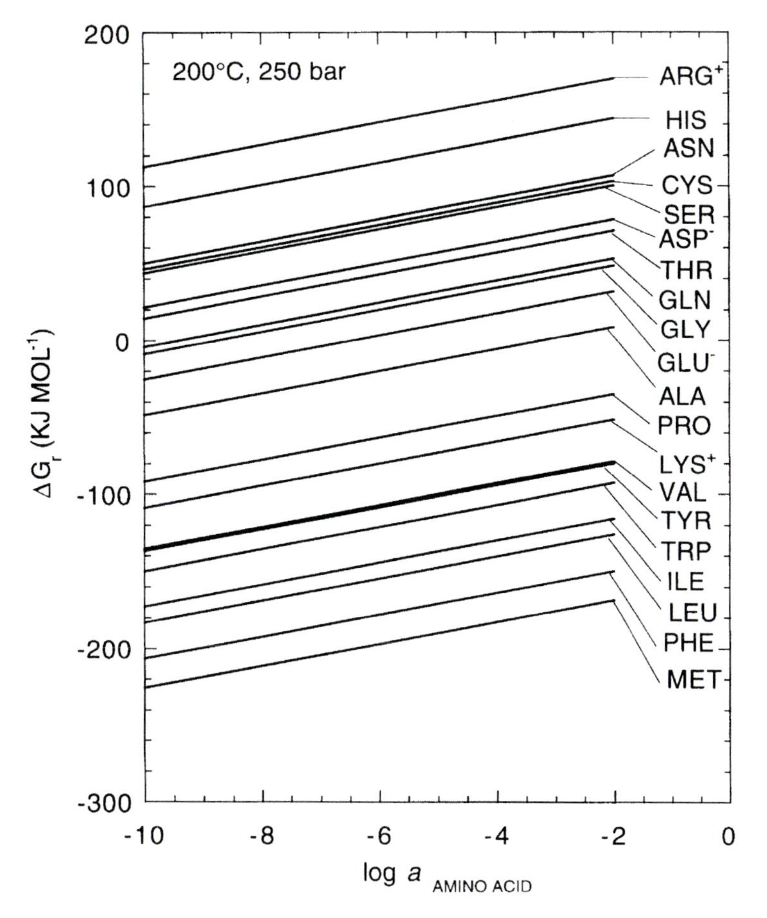

Fig. 3-12. Values of ΔG_r for the amino acid synthesis reactions given in table 3-5 at 200°C and 250 bar as a function of amino acid activity using the mixed hydrothermal solution with 'Ocean-3.'

CONCLUDING REMARKS

Calculations of the energetics of amino acid synthesis from CO_2 or CO, H_2, NH_4^+, and H_2S in hydrothermal solutions lend further support to the argument that aqueous organic synthesis can be tremendously favorable in geochemical conditions in and around submarine vent environments. The thermodynamic consequences of chemical disequilibria in zones of mixing between two fluids of different compositions show the types of synthesis processes that were possible in hydrothermal systems on the early Earth. Whether these processes were plausible can be tested in laboratory organic synthesis experiments under controlled and monitored conditions that resemble the natural systems of interest. A potentially fruitful experimental approach might involve the preparation and subsequent mixing, in the presence of mineral surfaces, of two fluids with compositions derived from data given in table 3-3. In order to expedite potentially sluggish reactions, experiments can be conducted at higher temperatures at which the increase in reaction rates may compensate for the somewhat lower energetic drive. Ideal temperatures, at least for amino acid synthesis from CO or CO_2 under early Earth conditions, appear to be between 100 and 200°C. Although synthesis of amino acids from CO relative to CO_2 does not release as much energy in the geochemical conditions considered here, the chemical instability of the C≡O triple bond may afford a nearly perfect target for studying the initiation of abiotic organic synthesis, and, by association, the emergence of metabolic and self-replicating biochemical systems.

Acknowledgments: We thank Misha Zolotov, Andrey Plyasunov, Panjai Prapaipong, Giles Farrant, Gary Olsen, and Harold Helgeson for many helpful discussions. Special thanks go to Tom McCollom and Mitch Schulte, because without their pioneering work on mixing and organic synthesis calculations this study could not have been completed. Financial support was provided by NSF-LExEn Grant OCE-9714288 and by NASA Grant NAG5-4002.

NOTES

1. The standard state convention used for aqueous species is unit activity in a hypothetical 1 molal solution referenced to infinite dilution at any temperature and pressure, and that for gases is unit fugacity of the hypothetical perfect gas at 1 bar and any temperature. The standard states for water and calcite are unit activity of pure H_2O and calcite, respectively, at any temperature and pressure.

2. The convention for the *apparent* standard Gibbs free energy of formation is to use the standard Gibbs free energy of formation from the elements at 25°C and 1 bar and integrate from there in temperature and pressure. Because the properties of the elements will cancel in any reaction, using apparent standard Gibbs free energies of formation at elevated temperatures and pressures makes it unnecessary to calculate the high temperature and pressure standard Gibbs free energies of the elements.

3. In contrast to changes in temperature, changes in pressure over the range of interest in this study contribute negligibly to the values of the equilibrium constants.

4. This is consistent with a present-day total concentration of Ca^{2+} in sea water equal to ~0.01 molal. Because of an activity coefficient of ~0.2 and complexation of Ca^{2+}, computed with the aid of computer program EQ3 (Wolery, 1992), the value of $a_{Ca^{2+}}$ is significantly less than the corresponding total molality of Ca^{2+}.

5. This is approximately six orders of magnitude greater than the concentration in present-day sea water and is perhaps similar to intracellular concentrations in extant microorganisms (see Amend and Shock, 1998).

6. This assumption will introduce only minimal error in mixed hydrothermal solutions unless the concentration of CO in hydrothermal fluids was substantially higher than in sea water.

REFERENCES

Amend J. P. and Helgeson H. C. (1997a) Calculation of the standard molal thermodynamic properties of aqueous biomolecules at elevated temperatures and pressures. Part 1. L-α-amino acids. *J. Chem. Soc. Faraday Trans.* **93**, 1927–1941.

Amend J. P. and Helgeson H. C. (1997b) Group additivity equations of state for calculating the standard molal thermodynamic properties of aqueous organic molecules at elevated temperatures and pressures. *Geochim. Cosmochim. Acta* **61**, 11–46.

Amend J. P. and Shock E. L. (1998) Energetics of amino acid synthesis in hydrothermal ecosystems. *Science* **281**, 1659–1662.

Baross J. A. and Hoffman S. E. (1985) Submarine hydrothermal vents and associated gradient environments as sites for the origin and evolution of life. *Orig. Life* **15**, 327–345.

Brandes J. A., Boctor N. Z., Cody G. D., Cooper B. A., Hazen R. M., and Yoder H. S., Jr (1998) Abiotic nitrogen reduction on the early Earth. *Nature* **395**, 365–367.

Canuto V. M., Levine J. S., Augustsson T. R., Imhoff C. L., and Giampapa M. S. (1983) The young Sun and the atmosphere and photochemistry of the early Earth. *Nature* **305**, 281–286.

Chyba C. F. (1989) Impact delivery and erosion of planetary oceans. *Nature* **343**, 129–132.

Corliss J. B., Dymond J., Gordon L. I., Edmond J. M., von Herzen R. P., Ballard R. D., Green K., Williams D., Bainbridge A., Crane K., and van Andel T. H. (1979) Submarine thermal springs on the Galapagos Rift. *Science* **203**, 1073–1083.

Corliss J. B., Baross J. A., and Hoffman S. E. (1981) An hypothesis concerning the relationship between submarine hot springs and the origin of life on Earth. *Oceanologica Acta* **4 suppl.**, 59–69.

de Duve C. (1995) The beginnings of life on Earth. *Am. Sci.* **83**, 428–437.

French B. M. (1964) Synthesis and stability of siderite, $FeCO_3$. 2. Progressive contact metamorphism of the Biwabik Iron Formation on the Mesabi Range, Minnesota. Ph.D. Thesis, The Johns Hopkins University, Baltimore, MD.

Grotzinger J. P. and Kasting J. F. (1993) New constraints on Precambrian ocean composition. *J. Geol.* **101**, 235–243.

Helgeson H. C., Delany J. M., Nesbitt W. H., and Bird D. K. (1978) Summary and critique of the thermodynamic properties of rock-forming minerals. *Amer. J. Sci.* **278A**, 1–229.

Helgeson H. C., Knox A. M., Owens C. E., and Shock E. L. (1993) Petroleum, oil field waters, and authigenic mineral assemblages: are they in metastable equilibrium in hydrocarbon reservoirs? *Geochim. Cosmochim. Acta* **57**, 3295–3339.

Holland H. D. (1984) *The Chemical Evolution of the Atmosphere and Oceans.* Princeton University Press.

Huber C. and Wächtershäuser G. (1997) Activated acetic acid by carbon fixation on (Fe,Ni)S under primordial conditions. *Science* **276**, 245–247.

Huber C. and Wächtershäuser G. (1998) Peptides by activation of amino acids with CO on (Ni,Fe)S surfaces: implications for the origin of life. *Science* **281**, 670–672.

Ingmanson D. E. and Dowler M. J. (1977) Chemical evolution and the evolution of the Earth's crust. *Orig. Life* **8**, 221–224.

Johnson J. W., Oelkers E. H., and Helgeson H. C. (1992) SUPCRT92: a software package for calculating the standard molal properties of minerals, gases, aqueous species, and reactions from 1 to 5000 bar and 0 to 1000°C. *Comp. Geosci.* **18**, 899–947.

Kasting J. F. (1993a) Early evolution of the atmosphere and ocean. In *The Chemistry of Life's Origins* (eds J. M. Greenberg, C. X. Mendoza-Gomez, and V. Pirronello), pp. 149–176. Kluwer Academic Publishers.

Kasting J. F. (1993b) Earth's early atmosphere. *Science* **259**, 920–926.

Kasting J. F. (1993c) Evolution of the Earth's atmosphere and hydrosphere. In *Organic Geochemistry* (eds M. H. Engel and S. A. Macko), pp. 611–624. Plenum Press.

Kasting J. F. and Ackerman T. P. (1986) Climatic consequences of very high carbon dioxide levels in the Earth's early atmosphere. *Science* **234**, 1383–1385.

Kasting J. F., Zahnle K. J. and Walker J. C. G. (1983) Photochemistry of methane in the Earth's early atmosphere. *Precambrian Res.* **20**, 121–148.

Levine J. S., Augustsson T. R., and Natarajan M. (1982) The prebiological paleoatmosphere: stability and composition. *Orig. Life* **12**, 245–259.

McCollom T. M. and Shock E. L. (1997) Geochemical constraints on chemolithoautotrophic metabolism by microorganisms in seafloor hydrothermal systems. *Geochim. Cosmochim. Acta* **61**, 4375–4391.

McCollom T. M. and Shock E. L. (1998) Fluid–rock interactions in the lower oceanic crust: thermodynamic models of hydrothermal alteration. *J. Geophys. Res.* **103**, 547–575.

Miller S. L. (1953) A production of amino acids under possible primitive Earth conditions. *Science* **117**, 528–529.

Miller S. L. (1957) The mechanism of synthesis of amino acids by electric discharges. *Biochim. Biophys. Acta* **23**, 480–489.

Miller S. L. and Urey H. C. (1959) Organic compound synthesis on the primitive Earth. *Science* **130**, 245–251.

Mojzsis S. J., Arrhenius G., McKeegan K. D., Harrison T. M., Nutman A. P., and Friend C. R. L. (1996) Evidence for life on Earth before 3,800 million years ago. *Nature* **384**, 55–59.

Ohmoto H. (1997) When did the Earth's atmosphere become oxic? *Geochem. News* **93**, 12–27.

Ohmoto H., Kakegawa T., and Lowe D. R. (1993) 3.4-Billion-year-old biogenic pyrites from Barberton, South Africa: sulfur isotope evidence. *Science* **262**, 555–557.

Pace N. R. (1991) Origin of life: facing up to the physical setting. *Cell* **65**, 531–533.

Rosing M. T. (1999) ^{13}C-depleted carbon microparticles in >3700-Ma sea-floor sedimentary rocks from West Greenland. *Science* **283**, 674–676.

Russell M. J. and Hall A. J. (1997) The emergence of life from iron monosulphide bubbles at a submarine hydrothermal redox and pH front. *J. Geol. Soc., Lond.* **154**, 377–402.

Sagan C. and Chyba C. (1997) The early faint sun paradox: organic shielding of ultraviolet-labile greenhouse gases. *Science* **276**, 1217–1221.

Schidlowski M. (1988) A 3,800-million-year isotopic record of life from carbon in sedimentary rocks. *Nature* **333**, 313–318.

Schopf J. W. (1993) Microfossils of the early Archean Apex Chert: new evidence of the antiquity of life. *Science* **260**, 640–646.

Schopf J. W. and Packer B. M. (1987) Early Archean (3.3-billion to 3.5-billion-year-old) microfossils from Warrawoona Group, Australia. *Science* **237**, 70–73.

Schulte M. D. and Shock E. L. (1995) Thermodynamics of Strecker synthesis in hydrothermal systems. *Orig. Life Evol. Biosph.* **25**, 161–173.

Shock E. L. (1988) Organic acid metastability in sedimentary basins. *Geology* **16**, 886–890.

Shock E. L. (1989) Corrections to "Organic acid metastability in sedimentary basins". *Geology* **17**, 572–573.

Shock E. L. (1990) Geochemical constraints on the origin of organic compounds in hydrothermal systems. *Orig. Life Evol. Biosph.* **20**, 331–367.

Shock E. L. (1992) Chemical environments of submarine hydrothermal systems. *Orig. Life Evol. Biosph.* **22**, 67–107.

Shock E. L. (1994) Application of thermodynamic calculations to geochemical processes involving organic acids. In *The Role of Organic Acids in Geological Processes* (eds M. Lewan and E. Pittman), pp. 270–318. Springer-Verlag.

Shock E. L. (1996) Hydrothermal systems as environments for the emergence of life. In *Evolution of Hydrothermal Ecosystems on Earth (and Mars?)* (eds G. R. Bock and J. A. Goode), pp. 40–60. John Wiley & Sons.

Shock E. L. and Helgeson H. C. (1988) Calculation of the thermodynamic and transport properties of aqueous species at high pressures and temperatures: correlation algorithms for ionic species and equation of state predictions to 5 kb and 1000°C. *Geochim. Cosmochim. Acta* **52**, 2009–2036.

Shock E. L. and Schulte M. D. (1998) Organic synthesis during fluid mixing in hydrothermal systems. *J. Geophys. Res.* (*Planets*) **103**, 28513–28527.

Shock E. L., Helgeson H. C., and Sverjensky D. A. (1989) Calculation of the thermodynamic and transport properties of aqueous species at high pressures and temperatures: standard partial molal properties of inorganic neutral species. *Geochim. Cosmochim. Acta* **53**, 2157–2183.

Shock E. L., McCollom T., and Schulte M. D. (1998) The emergence of metabolism from within hydrothermal systems. In *Thermophiles: the Keys to Molecular Evolution and the Origin of Life?*

(eds J. Wiegel and M. W. W. Adams), pp. 59–76. Taylor & Francis.

Shock E. L., Oelkers E. H., Johnson J. W., Sverjensky D. A., and Helgeson H. C. (1992) Calculation of the thermodynamic properties of aqueous species at high pressures and temperatures: effective electrostatic radii, dissociation constants and standard partial molal properties to 1000°C and 5 kbar. *J. Chem. Soc. Faraday Trans.* **88**, 803–826.

Shock E. L., Sassani D. C., Willis M., and Sverjensky D. A. (1997) Inorganic species in geologic fluids: correlations among standard molal thermodynamic properties of aqueous ions and hydroxide complexes. *Geochim. Cosmochim. Acta* **61**, 907–950.

Sleep N. H., Zahnle K. J., Kasting J. F., and Morowitz H. J. (1989) Annihilation of ecosystems by large asteroid impacts on the early Earth. *Nature* **342**, 139–142.

Stevenson D. J. (1983) The nature of the Earth prior to the oldest known rock record: the Hadean Earth. In *Earth's Earliest Biosphere–its Origin and Evolution* (ed. J. W. Schopf), pp. 32–40. Princeton University Press.

Von Damm K. L. (1990) Seafloor hydrothermal activity: black smoker chemistry and chimneys. *Ann. Rev. Earth Planet. Sci.* **18**, 173–204.

Von Damm K. L., Edmond J. M., Grant B., Measures C. I., Walden B., and Weiss R. F. (1985) Chemistry of submarine hydrothermal solutions at 21° N, East Pacific Rise. *Geochim. Cosmochim. Acta* **49**, 2197–2220.

Walker J. C. G. (1977) *Evolution of the Atmosphere.* Macmillan.

Walker J. C. G. (1983) Possible limits on the composition of the Archaean ocean. *Nature* **302**, 518–520.

Walker J. C. G. (1985) Carbon dioxide on the early Earth. *Orig. Life* **16**, 117–127.

Wetzel L. R. and Shock E. L. (2000). Distinguishing ultramafic- from basalt-hosted submarine hydrothermal systems by comparing calculated vent fluid compositions. *J. Geophys. Res. (Solid Earth)* **105**, 8319–8346.

Wöhler F. (1828) Über künstliche Bildung von Harnstoff. *Annal. Physik* **12**, 253.

Wolery T. (1992) *EQ3NR, A Computer Program for Geochemical Aqueous Speciation-Solubility Calculations: Theoretical Manual, User's Guide, and Related Documentation.* Lawrence Livermore National Laboratory.

Zahnle K. J., Kasting J. F., and Pollack J. B. (1988) Evolution of a steam atmosphere during Earth's accretion. *Icarus* **74**, 62–97.

4. Early pre- and post-biotic synthesis of alanine: an alternative to the Strecker synthesis

Jay A. Brandes, Robert M. Hazen, Hatten S. Yoder, Jr, and George D. Cody

The prebiotic synthesis of amino acids is considered a primary requirement for the origin of life. Much of the research in the origin-of-life field has therefore concentrated upon experiments designed to synthesize amino acids, either via electrical discharges (Miller, 1955), ultraviolet irradiation of solutions (Oparin, 1957), polymerization reactions under aqueous conditions (Peltzer et al., 1984; Stribling and Miller, 1987; Weber, 1998), or by Fischer Tropsch chemistry (Hayatsu and Anders, 1981). Although some success has been attained by each of these approaches, none has proved capable of synthesizing all of the 20 amino acids commonly employed by organisms, and most require conditions inhospitable to life. The most successful synthesis pathways involve the reaction and polymerization of aldehydes and cyanides (e.g., Strecker synthesis):

$$HCN + CH_2O \rightarrow \text{Amino acids} + \text{other nitrogen-containing organics.} \quad (4\text{-}1)$$

This reaction is capable of producing many nitrogen-containing compounds, including amino acids and nucleic acid subunits. The reaction is not, however, selective for specific amino or nucleic acid molecules. Furthermore, the presence of high concentrations of HCN, a powerful and highly reactive polymerizing agent, would seem to preclude the use of this pathway, by the earliest life-forms to produce amino acids. This is because HCN would react with, and ultimately destroy, the very catalysts necessary for maintenance of biochemical cycles. Thus, the fundamental question of how amino acids were synthesized by early life-forms remains unanswered.

Modern organisms employ specific enzyme-mediated biochemical pathways to control chemical reactions. The formation of amino acids proceeds via the reductive amination of α-keto acids produced initially in the citric acid cycle or elsewhere (e.g., α-ketoglutaric acid, pyruvic acid), or by modification of existing amino acids (e.g., cysteine from alanine). The enzymatic pathways are highly regulated to prevent wasteful production of unnecessary amino acids. The production of amino acids from α-keto acids involves the transfer of ammonia to the terminal carboxyl group of glutamate to form glutamine, followed by the amination and subsequent reduction of an α-keto acid to form the amino acid:

$$NH_3 + \text{glutamic acid} \rightarrow \text{glutamine:}$$
$$\text{glutamine} + \text{pyruvic acid} \rightarrow \text{glutamic acid} + \text{alanine.} \quad (4\text{-}2)$$

In some organisms (e.g., *Escherichia coli*) under ammonium-replete conditions, this reaction pathway is less complex. Under these circumstances, the glutamate/glutamine intermediate step is bypassed, and the α-keto acid is aminated and reduced in one step:

$$NH_3 + \text{pyruvic acid} \rightarrow \text{alanine.} \quad (4\text{-}3)$$

Of interest from a prebiotic chemistry standpoint is the observation that this reaction can take place in the absence of enzymes or external catalysts (Morowitz, 1992; Morowitz et al. 1996). It could be postulated that some of the biochemical pathways used by modern organisms were present in the first organisms (Morowitz, 1992). It follows logically that those reactions requiring little or no assistance from heterogeneous catalysts or enzymes and which produce a limited or controlled subset of possible reaction products are the most likely candidates for "relict" biochemical pathway status. The formation of alanine from pyruvic acid appears to meet these criteria, but this chemistry has been explored only under a very limited subset of conditions.

The goals of this work were to extend the conditions under which this reaction occurs in order to understand alanine synthesis and to understand the potential significance of this chemistry to prebiotic systems. Thus a series of experiments were undertaken with pyruvic acid, the simplest α-keto acid, and ammonium ions in aqueous solution under a variety of pH, temperature and pressure conditions.

MATERIALS AND METHODS

Samples were prepared from reagent-quality pyruvic acid and ammonium chloride. Most experiments were conducted with a solution containing 0.4 g pyruvic acid (purim grade; Fluka Chemical) and 1.0 g of a 4.7 M NH_4Cl solution. The pH of the solution was subsequently adjusted using a saturated solution of NaOH to the desired value, and finally all samples had distilled, deionized water added so that the total of NaOH solution plus water equaled 1.5 g. Dilution experiments were undertaken in the same manner, except that the volume of NH_4Cl solution and distilled, deionized water was increased to maintain NH_4^+ and pH conditions similar to those of the other solutions. Following initial sample preparation, 15 mg sample aliquots were placed in 10 mm (length) by 2 mm (diameter) gold (99.95% purity) tubes. The gold tubing was cleaned by sequential washes in boiling 10% HNO_3 and 10% HCl solutions, followed by distilled water washes and annealing at 500°C for 1 hr. One tube end was sealed by arc-welding prior to sample introduction. After sample introduction, solutions were placed in a glove bag and purged with N_2 for 20 min. The open tube ends were then crimp-sealed under N_2; this was followed by immersion in liquid N_2 and subsequent arc-welding of the tube end.

Atmospheric pressure experiments were conducted by placing the gold sample tubes within a larger glass tube (12 mm i.d. × 15 cm length) containing 0.05 ml H_2O to maintain internal pressure. The glass tube was sealed and the sample was placed in a convection oven for incubation. High-pressure experiments were conducted using an internally heated gas-pressure device (Yoder, 1950). Variations in temperature and pressure were controlled to better than 1°C and 101 kPa precision, respectively.

After incubation, sample tubes were washed in methanol, weighed, frozen in liquid N_2 and opened while frozen. Sample tube contents were extracted in either 1 ml methanol or 1 ml pH 2.0 HCl solution, followed by vortexing of sample contents for 30 seconds. Aliquots of sample were removed for both high-performance liquid chromatography (HPLC) and gas chromatography/mass spectrometry (GC–MS). Amino acid concentrations were determined using ion-exchange HPLC with post-column *o*-phthalaldehyde (OPA) derivitization (Serban et al., 1988). Mixed amino acid standards were also analyzed during each run. Typical uncertainties in yields were 5%, primarily because of errors in sample extraction and dilution. Sample aliquots for GC–MS analysis were derivitized using an acetyl chloride/isopropyl alcohol derivitization for the acidic moiety, followed by *N*-trifluoroacetic acid (*N*-TFA) derivitization for the amine moiety (Goodfriend 1991). Compounds were identified by GC–MS using a Hewlett Packard 6890 GC–MS system; either an HP-5 or a Supelco 1790 capillary column was used during experimental analysis.

RESULTS

Alanine synthesis

The formation of alanine from pyruvic acid requires both the reaction of the α-keto moiety with ammonia to form an intermediate, presumably an imine, followed by the reduction of that intermediate to form alanine. It is well known that acidic conditions promote decarboxylation reactions, which would increase the efficiency of pyruvic acid as a reducing agent. Previous work on alanine formation from pyruvic acid (Morowitz, 1992) has employed formic acid as a putative reducing agent, but it is clear from this work that no external reducing agent need be applied to the system. It is remarkable, however, how efficiently the reducing power is transferred to the imine intermediate at low pH values (figure 4-1). When one considers that the stoichiometry of the reaction is as follows:

$$2 \text{ pyruvic acid} + 1\, NH_4^+ \rightarrow 1 \text{ alanine} + 1 \text{ acetic acid} + H_2CO_3, \quad (4\text{-}4)$$

well over 50% of the pyruvic acid in these systems either ends up as alanine or is used as a reducing agent.

The synthesis of alanine from mixtures of pyruvic acid and ammonia was also rapid, reaching peak values at 100°C within 6–24 hr (figure 4-1). Synthesis was most rapid and efficient at pH values

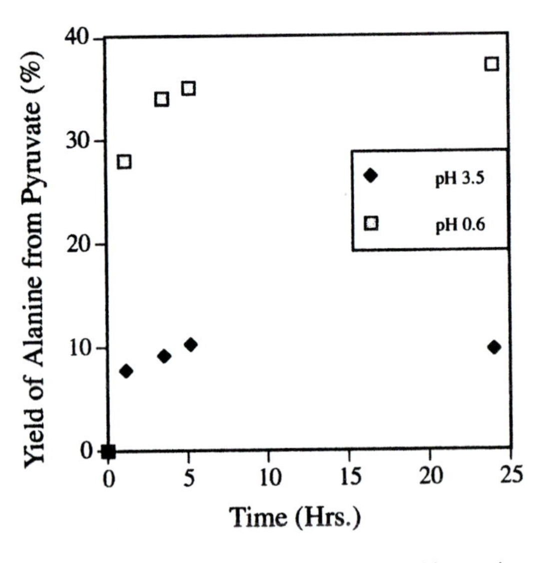

Fig. 4-1. Yield of alanine from pyruvic acid vs. time. Reaction conditions were 100°C and 0.1 MPa.

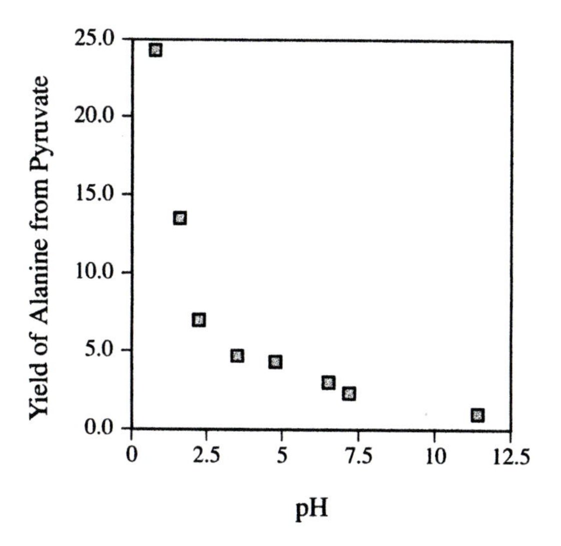

Fig. 4-2. Yield of alanine from pyruvic acid vs. pH. Reaction conditions were 100°C, 24 hr and 0.1 MPa.

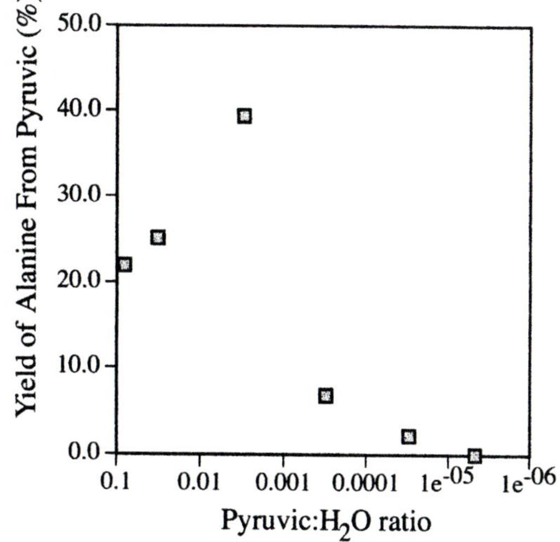

Fig. 4-4. Yield of alanine from pyuvic acid under different dilution conditions. Reaction conditions were 100°C, 24 hr, 0.1 MPa, 4.7 molar HN_4Cl and pH 0.7. Note: a 0.1 pyruvic acid:water ratio is equal to a concentration of 5.5 molar.

below 2, with conversion efficiencies of up to 40%. However, alanine concentrations plateaued after 6 hr.

The pH dependence of alanine synthesis is shown in more detail in figure 4-2. The reductive amination of pyruvic acid is very pH sensitive, with the highest conversion efficiencies being at pH 0.6–0.8 and dropping to below 10% above pH 2. No production of other identifiable amino acids was observed in the reaction mixtures. Experiments run at several pressures (figure 4-3) produced similar yields, with the exception that yields at pH = 2.95 were increased at the highest pressure.

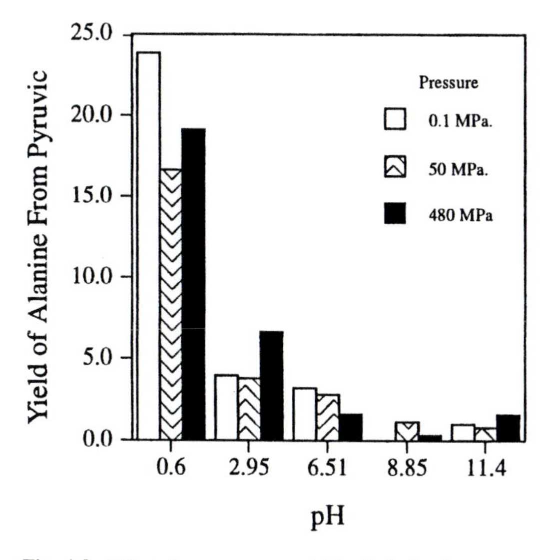

Fig. 4-3. Effect of pressure upon yields of alanine from pyruvic acid vs. pH. Reaction conditions were 100°C and 24 hr.

The influence of dilution of pyruvic acid upon alanine synthesis is shown in figure 4-4. Alanine yields increased slightly with increasing dilution until a pyruvic acid:water ratio of 1:100 (0.55 M) pyruvic acid was reached, at which point a sharp decline in yields was noted. Peak conversions (39%) of alanine were noted at pyruvic acid H_2O ratios of 1:100 .

GC–MS analyses

The efficiency of alanine synthesis at low pH values is also indicated by the lack of identifiable compounds other than alanine in the GC–MS trace (figure 4-5). However, the distribution of identifiable products from the reaction of pyruvic acid with ammonia shifted significantly with pH (figure 4-5). Below pH 2, the reaction products analyzable by GC–MS using isopropyl/*N*-TFA derivitization consisted entirely of alanine. However, above pH 2 another suite of products was observed in addition to alanine. Two major products, one eluting just after alanine at 12 min and the other at 15 min, were noted. Above pH 4, the predominant products shifted from alanine and smaller compounds to larger aromatic compounds, with one compound at 24 min dominating the system between pH 4 and pH 8.5. Above pH 8.5, the product mixture became more complex, with multiple compounds of aromatic character being observed.

The production of alanine is less efficient and the system becomes more complex as the pH increases. By pH 2.95, alanine comprised only 24% of the major products (by area). The two significant peaks, labeled Product A and Product B on figure

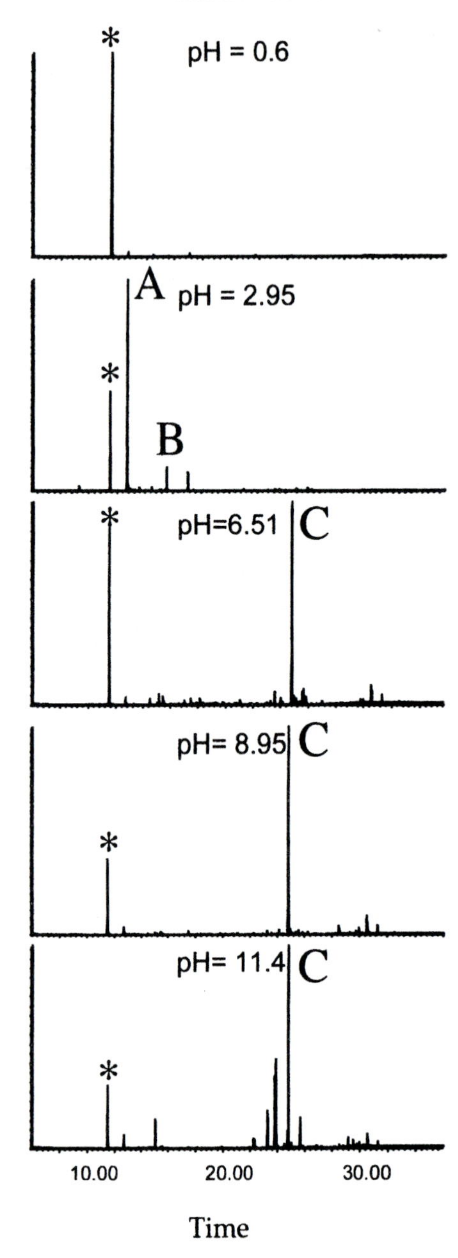

Fig. 4-5. Gas chromatography data from derivatized run products at different pH values. Reaction conditions: 100°C, 24 hr, and 50 MPa. The peaks representing alanine are denoted by asterisks; reaction products A, B and C are discussed in the text.

4-5, were identified (by fragmentation pattern) as aminated aldol condensation products. A proposed reaction scheme is shown in figure 4-6. Our work with pyruvic acid/water mixtures (G. D. Cody, unpublished data) has indicated that at intermediate pH values a primary reaction pathway for pyruvic acid is via aldol condensation. Once the aldol condensate is formed, the product is attacked either by ammonia or by another amine or amide. If attacked by ammonia, an imine is formed, which would then require reduction to the amine form to create a stable product. Also, the amine product can be attacked by acetic acid, which is present in high concentrations because of the decarboxylation of pyruvic acid, and this group may derivatize the amine as well (figure 4-6). This reaction pathway explains the lack, in Product A, of an $m/z = 69$ peak (expected from derivitization of the amine group by TFA), as well as a similarity of Product B's fragmentation pattern to that of alanine. Also these compounds or their precursors would be expected to hydrolyze to form alanine as well as other products (Streitwieser and Heathcock, 1985). The presence of bound (or otherwise underivatizable) alanine was tested by hydrolyzing an aliquot of the pH 2.95-run products in 6 N HCl for 15 min at 150°C. There was significantly more alanine present after hydrolysis (figure 4-7), supporting the hypothesis that products A and B were reduced aminated aldol condensates.

Further evidence for hydrolyzable products was obtained by examining the pressure dependence of alanine yields and reaction run products. The influence of pressure on the pyruvic acid/ammonia system was determined by comparing run products at 0.1, 50, and 480 MPa (figure 4-3). The general trend of decreasing yields above pH 0.6 was noted at all pressures, but at higher pressures the yield curve flattened out, with less production at the lowest pH and more observable alanine production at pH 2.95. It has been observed that the hydrolysis of protein bonds is accelerated by pressure (Qian et al., 1993); thus the increased yield at 480 MPa compared to lower pressures may be explained by hydrolysis of bound or underivatizable (by OPA) amino acid from compounds such as products A and B. Examination of the GC trace at pH 2.95 (figure 4-8) indicated that the yields of production compounds A and B were greatly reduced by increased pressure—a trend not noted in analysis of the other pH run products at pressure. It should be noted that overall yields, including the hydrolyzable fraction, were still greater in the high-pressure run than in low-pressure runs.

The pH 6.5 run again indicated a strongly pH-dependent change in the chemistry. Although alanine was still formed in significant quantity (figure 4-1), the primary product (Product C) was one with a much longer retention time, and the products noted in the pH 2.95 run were reduced to trace quantities. Product C was most closely related, in terms of mass spectra, to a family of nitrogen-containing aromatic carboxylic acids. We have observed the formation of aromatic compounds in the pyruvic acid/H_2O system as well (Cody et al., in press). The mass spectra from this compound indicated that that it contained a nitro-

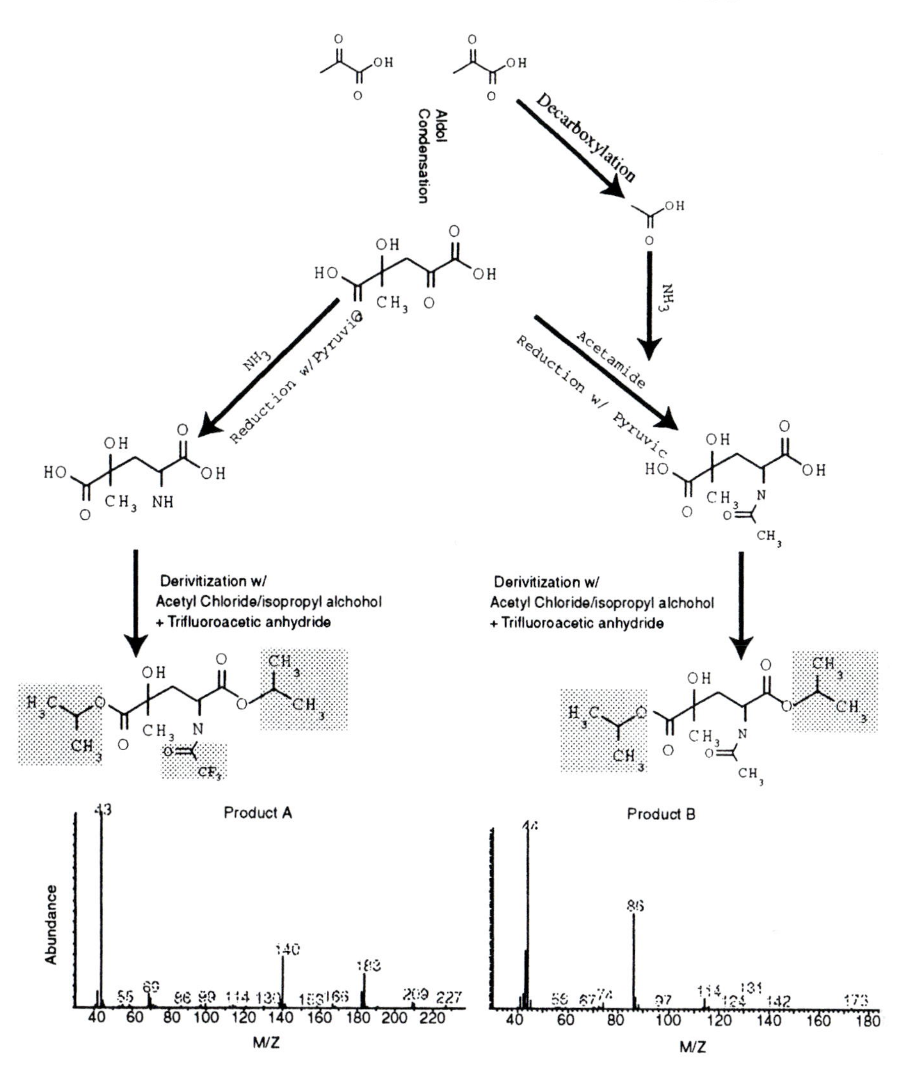

Fig. 4-6. Proposed reaction scheme and mass spectra for products "A" and "B" at pH 2.95. Shaded boxes represent functional groups added during derivatization.

gen moiety of either a derivatized primary amine (similar to Product A, and again lacking an $m/z = 69$ peak) or a 2 or 3° amine.

DISCUSSION AND CONCLUSIONS

Most prebiotic studies have concentrated upon the synthesis of compounds from simple precursors. Although the synthesis of ammonia in hydrothermal systems has been demonstrated (Brandes et al., 1998), the synthesis of pyruvic acid from simple one- and two-carbon precursors has not, to our knowledge, been reported. One hypothesis suggests that current biochemical pathways represent biochemical pathways of the earliest organisms, i.e., that core biochemical pathways have not been altered since the first life forms (Morowitz, 1992). If one accepts this hypothesis, then a reasonable corollary is that the chemical reactions that mimic biochemical pathways and that require the least specialized catalysis were the first to be utilized by life-forms. The results shown here indicate that the formation of amino acids from pyruvic acid and other α-keto acids fulfill this requirement. At low pH values, the synthesis of alanine is rapid, efficient and produces a minimum of side-products. Other proposed mechanisms employing cyano-containing compounds produce amino acids, but also produce β-amino acids as well as non-biological amines and acids (Stribling and Miller, 1987; Schesinger and Miller, 1983). The selectivity of the α-keto reaction,

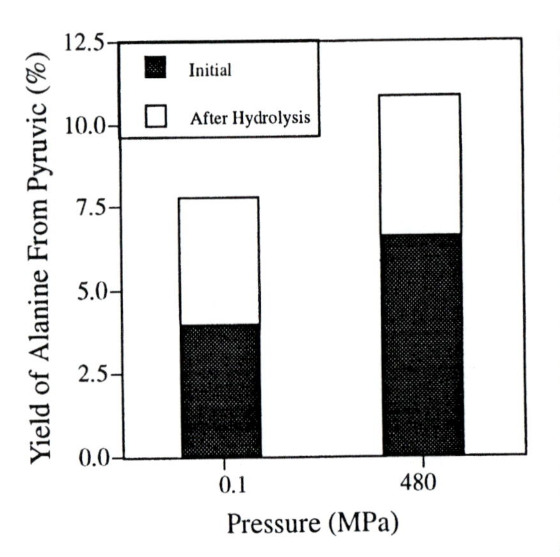

Fig. 4-7. Yield of alanine before and after hydrolysis at pH 2.95. Reaction conditions were 100°C and 24 hr.

from a biochemical perspective, is exactly what an early life form requires in order to target the production of valuable metabolic compounds. Given a source of α-keto acids, presumably from another primitive catalytic system, primitive life forms would have been able to produce pure α-amino acids relatively easily. Experiments with α-ketoglutarate and oxaloacetic acid have also shown that these compounds can produce specific amino acids as well, albeit in low yield (Morowitz et al., 1995; Maughan and Miller, 1999). Another possibility is the formation of α-keto acids via the decomposition of larger molecules (Cody et al., in press). Moreover, once formed, amino acids can act as nitrogen sources via transamination (Bishop et al., 1997). This work joins a body of research that suggests that the formation of amino acids in early biotic systems may have proceeded relatively easily provided that other pathways producing more reactive carbon compounds evolved first.

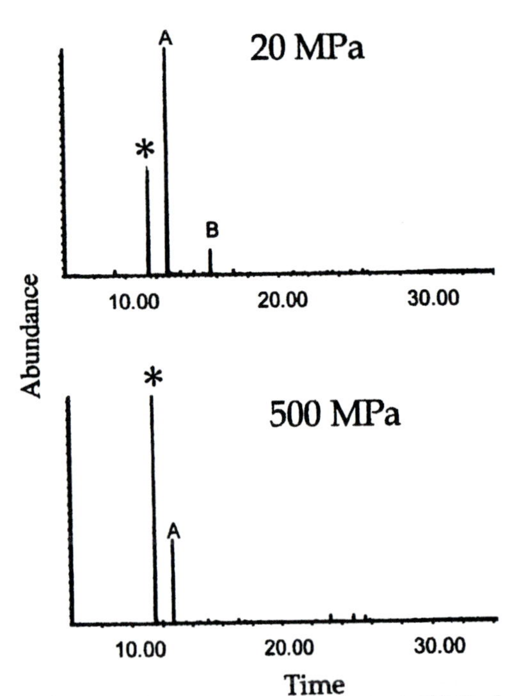

Fig. 4-8. Effect of pressure upon run products at pH 2.95 of pyruvic acid/ammonia reaction. Reaction conditions: 100°C and 24 hr.

On the basis of the results in this paper, it is possible to envision a primitive metabolic pathway quite different than that utilized by modern organisms. By simple control of the acidity of the reactive site, a primitive organism could (in theory) produce longer chain aldol condensation products such as products A and B. The underivatized forms of these compounds bear a resemblance to glutamic acid. Given that we see evidence for the formation of peptide-like bonds between acetate and the amine group (products A and C), it is quite possible that this chemistry could, with slight modification, yield both larger metabolic products and simple peptide or peptide-like products. Aldol-condensation chemistry is quite different from the CO_2-addition chemistry undertaken by modern photosynthetic and chemoautotrophic organisms, and could represent an early alternative in the formation of larger molecules. At higher pH values, aromatic compounds could be formed that could (in theory) take the place of phenylalanine or form the basis for the formation of aromatic-containing compounds. All of these reaction pathways may also be influenced by the presence of mineral surfaces (Hafenbradl et al., 1995). As the true nature of prebiotic chemistry remains uncertain, the existence of alternative chemical pathways within the framework of biochemistry require exploration with the eventual goal of understanding the possibilities contained in present-day metabolic pathways.

Acknowledgments: The authors wish to thank P. E. Hare for the use of his amino acid analyzer and for laboratory space, G. Goodfriend and D. von Endt for assistance with the amino acid analyses, R. Filley for assistance with the GC–MS analyses, and M. Teece and M. L. Fogel for helpful reviews of this manuscript.

REFERENCES

Bishop J. C., Cross S. T., and Waddell T. G. (1997) Prebiotic transamination. *Orig. Life Evol. Biosph.* **27**, 319–324.

Brandes J. A., Boctor N. Z., Cody G. D., Cooper B. A., Hazen R. M., and Yoder H. S. (1998) Abiotic nitrogen reduction on the early Earth. *Nature* **395**, 365–367.

Cody G. D., Brandes J. A., Hazen R. M., Morowitz H. J., and Yoder, Jr. H. S. (in press) The geochemical roots of archaic autotrophic carbon fixation: Implications from experiments in the system citric acid–H_2O–FeS–NiS at high pressures and moderate temperatures. *Orig. Life Evol. Biosph.*

Goodfriend G. A. (1991) Patterns of racemization and epimerization of amino acids in land snails over the course of the Holocene. *Geochim. Cosmochim. Acta* **55**, 293–302.

Hafenbradl D., Keller M., Wachtershauser G., and Stetter K. O. (1995). Primordial amino-acids by reductive amination of alpha-oxo acids in conjunction with the oxidative formation of pyrite. *Tetrahedron Lett.* **36**, 5179–5182.

Hayatsu R. and Anders E. (1981) Organic compounds in meteorites and their origins. *Top. Curr. Chem.* **99**, 3–37.

Liu R. and Orgel L. E. (1997) Oxidative acylation using thioacids. *Nature* **389**, 52–54.

Maughan Q. and Miller S. L. (1999). Does formate reduce alpha-ketoglutarate and ammonia to glutamate? *Orig. Life Evol. Biosph.* **29**, 355–360.

Miller S. L. (1955) Production of some organic compounds under possible primitive earth conditions. *J. Amer. Chem. Soc.* **77**, 2351–2361.

Morowitz H. J. (1992) *Beginnings of Cellular Life: Metabolism Recapitulates Biogenesis.* Yale University Press.

Morowitz H., Peterson E., and Chang S. (1996) The synthesis of glutamic acid in the absence of enzymes: implications for biogenesis. *Orig. Life Evol. Biosph.* **25**, 395–399.

Oparin A. (1957) *The Origin of Life on the Earth.* Academy of Sciences of the USSR.

Peltzer E. T., Bada J. L., Schlesinger S., and Miller S. L. (1984) The chemical conditions on the parent body of the Murchison Meteorite: some conclusions based upon amino, hydroxy, and dicarboxylic acids. *Adv. Space Res.* **4**, 69–74.

Qian Y., Engel M. H., Macko S. A., Carpenter S., and Deming J. W. (1993). Kinetics of peptide hydrolysis and amino acid decomposition at high temperatures. *Geochim. Cosmochim. Acta* **57**, 3281–3293.

Schlesinger G. and Miller S. L. (1983) Prebiotic synthesis in atmospheres containing CH_4, CO and CO_2. I. Amino acids. *J. Mol. Evol.* **19**, 376–383.

Serban A., Engel M. H., and Macko S. A. (1988) *Org. Geochem.* **13**, 1123–1129.

Streitwieser A., Jr, and Heathcock C. H. (1985) *Introduction to Organic Chemistry*, 3rd edn. Macmillan.

Stribling R. and Miller S. L. (1987) Energy yields for hydrogen cyanide and formaldehyde syntheses: the HCN and amino acid concentrations in the primitive ocean. *Orig. Life Evol. Biosph.* **17**, 261–273.

Weber A. L. (1998) Prebiotic amino acid thioester synthesis: thiol-dependent amino acid synthesis frfom formose substrates (formaldehyde and glycolaldehyde) and ammonia. *Orig. Life Evol. Biosph.* **28**, 259–270.

Yoder H. S. (1950) High–low quartz inversion up to 10,000 bars. *Trans. Amer. Geophys. Union* **31**, 821–835.

II. ISOTOPIC STUDIES OF AMINO ACIDS AND PROTEINS

5. A model for the effect of weaning on nitrogen isotope ratios in humans

Andrew R. Millard

Studies on the age of weaning in archaeological populations have increased in number and importance in recent years. A variety of factors have contributed to this: (1) the realization, in a number of areas of research, that the duration of breast-feeding is an important factor in the health of both mother and child; (2) increased research emphasis on the role and activities of women and children in past societies; and (3) a focus on using archaeological data as a proxy for evolutionary data on the "natural" or "optimum" duration of breast-feeding.

There are three main approaches to obtaining information on weaning from skeletal remains (Stuart-Macadam, 1995): (1) examining the pattern of occurrence of enamel hypoplasias; (2) using mortality curves; and (3) chemical analysis. Studies on the association of hypoplasias with weaning have been reviewed by Katzenberg et al. (1996), who concluded that most of them cannot demonstrate a causal link between weaning and the peak occurrence of hypoplasias. The use of mortality curves has been little used in archaeological work. Stuart-Macadam (1995) suggests that it might find wider application but it is likely to suffer from problems of trying to detect small changes in age-specific death rates using small sample sizes. At best this method can be considered experimental at the present time, with only the recent study of Herring et al. (1998) showing clear results.

The third method, chemical analysis, relies on the fact that an infant undergoes two trophic-level shifts. First, it moves from placental nourishment, where it has the same trophic level as the mother, to breast-feeding, where it feeds off its mother. Second, it moves down a level in moving from breast milk to the same foods as its mother. In principle this should be detectable in the bone chemistry in four ways.

(1) Changes in $\delta^{15}N$ as the infant's protein source changes (Fogel et al. 1989).
(2) Changes in strontium/calcium (Sr/Ca) as the infant's calcium source changes (Sillen and Smith, 1984).
(3) Changes in $\delta^{18}O$ as the infant's water source changes (Bryant and Froelich, 1996).
(4) Changes in $\delta^{13}C$ as the infant's carbon source changes (Katzenberg, 1993).

In practice, only the Sr/Ca and the $\delta^{15}N$ methods have been applied to archaeological populations, the former only in a couple of cases.

NITROGEN–ISOTOPE STUDIES OF WEANING

There have been a series of studies that have related $\delta^{15}N$ changes to weaning (Katzenberg, 1993; White and Schwarcz, 1994; Katzenberg et al., 1996; Fogel et al., 1997; Schurr, 1997) and almost as many reviews of the topic (Katzenberg et al., 1996; Stuart-Macadam, 1995; Schurr, 1998). These studies are based on principles first demonstrated by Fogel et al. (1989, 1997), i.e.

- that $\delta^{15}N$ in a mother's milk is elevated relative to her diet, and is comparable to the nitrogen-isotope composition of her body;
- that the suckling child shows a further elevation of $\delta^{15}N$ due to the trophic-level effect;
- that, consequently, the $\delta^{15}N$ of a child varies with age, because of suckling and subsequent weaning.

The first fully published study was that of Katzenberg (1993; Katzenberg and Pfeiffer, 1995) who studied two 19th century European cemeteries and one protohistoric Native American cemetery in Ontario. Initial study of the Native American site at MacPherson suggested a correlation of age and $\delta^{15}N$, and the other two cemeteries were explored to confirm the effect of age on $\delta^{15}N$. In all three cemeteries the $\delta^{15}N$ values for infants <2 and >2 were significantly different, and this was attributed to a weaning trophic-level shift. The $\delta^{15}N$ values of neonates and known still-births were similar to those of adults. The peak of $\delta^{15}N$ in all three cases occurred at approximately 1 year of age.

White and Schwarcz (1994) undertook a wide-ranging isotopic study of five sites in the Wadi Halfa area of Sudanese Nubia. They observed a significant negative correlation between $\delta^{15}N$ and age in the range 1–6 years, with a maximum enrichment of approx. 3‰ over the adult mean. They suggest that the gradual decrease in $\delta^{15}N$ with age represents a gradual decline in breast-feeding, rather than a discrete end to breast-feeding.

Fogel et al. (1997) report $\delta^{15}N$ analyses of two prehistoric populations. By comparing Archaic hunter-gatherers from the Tennessee Valley and Plains agriculturists from South Dakota, they sought to test the hypothesis that weaning would occur earlier in agricultural populations than in non-agricultural populations. No attempt was made to establish age of weaning, but the patterns of change of $\delta^{15}N$ were compared to discover if weaning patterns differed. The two populations were indistinguishable, contrary to the hypothesis.

Schurr (1997) studied the prehistoric agriculturalist Angel Site in the Ohio Valley, in an attempt to estimate weaning age from $\delta^{15}N$ and to compare the result with an estimate from mortality profiles. The $\delta^{15}N$ values peak at approx. 1.5 years of age. By fitting nonlinear equations to the data, an estimate of 1.25–2 years for the age of onset of weaning was obtained, with the suggestion that weaning was gradual rather than abrupt. Schurr also notes that the skeletal age estimates are the limiting factor in the precision of the data.

In an excellent study combining historical records, skeletal data and $\delta^{15}N$ measurements, Herring et al. (1998) have investigated the process of weaning in a 19th century sample from St Thomas's Anglican Cemetery in Belleville, Ontario. From the demographic data they conclude that weaning commenced at about 5 months, and from the $\delta^{15}N$ evidence they conclude that weaning was complete by about 14 months, although some individuals show evidence of much earlier completion weaning.

Interpreting Weaning Data from Bone Chemistry

The use of the term "weaning" varies widely in the literature, and it has been used loosely here, so far. Katzenberg et al. (1996) provide a useful overview of how it has been used and note, in particular, that weaning is a *process* and cannot be regarded as occurring at a specific age. Some studies have used the term weaning to indicate the onset of the process, others to indicate the *completion* of the process, and there is a tendency to try to estimate, or at least discuss, a single age of weaning. So, for example, Sillen and Smith (1984) were confident that weaning was complete by age 2–3 and this is the age that features in their discussions and many citations of their paper, whilst, on the other hand, studies using $\delta^{15}N$ usually estimate an age for the onset of weaning (e.g., Schurr, 1997). This has led Katzenberg et al. (1996) to state that "in contrast to Sr/Ca ratios, which change with the introduction of other foods, $\delta^{15}N$ values change due to the loss of breast milk in the diet. Therefore the methods are really measuring different events...." This is not entirely true, as both measures reflect the *process* of weaning (not an event), but with respect to different dietary components, so that the timing and tempo of change may differ between them, and will depend on the particular weaning foods used. Other studies (e.g., Katzenberg et al., 1993) refer to the age of maximum $\delta^{15}N$ to typify an age-at-weaning.

In order to estimate ages for the onset and completion of weaning it is necessary to have a model to compare with a data set. Schurr (1997) criticizes previous studies (Katzenberg et al., 1993; White and Schwarcz, 1994) for using simple linear models to evaluate age-related changes in $\delta^{15}N$; for the most part, however, the aim of the analyses in these early studies was simply to demonstrate that the age-related variation was significant, and not to evaluate weaning ages. In the absence of models, most studies have either simply noted that age-variation in $\delta^{15}N$ can be attributed to weaning, or, in the case of Fogel et al. (1997) tried to detect changes in weaning patterns without estimating ages.

In order to overcome these deficiencies, Schurr (1997) introduced an explicit mathematical model in order to estimate age of onset and relative rates of weaning. He suggests that a version of the Weibull Function (i.e., a saturating exponential) can be used to describe the changes in $\delta^{15}N$ in two stages:

$$\delta^{15}N(t) = \delta^{15}N_{max} - \Delta e^{-R_n t},$$

(from birth to onset of weaning): (5-1)

$$\delta^{15}N(t) = \delta^{15}N_{children} - \Delta e^{-R_w(t-t_s)},$$

(during weaning) (5-2)

where $\delta^{15}N(t)$ is $\delta^{15}N$ at age t, $\delta^{15}N_{max}$ is the maximum attainable $\delta^{15}N$ during breast-feeding, $\delta^{15}N_{children}$ is the average value of $\delta^{15}N$ in weaned children, Δ is the shift from the mother's diet to her milk, R_n and R_w are constants related to collagen synthesis rates and growth rates, with R_w also incorporating a component for the rate of change of diet during weaning, and t_s is the age at which weaning starts.

This model suffers from a number of deficiencies, the most serious one being embodied in equation 5-1: at birth there is an instantaneous shift in dietary $\delta^{15}N$ of magnitude Δ, and for the subsequent change in body-tissue $\delta^{15}N$ to follow a saturating exponential to a new value requires that body weight is constant and R_n represents only a protein-turnover rate. This is clearly unrealistic for an infant, which, in the few months after birth, has a more rapid rate of growth per unit of body weight than it will at any other time in its life (figures 73 and 74 of Tanner, 1989). Thus, rapid growth confounds the use of the Weibull Function and the constants cannot have the claimed physiological significance. Similarly, equation 5-2 is suitable for describing an instantaneous change of diet with constant body weight and is clearly inappropriate for describing the gradual change of diet in the process of weaning a growing infant. Schurr's model thus gives estimates of parameters that, at best, will allow relative comparison of populations but which cannot allow meaningful ages to be calculated.

A NEW MODEL

Any model of $\delta^{15}N$ changes with weaning must, therefore, take into account that weaning is a process which starts at one age (t_s) and ends at another (t_e), and that the infant will be growing. Growth over the period of interest (0–5 years) is well represented by a Jenss Curve (Berkey, 1982) and there are good modern data on infants (Freeman et al., 1995), which can be used as the basis of a model.

Some simplifying assumptions have to be made, notably those outlined below.

(1) It is assumed that non-milk food is sufficiently uniform in its mean $\delta^{15}N$ that it can be treated as constant in $\delta^{15}N$ for both mother and child. There may be a difference between the mother's diet and the weaning food, but we have no other handle on the likely $\delta^{15}N$ of weaning foods in the past. However, as will be seen later, the $\delta^{15}N$ of weaned children is usually slightly below that of adults in archaeological populations, as predicted for growing children on a diet with the same $\delta^{15}N$ as their parents. This suggests that the approximation holds reasonably well. A specialist weaning food, e.g., cereal gruel, will render the model less accurate.
(2) It is assumed that isotopic fractionation of nitrogen can be simplified to two steps: one is the conversion of food into body protein; and the other is nitrogen excretion. That the complexities of nitrogen-isotope biochemistry can be validly simplified in this way has been demonstrated by Schoeller (1999).
(3) It is assumed that the bone-collagen turnover is rapid enough to be considered comparable to whole-body protein turnover. This assumption may be justified on the grounds that whole-body protein turnover time in infants at 0.8–1.6 years (Fogel et al., 1997) is comparable to bone turnover of 100% in the first year of life (Vaughan, 1975).

Nitrogen-Isotope Mass Balance

The $^{15}N/^{14}N$-isotope ratio of a child (N_c) depends on the following.

(1) The rate of consumption of nitrogen as milk (M_m) and non-milk (M_n) foods, their isotope ratios (N_m and N_n respectively) and the fractionation in food absorption (α_c).
(2) The rate of excretion of nitrogen, primarily in urine, (M_u) and the fractionation in its production (α_u).
(3) The total nitrogen content of the child's body which is proportional to its weight as given by the Jenss Equation ($W \propto X + Zt + Se^{-Yt}$), where S, X, Y and Z are mathematical parameters with no physical meaning (Berkey, 1982).
(4) The relative proportions of milk and non-milk nitrogen in the child's food, described here by the proportion of non-milk food ($p = M_n/(M_n + M_m)$), which is a function of age.
(5) The ratio of dietary protein to body protein per unit time ($K = (M_n + M_m)/W$), which will vary with age.

The isotope mass balance equation is thus

$$\frac{d}{dt}(N_c W) = (M_m N_m + M_n N_n)\alpha_c - M_u N_c \alpha_u \qquad (5\text{-}3)$$

and the nitrogen mass balance is

$$\frac{dW}{dt} = M_n + M_m - M_u. \qquad (5\text{-}4)$$

Rearranging and substituting p and K gives

$$\frac{dN_c}{dt} = K[([1-p]N_m + pN_n)\alpha_c - N_c\alpha_u] + N_c(\alpha_u - 1)\frac{1}{W}\frac{dW}{dt}. \qquad (5\text{-}5)$$

Equation 5 has to be solved numerically when p and W vary with t. In the following calculations, a finite difference method implemented as a Microsoft Excel spreadsheet has been used to provide numerical solutions.

Implementation

In order to implement this model, a variety of parameters must be estimated or assumed, but the effect of their variation can be explored to give an idea of the sensitivity of the model to the assumptions.

Ambrose (1991) gives a survey of $\delta^{15}N$ fractionation in a variety of animals and suggests that body tissues are generally +3 to +4‰ from diet and urinary urea is −2 to −4‰ from diet. Here, for simplicity, the inter- and intraspecies variation is neglected and it is assumed that, at steady state, body tissues are +3.5‰ from diet and urine is −3‰ from diet. Consequently, consideration of the isotopic mass balance in the steady state shows that there must be a fractionation of −6.5‰ in producing urea from the general body reservoir of nitrogen, and thus, in order to maintain isotope balance in the simple two-step model adopted here, a fractionation of −3‰ in the absorption of food. Thus, values of $\alpha_u = 0.9935$ and $\alpha_c = 0.9970$ are adopted here.

A numerical model requires that values are specified for the $\delta^{15}N$ values of foods, although the shifts in $\delta^{15}N$ will be independent of the absolute values and depend only on the difference between $\delta^{15}N_m$ and $\delta^{15}N_n$. If the non-milk food is common to mother and child, and if the mother is producing milk with $\delta^{15}N$ one trophic-level shift above that food, then $\delta^{15}N_n = \delta^{15}N_m - 3.5‰$. If $\delta^{15}N_m$ is taken as 8‰, then $\delta^{15}N_n = 4.5‰$. Values of N_m and N_n (the absolute isotope ratios) are calculated using $(^{15}N/^{14}N)_{AIR} = 1/272$ (Cotton and Wilkinson, 1980) and the definition

$$\delta^{15}N = \left(\frac{(^{14}N/^{15}N)_{sample}}{(^{14}N/^{15}N)_{AIR}} - 1\right) \times 1000 \qquad (5\text{-}6)$$

The value of K and its variation with age may be derived from data on protein requirements of infants, scaled to represent likely levels of nourishment. FAO/WHO data on safe levels of protein intake are given by Taylor (1978). Taking into account that safe levels are 1.3 times average requirements (i.e., the mean plus twice the S.D. of 15%) and that protein is only 11% of a neonate's mass, and fitting a suitable, simple equation gives

$$K(\text{kg kg}^{-1}\ \text{a}^{-1}) = 3.57 - 0.64 \ln(t + 0.008) \qquad (5\text{-}7)$$

for *average* nutritional requirements.

W takes the form $W \propto X + Zt + Se^{-Yt}$, and, for convenience here, the median weight curve for U.K. boys in 1990 is used, with $X = 8.39$, $Y = 2.48$, $Z = 2.02$ and $S = 4.89$, unless otherwise indicated.

Finally, the form of $p(t)$ has to be decided. Before weaning $(t < t_s) p = 0$; after weaning, $(t > t_e) p = 1$. During weaning it describes the pattern of supplementation of milk protein by non-milk protein. For exploratory purposes here, four arbitrary models are investigated, as outlined below.

(1) In the linear model, non-milk protein replaces milk protein at a constant rate,

$$p = \frac{t - t_s}{t_e - t_s} \qquad (5\text{-}8)$$

(2) The parabolic model represents slow initial weaning and then rapid final weaning,

$$p = \left(\frac{t - t_s}{t_e - t_s}\right)^2 \qquad (5\text{-}9)$$

(3) The reverse parabolic model represents rapid initial weaning and then slow final weaning,

$$p = 1 - \left(\frac{t_e - t}{t_e - t_s}\right)^2 \qquad (5\text{-}10)$$

(4) The sigmoid model represents slow initial weaning, a period of rapid supplementation and then a slow final weaning from breast-milk,

$$p = \frac{1}{2} + \frac{1}{2}\sqrt[3]{\frac{2t - (t_s + t_e)}{t_e - t_s}} \qquad (5\text{-}11)$$

Values for t_e and t_s, the key parameters of interest, would be varied to fit the data in an archaeological investigation, but, for initial exploratory purposes here, values of $t_s = 0.5$ and $t_e = 2.5$ years will be taken as "typical" for pre-modern societies, on the basis of historical and ethnographic evidence (Stuart-Macadam, 1995).

EXPLORING THE MODEL

Before comparing the model with modern longitudinal studies and applying it to archaeological data sets, it is useful to explore the possible variation in some of the parameters and gauge their effect on the results.

Varying the Weaning Pattern

Figure 5-1 shows variation in $\delta^{15}N$ with age for the three different weaning patterns, and with all other parameters set as indicated above. It is noticeable that the pattern of weaning affects the pattern of $\delta^{15}N$ change, although it is possible that the three patterns could be made indistinguishable by varying the start and end dates of weaning or by the limitation of the current best precision of 0.2‰ in $\delta^{15}N$ measurements. In these scenarios it is worth noting that the maximum $\delta^{15}N$ values are reached 6–9 months after weaning has started, which means that peak $\delta^{15}N$ values do not necessarily represent the onset of weaning with any precision. Also, $\delta^{15}N$

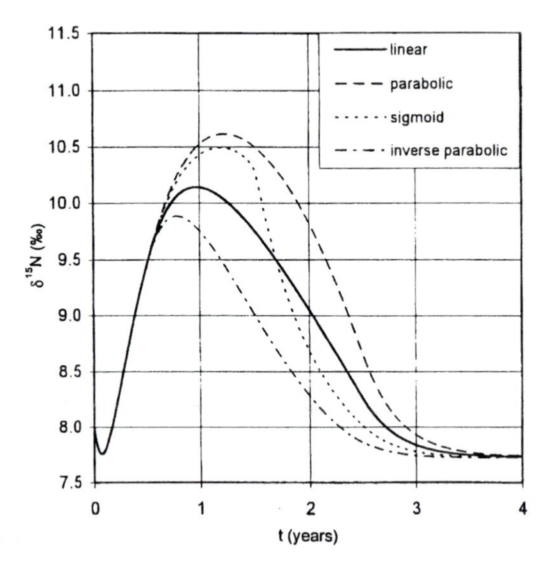

Fig. 5-1. Modeled variation of $\delta^{15}N$ with age for the four weaning patterns described in the text.

values continue to decline for up to a year after weaning has been completed, so the point of leveling out of the $\delta^{15}N$ values will be an overestimate of the end of weaning. However, changes in $\delta^{15}N$ will be more rapid in cases with higher protein intake than with the simple minimum nutritional requirements assumed here. The $\delta^{15}N$ of the fully weaned child settles at a level (7.75‰) below that of an adult (8‰), owing to its continued growth, as observed in several populations previously studied.

Varying Protein Intake

Figure 5-2 shows variation in $\delta^{15}N$ when protein intake varies from the mean requirement. The variation in protein intake is not great (±30% at 2 S.D.) but the variation in maximum $\delta^{15}N$ is almost 1‰ for that variation. Higher levels of protein intake (above minimum nutritional requirements) will lead to peaks in $\delta^{15}N$ similar to those observed in the various populations described above, i.e., 2.5–3.5‰ above adults.

Varying the Growth Curve

Calculations using curves for the 75th and 25th centiles and boys and girls growth curves give very little variation (not shown). This is not surprising as the main weight term in the equation governing $\delta^{15}N$ is $(1/W)(dW/dt)$. Thus it is not the weight of the infant that is important, but its growth per unit weight. Thus $\delta^{15}N$ is only weakly dependent on the value of W, and is more dependent on the form of W.

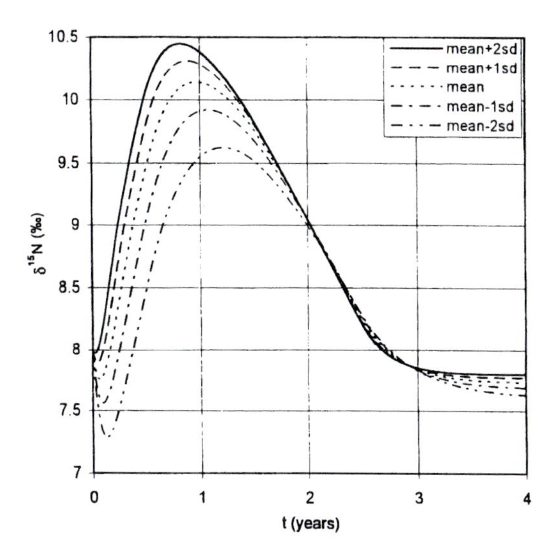

Fig. 5-2. Modeled variation of $\delta^{15}N$ with age as protein intake varies from the mean requirement. 1 S.D. = 15% (Taylor, 1978). A linear weaning pattern, and other parameters, are as described in the text.

COMPARISON WITH LONGITUDINAL STUDIES ON MODERN INFANTS

Very little work has been done on modern infants in comparison even with the half-dozen or so data sets on archaeological populations. Studies on three infants were available for consideration here. One is the data set of Fogel et al. (1997), derived from analyses of fingernail clippings from a mother and daughter over the first year of the child's life (for brevity, the child is referred to here as Baby F); the other two are derived from analyses of hair samples from mother–son (Baby D) and mother–daughter (Baby M) pairs (T. C. O'Connell, unpublished data). All of these infants differed in their manner of weaning.

Baby D

Baby D is the best characterized of the three children, as the approximate ages for his first non-milk food and final breast-feed are known, as well as his actual weight data. Figure 5-3 shows the data for Baby D and modeled changes for two different protein intakes (twice and three times the mean requirement). Other parameters used were $\delta^{15}N_m = 7.5$‰ (average of mother's hair), $\delta^{15}N_n = 4$‰, W given by $X = 6.74$, $S = -3.44$, $Y = 4.43$, $Z = 2.37$, and $t_s = 0.3$, $t_e = 1.17$ (known) with parabolic weaning. Using twice the mean protein requirement gives a better fit than three times the mean, but neither is ideal. Linear, reverse parabolic and sigmoid weaning (not shown) both lead to an earlier fall in $\delta^{15}N$ than the data show. It appears that either the parabolic weaning pattern overestimates the initial rate of weaning or the hair-sample ages are overestimated by 0.1–0.15 year.

Baby M

Figure 5-3 also shows data and modeled $\delta^{15}N$ values for Baby M prior to the onset of weaning. The same two protein intakes are shown, with other parameters as $\delta^{15}N_m = 9.8$‰ (mother's hair), W as the median girls weight curve (Child Growth Foundation, 1994; Freeman et al., 1995) with $X = 8.09$, $S = -4.79$, $Y = 2.21$, $Z = 2.03$. There is a good fit for the protein intake at twice the mean requirement, but not for three times the mean requirement, unless the ages are in error by about 0.1 year.

Baby F

The data of Fogel et al. (1997) are more extensive than those for babies D and M, but they suffer

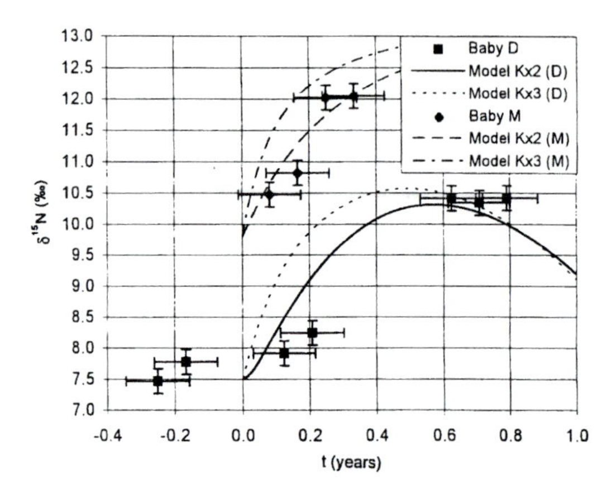

Fig. 5-3. Modeled and actual $\delta^{15}N$ variation for the hair of babies D and M. Length measurements of hair have been converted to ages by assuming it takes 2 weeks (0.04 year) for growth to emerge from the scalp, and that hair grows by 12 cm per year. *Model Kx2* was calculated with a protein intake of twice the mean requirement, and *Model Kx3* was calculated with three times that requirement. Other parameters as indicated in the text. Time is expressed relative to birth, with negative values indicating pre-natal data.

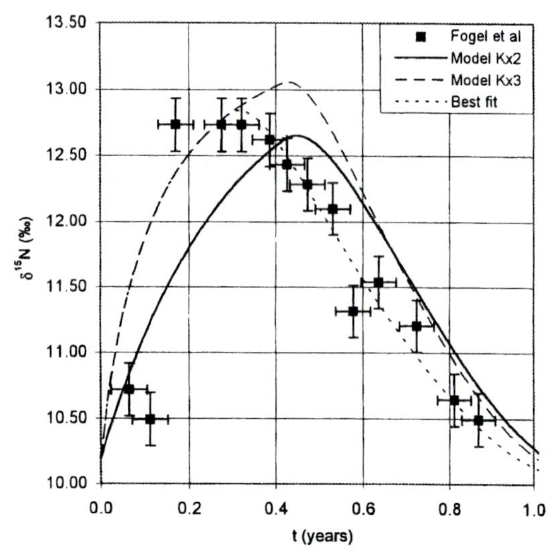

Fig. 5-4. Modeled and actual $\delta^{15}N$ variation for the nails of Baby F (data of Fogel et al., 1997). Ages of sampling have been converted to ages of formation of the nails by subtracting 1.6 months as described in the text. *Model Kx2* was calculated with a protein intake of twice the mean requirement, and *Model Kx3* was calculated with three times that requirement. *Best fit* is calculated with three times the mean protein requirement, and $t_s = 3.5$ months. Other parameters are as indicated in the text.

from a difficulty in estimating age at formation of the fingernail. Fogel et al. suggest that the nail is approx. 2–3 months old at sampling, but it is not clear if this is on the basis of their data, or other information. However, the data from their cross-sectional study (shown in their figure 5; figs 5 and 6 in the paper, but not the captions, are transposed) suggest a shorter interval. They show that some infants have fingernail $\delta^{15}N$ elevated above values that are the values of their mothers in clippings taken at 2 months, and possibly even at 1.8 months, whereas this is not the case for infants sampled at 0.8 and 1.2 months. The data from their longitudinal study are therefore used here with 1.6 months subtracted, to allow for the time between growth and sampling, and 0.5-month error bars show the uncertainty in estimation of age-at-formation.

Figure 5-4 shows age-corrected data and the model with the two different protein intakes. The other parameters are $\delta^{15}N_m = 10.1‰$ (average of mother's nails), and $\delta^{15}N_n = 6.6‰$; W is given by the median girl's weight curve as for Baby M. The start of weaning was 5 months and the end 13 months (M. L. Fogel, personal communication). Only a diet with two or more times the dietary requirement can account for the 2.5‰ rise in $\delta^{15}N$. However, the modeled curve shows a fall that is too late in comparison with the data, and a value earlier than the true one of $t_s = 3.5$ months has to be taken to fit the data (figure 5-4). The shape of the curve is best fitted with the reverse parabolic model.

Discussion of Modern Studies

All three modern infants studied show that protein intake estimated by nutritional requirement only is a severe underestimate of the protein obtained from breast-milk and weaning foods, and that 2–3 times the mean requirement is a better estimate of their true intake. On this basis, 2.5 times the mean is the value used to model archaeological populations below.

For 0.2 year after birth, all three infants show lower values than predicted. This could be due to an underestimation in the delay between growth and sampling, or it may be a real phenomenon. Baby M and Baby F then show a more rapid rise in $\delta^{15}N$ than can be accounted for by the model (there are no data for Baby D for this age). This may be because the model does not take account of a neonatal weight loss of 5–8% in the first few days (Akre, 1991). Although most of this is fluid, any reduction in total body protein by urea excretion will lead to increased $\delta^{15}N$. Breast-milk is also c.15% richer in nitrogen in the first month than later in lactation (Akre, 1991), which may mean that K should be higher soon after birth.

The difference in weaning pattern between babies D and F probably reflects the different methods of weaning of the two infants. Baby F was first weaned onto infant formula, a food designed to substitute, rather than supplement, breast-milk, so her initial weaning may have been more rapid than that of Baby D, who was never fed formula. The pattern of Baby D's weaning is thus more likely to approximate patterns used in the past before the advent of formula. For this reason, a parabolic weaning pattern is adopted in the modeling of archaeological populations below.

The data for Baby F show a severe underestimate of the timing of the fall in δ^{15}N with weaning. There are several possible causes for this, as outlined below.

(1) The δ^{15}N of the weaning food may have been overestimated, but realistic reductions in the value used in the model fail to account for the observed fall.
(2) The initial rate of weaning may have been faster than that estimated by the inverse parabolic pattern.
(3) Body protein is unlikely to be homogeneously mixed with incoming protein in the synthesis of new protein, as the model assumes; in adults, hair keratin responds to a dietary change in δ^{13}C within a few weeks (O'Connell and Hedges 1999), whereas body-protein turnover time is much greater than this. Hair and nails are therefore likely to show a more rapid response to dietary change than whole-body protein or bone collagen.

The model, designed for whole-body protein, is thus underestimating the rate of change observed in nails. When a fitted model is used, the beginning of weaning is placed too early: this suggests that the same may be true, albeit to a lesser extent, in archaeological populations.

Further testing and development of the model requires more δ^{15}N data sets covering the pre- *and* post-weaning periods, as well as the weaning period, and ideally are accompanied by information on infant growth and the ages of weaning onset and completion. If a method could be found to sample whole-body protein rather than newly synthesized protein, then that would be advantageous.

COMPARISON WITH ARCHAEOLOGICAL DATA

Two archaeological populations are compared with the model as exemplars of the potential and problems of the modeling approach. Both data sets are modeled with parabolic weaning, with W as the median weight curve for boys, and 2.5 times the mean required protein intake.

MacPherson Site

Katzenberg et al. (1993) report data from a protohistoric maize horticulturalist site at MacPherson, Ontario. Their data are shown in figure 5-5, together with δ^{15}N modeled with the parameters $\delta^{15}N_m = 11.6‰$ (adult average), $\delta^{15}N_n = 8.1‰$,

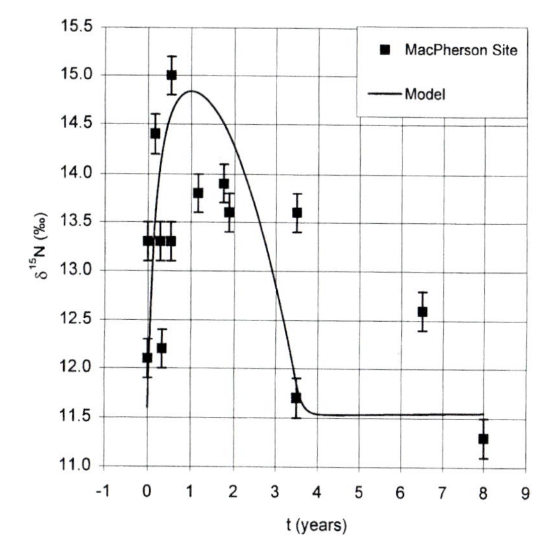

Fig. 5-5. Data on δ^{15}N of children from MacPherson, Ontario (Katzenberg et al., 1993), together with the model calculated with parameters as described in the text. Time is expressed relative to birth, with negative values indicating prenatal data.

$t_s = 0.5$, and $t_e = 3.5$ years. The fit of the curve to the points is not very good, mostly because of the scatter of points due to variation between individuals. Varying the parameters, e.g., with a $\delta^{15}N_c$ at birth of 12.2‰, or $t_e = 4$, makes little difference to the fit of the model to the data.

Angel Site

The data of Schurr (1997) on prehistoric maize farmers from the Ohio Valley in Indiana are plotted in figure 5-6 with modeled δ^{15}N values. The modeled curves use $\delta^{15}N_m = 8.3‰$ (adult average), $\delta^{15}N_n = 4.8‰$, $t_s = 0.5$, and $t_e = 3$ years. The fit of the model to the data is better than in the previous case, but is still limited by the scatter of points. The $\delta^{15}N_c$ at birth can be adjusted to 7.5‰ using the δ^{15}N evidence from neonates, but this makes little change to the fit of the curve. In Model 2, $\delta^{15}N_n$ is reduced to 7.8‰ on the basis of the neonates' and older children's δ^{15}N values, and this curve appears to give a better fit to the data; without an explanation as to why $\delta^{15}N_m$ should differ from the adult mean, however, it is difficult to justify changing it.

Also shown is the model of Schurr (1997), using the parameters he claims as best fit. It is clear that, in comparison to the model presented here, the Weibull Function gives an unrealistically slow increase in δ^{15}N after birth and overestimates the persistence of the increased δ^{15}N due to breastfeeding.

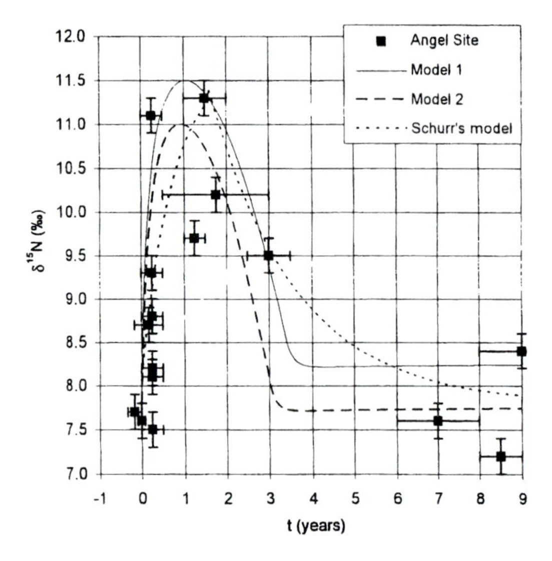

Fig. 5-6. Data on $\delta^{15}N$ of children from Angel Site, Indiana (Schurr, 1997), together with the model calculated with parameters as described in the text. *Model 1* takes $\delta^{15}N_m = 8.3‰$. *Model 2* takes $\delta^{15}N_m = 7.8‰$. *Schurr's model* was calculated from equations 5-1 and 5-2 using $t_s = 1.6$, $\delta^{15}N_{max} = 11.7‰$, $\delta^{15}N_{children} = 7.8‰$, $\Delta = 3.6‰$, $R_n = 1.33$ and $R_w = 0.5$, as given in Schurr (1997). Time is expressed relative to birth, with negative values indicating pre-natal data.

Discussion of the Modeling of Archaeological Data

The comparison of the (longitudinal) model with (pseudo-cross-sectional) archaeological data is hampered by interindividual variation, which can arise both from variations in the diets of the mothers and variations in weaning practice from child to child. This means that although there are clearly visible trends in archaeological data sets, the points are quite scattered and the fitting of curves to obtain population averages is difficult. More formal measures of the fit of individual curves might generate "best" fit values for the various parameters of the model, but the range of "good" fit values is likely to be quite large.

Another limitation of the archaeological data sets is that the rapid change in $\delta^{15}N_c$ in the first year of life, combined with imprecise estimates of age and the errors in $\delta^{15}N$ measurement, will always limit the precision of curve-fitting for the pre-weaning stage. Conversely, in most populations there is a period of reduced mortality in the 2–10-year age group, which means that there are few data points and limits to the precision of estimates of the end of weaning. Also, individuals from archaeological contexts represent deaths from unknown causes, and some of them may never have been breast-fed.

Despite these difficulties, the attempt to model the data sets here shows that the weaning period probably extended to 2.5–3.5 years in the first case and 3–4 years in the second case. These ages for the end of weaning are later than those suggested by the generality of ethnographic and historical evidence (Stuart-Macadam, 1995) but are by no means extreme. They are also older than the suggestions of the authors of the original studies suggested; the difference is only slight for the MacPherson site, where 2–3 years is suggested on the basis of historical evidence (Katzenberg et al., 1993) but the difference is much more significant in the case of the Angel site, where "sometime during the second year of life" is suggested based on the same isotopic evidence (Schurr, 1997).

CONCLUSIONS

The model developed here has been validated against the available modern longitudinal data sets, and appears realistic within the limits of accuracy of both model and data. To evaluate the model fully requires modern data on whole-body protein or a revision of the model to account for several "compartments" of protein within the body. In order to be usefully applicable to archaeological populations it will require large populations (as do demographic methods—Herring et al., 1998) which contain sufficient individuals >2 years old for the age of completion of weaning to be established. Larger populations also lend themselves to the possibility of calculating meaningful averages for a series of age classes (0–3 months, 3–6 months, etc.) containing a good number of individuals and comparing them to the model. This would reduce the problem of scatter demonstrated above, and would allow some sort of population average weaning pattern to be established.

It should also be possible to conduct longitudinal studies on archaeological individuals, using either hair from mummified infants or tooth dentine. Tooth dentine is laid down at the time of growth of the tooth and remains metabolically isolated thereafter. Deciduous teeth are formed from 5 months before birth to 3–4 years after birth, and permanent teeth are forming from birth (Hillson, 1986); thus, appropriate sampling may yield limited longitudinal data on an individual. A variant of this approach has recently been demonstrated for weaning in cattle by sampling bone from the region of the first molar and from the dental bud region to give samples with different formation ages (Balasse et al., 1997).

Another desirable archaeological study is one which combines historical records of weaning practices with $\delta^{15}N$ study of the same population, and thus allows testing of the deductions made by comparing a longitudinal model with population data.

Studies of weaning in archaeological populations are in their infancy. It is now time to wean them off simple qualitative analyses of the weaning process onto the more solid fare of quantitative estimates which can contribute to clarification of the archaeological, demographic, medical, and evolutionary questions outlined at the start of this Chapter. This model is a step in that direction.

Acknowledgments: I am grateful to the conference organizers for the opportunity to present this work, and to the Carnegie Institute of Washington and the University of Durham for funding my attendance at the conference. I am very grateful to Tamsin O'Connell for allowing me to use her unpublished data on modern mothers and babies. I also thank my wife and family, who have tolerated (and sometimes encouraged) discussions about breast-feeding and weaning. Reviewers' comments from Marilyn Fogel and Anne Katzenberg have greatly improved this contribution.

REFERENCES

Akre J (ed.) (1991) *Infant Feeding: the Physiological Basis. Supplement to Bulletin of the World Health Organization*, **vol. 67** (1989). World Health Organization.

Ambrose S. H. (1991) Effects of diet, climate and physiology on nitrogen isotope abundances in terrestrial foodwebs. *J. Archaeol. Sci.* **18**, 293–317.

Balasse M., Bocherens H., Tresset A., Mariotti A., and Vigne J.-D. (1997) Émergence de la production laitière au Néolithique? Contribution de l'analyse isotopique d'ossements de bovins archéologique. *C. R. Acad. Sci. Paris, Ser. II, Sci. Terre Planèt* **325**, 1005–1010.

Berkey C. S. (1982) Comparison of two longitudinal growth models for preschool children. *Biometrics* **38**, 221–234.

Bryant J. D. and Froelich P. N (1996) Oxygen isotope composition of human tooth enamel from Medieval Greenland: linking climate and society (Comment) *Geology* **24**, 477–478.

Child Growth Foundation (1994) Boys'/girls' growth charts. In *Personal Child Health Record.* British Paediatric Association.

Cotton F. A. and Wilkinson G. (1980) *Advanced Inorganic Chemistry: a Comprehensive Text.* Wiley.

Fogel M. L., Tuross N., and Owsley D. W. (1989) Nitrogen isotope tracers. *Carnegie Instn Washington Yrbk* **88**, 133–134.

Fogel M. L., Tuross N., Johnson B. J., and Miller G. (1997) Biogeochemical record of ancient humans. *Org. Geochem.* **27**, 275–287.

Freeman J. V., Cole T. J., Chinn S., Jones P. R. M., White E. M., and Preece M. A. (1995) Cross sectional stature and weight reference curves for the UK, 1990. *Arch. Disease Childhood* **73**, 17–24.

Herring D. A., Saunders S. R., and Katzenberg M. A. (1998) Investigating the weaning process in past populations. *Amer. J. Phys. Anth.* **105**, 425–439.

Hillson S. (1986) *Teeth.* Cambridge University Press.

Katzenberg M. A. (1993) Age differences and population variation in stable isotope values from Ontario, Canada. In *Prehistoric Human Bone: Archaeology at the Molecular Level.* (eds J. B. Lambert and G. Grupe), pp. 39–62. Springer Verlag.

Katzenberg M. A. and Pfeiffer S. (1995) Nitrogen isotope evidence for weaning age in a nineteenth century Canadian skeletal sample. In *Bodies of Evidence* (ed. A. L. Grauer), pp. 221–235. Wiley.

Katzenberg M. A., Saunders S. R., and Fitzgerald W. R. (1993) Age differences in stable carbon and nitrogen isotope ratios in a population of prehistoric maize horticulturists. *Amer. J. Phys. Anth.* **90**, 267–281.

Katzenberg M. A., Herring D. A., and Saunders S. R. (1996) Weaning and infant mortality: evaluating the skeletal evidence. *Yrbk Phys. Anth.* **39**, 177–199.

O'Connell T. C. and Hedges R. E. M. (1999) Investigations into the effect of diet on modern human hair isotopic values. *Amer. J. Phys. Anth.* **108**, 409–425.

Schoeller D. A. (1999) Isotope fractionation: why aren't we what we eat? *J. Archaeol. Sci.* **26**, 667–673.

Schurr M. R. (1997) Stable nitrogen isotopes as evidence for the age of weaning at the Angel site: a comparison of isotopic and demographic measures of weaning age. *J. Archaeol. Sci.* **24**, 919–927.

Schurr M. R. (1998) Using stable nitrogen isotopes to study weaning behavior in past populations. *World Archaeol.* **30**, 327–342.

Sillen A. and Smith P. (1984) Weaning patterns are reflected in strontium–calcium ratios of juvenile skeletons. *J. Archaeol. Sci.* **11**, 237–245.

Stuart-Macadam P. (1995) Breastfeeding in prehistory. In *Breastfeeding: Biocultural Perspectives* (eds P. Stuart-Macadam and K. A. Dettwyler), pp. 75–99. Aldine de Gruyter.

Tanner J. M. (1989) *Foetus into Man: Physical Growth from Conception to Maturity*, 2nd edn. Castlemead Publications.

Taylor T. G. (1978) *Principles of Human Nutrition.* Edward Arnold and Institute of Biology.

Vaughan J. (1975) *The Physiology of Bone.* Oxford University Press.

White C. D. and Schwarcz H. P. (1994) Temporal trends in stable isotopes for Nubian mummy tissues. *Amer. J. Phys. Anth.* **93**, 165–187.

6. Isotopic integrity of α-carboxyglycine (aminomalonate) in fossil bone based on ^{14}C data

Richard R. Burky, Donna L. Kirner, R. E. Taylor, P. E. Hare, and John R. Southon

Bone is an important source of data in paleontological and archaeological studies involving a wide range of research topics, including the reconstruction of terrestrial paleoenvironments (Behrensmeyer and Hill, 1980; Price 1989). For example, bone has been used to reconstruct evolving dietary patterns both for fossil hominids and for more recent human populations (Schoeninger, 1989). The changing subsistence regimes of various hominid populations are an important part of our understanding of the evolution of past human behavioral systems based on archaeological and paleoanthropological data. In such studies, it is important to document temporal relationships securely since a primary goal is understand how hominid diets have changed over time in diverse environments. However, this is only one of a number of important contexts for efforts to obtain accurate ^{14}C age determinations on bone (Taylor, 1982, 1987, 1991).

Dialogue concerning the reliability of ^{14}C-based age determinations on bone have occurred essentially from the beginning of ^{14}C research. Despite the great amount of attention given to the exclusion of contamination by isolation and purification of specific chemical and molecular fractions of bone, there continues to be a tradition of skepticism concerning the general reliability of ^{14}C values for bone (e.g., Brown, 1988; see Taylor, 1987, 1992; and Hedges and van Klinken, 1992 for literature).

Despite the widespread knowledge of problems with bone in ^{14}C work, for samples containing significant quantities of well-preserved collagen, it is now generally agreed by most radiocarbon specialists that appropriate physical and chemical pretreatment can, in the majority of cases, effectively isolate and purify the *in situ* residual collagen and that accurate ^{14}C age estimates can be obtained. Many laboratories now have a series of explicit acceptance/rejection criteria to determine which bones can and cannot be expected to yield accurate ^{14}C age specimens (Hedges and Law, 1989; Long et al., 1989; Taylor, 1992). For well-preserved bone samples, the isolation of a total amino acid fraction along with the use of chromatographic methods to remove humate compounds (Gillespie et al., 1986; Stafford et al., 1987, 1988; Long et al., 1989; van Klinken and Mook, 1990) have become widespread. Another approach adds an ultrafiltration step designed to exclude low-molecular-weight components under the assumption that most of the exogenous contamination will be contained in this fraction (Brown et al., 1988). Yet another strategy, originally developed to purify collagen for stable-isotope analyses, involves the use of collagenase, which preferentially isolates peptides of known length from the surviving collagen fragments (DiNiro and Weiner, 1988).

THE PROBLEM

In contrast to the various techniques generally employed successfully to purify collagen in cases in which it is retained in appropriate quantities, a continuing problem is the challenge of obtaining reliable ^{14}C-based age estimates on collagen-degraded bone. For many geographical regions, including many temperate (and most tropical) areas, some of the most interesting archaeologically related bones exhibit low ($<5\%$) or trace ($<1\%$) amounts of collagen. There is currently no consensus as to biogeochemical methods that can be used in these types of bones to distinguish indigenous amino acids, peptides, and other products of collagen diagenesis from external contamination.

Because of the difficulty of identifying and isolating indigenous organics and excluding exogenous ones in collagen-deficient bone, previous researchers have suggested targeting one or more major ($>1\%$) non-collagen component(s) of bone. Sometimes characterized as "matrix proteins," these components include osteocalcin, osteonectin and other phosphoproteins, proteoglycans, and sialo- and glycoproteins (Hauschka and Wiams, 1989). Other non-collagen components include blood proteins such as hemoglobin, serum albumin and the immunoglobins (Nelson et al., 1986; Masters, 1987; Long et al., 1989).

The UCR Radiocarbon Laboratory has previously been involved in examining the biochemical and isotopic integrity of osteocalcin ^{14}C values in fossil bone (Ajie et al., 1990, 1991, 1992). It was originally thought that several properties of osteocalcin offered the possibility that it would be a potentially useful protein in the ^{14}C dating of

TABLE 6-1 Aminomalonate (Am) content compared with other bone fractions. Total amino acids (AA) [total AA in acid hydrolysate], Gla, and Am content and compositional data for bones used in the study are listed in order of decreasing expected age. Total AA/acid, Gla, and Am content are expressed as percentages contained in our modern bone standard. Am values are in bold italics

Locality	Total AA/acid (% modern)[a]	Gla (% modern)[a]	***Am (% modern)***[a]	Gly/Glu[b]
1. Medicine Hat, Alberta, Canada	49	58	***2300***	2.8
2. Abri Pataud, Les Eyzies, Dordogne, France	30	38	***400***	4.4
3. La Brea Tar Pits, Los Angeles, California	79	38	***200***	4.5
4. Colorado Creek Mammoth, Alaska	109	42	***1400***	4.2
5. Burning Tree Mastodon, Allen County, Indiana	101	<1	***500***	5.4
6. Dent Mammoth, Colorado	28	34	***200***	4.5
7. Wizard Beach, Pyramid Lake, Nevada	93	52	***600***	4.3
8. Spirit Cave, Fallon County, Nevada	119	80	***300***	4.2

[a] Modern bone used as the standard for 100% total amino acids, Gla, and Am.
[b] Our modern bone standard exhibits a Gly/Glu ratio of 4.1 (Hare, 1980).

collagen-degraded fossil bones (Hauschka, 1977, 1980; Hauschka et al., 1989). Although early results were interpreted as potentially promising, recent studies have determined that in many fossil bones, osteocalcin and its characteristic amino acid, γ-carboxyglutamic acid (Gla), do not retain biogeochemical or isotopic integrity (Burky et al., 1998).

ISOLATION OF AMINOMALONATE IN FOSSIL BONE

The occurrence of aminomalonate in fossil bone was determined as a consequence of studies involving the isolation of Gla for ^{14}C analysis. The extraction of Gla was carried out using base hydrolyses followed by high-performance liquid chromatography (HPLC) separation on a cation-exchange resin. (Base hydrolyses are required since acid hydrolysis converts Gla to Glu.) An American Research Products Corporation (ARPCO) HPLC system and Bio-Rad analytical grade cation-exchange resin (AG50W0X8, 200–400 mesh, hydrogen form) was used. Gla was identified by comparing the position and timing of its elution peak with that of a commercial Gla standard (Sigma No. C3767) on an HPLC chromatogram. Additional verification of the identity of Gla was attempted by applying a thermal decarboxylation test (Hauschka et al., 1980).

Unexpectedly, it was found that both glutamic acid and glycine resulted from the decarboxylation of the eluted product. If only Gla were present in the elutant, decarboxylation would have been expected to yield only glutamic acid. By adjusting the conditions under which amino acids were eluted off our cation-exchange HPLC system, two peaks were resolved. When the elutants were individually decarboxylated, only one yielded glutamic acid. This is the result expected from the decarboxylation of Gla.

The decarboxylation of the second elutant yielded glycine. On the basis of previous work of Hauschka et al. (1980), it was proposed that the second peak contained α-carboxyglycine or aminomalonate (Am). To test this possibility, Am was produced by base hydrolysis of both diethyl aminonomalonate hydrochloride (Sigma No. D7144) and acetamidomalonic acid diethyl ester (Sigma No. A6384). The product was applied to our ion-exchange resin and eluted in single peak that had a reproducible characteristic retention time. No other peaks were observed. The second elutant obtained from the fossil bone samples exhibited the same characteristic retention time. The details of the chromatographic procedures employed are presented in Burky (1996). No mass spectrometric analysis of the elutant was attempted because of the amount of sample available.

TABLE 6-2 Am radiocarbon determinations compared with other fractions. AMS-based ^{14}C ages of total amino acid (total AA) obtained by acid- or base hydrolysis, Gla and Am fractions isolated from bone samples from archaeological and paleontological contexts. Am ^{14}C values are in bold italics

Site/Fraction	CAMS-	$\delta^{13}C$(‰)	^{14}C age (yr BP)
1. Medicine Hat, Alberta, Canada (UCR-1421)			
Total AA/acid	24195	−18[a]	46,990 ± 1170
	26944	−18.5	49,600 ± 2100
Total AA/base	22577	−18.1	> 50,000
	26934	−18[a]	42,310 ± 280
	26935	−18[a]	44,600 ± 300
	26936	−18[a]	51,920 ± 320
	26937	−18[a]	50,910 ± 380
	26938	−18[a]	52,080 ± 540
	28125	−18.1	47,920 ± 320
	28126	−18.1	44,290 ± 370
	28127	−18.1	42,840 ± 190
	28128	−18.1	42,100 ± 220
	28515	−18.1	43,910 ± 370
	28516	−18.1	42,220 ± 180
	28517	−18.1	41,160 ± 340
	29813	−18.1	42,210 ± 260
	29814	−18.1	40,950 ± 250
	29815	−18.1	44,270 ± 450
	29816	−18.1	46,470 ± 300
Gla	22587	−19[a]	28,220 ± 810
	24196	−18[a]	28,170 ± 340
	26939	−18.4	20,400 ± 140
	28117	−17.8	1,000 ± 70[b]
	28118	−19.3	27,640 ± 330
	28522	−17.3	25,160 ± 340
	28523	−18[a]	17,980 ± 150
Am	***24198***	***−18***[a]	***28,810 ± 190***
	26946	***−18***[a]	***32,420 ± 310***
	28119	***−18***[a]	***8,140 ± 70***[b]
	28120	***−18***[a]	***33,140 ± 610***
	28520	***−14.4***	***36,000 ± 1110***
	28524	***−12.9***	***34,700 ± 920***
2. Abri Pataud, Les Eyzies, Dordogne, France (UCR-3430)			
Total AA/acid	28518	−19.8	31,130 ± 480
Total AA/base	29807	−19[a]	31,070 ± 470
Gla	29796	−19[a]	15,430 ± 150
Am	***29804***	***−19***[a]	***14,440 ± 80***
3. La Brea Tar Pits, Los Angeles County, California (UCR-2470)			
Total AA/acid	28121	−19.3	15,600 ± 90
Total AA/base	29809	−19[a]	15,680 ± 80
Gla	29797	−19[a]	15,740 ± 100
Am	***29802***	***−19***[a]	***17,440 ± 110***
4. Colorado Creek Mammoth, Alaska (UCR-3441)			
Total AA/acid	28123	−18[a]	16,440 ± 90
Total AA/base	29808	−18[a]	16,260 ± 90
Gla	29799	−18[a]	15,570 ± 130
Am	***29805***	***−18***[a]	***16,330 ± 90***

Site/Fraction	CAMS-	$\delta^{13}C$(‰)	^{14}C age (yr BP)
5. Burning Tree Mastodon, Allen County, Indiana (UCR-3253)			
Total AA/acid	26949	−19[a]	10,960 ± 60
Total AA/base	26948	−19[a]	10,910 ± 50
Gla	26942	−18.2	10,130 ± 70
Am	***26941***	***−19***[a]	***10,530 ± 70***
6. Dent Mammoth, Colorado (UCR-3131)			
Total AA/acid	26950	−19[a]	10,840 ± 70
Total AA/base	28116	−13.7	11,000 ± 70
Gla	28113	−16.1	8810 ± 60
Am	***28114***	***−17.2***	***7660 ± 60***
7. Wizard Beach, Pyramid Lake, Nevada (UCR-3445)			
Total AA/acid	28124	−15.3	9250 ± 60
Total AA/base	29810	−15[a]	9200 ± 60
Gla	29798	−15[a]	9450 ± 100
Am	***29803***	***−15***[a]	***9240 ± 70***
8. Spirit Cave, Fallon County, Nevada (UCR-3260)			
Total AA/acid	28122	−15.4	9490 ± 60
Total AA/base	29806	−22[a]	9330 ± 60
Am	***29801***	***−22***[a]	***9870 ± 120***

[a] Estimated $\delta^{13}C$ value.
[b] Value seriously anomalous; excluded from summary of comparisons in table 6-3.

The presence of Am in base hydrolysates of fossil bone was not expected since Hauschka et al. (1980) had previously reported that an examination of various proteins and tissues had yielded no evidence of this amino acid.

AMINOMALONATE CONTENT IN FOSSIL BONE

We have examined eight samples of fossil bone in which Am along with Gla and total amino acid fractions were isolated (table 6-1).

In the bones examined, Gla tends to decrease in fossil bones in concert with the total amino acid concentration, which in general, reflects reductions in collagen content. The suggestion that there would be a preferential retention of Gla in fossil bone is not supported by these data. As expected, there is no correlation with age; rather, the Gla content and the total amino acid content as a percentage in our modern bone standard were, in general, highest in bones from environments characterized by constant cold and dry conditions, regardless of age. We lack data on effective soil pH at the sites from which the fossil bone is derived, but this also could be an important factor.

In contrast, Am concentration typically exceeds that which is characteristic of modern bone, sometimes, as in the case in a bone of Middle Pleistocene age from Medicine Hat, Alberta, Canada, by a factor of more than 20. This pattern suggests that Am is produced as a breakdown product, perhaps through several diagenetic mechanisms.

It should be noted that the measured concentration of Gla in our modern bone standard was 545 nmol/g, while the concentration of Am was measured as 438 nmol/g. King (1980) had previously reported a Gla concentration of 1100 nmol/g in modern bone. This is approximately a factor of 2 larger than our measurements. One possible explanation of this discrepancy is that King used a HPLC system and procedures in which Gla and Am coeluted in the same peak. Other factors, including incomplete hydrolysis, might also be partly responsible for the discrepancy.

AMINOMALONATE RADIOCARBON VALUES

We have obtained a suite of ^{14}C values on Am fractions isolated from a series of eight bone samples excavated from a series of localities ranging in age, previously estimated or directly determined by various means, from about 9.4 ka to >50,000 ka. From the same samples, total amino acid fractions obtained by both acid- and base hydrolysis were also isolated and ^{14}C analysis

TABLE 6-3 Summary of Am radiocarbon comparisons showing expected ages (ka) and measured ages (^{14}C yr BP) of different fractions isolated from bone samples from archaeological and paleontological contexts. Where duplicate measurements were undertaken, the maximum range in values are indicated with the number of duplicate ^{14}C determinations in parenthesis. Am ^{14}C values are in bold italics

		Measured ^{14}C age (yr × 10^3 BP)			
Locality	Expected age (ka)	Total AA	Gla	*Am*	Total AA-Am (yr × 10^3 BP)
1. Medicine Hat, Alberta, Canada	>50	46.9–49.6 (2)	18.0–28.1 (6)	***32.4–34.7 (5)***	14.5–14.9
2. Abri Pataud, Les Eyzies Dordogne, France	31.8	31.1	15.4	***14.4***	16.7
3. La Brea Tar Pits, Los Angeles County, California	15.5	15.6	15.7	***17.4***	−1.8
4. Colorado Creek Mammoth, Alaska	15.1	16.4	15.6	***16.3***	0.1
5. Burning Tree Mastodon, Allen County, Indiana	11.0	11.0	10.1	***10.5***	0.5
6. Dent Mammoth, Colorado	10.8	10.8	8.7	***7.6***	3.2
7. Wizard Beach, Pyramid, Lake, Nevada	9.5	9.2	9.5	***9.2***	0.0
8. Spirit Cave, Fallon County, Nevada	9.4	9.5	–	***9.9***	−0.4

undertaken (table 6-2). The ability to undertake these analysis was made possible by the development of methods to obtain ^{14}C measurements by accelerator mass spectrometry (AMS) technology on as little as 20 μg carbon (Kirner et al., 1995, 1996, 1997).

Despite problems with the reproducibility in total amino acid ^{14}C values for our bone blank (Medicine Hat), the age offsets reflected in the Gla and Am ^{14}C values indicate significant age anomalies reflected in these fractions (table 6-3). Assuming that contamination is modern, Gla extracts experienced a greater degree of contamination than was reflected in the Am ^{14}C values.

CONCLUSION

As part of our study of the isotopic characteristics of Gla in fossil bone, we have detected, for the first time in this sample type, the presence of Am. In the bones we studied, Am concentration exceeds that which is characteristic of our modern bone standard, suggesting that it is produced by one or more diagenetic process(es). Also, on the basis of the anomalous ^{14}C ages exhibited by the Am fractions, it appears that the isotopic integrity of Am, like Gla, can be significantly compromised.

Acknowledgments: The UCR Radiocarbon Laboratory is supported by the Gabrielle O. Vierra Memorial Fund, the Dean of the UCR College of Humanities, Arts and Social Sciences, and the UCR Academic Senate Intermural Research Fund. The Lawrence Livermore National Laboratory is funded by the United States Department of Energy under Contract W-7405-Eng-48. We acknowledge the dedicated laboratory assistance of Karen Selzer (UCR Radiocarbon Laboratory). We also wish to acknowledge the helpful assistance of those supplying the bone samples used in this study: Archie Stocker (Medicine Hat); J. van der Plicht, University of Groningen, The Netherlands (Abri Pataud); Roland Gangloff (Titakuk River and Colorado River Mammoth); George Jefferson, Page Museum, Los Angeles, California (La Brea Tar Pits); C. Vance Haynes, Jr, University of Arizona, Tuscson (Dent, BLM, BWD and Naco Mammoth samples); Donald Tuohy and Amy Dansie, Nevada State Museum, Carson City, Nevada (Wizard Beach and Spirit Cave); Gail Kennedy, University of California, Los Angeles, (Haverty) and David Stronanch, University of California, Berkeley [Monsul (Nineveh)]. This is contribution 98/6 of the Institute of Geophysics and Planetary Physics, University of California.

REFERENCES

Ajie H. O., Kaplan I. R., Slota P. J., Jr, and Taylor R. E. (1990) AMS radiocarbon dating of bone osteocalcin. *Nuclear Instr. Methods Physics Res.* **B52**, 433–437.

Ajie H. O., Hauschka P. V., Kaplan I. R., and Sobel H. (1991) Comparision of bone collagen and osteocalcin for determination of radiocarbon ages and paleodietary reconstruction. *Earth Planet. Sci. Lett.* **107**, 380–388.

Ajie H. O., Kaplan I. R., Hauschka P. V., Kirner D., Slota P. J., Jr, and Taylor R. E. (1992) Radiocarbon dating of bone osteocalcin: Isolating and characterizing a non-collagen protein. *Radiocarbon* **34**, 296–305.

Behrensmeyer A. K. and Hill A. P. (eds) (1980) *Fossils in the Making: Vertebrate Taphonomy and Paleoecology*. University of Chicago Press.

Brown F. H. (1988) Geochronometry. In *Encyclopedia of Human Evolution and Prehistory* (eds I. Tattersall, E. Delson, and J. Van Couvering), pp. 222–225. Garland Publishing.

Brown T. A., Nelson D. E., Vogel S. J., and Southon, J. R. (1988) Improved collagen extraction by modified Longin method. *Radiocarbon* **30**, 171–177.

Burky R. R. (1996) Radiocarbon dating archaeologically significant bone using gamma-carboxyglutamic acid (Gla) and alpha-carboxyglycine (aminomalonate). Ph.D. Thesis, University of California, Riverside, California, USA.

Burky R. R., Kirner D. L., Taylor R. E., Hare P. E., and Southon J. R. (1998) ^{14}C dating of bone using gamma-carboxyglutamic acid (Gla) and alpha-carboxyglycine (aminomalonate). *Radiocarbon* **40**, 11–20.

DeNiro M. J. and Weiner S. (1988) Chemical, enzymatic and spectroscopic characterization of "collagen" and teeth organic fraction from prehistoric bones. *Geochim. Cosmochim. Acta* **52**, 2197–2206.

Gillespie R., Hedges R. E. M., and Humm M. J. (1986) Routine AMS dating of bone and shell proteins. *Radiocarbon* **28**, 451–456.

Hare P. E. (1980) Organic geochemistry of bone and its relation to the survival of bone in the natural environment. In *Fossils in the Making* (eds K. Behrensmeyer and A. P. Hill), pp. 208–219. University of Chicago Press.

Hauschka P. V. (1977) Quantitative determination of gamma-carboxyglutamic acid in proteins. *Anal. Biochem.* **80**, 212–223.

Hauschka P. V. (1980) Osteocalcin: a specific protein of bone with potential for fossil dating. In *Biogeochemistry of Amino Acids* (eds P. E. Hare, T. C. Hoering, and K. King), pp. 75–82. Wiley.

Hauschka P. V. and Wiams F. H., Jr (1989) Osteocalcin–hydroxyapatite interaction in the extracellular organic matrix of bone. *Anatom. Rec.* **224**, 180–188.

Hauschka P. V., Henson E. B., and Gallop P. M. (1980) Quantitative analysis and comparative decarboxylation of aminomalonic acid, beta-carboxyaspartic acid, and gamma-carboxyglutamic acid. *Anal. Biochem.* **108**, 57–63.

Hauschka P. V., Lian J. B., Cole D. E. C., and Gundberg C. M. (1989) Osteocalcin and matrix Gla protein: vitamin K-dependent proteins in bone. *Physiol. Rev.* **69**, 990–1047.

Hedges R. E. M. and van Klinken G. J. (1992) A review of current approaches in the pretreatment of bone for radiocarbon dating by AMS. *Radiocarbon* **34**, 279–291.

Hedges R. E. M. and Law I. A. 1989 The radiocarbon dating of bone. *Appl. Geochem.* **4**, 249–253.

King K., Jr (1980) Gamma-carboxyglutamic acid in fossil bone. In *Biogeochemistry of Amino Acids* (eds P. E. Hare, T. C. Hoering and K. King, Jr), pp. 491–501. Wiley.

Kirner D. K., Taylor R. E., and Southon J. R. (1995) Reduction in backgrounds of microsamples for AMS ^{14}C dating. *Radiocarbon* **37**, 697–704.

Kirner D. L., Southon J. R., Hare P. E., and Taylor R. E. (1996) Accelerator mass spectrometry radiocarbon measurement of submilligram samples. In *Archaeological Chemistry: Organic, Inorganic, and Biochemical Analysis* (ed. A. V. Orna), pp. 434–442. American Chemical Society.

Kirner D. L, Burky R., Taylor R. E., and Southon J. R. (1997) Radiocarbon dating organic residues at the microgram level. *Nuclear Instrum. Meth. Physics Res.* **B123**, 214–217.

van Klinken G. J. and Mook W. G. (1990) Preparative high-performance liquid chromatographic separation of individual amino acids derived from fossil bone collagen. *Radiocarbon* **32**, 155–164.

Long A., Wilson A. T., Ernst R. D., Gore B. H., Hare P. E., and Tuross N. (1989) AMS radiocarbon dating of bones at Arizona. *Radiocarbon* **31**, 231–238.

Masters P. M. (1987) Preferential preservation of noncollagenous protein during bone diagenesis: Implications for chronometric and stable isotopic measurements. *Geochim. Cosmochim. Acta* **51**, 3209–3214.

Nelson D. E., Moreland R. E., Vogel J. S., Southern J. R., and Harington C. R. (1986) New radiocarbon dates on artifacts from the northern Yukon Territory: Holocene not upper Pleistocene in age. *Science* **232**, 749–751.

Price T. D. (ed.) (1989) *The Chemistry of Prehistoric Human Bone*. Cambridge University Press.

Schoeninger M. F. (1989) Reconstructing prehistoric human diet. In *The Chemistry of Prehistoric Human Bone* (ed. T. D. Price), pp. 38–67. Cambridge University Press.

Stafford T. W., Jr, Jull A. J. T., Brendel K., Duhamel R. C., and Donahue D. (1987) Study of bone radiocarbon dating accuracy at the University of Arizona NSF accelerator facility for radioisotope analysis. *Radiocarbon* **29**, 24–44.

Stafford T. W., Jr, Brendel D., and Duhamel R. C. (1988) Radiocarbon, ^{13}C, and ^{15}N analysis of fossil bone: removal of humates with XAD-2 resin. *Geochim. Cosmochim. Acta* **52**, 2257–2267.

Taylor R. E. (1982) Problems in the radiocarbon dating of bone. In *Nuclear and Chemical Dating Techniques: Interpreting the Environmental Record* (ed. L. A. Currie), pp. 453–473. American Chemical Society.

Taylor R. E. (1987) *Radiocarbon Dating: An Archaeological Perspective.* Academic Press.

Taylor R. E. (1991) Frameworks for dating the Late Pleistocene peopling of the Americas. In *The First Americans: Search and Research* (eds T. D. Dillehay and D. J. Meltzer), pp. 77–112. CRC Press.

Taylor R. E. (1992) Radiocarbon dating of bone: to collagen and beyond. In *Radiocarbon After Four Decades: An Interdisciplinary Perspective* (eds R. E. Taylor, R. Kra and A. Long), pp. 375–402. Springer-Verlag.

III. PRESERVATION AND DIAGENESIS OF AMINO ACIDS AND PROTEINS

7. Early diagenesis of particulate amino acids in marine systems

Richard G. Keil, Elizabeth Tsamakis, and John I. Hedges

The decay of organic matter is the primary factor that leads to diagenetic change in marine sediments. The process of organic-matter decay ultimately influences not only the amount and quality of organic matter preserved, but also the extent of remineralization of nutrients and the formation of authigenic mineral phases such as metal sulfides (e.g., Hedges and Keil, 1995). Because of the importance of these processes over geological time and the role sediments play in moderating and recording global change, extensive effort has gone into investigating early diagenesis in marine sediments. Amino acids represent a substantial portion of the organic-matter input to continental margin sediments (Lee, 1988), and amino acids extracted from a sedimentary system can provide information about the conditions during deposition and the extent to which the deposit has undergone decay (Henrichs, 1992, 1993).

Since amino acids are found in all living organisms, the presence or composition of amino acids in continental margin sediments does not typically reflect a specific source. Despite the presence of proteins with unique amino acid compositions (e.g., collagen), the bulk amino acid compositions of various organisms (plankton, bacteria, vascular plants, etc.) are quite similar (Lee, 1988; Cowie and Hedges, 1992b; Cowie et al., 1992). Biogenic production of amino acids in sediments by benthic organisms also contribute to the pool and further complicates any potential use of amino acids as source indicators (Keil and Fogel, in press). Of course, there are exceptions to this generality: amino acids bound in biogenic matrices (carbonates, silicates, and cell wall structures; e.g., Hecky et al., 1973; Shemesh et al., 1993) have the possibility of being preserved in sediments without undergoing appreciable change and potentially could be isolated and used as source indicators. However, for the bulk of the sedimentary amino acid pool, it appears that concentrations and compositions best reflect the diagenetic history of sediment.

Amino acid concentrations and compositions in marine sediments change during diagenesis, and these changes have been related to progress along a diagenetic continuum between fresh material (e.g., phytoplankton) and organic-depleted materials (e.g., oxidized deep-sea turbidites; Whelan and Emeis, 1992; Cowie and Hedges, 1994; Dauwe and Middelburg, 1998). Over a wide variety of time-scales, amino acids have been shown to be degraded faster than bulk organic matter (Henrichs, 1993; Wakeham and Lee, 1993; Harvey et al., 1995). This results in a depletion of amino acids with respect to the total organic matter of a sample. Two amino acid-based parameters decrease during diagenesis: the percentage of organic carbon identifiable as amino acid ($\%T_{AA}C$), and the percentage of total sedimentary nitrogen measurable as amino acid nitrogen ($\%T_{AA}N$). There are also amino acid diagenetic indicators that increase in relative magnitude during diagenesis. The relative abundance of non-protein amino acids generally increases during diagenesis, thus the ratio of a non-protein amino acid to its precursor, or the total mole percentage of non-protein amino acids in the amino acid pool, increases as materials become increasingly degraded (e.g., Cowie and Hedges, 1994). There are four common non-protein amino acids in margin sediments: β-alanine, γ-aminobutyric acid, α-aminobutyric acid, and ornithine. Beta-alanine is typically produced by the decarboxylation of the peptide-forming carboxyl group of aspartic acid, although it can also be produced during the degradation of uracil (Lee and Cronin, 1982). Similarly, γ-aminobutyric acid is produced by the decarboxylation of the peptide-forming carboxyl group of glutamic acid. In contrast, α-aminobutyric acid is formed via the dehydroxylation of threonine. Ornithine is not produced during diagenesis via decarboxylation reactions, but is an intermediate during the synthesis and degradation of arginine and other biomolecules. Ornithine is also used by some mollusks as an osmolite. Since the presence of ornithine cannot always be linked to a diagenetic process, it is sometimes omitted when distributions of non-protein amino acids are examined.

Much of the information about amino acid diagenesis has been gathered from degradation experiments (e.g., Harvey et al., 1995; Nguyen and Harvey, 1997, 2000), sampling of water-column particles, using sediment traps (e.g., Lee and Cronin, 1982; Lee, 1988; Lee et al., 1983; Lee and Wakeham, 1988; Wakeham and Lee, 1989, 1993) or examination of diagenetic signals down cores (e.g., Henrichs et al., 1984; Henrichs, 1987; Burdige and Martens, 1988; Haugen and Lichtentaler, 1991; Cowie and Hedges, 1992b). Despite the large number of individual studies, there have been few previous attempts to compile broad sedimentary data for gross, system-level evaluation of amino acid diagenesis under a wide range of depositional conditions (Dauwe et al., 1999). This chapter presents amino acid concentration and compositional data for a wide range of marine sediments, and relates the compositions to source materials and depositional settings. The goal of this compilation is to illustrate the fact that amino acids are a powerful tool for probing the depositional and degradative history of sedimentary samples.

METHODS

Variation in the techniques used by different researchers can make comparison of data difficult. For this chapter, we have compiled data collected mainly using reverse-phase high-performance liquid chromatography (HPLC)/*o*-phthalaldehyde (OPA) derivatization techniques. We have also included data quantified after separation of esterified amino acids by gas chromatography (GC). In general, much of the marine literature collected using cation-exchange chromatography is not consistent with the HPLC- or GC-based literature at the molecular level, and for this reason cation-exchange-based literature has not been presented here. Cation-exchange-based data are markedly depleted in basic amino acids and lacking in identified non-protein amino acids, especially for sediment samples. Also, cation-exchange-based data exhibit a much wider range of concentrations and mole percent compositions for both sediment samples and plankton endmembers.

The geographic focus of the data compilation is oceanic margin sediments. Although an attempt has been made broadly to cover a variety of marine sedimentary environments from fjord and shelf to deep-sea basin, this compilation does not necessarily represent a complete review of the literature. Table 7-1 lists the types of samples compiled and the sources of data. All of the data presented here are for modern (Holocene) sediments. Many articles in the literature could not be used for this compilation because the data were incomplete (most commonly, no compositional data are reported). On the basis of this literature review, it appears that there is a dearth of information on amino acid diagenesis in open ocean systems, from margins to the deep sea, relative to the many articles on enclosed basins or very shallow systems.

The data have been clustered into five groups: plankton samples (either cultures or size-specific net tows), sinking particles caught in sediment traps, high-carbon sediments deposited under oxygen-deficient water masses, shallow marine sediments (3000 m or less; most are from <300 m), and open-ocean sediments (slowly accumulating sediments and oxidized turbidite samples). Two additional groups, diatom frustules and bacterial peptidoglycan, were used to reconstruct a hypothetical marine-sediment sample (table 7-1). Several comparisons are made with an "average marine sediment" which is the average of the 75 individual sediment sample locations compiled for the shallow marine sediments category. This average marine sediment does not include deep-sea sediments or sediments overlain by oxygen-deficient water masses.

The data compiled were collected using modifications of a single analytical scheme. Hydrolysis was achieved either through conventional heating in 6 N HCl at 100–110°C for 20–24 hr or by rapid hydrolysis at elevated temperates (150–170°C) for 1–4 hrs. In all cases, an internal standard was added prior to hydrolysis to account for losses during hydrolysis and sample work-up. For a portion of the data (all references by Cowie and/or Keil), selective degradation of individual amino acids during hydrolysis and selective loss of compounds via post-hydrolysis sediment/amino acid interactions were accounted for using the charge-matched recovery standard technique of Cowie and Hedges (1992a). After hydrolysis, most samples were quantified using reverse-phase HPLC and pre-column derivatization with OPA (Lindroth and Mopper, 1979). Using this technique, 19 hydrolyzable amino acids (15 primary amines and 4 non-protein amino acids) can be extracted and quantified. Not all 19 amino acids are resolved and reported by the various investigators. Data for three regions were hydrolyzed in a similar fashion but quantified after esterification by GC (Whelan, 1977; Henrichs et al., 1984; Henrichs, 1987). This technique resolves 16 amino acids, including the secondary amines (proline and hydroxyproline) and one or two non-protein amino acids (γ-aminobutyric acid and β-alanine). Given the limited ability to compare proline and hydroxyproline with other data sets, these amino acids were not compiled for this study.

RESULTS AND DISCUSSION

Quantitatively, plankton and sinking particles are enriched in amino acids relative to sediments (figs 7-1 and 7-2). On average, $52 \pm 23\%$ of the organic carbon in plankton is chromatographically-identifiable amino acid carbon, and $77 \pm 16\%$ of the nitrogen is amino acid nitrogen (figure 7-1C–D). Subsequent depletion of amino acids relative to bulk carbon is an ongoing process, but two stages of degradation can markedly alter the bulk amino acid content of organic matter, i.e., conversion from plankton to sinking particle (Lee and Wakeham,

TABLE 7-1 Sample types and source literature

Sample type	Data source
plankton and bacteria	Cowie and Hedges (1992b); Colombo et al. (1998)
diatom frustules	Hecky et al. (1973); Shemesh et al. (1993); Cowie and Hedges (1996)
bacterial peptidoglycan	Koch (1990); Madigan et al. (1997)
sinking particles	Peru upwelling region; Lee and Cronin (1982)
	Panama Basin; Ittekkot et al. (1984a)
	Sargasso Sea; Ittekkot et al. (1984b)
	VERTEX I and II; Lee and Cronin (1984)
	Dabob Bay; Cowie and Hedges (1992b)
	Laurentian Trough; Colombo et al. (1998)
	EqPAC equator station; Wakeham et al. (1999)
semi-enclosed regions	Cape Lookout Bight; Burdige and Martens (1988)[a]
	Dabob Bay; Cowie and Hedges (1992b)
	Buzzards Bay; Henrichs (1987)
	North Sea; Dauwe and Middelburg (1998)
	Laurentian Trough; Colombo et al. (1998)
	Oslofjord; Haugen and Lichtentaler 1991)
continental margins	N.W. European Margin; Boski et al. (1998)
	Washington margin; Keil et al. (1998); Hedges et al. (1999); Keil and Fogel in press
	Deep Peru margin; Henrichs et al. (1984)[a]
	Deep Mexican margin; Keil et al. in press
oxygen-deficient regions	Peru upwelling region; Henrichs et al. (1984)[b]
	Pettaquamscutt River Estuary; Henrichs (1987)
	Saanich Inlet; Cowie et al. (1992)
	Mexican margin; Keil et al. in press
deep ocean	Deep Atlantic; Whelan (1977)
	Madiera Abyssal Plain turbidite-oxidized segment; Cowie et al. (1995)
	Porcupine Abyssal Plain, Atlantic Ocean; Horsfall and Wolf (1997)
	EqPAC equator station; Wakeham et al. (2000)

[a] Data for organic carbon and total nitrogen are from Haddad and Martens (1987).
[b] Data for organic carbon and total nitrogen are from Henrichs and Farrington (1984).

1988), and incorporation of the sinking particle into the sediment (Wakeham et al., 1997). Sinking particles are rapidly depleted in amino acids relative to plankton (figure 7-1), and particles caught in deep water sediment traps ($\geq$500 m) contain, on average, 24 $\pm$ 12% and 68 $\pm$ 9% of their carbon and nitrogen as amino acid. This is in contrast with sinking particles trapped near the surface of the water column, which have bulk concentrations essentially identical to those of plankton (Wakeham et al., 1997).

In contrast to either plankton sources or sinking particles, marine sediments are depleted of amino acids by a factor of 2 or more (figures 7-1 and 7-2), with open-ocean sediment being depleted by several orders of magnitude (Whelan, 1977). Continental margin sediments contain, on average, only 12 $\pm$ 10% of their carbon and 30 $\pm$ 12% of their nitrogen in amino acid form. While the mass-normalized amino acid content (mg/gdw) of a sediment is strongly linked to the organic carbon content of the sediments (figure 7-2A,B), there is no significant correlation between the carbon (or nitrogen) content of a continental margin sediment sample and the fraction of organic carbon (or nitrogen) composed of amino acids (figure 7-2C–D). For an organic-matter concentration range of almost a factor of 20 (e.g., 10–160 mg OC/gdw), the fraction of organic matter that is amino acid does not change markedly or in a systematic way (figure 7-2C,D). Carbon depocenters such as the Mexican margin (Hartnett et al., 1998) or the Peru upwelling region (Henrichs and Farrington, 1984), which can contain as much as 30% organic matter, have been hypothesized to contain organic matter that is less degraded and more similar in composition to source materials (Tyson, 1987). On the basis of bulk amino acid content (figures 7-1C,D and 7-2C,D), these sediments are clearly not enriched in amino acids, as one might expect if the organic matter were less

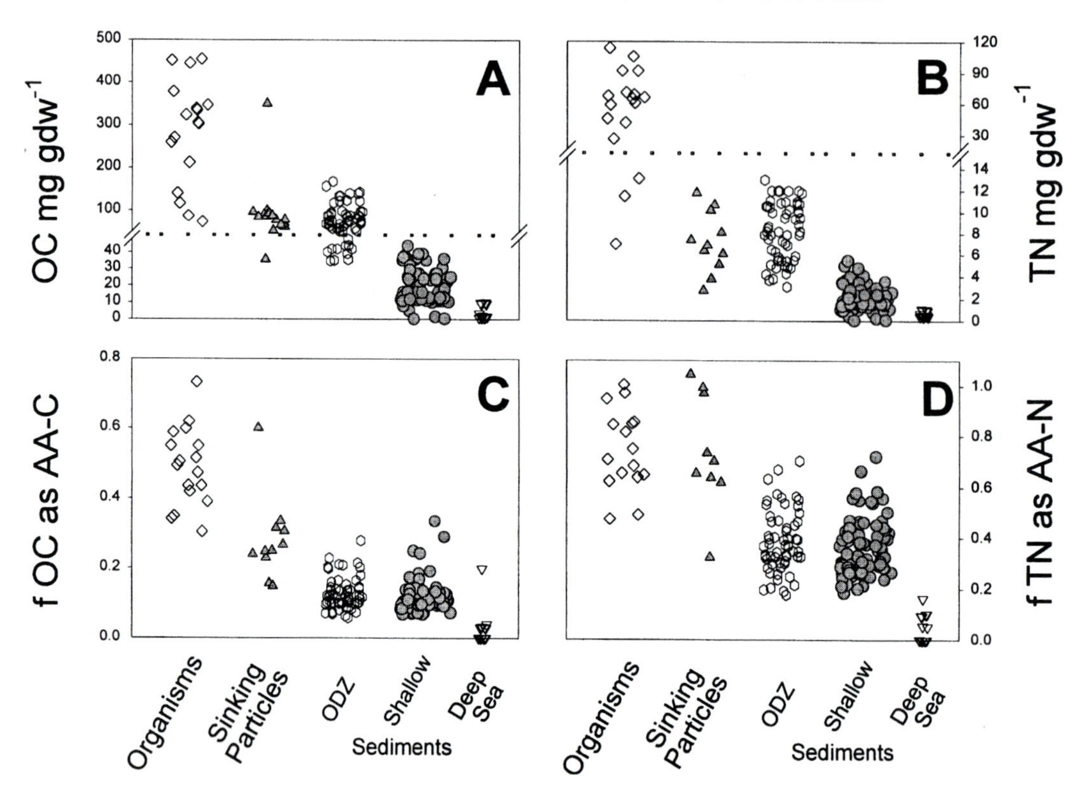

Fig. 7-1. Quantities of organic carbon, total nitrogen and amino acids as a function of sample classification. A. Organic carbon content. B. Total sedimentary nitrogen. C. The fraction of organic carbon that is amino acid carbon. D. The fraction of total nitrogen that is amino acid nitrogen. Dotted lines denote scale breaks. ODZ, Sediment-underlain oxygen-deficient waters; ◇, plankton sources; ▲, sinking particles; ⬡, ODZ samples; ●, shallow marine sediments; ▽, deep-sea sediments.

degraded. However, as discussed later, the amino acid composition of high-organic-matter sediments is different (i.e., less degraded) than that of many other sediments.

Compositional changes in amino acids are reflected at the compound-class and molecular levels. Relative to source materials, neutrally charged amino acids are slightly, but significantly, enriched in sinking particles (Student's t-test, $P > 0.005$; figure 7-3). Acidic and basic amino acids are depleted in most sediments (Student's t-test, $P < 0.001$ for acidics and 0.005 for basics; figure 7-3). This trend is robust across the literature data sets that report all three compound classes, and suggests that the changes are not analytical. However, we note an analytical issue that could potentially influence the magnitude of this trend. Losses of basic and acidic amino acids due to interactions with other hydrolysates or minerals can be an important sink for amino acids liberated during hydrolysis (Cowie and Hedges, 1992a), especially for basic amino acids. For the basic amino acids, there is a statistically significant positive offset (Student's t-test, $P < 0.001$) between samples for which compound-class-specific losses have been accounted and those for which a single (and usually neutral) amino acid was used to correct for losses across all compound classes. This is most easily observed in figure 7-3F, which shows small gaps in the oxygen-deficient zone (ODZ) and shallow-sediment data clusters. The size of the gap is larger than expected because of the selective loss upon hydrolysis of arginine, a significant component of the basic amino acids. However, even after adjustment for arginine (data not shown), there is still a statistically significant increase in the mole percent composition of basic amino acids in sediments for which selective losses during hydrolysis are accounted. Thus, future studies may need to consider careful evaluation of losses during sample hydrolysis in order to better constrain actual changes in composition in marine systems.

At the individual amino acid level, glycine, serine, threonine and arginine are preferentially present in sediment relative to sources (figures 7-4 and 7-5). Three of these amino acids (glycine, serine, and threonine) are present in excess abundance in silicious diatom frustules and their associated organic matrix (Hecky et al., 1973), suggesting that this material might be preferentially preserved

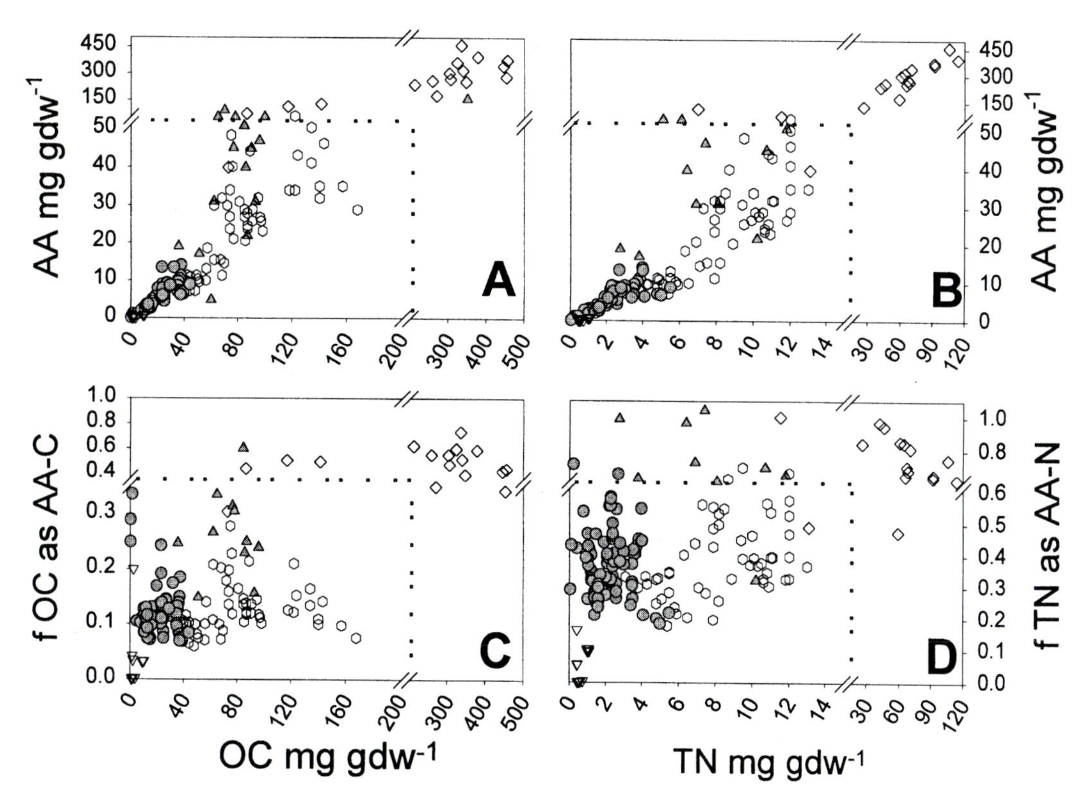

Fig. 7-2. Quantities of amino acids normalized to mass or organic matter content. A. Amino acid yields as a function of sample organic carbon content. B. Amino acid yields as a function of sample total nitrogen content. C. The fraction of organic carbon that is amino acid carbon as a function of sample organic carbon content, and D. the fraction of total nitrogen that is amino acid nitrogen as a function of sample total nitrogen content. Symbols and broken lines are as in figure 7-1.

in marine sediments. If the available data (table 7-1) are used as a reference, the proportion of diatom frustule-derived amino acids can be estimated by comparing the composition of amino acids in frustule-free planktonic materials, diatom frustules, and sediments. We used an iterative approach that mixes different proportions of frustule-free plankton biomass and diatom frustules together in an attempt to 'create' the amino acid composition of an average marine sediment based on these two sources. The model minimized the residual mole percent value for all protein amino acids. That is, it looked for the best overall fit without forcing any particular amino acid to fit perfectly. Using only frustules and plankton biomass, the best reconstruction of an average sediment required mixing of approximately 65% diatom frustules and 35% plankton materials, and yielded high residual mole percentages, especially for glycine. Since the fit for glycine was not particularly good, it is possible that there are other sources of glycine in addition to diatom frustules. One potential source of glycine is the peptidoglycan found in bacterial cell walls.

It is arguable as to whether bacterial biomass is a major component (Parkes et al., 1990, 1993; Harvey and Macko, 1997; Ransom et al., 1997) or an insignificant component (Rowe et al., 1991; Hartgers et al., 1994) of sedimentary organic matter. Sedimentary enrichment of glycine could potentially support the idea that bacterial biomass is important. Bacterial peptidoglycan can be replete in glycine, which in pentameric form acts as a bridge peptide (Morrison and Boyd, 1974; Koch, 1990). This peptide bridge is most abundant in Gram-positive bacteria, although it can also be found in some Gram-negative species (Madigan et al., 1997). Although there are several other amino acids that can be found in peptidoglycan, only two are typically quantified in the literature, i.e., D-alanine (the literature compiled here reports D-alanine as part of the D + L-alanine signal), and glutamic acid (Koch, 1990). Adding peptidoglycan (and whole bacterial cultures; Cowie and Hedges, 1992b) to the iterative model significantly increases the fit of the model and results in an average marine sediment being composed of 55% diatom frustules, 22% peptidoglycan, and 23% planktonic or

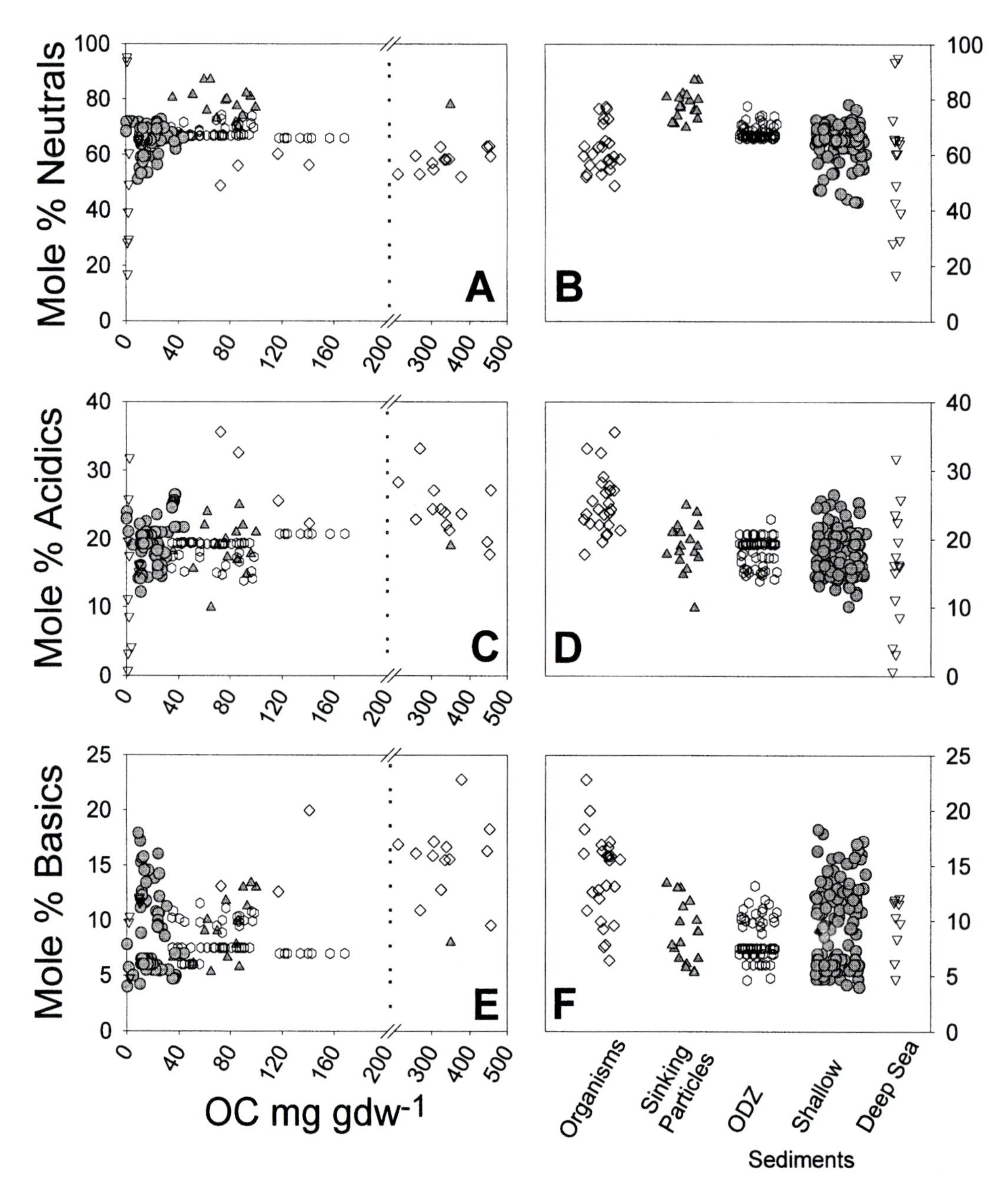

Fig. 7-3. Compositional changes in amino acid functionality between marine sources and sediment samples. A,B. Neutral amino acids, including serine, glycine, threonine, alanine, tyrosine, methionine, valine, phenylalanine, isoleucine, and leucine. C,D. Acidic amino acids including aspartic acid and glutamic acid. E,F. Basic amino acids, including histidine, arginine, and lysine. Dotted lines denote scale breaks. Symbols as for figure 7-1.

bacterial proteins (figure 7-5). The model output is sensitive to changes in the proportions of diatom frustule and peptidoglycan. Dominance by frustules is required in order to balance the alanine, serine, and threonine contents of the sediment. Conversely, the model is relatively insensitive to the composition of either planktonic or bacterial protein. For example, increasing the bacterial protein at the expense of the planktonic component allows better resolution of lysine and phenylalanine at the expense of the serine, aspartic acid, and alanine.

The presence of high mole percentages of glycine in sediments could be a good indicator of the presence of a significant bacterial contribution to the sedimentary amino acid pool, if it transpires that sedimentary bacteria contain significant glycine within their cell walls. If we consider that approximately 30% of the carbon in peptidoglycan is amino-acid-derived (glycine + alanine + glutamic acid), and use results from the mixing model, then the glycine, alanine, and glutamic acid content of sediments can be used to estimate the fraction of organic carbon that comes from peptidoglycan. This back-of-the-envelope calculation suggests that sedimentary organic carbon is composed of, on average, $\sim 2.6 \pm 0.9\%$ peptidoglycan. Working through this same calculation, assuming that 70% of the nitrogen in peptidoglycan is contained within these three amino acids, leads to the suggestion that $4.1 \pm 1.2\%$ of the nitrogen in marine sediments is in the form of peptidoglycan. While these estimates fit the average data, clearly the exact proportion of amino acids coming from various sources changes with location and diagenetic stage. Keil and Fogel (in press) used the stable carbon isotopic composition of amino acids to estimate the proportion of amino acids in Washington margin sediments that had been synthesized by bacteria. On the basis of the unique isotopic composition of valine, which is biosynthesized by bacteria using a prokaryote-specific pathway, they suggested that 15–40% of the valine in their samples was produced in the sediment by bacteria. Thus, while there may be considerable variability in the quantities of amino acids derived from various sources, bacterially derived amino acids are probably an important component in sediments. Finally, it is important to reiterate that the hypothesis of peptidoglycan as the source of the high glycine contents remains unsubstantiated at this time.

Although the exercise of mixing potential sources is informative, it does not completely reproduce the composition of an average sediment, and it does not take diagenesis into account. Relative to sources or any mixture thereof, the average marine sediment is enriched in arginine and depleted in glutamic acid, isoleucine, leucine, methionine, and perhaps lysine (figure 7-4). There are no proposed sources for the enrichment in arginine (blue-whale sperm is rich in arginine but most proteins are not; Morrison and Boyd, 1974). However, given arginine's strongly basic side-chain and its propensity to interact with negatively charged molecules or surfaces, its enrichment in sediments could be indicative of non-biological partitioning. Keil et al. (1998) observed enrichment of arginine (and all basic amino acids) in sediment-size fractions containing high clay contents; this suggests selective interactions between positively charged amino acids and negatively charged mineral surfaces.

The reason for the large depletion (factor of 2; figure 7-4) of glutamic acid in sediments is not known. With an argument similar to the electrostatic argument for arginine, it can be postulated that glutamic acid could be lost from sediments because it does not interact well with surfaces or other sites that might allow it to be preferentially retained in sediments. However, that argument is not necessarily strong because literature values for glutamic acid contain a glutamine component (the amine is lost during hydrolysis and thus for hydrolyzed samples glutamic acid = glutamic acid + deaminated glutamine), which could potentially counterbalance the acidic nature of glutamic acid. Also, aspartic acid, which might be expected to behave in the same way, does not show any depletion in sediments (figure 7-4). Finally, depletion of glutamic acid is not consistent with a significant peptidoglycan component in marine systems. Although the exact quantities of amino acids present in peptidoglycan are variable (other than the presence of D-alanine), glutamic acid is often reported as one component (Koch, 1990). If glutamic acid were generally present in the peptidoglycan of marine bacteria (this has not been evaluated), then glutamic acid should be enriched in sediments. Thus, as for arginine, the factors controlling the distribution of glutamic acid remain unclear.

Depletion of leucine and isoleucine in sediments may indicate that these amino acids are present in more labile proteins, and are thus preferentially degraded over other amino acids. The depletion of histidine and methionine may also indicate preferential utilization, although trends for these amino acids may simply reflect analytical difficulties. In addition to lysine (Lee, 1988), histidine and methionine are difficult to resolve and quantify by OPA-derivatization. If there are matrix effects associated with the detection and recovery of these amino acids, as suggested by Cowie and Hedges (1992a), the general depletion of histidine and methionine may be analytical rather than environmental. However, it is interesting to note that histidines in source materials, sinking particles, and high-organic-content sediments are all the same (figure 7-4). Only in low-organic-content sediments does the mole percentage of histidine drop.

Non-protein amino acids can account for as much as 2% of the organic carbon in sources (bacteria and plankton; figure 7-6A). However, as a mole percentage, non-protein amino acids are inconsequential in most source materials (figure 7-6B–F). Non-protein amino acids only become a significant component of the amino acid pool at very low organic matter concentrations, mostly in deep-sea sediments (figure 7-6). It has not been clearly established as to whether the large mole percent enrichments in non-protein amino acids observed in deep-sea sediments reflect selective preservation of non-protein amino acids present in source materials or whether there is significant production of

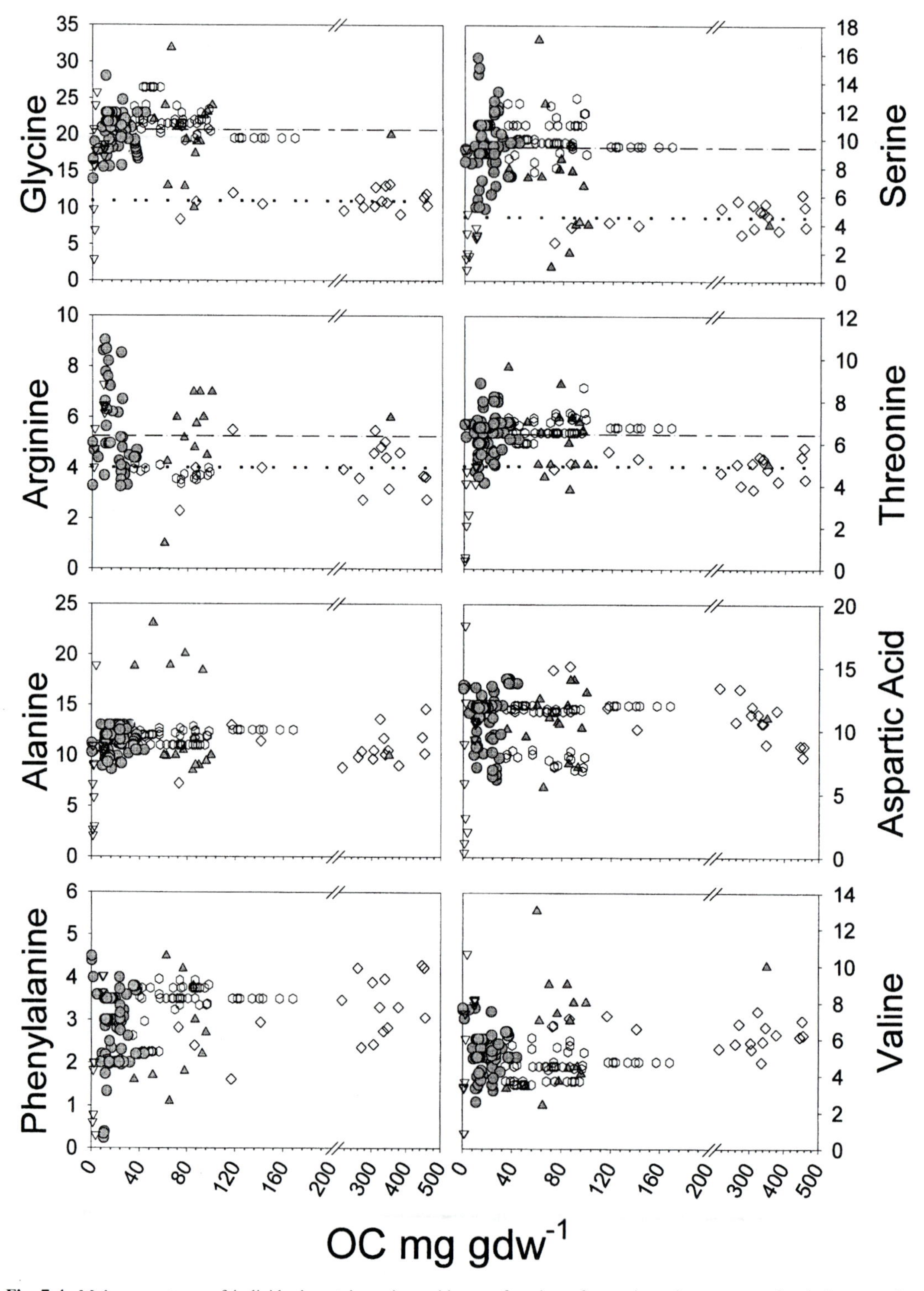

Fig. 7-4. Mole percentages of individual protein amino acids as a function of organic carbon content. Symbols are as for figure 7-1. For amino acids where there is a significant difference (Student's *t*-test, $P > 0.005$), the dotted lines mark the average composition of plankton + bacteria and the broken lines denote the average for continental margin sediments. Scale breaks are not denoted.

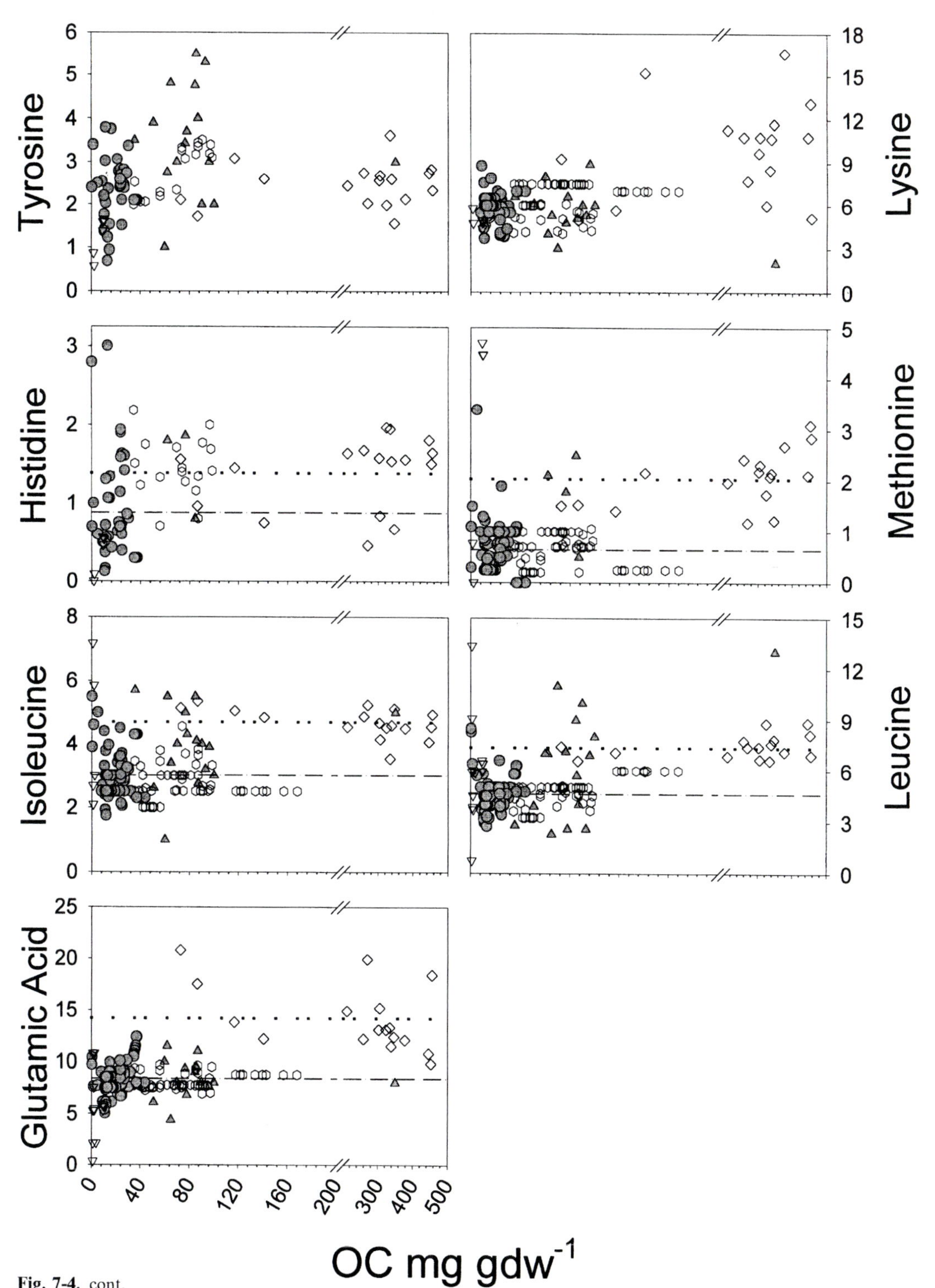

Fig. 7-4. cont.

these amino acids *in situ* (Whelan, 1977; Cowie and Hedges, 1994; Cowie et al., 1995). Non-protein amino acids are produced through all stages of amino acid diagenesis, including early stages such as when sinking particles first leave the euphotic zone (Lee and Cronin, 1982, 1984). It seems most likely that the non-protein amino acids preserved in sediments come from *in situ* production. As organic

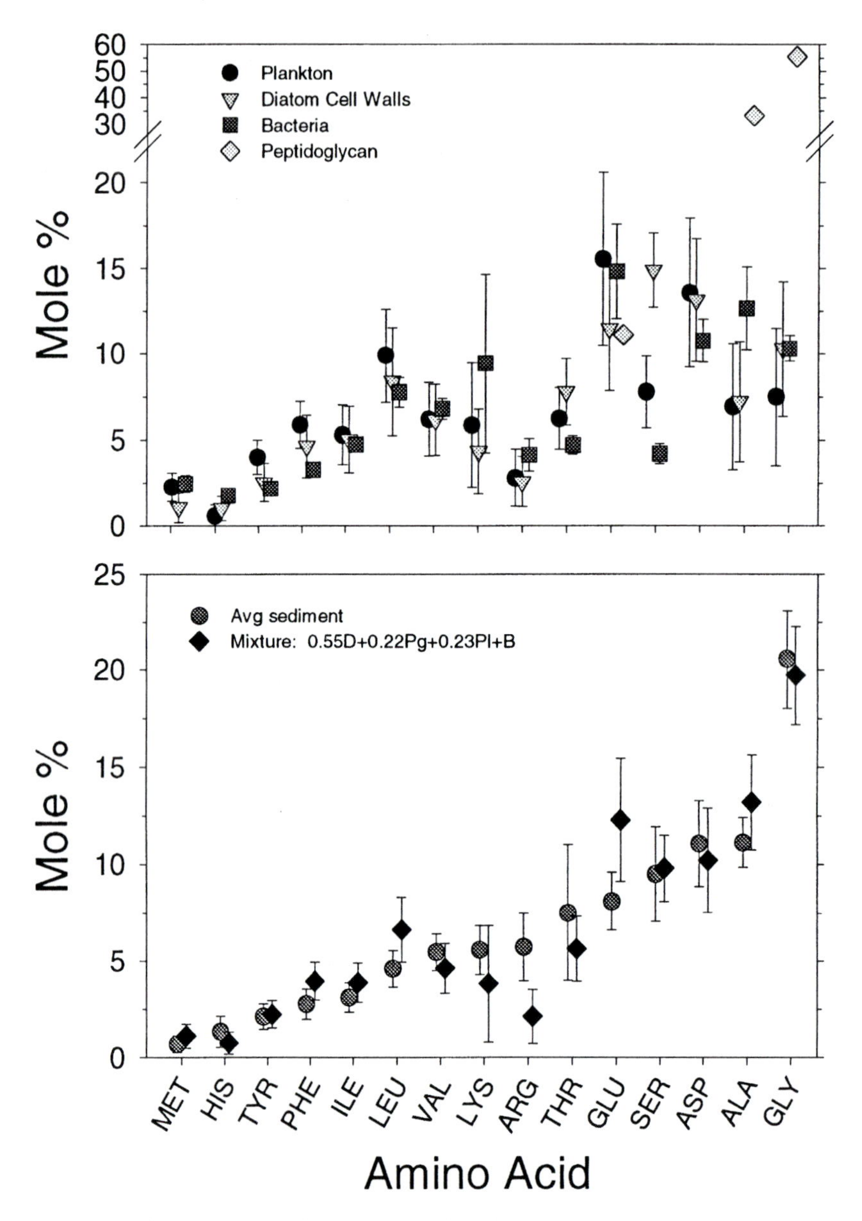

Fig. 7-5. Upper panel. Mole percent composition (±1 S.D.) of planktonic sources (diatom cell wall-free; $n = 7$), diatom cell walls ($n = 6$), bacteria (with associated peptidoglycan; $n = 3$) and average peptidoglycan ($n = 1$ average value from Koch, 1990). Lower panel. Average composition of shallow marine sediments (±1 S.D., $n = 75$) and the composition of a mixture that is 55% diatom cell wall (D), 22% peptidoglycan (Pg), and 23% plankton and bacterial (Pl + B) proteins.

carbon content decreases, the fraction of organic matter as amino acids decreases (figure 7-1C), yet the fraction of organic matter as non-protein amino acids decreases and then increases (figure 7-6A). If the original non-protein component of plankton was being selectively preserved, then there would be a linear inverse relationship between organic carbon content and the fraction of organic matter as non-protein amino acids. The curved shape in figure 7-6A indicates that as the overall amino acid pool is being consumed, a non-protein component is added at low organic carbon contents. A back-of-the-envelope calculation supports the hypothesis that *in situ* production dominates over preservation of planktonic material when organic contents of sediments are low. Assuming that 2% of the planktonic organic carbon is in the form of non-protein amino acid (the most conservative values from figure 7-6A), and that 23% of the organic carbon in sediments represents direct preservation of planktonic materials (result of iterative model; figure 7-5), then (at most) 0.5% (0.005 f OC;

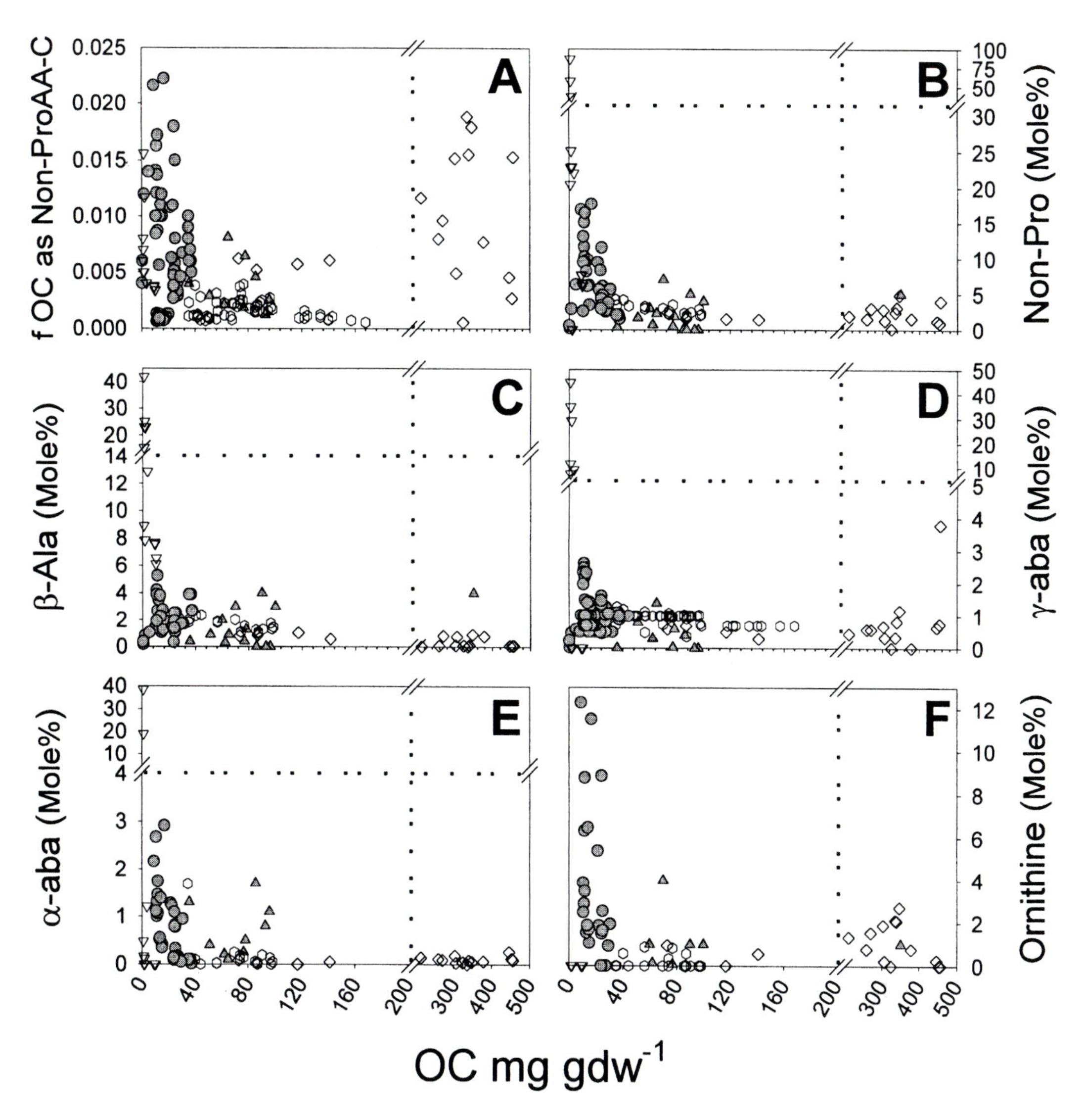

Fig. 7-6. Contribution of non-protein amino acids to sediments and sources as a function of organic carbon content. Symbols and dotted lines are as in figure 7-1. A. The fraction of organic carbon that is identifiable as non-protein amino acid carbon. B–F. Mole percent composition of non-protein amino acids, β-alanine, γ-aminobutyric acid, α-aminobutyric acid, and ornithine as a function of organic matter content.

figure 7-6A) of the non-protein amino acids in sediments can be directly derived from planktonic sources. Figure 7-6A illustrates that during diagenesis from source materials to organic carbon contents of ~75 mg OC/gdw, the non-protein amino acid pool can potentially be accounted for by selective preservation of planktonic source materials. When diagenesis proceeds further and organic carbon contents become less than ~75 mg OC/gdw, a substantial portion of the non-protein amino acid pool must be produced *in situ* (figure 7-6A).

In order to investigate the relationship between amino acid compositions and diagenesis in more detail, Dauwe and Middelburg (1998) and Dauwe et al. (1999) have used principal component analysis to categorize the changes in amino acid compositions. They have reported an algorithm that quantifies the changes in 14 amino acids (excluding lysine). The Dauwe Degradation Index ranks changes in individual amino acids in a general scheme similar to that shown for the overall data compilation: enrichment of threonine, arginine, and glycine, and depletion of isoleucine, leucine and histidine. Applying their algorithm to the to data compiled here illustrates that the amino acid composition of marine sediments is continually changing toward more degraded materials as a function of sedimentary organic matter content

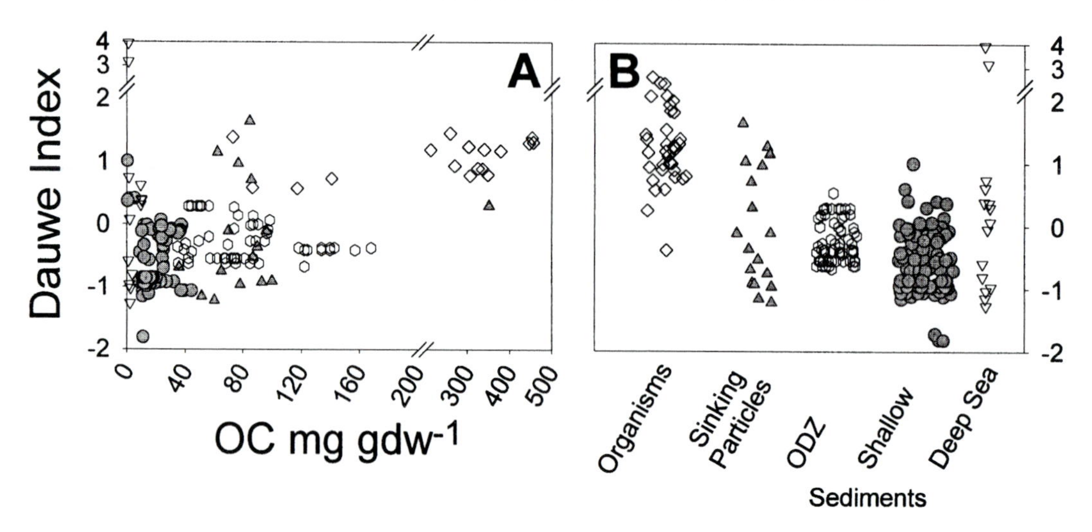

Fig. 7-7. The Dauwe Index of amino acid diagenetic stage (Dauwe and Middelburg, 1998) as a function of organic matter content (A) or sample classification (B). Abbreviations and symbols are as in figure 7-1.

(figure 7-7). The Dauwe Index is most sensitive to change during the early stages of diagenesis as biomass is transformed and transferred to sediments. This is complementary to the non-protein amino acids, which exhibit their greatest relative changes at the later stages of diagenesis. Interestingly, the diagenetic status of some fresh diatom frustules can fall into a "degraded" status (value <0.5) because frustules are enriched in three of the four amino acids that can drive the diagenetic indicator. Peptidoglycan can also have a relatively low value (−1.1; not plotted in figure 7-7) because of its enrichment of glycine. Thus, the index seems to incorporate the enhanced preservation of cell walls in its quantification of diagenetic state. The high index values for some deep-sea sediments reflect the fact that only some amino acids were detected (Whelan, 1977).

An interesting result observed in the data in figures 7-6 and 7-7 is that there are no sites of highly degraded amino acids (mole percent non-protein amino acids >5 or Dauwe Degradative Index <0.5) that correspond to organic carbon-rich sediments. In fact, sediments from both the Peru upwelling region (Henrichs et al., 1984) and the Mexican margin (Keil et al., in press) contain amino acids that better reflect sources rather than highly altered amino acid remains. Using the mixing model to derive a "high-organic-content" sediment with diatom cell walls, peptidoglycan, and source materials yields a mixture that is 0.25 diatom cell wall, 0.25 peptidoglycan and 0.50 plankton or bacterial protein. The relative abundance of glycine, threonine, and serine fixes the fractions of diatom versus peptidoglycan contributions. Interestingly, these sediments are not enriched in arginine (figure 7-4). As mentioned previously, these sediments do not contain more amino acids per unit organic carbon than other continental margin sediments. One possible interpretation of this may be that transit through the water column (even an oxygen-deficient one; Lee and Cronin, 1982; Lee and Wakeham, 1988) alters the amino acid content of the settling material, but the large-scale changes in composition which happen during diagenesis do not occur in high-carbon-content sediments. While no known organic-matter-rich sediments contain highly degraded amino acids, there are organic-poor sediments that contain relatively fresh amino acid compositions (figure 7-7; Dauwe and Middelburg, 1998).

Overall, the content and composition of amino acids in marine systems potentially provides a wealth of information about processing and diagenesis. Several interesting trends in the literature are validated when data are compiled, and some new aspects of amino acid geochemistry are discernable in the compiled data set. However, inconsistencies in methodologies and unquantified sources or processes contribute to molecular level compositions that cannot at this time be adequately explained. In order to harness the power of amino acid analysis more effectively in geochemical samples, we suggest that individual investigators attempt to quantify as many amino acids as possible, including the non-protein amino acids, and that source materials including sedimentary bacterial peptidoglycan be further investigated. Finally, we hope that processing indicators such as the Dauwe Degradative Index be further developed and used in order to relate amino acid diagenesis to other processes in sediments.

Acknowledgments: We are grateful to Ed Hare for the encouragement and advice he willingly dispensed during our many interactions with him.

R.G.K. specifically and joyfully remembers his first meetings with Ed and the calm manner in which Ed dispensed with Rick's nervousness and awe. We also thank Timothy Filley and Susan Ziegler for constructive comments on the manuscript. This research was supported by NSF grants OCE9402091 and OCE9711792 to R.G.K. This is Contribution No. 2210 from the School of Oceanography, University of Washington, Seattle.

REFERENCES

Boski T., Pessoa J., Pedro P., Thorez J., Dias J. M. A., and Hall I. R. (1998) Factors governing abundance of hydrolyzable amino acids in the sediments from the N.W. European continental margin. *Prog. Ocean.* **42**, 145–164.

Burdige D. J. and Martens C. S. (1988) Biogeochemical cycling in an organic-rich coastal marine basin. 10. The role of amino acids in sedimentary carbon and nitrogen cycling. *Geochim. Cosmochim. Acta* **52**, 1571–1584.

Colombo J. C, Silverberg N., and Gearing J. N. (1998) Amino acid biogeochemistry in the Laurentian Trough: vertical fluxes and individual reactivity during early diagenesis. *Org. Geochem.* **29**, 933–945.

Cowie G. L. and Hedges J. I. (1992a) Improved amino acid quanification in environmental samples: charge-matched recovery standards and reduced analysis time. *Mar. Chem.* **37**, 223–238.

Cowie G. L. and Hedges J. I. (1992b) Sources and reactivities of amino acids in a coastal marine environment. *Limnol. Ocean.* **37**, 703–724.

Cowie G. L. and Hedges J. I. (1994) Biochemical indicators of diagenetic alteration in natural organic matter mixtures. *Nature* **369**, 304–307.

Cowie G. L. and Hedges J. I. (1996) Digestion and alteration of the biochemical constituents of a diatom (*Thalassiosira weissflogii*) ingested by an herbivorous zooplankton (*Calanus pacificus*). *Limnol. Ocean.* **41**, 581–594.

Cowie G. L., Hedges J. I., and Calvert S. E. (1992) Sources and relative reactivities of amino acids, neutral sugars, and lignin in an intermittently anoxic marine environment. *Geochim. Cosmochim. Acta* **56**, 1963–1978.

Cowie G. L., Hedges J. I., Prahl F. G., and de Lange G. J. (1995) Elemental and major biochemical changes across an oxidation front in a relict turbidite: an oxygen effect. *Geochim. Cosmoch. Acta* **59**, 33–46.

Dauwe B. and Middelburg J. J. (1998) Amino acids and hexosamines as indicators of organic matter degradation state in North Sea sediments. *Limnol. Ocean.* **43**, 782–798.

Dauwe B., Middelburg J. J., Herman M. J., and Heip C. H. R. (1999) Particulate organic matter degradation in the ocean: changes in biochemical composition and reactivity are linked. *Limnol. Ocean.* **44**, 1809–1814.

Haddad R. I. and Martens C. S. (1987) Biogeochemical cycling in an organic-rich coastal marine basin. 9. Sources and accumulation rates of vascular plant-derived organic material. *Geochim. Cosmochim. Acta* **51**, 2991–3001.

Hartgers W. A., Damste J. S. S., Requejo A. G., Allan J., Hayes J. M., and de Leeuw J. W. (1994) Evidence for only minor contributions from bacteria to sedimentary organic carbon. *Nature* **369**, 224–227.

Hartnett H. E., Keil R. G., Hedges J. I., and Devol A. H. (1998) Influence on oxygen exposure time on organic carbon preservation in continental margin sediments. *Nature* **391**, 372–374.

Harvey H. R. and Macko S. A. (1997) Catalysts or contributors? Tracking bacterial mediation of early diagenesis in the marine water column. *Org. Geochem.* **26**, 531–544.

Harvey H. R., Tuttle J. H., and Bell J. T. (1995) Kinetics of phytoplankton decay during simulated sedimentation: changes in biochemical composition and microbial activity under oxic and anoxic conditions. *Geochim. Cosmochim. Acta* **59**, 3367–3377.

Haugen J.-E. and Lichtentaler R. (1991) Amino acid diagenesis, organic carbon and nitrogen mineralization in surface sediments from the inner Oslofjord, Norway. *Geochim. Cosmochim. Acta* **55**, 1649–1661.

Hecky R. E., Mopper K., Kilham P., and Degens E. T. (1973) The amino acid and sugar composition of diatom cell walls. *Mar. Biol.* **10**, 323–331.

Hedges J. I. and Keil R. G. (1995) Sedimentary organic matter preservation: an assessment and speculative synthesis. *Mar. Chem.* **49**, 81–115.

Hedges J. I., Hu F. S., Devol A. H., Hartnett H. E., Tsamakis E., and Keil R. G. (1999) Sedimentary organic matter preservation: a test for selective degradation under oxic conditions. *Amer. J. Sci.* **299**, 529–555.

Henrichs S. M. (1987) Early diagenesis of amino acids and organic matter in two coastal marine sediments. *Geochim. Cosmochim. Acta* **51**, 1–15.

Henrichs S. M. (1992) Early diagenesis of organic matter in marine sediments: progress and perplexity. *Mar. Chem.* **39**, 119–149.

Henrichs S. M. (1993) Early diagenesis of organic matter: the dynamics (rates) of cycling of organic compounds. In *Organic Geochemistry* (eds M. H. Engel and S. A. Macko), pp. 101–117. Plenum Press.

Henrichs S. M. and Farrington J. W. (1984) Peru upwelling region sediments near 15S. 1. Remineralization and accumulation of organic matter. *Limnol. Ocean.* **29**, 1–19.

Henrichs S. M., Farrington J. W., and Lee C. (1984) Peru upwelling region sediments near 15°S. 2. Dissolved free and total hydrolyzable amino acids. *Limnol. Ocean.* **29**, 20–34.

Horsfall I. M. and Wolf G. A. (1997) Hydrolysable amino acids in sediments from the Porcupine

Abyssal Plain, northeast Atlantic Ocean. *Org. Geochem.* **26**, 311–320.

Ittekkot V., Degens E. T., and Honjo S. (1984a) Seasonality in the fluxes of sugars, amino acids, and amino sugars to the deep ocean: Panama Basin. *D. S. Res.* **31**, 1071–1083.

Ittekkot V., Deuser W. G., and Degens E. T. (1984b) Seasonality in the fluxes of sugars, amino acids, and amino sugars to the deep ocean: Sargasso Sea. *D. S. Res.* **31**, 1057–1069.

Keil R. G., Tsamakis E., Giddings J. C., and Hedges J. I. (1998) Biochemical distributions among size-classes of modern marine sediments. *Geochim. Cosmochim. Acta* **62**, 1347–1364.

Keil R. G. and Fogel M. L. (in press). Stable carbon isotopic composition of amino acids along the Washington coast. *Limnol. Ocean.*

Keil R. G., Tsamakis E., and Devol A. H. (in press). Amino acid compositions and OC:SA ratios indicate enhanced preservation of organic matter in Pacific Mexican Margin sediments. *Geochim. Cosmochim Acta.*

Koch A. L. (1990) Growth and form of the bacterial cell wall. *Amer. Sci.* **78**, 327–341.

Lee C. (1988) Amino acid chemistry and amine biogeochemistry in particulate material and sediments. In *Nitrogen Cycling in Coastal Marine Environments* (eds T. H. Blackburn and J. Sorensen), pp. 126–141. Wiley.

Lee C. and Cronin C. (1982) The vertical flux of particulate nitrogen in the sea: decomposition of amino acids in the Peru upwelling area and the equatorial Pacific. *J. Mar. Res.* **40**, 227–251.

Lee C. and Cronin C. (1984) Particulate amino acids in the sea: effects of primary productivity and biological decomposition. *J. Mar. Res.* **42**, 1075–1097.

Lee C. and Wakeham S. G. (1988) Organic matter in seawater: biogeochemical processes. In *Chemical Oceanography*, vol. 9, pp. 49.1–49.44. Academic Press.

Lee C., Wakeham S. G., and Farrington J. W. (1983) Variations in the composition of particulate organic matter in a time-series sediment trap. *Mar. Chem.* **13**, 181–194.

Lindroth P. and Mopper K. (1979) High performance liquid chromatographic determination of subpicomole amounts of amino acids by precolumn fluorescence derivatization with *o*-phthaldialdehyde. *Anal. Chem.* **51**, 1667–1674.

Madigan M. T., Martinko J. M., and Parker J. (1997) *Brock: Biology of Microorganisms.* Prentice Hall.

Morrison R. T. and Boyd R. N. (1974) *Organic Chemistry*, 3rd edn. Allyn and Bacon, Inc.

Nguyen R. T. and Harvey H. R. (1997) Protein and amino acid cycling during phytoplankton decomposition in oxic and anoxic waters. *Org. Geochem.* **27**, 115–128.

Nguyen R. T. and Harvey H. R. (2000) Protein preservation during early diagenesis in marine waters and sediments. In *Nitrogen-containing Macromolecules in the Biosphere and Geosphere* (eds Stankiewicz and Van Bergen), pp. 561–587. Oxford University Press.

Parkes R. J., Cragg B. A., Fry J. C., Herbert R. A., and Wimpenny J. W. T. (1990) Bacterial biomass and activity in deep sediment layers from the Peru margin. *Phil. Trans. R. Soc. Lond.* **331**, 139–153.

Parkes R. J., Cragg B. A., Getliff J. M., Harvey S. M., Fry J. C., Lewis C. A., and Rowland S. J. (1993) A quantitative study of mircobial decomposition of biopolymers in Recent sediments from the Peru Margin. *Mar. Geol.* **113**, 55–66.

Ransom B., Bennett R. H., Baerwald R., and Shea K. (1997) TEM study of in situ organic matter on continental margins: occurrence and the "monolayer" hypothesis. *Mar. Geol.* **138**, 1–9.

Rowe G. T., Sibuet M., Demming J. W., Khripounoff A., Tietjen J., Macko S. A., and Theroux R. (1991) 'Total' sediment biomass and preliminary estimates of organic carbon residence time in deep-sea benthos. *Mar. Ecol. Prog. Ser.* **70**, 99–114.

Shemesh, A., Macko S., Charles C., and Rau G. (1993) Isotopic evidence for reduced productivity in the glacial southern ocean. *Science* **262**, 407–410.

Tyson R. V. (1987) The genesis and palynofacies characteristics of marine petroleum source rocks. *Geol. Soc. Spec. Pub.* **26**, 47–67.

Wakeham S. G. and Lee C. (1989) Organic geochemistry of particulate matter in the ocean: the role of particulates in oceanic sedimentary cycling. *Org. Geochem.* **14**, 83–96.

Wakeham S. G. and Lee C. (1993) Production, transport, and alteration of particulate organic matter in the marine water column. In *Organic Geochemistry* (eds M. H. Engel and S. A. Macko), pp. 145–169. Plenum Press.

Wakeham S. G., Lee C., Hedges J., Hernes P. J., and Peterson M. L. (1997) Molecular indicators of diagenetic status in marine organic matter. *Geochim. Cosmochim. Acta* **61**, 5363–5369.

Wakeham S. G., Lee C., and Hedges J. I. (2000) Fluxes of major biochemicals in the Equatorial Pacific Ocean. In *Biogeochemistry of Marine Organic Matter* (ed. N. Handa), pp. 363–379. Elsevier.

Whelan J. K. (1977) Amino acids in a surface sediment core of the Atlantic abyssal plain. *Geochim. Cosmochim. Acta* **41**, 803–810.

Whelan J. K. and Emeis K.-C. (1992) Sedimentation and preservation of amino compounds and carbohydrates in marine sediments. In *Organic Matter: Productivity, Accumulation, and Preservation in Recent and Ancient Sediments* (eds J. K. Whelan and J. W. Farrington), pp. 176–200. Columbia University Press.

8. Preservation of amino acids in museum herbarium samples

Mark A. Teece, Noreen Tuross, W. John Kress, Paul M. Peterson, George Russell, and Marilyn L. Fogel

The vast collections of plants preserved in herbaria throughout the world are a record of environmental change over the past two centuries. Plants adapt to changes in their growth conditions and environment in a variety of ways. Such biological and physiological responses are recorded in plant tissue, and therefore preserved plant tissue in herbaria may provide a record of such changes over the past two hundred years. Since the onset of industrialization, and burning of fossil fuels, in the 1850s, atmospheric concentrations of CO_2 have increased substantially (Keeling et al., 1996). The response of plants to increased CO_2 levels has been studied using herbarium specimens. For example, such increases affect stomatal density (Woodward, 1987), and water-use efficiency (Penuelas and Azcon-Bieto, 1992). In addition, the $\delta^{15}N$ of preserved plant specimens was employed as an indicator of long-term responses to changes in the availability of water, carbon, and nitrogen since 1920 (Penuelas and Estiarte, 1997).

The use of preserved plant specimens is based on the assumption that the biological and biochemical signal preserved in such samples has not been affected by the storage and/or conservation processes used. Although the macro-integrity of plant specimens is intact, i.e., whole plants, leaves and roots are easily recognized, little is known about the preservation of the organic material of the plants. In order to utilize the signal preserved in the organic matter of these specimens, for environmental and other studies, the preservation of the major biochemicals has to be addressed. We measured the concentration of hydrolyzable amino acids in specimens (dating back to 1849) of two plant species preserved in the U.S. National Herbarium to determine whether storage under these conditions results in the preservation of amino acids.

MATERIALS AND METHODS

The two species of grasses studied were *Bouteloua hirsuta* and *Pseudoroegneria spicata*. *B. hirsuta* Lag. (Hairy grama) is a native, warm-season, densely tufted perennial that is drought-tolerant and widely distributed throughout the Great Plains and Southwest of the U.S.A. (Stubbendieck et al., 1986; Cronquist et al., 1977). *P. spicata* (Bluebunch wheatgrass) is a perennial, cool-season, drought-resistant wheatgrass that is distributed throughout a broad elevational band in the western U.S.A. (Wright and Bailey, 1982). Leaves of both species had been stored in the U.S. National Herbarium at the National Museum of Natural History in Washington, DC. The herbarium samples of *B. hirsuta* were collected in Texas, U.S.A., between 1849 and 1992, and samples of *P. spicata* originated from N.W. U.S.A. (Idaho, Wyoming, Nevada and Colorado), from 1874 to 1997. Leaf tissue was carefully removed from similar terminal positions and inspected for damage prior to analysis.

Herbarium sheets were visually inspected for damage or signs of chemical treatment, as conservation practices have changed over the past two centuries. For example, some older collection protocols suggested the use of chemicals to preserve plant material. Plant specimens were treated with chemicals such as ethanol, formaldehyde, camphor, naphthalene, and mercuric chloride, to prevent insect damage and enhance plant preservation (Womersley, 1981). In the middle years of the 20th century, dichlorodiphenyltrichloroethane (DDT) and its derivatives were used to control pests, and the 1960s and 1970s were periods of large-scale prophylactic treatment of museum specimens (Child and Pinnager, 1992). Such techniques are now discouraged, and are only used in the preservation of some tropical plant species, whose tissues are extremely susceptible to insect attack (Fosberg and Sachet, 1965; Rao and Sharma, 1990). No records of chemical treatment of the plant specimens removed for the present study were found. Visual inspection of the specimens and the associated herbarium sheets also indicated that these samples had not been treated with chemicals during their conservation and storage.

Dried plant samples were placed in glass test tubes, which were immersed in liquid nitrogen, and then ground to a fine powder using a ground-glass pestle; the powder was then stored in glass vials. An aliquot of the finely ground plant tissue was hydrolyzed in 6 N HCl (Pierce Chemicals) for 20 hrs at 110°C. The optimum conditions for hydrolysis, namely temperature and duration of heating, to maximize yields of amino acids were taken from Cowie and Hedges (1992). Hydrolysis was not performed under a nitrogen atmosphere and therefore losses of specific amino acids may have occurred because of oxidation (Engel and Hare, 1985). Concentrations of individual amino acids were measured using high-performance liquid chromatography (HPLC) with fluorescence detection according to the methods of Hare et al. (1991). This method does not detect the amino acids tryptophan or proline; cysteine and histidine are not well quantified. As a rule, all amino acids yields typically exhibited sample mean deviations of 8% or less, and less than 5% (approx. ±5 pmol/μg dry wt for Asx, Glx, Gly, Ala, Leu) for the major components. The carbon and nitrogen content of plant specimens was measured on a Carlo Erba EA-1108 elemental analyzer. The typical error of measurement at the low concentrations of nitrogen found in these samples was ±0.5%.

RESULTS AND DISCUSSION

The relative distribution of hydrolyzable amino acids present in *B. hirsuta* (figure 8-1A, table 8-1) and *P. spicata* were similar (figure 8-1B, table 8-2). Furthermore, the relative distribution was similar to that observed in seagrasses and mangroves (Zieman et al., 1984). In all cases, the aliphatic amino acids alanine, glycine, and leucine and the acidic amino acids aspartic and glutamic acid were the most abundant. Lower concentrations of the aromatic amino acids tyrosine and phenylalanine and the basic components lysine and arginine were observed. Concentrations of methionine in the two grasses in the present study were low, and resulted from oxidation of this component under the hydrolysis conditions used (Engel and Hare, 1985).

The concentrations of acidic amino acids Asx and Glx are a composite signal (i.e., Asx = asparagine + aspartate; Glx = glutamine and glutamate), as hydrolysis of the amide group of the acidic amino acids, namely asparagine and glutamine, to their corresponding acids occurs during the acid hydrolysis step. The high concentration of aspartic acid measured in these plants probably arises from the amide derivative asparagine, which is the major transport and storage compound for nitrogen found in many plants (Rabe, 1990; Azevedo et al., 1997).

The absolute concentration of amino acids in specimens of *B. hirsuta* exhibited an overall decrease over time, whereas specimens of *P. spicata* showed no such trend (figure 8-1). Specimens of *B. hirsuta* collected between 1849 and 1936 contained the lowest concentrations of amino acids and no substantial difference was observed between these samples in terms of concentration or relative distribution. The consistently lower concentrations of amino acids may result from a change in museum conservation practices after 1936. As noted, however, these specimens showed no visual signs of chemical treatment, and no written records were found to indicate that the U.S. Herbarium used chemical treatments during this period to preserve plant material. Within experimental error, the carbon and nitrogen concentrations of these specimens were not significantly different from those of more recently collected samples (table 8-1). Furthermore, the amino acid content of two specimens of *P. spicata* collected in 1849 and 1882, and similarly stored in the U.S. National Herbarium, were not consistently lower than those of samples collected more recently (figure 8-1B). It is therefore more likely that the lower concentrations of amino acids in these specimens of *B. hirsuta* are a result of biological variability rather than differences in conservation practices prior to 1936. The lower concentrations could be related to changes in plant metabolism. The recent study of herbarium specimens by Penuelas and Estiarte (1997) attributed changes in the concentration of bulk carbon and nitrogen content to changes in plant physiology as a result of human activities.

With increasing storage time, the distribution of amino acids in both species of grass remained constant. All specimens of *P. spicata* analyzed displayed a similar distribution and concentration of amino acids, which did not vary as a function of storage time (figure 8-1B). The distributions of amino acids in the specimens of *B. hirsuta* analyzed were identical, and even the specimens collected prior to 1936 exhibited the same distribution (figure 8-1A). Furthermore, within the errors of measurement, the carbon and nitrogen composition (% by weight) of specimens of both species did not change markedly with increased storage time. These observations indicate that degradation of amino acids was not occurring under these conditions. These results further suggest that the degradation of major proteins was not a significant process. To establish whether proteins are preserved intact under these conditions, additional techniques such as sodium dodecyl sulphate/polyacrylamide gel electrophoresis (SDS–PAGE) and enzyme-linked immunosorbent assay (ELISA) are required (e.g., Fogel and Tuross, 1999). Proteins are among the more labile plant biochemicals (Fogel and Tuross, 1999), and their apparent resistance to decomposition under herbarium conditions strongly indicates that plant material preserved under such conditions is a reasonable proxy for freshly collected material.

FUTURE STUDIES OF HERBARIUM SPECIMENS

The study of the biochemical adaptation and response of plants to ecological factors is the subject of chemical ecology. A general characteristic of plants subjected to stress is the increased level of

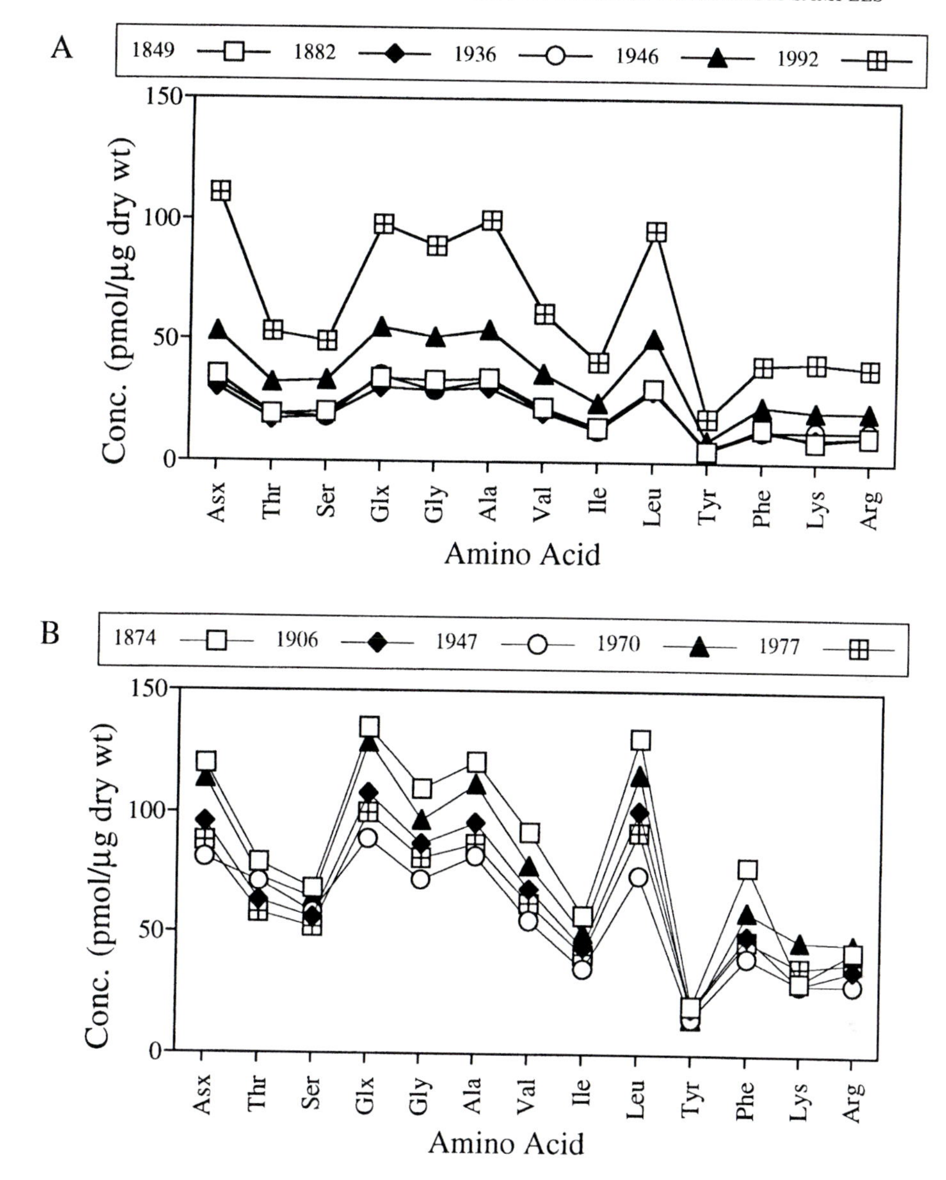

Fig. 8-1. Concentrations of hydrolyzable amino acids in herbarium specimens of (A) *B. hirsuta* and (B) *P. spicata.*

total free amino acids, and in many cases this is accompanied by reduced rates of protein synthesis (for a review, see Rabe, 1990). Plants respond to low temperature by accumulating sugars and increasing the proportion of unsaturated fatty acids in the chloroplast membrane, whereas higher temperatures result in the production of heat-shock proteins (Harborne, 1997). The plant response to elevated CO_2 levels affects growth rate (Kirschbaum, 1994), concentrations of chlorophyll and lipid in the chloroplast (Sgherri et al., 1998), and the relative abundance of fatty acids in surface waxes (Williams et al., 1998). Measurement of such changes in the organic matter composition of plants stored in herbaria can be extremely useful in indicating changes in environmental stress over time.

We have shown that plant material preserved in herbaria retains the amino acid signal (this study) and also the lipid signal (M. A. Teece, unpublished data) of the original plant, and therefore any biochemical or biological responses to stress should also be preserved. DNA was amplified from herbarium specimens dating back to 1940, indicating that even DNA is preserved under these conditions (Karen et al., 1997). The preservation of protein, lipid, and DNA in herbarium

TABLE 8-1 Concentration (residues per 1000) of amino acids, total hydrolyzable amino acid (THAA; pmol/μg dry wt) concentrations and elemental composition of herbarium samples of the C_4 grass *B. hirsuta*. (wt% and atomic C:N)

Amino acid	1849	1882	August 1936	July 1946	Nov. 1952[b]	Sept. 1963[b]	October 1965[b]	August 1981	October 1992
Asx	125	116	116	114	112	105	108	110	132
Thr	68	66	69	69	112	111	108	74	63
Ser	71	70	65	71	43	41	42	65	58
Glx	121	116	127	118	128	126	126	128	116
Gly	117	112	105	109	97	92	94	94	106
Ala	121	116	120	116	117	113	118	111	119
Val	78	78	76	77	77	77	77	76	72
Met	4	8	7	6	0	2	0	0	8
Iso	50	50	47	52	50	51	51	49	49
Leu	107	112	109	109	110	111	111	107	114
Tyr	18	23	18	19	18	23	19	19	21
Phe	50	50	47	49	50	51	51	67	47
Lys	32	39	47	45	45	51	51	51	49
Arg	39	43	47	45	41	47	45	48	46
THAA	281	258	276	466	556	611	692	876	843
wt.% C	46.1	42.6	45.4	42.3	NM	NM	NM	40.9	44.8
wt.% N[a]	1.0	0.4	0.5	1.0	NM	NM	NM	1.1	2.5
at. C:N	56	115	106	50	NM	NM	NM	45	21

[a] Error of measurement at low concentrations ±0.5%.
[b] NM, Not measured.

TABLE 8-2 Amino acid concentration (residues per 1000), total hydrolyzable amino acid (THAA; pmol/μg dry wt) concentrations and elemental composition of specimens of *P. spicata* preserved in the U.S. National Herbarium (wt% and atomic C:N)

Amino acid	July 1874[b]	August 1906	July 1947	July 1970	July 1977[b]
Asx	111	113	111	114	110
Thr	73	74	97	73	73
Ser	63	66	79	64	65
Glx	125	127	122	129	125
Gly	101	102	99	97	101
Ala	112	113	112	112	109
Val	85	80	75	78	78
Met	0	0	0	0	0
Iso	53	52	48	50	51
Leu	121	119	101	116	115
Tyr	18	22	21	15	23
Phe	72	58	55	59	59
Lys	28	34	40	47	45
Arg	40	41	40	46	48
THAA	1084	851	730	1000	800
wt.% C	NM	39.1	45.3	40.3	NM
wt.% N[a]	NM	1.5	2.0	1.2	NM
at. C:N	NM	30	26	39	NM

[a] Error of measurement at low concentrations ±0.5%.
[b] NM, Not measured.

specimens suggests that plant specimens preserved in museums and herbaria around the globe offer a unique resource for studying the effect of human practices since the early 19th century. Indeed, the change in the atmospheric CO_2 levels resulting from industrialization was recorded in the stable carbon isotope composition of preserved plant tissue dating back to 1849 (M. A. Teece, unpublished data). A recent study (Herpin et al., 1997) utilized herbarium moss samples to document the effect of air pollution resulting from industrialization. As herbarium specimens are geographically and temporally constrained, they should provide an enormous information resource for these environmental effects both on a local as well as a global scale. Future analyses of herbarium specimens may provide valuable information on the effects of an even greater range of human practices such as deforestation, the introduction of domesticated livestock, and the use of pesticides and insecticides on ecosystems around the world.

Acknowledgments: The authors would like to thank Jill Russ for amino acid analyses, and Sue Ziegler and Henry Fricke for comments on an earlier version of the manuscript.

REFERENCES

Azevedo R. A., Arruda P., Turner W. L., and Lea P. J. (1997) The biosynthesis and metabolism of the aspartate derived amino acids in higher plants. *Phytochemistry* **46**, 395–419.

Child R. and Pinnager D. (1992) The inefficient use of insecticides in museums. In *Life After Death: the Practical Conservation of Natural History Collections*, pp. 15–17. United Kingdom Institute for Conservation of Historic and Artistic Works of Art.

Cowie G. L. and Hedges J. I. (1992) Improved amino acid quantification in environmental samples: charge-matched recovery standards and reduced analysis time. *Mar. Chem.* **37**, 223–238.

Cronquist A., Holmgren A. H., and Holmgren N. H. (1977) *Intermountain Flora: Vascular Plants of the Intermountain West, U.S.A.* Columbia University Press.

Engel M. H. and Hare P. E. (1985) Gas–liquid chromatographic separation of amino acids and their derivatives. In *Chemistry and Biochemistry of the Amino Acids* (ed. G. C. Garrett), pp. 462–479. Chapman and Hall.

Fogel M. L. and Tuross N. (1999) Transformations of plant biochemicals to geological macromolecules during early diagenesis. *Oecologia* **120**, 336–346.

Fosberg F. R. and Sachet M. H. (1965) *Manual for Tropical Herbaria. Regnum Vegetabile*, vol. 39. International Bureau for Plant Taxonomy and Nomenclature.

Harborne J. B. (1997) Biochemical plant ecology. In *Plant Biochemistry* (eds P. M. Dey and J. B. Harborne), pp. 503–516. Academic Press.

Hare P. E., Fogel M. L., Stafford T. W., Jr, and Hoering T. C. (1991) The isotopic composition of carbon and nitrogen in individual amino acids isolated from modern and fossil proteins. *J. Arch. Sci.* **18**, 277–292.

Herpin U., Markert B., Weckert V., Berlekamp J., Friese K., Siewers U., and Lieth H. (1997) Retrospective analysis of heavy metal concentrations at selected locations in the Federal Republic of Germany using moss material from a herbarium. *Sci. Tot. Environ.* **205**, 1–12.

Karen O., Hogberg N., Dahlberg A., Jonsson L., and Nylund J. E. (1997) Inter- and intraspecific variation in the ITS region of rDNA of ectomycorrhizal fungi in Fennoscandia as detected by endonuclease analysis. *New Phytol.* **136**, 313–325.

Keeling R. F., Piper S. C., and Heimann M. (1996) Global and hemispheric CO_2 sinks deduced from changes in atmospheric O_2 concentration. *Nature* **381**, 218–221.

Kirschbaum M. U. F. (1994) The sensitivity of C3 photosynthesis to increasing CO_2 concentration: a theoretical analysis of its dependence on temperature and background CO_2 concentration. *Plant, Cell Environ.* **17**, 747–754.

Penuelas J. and Azcon-Bieto J. (1992) Changes in leaf $\Delta^{13}C$ of herbarium plant species during the last 3 centuries of CO_2 increase. *Plant, Cell Environ.* **15**, 485–489.

Penuelas J. and Estiarte M. (1997) Trends in plant carbon concentration and plant demand for N throughout this century. *Oecologia* **109**, 69–73.

Rabe E. (1990) Stress physiology: the functional significance of the accumulation of nitrogen-containing compounds. *J. Hort. Sci.* **65**, 231–243.

Rao R. R. and Sharma B. D. (1990) *A Manual for Herbarium Collections.* Botanical Survey of India.

Sgherri C. L. M., Quartacci M. F., Menconi M., Raschi A., and Navari-Izzo F. (1998) Interactions between drought and elevated CO_2 on alfalfa plants. *J. Plant Physiol.* **152**, 118–124.

Stubbendieck J., Hatch S. L., and Hirsch K. J. (1986) *North American Range Plants.* University of Nebraska Press.

Williams M., Robertson E. J., Leech R. M., and Harwood J. L. (1998) Lipid metabolism in leaves from young wheat (*Triticum aestivum* cv. Hereward) plants grown at two carbon dioxide levels. *J. Exp. Bot.* **49**, 511–520.

Womersley J. S. (1981) *Plant Collecting and Herbarium Development.* FAO Plant Production and Protection Paper 33, Food and Agriculture Organization of the United Nations.

Woodward F. I. (1987) Stomatal numbers are sensitive to increases in CO_2 from pre-industrial levels. *Nature* **327**, 617–618.

Wright H. A. and Bailey A. W. (1982) *Fire Ecology: United States and Southern Canada.* Wiley.

Zieman J. C., Macko S. A., and Mills A. L. (1984) Role of seagrasses and mangroves in estuarine food webs: temporal and spatial changes in stable isotopic composition and amino acid content during decomposition. *Bull. Mar. Sci.* **35**, 380–392.

9. Biogeochemical diagenesis in recent mammalian bones from saline lakes in western Victoria, Australia

Bonnie A. B. Blackwell, William M. Last, and N. W. Rutter

Amino acid racemization (AAR) dating in bones and teeth requires that their AAR rates rely solely on sedimentary temperature and the time since deposition (e.g., Bada, 1985). Tests using well-dated bones and teeth, however, have implicated diagenetic alteration as an important parameter in AAR under natural conditions (Blackwell et al., 1990). Hare (1974, 1980) showed experimentally that water availability limited AAR, whereas pH alters AAR rates (Baum and Smith, 1986; Blackwell, 1987). Paleontological wisdom holds that complete bone remineralization requires many thousands of years (Goffer, 1980). In subaerial, semiarid environments, however, bone rarely survives even five years of exposure (Behrensmeyer, 1978; Tuross et al., 1989) before erosion reduces it to small, taxonomically unidentifiable shards. Yet, in the fossil record, numerous identifiable fossils have persisted long enough to be remineralized, and some even preserve recognizable collagen.

To examine how rapid mineralization affects AAR, recently deposited remineralized bovid fossils and their surrounding sediment from four saline lakes in western Victoria, Australia, were examined. AAR, mineralogical, scanning electron microscopic, neutron activation analysis (NAA) and other chemical analyses show that the geologically instantaneous secondary mineralization in the fossils reflected the sedimentary secondary mineralization, and that conditions associated with fossilization resulted in very rapid and dramatic changes in amino acid compositions and D/L ratios.

THE SALINE LAKES IN WESTERN VICTORIA

During the late Cenozoic in western Victoria, Australia, frequent volcanism created an olivine tholeiite and alkali olivine basalt plain 40,000 km^2 in area, interrupted by lava ridges and numerous eruption points (figure 9-1; Clark and Cook, 1988; Douglas and Ferguson, 1988). Blanketed by black chernozems, the gently rolling terrain experiences a Mediterranean climate (table 9-1). In summer, high evaporation rates cause annual evaporation/precipitation ratios ranging from >1.0 to 2.0. Regionally, the three main aquifers reside in the Quaternary volcanic units, the late Tertiary limestone, and the late Cretaceous/early Tertiary quartzose sandstone (Gill, 1987). Under poorly developed drainage conditions, numerous, mainly saline, lakes occupy topographically closed basins ranging in size from Lake Corangamite, Australia's largest permanent lake, to small ephemeral ponds (Jenkin, 1988). Most smaller lakes occupy local depressions, whereas some larger playas have formed behind lava dams. Many volcanic cones and eruption centers contain small, but often deep, maar lakes. Because of the high evaporation rates, lake depths and areal extents vary dramatically during the year. Regional groundwater contains 1000–3000 mg/L total dissolved solids (TDS) with $Na^+ \gg Mg^{2+} > K^+ > Ca^{2+}$ and $Cl^- \gg SO_4^{2-} > HCO_3^- > CO_3^{2-}$ (Williams and Buckney, 1976).

The Maar Lakes: East and West Basins

East and West Basins, two small adjacent maar lakes near Lake Corangamite (figure 9-1; tables 9-1 and 9-2), contain permanent saline alkaline water dominated by Na^+ and Cl^-, with highly productive infauna (Timms, 1972; Timms and Brand, 1973; Last and De Deckker, 1990). West Basin has a chemocline at a depth of 3.5 m, but East Basin is presently monomictic (Last and De Deckker, 1990).

Besides the meromixis in West Basin, both lakes can develop seasonal thermal stratification complicating the modern sedimentological facies (Last and De Deckker, 1990). Many indurated, algally generated, evaporitic carbonate hardgrounds and stromatolites, exhibiting complex fabrics, geometries, and mineralogies, ring the lakes to depths near 1 m. In addition to dolomite and protodolomite, hydromagnesite and magnesite lithify hardgrounds in West Basin, while calcite and monohydrocalcite also occur in East Basin (table 9-3; Last, 1992). Kutnahorite, siderite, protohydromagnesite, and aragonite all occur as major local cements. The Holocene limestones grade basinward into unconsolidated coarse clastics and laminated organic-rich clays. Dolomite, calcite, hydromagnesite, magnesite, monohydrocalcite, and pyrite all form primary endogenic or authigenic precipitates in the modern deep-water offshore sediment (Gell et al., 1994).

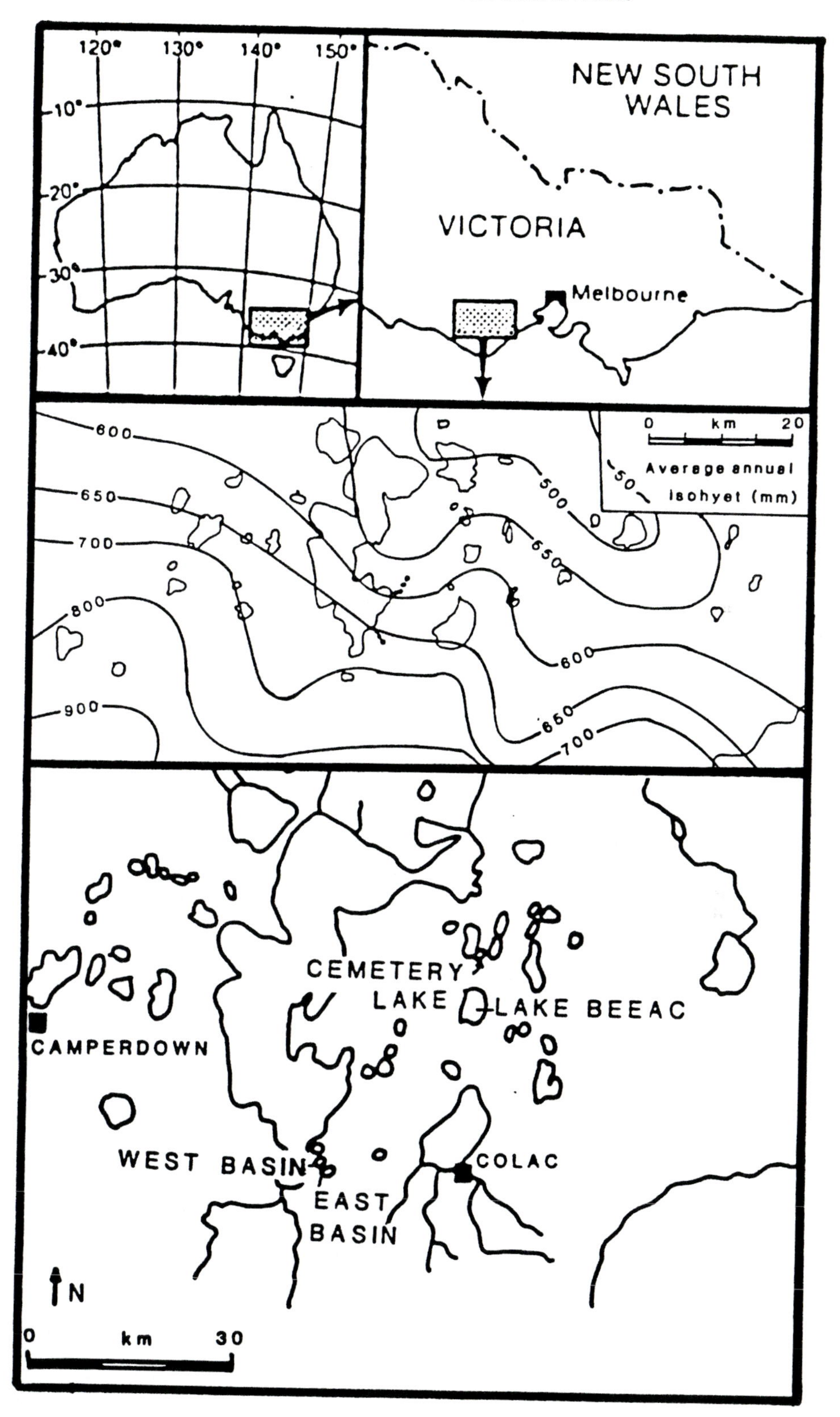

Fig. 9-1. The four saline lakes near Colac, Western Victoria, Australia. East and West Basins are permanent maar lakes, but the playas, Lake Beeac and Cemetery Lake, can dry up completely.

TABLE 9-1 Climatic conditions in the saline lakes and comparative sites

Site	Mean annual precipitation (mm/yr)	Mean annual temperature (°C)	Diagenetic temperature (°C)
Colac, Victoria	750	15	15
East and West Basin Lakes, Victoria	750	15	15
Lake Beeac, Victoria	575	15	15
Cemetary Lake, Victoria	550	15	15
La Chaise-de-Vouthon Cave, France	1600	14	7

The Playas, Lake Beeac and Cemetery Lake

Lake Beeac, a large playa north of Colac, occupies a topographically closed basin with no associated perennial streams, but with several small ephemeral creeks (figure 9-1, table 9-2). Cemetery Lake, a much smaller (but similar) playa 8 km north of Lake Beeac, has no associated perennial streams. Annually, their water levels and salinities vary greatly (De Deckker and Last, 1988; Last and De Deckker, 1990, 1992). Water depths peak to about 1 m in winter. Although Lake Beeac normally retains some water year round, Cemetery Lake usually dries up completely by late summer.

Diffuse shallow subsurface inflow from the volcanic ridges immediately to the west dominates Lake Beeac's hydrologic budget (Last and De Deckker, 1992). The saline evaporitic conditions, high alkalinities, and elevated Mg/Ca ratios all favor Mg-carbonate precipitation. In both basins, magnesite, high-Mg calcite, and dolomite occur as primary precipitates in the modern sediment (De Deckker and Last, 1988, 1989; Last, 1990), but only Lake Beeac experiences dolomite whitings (P. De Deckker, personal communication, 1987). Late Quaternary sediment from Lake Beeac contains approximately 1.3 m of fine grained, generally nonstratified, nonfossiliferous, calcareous clay and silty clay. Cemetery Lake has not been investigated stratigraphically.

METHODOLOGY

Recently deposited bones and teeth, and the sediment surrounding them, were collected from the four saline lakes (tables 9-2 and 9-3) in the fall, when water levels were near their lowest, thus exposing samples that, at other times, would have been submerged. Because all but two samples (a bovid rib and a juvenile tibia) were positively identified as being from *Bos taurus* (the domestic cow), a species first imported to Australia in 1788, the samples cannot exceed 200 ± 10 years in age.

After collection, samples were immediately doubly sealed in airtight bags to prevent salt crystallization due to desiccation, and covered in black plastic bags to slow algal growth.

TABLE 9-2 The saline lakes in the study

Character	East Basin	West Basin	Lake Beeac	Cemetery Lake
lake type	maar	maar	playa	playa
surface area	$<1\ km^2$	$<1\ km^2$	$\sim 7\ km^2$	$\sim 1\ km^2$
watershed area	$\sim 1\ km^2$	$\sim 1\ km^2$	$25\ km^2$	$\sim 2\ km^2$
water depth[a]	4.5 ± 0.3 m	5.8 ± 0.2 m	$1.0^{+0.5}_{-0.5}$ m	$\sim 1.0^{+0.5}_{-0.5}$ m
completely dried	no	no	~1 yr in 10–20	most years
thermal stratification	seasonally	seasonally	no	no
chemical stratification	monomictic	meromictic	monomictic	monomictic
chemocline depth	–	3.5 m	–	–
brine concentration[b]	50 ppt TDS	mixo: 75 ppt TDS mono: 89 ppt TDS	1–3 g/L TDS	1–3 g/L TDS
surface pH	~8	~8.2	>8.5	>8.5
dominant ionic composition	Na^+, Cl^-	Na^+, Cl^-	$Na^+ \gg Mg^{2+} > K^+ > Ca^{2+}$ $Cl^- \gg SO_4^{2-} > HCO_3^- > CO_3^{2-}$	

[a] Typical years.
[b] mixo = mixolimnion; mono = monolimnion.

TABLE 9-3 Sample and sediment mineralogy in Australian saline lakes. Abbreviations: lo = low; hi = high; exp = exposed; subm = submerged; amorf = amorphous; sed = sediment

Sample	Conditions	Depth below surface[a] (cm)			Minerals[b]
		Hi H_2O	Lo H_2O	Sed	
A. Cemetery Lake[c]					
87001 bone	exp	30	−120	−5–2	apatite, halite, quartz, clays, dolomite
87002 bone	exp	30	−120	−5–2	halite, apatite, pyrite, dolomite, clays
surface gel[d]	subm	100	−50	0–1	clays, quartz, feldspars, magnesite, dolomite
sediment[d]	subm	105	−45	4–6	clays, magnesite, domolite, quartz (tr), feldspar (tr), calcite (tr)
B. Lake Beeac					
87004 bone	part subm	88–100	−12–0	−2–10	apatite, siderite, dolomite, hi-Mg calcite, aragonite, protodolomite
surface gel[e]	subm	110–112	10–12	0–2	clays, magnesite, dolomite
sediment[e]	subm	115–120	15–20	5–10	clays, magnesite, dolomite, quartz, feldspar (tr)
C. East Basin					
87003 bone	exp	30	−30	−5–2	apatite, quartz, clays, monohydrocalcite, dolomite
boundstone	exp	75–90	15–30	0–5	dolomite, monohydrocalcite, protodolomite
boundstone	subm	85–135	25–75	0–5	monohydrocalcite, dolomite, protodolomite, hydromagnesite
sediment[e]	subm	75–80	15–20	5–10	dolomite, protodolomite, lo-Mg calcite, clays, feldspars, amorf Fe oxides, monohydrocalcite
D. West Basin					
87005 bone	exp	20	−20	−5–2	apatite, siderite, calcite, hydromagnesite
87006A bone	subm	50	10	−5–2	apatite, magnesite, pyrite, hydromagnesite, clays
87006B bone	subm	50	10	−5–2	halite, siderite, apatite, quartz
87009A buccal bone, dentine	subm	60	20	4–6	siderite, magnesite, apatite, aragonite, kutnahorite, hydromagnesite, vivianite
87009B lingual bone	subm	50	10	0–2	quartz, clays, feldspars, dolomite, hi-Mg calcite, magnesite, siderite
87010 bone	subm	45	5	0–2	siderite, apatite, pyrite, dolomite
87011 bone	subm	60–78	20–38	−8–10	
87011A proximal bone	subm	72–74	22–24	−6–−4	siderite, apatite, halite, dolomite, amorf Fe oxides
87011B distal bone	subm buried	74–76	34–36	6–8	apatite, siderite, pyrite, halite, amorf Fe oxides
boundstone	exp	30–35	−10–−5	5–15	dolomite, protodolomite
boundstone	subm	65–165	25–125	5–10	hydromagnesite, magnesite, protodolomite, dolomite, kutnahorite, aragonite, siderite
sediment[d]	subm	90–100	50–60	1–4	dolomite, lo-Mg calcite, clays, feldspar, amorf Fe oxides, magnesite, hydromagnesite

[a] Negative numbers indicate distances above.
[b] In order of abundance, most common first. tr = trace.
[c] Cemetary Lake = "Lake near Cemetery" (De Deckker and Last, 1988).
[d] Last and De Deckker (1990).
[e] De Deckker and Last (1988).

Preliminary identifications were made in the field, but confirmed later in the laboratory. Simultaneously, surrounding sediment was sampled and doubly sealed in airtight bags. The samples were stored at 25°C to minimize mineralogical changes among the evaporate minerals. Samples were selected for analysis to represent a range of weathering, infaunal infestation, diagenetic alteration, and secondary mineralization. Each was analyzed by several techniques to determine its diagenetic alteration, biogeochemistry, and mineralogy. Subsamples for AAR and mineralogical analyses were removed within 2–5 d after collection. AAR subsamples were then stored at ca. 5°C in the dark to minimize further algal growth and were dissolved for analysis within 5–7 d. Although Behrensmeyer's (1978) bone diagenesis stages do not incorporate considerations of secondary mineralization, the time since deposition was estimated from comparative diagenetic alteration.

Diagenetic Alteration and Mineralogy

Immediately after each bag was opened, the bone was photographed, visually examined, and speciated and typed. After slicing, each new surface was photographed and visually re-examined. Visual, petrographic, microscopic, and scanning electron microscopic (SEM) examination of surfaces and internal sections revealed evidence of physical destruction, secondary mineralization, residues, stains, cellular structure, and infaunal infestation. Algal filaments were removed using a binocular microscope and then mounted on glass slides for identification. Sedimentary and fossil mineralogy were determined by SEM, microprobe, thin-section petrology, and X-ray diffraction (XRD). An automated search–match program (Marquart, 1986) aided mineral identification, and mineral percentages were estimated using the methods in Last and De Deckker (1990). Fourteen trace elements were analyzed by instrumental and delayed neutron-counting neutron activation analysis (NAA).

Amino Acid Chemistry

Amino acid concentrations and D/L ratios were analyzed by gas chromatography and automated amino acid analysis (Blackwell, 1987). In replicate analyses of bones and teeth, the most reproducible AAR ratios resulted from alanine, leucine, glutamic acid, and aspartic acid, whereas glycine, proline, alanine, and glutamic and aspartic acids yield the most reproducible concentration data (Rutter and Blackwell, 1995). Though species-independent, AAR ratios and concentrations in ungulate tissues do depend on tissue type and diagenetic alteration (Blackwell, 1987; Blackwell et al., 1990). As discussed below, however, tissue fractions may have also contained infauna and possibly dissolved amino acids in the porosity elements.

Data were compared with those for the same tissues and species from fresh tissues and La Chaise-de-Vouthon (France) cave fossils dating at 100–200 ka (table 9-2; Blackwell et al., 1990). Despite similar modern mean annual temperatures, diagenetic temperatures at the lakes are twice, and annual precipitation half, those at La Chaise-de-Vouthon. Periodic glacial advances since 200 ka significantly reduced the annual temperatures and precipitation affecting La Chaise-de-Vouthon.

BONE DIAGENESIS AND PRESERVATION IN WESTERN VICTORIA SALINE LAKES

Several Australian samples were cemented in recently formed beachrock whose very young age was confirmed by the *in situ* anthropogenic debris and ^{14}C (Last and De Deckker, 1990). The rapid sedimentation rates on the lakeshores (Plate 1A) suggest that the samples had been deposited not more than five years, or perhaps only months, before collection, which agrees with bone-preservation data from the Serengeti, which has a comparable climate (Behrensmeyer, 1978).

For the Australian fossils, average subaerial survival time is probably considerably less than that on the Serengeti grasslands, because continual salt spray and higher spring water levels cause salt brines to penetrate the pores. With repeated wetting and drying, salts crystallize from these brines within bones' porosity elements, forcing the mineral crystals apart. All bones near the saline lakes exhibited abundant evidence for salt crystallization. This commonly affected the thinnest bone most dramatically, especially that near the mandibular and maxillar alveolar margins, the articular surfaces on the joints, edges near the foramen or defects, such as cracks caused by trampling, algal borings, or gnawed areas. In thicker compact bone, "green bone" fractures formed parallel to the collagen bundles, because of salt crystallization and desiccation, while the outer lamellar bone often flaked off as plaquettes. Although few experiments have determined bone-destruction rates in saline lakes, anecdotal descriptions by local inhabitants, together with our data, suggest that most bones break down within 1–2 years if subaerially exposed and disappear within 5 years if not buried. Teeth survived longer, but were often separated from the jaws within 1–2 years. To survive fossilization as an identifiable bone in Victoria, the sample must be submerged or buried within 2–3 years. If submerged rapidly without suffering total weathering, however, the "survival" rate was perhaps as high as 75%. In fossils <200 years old, secondary carbonate and iron-rich mineralization had infilled nearly all porosity, helping to ensure preservation.

Algal Infestation

Bones experience infaunal boring by numerous organisms (Behrensmeyer, 1978; Stout, 1978; White and Hannus, 1983), including fungi, bacteria, insect larvae, and algae. On even the freshest Australian bones, green and blue-green algae (cyanobacteria) ubiquitously colonized the outer surfaces (Plates 1 and 2). The more weathered the bone, the more algae had bored through the compact bone in any fossil. Algae had colonized the medullary cavities near the upper surface, the cancellous bone, and other porosity elements in any fossil, whether in sunlit oxygenated water or subaerially exposed for more than a few weeks (Plates 1B,C and 2C,D).

Although these algae are not symbiotic, their general location within the bone relative to the external light source mimics that seen for *Zooxanthellae* in hermatypic corals. The algae colonized a band in the cancellous bone or medullary cavities approximately 2–5 cm thick below the lamellar bone (i.e., ~1–3 cm below the bone surface) on the sides that were exposed to light (Plates 1B, 2C, D, figure 9-2). This "diffuse light" zone provides ideal conditions for algal growth, since sufficient light penetrates to permit growth, but the protected location ensures a more reliable moisture supply than would be available on the exposed hardgrounds, especially during periods of lower precipitation. In areas where the lamellar bone was thinner, light penetrated further into the bone, allowing the algae to occupy more cancellous bone. Algae did not occur in the half of the bone that was closest to the sediment where, apparently, insufficient light penetrated. Bones collected during dry periods displayed desiccated algal filaments attached to their outer surfaces; inside, however, there were live green filaments in the intertrabecular spaces within the cancellous bone. Only severely desiccated bones with cracks that exposed and dried the cancellous bone or medullary cavities contained desiccated algae in the diffuse light zone. When rehydrated, however, the algae flourished. Storage of algally infested bone samples in complete darkness for >1 year did not kill the algae. When re-exposed to light, the algae began to grow again within 1–2 hours.

Fossil Mineralogy

Without exception, all fossils from the saline lakes had suffered some remineralization or secondary mineralization infilling their porosity (Plate 2, table 9-3). Among those found submerged, all had experienced >50% remineralization of their hydroxyapatite. Because the XRD protocol could not differentiate between the apatite minerals common in calcified tissues, "apatite" (table 9-3) implies any one (or some combination) of hydroxyapatite [HAP, $Ca_{10}(PO_4,CO_3,OH)_6(OH,F,Cl)_2$], carbonate-HAP, and calcium-deficient HAP, typical in fresh bones (Dallemagne, 1964; Neuman, 1980), as well as the common bone-replacement minerals francolite, uranylapatite, vivianite, and Mg-, Ba-, K-, or Sr-apatites (Denys et al., 1996; Hubert et al., 1996; Michel et al., 1996). Replacement minerals, which can contain higher trace-element concentrations than does fresh HAP (Farnum et al., 1995), were identified using the SEM and visual determinations, but the amorphous or very small crystal sizes in some secondary cements hampered identification.

In areas where sedimentary calcium carbonates are currently precipitating, the fossils also preserved calcium carbonates (table 9-3). Those collected from reducing sedimentary environments had typical reduced iron minerals. Common secondary minerals included halite, dolomite, aragonite, protodolomite, calcite, high-Mg-calcite, monohydrocalcite, siderite, magnesite, hydromagnesite, uranylapatite and other apatites, but secondary vivianite, kutnahorite, and fluoroapatite also occurred. The concentrations of Fe, Mg, U, F, Cl, Ca, and Na all increased during secondary mineralization. Initially during fossilization, secondary mineralization concentrated in porosity elements (i.e., along dentine tubules or cracks, enamel lamellae or growth lines, or in intertrabecular spaces, cracks, borings, or canals in the bone). Secondary mineralization was favored wherever solutions bearing essential dissolved chemical species could penetrate. High concentrations of organic matter lining these cavities could complex the ions and

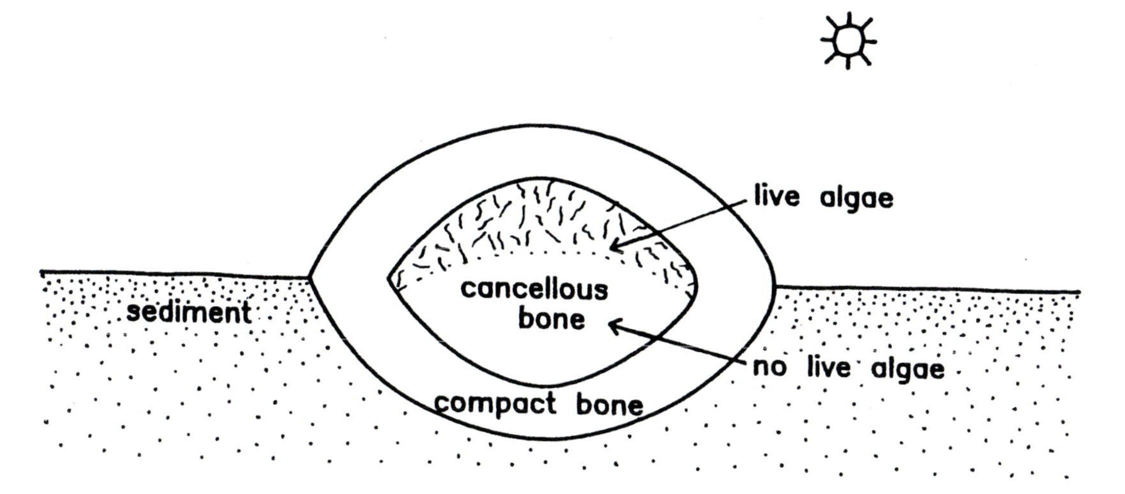

Fig. 9-2. The cryptic habitat for algae infesting bones. In bones lying on, or partially submerged in, the sediment, algae infest the diffuse light zone.

probably catalyzed some reactions, as could the microfauna living therein.

Bones with Little Diagenesis

All bones with little diagenesis occurred well below the high water mark, but had been completely or partially subaerially exposed at collection (table 9-3). Still in Stage 1 (cf. Behrensmeyer, 1978) and probably not deposited more than a few months to a year before collection, all showed little evidence for diagenesis or alteration, except for the ubiquitous exterior algal colonization and salt cracks, especially near the alveolar margins and foramen. All did have secondary mineralization in porosity elements.

In both mandibles from Cemetery Lake (87001 and 87002; tables 9-3, 9-4), clays and carbonates dominated the gray detrital "sediment" in the medullary cavity, as they did in the adjacent sediment and the lacustrine sediment generally (De Deckker and Last, 1988). In East Basin, the mandible 87003 contained quartz, clays, monohydrocalcite, and dolomite. In nearby boundstones, dolomite, protodolomite, monohydrocalcite, and hydromagnesite occur (Last and De Deckker, 1990), while unconsolidated sediment surrounding the bone contained those same carbonates, low-Mg calcite, clays, and amorphous iron oxides. Despite the spongy marrow in 87005, which suggests deposition in West Basin <1–2 months before collection (Plate 1A), it already contained secondary siderite, calcite, and hydromagnesite. The beachrock cemented to 87005 contained hydromagnesite, dolomite, protodolomite, and calcite.

The quartz, protodolomite, dolomite, and clays in the bones occurred as either detrital or secondary minerals, whereas halite and pyrite were probably secondary. Any detrital minerals could have entered through the foramen, cracks around the alveolus, or insect borings. Whereas blood residues do occur in archaeological bones (Maat, 1991, 1993), pyrite is very common in older fossils (Hubert et al., 1996; Martill and Unwin, 1997). Possibly, the pyrite seen in 87002 derived its iron from hemoglobin in the blood residues, although bacterial action may have precipitated pyrite directly from brines in the medullary cavity.

Bones with Significant Diagenesis

All submerged bones from West Basin showed significant diagenesis, as did 87004, found partially submerged in Lake Beeac. With moderate to significant cracking and compact bone loss, all had reached late Stage 1 (tables 9-3 and 9-4) and experienced significant remineralization and secondary mineralization. Those with less weathering, like 87004, 87010, and 87011, were probably not exposed subaerially for long, before total submergence stopped most bone weathering. Surprisingly, the teeth in 87009 had been completely dissolved to the gumline, yet the underlying bone and dentine showed no effects, although the surface exposed to the water was badly cracked and had lost cortical bone. While algae colonized all these bones on the exterior, all but sample 87011 had algal infestation within the medullary cavities or in the "diffuse light zone" (Plates 1B–D, 2C,D, figure 9-2). When sliced open, the wet cancellous bone in 87004 hosted live algal filaments, as did that of sample 87006 and that of sample 87010. On 87006, long algal filaments completely coating the outer surface also hosted ostracode eggs (Plate 1D).

In 87004, a pearly gray, poorly crystallized, secondary mineral mixture with co-occurring high-Mg calcite, aragonite, and dolomite infilled most intertrabecular and all other porosity, as did a similar cement with halite, siderite, and dolomite in the vertebral body of 87006 (tables 9-3 and 9-4, Plate 2). Sample 87006 also had a well-crystallized dark-gray mineral layer, rich in siderite, at the cortical/cancellous bone interface. Although not visible in 87010, siderite, pyrite, dolomite, protodolomite, and calcite (all probably secondary) were identified by XRD analysis. Within West Basin, dolomite and protodolomite dominate all boundstones, while submerged ones additionally have hydromagnesite, magnesite, kutnahorite, aragonite, and siderite (Last and De Deckker, 1990). Unconsolidated sediment surrounding the bones contained dolomite, high-Mg calcite, clays, feldspars, magnesite, iron hydroxides, and hydromagnesite. In Lake Beeac generally, and immediately adjacent to 87004, protodolomite and dolomite precipitation currently dominates the sedimentation, whereas magnesite, clays, and quartz are less common (De Deckker and Last, 1988).

Two bones had unusual Fe–Mg-rich mineralization (tables 9-3D and 9-4D). Found near an anthropogenic metal deposit replete with discarded sinks and appliances, 87009 was cemented to the boundstone by a hard brown pelloidal cement, beneath which the adjacent bone was completely altered by Fe-rich organic stains. Black carbonate-rich secondary mineralization partially filled the medullary cavity and invasively mineralized the dentine along the neonatal line, cracks, and tubules (Plate 2A,B). The adjacent boundstones, the cement, and the bone all had considerable kutnahorite, which was unknown anywhere except in ocean sediment below 2000 m (Mucci, 1988) until its discovery in West Basin (Last and De Deckker, 1990). The conditions able to precipitate kutnahorite in the sediment and fossil, and dissolve enamel, without dissolving bone or carbonate-rich cements, remain a mystery. Standing vertically, 87011 had its distal end submerged, and the proximal end buried 10 cm in sediment. On the distal end exposed to oxygenated water, siderite, red (oxidized) amorphous Fe-oxides and organometallic complexes had remineralized the cortical bone on both the exterior and medullary cavity surfaces, while on the proximal end, pyrite, siderite, black (reduced) amorphous Fe-oxides and organometallic complexes had almost completely altered all the cortical and cancellous bone. In the diaphysis, the black mineralization progressively thinned away from the medullary cavity and proximal end, while the red mineralization thickened toward the distal end, leaving some pristine bone in the diaphysis and at the distal end. This strange pattern may indicate that the interface between the oxidized and reduced water had recently migrated, that the bone had sunk gradually into the sediment, or that the medullary cavity can maintain oxidizing conditions longer than adjacent sediment. The strongly reducing sediment here, or bacterial action, may have caused pyrite to precipitate in the bone.

Whether the dolomite in all these bones entered as detritus through the foramen and cracks or precipitated a secondary cement is uncertain, but any siderite, pyrite, kutnahorite, and aragonite appeared to be secondary. Although halite is probably a secondary precipitate, the quartz, feldspars, and clays are probably detrital.

BIOGEOCHEMICAL DIAGENESIS IN BONES AND TEETH

Within the bones and teeth, the type of biogeochemical diagenesis correlated with other factors, including algal infestation, amino acid leaching or enrichment, secondary mineralization, and sediment type. In the discussion below, amino acid "concentrations" imply "relative concentrations." Any effects in "whole tooth" reflect changes primarily in the dentinal amino acids, which dominate such samples, because they average 75% dentine, and enamel (~20%) contains minimal amino acids (Blackwell, 1987). The limited sample set, combined with the large numbers of variables that influenced the amino acid geochemistry, made statistical analysis difficult; the situation was further complicated by the need to obtain multiple splits of a subsample for different analyses. Within the tissue samples, mineralogy varied considerably over very small distances, but AAR protocols precluded powdering the samples before removal of subsamples for mineralogical or trace-element analyses. Therefore, mineralogy in the AAR subsamples may have differed somewhat from the immediately adjacent subsamples in which the mineralogy was determined. This analytical problem must be solved before the trends described here can be confirmed with any precision.

Amino Acid Racemization and Compositional Changes

Given that the Australian teeth and bones were <200 years old, and probably <1–3 years old, their AAR ratios should not have exceeded those induced by the analysis in fresh tissues ("fresh sample background ratios"). Nonetheless, some had anomalously high AAR ratios (figures 9-3 to 9-8), which agrees with rapid organic degradation seen in bones with mechanical destruction at or beyond Stage 2 (Tuross et al., 1989). While others had AAR ratios significantly below background, most had relative amino acid concentrations that deviated from those expected for the fresh tissues.

Roughly half the tissues had alanine D/L ratios >0.02 (figure 9-3A), the background value for fresh samples, while several had D/L ratios exceeding 0.20, a value normally attained in bone racemization at 25°C for >50 ky! Although summer temperatures in Victoria can reach 45–50°C, this alone cannot explain the extreme D/L ratios. Alanine D/L ratios were elevated only in samples in which the alanine concentrations normalized against total alanine, proline, and aspartic acid concentration exceeded 70 ± 2%. Below this threshold, D/L ratios averaged 0.01. Conditions favoring increased alanine concentrations are associated with high D/L ratios.

For valine, proline, and glutamic and aspartic acid, however, conditions causing their depletion favored their high D/L ratios. All bones with glutamic acid concentrations (relative to glutamic acid + alanine + proline) below the 10–12% threshold had glutamic acid D/L ratios >0.05—the fresh tissue background (figure 9-3B). The five cementum samples (including that from La Chaise-de-Vouthon) showed a strong inverse proportionality ($R^2 > 0.98$) between their aspartic acid D/L ratios and their aspartic acid concentrations relative to proline, and aspartic and glutamic acids (figure 9-4). In whole tooth and cementum, glutamic acid concentrations, although reduced, did not fall below the threshold, and none had a significantly higher D/L ratio. For proline in teeth, a strong negative correlation existed between relative proline concentrations and D/L ratios ($R^2 = 0.88$–0.92; figure 9-5B). In both tooth and bone, valine D/L ratios as high as 0.54 occurred. Those that exceeded fresh tissue background values (0.005–0.02) occurred only when the relative valine concentration normalized against alanine, valine, and proline dropped below a threshold at 4–5% (figure 9-6). More data are needed to ascertain if phenylalanine losses are associated with elevated racemization (figure 9-7). The high D/L ratios occurring here are not correlative among the amino acids, but occur under specific conditions (algal infestation, secondary mineralization, sedimentary geochemistry) for each case (figures 9-3 to 9-10).

Algal Infestation

In six sample sets, comparative AAR analyses from algally infested and adjacent non-infested tissue

TABLE 9-4 Diagenetic features in Australian saline lake bones and teeth. Abbreviations: diag = diagenetic; ext = exterior; ant = anterior; prox = proximal; intertrab = interbecular spaces; med cav = medullary cavity; alv = alveolar margins; cort = cortical; canc = cancellous; mand for = mandibular foramen; adj = adjacent to; exp = exposed; penetr = penetrating into; minlzn = mineralization; alter'n = alteration; 2° = secondary; filam = filaments; cov = coverage; DLZ: diffuse light zone; res = residue; dk = dark

Sample number	Species	Type, subsample	Diagenetic stage[a]	Deposition time[b]	Surface strains, fauna, residues, cements[c]	Cracks and borings	Bone loss	Infilling porosity
A. Cemetery Lake[4]								
87001	adult *B. taurus*	mandible, teeth	1	≤1 yr	buried ext: orangy-brown stain, 0.1 mm thick; 20–25% algal cov	salt cracks near alv and mand for, not penetr med cav	none	med cav: gray detritus, clays, carbonate[3]
87002	juvenile *B. taurus*	mandible, teeth	1	≤0.5–1 yr	buried ext: orangy-brown stain, 0.1 mm thick; ~20% algal cov med cav: dried blood res	none	none	med cav: gray detritus, clays, carbonate[3]
B. Lake Beeac								
87004	adult *B. taurus*	inominate (hemi-pelvis)	early 2	1–3 yr	ext: slightly bleached; ~20% algal cov	cracks not penetr through cort bone	iliac crest: 30% cort bone lost; canc bone exp	intertrab: 95% 2° minlzn[c]; **DLZ**: algal filam
C. East Basin								
87003	adult *B. taurus*	mandible, teeth	1	≤1 yr	ext: orangy-brown stain, partial cov; 20–25% algal cov	salt cracks near alv and mand for, not penetr med cav	none	none
D. West Basin								
87005	adult *B. taurus*	mandible, teeth	early 1	0.1–0.2 yr	no stains; ~20% algal cov, filam <1 mm	none	none	marrow still present
87006	*B. taurus*	vertebra	late 1	1–2 yr	ext: 100% algae cov, all surfaces; ostracode eggs in algae			intertrab: 100% 2° minlzn[c], **DLZ**: algal filam
87006A		spinous process			ext: no visible stains; algal filam ~2 cm	none	dorsal crest: abraded	

87006B		body			ext: no visible stain; algal filam ~5 cm cort/canc bone interface: grey 2° minlzn[c], 2 mm thick	algal filaments penetr canc bone		
87010	Bovida	rib	early 2	1–5 yr	exposed ext: orangy-brown stain; 60–70% algal cov buried ext: dk red stain, 20% algal cov	distal end: cracks penetr med cav	distal end: eroded; prox end: abraded	DLZ: algal filam
87009	adult *B. taurus*	mandible, teeth	late 2–early 3	1–3 yr		salt cracks, green bone fractures		med cav: 50% black minlzn[c]
87009A		buccal surface			ext: brown-black stain adj mineral cement[c]; cement 1–2 cm thick; <5% algal cov med cav: 2° minlzn[c]	cracks penetr med cav; insect borings	obscured by cement	foramen & canals: ~90% 2° minlzn[c]; DLZ: algal filam
87009B		lingual surface			ext: red stain, partial cov, surficial only; 40–50% algal cov med cav: none	cracks penetr med cav; insect borings	plaquettes lost; ramus broken off	foramen & canals: ~10% 2° minlzn[c]
87009D		premolar dentine			roots: 2° invasive alteration[c] along neonatal line, cracks, tubules	roots: desiccation cracks	all enamel & dentine above gum dissolved	tubules & neonatal line: ~50% 2° minlzn[c]
87011	Bovida	tibia	1	1–2 yr		none	none	
87011A		distal end: exp water			ext: red 2° minlzn[c], 1 cm thick; 40–50% algal cov med cav: red 2° minlzn[c], 3–5 mm thick			intertrab: <10% 2° minlzn[c]; DLZ: algal filam
87011B		prox end: buried			all bone: black 2° minlzn[c]			intertrab: <10% 2° minlzn[c]
87011C		diaphysis: buried			ext: black 2° minlzn[c], 1.0–1.2 cm thick med cav: red 2° minlzn[c], 5–7 mm thick			med cav: none

[a] Diagenetic stage as described by Behrensmeyer (1978).
[b] estimated time since deposition in sediment.
[c] see table 9-2 for mineralogies of cements, secondary minerals, and infilling sediment.
[d] Cemetery Lake = "Lake near Cemetery" in De Deckker and Last (1988).

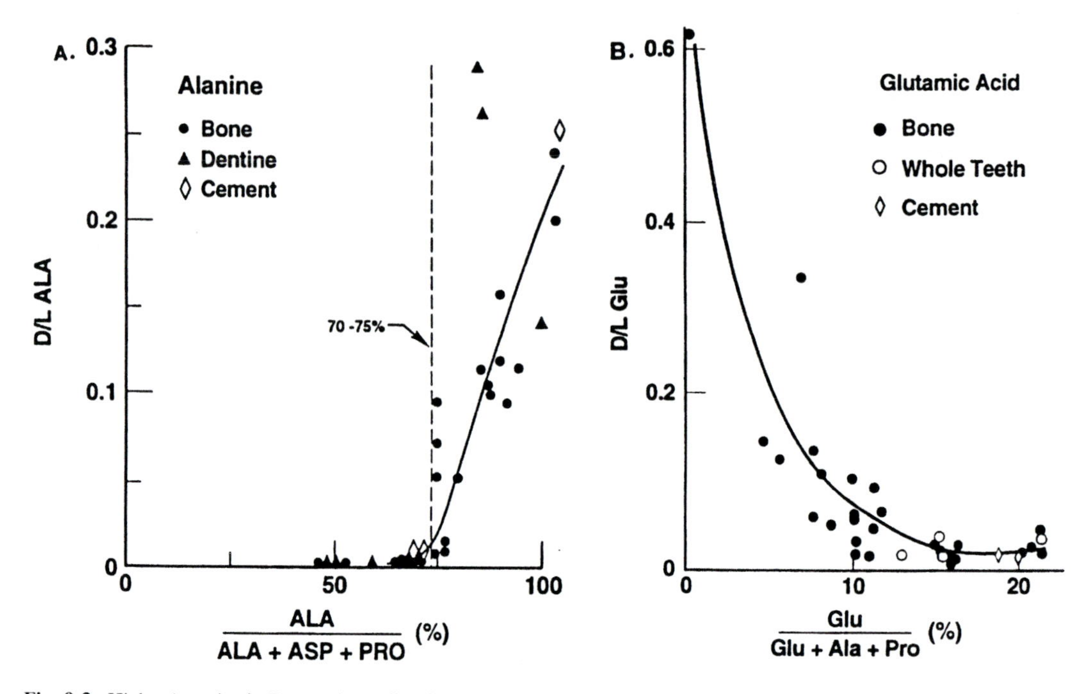

Fig. 9-3. High D/L ratios in Recent Australian fossils. For samples in which amino acid concentrations passed the threshold values, D/L ratios exceeded the fresh tissue backgrounds. A. When alanine concentrations relative to alanine, aspartic acid and proline exceeded $70 \pm 2\%$, alanine D/L ratios rose significantly above 0.02. B. When glutamic acid concentrations relative to glutamic acid, alanine and proline dropped below 15%, glutamic acid D/L ratios significantly exceeded 0.05.

showed significantly different relative amino acid concentrations (figure 9-9). Algal infestation in all tissues correlated with reduced phenylalanine, aspartic acid, and glutamic acid concentrations, and increased leucine, valine, and proline concentrations (figures 9-3–9-8). In algally infested cementum, glutamic acid also decreased relative to aspartic acid. Algally infested dentine and bone showed slight, but variable, changes in leucine relative to valine, but changes in aspartic and glutamic acid concentrations fell within the random error. Although the uninfested tissues had no visible algal filaments or cells, their surfaces or pores may have had some adhering, dissolved, algally derived amino acids. The close similarity between concentrations in fresh tissues and most uninfested Australian tissues suggests that the latter suffered little algal contamination. Given the small sample sets, the differences among the tissue types may reflect random variation.

Algally infested samples also had unusual racemization ratios. Teeth with algae but no dolomite showed very reduced D/L aspartic acid ratios compared to fresh sample backgrounds (figure 9-4), whereas tissues with algae but no secondary Fe–Mg-minerals showed significantly reduced valine and phenylalanine D/L ratios compared to fresh samples (figures 9-6 and 9-7). All algally infested samples with secondary dolomite, particularly cancellous bone, had higher proline D/L ratios than fresh samples, up to 0.22, with slightly to moderately reduced relative proline concentrations (figure 9-5). The highest leucine D/L ratios occurred in samples having algal infestation and some combination of secondary siderite, pyrite, kutnahorite, and vivianite, but no dolomite (figure 9-8). Since glutamic acid concentrations in algally infested tissues dropped below the threshold value, algae may be a major contributor to, or indicator for, the high glutamic acid D/L ratios seen in fossils here. Therefore, algal infestation correlated with dramatically altered amino acid concentrations, coupled with altered D/L ratios for some amino acids, depending on the secondary minerals present.

During sample preparation, each tissue was sonicated with dilute HCl(aq) and then repeatedly rinsed with water. Microscopic examination prior to dissolution confirmed that all easily identified filaments and their fragments had been removed from the surfaces. Moreover, HCl(aq) should have removed any free amino acids from large porosity elements, including any algal borings. Whilst we cannot totally discount contamination by algal filaments as a cause, the dramatic shift in amino acid composition may result from one or more effects:

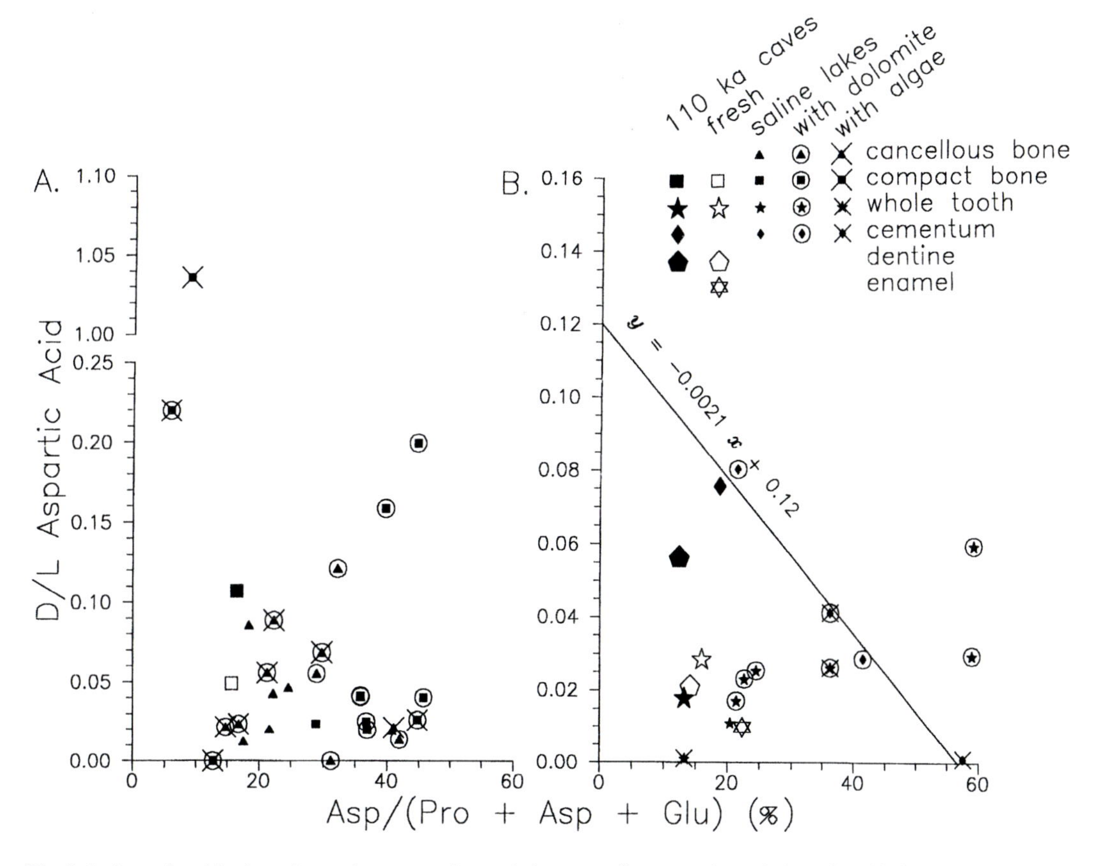

Fig. 9-4. Aspartic acid D/L ratios and concentrations relative to proline, aspartic and glutamic acids in Recent Australian fossils. Dolomitized tissues without algae show increased concentrations compared to fresh samples. A. A bone with algae, secondary siderite, kutnahorite, and vivianite had a D/L ratio >1.0! B. In teeth, samples with algae but no dolomite showed very reduced D/L ratios compared to fresh sample backgrounds. In cementum, aspartic acid D/L ratios correlated inversely with relative concentration ($R^2 > 98\%$). A loose correlation between D/L ratios and concentration occurs for dolomitized whole tooth.

1. The algae might have digested phenylalanine, glutamic acid, and aspartic acid from the collagen.
2. Algal borings increased porosity, and phenylalanine, glutamic and aspartic acid leaching.
3. The algal presence changed the microenvironmental conditions, promoting amino acid leaching, protein degradation, secondary carbonate or Fe–Mg-rich mineral precipitation.

Although they strongly resemble those in the lacustrine stromatolitic (microbolitic) mounds, the algal species have not been identified. Consequently, their bio- and geochemistry are poorly understood, but their effect is probably complex. Undoubtedly, the algae have altered the pH and Eh within the tissues through respiration, photosynthesis, and bacterial decay of their filaments; this could account for all of the amino acid variation (Bada, 1972; Blackwell, 1987). Although algal borings increased the tissue porosity, secondary mineralization reduced the porosity in samples with the highest filament densities, hinting at a causative link between algal infestation and secondary mineralization. Algae, or effects from their decay, seem to add to infested, dolomitized tissue, proline that is slightly more racemized than that in uninfested tissues and less racemized leucine to infested tissue with Fe–Mg-mineralization. All of the heavily infested tissue also had secondary mineralization, either from dolomite or Fe–Mg-rich minerals. Irrespective of whether stromatolitic algal filaments scavenge their carbonate as discrete particles from an abiotic carbonate particulate rain or instead actively precipitate the biogenic carbonate from solution because of photosynthetic or respiratory chemical effects, the algae in the bones could exploit same mechanism. Whilst achieving a significant particulate rain would be difficult, the same condi-

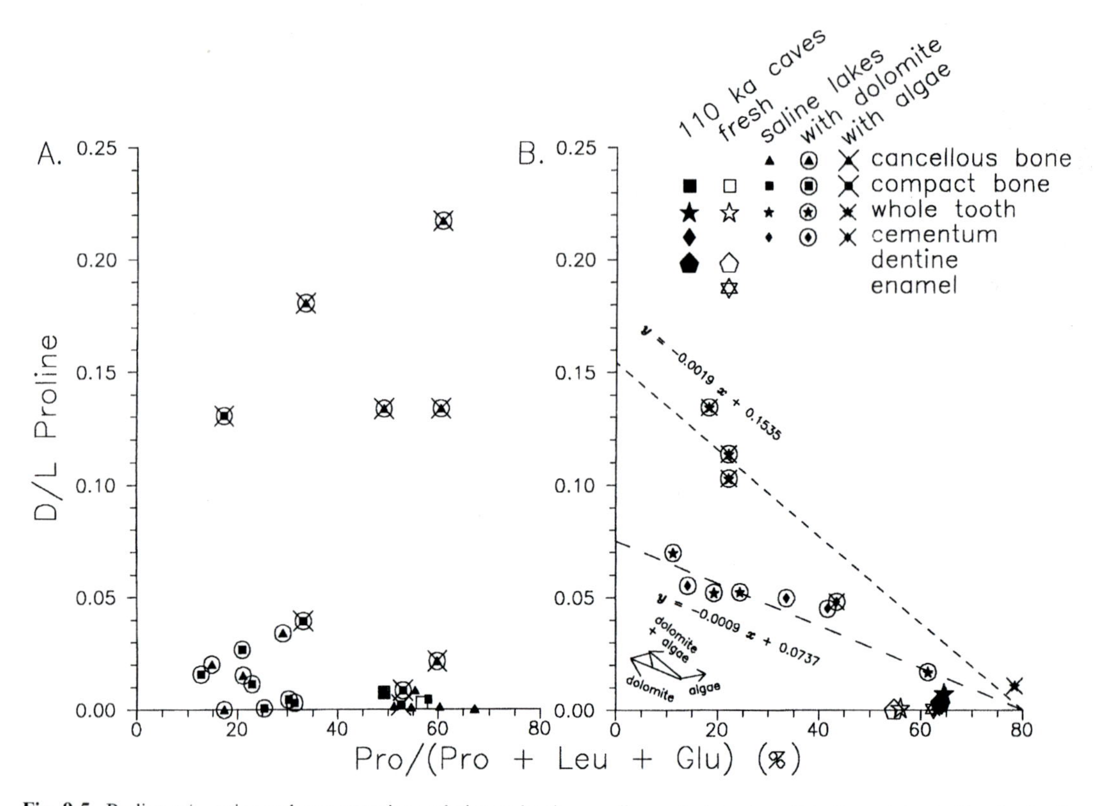

Fig. 9-5. Proline D/L ratios and concentrations relative to leucine, proline and glutamic acid in Recent Australian fossils. In secondarily dolomitized samples without algae, proline concentrations relative to proline, leucine and glutamic acid never exceeded those in fresh tissues. All dolomitized tissues with algae had significantly higher D/L ratios compared to fresh tissues and dolomitized tissues lacking algae. A. Dolomitized bone lacking algae had significantly reduced proline concentrations, whereas some algally infested, dolomitized bone had elevated concentrations. B. In dolomitized teeth, higher D/L ratios correlated with reduced proline concentrations, at $R^2 = 92\%$, for those with, and at $R^2 = 88\%$, for those without, algae. One with algae but no dolomite had a higher proline concentration than fresh samples.

tions that cause abiotic precipitation in the water column could occur intertrabecularly. Whatever the mechanism, the amino acid chemistry changed strikingly.

Bacterial Infestation?

Alanine and glutamic acid in many samples analyzed here, along with other amino acids in a few samples, had particularly elevated D/L ratios (figures 9-3–9-9). In some bacteria, peptidoglycans in cell walls contain D-amino acids, especially alanine and glutamic acid (Friedman, 1999). If due to bacteria, the elevated D/L ratios would not be generally correlative among the various amino acids, but would depend on the various bacterial species present and their abundance. If, however, racemization were occurring *in situ*, a positive correlation among the D/L ratios should result, with the amino acids that racemize more rapidly, such as aspartic acid, showing greater D/L ratios. In the data given here, the D/L ratios do not correlate among amino acids (figures 9-3–9-8). Although the high ratios occur in alanine (figure 9-3A) and aspartic acid (figure 9-4A), several samples had high valine D/L ratios, even though valine racemizes very slowly. These observations all support, but cannot definitively prove, that bacterial infestation during early diagenesis may cause the elevated D/L ratios and altered amino acid concentrations. Other potential sources include fungi and insects (Friedman, 1999). Bacteria have also been implicated in secondary mineralization (Ehrlich, 1996), which could then cause these altered amino acid compositions to be locked into fossilized tissues.

Secondary Mineralization

Precipitation of secondary dolomite and other carbonates in the Australian fossils was associated with significant changes in some amino acid compositions and D/L ratios. All samples with the highest valine D/L ratios (>0.03) had secondary

PLATE 1. Rapid beachrock formation, fossil preservation, infaunal colonization and infestation. Infaunal colonizations and infestations occur ubiquitously in bones from the Australian saline lakes. Within a few months, algal colonization and infestation increase from 0 to >80%. Scale bars are in mm. **A.** Firmly cemented into the beachrock (boundstone) at the West Basin waterline, this *B. taurus* pelvis displays no green bone fracturing, no plaquette formation, little cortical bone loss (<5%, all from articular surfaces), and most importantly, only ~20% algal colonization, cf. B–D and Plate 2C–D. **B.** In the rib, 87010, from West Basin, filamentous green algae (c) colonize the cancellous bone in a band approximately 2–5 cm thick about 2.5–3.0 cm below the exterior surface exposed to the light (a), but not similar locations on the side buried in the sediment (b). **C.** Another view of 87010, with algae (c) in the "diffuse light zone." **D.** An algal mat (c) completely coats 87006, a *B. taurus* vertebra from West Basin. The round red globules (e) are ostracode eggs living in the algal mat (see Plate 2C–D).

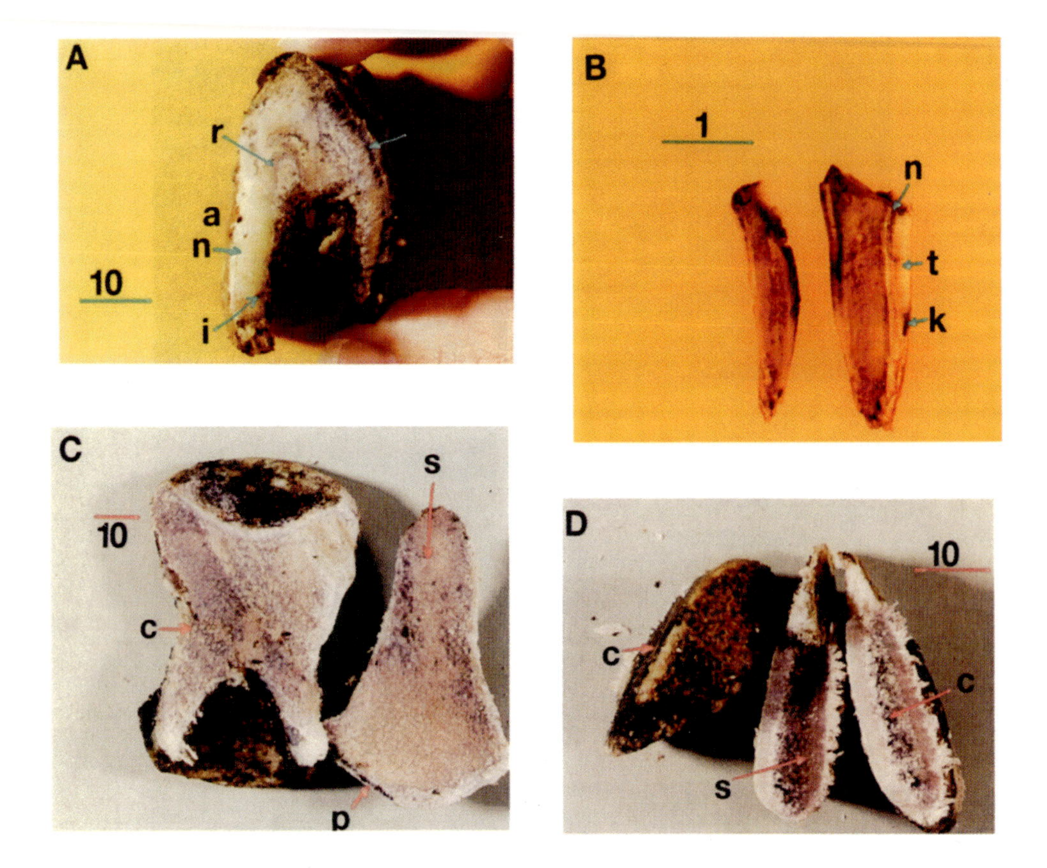

PLATE 2. Secondary mineralization in fossils from West Basin. Dolomite, calcite, aragonite, and several Fe- and Mg-rich minerals are common secondary minerals. **A.** In 87009, a *B. taurus* mandible, the enamel and dentine above the gumline were completely dissolved, without affecting the bone and dentine below. On the side abutting the sedimentary cement (b), secondary siderite, magnesite, kutnahorite, and vivianite invasively fill the porosity elements, highlighting the neonatal line (n), the dentinal tubules (t), the cementum around the tooth root (r), cracks, and insect borings (i). On the exposed side (a), dolomite, magnesite, high-Mg calcite, and siderite occur as secondary minerals. **B.** In the 87009 tooth root, Fe–Mg-rich minerals mark the neonatal line (n), the dentine tubules (t), and a crack (k) formed by salt crystallization. **C.** Within the vertebral body of 87006, secondary siderite and dolomite infill the intertrabecular spaces (s). Algal filaments (c) live in the diffuse light zone. Salt crystallization has loosened a bone plaquette (p). **D.** In the 87006 spinous process, secondary magnesite, pyrite, hydromagnesite, and dolomite infill the cancellous intertrabecular spaces (s). Algal filaments (c) inhabit the diffuse light zone and colonize the surface.

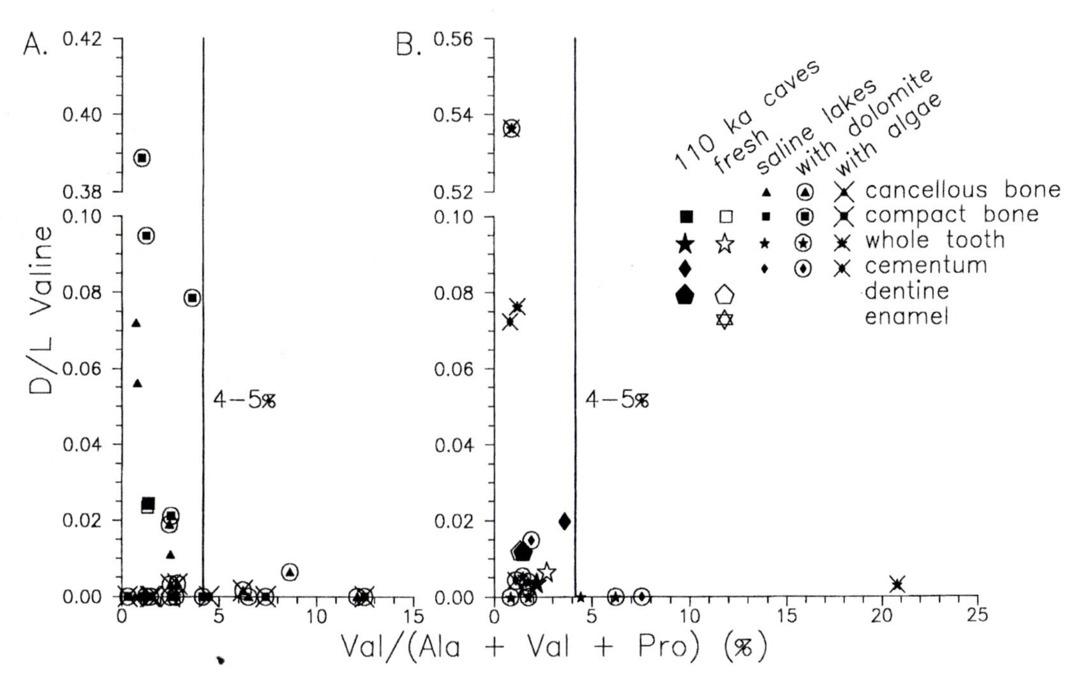

Fig. 9-6. Valine D/L ratios and concentrations relative to alanine, valine and proline in Recent Australian fossils. High D/L ratios occurred if valine concentrations relative to alanine, valine and proline dropped <4–5%. Tissues with the highest D/L ratios all had secondary monohydrocalcite, high-Mg or low-Mg calcite, while those with concentrations >5% contained secondary siderite, pyrite or vivianite. A. In algally infested bone, D/L ratios were significantly lower than those in fresh bone. B. Many, but not all, teeth with algae showed D/L ratios that were lower than those of the fresh sample background.

high-Mg calcite, low-Mg calcite, or monohydrocalcite (figure 9-6), but the converse was not true. Compared to fresh sample backgrounds, dolomitized tooth and bone had reduced proline concentrations normalized to proline, leucine, and glutamic acid (figure 9-5), while many had higher proline D/L ratios. Among dolomitized tooth samples with or without algae, D/L ratios correlated inversely with proline concentrations ($R^2 = 0.88$–0.92). Except for some algally infested samples, most dolomitized tooth and bone had significantly higher relative phenylalanine, leucine, and aspartic acid concentrations than fresh tissues, while some, but not all, also showed significantly increased D/L ratios, many equaling or exceeding those in 110 ka samples (figures 9-4, 9-7, and 9-8). Dolomitized samples had enriched alanine and high D/L alanine ratios (figures 9-3 and 9-10). When environmental conditions favor dolomitization in bone, therefore, high proline D/L ratios should also occur, as should relative leucine and aspartic acid enrichment, possibly with elevated aspartic acid, leucine, and phenylalanine D/L ratios. Algal infestation should heighten the effects for proline.

Iron- and Mg-rich mineralization correlated with other effects. An algally infested bone with secondary siderite, vivianite, and kutnahorite had an aspartic acid D/L ratio >1.0 (figure 9-4A), whereas a bone and a tooth with the same minerals and algae showed extremely elevated phenylalanine D/L ratios (figure 9-7). The highest D/L leucine ratios coupled with increased leucine concentrations (normalized to proline, leucine, and glutamic acid) occurred in samples having algal infestation and some combination of secondary siderite, pyrite, kutnahorite, and vivianite, but no dolomite (figure 9-8). Tooth samples having algae, siderite, and no dolomite had higher proline concentrations than fresh samples, but had unremarkable D/L ratios (figure 9-5). All samples with valine relative concentrations >5%, and hence, unremarkable D/L ratios, had secondary siderite, pyrite, or vivianite (figure 9-6), but not all samples with those minerals had elevated concentrations. Therefore, conditions favoring secondary Fe- and Mg-rich mineralization and algal infestation can also favor high D/L phenylalanine and leucine ratios, as well as enriched leucine concentrations in all tissues, and high aspartic acid D/L ratios in bone, and enriched proline in tooth. Such mineralization correlated with enriched valine in samples without algal infestation.

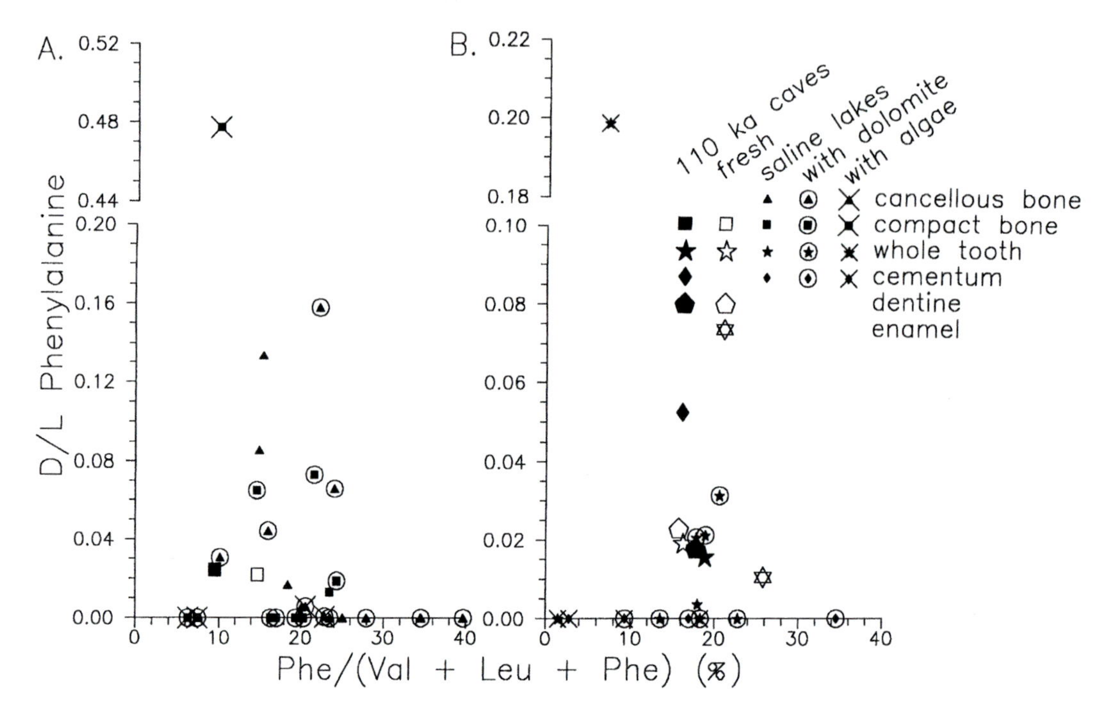

Fig. 9-7. Phenylalanine D/L ratios and concentrations relative to valine, leucine and phenylalanine in Recent Australian fossils. Each sample containing secondary kutnahorite or vivianite and algae showed extremely elevated D/L ratios and reduced phenylalanine concentrations relative to valine, leucine, and phenylalanine. Compared to fresh tissues, all those with algae lacking Fe-rich minerals showed lower D/L ratios. Most dolomitized tissues without algae had increased relative concentrations. A. Compared to fresh bone backgrounds, some dolomitized bone had high D/L ratios. B. Compared to fresh tissues, no teeth containing algae had significantly higher relative phenylalanine concentrations, and cementum concentrations were significantly lower.

Sedimentary Mineralization and Geochemistry

All of the Australian sediment samples had relatively less alanine and glutamic acid but more glycine and proline than the adjacent fossils. Relative alanine/glycine, aspartic acid/glycine, and leucine/glycine or proline/glycine ratios correlated with sediment type (figure 9-10). In all tissues from oxidized sediment precipitating calcium carbonates, especially dolomite, alanine was significantly enriched, whereas aspartic acid, leucine, and proline were depleted compared to fresh samples ($R^2 = 0.99$). In tissues from reduced sediment depositing Fe–Mg-rich minerals, compact bone had moderately enriched, but whole teeth had slightly reduced, alanine compared to modern samples. Teeth also had slightly higher aspartic acid relative to proline and alanine. Although all cancellous bone had high alanine/glycine ratios, the differences among those from different sedimentary geochemistries, though present, were less obvious.

CONCLUSIONS

Because of the average bone survival times of 2–3 years in the saline lakes of Western Victoria, fossilization of identifiable bones there requires rapid submergence or burial. In subaerially exposed bone, physical destruction, mainly due to salt crystallization, caused green bone fractures and removed teeth and bone plaquettes, starting in the thinnest compact bone. Whether limited to surface colonization, or completely infesting the medullary cavity and/or the diffuse light zone, some algae colonized all defleshed tissue except enamel. Secondary mineralization, especially by calcium, magnesium, and iron carbonates, apatite-group minerals, and pyrite, however, counters weathering. During initial fossilization in submerged tissues, secondary mineralization is concentrated in porosity elements and on exposed surfaces. High organic-matter concentrations may have helped to complex ions, catalyze reactions, or provide nutrients for microorganisms that had the same effects. Not only can all tissues be completely remineralized after a few years of submergence or burial, their organic geochemistry can be radically altered within the same time frame.

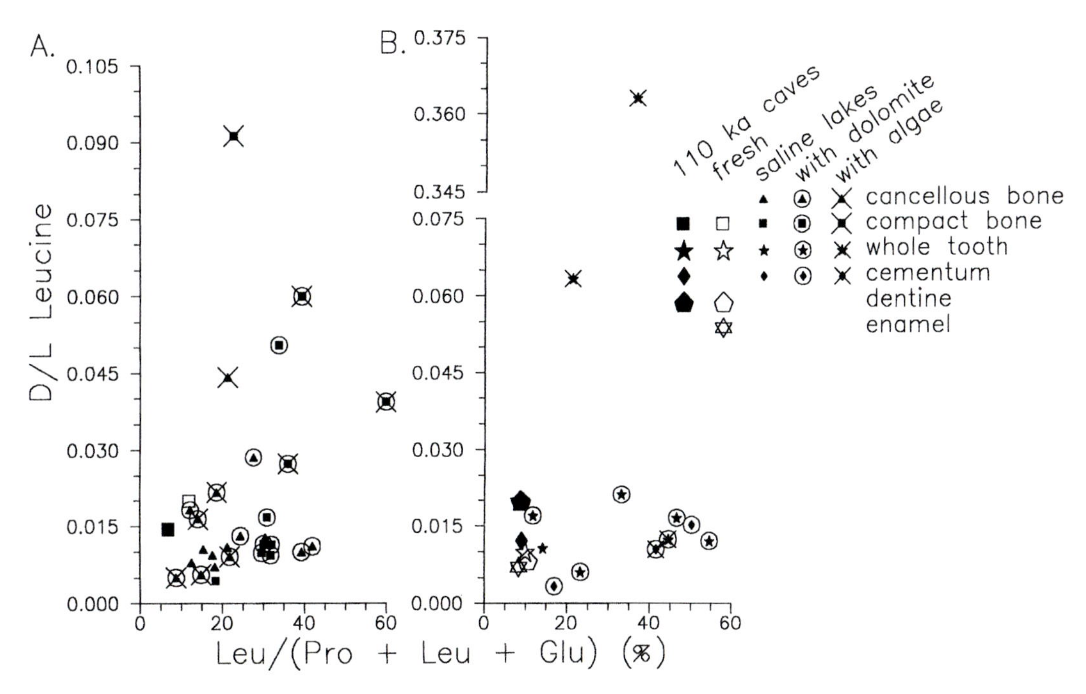

Fig. 9-8. Leucine D/L ratios and concentrations relative to proline, leucine and glutamic acid in Recent Australian fossils. Compared to fresh tissues, tissues with algae but no dolomite showed high racemization ratios, with moderate to dramatic increases in leucine concentrations relative to proline, leucine, and glutamic acid. Some secondarily dolomitized tissues had D/L ratios equaling those accrued by 110 ka cave fossils. A. Bone with secondary dolomite often showed increased leucine concentrations. B. All dolomitized teeth had leucine concentrations greater than those in fresh samples.

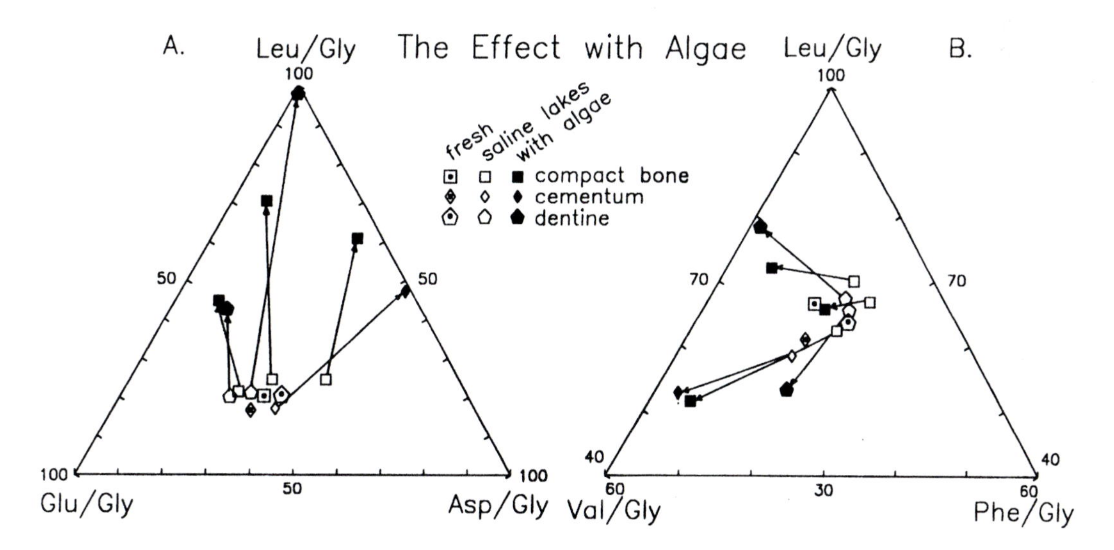

Fig. 9-9. Amino acid concentrations in algally infested Recent Australian bone and tooth tissues. In adjacent subsamples, algally infested subsamples have elevated leucine and valine concentrations compared to phenylalanine, aspartic and glutamic acids. Arrows point from the subsample *without* algae to that *with* algae present: A. Leu/Gly vs. Glu/Gly vs. Asp/Gly. B. Leu/Gly vs. Val/Gly vs. Phe/Gly.

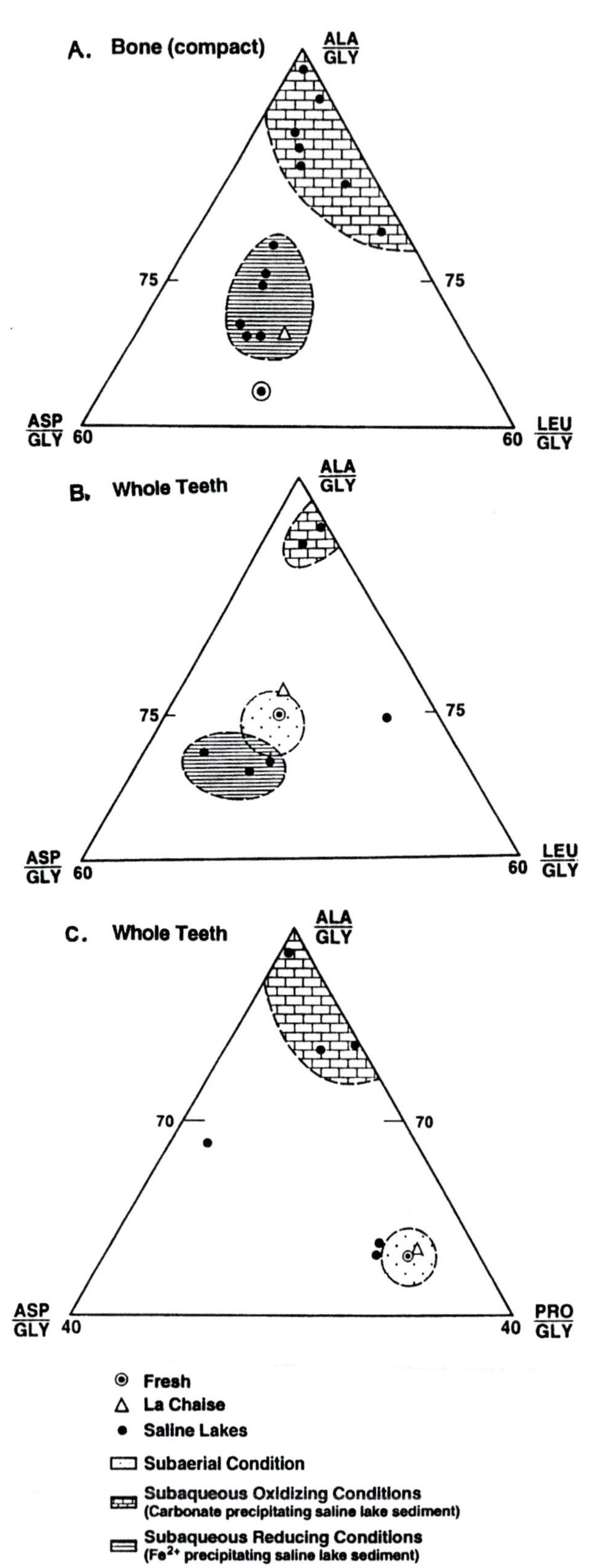

Fig. 9-10. Relationships between amino acid and sedimentary geochemistry in Australian fossils. On ternary plots, relative amino acid compositions differentiate the sediments from which the samples were collected. A. Ala/Gly vs. Asp/Gly vs. Leu/Gly in compact bone. B. Ala/Gly vs. Asp/Gly vs. Leu/Gly for whole tooth (~70 vol% dentine, 25 vol% enamel, 5 vol% cementum, dominated by dentinal amino acids). C. Ala/Gly vs. Asp/Gly vs. Pro/Gly in whole tooth.

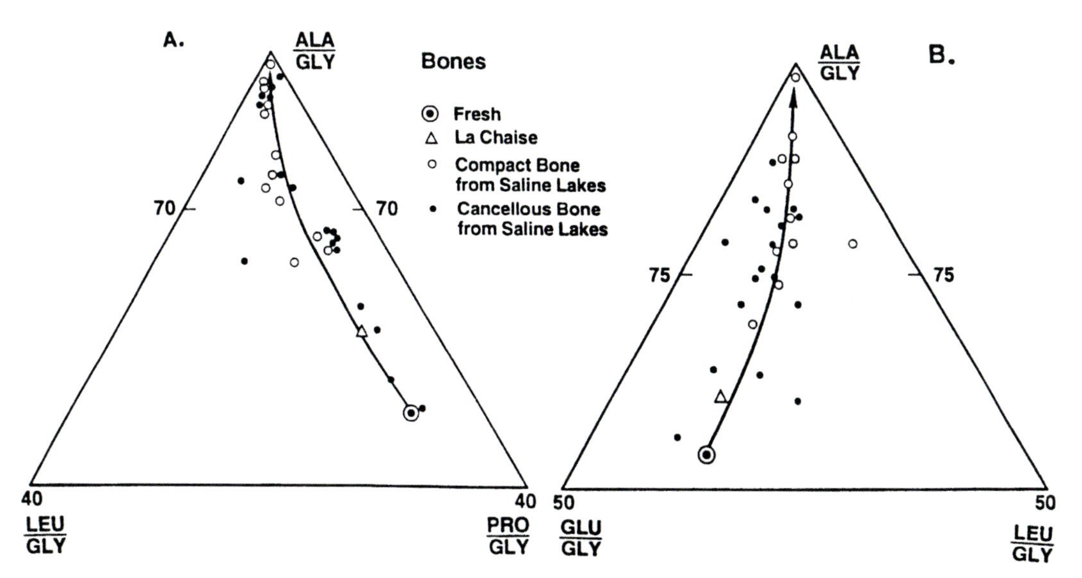

Fig. 9-11. Possible pathways for amino acid compositional variation during bone fossilization under oxidizing conditions. A. Alanine compared to leucine and glutamic acid. B. Alanine compared to leucine and proline.

As secondary minerals were precipitated and algae and bacteria thrived, related geochemical conditions apparently caused significant disruption in amino acid geochemistry. As bone began to fossilize with increasingly more secondary carbonate mineralization under oxidizing conditions, it gained alanine relative to proline, leucine, glutamic acid, and aspartic acid (figure 9-11). Possibly because the greater porosity in cancellous bone promotes more pervasive leaching, the trend in compact bone was more defined. Crystallinity studies (e.g., Tuross et al., 1989; Person et al., 1996; Stuart-Williams et al., 1996) have demonstrated that bone-apatite alteration is closely linked with degradation of organic matter. Geologically instantaneous fossilization, in which both the mineral and collagen matrices undergo diagenetic alteration, occurs in saline lakes, and may be responsible for most of the bones preserved in the geological record.

Neither temperature nor time can be the major controlling factor in the diagenesis of these samples. Wide variation in D/L ratios occurred in samples less than 5–10 m apart. Several samples had D/L ratios exceeding those in 100–200 ka tissues, and one bone had a completely racemized aspartic acid ratio. Since aspartic acid D/L ratios <0.1 may indicate that pristine DNA is present (Poinar et al., 1996), the high level of racemization suggests that some fossils here, though modern, would be unsuitable for gene sequencing. Instead, D/L ratios and compositions here vary in relation to algal (and possibly bacterial) infestation, amino acid leaching and enrichment, secondary tissue mineralization, and sedimentary geochemistry. It remains to be determined as to whether these factors are direct causal agents or are themselves by-products of biogeochemical conditions that simultaneously caused the amino acid effects. A strongly circumstantial, if not causal, link exists between algal infestation and secondary mineralization in the tissues. It remains to be determined as to whether bacterial or other infaunal infestation has caused the secondary mineralization, high D/L ratios, and altered amino acid compositions.

During early diagenesis, D-amino acids added by bacteria or other infauna would result in anomalously old, and inaccurate, AAR ages. Such infaunal infestation, especially by bacteria, is ubiquitous during fossilization, and the fact that secondary mineralization can lock these amino acids into the tissues probably dooms any attempt to use AAR for dating bones or teeth, unless these extraneous amino acids can be completely isolated from those derived solely from the original tissues. The patterns of elevated D/L ratios and altered concentrations in the Australian samples, however, suggests that the evaluation of D/L ratios and concentrations for several amino acids may allow bone samples to be screened for potential bacterial, algal, or other infaunal D-amino acid contamination. Samples showing concordance among their D/L ratios, and having higher ratios for those amino acids that racemize more quickly, could be deemed free of significant bacterial contamination.

The precise pathway of secondary tissue remineralization and organic alteration depends strongly upon the sedimentary mineralogy and geochemistry of the area in which the fossil resides. Unlike sediment, in which earlier generations of soluble secondary mineral precipitates may redissolve during the deposition of later generations, successive secondary mineralizations may be preserved within bone, dentine, and cementum, as the porosity infills (especially in the intertrabecular spaces). There, the amino acid geochemistry alone may serve to identify the sedimentary geochemistry and mineralogy at the time of initial fossilization, even if the fossil has been reworked and all other traces of its "fossilization habitat" have been removed.

Given the diverse nature of sediment and water chemistries in saline lakes, this study of fossils from four lakes could explore neither all the possible variations in fossilization processes nor all the paleoenvironmentally important indicators. While most, if not all, of the results may be generally applicable, confirmation must await analyses from saline lakes in other regions. Future studies should address the issue of cause vs. effect for algal and bacterial contamination, secondary mineralization, amino acid leaching and enrichment.

Acknowledgments: Grants from the National Sciences and Engineering Research Council, Canada (NSERC PDF to B.A.B.B., 1987–1989; NSERC operating grants to W.M.L., N.W.R.), l'Université du Québec à Montréal (to B.A.B.B.), and NSF (to B.A.B.B.) partially supported this work. The Geography Department, Monash University, provided field equipment and laboratory facilities. NAA analyses were performed at the McMaster Nuclear Reactor (facilitated by A. Pidruczny). B. Wright, K. Goodger, and G. Lyons assisted in the analyses. M. Lauthier, F. Dimitrov, J. Whorwood, J. I. Blickstein, S. Berman, and L. Provencher helped to prepare the manuscript. Glenn Goodfriend and two others provided constructive reviews.

REFERENCES

Bada J. L. (1972) Kinetics of racemization of amino acids as a function of pH. *J. Amer. Chem. Soc.* **94**, 1371–1373.

Bada J. L. (1985) Amino acid racemization dating of fossil bones. *Ann. Rev. Earth Planet. Sci.* **13**, 241–268.

Baum R. and Smith G. G. (1986) Systematic pH study on the acid-catalyzed and base-catalyzed racemization of free amino acids to determine the 6 constants, one for each of the 3 ionic species. *J. Amer. Chem. Soc.* **108**, 7325–7327.

Behrensmeyer A. K. (1978) Taphonomic and ecological information from bone weathering. *Paleobiology* **4**,150–162.

Blackwell B. A. (1987) *Problems in Amino Acid Racemization Dating Analyses: Bones and Teeth from the Archeological Sites Lachaise and Montgaudier (Charente, France).* Ph.D. thesis, University of Alberta, Edmonton, 2 vols., 650 pp.

Blackwell B. A., Rutter N. W., and Debénath A. (1990) Amino acid racemization analysis of mammalian bones and teeth from La Chaise-de-Vouthon (Charente), France. *Geoarchaeology* **5**, 121–147.

Clark I. and Cook B. (eds) (1988) *Victoria Geology Excursion Guide*. Australian Academy of Science.

Dallemagne M. J. (1964) Phosphate and carbonate in bone and teeth. In *Bones and Teeth, Proceedings of the 1st European Symposium (Oxford, 1963)* (ed. H. J. J. Blackwood), pp. 171–174. Pergamon Press.

De Deckker P. and Last W. M. (1988) A newly discovered region of modern dolomite deposition in western Victoria, Australia. *Geology* **16**, 29–32.

De Deckker P. and Last W. M. (1989) Modern dolomite in continental evaporitic playa lakes in Western Victoria, Australia. *Sed. Geol.* **64**, 223–238.

Denys C., Williams C. T., Dauphin Y., Andrews P., and Fernadez Jalvo Y. (1996) Diagenetic changes in Pleistocene small mammal bones from Olduvai Bed I. *Palaeogeog. Palaeoclimat. Palaeoecol.* **126**, 121–134.

Douglas J. G. and Ferguson J. A. (eds) (1988) *Geology of Victoria*. Geological Society of Australia.

Ehrlich H. L. (1996) *Geomicrobiology*, 3rd edn. Marcel Dekker.

Farnum J. F., Glascock M. D., Sandford M. K., and Gerritsen S. (1995) Trace elements in ancient human bone and associated soil using NAA. *J. Radioanalyt. Nuclear Chem.* **196**, 267–274.

Friedman M. (1999) Chemistry, nutrition, and microbiology of D-amino acids. *J. Agric. Food Chem.* **47**, 3457–3479.

Gell P. A., Barker P. A., De Deckker P., Last W. M., and Jelicic L. (1994) The Holocene history of West Basin Lake, Victoria, Australia. *J. Paleolimnol.* **12**, 235–259.

Gill B. (1987) *Preliminary Hydrogeological Review of Salinity Problems in the Barwon/Corangamite Region.* Rural Water Commission of Victoria.

Goffer Z. (ed.) (1980) *Archeological Chemistry* (Chemical Analysis, vol. 55). Wiley.

Hare P. E. (1974) Amino acid dating of bone: the influence of water. *Carnegie Instn Wash. Yrbk* **73**, 576–581.

Hare P. E. (1980) Organic geochemistry of bones, and its relation to the survival of bone in the natural environment. In *Fossils in the Making* (eds A. K. Behrensmeyer and A. P. Hill), pp. 208–219. Chicago Univ. Press.

Hubert J. F., Panish P. T., Chure D. J., and Prostak J. S. (1996) Chemistry, microstructure, petrology, and diagenetic model of Jurassic dinosaur bones,

Dinosaur National Monument, Utah. *J. Sed. Res.* **66**, 531–547.

Jenkin J. J. (1988) Geomorphology. In *Geology of Victoria* (eds J. G. Douglas and J. A. Ferguson), p. 403–426. Geological Society of Australia.

Last W. M. (1990) Lacustrine dolomite: an overview of modern Holocene and Pleistocene occurrences. *Earth Sci. Rev.* **27**, 221–263.

Last W. M. (1992) Petrology of modern carbonate hardgrounds from East Basin Lake, a saline maar lake, Southern Australia. *Sed. Geol.* **81**, 215–229.

Last W. M. and De Deckker P. (1990) Modern and Holocene carbonate sedimentology of two saline volcanic maar lakes, southern Australia. *Sedimentology* **37**, 967–981.

Last W. M. and De Deckker P. (1992) Paleohydrology and paleochemistry of Lake Beeac, a saline playa in southern Australia. In *Aquatic Ecosystems in Semi-arid Regions. National Hydrological Research Institute Symposium Series*, vol. 7 (eds R. D. Robarts and M. L. Bothwell), pp. 63–74. Environment Canada.

Maat G. J. R. (1991) Ultrastructure of normal and pathological fossilized red blood cells compared with pseudo-pathological biological structures. *Int. J. Osteoarchaeol.* **1**, 209–214.

Maat G. J. R. (1993) Bone preservation, decay and its related conditions in ancient human bones from Kuwait. *Int. J. Osteoarchaeol.* **3**, 77–86.

Marquart R. G. (1986) UPDSM: mainframe search–match on an IBM-PC. *Powder Diffractometry* **1**, 34–36.

Martill D. M. and Unwin D. M. (1997) Small spheres in fossil bones: blood corpuscles or diagenetic products? *Palaeontology* **40**, 619–624.

Michel V., Ildefonse P., and Morin G. (1996) Assessment of archaeological bone and dentine preservation from Lazaret Cave (Middle Pleistocene) in France. *Palaeogeog. Palaeoclimatol. Palaeoecol.* **126**, 109–119.

Mucci A. (1988) Manganese uptake during calcite precipitation from seawater: conditions leading to the formation of a pseudo-kutnahorite (*sic*). *Geochim. Cosmochim. Acta* **52**, 1859–1868.

Neuman W. F. (1980) Bone material and calcification mechanisms. In *Fundamental and Clinical Bone Physiology* (ed. M. R. Urist), pp. 83–107. Lippencott.

Person A., Bocherens H., Mariotti A., and Renard M. (1996) Diagenetic evolution and experimental heating of bone phosphate. *Palaeogeog. Palaeoclimatol. Palaeoecol.* **126**, 135–149.

Poinar H. N., Hoss M., Bada J. L., and Paabo S. (1996) Amino acid racemization and the preservation of ancient DNA. *Science* **272**, 864–866.

Rutter N. W. and Blackwell B. A. (1995) Amino acid racemization dating. In *Dating Methods of Quaternary Deposits* (eds N. W. Rutter and N. R. Catto). Geological Association of Canada, *GEOtext* **2**, 125–164.

Stout S. P. (1978) Histological structure and its preservation in ancient bone. *Curr. Anthropol.* **19**, 601–604.

Stuart-Williams H. L. Q., Schwarcz H. P., White C. D., and Spence M. W. (1996) The isotopic composition and diagenesis of human bone from Teotihuacan and Oaxaca, Mexico. *Palaeogeog. Palaeoclimatol. Palaeoecol.* **126**, 1–14.

Timms B. V. (1972) A meromictic lake in Australia. *Limnol. Oceanogr.* **17**, 918–922.

Timms B. V. and Brand G. W. (1973) A limnological survey of the Basin lakes, Nalangil, Western Victoria, Australia. *Bull. Austral. Soc. Limnol.* **5**, 32–40.

Tuross N., Behrensmeyer A. K., Eanes E. D., Fisher L. W., and Hare P. E. (1989) Molecular preservation and crystallographic alterations in a weathering sequence of wildebeest bones. *Applied Geochem.* **4**, 261–270.

White E. M. and Hannus L. A. (1983) Chemical weathering of bone in archeological soils. *Amer. Antiquity* **48**, 316–322.

Williams W. D. and Buckney R. T. (1976) Stability of ionic proportions in five salt lakes in Victoria, Australia. *Austral. J. Marine Freshwater Res.* **27**, 367–377.

10. Characterization of ultrastructural and biochemical characteristics of modern and fossil shells

L. L. Robbins, S. Andrews, and P. H. Ostrom

How are the diagenetic products of ancient shell-matrix organic matter related to the original organic material of the organism? Ultrastructural and molecular data from shell proteins provide multiple lines of evidence for addressing this question and describing the diagenetic pathways of fossil biomolecules. By isolating and characterizing shell organic matter, we can begin to understand how individual proteins degrade over time. The complexity of diagenetic reactions, however, has impeded our understanding of the mechanisms involved in the diagenesis of proteins. Despite this complexity, accurate paleobiochemical interpretations require a better comprehension of the diagenetic history of proteinaceous material in carbonate shells. This is particularly important in paleoenvironmental studies based on isotopic analysis of ancient organic proteins.

Shells of a modern and fossil gastropod, *Polinices duplicatus,* and a bivalve, *Mercenaria mercenaria,* were analyzed and compared for shell permeability, proteins, amino acids, and stable carbon and nitrogen isotopic compositions as a means of clarifying diagenetic resistance and protein indigeneity in fossils and to assess isotopic variation among modern shells from different environments. The two classes of mollusks we focus on provide useful comparisons because gastropods have a mechanically stronger structure than bivalves (Andrews et al., 1985) and both bivalves and gastropods contain similar biochemical components, which allows evaluation of the pathways of degradation. In this study, a subsample of each shell was measured for permeability, to facilitate evaluation of the potential transport of molecules through the mineral matrix. We tested the hypothesis that shells with greater permeability have a greater chance of *in situ* degradation. To determine the effects of permeability on protein preservation, the yield of organic material was compared with the permeability. Although such an approach is simplistic, to our knowledge no previous studies relate permeability to chemical preservation. We also heated samples, to simulate diagenetic alteration and to model one pathway of diagenesis. Carbon isotope data are presented for the total organic matrix of modern shells from different ecosystems, to determine if they are indicative of environmental conditions. To investigate diagenetic effects that might obscure these environmental interpretations in the fossil record, $\delta^{13}C$ and $\delta^{15}N$ values of individual protein fractions of modern and fossil shells were also measured.

BACKGROUND

Shell Ultrastructure

Shell mineralogy and ultrastructure play key roles in the organic preservational potential in a fossil because they influence the extent to which the shell behaves as a closed system. *Mercenaria* is a member of the superfamily Veneracea which contains an outer and inner aragonitic prismatic structure bounding a laminar nacreous layer (Carter, 1990). The prismatic layer contains parallel prisms oriented vertically, with the long axes perpendicular to the shell surface, whereas the aragonitic nacreous layer contains lamellar or sheet-like structures that form one layer at a time (Carter, 1990).

Polinices, a member of the family Naticidae, has an aragonitic shell with a non-nacreous crossed lamellar structure (homologous to the prismatic structure of the bivalve), composed of two mineralogically distinct alternating layers of lamellae (Carter, 1990). The first-order lamellae are rod-shaped and oriented perpendicular to the shell surface. The second-order lamellae are differentiated by the dip angle away from the shell margins.

Researchers have assumed that the permeability of a modern shell plays an influential role in the decrease in organic material during diagenesis (Andrews et al., 1985; Curry, 1987, 1988), but no baseline data exist to support this assumption. For example, Andrews et al. (1985) noted the impermeable nature of the molluscan shell matrix and its resistance to losses of primary amino acids and incorporation of secondary amino acids. This observation suggests that a comparison of permeability data from different mineral matrices may provide ways of comparing intergeneric diagenesis.

The location of the organic matrix within the ultrastructure of a shell is also important to its preservational potential. Although the specific locations of the organic components vary between taxa, the matrix is located between, and enclosed in, the carbonate crystals of both *Polinices* and

Mercenaria (Crenshaw, 1972; Weiner et al., 1983; Wheeler et al., 1988; Robbins et al., 1993), suggesting that at least a subset of the organic matrix may be potentially preserved. Not only does the entombed organic material contribute to the mechanical properties of the shell (Weiner and Traub, 1984), but the ultrastructural design of the shell, coupled with the location of the organic matter, will ultimately influence the state of preservation of the material. For example, organic matter contained within nacreous structures is typically present in thick, continuous sheets and is often exposed to attack by hydrolyzing or oxidizing solutions (Hudson, 1967). In contrast, intricate ultrastructures of complex crossed lamellae containing thin organic sheaths are less likely to be attacked (Hudson, 1967). Consequently, gastropods consisting of the complex crossed lamellar structure are less prone to degradation than bivalves (that have the prismatic structure). Researchers have shown that well-protected macromolecular proteins can escape destruction, and at least part of their original structure can be preserved over long periods (Collins et al., 1991; Muyzer et al., 1992; Robbins et al., 1993). These proteins, trapped between and within crystals, include soluble matrix (SM) proteins and insoluble matrix (IM) proteins and constitute 0.01–0.1 weight percent of the mineral (Addadi and Weiner, 1989; Albeck et al., 1993).

The geochemistry of *Mercenaria* has been extensively studied by many researchers (e.g., Abelson, 1954, 1955, 1956; Hare and Abelson, 1964; Degens et al., 1967; Hare, 1969; Vallentyne, 1969; Akiyama, 1971; Crenshaw, 1972; Weiner and Hood, 1975; Hare and Hoering, 1977; Kennish, 1980; Andrews et al., 1985; Serban et al., 1988; Muyzer et al., 1988; Engel et al., 1994; Silfer et al., 1994; Robbins and Ostrom, 1995). Wheeler et al. (1988) showed that matrix fractions can range from many low-molecular-weight components within gastropod shells to a single, high-molecular-weight glycoprotein within *M. mercenaria*. Other researchers have described additional protein and sulfated glycoprotein fractions (Crenshaw, 1972; Muyzer et al., 1988; Engel et al., 1994; Silfer et al., 1994; Robbins and Ostrom, 1995).

Although *Polinices* has not been studied as widely as *Mercenaria*, several investigations exist in the literature (Degens et al., 1967; Ghiselin et al., 1967; Ostrom et al., 1994; Robbins and Ostrom, 1995). Previous experiments performed on *Polinices* showed that the shell-matrix protein solution was entirely soluble in water (no IM).

Isotope values of shell organic material have not been widely used as paleoecological indicators. However, efforts to evaluate diagenesis using isotopic analyses have been conducted. Robbins and Ostrom (1995) recognized that modern *Polinices* SM was composed of two major protein fractions containing individual amino acids that differed in isotopic composition. Substantial differences were observed between the isotope data for modern protein amino acids and those of fossil specimens (Robbins and Ostrom, 1995). Fundamentally, this and other studies suggested that isotopic analyses of both carbon and nitrogen represent valuable tools for tracing specific compounds through biosynthesis and early diagenesis and, ultimately, into the fossil record (Macko et al., 1983; Serban et al., 1988; Qian, 1993).

METHODS

Sample Collection and Preparation

Fossil *M. mercenaria* (MF) and *P. duplicatus* (PF) samples were collected from the Pinecrest Shell Beds in Sarasota, Florida (3.5–2.0 Ma). The investigation of fossils collected at one locality reduced the variability of the diagenetic environment and the thermal history of the fossils (Abelson, 1954). In addition, these shell beds consist of unconsolidated shell hash, eliminating the problem of diagenetic processes that occur with cementation. Pristine shells separated from the sediment matrix were used. Living *M. mercenaria* (MM) samples were collected from the east side of Burgess Bay near Cape Coral, Florida, and Fort Desoto, Florida. Modern *P. duplicatus* (PM) samples were collected from Essex, Massachusetts. Modern samples were stored at −70°C until processed.

The entire gastropod shell and single valves of the bivalves constituted a sample. Shells were scrubbed using a brush and rinsed in triple-distilled water ($3 \times DH_2O$) and shell ultrastructure was analyzed in subsamples using scanning electron microscopy (SEM). Samples were then soaked in 5% NaOCl to remove any remaining sediment and contaminants, thoroughly rinsed with $3 \times DH_2O$ and air-dried prior to being crushed by mortar and pestle and subsequently ground to a powder. Samples were decalcified in 1 M HCl or 10% (w/v) ethylenediaminetetraacetic acid (EDTA, pH 8) containing 0.1% sodium azide at 4°C. After all the carbonate had been dissolved, the sample was centrifuged at 12,500 rpm for 25 min to concentrate the acid-insoluble organics. The supernatant solution containing the acid-soluble matrix (SM) proteins was then exhaustively dialyzed against $3 \times DH_2O$ using an Amicon ultrafiltration device (molecular weight cut-off, 10,000 Da) to remove excess salts and free amino acids. After dialysis, the SM was lyophilized and stored at −70°C prior to analysis.

Reverse-phase high-performance liquid chromatography

The SM was separated into fractions by reverse-phase high-performance liquid chromatography (RPHPLC) equilibrated with 0.1% trifluoroacetic acid (TFA) in $3 \times DH_2O$ (Buffer A). Separation was done with a Vydac Protein C_4 column

(Vydac, Separation Group) at a flow rate of 5 ml/min at 25°C through a variable UV detector (280 nm). The proteins were eluted using increasing concentrations of acetonitrile to 95% acetonitrile (containing 0.1% TFA; (Buffer B) with a gradient of 0–5 min at 0% Buffer B and 15–40 min at 100% Buffer B. All fractions for amino acid analysis were manually collected into test tubes and lyophilized. Protein fractions for analysis of $\delta^{13}C$ and $\delta^{15}N$ were collected into ballasts, rotoevaporated, transferred into quartz tubes and subsequently lyophilized. Bovine serum albumin (BSA) was used as a control to determine if isotopic fractionation occurred during chromatographic separation of proteins (Ostrom et al., 1994). After an initial run designed to obtain the retention time of individual protein fractions, samples were collected. The column was acid-washed for 40 min between samples by using a 1:4 0.1 N HNO_3:isopropanol solution. After cleaning of the column, a blank was run and column eluent was always collected at exactly the same times.

$\delta^{13}C$ and $\delta^{15}N$ of RPHPLC Fractions

Hydrophilic and hydrophobic fractions from *Mercenaria* and *Polinices* organic matter were analyzed for their $\delta^{13}C$ and $\delta^{15}N$ values. The analysis of $\delta^{13}C$ and $\delta^{15}N$ compositions of bulk protein material was performed on a PRISM (Micromass) isotope-ratio mass spectrometer subsequent to cryogenic purification. The standard deviation of $\delta^{13}C$ values for proteins is typically 0.1 (Ostrom et al., 1994).

Amino Acid Analysis

Bulk polypeptides and fractions containing polypeptides collected from RPHPLC were hydrolyzed with 200 µl 6 N HCl at 110°C *in vacuo* for 24 hr on a Waters Pico-Tag workstation and redried. Hydrolyzed samples were analyzed for amino acids at the Protein Chemistry Core Laboratory at the University of Florida, using a Beckman System 6300 high-performance amino acid analyzer.

Permeability

Shell permeability was measured on the two species as a reflection of the possible fluid flow rates that would contribute to diagenetic alteration. A piece of shell material from *Mercenaria* and *Polinices* was drilled from specific areas on the specimen, using a Dremel Moto-Tool. The permeability was then determined for samples by the process of microflow permeability (Porous Materials, Inc.). In this method, the sample is sealed in a chamber of known (calibrated) volume. Pressurized gas is then applied to one side of the sample and the pressure on the other side of the sample is measured. If flow is occurring through the sample, the pressure on the downgradient side will increase proportionately. Estimates of permeability were based on ASTM Method D1434-82. A linear regression was performed on the pressure-versus-time data to yield the slope of the line (torrA/sec).

Heating Experiments

Temperature studies were performed on the modern specimens to simulate diagenesis. By subjecting the shell material to elevated temperatures for specific periods of times, the decay process can be accelerated, degradation products measured, and pathways of diagenesis inferred under these model conditions. Modern samples were placed in a beaker with distilled water, with an adjusted pH of ~7.6 and heated at 100°C for 1 week. After heating, shell samples were decalcified and analyzed for protein and amino acid composition.

RESULTS

Proteins and Amino Acids

A comparison between protein chromatograms of modern and fossil specimens of *Mercenaria* (figure 10-1A, B) and *Polinices* (figure 10-2A, B) showed that the concentration of the hydrophobic fraction increased in the fossil materials. Note that some protein peaks may appear off scale because each chromatogram represents an injection of the same amount of material. This was done intentionally so that a comparison among chromatograms would be representative of differences in the relative abundance of proteins among samples rather than injection amounts. Amino acids were measured on the two major protein-containing fractions and the results for *Mercenaria* and *Polinices* for these fractions are shown in figures 10-1C and 10-2C, respectively.

In both protein fractions of modern *Mercenaria,* aspartic acid and glycine were the predominant amino acids together accounting for nearly 40% or more of the total amino acid composition (figure 10-1C). In comparisons of the individual protein fractions of modern *Mercenaria*, aspartic acid decreased while serine and glycine increased in the hydrophilic fraction, and basic amino acids (phenylalanine, histidine, and lysine) increased in the hydrophobic fraction of fossil *Mercenaria* (figure 10-1C). In comparison to modern *Polinices*, aspartic acid, threonine, serine, glutamic acid, phenylalanine, and lysine decreased, whereas alanine and valine increased in the fossil hydrophilic fraction (figure 10-2). The most apparent differences in the hydrophobic fraction of fossil *Polinices* samples relative to modern samples are increases in glutamic acid, glycine, alanine, and valine and decreases in aspartic acid, phenylalanine, and lysine.

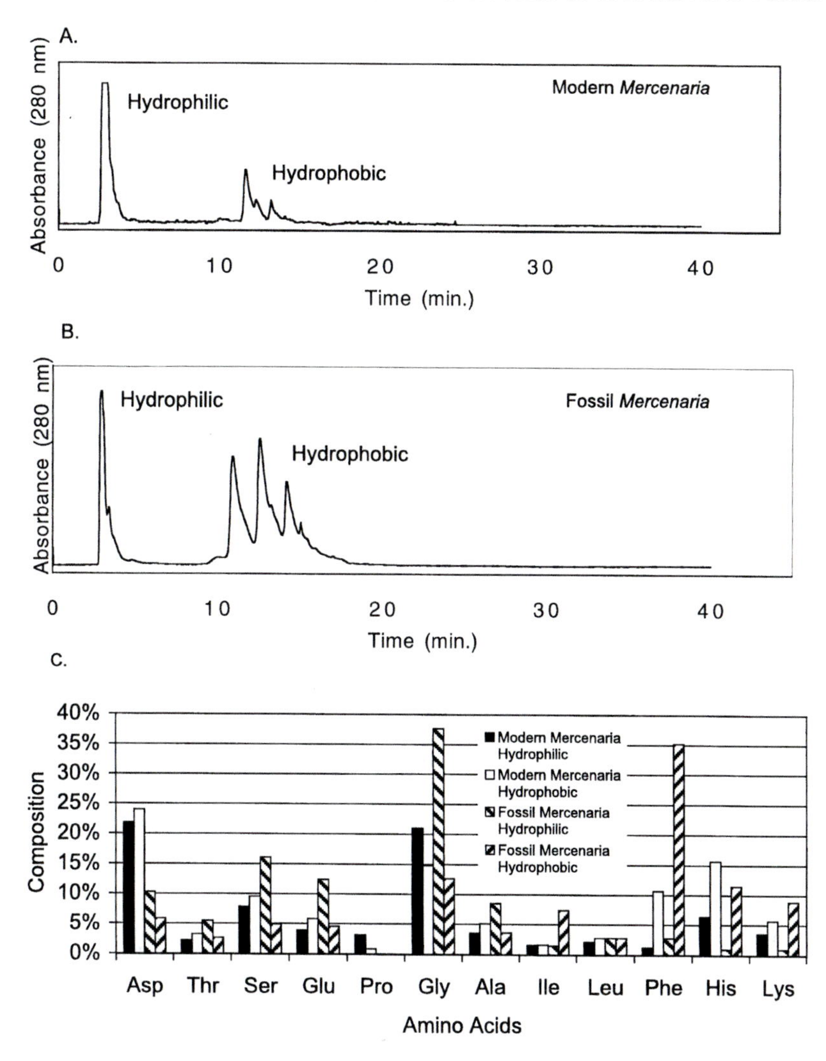

Fig. 10-1. A. Chromatogram of protein separations of modern *M. mercenaria*, revealing two major fractions. B. Chromatogram of protein separations of fossil *M. mercenaria*. C. Amino acid composition of hydrophilic and hydrophobic fractions of modern and fossil *Mercenaria* samples. Note that minor amino acids (< 2%) are not plotted.

Heated Samples

Chromatograms of heated *Mercenaria* and *Polinices* shell proteins and their amino acid compositions are shown in figures 10-3 and 10-4, respectively. The heated *Mercenaria* sample (figure 10-3A) showed a larger hydrophobic fraction and smaller hydrophilic fraction compared to the modern shell counterpart (figure 10-1A). In comparison to the modern shell, the hydrophobic fraction from heated *Polinices* samples also appears to be composed of at least three proteins (figure 10-2A). However, unlike the modern shell this fraction comprised over 94% of the total SM in the heated sample. The predominant amino acids in the hydrophilic fraction of heated *Mercenaria* and *Polinices* samples were glycine, and proline, or glycine and glutamic acid, respectively, while the hydrophobic fraction showed glutamic acid, proline, and glycine as the dominant amino acids in *Mercenaria* and *Polinices*, respectively.

Stable Isotopes

Comparisons of $\delta^{13}C$ values for the total organic fractions of modern shells indicate ecosystem and trophic level differences among gastropods (table 10-1). Data obtained from modern and fossil *Polinices* shells decalcified with EDTA indicated extremely low C isotope values (−28 to −37‰) relative to those of shells decalcified with HCl and likely reflect contamination from ^{13}C-depleted EDTA ($\delta^{13}C = -43.3‰$) (tables 10-2 and

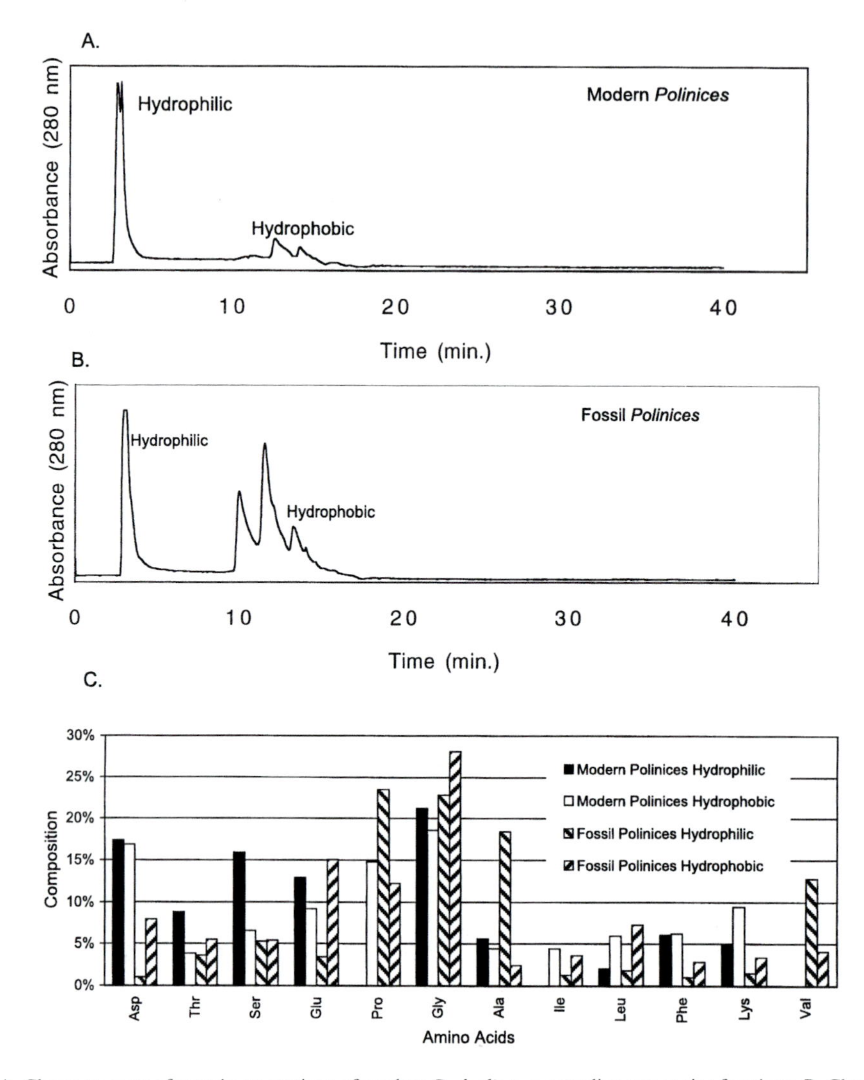

Fig. 10-2. A. Chromatogram of protein separations of modern *P. duplicatus*, revealing two major fractions. B. Chromatogram of protein separations of fossil *Polinices* samples. C. Amino acid composition of hydrophilic and hydrophobic fractions of modern and fossil *Polinices* samples. Note that minor amino acids (< 2%) are not plotted.

10-3). Consequently, the proteins, although dialyzed exhaustively and separated by HPLC, are not entirely free of EDTA. Traces of EDTA have previously been observed on shell-matrix proteins after dialysis (Johnson, 1995; Albeck et al., 1996).

Carbon-isotope values for *Polinices* (table 10-3), indicate differences of 1.5‰ and 0.6‰

TABLE 10-1 $\delta^{13}C$ values (in ‰) for total shell organic hydrolyzate for *Strombus* and *Polinices*. Specimens came from distinctively different environments: a Florida seagrass bed and a temperate sandy mud estuary in Massachusetts

Specimen	Seagrass (‰)	Estuary (‰)
Polinices (carnivore)	−5.3	−24.5
Strombus (herbivore)	−16.3	N/A

TABLE 10-2 Isotope results for *Polinices* samples decalcified with EDTA

	$\delta^{13}C$ (‰)		$\delta^{15}N$ (‰)	
Fraction	Modern	Fossil	Modern	Fossil
hydrophilic	−37.5	−38.8	−0.09	1.06
hydrophobic	−28.6	−29.7	10.7	12.6

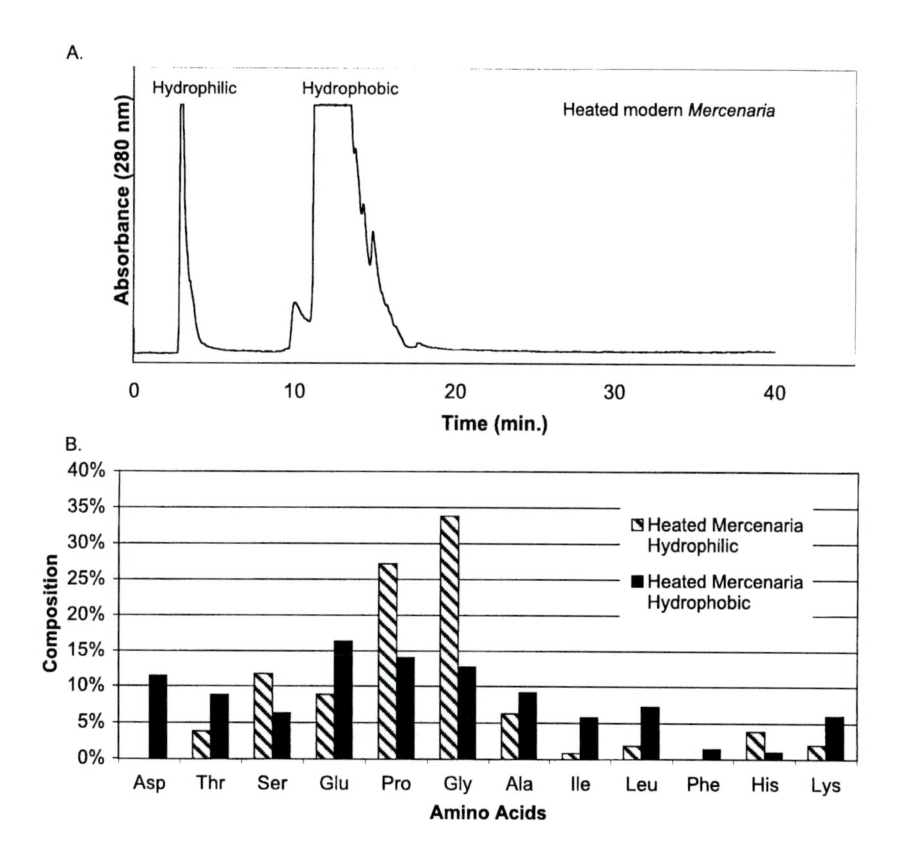

Fig. 10-3. A. Chromatogram of separation of hydrophilic and hydrophobic protein fractions from heated *Mercenaria* samples. B. Amino acid composition of protein fractions from heated *Mercenaria* samples. Note that only amino acids above 2% were plotted.

between the hydrophilic and hydrophobic fractions of the modern and fossil shells, respectively. In both cases, the hydrophilic fraction was slightly depleted in ^{13}C relative to the hydrophobic fraction. Nitrogen-isotope values for the hydrophobic fractions differed by +7.4‰ and −1.4‰ relative to those of the hydrophilic fractions for the modern and fossil shell, respectively. In addition, the C- and N-isotope values of the fossil protein fractions are lower than those of the corresponding fraction in the modern mollusk.

Permeability

Although SEM photomicrographs (figure 10-6; Andrews, 1998) support the hypothesis that gastropods have a less-permeable shell structure, preli-

TABLE 10-3 Averages of isotope data for modern and fossil *Polinices* and *Mercenaria*

	Polinices[a]				*Mercenaria*[b]			
	δ^{13}C(‰)		δ^{15}N(‰)		δ^{13}C(‰)		δ^{15}N(‰)	
Fraction	Modern	Fossil	Modern	Fossil	Modern	Fossil	Modern	Fossil
hydrophilic	−22.6	−28.8	2.2	−1.7	−38.0	−37.5	0.96	0.67
hydrophobic	−21.1	−28.2	9.6	−3.1	−21.0	−27.5	5.0	3.7

[a] HCl-treated.
[b] EDTA-treated.

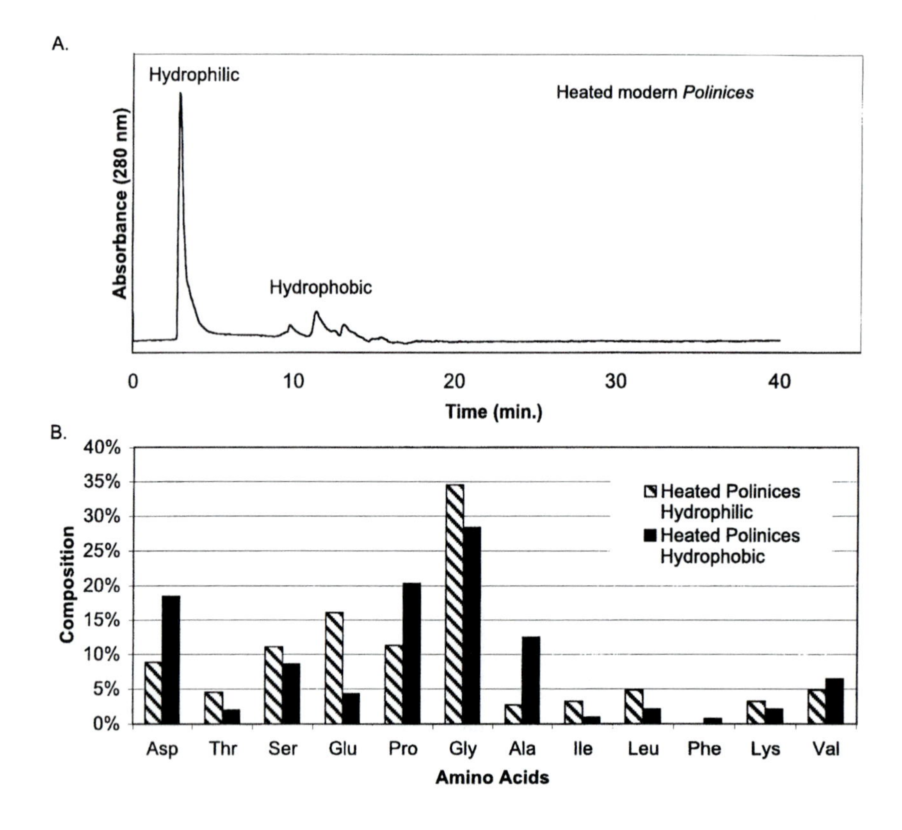

Fig. 10-4. A. Chromatogram of separation of hydrophilic and hydrophobic protein fractions from heated *Polinices* samples. B. Amino acid composition of protein fractions from heated *Polinices* samples. Note that only amino acids above 2% were plotted.

minary results of the microflow permeability analysis (figure 10-5) indicate no significant trends between modern *Mercenaria* and *Polinices*. As expected, intraspecies comparisons indicate that the permeability of fossil specimens was higher and more variable than that of their modern counterparts. For example, fossil *Mercenaria* and *Polinices* permeability differed by 2 and 6 orders of magnitude, respectively.

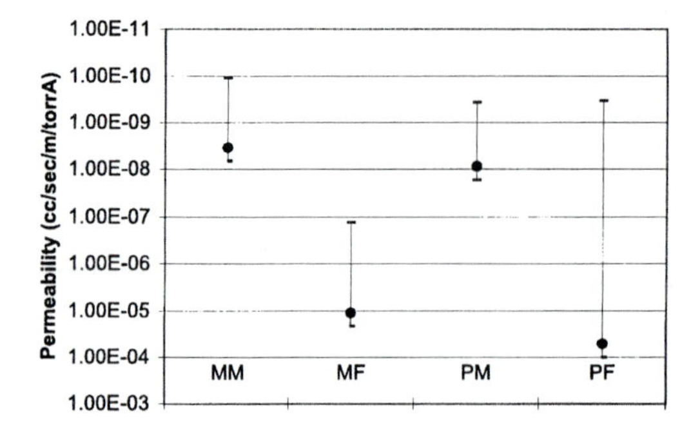

Fig. 10-5. Plot of shell permeability for modern *Mercenaria* (MM), fossil *Mercenaria* (MF), modern *Polinices* (PM), and fossil *Polinices* (PF) samples. Bars record minimum, average and maximum values on a log scale.

DISCUSSION

Although the total organic shell matrices in *Mercenaria* and *Polinices* are composed of IM and SM, only the SM was analyzed in this study because the IM is difficult to characterize using common analytical techniques which depend upon separating molecules from solution. Within our samples, the insoluble portions of *Polinices* and *Mercenaria* comprised only a small proportion of the shell, approximately 0.3% and 1.0% of the total weight, respectively. The SM comprised an average of 0.84% and 0.06% of modern total shell material for *Polinices* and *Mercenaria*, respectively (table 10-4). During preparation it was noted that several *Polinices* samples contained only SM and entirely lacked any IM. Although at this time there is no exact explanation for specimens demonstrating different characteristics, this was also observed by Robbins and Ostrom (1995).

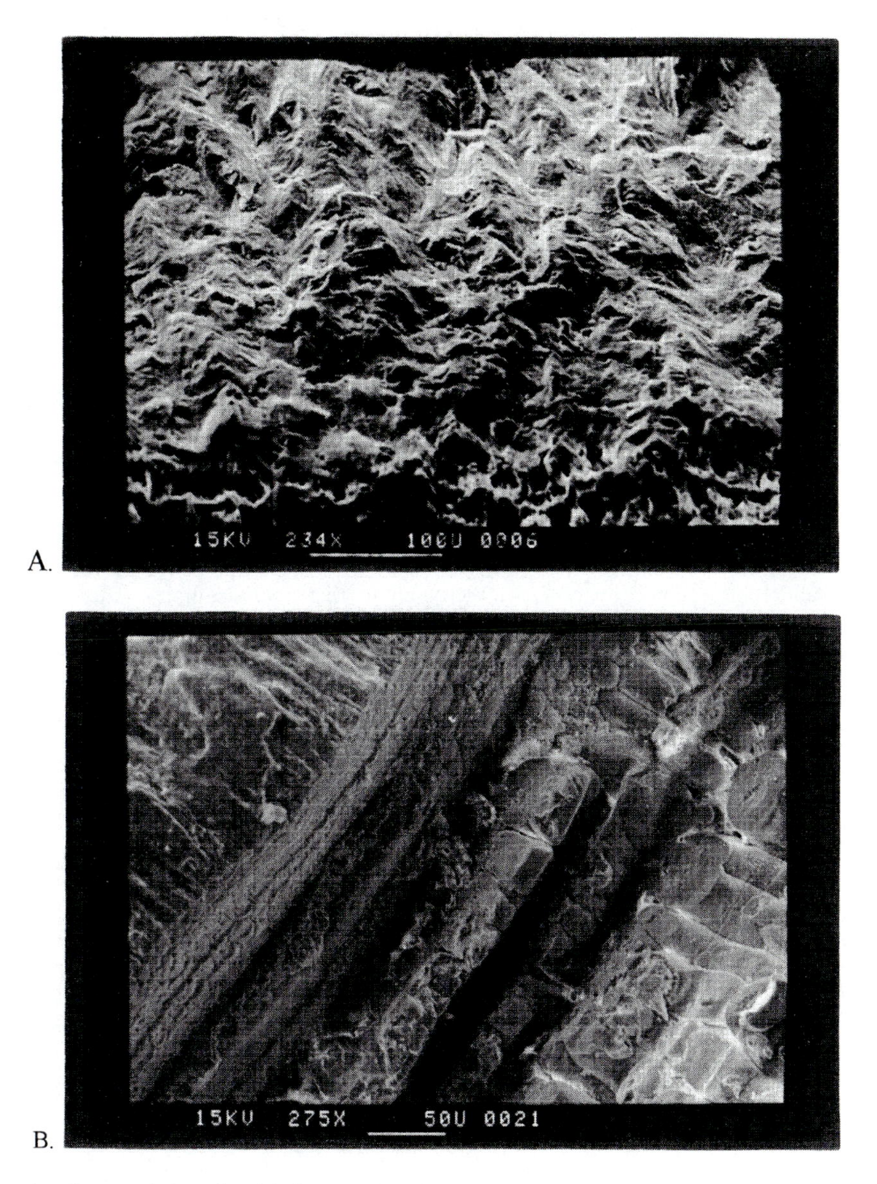

Fig. 10-6. Scanning electron micrographs of shell matrix of modern *Polinices* sp. (A), showing external nacre, having an intact non-porous structure, and modern *Mercenaria* sp. (B), showing an external surface of well-defined shell ridges and interridges, which could act as paths of fluid flow.

TABLE 10-4 Comparison of organic matrices of modern and fossil *Mercenaria* and *Polinices* samples

Sample	SM content (% of total shell)	Content (%) of SM		IM content (% of total shell)
		Hydrophilic	Hydrophobic	
modern *Mercenaria*	0.06	80	20	1.0
fossil *Mercenaria*	0.39	60.7	39.3	0.2
modern *Polinices*	0.21	75.7	24.3	0.3
fossil *Polinices*	0.84	50.5	49.4	0.0

Protein-fraction percentages were calculated as weight percents, using total absolute values measured in both hydrophilic and hydrophobic fractions.

Fossil samples of *Mercenaria* and *Polinices* (in cases where IM was present) demonstrated less IM, as a percentage of the total, than the modern samples, and showed a concurrent increase in SM. This suggests that the proportion of SM increased as the IM decomposed over time (Abelson, 1955; Hare, 1969; Akiyama, 1971), as has been shown with artificial diagenesis (Totten et al., 1972). Very little is known about how the IM solubilizes over time. Perhaps the increase in the hydrophobic fraction relative to the hydrophilic fraction in fossils and heated shells is a consequence of degradation products solubilized from the IM. Relative to *Mercenaria*, the observation of *Polinices* did not show as large an increase in the hydrophobic fraction during the heating experiment is likely because this species contains little or no IM.

Variation in the relative abundance of hydrophilic and hydrophobic protein fractions of shells is related to changes in amino acid abundances and distribution. The amino acid composition of the fractions demonstrate matrix-protein signatures (figure 10-1C and 10-2C) similar to those of other modern and fossil shells (Weiner and Lowenstam, 1980; Serban et al., 1988; Robbins et al., 1993). The aspartic acid content was greater than 5% in fossil hydrophobic fractions. In fossil shells, an abundance of aspartic acid, also typical of modern shells, has often been cited as reflecting good preservation (Akiyama, 1971), but clearly this criterion can be ambiguous. This is because, as demonstrated by the permeability data, the fossil shell is not a closed system and thus peptides and amino acids can infiltrate the shell matrix. The preferential decrease in the percentage of hydrophilic amino acids and the subsequent increase in the hydrophobic amino acids in the fossil fractions relative to the modern fractions in both species provide evidence for diagenetic alteration of the original total organic matter (Mitterer, 1993). Specifically, in comparison to modern *Mercenaria,* there is an increase in basic amino acids in the hydrophobic fractions of the fossils. Similarly, relative to modern *Polinices*, large increases in proline and alanine occur in the hydrophobic fraction of fossil *Polinices*. We also note, however, that changes in the relative abundance of amino acids in either fraction could be a function of biochemical reactions that gradually break down some of the amino acids in other compounds (e.g., proteoglycans) that are not detected by the amino acid analyzer (Mitchell and Curry, 1997). Moreover, the observed differences in the amino acid compositions among the samples are related to differences in the pathway and extent of diagenesis unique to each fossil. Although specific pathways for individual amino acids are difficult to characterize, we suggest that there are at least two major separate processes that affect the fossil SM: (1) the breakdown of the hydrophilic fraction and its subsequent incorporation into the hydrophobic fractions; and (2) the solubilization of the IM and its incorporation into the more hydrophobic component. These processes are also possible for *Polinices*, although, as discussed, individuals of modern *Polinices* specimens do not consistently possess an IM, and therefore may not exhibit its incorporation into the hydrophobic SM.

Diagenetic alteration of organic matter in shells is a function of mineral matrix structure and associated porosity. *Polinices* demonstrates a crossed lamellar shell structure that contrasts with the prismatic and nacreous structure found in *Mercenaria* (figure 10-6). Although no significant differences in permeability were found between species in modern shell samples (figure 10-5), large increases in permeability were observed in the fossil shells. We suggest that, as in recent studies of fossil-bone proteins, quantification of permeability may be an important step in our understanding of the loss of shell organic matter over time (Nielsen-Marsh and Hedges, 1999). Consistent with this perspective, it is important to emphasize that modern shell samples showed a wide range of values and, accordingly, that fossil samples reflect an even wider range. These preliminary data provide another line of empirical evidence demonstrating that variability expressed by a suite of modern samples is magnified in the fossil shells.

Clearly, environmental conditions can influence diagenesis, and stable isotope ratios have a long-established history for use in distinguishing the environmental origins of a variety of biological and geological materials (Degens, 1989; Peterson and Fry, 1989; Ostrom and Macko, 1992). Therefore, isotope values offer an insight into the diagenetic history of fossils. Our isotope data are consistent with this concept and show that both environment and trophic level are important determinants of the isotopic signatures of shell organic matter (table 10-3). The total organic fraction of *Polinices* shells from a Florida seagrass bed are enriched in ^{13}C relative to *Polinices* shells from a Massachusetts estuary. The high isotope values of shells from Florida probably reflect the influence of seagrass carbon that is typically enriched in ^{13}C relative to other sources of marine organic matter (Harrigan et al., 1989). Currently, isotopic analysis of shell proteins remains an under-utilized tool in paleoecology. Data reported here for the total organic fraction of shells suggest the potential for this approach if the effects of diagenetic alteration do not obscure the original isotopic composition of the once-living organism. This possibility could be assessed through comparison of fossils from environments predicted to be isotopically distinct or by evaluating the isotopic effects associated with diagenesis. The latter approach was implemented here in a preliminary isotopic comparison of individual protein fractions between a modern and a fossil shell.

Initial isotope data on individual proteins were obtained from EDTA-decalcified shells (table

10-2). The results for an individual show a marked depletion in $\delta^{13}C$ values that are consistent with prior reports of incomplete removal of EDTA during dialysis or RPHPLC (Johnson, 1995). This also indicates that EDTA is a poor choice as a demineralizing agent (Johnson, 1995) in paleobiochemical studies.

The hydrophobic and hydrophilic proteins of the modern shell differ in their $\delta^{13}C$ and $\delta^{15}N$ values. Factors that govern the observed isotope values of an individual molecule include the signature of its ultimate precursor, the size of the substrate reservoir, the degree of fractionation during rate-limiting steps and at major branch points of metabolic pathways, the diffusion of materials across cell boundaries, and the number of reactions in which the compound is involved (Macko et al., 1987). Currently, the effects of such factors on the isotope values of shell proteins have not been quantified.

The $\delta^{13}C$ and $\delta^{15}N$ values of proteins from fossil shells demineralized with HCl are lower than those of corresponding proteins in the modern specimens. Such differences could result from a number of environmental factors such as differences in the isotopic value of the predominant vegetation type at the base of the food web of the modern and fossil shells and/or isotopic discrimination during diagenesis. The low $\delta^{15}N$ values associated with fossil *Polinices* (−1.7−−3.1‰) are a particularly salient feature of the data. Nitrogen-isotope values less than 0‰ can be associated with nitrogen-fixing algae (Minagawa and Wada, 1986; Macko et al., 1984) but, given the expected increase in ^{15}N with trophic level (Harrigan et al., 1989), the values probably would not remain this low for a carnivore such as *Polinices*. Alternatively, low C and N isotope values could be associated with reincorporation of ^{15}N- and ^{13}C-depleted degradation products from the IM during condensation reactions associated with diagenesis.

CONCLUSIONS

We have applied multiple approaches to evaluate and assess pathways of protein diagenesis. Preliminary data show some similarity between diagenetic and degradative trends in organic components measured in heated shells and fossil samples. However, our ability to assess the effects of diagenesis on the isotopic composition of total shell organic material is limited without data on IMs and non-protein pools of organic material, such as carbohydrates and lipids. It is likely that the interactions of shell-matrix proteins and their hydrolytic products with the other organic compounds, and the effect of the carbonate matrix, substantially complicate isotopic fractionation during diagenesis (Qian, 1993). Stable isotopes on the organic material not only could provide paleoecologic information of fossil specimens, but also may ultimately elucidate any diagenetic overprints.

REFERENCES

Abelson P. H. (1954) Organic constituents of fossils. *Carnegie Instn Wash. Yrbk* **53**, 97–101.

Abelson P. H. (1955) Organic constituents of fossils. *Carnegie Instn Wash. Yrbk* **54**, 107–109.

Abelson P. H. (1956) Paleobiochemistry. *Scient. Amer.* **195**, 83–92.

Addadi L. and Weiner S. (1989) Stereochemical and structural relations between macromolecules and crystals in biomineralization. In *Biomineralization: Chemical and Biochemical Perspectives* (eds S. Mann, J. Webb, and R. J. P. Williams), pp. 133–156. VCH Verlagssgesellschaft.

Akiyama M. (1971) The amino acid composition of fossil scallop shell proteins and non-proteins. *Biomin. Res. Rep.* **3**, 65–70.

Albeck S., Aizenberg J., Addadi L., and Weiner S. (1993) Interactions of various skeletal intracrystalline components with calcite crystals. *J. Amer. Chem. Soc.* **115**, 11691–11697.

Albeck S., Weiner S., and Addadi L. (1996) Polysaccharides of intracrystalline glycoproteins modulate calcite crystal growth in vitro. *Chem. Eur. J.* 278–284.

Andrews S. (1998) Comparative biogeochemistry of modern, fossil, and artificially aged molluscs using protein, amino acid, stable isotopic and ultrastructural methods. M.S. Thesis, University of South Florida, Tampa.

Andrews J. T., Miller G. H., Davies D. C., and Davies K. H. (1985) Generic identification of fragmentary Quaternary molluscs by amino acid chromatography: a tool for Quaternary and paleontological research. *Geol. J.* **20**, 1–20.

Carter J. G. (1990) *Skeletal Biomineralization: Patterns, Processes, and Evolutionary Trends*, Vol. I. Van Nostrand Reinhold.

Collins M. J., Muyzer G., Westbroek P., Curry G. B., Sandberg P. A., Xu S. J., Quinn R., and Mackinnon D. (1991) Preservation of fossil biopolymeric structures: conclusive immunological evidence. *Geochim. Cosmochim. Acta* **55**, 2253–2257.

Crenshaw M. A. (1972) The soluble matrix from *Mercenaria mercenaria* shell. *Biomineralisation* **6**, 6–11.

Curry G. B. (1987) Molecular paleontology: new life for old molecules. *Trends Ecol. Evol.* **2**, 161–165.

Curry G. B. (1988) Amino acids and proteins from fossils. In *Molecular Evolution and the Fossil Record: Short Courses in Paleontology*, no. 1 (eds B. Runnegar and J. W. Schopf), pp. 20–33. The Paleontological Society, University of Tennessee.

Degens E. T. (1989) *Perspectives on Biogeochemistry*, Springer Verlag.

Degens E. T., Spencer D. W., and Parker R. H. (1967) Paleobiochemistry of molluscan shell proteins. *Comp. Biochem. Physiol.* **20**, 553–579.

Engel M. H., Goodfriend G. A., Qian Y., and Macko S. A. (1994) Indigeneity of organic matter in fossils: a test using stable isotope analysis of amino acid enantiomers in Quaternary mollusk shells. *Proc. Natl. Acad. Sci. U.S.A.* **91**, 10475–10478.

Ghiselin M. T., Degens E. T., Spencer D. W., and Parker R. H. (1967) A phylogenetic survey of molluscan shell matrix proteins. *Breviora* **262**, 1–35.

Hare P. E. (1969) Geochemistry of proteins, peptides and amino acids. In *Organic Geochemistry, Methods and Results* (eds G. Eglinton and M. T. J. Murphy), pp. 438–463. Springer-Verlag.

Hare P. E. and Abelson P. H. (1964) Proteins in mollusk shells. *Carnegie Instn Wash. Yrbk.* **63**, 267–270.

Hare P. E. and Hoering T. C. (1977) The organic constituents of fossil mollusc shells. *Carnegie Instn Wash. Yrbk* **76**, 625–631.

Harrigan P., Zieman J. C., and Macko S. A. (1989) The base of nutritional support for the gray snapper (*Lutjanus griseus*): an evaluation based on a combined stomach content and stable isotope analysis. *Bull. Mar. Sci.* **44**, 65–77.

Hudson J. D. (1967) The elemental composition of the organic fraction, and the water content, of some recent and fossil mollusc shells. *Geochim. Cosmochim. Acta* **31**, 2361–2378.

Johnson B. (1995) The stable isotope biogeochemistry of Ostrich Eggshell and its application to late Quaternary Paleoenvironmental reconstructions in South Africa, Ph. D., University of Colorado, Boulder.

Kennish M. J. (1980) Shell microgrowth analysis: *Mercenaria mercenaria* as a type example for research in population dynamics. In *Skeletal Growth of Aquatic Organisms: Biological Records of Environmental Change* (eds D. C. Rhoads and R. A. Lutz). Plenum Press.

Macko S. A., Estep M. L. F., Hare P. E., and Hoering T. C. (1983) Stable nitrogen and carbon isotopic composition of individual amino acids isolated from cultured microorganisms. *Carnegie Instn Wash. Yrbk* **82**, 404–409.

Macko S. A., Entzeroth L., and Parker P. L. (1984) Regional differences in nitrogen and carbon isotopes on the continental shelf of the Gulf of Mexico. *Naturwissenschaften.* **71**, 374–375.

Macko S. A., Estep M. F., Hare P. E., and Hoering T. C. (1987) Isotope fractionation of nitrogen and carbon in the synthesis of amino acids by microorganisms. *Chem. Geol.* **65**, 79–92.

Minigawa M. and Wada E. (1986) Nitrogen isotope ratios of red tide organisms in the East China Sea: a characterization of biological nitrogen fixation. *Mar. Chem.* **19**, 245–259.

Mitchell L. and Curry G. B. (1997) Diagenesis and survival of intracrystalline amino acids in fossil and recent mollusc shells. *Palaeontology* **40**, 855–874.

Mitterer R. M. (1993) The diagenesis of proteins and amino acids in fossil shells. In *Organic Geochem.* (eds M. H. Engel and S. A. Macko), pp. 739–753. Plenum.

Muyzer G., Westbroek P., and Wehmiller J. F. (1988) Phylogenetic implications and diagenetic stability of macromolecules from Pleistocene and recent shells of *Mercenaria mercenaria* (Mollusca, Bivalvia). *Histor. Biol.* **1**, 135–144.

Muyzer G., Sandberg P. A., Knapen M., Vermeer C., Collins M. J., and Westbroek P. (1992) Preservation of the bone protein osteocalcin in dinosaurs. *Geology* **20**, 871–874.

Nielsen-Marsh C. M. and Hedges, R. E. M. (1999). Bone porosity and the use of mercury intrusion porosimetry in bone diagenesis studies. *Archaeometry* **41**, 165–174.

Ostrom N. E. and Macko S. A. (1992) Sources, cycling, and distribution of water column particulate and sedimentary organic matter in northern fjords and bays: a stable isotope study. In *Organic Matter: Productivity, Accumulation and Preservation in Recent and Ancient Sediments* (eds J. K. Whelan and J. W. Farrington), pp. 55–81. Columbia University Press.

Ostrom P. H., Zonneveld J., and Robbins L. L. (1994) Organic geochemistry of hard parts: assessment of isotopic variability and indigeneity. *Palaeogeog., Palaeoclim., Palaeoecol.* **107**, 201–212.

Peterson B. J. and Fry, B. (1989) Stable isotopes in ecosystem studies. *Ann. Rev. Ecol. Syst.* **18**, 293–320.

Qian Y. (1993) Kinetic aspects of the diagenesis of organic compounds and the associated kinetic isotope fractionations, Ph. D. Thesis, University of Oklahoma, Norman, pp. 141–226.

Robbins L. L. and Ostrom P. H. (1995) Molecular isotopic and biochemical evidence of the origin and diagenesis of shell organic material. *Geology* **23**, 345–348.

Robbins L. L., Muyzer G., and Brew K. (1993) Macromolecules from living and fossil biominerals: Implications for the establishment of molecular phylogenies. In *Organic Geochemistry* (eds M. H. Engel and S. A. Macko), pp. 799–816. Plenum.

Serban A., Engel M. H., and Macko S. A. (1988) The distribution, stereochemistry and stable isotopic composition of amino acid constituents of fossil and modern mollusk shells. *Organic Geochem.* **13**, 1123–1129.

Silfer J. A., Qian Y., Macko S. A., and Engel M. H. (1994) Stable carbon isotope composition of individual amino acid enantiomers in mollusc shell by GC/C/IRMS. *Organic Geochem.* **21**, 603–609.

Totten D. K., Davidson F. D., and Wyckoff R. G. (1972) Amino acid composition of heated oyster shells. *Proc. Natl Acad. Sci. U.S.A.* **69**, 784–785.

Vallentyne J. R. (1969) Pyrolysis of amino acids in Pleistocene *Mercenaria* shells. *Geochim. Cosmochim. Acta* **33**, 1453–1458.

Weiner S. and Hood L. (1975) Soluble proteins of the organic matrix of mollusk shells: a potential template for shell formation. *Science* **190**, 987–988.

Weiner S. and Lowenstam H. A. (1980) Well-preserved fossil mollusk shells: characterization of mild diagenetic processes. In *Biogeochemistry of Amino Acids* (eds P. E. Hare, T. C. Hoering, and K. King, Jr), pp. 95–114. Wiley.

Weiner S. and Traub W. (1984) Macromolecules in mollusc shells and their function in biomineralization. *Phil Trans. Roy. Soc. Lond., Ser. B* **304**, 425–434.

Weiner S., Traub W., and Lowenstam H. A. (1983) Organic matrix in calcified exoskeletons. In *Biomineralization and Biological Metal Accumulation* (eds P. Westbroek and E. W. de Jong), pp. 205–224. D. Reidel Publishing Company.

Wheeler A. P., Rusenko K. W., Swift D. M., and Sikes C. S. (1988) Regulation of *in vitro* and *in vivo* $CaCO_3$ crystallization by fractions of oyster shell organic matrix. *Marine Biol.* **98**, 71–80.

11. Amino acid racemization in biominerals: the impact of protein degradation and loss

Matthew J. Collins and Michael S. Riley

Amino acid racemization[1] (AAR) presents a conundrum. Ever since Hare (1971), it has been appreciated that there is a clear interrelationship between AAR and protein diagenesis[2] (e.g., Wehmiller, 1980; Mitterer, 1993). Experimental studies of diagenesis have revealed marked differences in the rates of hydrolysis of peptide bonds, racemization and decomposition of constituent amino acids and the loss of organic fractions from different biominerals. Yet observed rates of racemization in fossils (with the exception of aspartic acid) display surprisingly little variation between amino acid, genera or biomineral type. This minimal variation enables AAR to be treated as a black box, in which the underlying processes can be ignored.

The black-box approach, however successful, leaves many questions unanswered, exposes results to challenge from those who query interpretations based upon AAR, and is perhaps partly responsible for the limited impact made by the approach in the wider archaeological and geological community. Recent reviews of protein diagenesis (Bada, 1991; Mitterer, 1993; Wehmiller, 1993; Bada, 1998; Collins et al., 1998a; Bada et al., 1999) permit a renewed consideration of the relationship between diagenesis and racemization. It is therefore to the process of racemization that we turn first and consider how the increasingly complex pattern of kinetics from free amino acids via proteins to the organic fraction of biominerals brings into play, first, a consideration of protein structure and, subsequently, biomineral ultrastructure and organization.

RACEMIZATION

Racemization of Free Amino Acids

There is broad agreement over the relative rates and rate constants for the racemization of free amino acids (k^F_{DL}); differences between studies are in large part due to variations in the pH conditions used (Bada and Shou, 1980). Amino acids with aliphatic side chains display a 4–6-fold variation in rate at pH 7.6, between the most rapid (Phe) and the slowest (Val) (e.g., Bada and Shou, 1980; Smith and Sol, 1980). At the same pH, Wonacott (1979) reports a variation in rate of over 150-fold between Ser and Glu. At pH 9, Liardon and Ledermann (1986) observed a 90-fold variation in rate between Ser and Val.

The variation in relative rates of racemization and their pH dependence is broadly in agreement with the mechanism proposed by Neuberger (1948), who suggested that under basic conditions racemization proceeds via abstraction of H^+ and the formation of a planar carbanion. This is illustrated by the correlation between relative rates of racemization and the electron-withdrawing capacity of the side-chain (estimated from the Taft inductive constant); the more electron-withdrawing the side-chain, the greater will be the stabilization of the carbanion and thus the more readily is this formed (Bada and Shou, 1980; Smith and de Sol, 1980; Liardon and Ledermann, 1986). Unfortunately, racemization of free amino acids has little bearing upon observed racemization in fossil biominerals.

Protein Decomposition and Racemization Rate

Unlike their free counterparts, relative rates of racemization between constituent amino acids in biominerals display surprisingly little variation (e.g., Lajoie et al., 1980; Goodfriend and Meyer, 1991; but see Csapó et al., 1990). Work by Steinberg and Bada (1981, 1983), Mitterer and Kriausakul (1984), Moir and Crawford (1988), Smith and Baum (1987), Gaines and Bada (1988) and Sepetov et al. (1991) have highlighted the importance of diketopiperazine (DKP), formed during decomposition, in the kinetics of racemization of dipeptides and proteins. Overall rates of racemization are more than ten times faster in dipeptides than for free amino acids, whereas in DKP itself, rates are estimated to be over two orders of magnitude greater.

Although studies at basic pH do reveal some evidence of racemization of internal residues (e.g., Fridkin et al., 1970), there is little evidence for significant racemization within polypeptides at neutral pH. Two lines of evidence support this contention: firstly, limited racemization in high-molecular-weight fractions (e.g., Kimber and Griffin 1987; Kimber and Hare, 1992; Kaufman and Miller, 1995) and melanoidins (Rafalska et al., 1991); second, the unusual behavior of Asx (Asp plus Asn).

Lajoie et al. (1980) observed a very significant positive intercept in the slope of D/L Leu vs. D/L Asx, because there is significant racemization of *bound* Asx residues (e.g., Geiger and Clarke, 1987; Collins et al., 1999, for review). The absence of a positive offset in any other amino acid (except occasionally Pro[3]) when plotted against D/L Leu suggests that in all other cases, bound residues racemize equally slowly if at all. The variation in relative racemization rates of both free amino acids and dipeptides is not apparent in biominerals, implying that the rate-limiting step in the observed racemization rates must be sought elsewhere.

Modelling Racemization

It is common to report estimated rate constants for racemization in biominerals by treating the net racemization observed as a single reaction with its own rate constant; we will prefix estimates of this type k_{DL}^{obs} to distinguish them from underlying rate constants. The relationship between k_{DL}^{obs} and the various underlying rates (i.e., a series of rate constants for hydrolysis, racemization, and decomposition) is potentially very complex. The more that is understood regarding the factors influencing racemization, namely the potential for non-equilibrium ratios (Moir and Crawford, 1988; Fujii et al., 1994), the impact of conformation (Fujii et al., 1994; van Duin and Collins, 1998) and the wide range of rate constants observed for free amino acids and dipeptides (above), the more surprising is the relatively small variation in the rates and simplicity in the patterns of racemization (although size fractions do display more convoluted kinetics; Kimber and Griffin, 1987). The approximation of early stages of racemization in biomineral proteins to first-order reversible kinetics means that k_{DL}^{obs} is widely reported as the rate of racemization. In recent years, mathematical transformations (Mitterer and Kriausakul, 1989; Murray-Wallace and Kimber, 1993) or polynomial functions (Gillard et al., 1991; Wehmiller, 1993) have been used to represent the net 'rate' over a wide range of D/L values but such approaches ignore the underlying reactions.

In order to utilize k_{DL}^{obs} safely, it is necessary to come to terms with the processes that influence it; these processes can be broadly termed 'protein diagenesis'. It is this approach developed by Kriausakul and Mitterer (1980a, b), Mitterer and Kriausakul (1984) and Wehmiller (1980, 1982, 1990, 1993), which has provided the most widely accepted conceptual model to explain the nonlinear kinetics (other models exist, e.g., Julg, 1984; Saint-Martin, 1991). We have already reviewed evidence demonstrating that the rate of racemization varies depending upon the state of protein degradation, namely (1) slow racemization of peptide-bound residues (2) fast racemization of either (a) terminal or (b) DKP residues, and (3) moderate rates of racemization of free amino acids. The conceptual models of Mitterer and Kriausakul and of Wehmiller rationalize the overall racemization rate in terms of protein degradation, at its simplest in three stages (figure 11-1), as outlined below.

1. Initially, racemization is slow because the overwhelming majority of amino acids are in the form of peptide-bound residues in long polypeptide chains.
2. Over time, as hydrolysis of peptide bonds leads to the solubilization of insoluble proteins and the generation of oligomers, dimers, and DKP, the rate of racemization is accelerated.[4]
3. Finally, the k_{DL}^{F} becomes the controlling factor, reflected in a declining rate of k_{DL}^{obs}.

This conceptual model, hereafter referred to as the "three-box model", is inadequate to explain the observations used to derive it. The example of nonlinear racemization used by Kriausakul and Mitterer (1980b) of *Mercenaria* heated at 152°C displays a final rate of epimerization less than half the rate of free isoleucine (Ile) from the same study (figure 11-2). In the simple three-box model, the slowest possible k_{DL}^{obs} is k_{DL}^{F}, as is observed for heated peptides and proteins *in vitro* (Kriausakul and Mitterer, 1980b). However, an adequate explanation of epimerization of Ile in *Mercenaria* and, by implication, in other biominerals, has to account for an k_{DL}^{obs} in the latter stages of diagenesis which is slower that k_{DL}^{F}.

INADEQUATE MODELS OF RACEMIZATION

Two processes may account for the slower-than-anticipated rate of racemization: (1) slow hydrolysis of a residual peptide-bound fraction or (2) preferential loss of free (highly racemized) amino acids. Each of these two possibilities will now be considered.

Hydrolysis

In a previous paper (Riley and Collins, 1994) we modeled a polypeptide comprising P_0 amino acid residues in terms of random chain scission, leading to an initial increase in *N*-termini and an overall decrease in the mean length of residual polypeptides, ultimately leading to the liberation of free amino acids (equation 11-1). The weight fraction, γ_x, of x-mers at time t, is given by

$$\gamma_x = \begin{cases} \alpha \dfrac{x}{P_0}(1-\alpha)^{x-1}[2+(P_0-x-1)\alpha] & \text{for } x < P_0 \\ (1-\alpha)^{P_0-1} & \text{for } x = P_0 \end{cases} \qquad (11\text{-}1)$$

where α is the probability of bond breakage given by $\alpha = 1 - \exp(-kt)$, and k is the rate of hydrolysis (Riley and Collins, 1994).

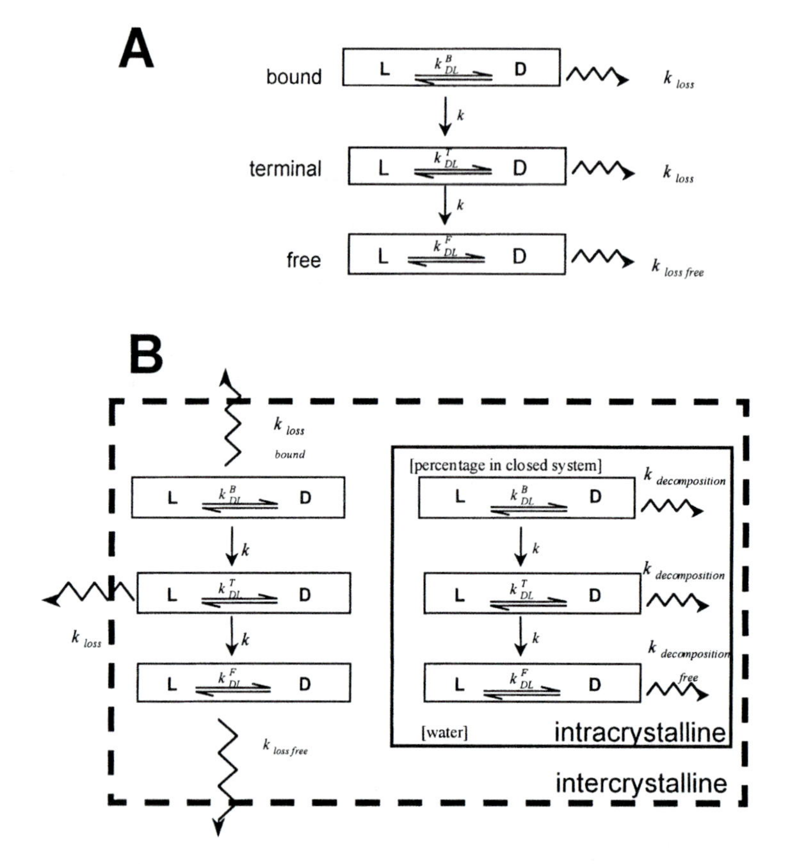

Fig. 11-1. Alternative models of amino acid racemization. The basic model (A) represents the original three-box model of Kriausakul and Mitterer (1980b), Mitterer and Kriausakul (1984) and Wehmiller (1980, 1982, 1990, 1993). This is modified (B) to include both an intercrystalline open system (left-hand side) and an intracrystalline fraction (right-hand side). Only the former can be lost from the biomineral; the latter will only become depleted in concentration via decomposition reactions. Rate constants are shown in italics. The rates will depend upon the concentrations of the parameters defined at starting conditions.

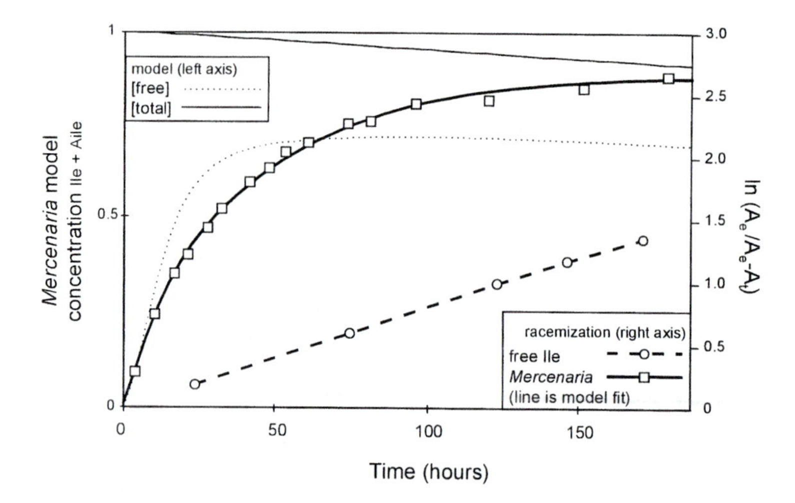

Fig. 11-2. Comparison of data for epimerization of *Mercenaria* samples and free Ile (pH 7.6, 152°C; data from Kriausakul and Mitterer, 1980b). Note that the latter rate of epimerization in *Mercenaria* is only one third that of free Ile, a result inconsistent with the prevailing explanation. The model fit of the *Mercenaria* racemization data uses the three-box two-pool model (figure 11-1B), note the predicted decline in both total and free amino acid concentrations in the latter part of the experiment.

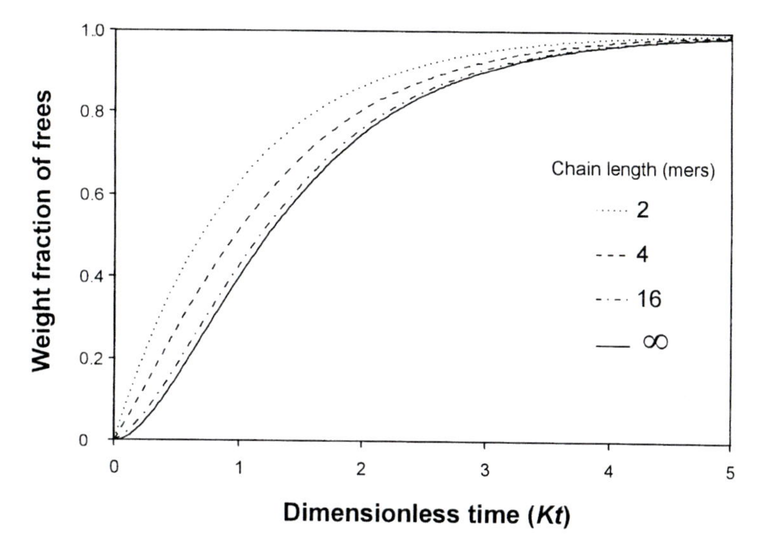

Fig. 11-3. Rate of release of free amino acids, assuming random chain scission; note the increasing lag observed within increasing length of polymer (P_0) from $P_0 = 2$ to ∞. The output is based on the polymer degradation model of Riley and Collins (1994).

In the random scission model (equation 11-1) we observe an initial lag in the rate of release of amino acids (figure 11-3), because the model assumes that all scission events occur randomly along the chain. In the case of oligopeptides, the number of random scission events required to generate a free amino acid increases with oligopeptide length, although the lag time remains stable above $P_0 \sim 16$ (figure 11-3). The lag would be exaggerated if the observed decrease in the rate of (acid- and base-catalyzed) peptide bond hydrolysis with reduction in polymer length (e.g., of Gly_n peptides; Hammel and Glasstone, 1954) were extrapolated to neutral pH. No lag is apparent in those few studies which report on the increase in the proportion of free amino acids over time (e.g., King, 1980; Müller, 1984; Goodfriend et al., 1992; Miller et al., 1992), although, with the exception of Miller et al. (1992), this may be due to a lack of sampling density.

For the interpretation of racemization kinetics it is important to establish the dominant mechanism of chain scission. At extreme pH, peptide-bond hydrolysis is strongly influenced by the amino acid side-chain. In particular, the approach of H^+ ions is hindered by bulky hydrocarbon side-chains (Hill, 1965) and the electrostatic repulsion of the terminal amino group (Hammel and Glasstone, 1954). At neutral pH there are surprisingly few data on rates and mechanisms of chain scission, with the notable exception of bonds involving aspartyl and asparingyl residues (e.g., Oliyai and Borchardt, 1993). In the latter case, decomposition of aspartyl and asparinglyl residues by formation of a cyclic succinimide (Geiger and Clarke, 1987) may result in chain cleavage. In an investigation of decomposition of tripeptides and a hexapeptide at neutral pH, Steinberg and Bada (1983) observed that the major pathway of chain scission was via the formation of DKP; see also Sepetov et al. (1991). The DKP subsequently decomposes to the two constituent amino acids. Thus aminolysis via DKP can potentially reduce the lag time to generate free amino acids, but the kinetics would appear to obey a rate law intermediate between first and zero-order, as the rate is dependent upon the attack of the *N*-terminal amino group on the adjacent amide carbonyl group (Sepetov et al., 1991). Rates of hydrolysis at neutral pH do appear to be more rapid than predicted if acid- or base-catalyzed. Kahne and Still (1988) suggested that the release of peptide bound Phe at neutral pH was the result of "water-catalyzed" hydrolysis; a similarly accelerated rate of gelatinization at neutral pH was observed by Rudakova and Zaikov (1987) (see figure 5 of Collins et al., 1995).

Assuming random scission, the generation of free amino acids can be characterized by

$$-\ln[1 - \sqrt{\gamma_1}] = kt \tag{11-2}$$

provided that P_0 is sufficiently large. The maximum absolute error in using equation 11-2 is $1/(2P_0)$ and so when $P_0 > 50$, (i.e., the size of the very smallest biomineral proteins, e.g., osteocalcin), the error is less than 1%.

If decomposition is by random scission, plotting the left-hand side of equation 11-2 should yield a straight line through the origin with slope k (figure 11-4) (for derivations see appendix 11-1). The rate

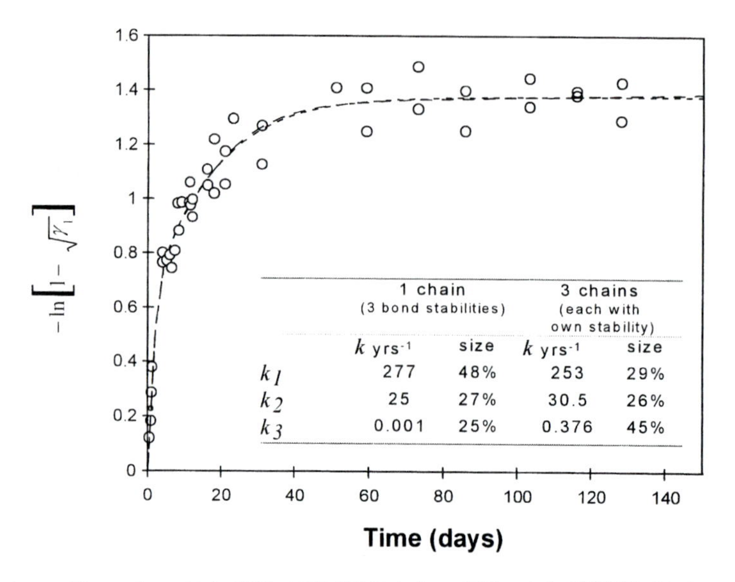

Fig. 11-4. The release of free amino acids in OES at 142.5°C (data from Miller et al., 1992) fitted to two versions of the polymer degradation model (for details see text); the outputs are almost indistinguishable. Estimates of the two models are given in the inset table.

of hydrolysis determined by this method will be larger than that estimated as just the rate of increase of free amino acids with time even if all peptide bonds are of equal stability. If the rate of increase of frees (k^{obs}) is estimated from a plot of the number of frees versus time over the time period $kt = 0$ to $kt = T$, for which scattered data might be understandably approximated by a straight line, the relationship between k^{obs} and k is given by

$$k = \frac{T}{[1 - \exp(-T)]^2} k^{obs}. \quad (11\text{-}3)$$

For the range of likely values of T, this means that k will be between 2.5–3.5 times larger than k^{obs}.

Plotting of the data set of Miller et al. (1992) yields a curve with a minimum of two breaks of slope (figure 11-4). In order to explore this further, we fitted two alternative versions of the polymer degradation model to the data. Both versions assume that there are three rate constants of hydrolysis of peptide bonds—one rapid (k_1), one intermediate (k_2), and one slow (k_3); in the latter case, because of the scatter of the data and the short time-scales relative to the rate, our estimate represents an approximation to an upper bound for k_3. In the first model (the three-oligomer model), there are three separate peptide chains ($P_0 > 50$), each having its own rate of hydrolysis. The weight fraction of free amino acids is given by

$$\gamma_1 = \sum_{i=1}^{3} \lambda_i (1 - \exp[-k_i t])^2 \quad (11\text{-}4)$$

where λ_i is the fraction of bonds which hydrolyse at rate k_i. In the second, (random) polymer model, peptide bonds have one of three different rate constants, which are randomly distributed along a single polymer. In this case the weight fraction is given by

$$\gamma_1 = \left(\sum_{i=1}^{3} \lambda_i (1 - \exp[-k_i t]) \right)^2 \quad (11\text{-}5)$$

It is possible to fit each model to the data and estimate the rate constant for each of the three rates of hydrolysis and the relative fraction each contributes to the total number of peptide bonds. Unfortunately, the outputs of the models are too similar to distinguish between them using the available data. The models could be distinguished if the concentration of dipeptides were known.

The data transformation provides support for the suggestion of Miller et al. (1992) that the pattern of hydrolysis is explicable in terms of variable peptide-bond stability (figure 11-4). The difference between k_1 and the upper bound of k_3 estimated by this approach is at the limit of comparative rates of (acid- and base-catalyzed) hydrolysis of different sequence combinations (Hill, 1965). The initial lag predicted by the polymer model is not seen in the untransformed data because k_1 is so rapid that it would require very high sampling density to be revealed. For ostrich eggshell (OES) proteins, our estimate of k_1 is eight times larger than the k value reported by Miller et al. (1992) (figure 11-4). Furthermore, because of the rapid initial loss of total amino acids described below, the observed concentration of free amino acids may be depleted and the real value of k_1 will be higher still.

In our interpretation, the k_{DL}^{obs} of Kriausakul and Mitterer (1980b) (figure 11-2) is slower than k_{DL}^{F} in the latter stages of diagenesis, because k_3 is slower than k_{DL}^{F} and therefore k_3 becomes the rate-limiting step. This could be tested by monitoring the rate of racemization and hydrolysis of purified biomineral proteins *in vitro* over extended periods of time and discerning whether the rate of racemization falls below the rate of racemization of free amino acids (cf. Kriausakul and Mitterer, 1980b). If racemization is never slower than k_{DL}^{F} (e.g., proteins and peptides reported by Kriausakul and Mitterer, 1980b), it would imply that the model is adequate to describe the diagenesis of proteins *in vitro* but not within biominerals themselves. Such an experiment would also be of interest in the case of Ile to determine whether the epimerization equilibrium constant was the same for the free amino acid, isolated and intact biomineral proteins, because the pool of slowly hydrolysing Ile observed in biominerals has the effect of apparently lowering the equilibrium constant.[5]

One problem with the above interpretation is that the claimed activation energies of most racemization reactions [96–135 kJ/mol (=23–32 kcal/mol); Bada, 1972; Smith and Evans, 1980; Mitterer, 1975; Miller, 1985; Miller et al., 1992, 1997; McCoy, 1987; Goodfriend and Meyer, 1991] are apparently higher than hydrolysis [83–99 kJ/mol =20–24 kcal/mol; Hare, 1976; Kriausakul and Mitterer, 1980a; Saban et al., 1992; Qian et al., 1993]. If this is indeed so, then racemization patterns observed in laboratory studies would have little relevance to fossil biominerals, as in the latter case k_{Hyd} will be an order of magnitude faster relative to k_{DL}^{obs} than in the laboratory. However, estimates of hydrolysis at extreme pH or in solution may not be relevant to proteins in biominerals.

Temperature Dependence of Hydrolysis

Miller et al. (1992) estimate an activation energy (E_a) for hydrolysis in OES by comparing the initial rate of release of free amino acids at high temperatures in laboratory experiments and at low temperatures from archaeological sites in southern Africa. The E_a of Leu hydrolysis in OES is calculated by Miller et al. (1992) to be 114 kJ/mol (27 kcal/mol), well outside the normal range reported above. Although the values for k estimated by Miller et al. will be low, errors introduced by using their approach are systematic and will not significantly impact upon the estimate of E_a. We have made our own estimates of E_a by determining k (using equation 11-2) for foraminifera (1.4–2.7°C; King, 1980; Müller, 1984) and by laboratory diagenesis of OES (142.5°C; Miller et al., 1992). The E_a for both the fast and medium k estimates, derived from optimization, are similar to those reported by Miller et al. (1992), albeit using two totally different biominerals.

There has been little investigation of either the temperature- or pH dependence of the diketopiperazine (DKP) aminolysis, and none at all of intact proteins (except inadvertently in racemization studies). The cyclization rate of DKP is much more rapid than hydrolysis at neutral pH (Gaines and Bada, 1988; Capasso et al., 1998) and has a lower activation energy [74 kJ/mol (=18 kcal/mol); Capasso et al., 1998]. Kriausakul and Mitterer (1980a) reported activation energies for the *hydrolysis* of the Ile–Gly dipeptide of 87.5 kJ/mol (21 kcal/mol) and for *racemization* of 96.6 kJ/mol (23 kcal/mol, pH 7.6, 0.01 M phosphate), which, since the latter is probably via DKP, suggests values similar to those for peptide bonds.

A possible explanation of the higher activation energy of hydrolysis in biominerals is the impact of conformation—a feature that would not be relevant to the hydrolysis of dipeptides (Kriausakul and Mitterer, 1980a; Qian et al., 1993) or proteins at extreme pH (Hare, 1976; Saban et al., 1992). The rates of peptide-bond hydrolysis have been observed to be lowered by structural constraints (Müller and Heidemann, 1993). Increasing temperature leads to greater conformational freedom; the effect will be to elevate E_a relative to reactions involving unconstrained sequences (such as oligopeptides *in vitro*). This will be particularly significant if the mechanism of decomposition involves a significant distortion of the peptide backbone (e.g., van Duin and Collins, 1998). The decomposition rate of osteocalcin monitored immunologically displayed a greater temperature dependence in bone than in free solution (Collins et al., 1998b). If hydrolysis has a temperature dependence in biominerals approaching that of k_{DL}^{obs}, then laboratory studies are relevant to the patterns of racemization in fossils.

"Leaching"

The second process that can dampen the rate of apparent racemization (figure 11-2) is loss of free amino acids. Loss of the highly racemized free pool will have the effect of increasing the relative proportion of (less-racemized) residues thereby dampening k_{DL}^{obs}. There are a number of ways in which amino acids could be lost: (1) condensation via the reaction of amide or amino groups with reducing sugars; (2) decomposition (deamination and decarboxylation; see Bada, 1991, for review); or (3) migration from the biomineral. Ile is thought to be one of the most stable amino acids and decomposition therefore seems to be an unlikely explanation.

It is rarely possible to distinguish between these three processes (Wehmiller, 1980, 1990), but in an interesting study Kimber and Griffin (1987) report a rapid decline in absolute amounts of Val in heated *Anadara* (bivalve mollusk) samples (their figure 6) *despite* the fact that the analysis was conducted in sealed glass vials and the whole tube contents were analyzed. The pattern of decline is similar to that for Leu, reported by Miller et al. (1992) for OES, which is also believed to be a closed system (Brooks et al., 1990). Similar rapid rates of initial loss of amino acids are reported in fossil foraminifera (King, 1980; Müller, 1984), although the decline in total concentration was originally attributed to dissolution of the foraminiferal tests.

Simple mathematical functions have difficulty in describing this loss[6], which is reminiscent of the so-called "3G" model of degradation of organic matter in sediments (Berner, 1979). The choice of three pools is somewhat arbitrary, being the simplest reduction of a system with multiple rates to produce a satisfactory pattern of loss. We have already shown that a three pool model is consistent with the data (figure 11-4) and it seems unlikely that the two phenomena, hydrolysis and loss, are unrelated.

The temperature-dependent *relative* rates of decomposition of this rapid initial phase between the fossil and high temperature data sets [137 kJ/mol (=33 kcal/mol)] are consistent with decarboxylation (Ala 131.5 kJ/mol [31 kcal/mol]; Conway and Libby 1958; Phe, 131.1 kJ/mol, Vallentyne, 1964), but the *absolute* rates are between 2 and 5 orders of magnitude faster.

The rate of amino acid decomposition is accelerated in the presence of oxygen (Conway and Libby, 1958), but even assuming that all the water within the biomineral (Gaffey, 1988) was fully oxygenated and chemically available, this would account for the decomposition of no more than 1% of the total protein. A second possibility is that the initial loss of concentration is an artefact of analysis. Miller (1985) reported that soluble high-molecular weight material sorbed to glass walls during demineralisation and was therefore lost to the analysis. The decomposition of the insoluble matrix to soluble polypeptides will increase the likelihood of loss via this process. However, subsequent decomposition to smaller peptides would increase the reversibility of sorption (e.g., Lyklema, 1986) and thus increase the apparent concentration again.

An alternative explanation for initially rapid loss is melanoidin formation (e.g., Qian et al., 1995). Vallentyne (1964) observed accelerated decomposition of amino acid in the presence of carbohydrates, and the rates are consistent with data reported by Collins et al. (1992) for Maillard reactions between bovine serum albumin (BSA) and $(\text{glucose})_n$. There is plentiful carbohydrate in mollusk shells, both as glycosylated residues and as β-chitin, and in OES (as proteoglycan). Hydrolysis of proteins and polysaccharides liberates reactive reducing sugars and amino groups; therefore the rate of loss of amino acids from the analytical window will mirror the rate of hydrolysis. This idea could be tested by monitoring total nitrogen as well as total amino acids; if the Maillard Reaction is responsible, nitrogen content will remain constant at the expense of total amino acids and reducing sugars.

Although condensation is a candidate for the initial loss of amino acids (see figure 11-7 later) this process would not explain the decline in k_{DL}^{obs} seen in the latter part of the *Mercenaria* curve (figure 11.4). Decarboxylation of free amino acids would appear to be too slow to remove free amino acids selectively in the time-scales required. The most probable explanation for analytical loss in the latter stages of diagenesis is diffusion (so-called "leaching") of free amino acids from the shell.

Diffusive Flux

The diffusive flux of decomposition products from the biomineral will be governed by: (1) the diffusion coefficient, D, of the compounds (which will increase with the decrease in molecular radius and the concomitant increase in charge density arising from hydrolysis of polypeptides to free amino acids); (2) the dimensions and tortuosity of the pore system; (3) the presence of residual polymer matrix, which acts to both obstruct and sieve solute molecules; (4) concentration of solute and polymer; (5) surface effects leading to sorption of the solute on to the polymer matrix or pore walls; and (6) temperature.

The diffusion coefficient for Leu in water is 11 times greater than the globular protein BSA and 1.18 times greater than that of Leu–Gly (Longsworth, 1952), meaning that even in the case of water-filled pores, amino acids will be more rapidly lost than larger proteins (figure 11-5). The effect is accentuated because the pore system in the biomineral is generated by the degradation of the organic matrix itself, a system somewhat analogous to the migration of hydrocarbons within a source rock (Stainforth and Reinders, 1989). The organic matrix within the pore system will both (1) obstruct (thereby increasing the path length for diffusive transport) and (2) sieve, the solute molecules (Amsden, 1998). The restriction on diffusive movement will be greatest in the early stages of diagenesis when most solute molecules will be large (oligopeptides) and the polymer matrix nearly intact. The net effect of all these factors will be a selective loss of the smallest solute molecules (free amino acids) which are the most highly racemized (e.g., Roof, 1997).

The disparity between the temperature dependence of this diffusive flux relative to racemization and hydrolysis is just as problematic as the preferential loss of racemized amino acids. Estimates of the temperature dependence of racemization and hydrolysis range between 80 and 120 kJ/mol, but Longsworth (1954) estimated E_a of D for amino acids and proteins between 0 and 37°C as lying

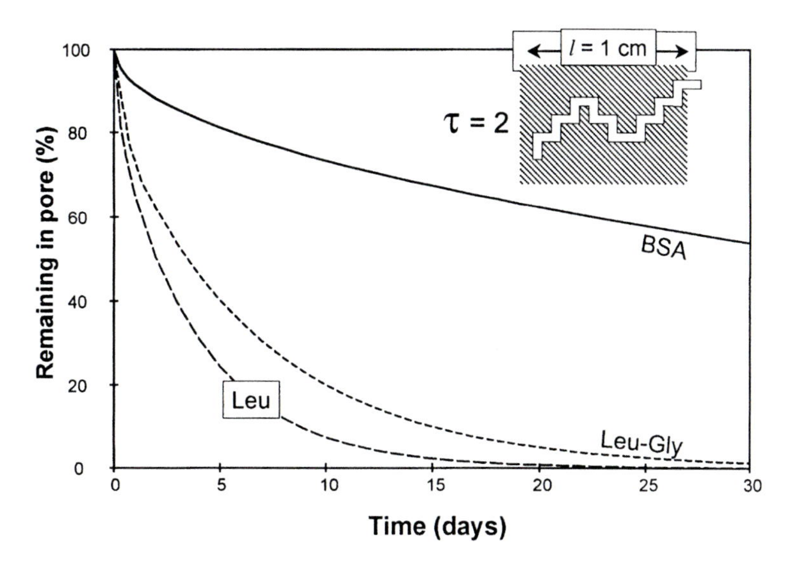

Fig. 11-5. Estimate of relative time taken to diffuse a protein (BSA), dipeptide (Leu–Gly) and amino acid (Leu) out of a simple blind-ended pore system (inset) across a shell thickness of 1 cm with a tortuosity of 2 at 1°C. The estimates are made using:

$$c_{av}(t) = \frac{8C_0}{\pi^2}\sum_{n=0}^{\infty}\frac{1}{(2n+1)^2}\exp\left[-(2n+1)^2\pi^2\frac{D}{b^2}t\right]$$

where D is the diffusion coefficient (data from Longsworth, 1952), b is the shell thickness, C_0 is the initial concentration of free amino acids (Carslaw and Jaeger, 1959). The estimated time for diffusion is somewhat arbitrary because the diffusion coefficients are established for dilute solutions; the model assumes diffusion into an infinite solution of low viscosity from a large diameter pore system and no surface effects. For further information on tortuosity see Boudreau (1997).

between 19.3 and 20.7 kJ/mol. Oelkers (1991) used a modified Arrhenius-type equation (11-6) to calculate diffusion coefficients (D) for aqueous species (including one amino acid, Gly) at the elevated temperatures:

$$D = A\exp\left[-\frac{E_a + a(T-228)^{-1.5}}{RT}\right] \qquad (11\text{-}6)$$

where A is a pre-exponential factor (for Gly $352.85 \times 10^{-5}\,\mathrm{cm^2/s}$), T is the temperature in Kelvin, a is the species-specific empirical constant (for Gly 346.57 kJ/mol $\mathrm{K}^{1.5}$), E_a is the apparent activation energy (13.93 kJ/mol) and R is the gas constant (8.31451 J/mol/K). From equation 11-6, it can be seen that D displays decreasing temperature dependence with increasing temperature (Oelkers, 1991). The calculated D values for Gly at 20 and 142.5°C are $9.33 \times 10^{-6}\,\mathrm{cm^2/s}$ and $62.5 \times 10^{-6}\,\mathrm{cm^2/s}$, respectively.

The flux of Gly from a biomineral via diffusion through water filled pores will be only seven times faster in the laboratory (142.5°C) than in the field (20°C), whereas the corresponding rates of hydrolysis and racemization are 5–6 orders of magnitude greater. Activation energies for diffusion within polymer matrices are higher and increase with solute size and polymer (and cross-link) concentration (Crank and Park, 1968). Thus, during earlier stages of diagenesis the flux will be more temperature dependent than when the matrix is more degraded. However, activation energies for diffusion in semicrystalline polymers 50 kJ/mol are much lower than those of either hydrolysis or racemization.

The selective diffusive flux of racemized (free) amino acids helps to explain an otherwise remarkable disparity in apparent rates of racemization in laboratory studies. The k_{DL}^{obs} (actually k_{AI}) for the bivalve *Anadara* reported by Kimber and Griffin (1987; 1.86×10^{-6}/sec, figure 11-2), is more than one order of magnitude greater than the value reported by Goodfriend and Meyer (1991) for the gastropod *Trochoidea* (1.02×10^{-7}/sec) even after correction for the temperature difference (3.5°C) between the two experiments; such differences are much greater than any intergeneric effect (Lajoie et al., 1980). A difference in k_{DL}^{obs} as large as this can be modeled if the flux of free amino acids is significantly different (figure 11-6). Details of the experimental methods suggest that this was the

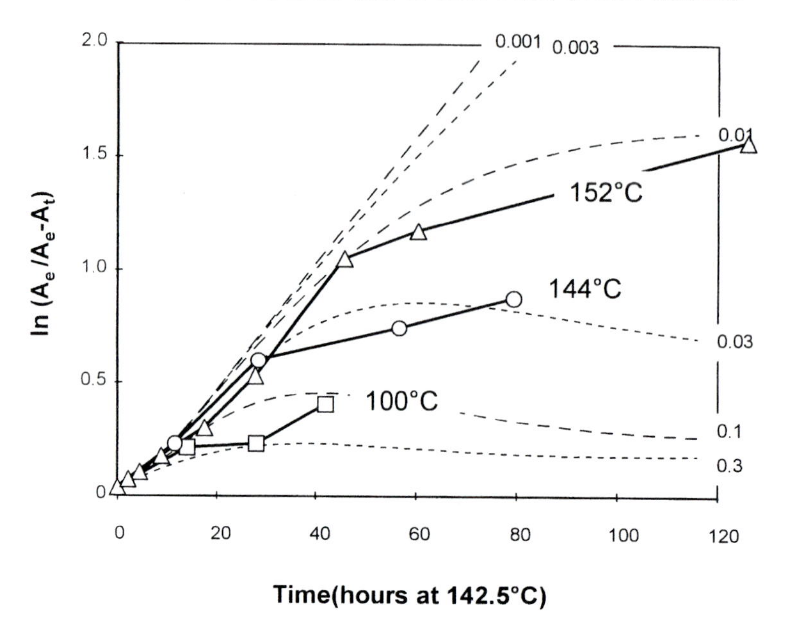

Fig. 11-6. One possible impact of loss of free amino acids on apparent rate of racemization, (data for *Hiatella arctica;* Miller and Hare, 1980). Optimized rates for leaching (times compensated to 142.5°C using a consensus E_a of 119.5 kJ/mol—derived from E_a for free Ile, (Bada, 1972; Smith and Evans, 1980) and biominerals (Mitterer, 1975; Miller and Hare 1980, Miller, 1985; Miller et al., 1992, 1997; McCoy, 1987). Note that in this model, as the rate of loss increases relative to k_{DL} (strictly AI) with decreasing temperature, both the linear phase and the initial rate of k_{DL}^{obs} decrease. The system is modeled as a three-box, one-pool system (figure 11-1A), $k = 0.05$ hr^{-1}, $k_{DL}^{I} = 0.00001$ hr^{-1}, $k_{DL}^{T} = 0.02$ hr^{-1}, $k_{DL}^{F} = 0.001$ hr^{-1}, and D ranges from 1 to 0.001.

case. Kimber and Griffin (1987) analyzed the total reaction vessel but Goodfriend and Meyer (1991) sonicated and rinsed the shell fragments prior to analysis.

Differences in diffusive flux have potentially serious implications for estimates of reaction rates in open systems. Because of the relatively small differences in D with increasing temperature, the total loss of amino acids will be greater for low-temperature experiments in which longer incubations are required to observe racemization—an effect accentuated by using pulverised or fragmented material. The effect is to reduce both the extent of the so-called "linear-phase" of the racemization reaction and k_{DL}^{obs} (due to the loss of more highly racemized amino acids), a process we term "flux dampening". At higher temperatures, the effect will be minimized as the experiments are shorter and there is a smaller overall loss of amino acids; the reaction will appear linear for a greater proportion of the experiment. The modeled impact imparted by diffusive flux upon k_{DL}^{obs} is dependent upon the interplay within the three-box model (figure 11-1), notably between the relative rates of hydrolysis and k_{DL}^{T} and the flux of the terminal/DKP box. Flux dampening will systematically lower k_{DL}^{obs} in lower-temperature experiments, particularly if $k > k_{DL}^{T}$ and/or flux is significant for the terminal/DKP fraction as well as for free amino acids. In these instances (figure 11-6) is significant, thereby overestimating the activation energy, in the case of *Hiatella* (illustrated in figure 11-6) by at least 10%. Analysis of the total experiment (i.e., residual and leached amino acids) will overcome the effect of flux dampening (e.g., Kimber and Griffin, 1987) and should lead to a more accurate estimation of activation energy. However, the true activation energy will very seriously overestimate the apparent rate of racemization in fossil shells, in which the diffusive flux will be most extreme.

This interpretation would predict that estimates of activation energy of racemization in biominerals are unreliable if the experiments have not controlled for loss. One way in which to test the hypothesis is to plot total and free amino acid concentrations (normalized to initial concentration) against total D/L (or alloisoleucine/isoleucine). In biominerals that appear to have little or no loss, there is a surprising degree of correspondence between high- and low-temperature data from very different studies, all of which plot in surprisingly tight fields (figure 11-7). In systems in which loss is significant (such as mollusk shells) it is anticipated that the loss of highly racemized free amino acids will skew (i.e., lower) the total D/L value and simultaneously lead to a loss of total concentration. The net effect will

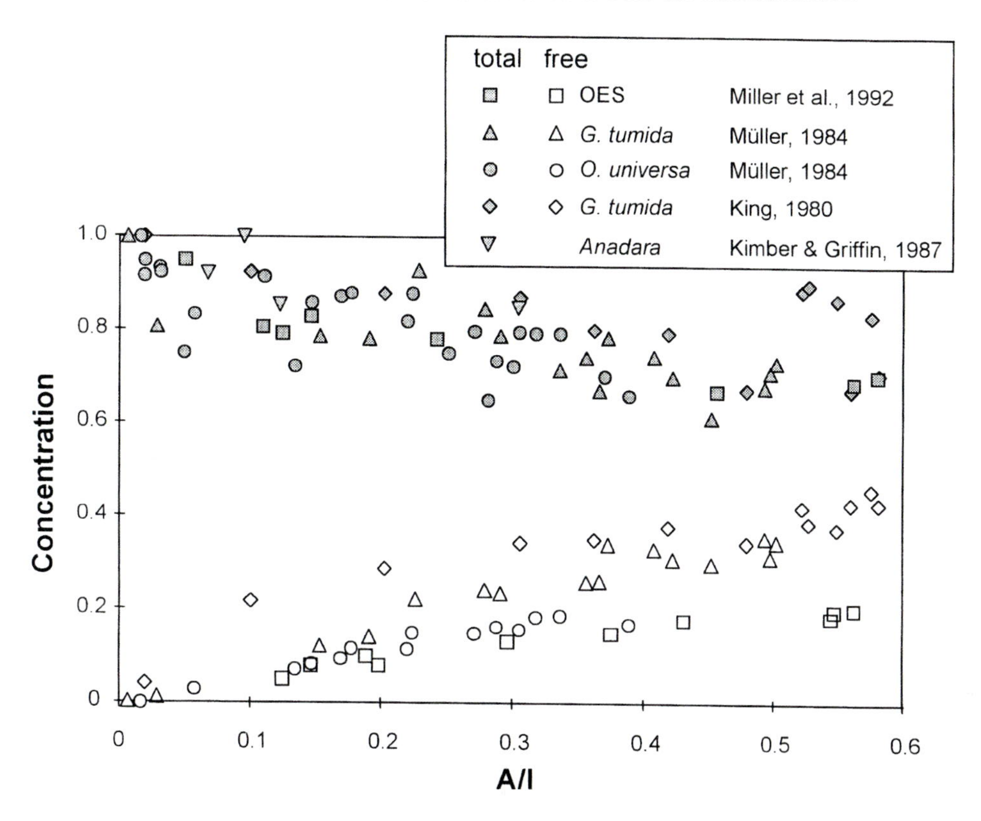

Fig. 11-7. Total and free concentration of Leu (OES at 142.5°C; Miller et al., 1992), Val (*Anadara* at 110°C; Kimber and Griffin, 1987), Aile + Ile (fossil foraminifera at 2.7°C; Müller 1984) and total AA (fossil foraminifera at 1.7°C; King, 1980) plotted against isoleucine epimerization. All concentration data (including free amino acids) are normalized to the highest total value reported. Note the narrow overlapping range of concentrations of both total and free observed in all studies. In the case of *Anadara*, total analysis of the sealed experimental tube reveals that loss must be due to diagenetic reactions. In the case of OES the lower concentration of free Ile relative to Ile may be attributed to diffusive flux of free amino acids. The patterns observed in foraminifera are similar and this is suggested as evidence that the test is behaving as a closed system (having lost the matrix by diffusive flux very early during diagenesis).

be a much steeper apparent fall in total concentration (as observed by Corrado et al., 1986; Roof, 1997; figure 11-8), and a concomitant underestimation of the racemization rate.

THE EXTENT OF LEACHING AND ITS POSSIBLE RELATIONSHIP TO ULTRASTRUCTURE

Patterns of Loss

The problem of "leaching" has been long recognized and the impact that the loss of free amino acids has on apparent racemization kinetics in fossils has been discussed exhaustively. The debate on the impact of leaching on bone dating (Hare, 1980), the nonlinear kinetics of shells (Masters and Bada, 1978; Bada, 1982), and the integrity of foraminfera (King, 1980) and OES (Brooks et al., 1990) will not be rehearsed again here. The term itself is problematic, as it implies removal by a percolating fluid. In long tortuous pore systems, diffusion will be the rate-limiting step and, as pore systems narrow, this will be replaced by molecular-scale processes.

Johnson and Miller (1997) reported that under conditions of continuous "leaching", bone lost 90% of its amino acids, mollusk shells lost 60% and OES lost only 10% (Brooks et al., 1990); these observed differences are consistent with patterns of long-term diagenesis. OES apparently retains a high proportion of its amino acids, whereas bone loses collagen rapidly following gelatinization (Hare, 1980). In mollusks, the pattern, summarized by Wehmiller (1980) is more complex: (1) a rapid decline in total amino acid abundance during early diagenesis is concomitant with a rapid rise in free amino acids; (2) in the later stages of diagenesis, a "plateau" of very gradually declining concentration (40% ± 10% of the initial total), of which free amino acids account for 70% ± 20% (e.g. figure 1 of Miller and Hare, 1980) is noted. A residual fraction

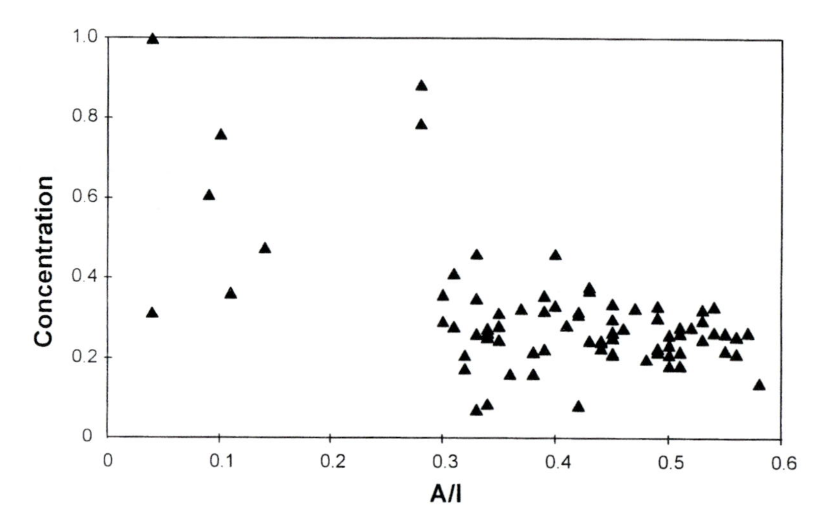

Fig. 11-8. Total concentration of Aile + Ile plotted against A/I for *Mulinia lateralis* from Charleston (S. Carolina), normalized to the highest concentration reported from beach material (data from Corrado et al., 1986). Despite that fact that this study is of a single mollusk species from a single geographic region, the scatter is much greater than that observed between all of the different studies illustrated in figure 11-7 (outlined). As predicted, the decline in total concentration is more rapid than for the apparently closed systems illustrated in Fig. 11-7.

apparently extends into older fossils, in which it is detected as organic carbon (Hudson, 1967), volatile fatty acids (Hoering, 1980), or hydrocarbons (Thompson and Creath, 1966).

Ultrastructure

The differences in the patterns of diagenesis can be related to the structure and organization of each biomineral. Bones are essentially composed of an insoluble protein (collagen) which has been overgrown with fine, minute crystals of apatite 0.02 μm long and 0.005 μm wide. Bones are 63% (by volume) hydrated crystalline collagen, which, during diagenesis, undergoes a phase change to gelatin (Nielsen-Marsh et al., 2000). Being a living tissue, bone has a highly organized "internal plumbing": canniliculi connect osteocyte lacunae to the Haversian system, so the diffusive path lengths for gelatin will be short and the diffusive flux high (Hare, 1980; Von Endt, 1980; Nielsen-Marsh and Hedges 1999). Only in cases in which there is no solution to remove the products does bone contain a high proportion of soluble decomposition products (Iacumin et al., 1996) that are highly racemized (e.g., Kimber and Hare, 1992). We have argued that the propensity for loss limits the scope for racemization analysis of the total protein (Collins et al., 1999; Ritz-Timme et al., 2000).

Eggshell is composed of larger crystallites than bone: in the palisade layer (i.e., the shell proper) they form bricks 1 μm in width and 0.3 μm high within an organic "mortar" (Arias et al., 1993). This organic matrix is 70% protein (e.g., Hinke et al., 1995) but also contains a high proportion of polysaccharide (11%; Baker and Balch, 1962), the protein having an amino acid composition similar to that of the core protein of the chondroitin sulphate proteoglycan of cartilage (Carrino et al., 1997). The organization of the palisade layer means that it is resistant to gas exchange (Panginelli, 1980), and it is therefore cross-cut by a system of pores that enhance the conductance of gases. The density and distribution of these pores vary from species to species because of packing of the calcite columns (Tullet, 1975). OES has a low conductance (Christensen et al., 1996), which reflects the demands of a large precocial embryo. The organization and microscopic scale of the "bricks and mortar" structure of the palisade layer and the presence of proteoglycans will increase tortuousity and (via Maillard condensation) reduce the pore size, thereby reducing diffusive flux. Miller et al. (Chapter 13 of this volume) argue that much of the protein in OES is intracrystalline.

Mollusk shells display a wider variety of microstructures than are observed in either bone or eggshell, and their crystallites are generally much larger. The organic matrix between crystallites comprises an insoluble framework of silk-fibroin-like protein and β-chitin, to which are sorbed acidic macromolecules (the latter controlling mineral formation; Falini et al., 1996). Some of the water-soluble proteins become entrapped within the mineral itself, which has been demonstrated in echinoderm sterome to alter the minerals mechanical properties (Berman et al., 1988). These "intracrystalline pro-

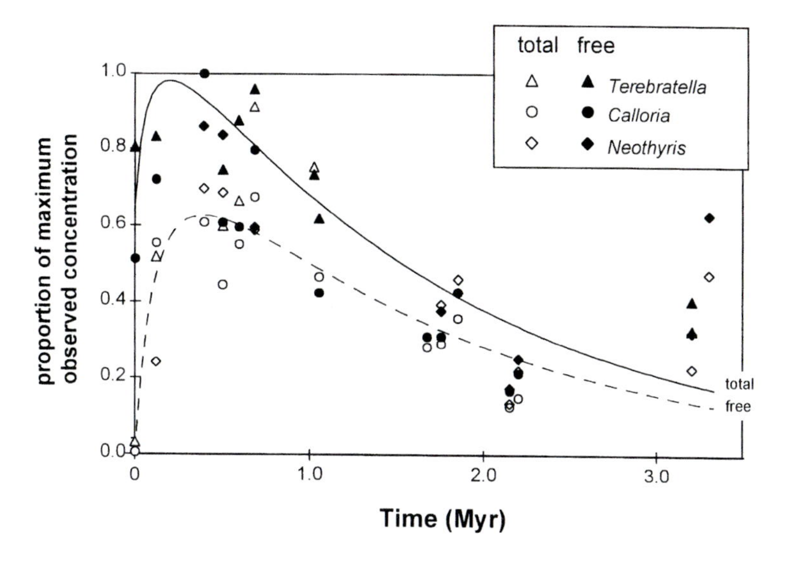

Fig. 11-9. Total and free amino acids from three genera of brachiopod shells interpreted as a closed system (e.g., figure 11-2B; data from Walton, 1998). Parameters: $k = 3 \times 10^{-13}$/sec; [water], 0.7 molecules per peptide bond (closed system), $k_{decomposition} = 2 \times 10^{-14}$/sec^{-1}, total concentration 130% of maximum soluble fraction observed in modern shells of which 50% is originally insoluble.

teins" (an operational term to describe those proteins isolated by exhaustive chemical oxidation of crystallites or powdered shell) have been reported in echinoderms (Berman et al., 1988), mollusks (Crenshaw, 1972; Sykes et al., 1995), and brachiopods (Collins et al., 1991).

In the prismatic layer of *Pinna*, loss of the organic matrix results in the formation of slit-shaped pores with a height of approximately 1 μm (Keller, 1981) (i.e., as large as the *longest* dimension of crystallites in the palisade layer of eggshell). Using bacterial enzymolysis of polished sections, Keller (1981) and Frerotte et al. (1983) were able to demonstrate organic inclusions *within* the crystallites of *Pinna* and *Crassostrea*, respectively. The (intracrystalline) organic dense layers within the prismatic layer are less than 5 nm in width (i.e., of the same order as the dimensions of proteins, BSA, 4.4 × 4.0 × 12.0 nm; Squire et al., 1968). Degradation of the matrix will generate an extensive, albeit tortuous, pore system.

The controlling factor in the apparent rate of leaching is not the increase in size of mineral elements (bone < eggshells < mollusk shells) but the dimensions and tortuosity of the pore systems within the biomineral (including the pores generated by the degrading organic matrix). Comments such as "the carbonate matrix ... approximates a closed system capable of retaining free amino acids and peptide chains" (Miller and Hare 1980), and "in most fossil shells [leaching] is generally accepted to be minimal" (Kimber and Griffin, 1987), suggest that, in the field of racemization, the importance of ultrastructure has been largely ignored.

The reason for the plateau observed in mollusks is that, unlike bone (De Niro and Weiner, 1988), a significant proportion of the proteins are intracrystalline. Walton (1998) has already demonstrated that amino acids remain entrapped within the large crystallites of the brachiopod secondary layer for over 3 Myr in temperate latitudes (figure 11-9). The pore system generated by the degrading intracrystalline fraction is of a very fine (molecular) scale and diffusive flux will be very low.

Closed Systems

Figure 11-7 plots the concentrations of total and free amino acids as a function of alloisoleucine/isoleucine (A/I) for a wide variety of "closed" biomineral systems. The similarity between all these systems is striking; free amino acid concentrations rise steadily whereas the total concentration of amino acids slowly declines. The figure appears very different from plots for mollusk shells (figure 11-8, see also figure 3 of Sykes et al., 1995, and figure 4 of Roof, 1997), wherein the total concentration falls sharply with increasing A/I (the proportion of free amino acids was not reported). A/I is much less useful as a proxy for state of degradation in mollusks as the value will be dependent upon the proportion of free amino acids retained, which can vary not only between shells but between different regions of the same shell (Brigham, 1983). The problem that this loss of free amino acids poses is well known and has led to the identification of OES as the material of choice for dating (e.g., Brooks et al., 1990).

If OES is indeed closed, then it would be the ideal material in which to exploit racemization reactions. We suspect, on the basis of eggshell ultrastructure, that high-temperature laboratory experiments (Brooks et al., 1990) will underestimate the flux, because of the low E_a of diffusion and the narrow, tortuous pore system in the brick and mortar structure of the palisade layer. If the system were not closed we would predict that overall concentrations of amino acids would decrease with increasing A/I values more in natural samples than in artificially heated samples. The similarity in plots of free and bound A/I in heated and fossil OES samples (Miller et al., Chapter 13 of this volume) provide strong evidence that the system is indeed closed. The claim for a closed system could be tested further by exhaustive bleach treatment of powdered OES (cf. Sykes et al., 1995; Walton, 1998) and also by investigating the pore-size distribution of the OES and molluscan shells using Hg intrusion porosimetry, which has already been successfully applied to bone. (Nielsen-Marsh and Hedges, 1999).

TOWARD RACEMIZATION REACTIONS OF CLOSED SYSTEMS

Closed systems such as OES not only display more tractable kinetics, but are also easier systems in which to simulate diagenesis in the laboratory. Ideally, the system will be composed of large, easily isolated elements of a diagenetically stable mineral. Possible candidates include echinoderm sterome, molluscan nacreous and prismatic layers, brachiopod secondary-layer fibers, vertebrate enamel, but not bone or dentine. The loss of free amino acids from the intracrystalline fraction will be much slower than that from the intercrystalline organic matrix (e.g., Walton 1998) and should therefore provide a more reliable pool of organic matter for racemization analysis. The isolation of this fraction is straightforward, merely requiring prolonged exposure of finely ground shell powders to NaOCl (Sykes et al., 1995; see also Stathoplos and Hare, 1993; Stern et al., 1999). In a previous study (Sykes et al., 1995), we demonstrated the usefulness of this approach in removing contamination from fossil mollusk shells. Foraminiferal tests have yielded very consistent patterns of diagenesis (King and Neville, 1977; King, 1980; Müller, 1984). The relationships between total A/I and the concentrations of free and total amino acids are similar to those in other "closed" systems and may be the result of the inadvertent analysis of an intracrystalline fraction. The small size of the test and the slow rate of epimerization in cold deep-sea sediments leads to the rapid loss of the intercrystalline matrix before extensive epimerization has occurred.

Characterization of the Intracrystalline Fraction

A further refinement of racemization dating would be to establish the primary structure around residues of interest, as the sequence will influence rates of hydrolysis (Hill, 1965), DKP formation, and concomitant racemization (e.g., Kriausakul and Mitterer 1978; Smith and de Sol, 1980; Smith and Baum, 1987; Gaines and Bada, 1988; Sepetov et al., 1991). One prevailing feature of biominerals other than bone has been a general lack of sequence information revealed using conventional sequencing techniques (e.g., Robbins and Brew, 1990; Cusack et al., 1992). However, molecular biochemical approaches are beginning to yield long sequences of considerable interest and importance to those investigating protein diagenesis (e.g., Shen et al., 1997; Sudo et al., 1997; Sarashina and Endo, 1998).

Sudo et al. (1997) have sequenced two matrix proteins from the mollusk *Pinctada*, both of which contain a sequence found in the nacreous layer. The larger of the two proteins (60 kDa) has 11 poly(Ala) blocks and is very similar in bulk composition not only to the insoluble fraction (ISF) from the nacreous layer of *Pinctada* (and other shells that are predominately nacreous, e.g., *Nautilus*, figure 11-10A), but also to silk proteins from spider dragline (Guerette et al., 1996). The polyalanine blocks are believed to form crystals of densely packed antiparallel β-sheet. The smaller Gly-rich protein has a composition similar to the ISF of the prismatic layer of *Pinctada*. These silk-like proteins combine with β-chitin to form the framework of the organic matrix (Levi et al., 1998).

The water-soluble fractions of many biominerals are rich in Asx and Ser (e.g., the mollusk *Patinopecten*, Akiyama, 1980; figure 11-10B); but a recent partial sequence from the foliated shell of the mollusk *Patinopecten* (MSP-1; Sarashina and Endo, 1998), though rich in Ser and Asp does not have the anticipated $(\text{Asp-Gly})_n$ motifs. Despite the absence in MCP-1 of the $(\text{Asp–Ser})_n$ motif seen in phosphophoryns (e.g., Ritchie and Wang, 1996), there appears to be surprising *functional* homology between molluscan and bone biomineral proteins (Atlan et al., 1997).

The most striking aspects of these data are (1) the similarity between the sequences and bulk composition (figure 11-10B), suggesting that these proteins (or members of the same gene family) represent the bulk of the biomineral content, and (2) the concomitant paucity of Ile sequence combinations (if the organic matrix comprises only a few proteins). However, a mucein-like protein found in the nacreous layer of *Pinna* does contain higher levels of Ile (F. Marin, personal communication) and not all are found in tandem repeat regions. In the case of *Pinctada*, the silk-like protein from the nacre contains only four Ile residues (GIR, DIK, QID, and PIR), whereas there is none in the full sequence of MSP-1 of *Patinopecten* (I. Sarashina

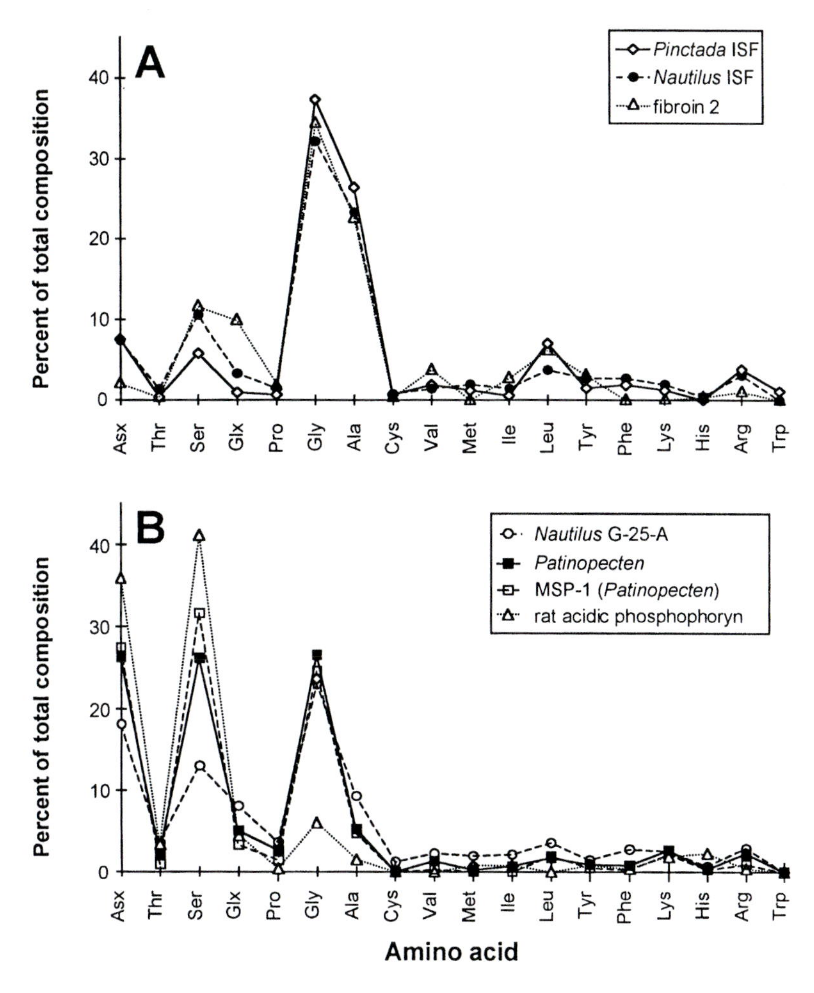

Fig. 11-10. Comparison of amino acid compositions from selected biominerals and recent sequences. The insoluble fraction of *Nautilus margaritacea* is compared with fibroin 2 from spider dragline (Guerette et al., 1996) and the Ala- and Gly-rich protein present in *Pinctada* nacre (Sudo et al., 1997). The soluble fraction isolated from *Nautilus* (Weiner and Lowenstam, 1980) and *Patinopecten* (Akiyama, 1980) is compared with the partial sequence MSP-1 from *Patinopecten* (Sarashina and Endo, 1998) and rat phosphophoryn (Ritchie and Wang, 1996).

and K. Endo, personal communication). Modeling of Ile-epimerization kinetics in molluscan shells may prove simpler than for abundant amino acids such as Asx in bone (e.g., Collins et al., 1999; Waite and Collins, 2000).

Three Rates or Two Pools?

We have suggested above that the pattern of release of free amino acids in OES (and also in foraminifera) can be accounted for by 3 pools with rates of hydrolysis differing by three orders of magnitude. However, from the perspective of molluscan skeletal ultrastructure, there appear to be two groups of proteins—the silk-like insoluble matrix and the (predominantly acidic) soluble proteins. There are two sources of proteins within biominerals, namely the inter- and intracrystalline fractions. On the basis of limited evidence (e.g., Crenshaw, 1972) the soluble protein dominates the intracrystalline matrix, the silk-like insoluble protein the intercrystalline matrix.

An alternative interpretation of the hydrolysis data would be that there are two pools rather than three rates of hydrolysis: (1) an open pool (i.e., the intercrystalline matrix) bathed in an excess of water and decaying rapidly with pseudo-first-order kinetics; and (2) a closed (intracrystalline) pool with restricted water displaying second-order kinetics and coming to a halt if there is insufficient water (cf. Hare, 1971; Bada et al., 1994; Nardi et al., 1994). The concentration of water in biomineral carbonates (Gaffey, 1988) is sufficiently high to decompose all the peptide bonds of the organic matrix, but much of the water may be in locations such as fluid inclusions or bonded to mineral surfaces, and thus not available for hydrolysis.

Evidence for this two-pool model is sparse and mostly obtained by inference, which is outlined below.

1. Initial rates of hydrolysis are consistent with experimental data (e.g., Kahne and Still, 1988), implying that subsequent rates are unexpectedly slow, but would be consistent with a closed system with restricted water.[7]
2. The presence in long-term data sets (Miller at al., 1992; Walton, 1998) of a residual bound fraction suggests either that there are some unusually stable peptide bonds or that there is a restrictive pool of water. Selective loss of free amino acids could lead to an apparent increase in the bound fraction, as reported for Glx (Walton, 1998).[8] Nevertheless, this cannot account for all the bound amino acids in the intracrystalline fraction of brachiopod shells after 3 Myr (Walton, 1998).
3. The presence of a plateau concentration of residual amino acids would not be predicted by the chain scission model with three rate constants of hydrolysis (see above), as peptide-bound residues would eventually diffuse out of the pore system, albeit more slowly than free amino acids (figure 11-5).
4. Volatile fatty acids (Hoering, 1980) and light hydrocarbons (Thompson and Creath, 1966) in Mesozoic and Palaeozoic fossils have an abundance and distribution consistent with amino acids of the intracrystalline fraction; in an open system these compounds would not be retained.

In terms of racemization, a two-component three-box model of diagenesis (figure 11-1B) is simpler than a modified three-box model with multiple rates of hydrolysis; it is more readily interpreted in terms of ultrastructure (i.e., an intra- and an intercrystalline fraction) and more capable of simulating the observed patterns of loss; in most other ways, however, the outputs from both models are the same.

Racemization versus Hydrolysis

We suspect that errors introduced into the determination of k and k_{DL}^{obs} in biominerals have had the effect of underestimating the temperature dependence of the former and overestimating that of the latter. We are surprised, in the case of Ile epimerization, that relative rates of k and k_{DL}^{obs} between different biominerals are so similar, particularly when any observed differences can be accounted for in part by the higher apparent rates observed in systems in which the loss of amino acids is limited (i.e., OES, foraminifera). It is tempting to speculate that the reason for the similarity in the E_a of racemization and of hydrolysis is due to the fact that k_{DL}^{obs} is a function of k (i.e., the original three-box model). If racemization rates are significantly faster in the second compartment of the three-box model (i.e., *N*-terminal and or DKP), then generation of reactive positions (hydrolysis, aminolysis) has the potential to be the rate-determining step. In this interpretation (which needs to be explored experimentally), the failure of biominerals to conform to the three-box model is due to the presence of both open and closed systems within the same material. The pattern of observed racemization and release of free amino acids is the result of an organic matrix displaying a pseudo-first-order rate and an intracrystalline fraction approximating to a second-order rate law. The complication introduced by the variable loss of the open pool will preclude meaningful interpretation unless the loss is carefully monitored, but it could be avoided by selecting the intracrystalline fraction. Racemization kinetics are complex enough without having to interpret results obtained from leaky reaction vessels!

CONCLUSIONS

Racemization is more predictable in closed systems, as the reaction vessels are not leaking away the products of diagenesis. It remains unclear as to whether OES is a closed system or one which loses amino acids slowly, although the ultrastructure would suggest the latter. Most of the protein in bivalve mollusk shells is associated with the organic matrix, which will degrade and diffuse out of the shell along the boundaries between the crystallites left by the decomposing organic matrix (cf. Collins, 1986). The prediction of racemization in such a leaky system is difficult as any loss will deplete Aile more than Ile (as observed by Roof, 1997). It is possible to argue that intergeneric variation in racemization rate (which must exist on theoretical grounds because of differences in primary and secondary structure) is in part due to differences in flux.

In a closed system, all products will be trapped, the kinetics will be nonlinear because of the retention of all the free amino acids, but the extent of deviation from linearity will be predictable. From a modeling standpoint, the advantages of a closed system are clear—the kinetics will ultimately be tractable. In an open system, the loss of amino acids with unpredictable degrees of racemization limits the scope for predicting the kinetics. Furthermore, if the system were genuinely closed, the decomposition products would also be trapped and their concentration and distribution could be used as a form of internal validation of dated material. Miller and Hare (1980) have already suggested that the high apparent activation energy for the decomposition of Thr (and also Ser: Bada, 1971; Bada et al., 1978; Belluomini, 1981) could be used to provide simultaneous equations with unique solutions for time and temperature or to identify anomalies in the thermal history (e.g., cooking events).

Despite all of the potential limitations highlighted above perhaps the most remarkable conclusion (as highlighted by other chapters in this volume) is how robust current methods are, even in the case of mollusk shells. This may be because (1) following hydrolysis, free amino acids rapidly diffuse out of the shells and therefore the observed D/L ratios are not complicated by extreme variations in this pool and rates of diffusion are so rapid that the concentration of free amino acids retained within the shell is low, or (2) that the intercrystalline fraction decomposes so rapidly that (in most cases) only the intracrystalline faction is analyzed (as appears to be the case in *Cepaea* sp; figure 2 of Sykes et al., 1995). Successes with OES (e.g., Miller et al., 1997) demonstrate the advantages of analyzing a closed system, advantages that could be extended to mollusks and other biominerals if an intracrystalline fraction could be isolated for AAR analysis.

Acknowledgments: Many people have helped to compile and refine the information for this Chapter. In particular, we would like to thank Frederic Marin and Kashioshi Endo (molluscan shell proteins), Max Hinke (eggshell structure and composition), Glenn Goodfriend, Giff Miller and Darrel Maddy (racemization), not forgetting the wonderful support of those at the Newcastle Research Group, notably Adri van Duin (racemization of DKP and free amino acids), Jason Dale and Steve Larter (high-temperature diffusion modeling), and Christina Nilesen-Marsh and Andy Aplin (the intricacies of porosimetry and tortuous flow). Any errors or omissions are entirely the fault of the authors. This work was funded by NERC, through the Ancient Biomolecules Initiative (NERC GST/02/1017), the Royal Society, and EU ENV4-CT98-0712 awards to M.J.C.

NOTES

1. Strictly speaking for diastereomic amino acids (Ile, Thr, Hyp) racemization is more properly termed epimerization, but the term is used here to encompass both reactions.

2. Here protein diagenesis is broadly defined to encompass all of the chemical transformations that ultimately yield to the formation of light hydrocarbons, CO_2 and NH_3.

3. The offset of Pro observed by Lajoie et al. (1980) in three of four mollusk species is of interest. Smith and de Sol (1980) observed high rates of racemization in Pro-containing dipeptides. The Pro–Gly DKP is commonly reported following hydrolysis, and Pro-hosted DKPs are common products of protein pyrolysis (Stankevitzch et al., 1996).

4. A more complex four-box model variant distinguishes between fast-racemizing terminal residues and DKP (Mitterer, 1993).

5. One implication of a residual pool with a very slow hydrolysis/epimerization rate is that the apparent equilibrium constant will be lower than the true value; indeed, if decomposition or flux of the free amino acids occurs the ratio will attain a maximum value and will then appear to decline slowly (e.g., Miller et al., 1992). In preliminary modeling experiments using the data of Miller et al. (1992), if it is assumed the rate of racemization in the large residual bound fraction is only 1% of the rate for free amino acids, then the optimized Aile/Ile equilibrium ratio is 1.4, not the observed value of 1.28. Equilibrium ratios of 1.4 are reported in simple or open systems (e.g., Miocene sediments; King and Hare, 1972: alkaline (but not aqueous) solution; Hare and Mitterer, 1967: and aqueous solutions, Smith et al., 1978). In biominerals which may have this residual bound fraction, equilibrium constants are generally reported to be lower (Miocene foraminifera: King and Hare, 1972; *Mercenaria*: Hare, 1969; bone: Bada, 1972; Dungworth et al., 1974; ostrich eggshell: Miller et al., 1992), as it is in strong mineral acid (Nakaparksin et al., 1970). With extended diagenesis the equilibrium constant of Ile epimerization is complicated by degradation of Ile and the possibility of isomerization at the β-carbon (Bade et al., 1986).

6. Miller et al. (1992) provide a polynomial fit to the rate of loss of amino acids, which subsequently predicts an *increase* in concentration (albeit after 20 years at 142.5°C). Similarly, the best-fit logarithmic decay function [(total conc) $= -0.114\ln(t) + 0.165$, where t is time in years, $R^2 = 0.953$] predicts negative concentration after less than 5 years.

7. It is probable that the watersoluble (Asx- and Ser-rich) intracrystalline fraction is more prone to hydrolysis than in the insoluble (Ala and Gly-rich) silk-fibroin-like organic matrix which forms the major component of the intercrystalline system.

8. The concentration of free Glx (i.e., Glu + Gln) is anticipated to be low because free Glu forms pyroglutamic acid (Wilson and Cannan, 1937), a fraction which would contribute racemic Glu to the total hydrolysate. Lactam formation is also predicted to produce an "inversion" in free Glx (e.g., Kimber and Griffin, 1987, figure 2, *Anadara trapezia*). Modeling of the inversion suggests a strikingly low pool concentration of free Glx. Although the concentration of free Glx was not reported by Kimber and Griffin (1987), lactam formation would provide an explanation for the absence of reported D/L ratios for free Glx in the later stages of the experiment.

REFERENCES

Akiyama M. (1980) Diagenetic decomposition of peptide linked serine residues in the fossil scallop shells. In *Biogeochemistry of Amino Acids*. (eds P. E. Hare, T. C. Hoering, and K. King, Jr), pp. 115–120. Wiley.

Amsden B. (1998) Solute diffusion within hydrogels. Mechanisms and models. *Macromolecules* **31**, 8382–8395.

Arias J. L., Fink D. J., Xiao S. Q., Heue A. H., and Caplan A. I. (1993) Biomineralization and eggshells—cell-mediated acellular compartments of mineralized extracellular matrix. *Intl. Rev. Cytol.*, **145**, 217–250.

Atlan G., Balmain N., Berland S., Vidal B., and Lopez E. (1997) Reconstruction of human maxillary defects with nacre powder: histological evidence for bone regeneration. *C. R. Acad. Sci. Paris III* **320**, 253–258.

Bada J. L. (1971) Kinetics of the non-biological decomposition and racemization of amino acids in natural waters. *Adv. Chem. Ser.* **106**, 309–331.

Bada J. L. (1972) The dating of fossil bones using the racemization of isoleucine. *Earth Planet. Sci. Lett.* **15**, 223–228.

Bada J. L. (1982) The racemization of amino acids in nature. *Interdis. Sci. Rev.* **7**, 30–460.

Bada J. L. (1991) Amino-acid cosmogeochemistry. *Phil. Trans. Roy. Soc. Lond. Ser. B.* **333**, 349–358.

Bada J. L. (1998) Biogeochemistry of organic nitrogen compounds. In *Nitrogen-containing Macromolecules in the Bio- and Geosphere. ACS Symp. series 707* (eds B. A. Stankiewicz and P. F. van Bergen), pp. 64–73.

Bada J. L. and Shou M-Y. (1980) Kinetics and mechanism of amino acid racemization in aqueous solution and in bones. In *Biogeochemistry of Amino Acids.* (eds P. E. Hare, T. C. Hoering, and K. King, Jr), pp. 235–255. Wiley.

Bada J. L., Shou M-Y., Man E. H., and Schroeder R. A. (1978) Decomposition of hydroxy amino acids in foraminiferal tests: kinetics, mechanism and geochronological implications. *Earth Plan. Sci. Lett.* **41**, 67–76.

Bada J. L., Zhao M., Steinberg S., and Ruth E. (1986) Isoleucine stereosomers on the Earth. *Nature* **319**, 314–316.

Bada J. L., Wang X. S., Poinar H. N., Pääbo S., and Poinar G. O. (1994) Amino acid racemization in amber-entombed insects—implications for DNA preservation. *Geochim. Cosmochim. Acta* **58**, 3131–3135.

Bada J. L., Wang X. S., and Hamilton H. (1999) Preservation of key biomolecules in the fossil record: current knowledge and future challenges. *Phil. Trans. Roy. Soc. Ser. B.* **354**, 77–87.

Baker J. R. and Balch D. A. (1962) A study of the organic material of hen's-egg shell. *Biochem. J.* **82**, 352–361.

Belluomini G. (1981) Direct aspartic acid racemization dating of human bones from archaeological sites of central southern Italy. *Archaeometry* **23**, 125–137.

Berman A., Addadi L., and Weiner S. (1988) Interactions of sea urchin skeleton macromolecules with growing calcite crystals: a study of intracrystalline proteins. *Nature* **331**, 546–548.

Berner R. A. (1979) A rate model for organic matter decomposition during bacterial sulphate reduction in marine sediments. In *Biogéochemie de la Matiére Organique l'Interface Eau–sédiment Marin,* pp. 36–42. CNRS Paris.

Boudreau B. P. (1997) *Diagenetic Models and their Implementation.* Springer-Verlag.

Brigham J. K. (1983) Intrashell variations in amino acid concentrations and isoleucine epimerization ratios in fossil *Hiatella arctica. Geology* **11**, 509–513.

Brooks A. S., Hare P. E., Kokis J. E., Miller G. H., Ernst R. D., and Wendorf F. (1990) Dating Pleistocene archaeological sites by protein diagenesis in ostrich eggshell. *Science* **248**, 379–385.

Capasso S., Vergara A., and Mazzarella L. (1998) Mechanism of 2,5-dioxopiperazine formation. *J. Amer. Chem. Soc.* **120**, 1990–1995.

Carrino D. A., Rodriguez J. P., and Caplan A. I. (1997) Dermatan sulfate proteoglycans from the mineralized matrix of the avian eggshell. *Connective Tissue Res.* **36**, 175–193.

Carslaw, H. S. and Jaeger, J. C. (1959) *Conduction of Heat in Solids.* 2nd edn Oxford University Press.

Christensen V. L., Davis G. S., and Lucore L. A. (1996) Eggshell conductance and other functional qualities of ostrich eggs. *Poultry Sci.* **75**, 1404–1410.

Collins M. J. (1986) Post mortality strength loss in shells of the Recent articulate brachiopod *Terebratulina retusa* (L.) from the west coast of Scotland. *Biostrat. Palaeozoic* **4**, 209–219.

Collins M. J., Muyzer G., Curry G. B., Sandberg P. A., and Westbroek, P. (1991) Macromolecules in brachiopod shells: characterisation and diagenesis. *Lethaia* **24**, 387–397.

Collins M. J., Westbroek P., Muyzer G., and De Leeuw J. W. (1992) Experimental evidence for condensation reactions between sugars and proteins in carbonate skeletons. *Geochim. Cosmochim. Acta* **56**, 1539–1544.

Collins M. J., Riley M., Child A. M., and Turner-Walker G. (1995) A basic mathematical simulation of the chemical degradation of ancient collagen. *J. Archaeol. Sci.* **22**, 175–183.

Collins M. J., Waite E. R., and van Duin A. T. C.(1999) Predicting protein decomposition; the case of aspartic acid racemization kinetics. *Phil. Trans. Roy. Soc. Ser. B* **354**, 51–64.

Collins M. J., Walton D., and King A. (1998a) The geochemical fate of proteins. In *Nitrogen-containing Macromolecules in the Bio- and Geosphere. ACS Symp. Series 707,* (eds B. A. Stankiewicz and P. F. van Bergen), pp. 74–87.

Collins M. J., Child A. M., van Duin A. T. C., and Vermeer C. (1998b) Ancient osteocalcin; the most stable bone protein? *Ancient Biomolecules* **2**, 223–238.

Conway D. and Libby, W. F. (1958) The measurement of very slow reaction rates; decarboxylation of alanine. *J. Amer. Chem. Soc.* **80**, 1077–10840.

Corrado J. C., Weems R. E., Hare P. E., and Bambach R. K. (1986) Capabilities and limitations of applied aminostratigraphy, as illustrated by analyses of *Mulinia lateralis* from the late Cenozoic marine beds near Charleston, South Carolina. *South Carolina Geology* **30**, 19–46.

Crank J. and Park G. S. (1968) *Diffusion in Polymers*. Academic Press.

Crenshaw M. A. (1972) The soluble matrix from *Mercenaria mercenaria* shell. *Biomineralisation* **6**, 6–11.

Csapó J., Csapó-Kiss Z., Kolto L., and Papp I. (1990) Age determination based on amino acid racemization: a new possibility. *Archaeometry* **90**, 627–635.

Cusack M., Curry G. B., Clegg H., and Abbott G. D. (1992) An intracrystalline chromoprotein from red brachiopod shells: implications for the process of biomineralization. *Comp. Biochem. Physiol.* **102B**, 93–95.

DeNiro M. J. and Weiner S. (1988) Organic matter within crystalline aggregates of hydroxyapatite—a new substrate for stable isotopic and possibly other biogeochemical analyses of bone. *Geochim. Cosmochim. Acta* **52**, 2415–2423.

van Duin A. C. T. and Collins, M . J. (1998) The effects of conformational constraints on aspartic acid racemization. *Org. Geochem.* **29**, 1227–1232.

Dungworth G., Vinken N. J., and Schwartz A. W. (1974) Racemization of aliphatic amino acids in fossil collagens of Pleistocene, Pliocene and Miocene ages. In *Advances in Organic Geochemistry* (eds B. Tissot and F. Bienner), pp. 689–700. Editions Technip Pubs.

Falini G., Albeck S., Weiner S., and Addadi L. (1996) Control of aragonite or calcite polymorphism by mollusk shell macromolecules. *Science* **271**, 67–69.

Frerotte B., Raguideau A., and Cuif J. P. (1983) Action of bacterial cultures on the shell of *Crassostrea gigas* (Thunberg). Ultrastructural data regarding the mineral components. *C. R. Acad. Sci. Paris II* **297**, 383–388.

Fridkin M., Wilcheck M., and Sheinblatt M. (1970) NMR studies of H-D exchange of α-CH group of amino residues in peptides. *Biochem. Biophys. Res. Commun.* **38**, 458–463.

Fujii, N., Satoh K., Harada K., and Ishibashi Y. (1994) Simultaneous stereoinversion and isomerization at specific aspartic acid residues in alpha-a-crystallin from human lens. *J. Biochem.* **116**, 663–669.

Gaffey S. J. (1988) Water in skeletal carbonates. *J. Sed. Petrol.* **58**, 397–414.

Gaines S. M. and Bada J. L. (1988) Aspartame decomposition and epimerization in the diketopiperazine and dipeptide products as a function of pH and temperature. *J. Org. Chem.* **53**, 2757–2764.

Geiger T. and Clarke, S. (1987) Deamidation, isomerization, and racemization at asparaginyl and aspartyl residues in peptides. Succinimide-linked reactions that contribute to protein degradation. *J. Biol. Chem.* **262**, 785–794.

Gillard R. D., Hardman S. M., Pollard A. M., Sutton P. A., and Whittaker D. K. (1991) Determinations of age at death in archaeological populations using the D/L ratio of aspartic acid in dental collagen. *Archaeometry* **90**, 637–644.

Goodfriend G. A. and Meyer V. R. (1991) A comparative study of the kinetics of amino-acid racemization epimerization in fossil and modern mollusk shells. *Geochim. Cosmochim. Acta* **55**, 3355–3367.

Goodfriend G. A., Hare P. E., and Druffel E. R. M. (1992) Aspartic acid racemization and protein diagenesis in corals over the last 350 years. *Geochim. Cosmochim. Acta* **56**, 3847–3850.

Guerette P. A., Ginzinger D. G., Weber B. H., and Gosline J. M. (1996) Silk properties determined by gland-specific expression of a spider fibroin gene family. *Science* **272**, 112–115.

Hammel E. F, Jr and Glasstone S. (1954) Physicochemical studies of the simpler polypeptides. III. The acid- and base-catalyzed hydrolysis of di-, tri-, tetra-, penta- and hexaglycine. *J. Amer. Chem. Soc.* **76**, 3741–3745.

Hare P. E. (1969) Geochemistry of proteins, peptides and amino acids. In *Organic geochemistry: Methods and Results* (eds G. Eglinton and M. T. J. Murphy), pp. 438–463. Springer-Verlag.

Hare P. E. (1971) Effect of hydrolysis on racemization rate of amino acids, *Carnegie Instn Wash. Yrbk* **70**, 801–806.

Hare P. E. (1976) Relative reaction rates and activation energies for some amino acid reactions, *Carnegie Instn Wash. Yrbk* **70**, 256–258.

Hare P. E. (1980) Organic geochemistry of bone and its relation to the survival of bone in the natural environment. In. *Fossils in the Making* (eds A. K. Behrensmeyer and A. P. Hill), pp. 208–219. University of Chicago.

Hare P. E. and Mitterer R. M. (1967) Non protein amino acids in fossil shells. *Carnegie Instn Wash. Yrbk* **65**, 362–364.

Hill R. L (1965) Hydrolysis of proteins. *Adv. Prot. Chem.* **20**, 37–107.

Hincke M. T., Tsang C. P. W., Courtney M., Hill V., and Narbaitz R. (1995) Purification and immunochemistry of a soluble matrix protein of the chicken eggshell (ovocleidin-17). *Calcif. Tiss. Intl.* **56**, 578–583.

Hoering T. C. (1980) The organic constituent of fossil mollusc shells. In *Biogeochemistry of*

Amino Acids. (eds P. E. Hare, T. C. Hoering, and K. King, Jr.), pp. 193–201. Wiley.

Hudson J. D. (1967) The elemental composition of the organic fraction, and the water content of some recent and fossil mollusc shells. *Geochim. Cosmochim. Acta* **31**, 2361–2378.

Iacumin P., Bocherens H., Mariotti A., and Longinelli A. (1996) An isotopic palaeoenvironmental study of human skeletal remains from the Nile Valley. *Palaeogeogr. Palaeoclimatol. Palaeoecol.* **126**, 15–30.

Johnson B. J. and Miller G. H. (1997) Archaeological applications of amino acid racemization. *Archaeometry* **39**, 265–287.

Julg A. (1984) The problem of enantiomers: support for a new interpretation of quantum mechanics. *Croatia Chemica Acta* **57**, 1497–1507.

Kahne D. and Still W. C. (1988) Hydrolysis of a peptide bond in neutral water. *J. Amer. Chem. Soc.* **110**, 7529–7534.

Kaufman D. S. and Miller G. H. (1995) Isoleucine epimerization and amino-acid composition in molecular-weight separations of Pleistocene *Genyornis* eggshell. *Geochim. Cosmochim. Acta* **59**, 2757–2765.

Keller J. P. (1981) Le degagement du materiel mineral des tests d'invertebres (Bivalves) par proteolyse enzymatique de la trame organique. *Géobios* **14**, 269–273.

Kimber R. W. L. and Griffin C. V. (1987) Further evidence of the complexity of the racemization process in fossil shells with implications from amino acid racemization dating. *Geochim. Cosmochim. Acta* **51**, 839–846.

Kimber R. W. L. and Hare P. E. (1992) Wide range of racemization of amino-acids in peptides from human fossil bone and its implications for amino acid racemization dating. *Geochim. Cosmochim. Acta* **56**, 739–743.

King K., Jr (1980) Applications of amino acid biogeochemistry for marine sediments. In *Biogeochemistry of Amino Acids.* (eds P. E. Hare, T. C. Hoering, and K. King, Jr.), pp. 377–391. Wiley.

King K., Jr. and Hare P. E. (1972) Amino acid composition of planktonic foraminitera: a palaeobiochemical approach to evolution. *Science* **175**, 1461–1463.

King K., Jr and Neville C. (1977) Isoleucine epimerization for dating marine sediments: Importance of analyzing monospecific foraminiferal samples. *Science* **195**, 1333–1335.

Kriausakul N. and Mitterer R. M. (1978) Isoluecine epimerization in peptides and proteins: Kinetic factors and application to fossil proteins. *Science* **201**, 1101–1104.

Kriausakul N. and Mitterer R. M. (1980a) Comparison of isoleucine epimerization of isoleucine in a model peptide and fossil protein. *Geochim. Cosmochim. Acta* **44**, 753–758.

Kriausakul N. and Mitterer R. M. (1980b) Some factors affecting the epimerisation of isoleucine in peptides and proteins. In *Biogeochemistry of Amino Acids.* (eds P. E. Hare, T. C. Hoering, and K. King, Jr.), pp. 283–296. Wiley.

Lajoie K. R., Wehmiller J. F., and Kennedy G. L. (1980) Inter- and intrageneric trends in apparent racemization kinetics of amino acids in Quaternary mollusks. In *Biogeochemistry of Amino Acids.* (eds P. E. Hare, T. C. Hoering and K. King, Jr.), pp. 305–340. Wiley.

Levi Y., Albeck S., Brack A., Weiner S., and Addadi L. (1998) Control over aragonite crystal nucleation and growth: an *in vitro* study of biomineralization. *Chemistry—A European Journal* **4**, 389–396.

Liardon R. and Ledermann S. (1986) Racemization kinetics of free and protein-bound amino acids under moderate alkaline treatment. *J. Agric. Food. Chem.* **34**, 557–565.

Longsworth L. G. (1952) Diffusion measurements at 1°, of amino acids, peptides and sugars. *J. Amer. Chem. Soc.* **58**, 770–773.

Longsworth L. G. (1954) Temperature dependence of diffusion in aqueous solutions. *J. Phys. Chem.* **58**, 770–773.

Lyklema J. (1986) How polymers adsorb and affect colloid stability stability, flocculation, sedimentation and consolidation. In *Flocculation, Sedimentation and Consolidation*: Proceedings of the Engineering Foundation Conference, Sea Island Georgia, (eds B. M. Moudgil and P. Somasundaran), pp. 3–21. United Engineering Trustees Inc.

McCoy W. D. (1987) The precision of amino acid geochronology and palaeothermometry. *Quat. Sci. Rev.* **6**, 43–54.

Masters P. M. and Bada J. L. (1978) Racemization of isoleucine in fossil molluscs from Indian middens and interglacial terraces in Southern California. *Earth Plan Sci. Lett.* **37**, 173–183.

Miller G. H. (1985) Aminostratigraphy of Baffin Island shell bearing deposits. In *Quaternary Environments, Eastern Canadian Artic, Baffin Bay and Western Greenland* (ed. J. T. Andrews), pp. 394–427B. Allen and Unwin.

Miller G. H. and Hare P. E. (1980) Amino acid geochronology: integrity of the carbonate matrix and potential of molluscan fossils. In *Biogeochemistry of Amino Acids.* (eds P. E. Hare, T. C. Hoering, and K. King, Jr.), pp. 415–443. Wiley.

Miller G. H., Beaumont P. B., Jull A. J. T., and Johnson B. J. (1992) Pleistocene geochronology and palaeothermometry from protein diagenesis in ostrich eggshells: implications for the evolution of modern humans. *Phil. Trans. Roy. Soc. Lond. Series B—Biol. Sci.* **337**,149–157.

Miller G. H., Magee J. W., and Jull A. J.T. (1997) Low-latitude glacial cooling in the Southern

Hemisphere from amino-acid racemization in emu eggshells. *Nature* **385**, 241–244.

Miller G. H., Hart C. P., and Roark E. B (2000). Isoleucine epimerization in eggshells of the flightless Australian birds *Genyornis* and *Dromaius*. In *Perspectives in Amino Acid and Protein Geochemistry* (eds G. A. Goodfriend, M. J. Collins, M. L. Fogel, S. A. Macko, and J. F. Wehmiller), pp. 16–18. Oxford University Press.

Mitterer R. M. (1975) Ages and diagenetic temperatures of Pleistocene deposits of Florida based on isoleucine epimerization in *Mercenaria*. *Earth Plan Sci. Lett.* **28**, 275–282.

Mitterer R. M. (1993) The diagenesis of proteins and amino acids in fossil shells. In *Organic Geochemistry* (eds M. H. Engel and S. A. Macko), pp. 739–752. Plenum.

Mitterer R. M. and Kriausakul N. (1984) Comparison of the rates and degrees of isoleucine epimerization in dipeptides and tripeptides. *Org. Geochem.* **7**, 91–98.

Mitterer R. M. and Kriausakul N. (1989) Calculation of amino acid racemization ages based on apparent parabolic kinetics. *Quat. Sci. Rev.* **8**, 353–357.

Moir M. E. and Crawford R. J. (1988) Model studies of competing hydrolysis and epimerization of some tetrapeptides of interest in amino acid racemization studies in geochronology. *Can. J. Chem.* **66**, 2903–2913.

Müller P. J. (1984) Isoleucine epimerization in Quaternary planktonic foraminifera: effects of diagenetic hydrolysis and leaching and Atlantic–Pacific intercore correlations, *"Meteor" Forschungsergerebnisse*, Reihe C. **38**, 25–47.

Müller H. T. and Heidemann E. (1993) Untersuchung der Gesetzmässigkeiten für den Säureabbau von Hautkollagen und Identifizierung der Kollagenspaltsstellen beim sauren Gelatineprozess. *Das Leder* **44**, 69–79.

Murray-Wallace C. V. and Kimber R. W. L. (1993) Further evidence for apparent parabolic racemization kinetics in Quaternary mollusks. *Australian J. Earth Sci.* **40**, 313–317.

Nakaparksin S., Gil-Av E., and Oró J. (1970) Study of the racemization of some neutral α-amino acids in acid solution using gas chromatographic techniques. *Anal. Biochem.* **33**, 374–382.

Nardi S., Binda P. L., Scudeler Bacelle L., and Concheri G. (1994) Amino acids of Proterozoic and Ordovician sulphide coated grains from western Canada: record of biologically mediated pyrite precipitation. *Chem. Geol.* **111**, 1–15.

Neuberger A. (1948) Stereochemistry of amino acids IV. *Adv. Prot. Chem.* **4**, 297–383.

Nielsen-Marsh C. M. and Hedges, R. E. M. (1999) Porosity and the use of mercury intrusion porosimetry in bone diagenesis studies. *Archaeometry* **41**, 165–174.

Nielsen-Marsh C. N., Gernaey A.. M., Turner-Walker G., Hedges R. E. M., Pike A. G., and Collins M. J. (2000) Chemical diagenesis of the protein and mineral fractions of bone. In *Osteology Current Practice and Future Prospects* (eds M. Cox and S. Mays), pp. 439–454.

Oelkers E. H. (1991) Calculation of diffusion coefficients for aqueous organic species at temperatures from 0 to 350°C. *Geochim. Cosmochim. Acta* **55**, 3515–3529.

Oliyai C. and Borchardt R. T. (1993) Chemical pathways of peptide degradation 4. Pathways, kinetics, and mechanism of degradation of an aspartyl residue in a model hexapeptide. *Pharm. Res.* **10**, 95–102.

Panginelli C. V. (1980) The physics of gas exchange across the avian eggshell. *Amer. Zool.* **20**, 329–342.

Qian Y. R., Engel M. H., Macko S. A., Carpenter S., and Deming J. W. (1993) Kinetics of peptide hydrolysis and amino-acid decomposition at high-temperature. *Geochim. Cosmochim. Acta* **57**, 3281–3293.

Qian Y. R., Engel M . H., Goodfriend G. A., and Macko S. A. (1995) Abundance and stable carbon-isotope composition of amino-acids in molecular-weight fractions of fossil and artificially aged mollusk shells. *Geochim. Cosmochim. Acta* **59**, 1113–1124.

Rafalska J. K., Engel M. H., and Lanier W. P. (1991) Retardation of racemization rates of amino acids incorporated into melanoidins. *Geochim. Cosmochim. Acta* **55**, 3669–3675.

Riley M. S. and Collins M. J. (1994) The polymer model of collagen degradation. *Poly. Degrad. Stab.* **46**, 93–97.

Ritchie H. H. and Wang L. H. (1996) Sequence determination of an extremely acidic rat dentin phosphoprotein. *J. Biol. Chem.* **271**, 21695–21698.

Ritz-Timme S., Shutz H. W., and Collins M. J. (2000) Commentary on Ohtani S., Matsushima Y., Kobayashi Y., and Kishi K. Evaluation of aspartic acid racemization ratios in the human femur for age estimation. *J. Forensic Sci.* 1998; **43**: 949–953; *J. Forensic Sci* **44**, 874–875.

Robbins L. L. and Brew K. (1990) Proteins from the organic matrix of core-top and fossil planktonic foraminifera. *Geochim. Cosmochim. Acta* **54**, 2285–2292.

Roof S. (1997) Comparison of isoleucine epimerization and leaching potential in the molluskan genera *Astarte*, *Macoma*, and *Mya*. *Geochim. Cosmochim. Acta* **61**, **24**, 5325–5333.

Rudakova T. E. and Zaikov G. E. (1987) Degradation of collagen and its possible applications in medicine. *Poly. Degrad. Stab.* **18**, 271–291.

Saban M. D., Scott J. D., and Cassidy R. M. (1992) Collagen proteins in electrorefining—rate constants for glue hydrolysis and effects of molar mass on glue activity. *Metallurg Trans. B—Process Metallurgy* **23**, 125–133.

Saint-Martin B. (1991) Influence of the intramolecular interactions on the amino acid racemization inside the fossil proteins, *C.R. Acad. Sci. Paris* **312**, 485–491.

Sarashina I. and Endo K. (1998) Primary structure of a soluble matrix protein of scallop shell: implications for calcium carbonate biomineralization. *Amer. Min.* **83**, 1510–1515.

Sepetov N. F., Krymsky M. A., Ovchinnikov M. V., Bespalava Z. D., Isakova O. L., Soucek M., and Lebl M. (1991) Rearrangement, racemization and decomposition of peptides in aqueous solutions. *Peptide Res.* **4**, 308–313.

Shen X., Belcher A. M., Hansma P. K., Stucky G. D., and Morse D. E. (1997) Molecular cloning and characterization of lustrin A, a matrix protein from shell and pearl nacre of *Haliotis rufescens. J. Biol. Chem.* **272**, 32472–32481.

Smith G. G. and Baum R. (1987) 1st-order rate constants for the racemization of each component in a mixture of isomeric dipeptides and their diketopiperazines, *J. Org. Chem.* **52**, 2248–2255.

Smith G. G. and Evans R. C. (1980) The effect of structure and conditions on the rate of racemization of free and bound amino acids. In *Biogeochemistry of Amino Acids.* (eds P. E. Hare, T. C. Hoering, and K. King, Jr.), pp. 257–282. Wiley.

Smith G. G. and de Sol B. S. (1980) Racemization of amino acids in dipeptides shows COOH > NH_2 for non-sterically hindered residues. *Science* **207**, 765–767.

Smith G. G., Williams K. M., and Wonnacott D. M. (1978) Factors affecting the rate of racemization of amino acids and their significance to geochronology. *J. Org. Chem.* **52**, 2248–2255.

Squire P. G., Moser P., and O'Konski C. T. (1968) The hydrodynamic properties of Bovine Serum Albumin Monomer and Dimer. *Biochemistry* **7**, 4261–4265.

Stainforth J. G. and Reinders J. E. A. (1989) Primary migration of hydrocarbons by diffusion through organic matter networks, and its effects on oil and gas generation. *Org. Geochem.* **16**, 61–74.

Stankevitzch B. A., van Bergen P. F., Duncan I. J., Carter J., Briggs D. E. G., and Evershed R. P. (1996) Recognition of chitin and proteins in invertebrate cuticles using analytical pyrolysis–gas chromatography and pyrolysis–gas chromatography/mass spectrometry. *Rapid. Commun. Mass Spec.* **10**, 1747–1757.

Stathoplos L. and Hare P. E. (1993) Bleach removes labile amino acids from deep sea planktonic foraminiferal shells. *J. Foram. Res.* **23**, 102–107.

Steinberg S. M. and Bada J. L. (1981) Diketopiperazine formation during investigations of amino acid racemization in dipeptides. *Science* **213**, 544–545.

Steinberg S. M. and Bada J. L. (1983) Peptide decomposition in the neutral pH region via the formation of diketopiperazines. *J. Org. Chem.* **48**, 2295–2298.

Stern B., Abbott G. D., Collins M. J., and Armstrong H. A. (1999) Development and comparison of different methods for the extraction of biomineral associated lipids. *Ancient Biomolecules* **3**, 321–334.

Sudo S., Fujikawa T., Nagakura T., Ohkubo T., Sakaguchi K., Tanaka M., Nakashima K., and Takahashi T. (1997) Structures of mollusc shell framework proteins. *Nature* **387**, 563–564.

Sykes G. A., Collins M. J., and Walton D. I. (1995) The significance of a geochemically isolated (intracrystalline) organic fraction within biominerals. *Org. Geochem.* **23**, 1059–1066.

Thompson R. R. and Creath W. B. (1966) Low molecular weight hydrocarbons in Recent and fossil shells. *Geochim. Cosmochim. Acta* **30**, 1137–1152.

Tullet S. G. (1975) Regulation of avian eggshell porosity. *J. Zool.* (*Lond.*) **177**, 339–348.

Vallentyne J. R. (1964) Biogeochemistry of organic matter II: Thermal reaction kinetics and transformation products of amino compounds. *Geochim. Cosmochim. Acta* **28**, 157–188.

Von Endt D. W. (1980) Protein hydrolysis and amino acid racemization in sized bone. In *Biogeochemistry of Amino Acids.* (eds P. E. Hare, T. C. Hoering, and K. King, Jr.), pp. 297–304. Wiley.

Waite E. R. and Collins M. J. (2000) The interpretation of aspartic acid racemization of dentine proteins. In *Perspectives in Amino Acid and Protein Geochemistry* (eds G. A. Goodfriend, M. J. Collins, M. L. Fogel, S. A. Macko, and J. F. Wehmiller), pp. 182–194. Oxford University Press.

Walton D. (1998) Degradation of intracrystalline proteins and amino acids in fossil brachiopods. *Org. Geochem.* **28**, 389–410.

Wehmiller J. F. (1980) Intergeneric differences in apparent racemization kinetics in mollusks and formainifera: implications for models of diagenetic racemization. In *Biogeochemistry of Amino Acids.* (eds P. E. Hare, T. C. Hoering, and K. King, Jr.), pp. 341–355. Wiley.

Wehmiller J. F. (1982) A review of amino acid racemization studies in Quaternary molluscs: stratigraphic and chronological applications in coastal and interglacial sites, Pacific and Atlantic coasts, US, UK, Baffin Island and Tropical Islands. *Quat. Sci. Rev.* **1**, 83–120.

Wehmiller J. F. (1990) Amino acid racemization: applications in chemical taxonomy and chronostratigraphy of Quaternary fossils. In *Skeletal Biomineralization.* (ed. J. G. Carter), vol. I, pp. 583–608. Van Nostrand Reinhold.

Wehmiller J. F. (1993) Applications of organic geochemistry for quaternary research; aminostratigraphy and aminochronology. In *Organic Geochemistry* (eds M. H. Engel and S. A. Macko), pp. 755–783. Plenum.

Weiner S. and Lowenstam H. A. (1980) Well-preserved fossil mollusk shells: characterization of mild diagenetic processes. In *Biogeochemistry of Amino Acids*. (eds P. E. Hare, T. C. Hoering, and K. King, Jr.), pp. 95–113. Wiley.

Wilson H. and Cannan R. K. (1937) The glutamic acid–pyrrolidone carboxylic acid system. *J. Biol. Chem.* **119**, 309–331.

Wonacott D. M. (1979) *The application of gas chromatography to the resolution of enatiomeric mixtures of amino acids*. Ph.D. Thesis, Utah State University, Logan, UT, USA.

APPENDIX: DETAILS OF CALCULATIONS

Analysis of the shape of the "free amino acids vs time curve"

For a chain molecule comprising P_0 amino acid residues, subject to random scission by hydrolysis, the weight fraction, γ_x, of x-mers at time t, is given by

$$\gamma_x = \begin{cases} \alpha \dfrac{x}{P_0}(1-\alpha)^{x-1}[2+(P_0-x-1)\alpha] & \text{for } x < P_0 \\ (1-\alpha)^{P_0-1} & \text{for } x = P_0 \end{cases} \qquad \text{(11-A1)}$$

where $\alpha = 1 - \exp(-kt)$ in which k is the rate of hydrolysis (Riley and Collins, 1994).

It follows by putting $x = 1$ that the weight fraction of free amino acids is given by

$$\gamma_1 = \frac{\alpha}{P_0}[2+(P_0-2)\alpha] \qquad \text{(11-A2)}$$

Differentiating twice with respect to t and then putting $t = 0$ gives

$$\left.\frac{d^2\gamma_1}{dt^2}\right|_{t=0} = \frac{2k^2}{P_0}(P_0-3) \qquad \text{(11-A3)}$$

so the curve of γ_1 against time is concave upwards at the origin provided that $P_0 > 3$.

Calculation of the rate constant, k, from the weight fraction of free amino acids

As in equation (11-A2), the weight fraction of free amino acids is given by

$$\gamma_1 = \frac{\alpha}{P_0}[2+(P_0-2)\alpha] \qquad \text{(11-A4)}$$

Case 1 ($P_0 = 2$)
Putting $P_0 = 2$ in equation (11-A2) gives

$$\gamma_1 = \alpha = 1 - \exp(-kt) \qquad \text{(11-A5)}$$

from which

$$-\ln[1-\gamma_1] = kt \qquad \text{(11-A6)}$$

and so plotting the left-hand side against t should produce a straight line through the origin with slope k.

Case 2 ($P_0 > 2$)

Provided that $P_0 > 2$, equation (11-A4) is a quadratic equation in α which can be solved to give

$$\alpha = \frac{-1+\sqrt{1+(P_0-2)P_0\gamma_1}}{P_0-2} \qquad \text{(11-A7)}$$

and since $\alpha = 1 - \exp(-kt)$, the relationship can be re-written

$$\frac{-1+\sqrt{1+(P_0-2)P_0\gamma_1}}{P_0-2} = 1 - \exp(-kt) \qquad \text{(11-A8)}$$

from which

$$-\ln\left[\frac{P_0-1-\sqrt{1+(P_0-2)P_0\gamma_1}}{P_0-2}\right] = kt \qquad \text{(11-A9)}$$

Thus, for $P_0 > 2$, plotting the left hand side of equation (11-A9) against t should produce a straight line through the origin with slope k.

Analysis for large values of P_0

The result in equation (11-A9) provides a method for calculating the rate of hydrolysis provided the initial length of the molecules in the sample is known. In many applications, either this length will not be known or there will be an initial distribution of lengths. In this section, we first consider the theoretical case in which molecules are infinitely long, and then estimate how long a molecule must be in reality for the infinite analysis to be applicable.

In the limit as P_0 tends to infinity equation (11-A10) becomes

$$-\ln[1-\sqrt{\gamma_1}] = kt \qquad \text{(11-A10)}$$

and so if P_0 is large (but perhaps unknown) plotting the left-hand side of equation (11-A10) against t should produce a straight line through the origin with slope k.

How large is "large"?
The weight fraction of free amino acids is given by

$$\gamma_1 = \frac{\alpha}{P_0}[2 + (P_0 - 2)\alpha] \qquad (11\text{-A11})$$

Letting P_0 tend to infinity gives

$$\gamma_1^\infty = \alpha^2 \qquad (11\text{-A12})$$

The error, ε, which results from assuming that the molecule is infinitely long when its true length is P_0 is

$$\begin{aligned}\varepsilon &= \gamma_1^\infty - \gamma_1 \\ &= -\frac{2}{P_0}\alpha(1 - \alpha)\end{aligned} \qquad (11\text{-A13})$$

which is never greater than zero and has its maximum absolute value when $\alpha = 1/2$. Thus the maximum absolute error in assuming the molecule is infinite is

$$|\varepsilon|_{\max} = \frac{1}{2P_0} \qquad (11\text{-A14})$$

which occurs at time $t = (1/k)\ln 2$.

So the maximum absolute error in the weight fraction of free amino acids, assuming that P_0 is infinite when in fact $P_0 = 50$, is only 1% of the total weight. It follows that the weight fraction of free amino acids at any time is not very sensitive to P_0 provided P_0 is large enough, say > 50. The implications are, first, that in most cases P_0 does not have to be known exactly, and second, that the asymptotic formula will remain a good approximation to reality even if there is a spread of initial values of P_0 in the sample under examination, provided, of course, that the initial distribution is not heavily skewed towards very low values of P_0.

IV. AMINO ACID RACEMIZATION KINETICS

12. Amino acid racemization in ostracodes

Darrell S. Kaufman

Improvements in chromatographic techniques during the last decade have enabled a significant reduction in the sample-size requirements for amino acid enantiomeric (D/L) separations. Kaufman and Manley (1998) developed a reverse-phase high-performance liquid chromatography (HPLC) procedure that improved detection capabilities by at least an order of magnitude compared to conventional ion-exchange technology, which has been used previously for most amino acid geochronological investigations and is presently used in many laboratories. These small sample sizes afford new opportunities to use microfossils for amino acid geochronology and paleothermometry.

Ostracodes (figure 12-1) are millimeter-size microcrustaceans that secrete a carbonaceous exoskeleton composed of a laminated chitin–protein complex (e.g., Bate and East, 1972). They inhabit a wide range of aquatic depositional settings. The potential for more continuous amino-geochronological time series is therefore greater than is possible with aquatic mollusks, which are generally restricted to marginal and shallow-water habitats. Because ostracodes are small, they are often present in sufficient abundance in sediment cores, allowing amino acid geochronology to be integrated into the study of lake and marine cores. Samples from these settings are well suited to amino acid geochronology because subaqueous sediments accumulate more continuously than in subaerial environments, the temperature is relatively stable over long intervals, and cores can often be dated using other techniques. This, in turn, provides a means of calibrating the rate of racemization so that amino acids can be used to derive the ages of undated sections of the core.

Because of the relatively high concentration of amino acids in ostracode shells (see below), and because of the low detection limits of the new reverse-phase technique, reliable results can be obtained on a single Pleistocene ostracode shell. Ten individuals (<0.1 mg) are sufficient for deriving a sample mean D/L ratio. This is an order of magnitude less material than is required for accelerator mass spectrometric (AMS) ^{14}C dating and presents a new opportunity for dating deposits that are organically poor, or beyond the applicable range of the ^{14}C technique (>40 ka). Finally, amino acid geochronology is not subject to the effects of hard water and reservoir ages, or to changes in the isotopic composition of the atmosphere, which hampers ^{14}C dating in some cases.

In addition to reducing the sample-size requirements, another advantage of the reverse-phase procedure over the conventional ion-exchange technology is that it separates enantiomers of multiple amino acids. Because each amino acid racemizes at a different rate, the range of reaction rates can be exploited by choosing the one that offers the optimal geochronometer for the time and temperature range of interest. Second, leaching and perhaps other diagenetic reactions are more likely to affect diastereomer pairs such as alloisoleucine/isoleucine differentially (e.g., Roof, 1997), whereas enantiomers (e.g., D- and L-aspartic acid) have identical chemical properties and thus are more likely to yield accurate results (Goodfriend et al., 1996). A further advantage of analyzing the extent of racemization in several amino acids, rather than one, is that the others can be used for cross checks to identify shells that yield unreliable results. Because the D/L ratios among the different amino acids tend to be strongly covariant (e.g., Goodfriend, 1991), divergence in these trends might be used to diagnose excessive leaching or contamination, both of which undermine the accuracy of the technique and contribute to high intersample variation (e.g., Müller, 1984). The ability to recognize and exclude aberrant samples should improve the resolving capacity of amino acid geochronology.

The purpose of this study is to develop the use of racemization, particularly in aspartic (Asp) and glutamic (Glu) acids, in ostracodes for amino acid geochronology. To assess the integrity of the new technique, I measured the extent of racemization in both fossil and laboratory-heated shells. Laboratory heating experiments were used to evaluate the reaction rates of amino acids in ostracodes over a range of temperatures (80–140°C) and D/L ratios (~0.1–0.6 in Asp). These data, combined with analysis of six ^{14}C-dated latest Quaternary ostracodes, were used to develop an age equation based on racemization of Asp and Glu in the ostracode genus *Candona*, one of the most common taxa in freshwater environments. The equations for the age and the associated age uncertainty were solved using a Monte Carlo simulation that incorporated the inter-related

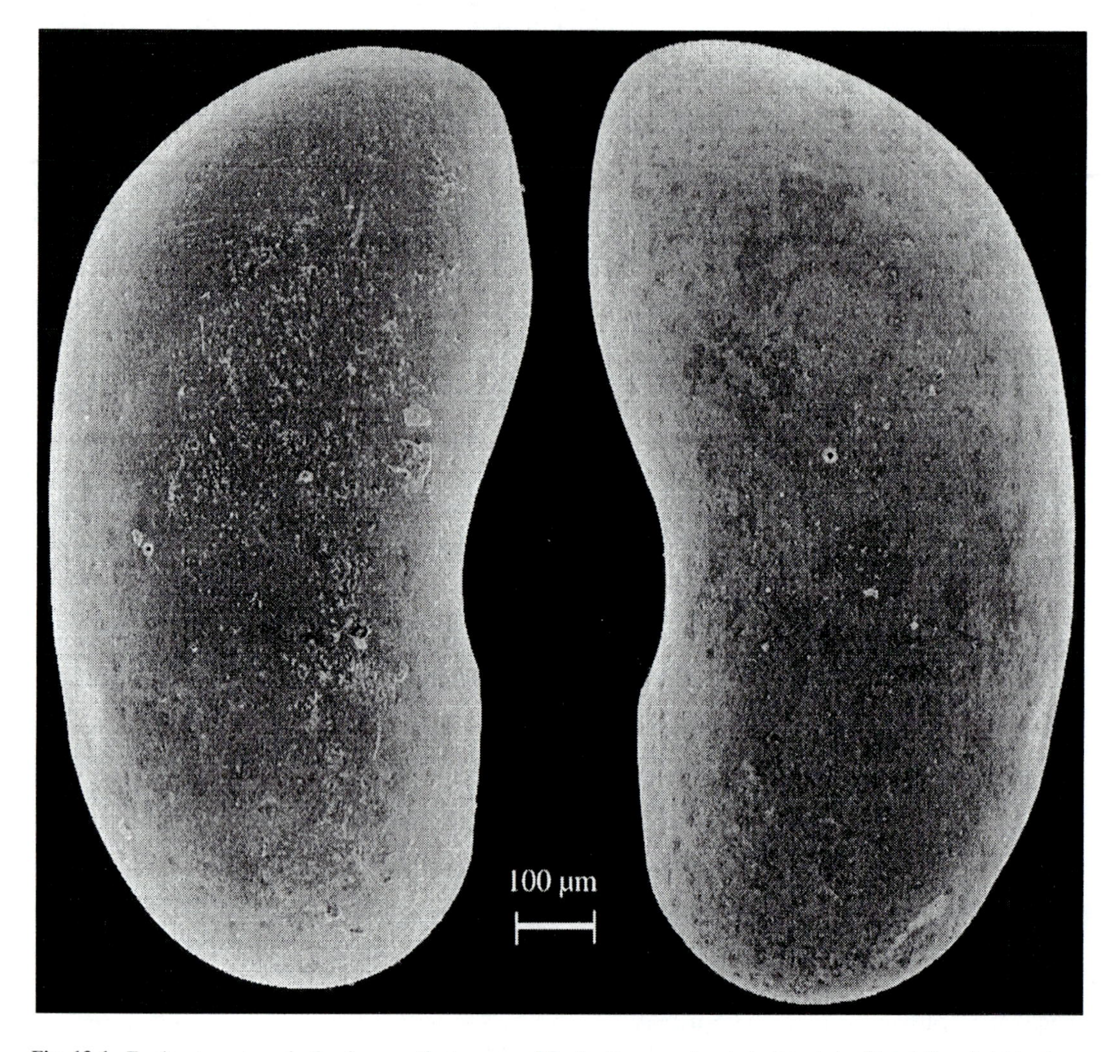

Fig. 12-1. Freshwater ostracode *Candona* sp., the taxon used for heating experiments in this study. Male left and right valves. Scale bar is 100 μm.

uncertainties in each of the principal variables used to derive the age estimate. Finally, the data from the heating experiments are compared with down-core analyses of ostracodes from the well-dated Burmester core from Utah.

This work follows the pioneering effort by McCoy (1988), who analyzed the extent of isoleucine epimerization in ostracodes from several stratigraphic sections with independent age control to evaluate the potential of ostracodes for aminostratigraphic purposes. He concluded that racemization in fossil ostracode shells had potential for geochronological and paleotemperature studies, but that more study of the kinetics of amino acid racemization was needed to assess the suitability in detail.

METHODS

Sample Material

Most of the ostracodes analysed in this study were collected in a single dredge sample of lake-bottom sediment from a ~30 m water depth in Bear Lake, Utah. The upper ~5 cm of mud contained abundant ostracode shells, which were collected in bulk. The sample included both dark gray and translucent white ostracode shells, but no live ostracodes were found. A cleaned subsample (see below) of translucent white *Candona* shells weighing 19.9 mg was submitted for AMS ^{14}C analysis. The results show that the average age of the bulk ostracode sample is ~1000 cal yr BP (table 12-1).

TABLE 12-1 AMS ^{14}C ages, amino acid (aspartic and glutamic acids) D/L ratios in *Candona* and site temperatures used to calculate forward rate constants for Arrhenius parameters

Lab. no. (UAL)	Field ID	Location	MAT (°C)	^{14}C lab. no.	Material	^{14}C age (yr BP)	Cal age[a] (yr BP)	Asp D/L[b]		Glu D/L		n
								mean	1 S.D.	mean	1 S.D.	
2753	30 m dredge	Bear Lake, Utah	4.6	CAMS-44856	ostracodes	1090 ± 190	970 ± 225	0.068	0.008	0.022	0.004	6
2758	DK96-06C	Lifton, Idaho	4.8	WW-1566	charcoal	5530 ± 50	6345 ± 55	0.149	0.009	0.035	0.003	15
2749	AL795: 14-1B	Deseret, Utah	9.5	AA-19050[c]	gastropod	1450 ± 50	1325 ± 30	0.153	0.011	0.041	0.007	6
2750	ID93-23D	Lifton, Idaho	4.8	OS-20383	gastropod	6750 ± 60	7570 ± 40	0.161	0.013	0.036	0.008	15
2621	DK97-01E	Fielding, Utah	7.5	AA-19064[c]	gastropod	12410 ± 110	14530 ± 210	0.213	0.009	0.053	0.006	9
2752	AL795: 11-2A	Salt Lake City, Utah	11.2	AA-19047[c]	gastropod	12505 ± 105	14660 ± 215	0.255	0.013	0.073	0.008	13

[a] Calibrated ages from Stuiver and Reimer (1993).
[b] Mean and standard deviation of "n" subsamples; each subsample composed of a single shell.
[c] From Light (1996); all other ^{14}C ages reported here for the first time.

Samples from the Burmester core were collected from material archived at the University of Utah. The 307-m core was drilled by Eardley et al. (1973) near the south shore of Great Salt Lake (latitude = 40° 41′N; longitude = 112° 27′W). The core is relatively well dated using paleomagnetic stratigraphy (Eardley et al., 1973) and tephrochronology (Williams, 1994). Its lithostratigraphy and paleoenvironmental interpretation has recently been re-examined (Oviatt et al., 1999).

Heating Procedure

All of the heated ostracodes were taken from the concentrated 1 ka sample. Aliquots of ~30 mg were combined in sterilized ampules with 1.1 mg sterilized sand (sieved to exclude grain sizes overlapping with the ostracode diameter, for easier retrieval) and moistened with 0.25 ml purified H_2O. The ampoules were flame-sealed and heated for up to 100 d at 80, 110, 120, or 140°C in a forced-convection oven calibrated using a certified thermocouple and meter. Ostracodes were then cleaned and prepared for analysis as described below.

Sample Preparation

The ostracode-cleaning procedure used in this study was modified from that of Forester et al. (1994). The shells were concentrated by wet-sieving bulk sediment at 250 μm using cold, distilled water. Whole ostracode shells were hand-picked by genus (or species, where sufficiently abundant) into a 50 ml glass beaker and immersed in 40 ml reagent-grade purified water. The horn of an ultrasonic cleaner was inserted into the beaker and operated at low intensity for 60 min to remove adhering clay particles. The shells were transferred to a clean 50 ml glass beaker and immersed in 10 ml 3% H_2O_2 for 2 hr, rinsed with purified H_2O, then air-dried under laminar flow. Repeated analyses of samples cleaned in peroxide showed a lower variation in amino acid composition among subsamples than those that had not been treated. Ostracodes were then hand-picked a second time to separate the best-preserved, thoroughly cleaned, whole individuals. These were placed in sterilized, conical bottomed microreaction vials for hydrolysis.

Sediment samples from the Burmester core required an additional step to disaggregate the consolidated mud and to remove adhering clay particles from the ostracode shells. Between 4 and 47 g (average = 33 g) raw sediment was placed into a 600 ml stainless steel beaker with 0.5 L freshly boiled purified water and allowed to cool to room temperature. Calgon (1.0–1.5 mg) was added to the mixture. The solution was stirred and frozen overnight. Samples were then thawed at room temperature and the sediment allowed to disaggregate for at least 24 hr prior to wet-sieving and further cleaning as described above. Ostracodes from some of the Burmester core samples analyzed in this study were cleaned and concentrated at the laboratory of the U.S. Geological Survey.

For the heated material, as few as 2 and as many as 17 (average = 6) monospecific *Candona* sp. (unnamed species; R. M. Forester, personal communication; figure 12-1) subsamples were prepared from either one or two ampoules. For the fossils from the Burmester core, two to six monogeneric, and in some cases, monospecific, subsamples were prepared from each of nine stratigraphic levels. Each subsample comprised approx. 10–40 individual shells and weighed 0.1–0.2 mg. Following the analysis of these samples, new refinements to the analytical technique have enabled accurate results to be obtained from a single ostracode shell (see below). Samples analyzed with the newer, single-shell procedure include the modern (live-collected) ostracodes and the ^{14}C-dated late Quaternary samples.

Subsamples comprising multiple shells of either heated or fossil valves were dissolved in 10 μl 7 M HCl; single-shell subsamples were dissolved in 6 μl 6 M HCl. Both were sealed under N_2 and heated at 110°C. Laboratory-heated and ^{14}C-dated samples were hydrolyzed for 6 hr to minimize the induced racemization; samples from the Burmester core were hydrolyzed for 22 hr for more complete recovery of amino acids (Kaufman and Manley, 1998). The hydrolysates were evaporated to dryness *in vacuo,* then rehydrated in 0.01 M HCl with 1.5 mM sodium azide (to inhibit bacterial growth) and 0.03 mM L-*homo*-arginine (L-*h*Arg) using a ratio of 0.02 ml/mg of shell, or a total of 4 μl for single shells, which were too small to weigh. The synthetic amino acid L-*h*Arg served as an internal standard for quantifying the concentration of amino acid enantiomers. The hydrolyzed solutions of multiple-shell subsamples were transferred using sterilized glass pipettes to 2 ml autoinjector vials containing 200 μl glass inserts.

Analytical Procedure

The chromatographic instrumentation and procedure used to separate and quantify the abundance of amino acid enantiomers is presented in detail by Kaufman and Manley (1998). Briefly, the analyses were performed on an integrated Hewlett-Packard HP1100 liquid chromatograph equipped with a quaternary pump and vacuum degasser, an autoinjector and autosampler, and a programmable fluorescence detector. The method employed pre-column derivatization with *o*-phthaldialdehyde (OPA) together with the chiral thiol, *N*-isobutyryl-L-cysteine (IBLC), to yield fluorescent diastereomeric derivatives of chiral primary amino acids. The derivatization was performed on-line prior to each injection using the automated features of the autoinjector. The OPA/IBLC derivatizing reagent was drawn into the off-line sample loop sequentially both before and after the sample solution. The sample and derivative were mixed by a motorized plunger, then injected on to the reverse-phase column (250 × 4 mm i.d.) packed with a C_{18} stationary phase (Hypersil BDS, 5 μm at 25°C). A linear gradient was programmed for three mobile-phase channels: Eluent A was 23 mM sodium acetate with 1.5 mM sodium azide, adjusted to pH 6.00 with 10% acetic acid. Eluents B and C were 100% HPLC-grade methanol and acetonitrile, respectively. The initial condition was 95% A and 5% B, grading to 50% B and 2% C after 75 min. Detection was by fluorescence (230 nm excitation and 445 nm emission). HP CHEMSTATION computer software performed the integration of the fluorescence signal and controlled the eluent gradient and automated sample derivatization and injection program. Most subsamples (hydrolysates) were analyzed at least twice using HPLC; the averaged results (based on area counts) were then averaged among the multiple subsamples to derive the grand mean and standard deviations reported in this study.

Improved Detection

Recently, the Amino Acid Geochronology Laboratory at Northern Arizona University has improved its analytical procedure to reduce further the minimum required sample size. Multiple injections during HPLC are now possible using a single Pleistocene ostracode shell (~ 5 μg shell per injection), with detection in the sub-picomole range. The D/L ratios measured in the single shells are indistinguishable from those in multiple-shell preparations. The improved analytical capability results from two simple modifications to the original procedure (Kaufman and Manley, 1998).

1. The waste associated with transfer of the sample solutions following hydrolysis is eliminated; the laboratory now uses a single microreaction vial suitable for both the hydrolysis step and the autoinjector (Kontes vials, PN-60710-110, with National Scientific solid caps, PN-B7815-8).
2. The detector response is improved (a narrow-bore column is used).

A 250 × 3 mm column packed with 5 μm Hypersil BDS (Batch no. 5/120/4772) and operated with a flow rate of 0.56 ml/min and gradient conditions specified above, performs the separation exceptionally well.

Precision

Kaufman and Manley (1998) evaluated the precision of the reverse-phase HPLC procedure used in this study. They found that the best and most consistently resolved amino acids are the enantiomers of aspartic acid (Asp[1]), glutamic acid (Glu[1]), serine (Ser), alanine (Ala), and valine (Val), although not all are equally well resolved in every analysis of every sample. On the basis of repeated analyses of a synthetic mixture of racemic amino acids, the analytical precision averages about 1.6% for 10 amino acid pairs. In fossil mollusks, which are biogeochemically more complex, analyzed over a longer interval (4–6 weeks), the reproducibility averages 7% across four Interlaboratory Comparison (ILC; Wehmiller, 1984) standards. Asp and Glu D/L ratios are the most consistently well resolved and reproduced, with analytical variations of 2 and 3%, respectively. D/L ratios measured in ILC standards by the reverse-phase procedure are listed in Kaufman and Manley (1998).

Analysis of synthetic racemic standards shows that the derivatized D and L-enantiomers have slightly different fluorescent properties. To compensate for this bias, the average D/L ratio calculated from multiple analyses of a racemic standard (i.e., the fluorescence factor) was applied to each sample D/L ratio. The fluorescence factors for Asp and Glu are 0.96 and 1.08, respectively (Kaufman and Manley, 1998). No attempt was made to determine whether the fluorescence factor derived from the racemic mixture is the same for one derived from a mixture with a lower D/L ratio.

RESULTS

Amino Acid Composition and Concentration

Chromatograms (figure 12-2) of fossil *Limnocythere* from the Burmester core illustrate the typical analytical performance of the reverse-phase technique, and the differences in amino acid concentrations and D/L ratios between samples of different ages. Eight amino acid pairs are separated with baseline resolution, although D-Val and D-Phe commonly had interfering peaks and are not reported in this study. The chromatograms illustrate the expected decrease in amino acid concentration and increase in D/L ratios with increasing age.

The chromatograms of the fossil samples are mimicked by those from ostracodes heated progressively in the laboratory. Asp and Gly (glycine, which coelutes with D-threonine) are the most abundant amino acids, followed by Glu and Ala, then Ser, Leu, and Ile in decreasing order (figure 12-3). On the basis of five analyses (UAL-2372), live-collected *Candona* specimens contain an amino acid concentration (a total of ~235 nmol/mg for these six amino acids; data not shown in figure 12-3) that is an order of magnitude higher than that of the unheated ostracodes from the 1 ka Bear Lake dredge sample (which contained a total of ~34 nmol/mg for these six amino acids). For each temperature, heated *Candona* sp. shows an initially high concentration of amino acids followed by an exponential decrease with increased heating time (figure 12-3). For example, after ~7 d

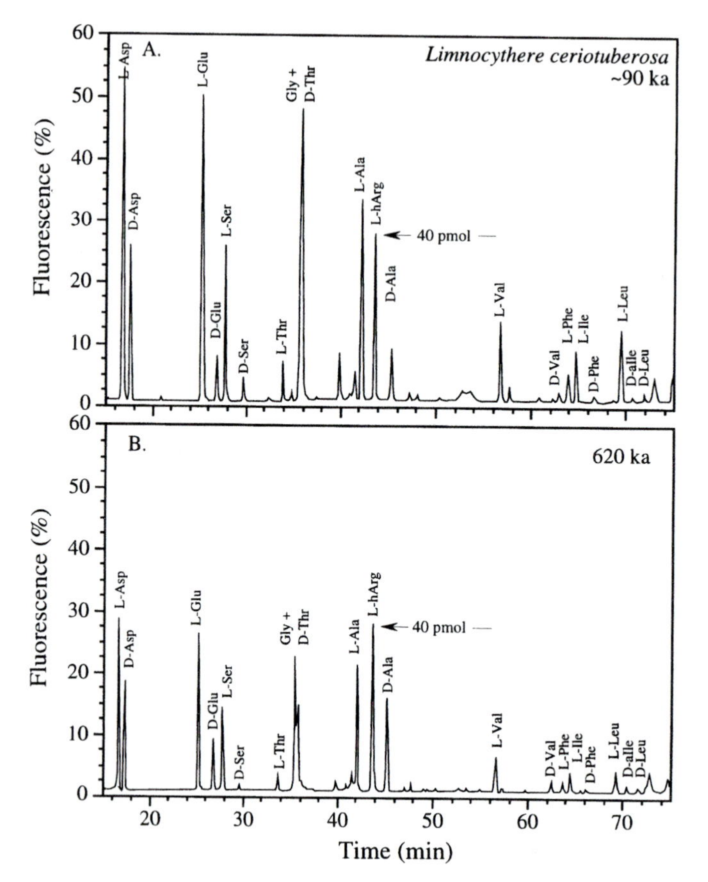

Fig. 12-2. Typical chromatograms showing the relative differences in amino acid D/L ratios and concentrations for two different ages of ostracodes. Samples are (A) ~90 ka *Limnocythere ceriotuberosa* and (B) 620 ka *L. ceriotuberosa* from the Burmester core, Utah. See text for analytical conditions.

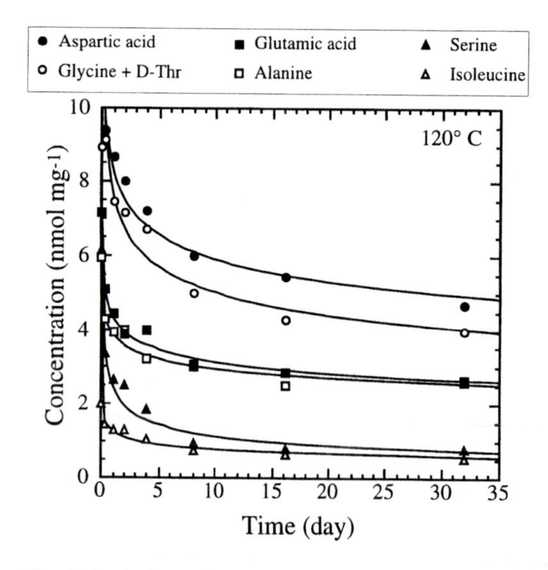

Fig. 12-3. Amino acid concentration in 1 ka ostracode shells (*Candona* sp.) from Bear Lake dredge sample heated at 120°C for up to 32 d. Lines are exponential fits.

at 120°C, 30–50% of the initial (1 ka shells) amino acid concentration is lost from the shell. The amino acid concentration then attains a relatively stable level (total = ~12 nmol/mg), with little additional decrease with increased heating time.

The relative proportions of amino acids and the rate at which they decrease with increased heating time are similar to those of the gastropod genus *Stagnicola* (D. Kaufman, unpublished data). The principal difference between snails and ostracodes is their amino acid concentration. For a given heating duration or fossil age, the concentration of amino acids in shells of the ostracode genus *Candona* is 5–9 times greater than that in the shells of the gastropod *Stagnicola* collected from the same stratigraphic horizon (figure 12-4) or heated for the same length of time.

Amino Acid Racemization

The extent of amino acid racemization (D/L ratio) in *Candona* sp. increases with increased heating time in five of the six amino acids that were consistently resolved and monitored (Asp, Glu, Ala, Val,

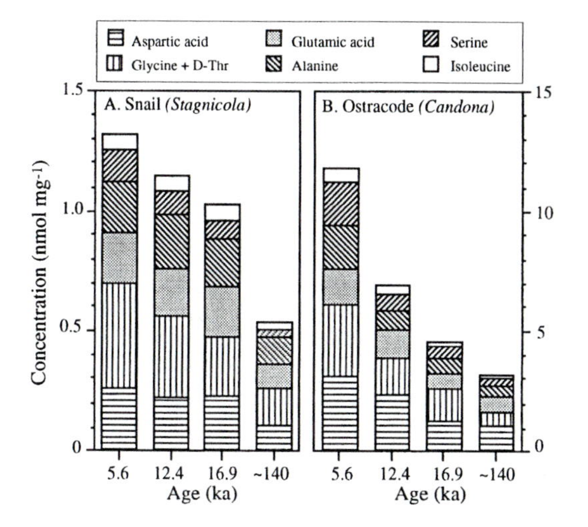

Fig. 12-4. Amino acid composition of fossil (a) snail (*Stagnicola*) and (b) ostracode (*Candona*) shells from the same dated stratigraphic horizons in the Bonneville basin, Utah. Ages are ^{14}C ka. Note the order-of-magnitude difference between the concentration scales for the two taxa.

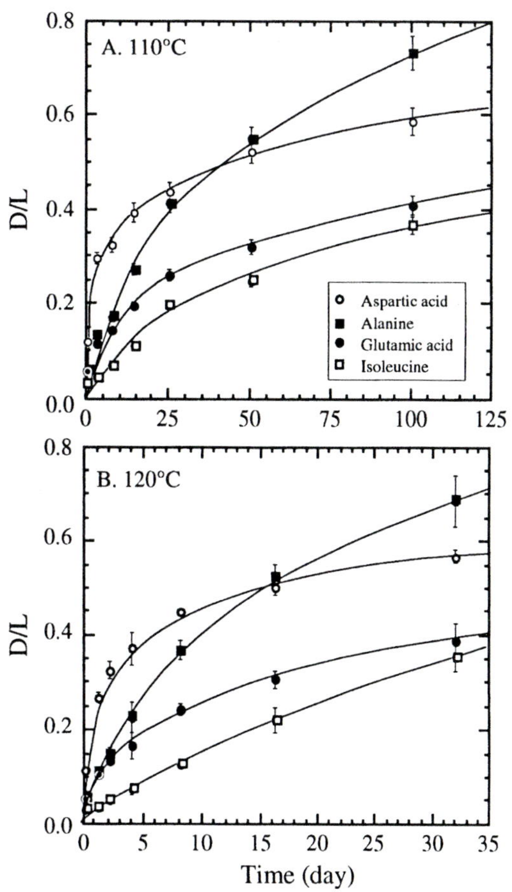

Fig. 12-5. D/L ratios in four amino acids in 1 ka ostracode shells (*Candona* sp.) from Bear Lake dredge sample heated at (A) 110°C and (B) 120°C. Error bars are ±1 S.D. of mean D/L ratio measured in 3–17 subsamples. Curved lines are fitted by hand for visual purposes. "Isoleucine" is the ratio of D-alloisoleucine to L-isoleucine.

Ile). Only Ser exhibits a reversal, with D/L increasing up to ~0.35, then decreasing upon increased heating at all temperatures. Here, I focus on a subset of the data that includes four amino acids and two temperatures (figure 12-5, table 12-2). D/L ratios in Asp, Glu, Ala, and Ile show the expected convex-upward trend, with Asp and Ala racemizing at a higher rate than Glu and Ile. Of all amino acids, Asp exhibits the most rapid rate of racemization during the initial stages of heating, followed by the strongest convexity. This characteristic of Asp is well documented in other carbonate fossils (e.g., Goodfriend, 1992; Goodfriend et al., 1996; Manley et al., Chapter 16). At a D/L ratio of ~0.42, the rate and extent of racemization in Ala exceeds that of Asp. Ile epimerizes somewhat more slowly than Glu racemizes over the range of temperatures and times in this experiment.

Equations that relate time and temperature to D/L were calculated for Asp and Glu racemization, two amino acids that are among the most abundant enantiomers, are the best resolved chromatographically, elute during the first 30 min of the sample run, and span most of the range of racemization rates. Choosing a mathematical model to describe the shape of the D/L versus time curves is not straightforward. A theoretical model of first-order reversible kinetics is known to overestimate the rate of racemization beyond the initial phases of the reaction (e.g., Mitterer and Krausikal, 1989), especially for Asp (e.g., Goodfriend and Meyer, 1991). Instead, I chose an empirical power-law model based on a general expression that approaches an asymptote value of 1.0 (see Manley et al., Chapter 16 for a discussion of this and other transformation models):

$$(1 + K')kt + C = [(1 + \text{D/L})/(1 - K'\text{D/L})]^n \qquad (12\text{-}1)$$

where: K' = the reciprocal of the equilibrium constant (1.0); k = forward rate constant for a given temperature (yr^{-1}); t = time (yr); C = a constant equivalent to the right side of the equation evaluated at $t = 0$. For the heated shells from the dredge sample, C was derived by measuring D/L ratios in unheated shells from the same collection; in the case of the ^{14}C-dated samples, C was calculated using D/L ratios measured in live-collected shells (table 12-2). n = the exponent that produced the best linearization of the data over the range of temperatures and D/L ratios evaluated.

TABLE 12-2 Aspartic acid (Asp) and glutamic acid (Glu) D/L ratios in *Candona* sp. shells from 1 ka Bear Lake dredge sample heated at four temperatures for up to 100 days

Lab. no. (UAL)	Temp (°C)	Time (yr)	D/L Asp[a] mean	D/L Asp[a] 1 S.D.	D/L Glu[a] mean	D/L Glu[a] 1 S.D.	*n*
2360	140	0.01370	0.596	0.011	0.366	0.020	6
2349, 2376–78	140	0.00822	0.543	0.019	0.295	0.034	14
2371	140	0.00548	0.504	0.023	0.261	0.023	4
2382–2384	140	0.00356	0.459	0.018	0.202	0.022	8
2357, 2373–75	140	0.00219	0.409	0.023	0.167	0.029	17
2359	140	0.00082	0.325	0.002	0.130	0.005	6
2387, 2389	140	0.00055	0.278	0.021	0.087	0.008	6
2396	120	0.08767	0.575	0.014	0.342	0.037	5
2330–33	120	0.04438	0.504	0.016	0.266	0.019	12
2395, 2403	120	0.02192	0.451	0.007	0.207	0.011	6
2394, 2402	120	0.01096	0.369	0.034	0.137	0.029	5
2393, 2401	120	0.00548	0.322	0.018	0.107	0.008	6
2392, 2400	120	0.00274	0.262	0.009	0.080	0.005	7
2289	110	0.27397	0.589	0.002	0.357	0.025	3
2365	110	0.13699	0.521	0.021	0.275	0.026	5
2364	110	0.06849	0.435	0.009	0.218	0.008	4
2363	110	0.03836	0.388	0.014	0.159	0.012	5
2366	110	0.01918	0.316	0.019	0.112	0.015	4
2362	110	0.00822	0.286	0.006	0.085	0.006	3
2446	80	0.27397	0.285	0.007	0.080	0.005	5
2369	80	0.13699	0.218	0.016	0.056	0.008	5
2368	80	0.06849	0.188	0.004	0.053	0.002	2
2367	80	0.02740	0.160	0.008	0.044	0.001	3
2388, 2390	80	0.01370	0.136	0.005	0.038	0.003	5
2646, 2653, 2656	live-collected[b]		0.041	0.002	0.017	0.003	24

[a] Mean and standard deviation of "*n*" subsamples; subsamples composed of multiple shells, except for live-collected, which were single shells.
[b] *Candona subacuminata* and *Candona sigmoides*.

To determine n, coefficients of determination (r^2) for least-squares linear regressions of $[(1 + \mathrm{D/L})/(1 - \mathrm{D/L})]^n - C$ on t were calculated for each of the four temperatures by incrementing n in intervals of 0.1. For these regressions, the y-intercept was forced at zero, the known initial value following subtraction of the initial transformed D/L value for shells prior to heating (i.e., C in equation 12-1). Furthermore, only those D/L ratios <0.60 for Asp and <0.37 for Glu were used in this analysis so that the range of D/L ratios for each amino acid was similar among temperatures, except 80°C, for which D/L ratios did not attain these values during the duration of the experiment. An n value of 3.6 produced the highest mean r^2 (0.98) among the four temperatures for Asp; an n value of 3.8 produced the highest mean r^2 (0.94) for Glu (figure 12-6). The $[(1 + \mathrm{D/L})/(1 - \mathrm{D/L})]^n$ transformation linearized the D/L ratios about equally well for each of the four temperatures (table 12-3), lending confidence to the use of this transformation to correctly model the form of racemization rates in fossils at ambient temperatures. The forward rate constant (k) for each of the four temperatures was then calculated as one-half of the slope of the least-squares regression of the transformed data (table 12-3).

In addition to data from the four high-temperature heating experiments, the evaluation of racemization kinetics was extended to ambient temperatures using unheated, fossil *Candona* shells from six AMS ^{14}C-dated sites in the Bonneville basin (table 12-3). The sample ages range from 1.0 to 14.6 cal ka; they were all deposited following the transition from full-glacial to late-glacial climate in the Bonneville basin (Thompson et al., 1993). The fall of Lake Bonneville from the Provo shoreline prior to the deposition of the oldest shells (Light, 1996) suggests that they experienced little, if any, glacial conditions. The average temperature experi-

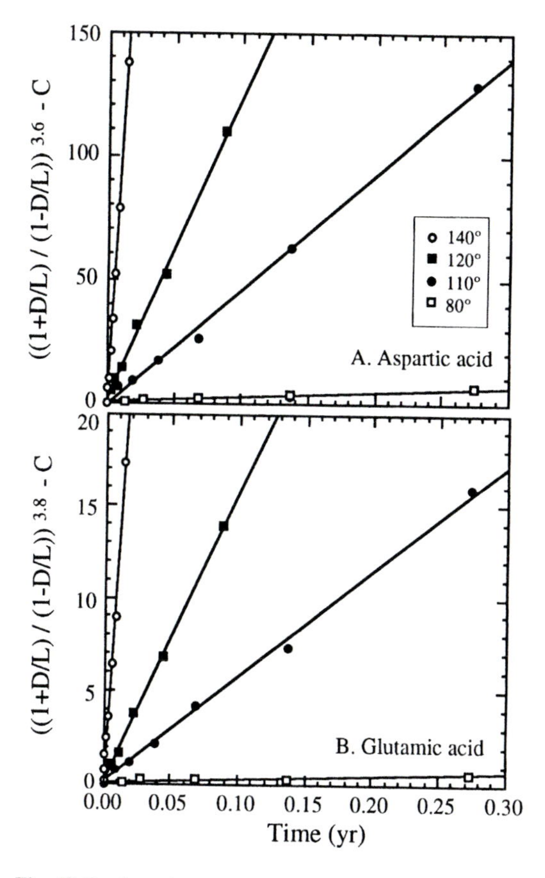

Fig. 12-6. D/L ratios measured in (A) aspartic acid and (B) glutamic acid in 1 ka ostracode shells (*Candona* sp.) from Bear Lake dredge sample heated at four temperatures. D/L data listed in table 12-2 have been transformed using equation 12-1 to achieve best-fit linear trends across the range of temperatures. The slope of the regression lines = 2k, where k = the forward rate constant (table 12-3).

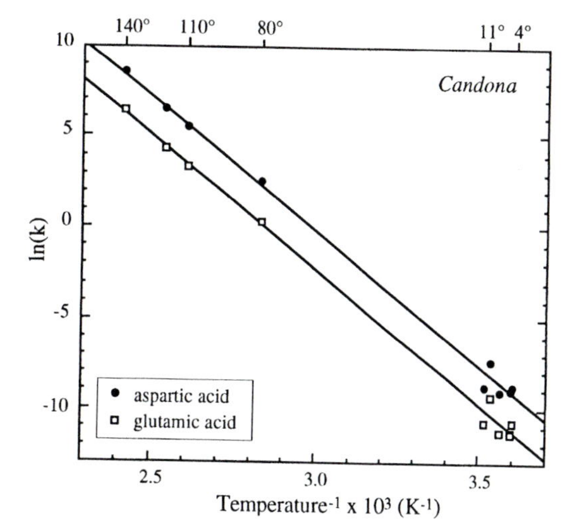

Fig. 12-7. Arrhenius plot for aspartic and glutamic acids in laboratory-heated (*Candona* sp.) and ^{14}C-dated ostracode (*Candona*) shells; data listed in table 12-3.

enced by the ^{14}C-dated fossils, all of which were deeply buried (> 1 m), was approximated by the current mean annual air temperature (MAT) recorded at nearby climate stations (e.g., Miller, 1985) (data courtesy of Utah State University Climate Center). The mean annual bottom-water temperature at Bear Lake (4.6°C; Wurtsbaugh and Luecke, 1995) was used for the 1 ka sample dredged from the lake. The forward rate constant (k) for each fossil shell sample was calculated analogously to the heated samples by dividing the net racemization (transformed D/L of the fossil ostracodes minus the transformed D/L of modern ostracodes) by twice the calibrated calendar-year sample age.

To derive an equation for sample age, the expression for k in the rate equation 12-1 was substituted into the Arrhenius equation to derive a single expression that relates k to both time and temperature. The Arrhenius equation is:

$$k = \mathrm{A}e^{(-\mathrm{Ea}/\mathrm{RT})} \tag{12-2}$$

where: A is the frequency factor; E_a = activation energy (kcal/M), R = the gas constant (0.001987 kcal/K/M); T = effective temperature (K). Substitution yields the age equation:

$$t = ([(1 + \mathrm{D/L})/(1 - \mathrm{K'D/L})]^n - \mathrm{C}) / [(1 + \mathrm{K'})\mathrm{A}e^{(-Ea/RT)}] \tag{12-3}$$

Here, C is the transformed D/L measured in modern shells following hydrolysis.

An Arrhenius plot was used to solve for the parameters E_a and A (figure 12-7). The plot shows the linear relationship between the natural log of k and the reciprocal of the absolute temperature. Fitting the data from the laboratory-heated and fossil samples to a linear regression shows that rate constants calculated for the ^{14}C-dated fossils are concordant with the values measured in the heated shells ($r^2 = 0.996$ for both Asp and Glu). In detail, the regression line is bracketed by the ^{14}C-derived data, with the Holocene samples positioned above the line and the latest Pleistocene samples below it. This probably indicates that, as expected, the current MAT that was used to approximate the average postdepositional temperature for the fossils is slightly too low for the Holocene and slightly too high for the late Pleistocene samples.

A Monte Carlo simulation was used to determine the best-fit regression for the Arrhenius plot and its associated uncertainty. The simulation takes into

TABLE 12-3 Data used to calculate the Arrhenius parameters for aspartic acid (Asp) and glutamic acid (Glu) racemization in *Candona*

Temp (°C)	T^{-1} (K^{-1})	Asp slope[a]	Asp r^2	Asp 1 S.E.[b]	Glu slope	Glu r^2	Glu 1 S.E.
Laboratory-heated samples							
140	2.420E-03	9.899E+03	0.998	106	1201	0.989	34
120	2.540E-03	1.250E+03	0.996	24.7	157.4	0.998	2.0
110	2.610E-03	4.645E+02	0.997	7.7	57.2	0.998	0.8
80	2.830E-03	2.500E+01	0.938	1.83	2.60	0.770	0.34
AMS ^{14}C-dated samples							
11.2	3.517E-03	3.20E-04			4.13E-05		
9.5	3.538E-03	1.27E-03			1.73E-04		
7.5	3.563E-03	2.35E-04			2.43E-05		
4.8	3.598E-03	2.48E-04			2.28E-05		
4.8	3.598E-03	2.51E-04			2.58E-05		
4.6	3.600E-03	3.02E-04			4.38E-05		
Arrhenius parameters							
E_a (kcal/mol)		29.5 ± 0.2			29.5 ± 0.2		
E_a (kJ/mol)		123.4 ± 0.8			123.4 ± 0.8		
ln A (yr^{-1})		44.2 ± 0.2			42.3 ± 0.2		

[a] Slope of least-squares linear regression of $[(1 + D/L)/(1 - D/L)]^n - C$ on t; twice the rate constant (yr^{-1}).
[b] S.E. = Standard error of slope of least-squares linear regression.

account the error associated with each forward rate constant and temperature used in the linear regression. The uncertainty assigned to each k value was the standard error of the regression of $[(1 + D/L)/(1 - K'D/L)]^n - C$ on t (table 12-3). For the fossil samples, I used 5% as the estimated uncertainty in k, approx. twice the ±1 S.D. analytical error associated with the ^{14}C age itself. I further assumed a ±1 S.D. uncertainty in the absolute temperature of 0.25°C for the heated samples, about twice the manufacturer's specification. For the ^{14}C-dated samples, I assumed that the average temperature for the shells was probably within 4°C of the current MAT. In other words, that there is a ~95% chance that the actual temperature is within ±2°C (2 S.D.); therefore, the estimated ±1 S.D. is 1°C (cf. McCoy, 1987). Finally, I used a ±1 S.D. value of 0.5°C for the dredge sample taken from below the thermocline of Bear Lake. The Monte Carlo simulation (programmed using a standard spreadsheet) randomly drew data from normal probability distributions generated from the means and standard deviations of all k and T values (cf. Goodfriend and Meyer, 1991). The regression for the Arrhenius plot was then calculated for each of 5000 random draws of ten k, T pairs (additional draws have no significant effect on the outcome). The mean and standard deviation of the slope and intercept were derived from these iterations. On the basis of the Monte Carlo simulation, the ±1 S.D. of the mean for both parameters are <1% for both Asp and Glu racemization. The slopes of the regression lines, after multiplying by R, yield the same E_a value of 29.5 ± 0.2 kcal/mol (123.4 ± 0.8 kJ/mol) for both Asp and Glu racemization in *Candona* (table 12-3). The intercepts yield lnA values of $44.2 \pm 0.2\,yr^{-1}$ for Asp and $42.3 \pm 0.2\,yr^{-1}$ for Glu. These values are somewhat lower than those calculated by Manley et al. (this volume) using an equivalent procedure for Asp racemization in the mollusc genera *Mya* (30.1 kcal/mol (125.9 kJ/mol); 46.9 yr^{-1}) and *Hiatella* (30.2 kcal/mol (126.4 kJ/mol); 46.8 yr^{-1}), which racemize at a higher rate than ostracodes.

The Arrhenius parameters calculated here are applicable over a limited range of D/L ratios. They are based on a model designed to linearize the racemization rate for D/L ratios ranging up to 0.60 in Asp and to 0.37 in Glu, and therefore are applicable only to samples with D/L ratios within this range. Although the model aptly describes the overall forward rate constant over this range, other models will be required to more accurately describe the shape of the racemization curve during the initial (<0.2 in Asp) and the late (>0.5 in Asp) stages of the reaction. This is especially true for the initial stages of the reaction because, in this study, only the 85°C heating experiment adequately captured the early phase of the reaction. For the other temperatures, the lowest D/L ratio measured was >0.26 for Asp and >0.08 for Glu.

Age Uncertainty

The Monte Carlo simulation was further used to evaluate the precision of ages derived using equation 12-3, and to assess the relative contribution of each variable to the overall uncertainty. This sensitivity analysis uses the simulation to propagate the effects of nonlinear errors through to the resulting age estimate. Five thousand random draws were again made from normally distributed populations of the following: (1) the activation energy (±1 S.D. as calculated based on the Monte Carlo simulation described above); (2) the frequency factor (±1 S.D. as calculated above); (3) the initial D/L ratio (±5%); (4) the D/L ratio of a hypothetical sample (± the variability among multiple subsamples, which, for this study, averages ~5% for Asp and ~10% for Glu); and (5) the effective temperature (evaluated for different scenarios). Because the slope and the intercept of the Arrhenius regression are strongly correlated ($r = 0.98$ for both the Asp and the Glu data generated by the Monte Carlo simulation of the Arrhenius plot regression), each pair of values was chosen randomly from their respective distributions using *the same* probability value. This procedure explicitly models the interdependency between E_a and A by correctly simulating the compensating effect in the age equation that the error in one variable has on the other (e.g., Miller, 1985; McCoy, 1987). The age (equation 12-3) was evaluated for each of the 5000 iterations, from which the mean age and standard deviation were calculated.

The cumulative age error can be assessed for a range of D/L ratios and temperatures using the Monte Carlo simulation. For this analysis, I focus on the age equation for Asp because the intershell variation in D/L ratios is lower than for Glu. First, I assume that the ±1 S.D. uncertainty in the average postdepositional temperature is ±3°C. For example, I might assume that the average Quaternary temperature is 10±3°C lower than the present value, or (in other words) that there is a 95% probability that the actual temperature is between 4 and 16°C lower than the present value. This value is a reasonable first approximation for Pleistocene samples. Applying the uncertainties in E_a, A, and D/L listed above, along with a temperature uncertainty of ±3°C results in an overall age error ranging from ~ ±65% of the sample age for Asp D/L ratios <0.3 to ~ ±75% for samples with Asp D/L ratios of 0.6 (figure 12-8)[2]. The dominant sources of error are the uncertainty in the D/L ratio and the temperature. Although the error in the sample D/L ratio has little influence on the overall age uncertainty for D/L ratios below ~0.2, at higher D/L ratios the accuracy of the D/L ratio becomes proportionally larger. For example, if the error in the sample D/L ratio were doubled to ±10%, as is typical of intershell variation for Glu D/L, then the estimated age error would increase to > ±100% of the sample age for

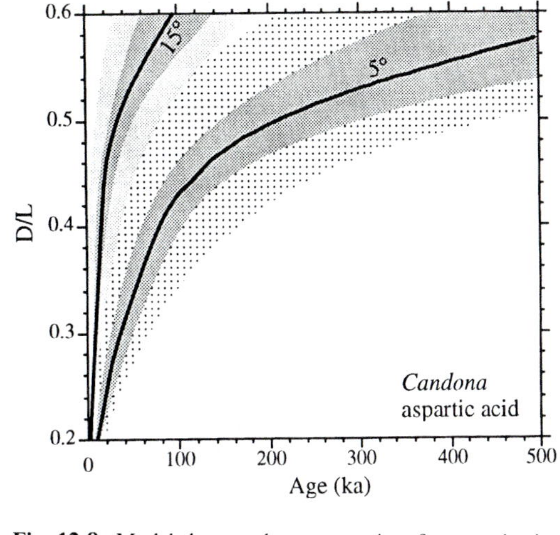

Fig. 12-8. Modeled age and age uncertainty for racemization of aspartic acid in *Candona* at 5 and 15°C. Curved lines show the solution to equation 12-3 using the values determined in this study. The inner dark gray bands represent the integrated ±1 S.D. uncertainty in age, assuming a ±1°C S.D. uncertainty in the temperature; lighter outer bands show the age error associated with a ±3°C temperature uncertainty. Error bands also incorporate uncertainties in the Arrhenius parameters (~1%) and D/L ratios (5%). See text for discussion of the Monte Carlo simulation and other procedures used in the analysis.

Asp D/L ratios >0.5. Conversely, if the error were reduced to ±3%, then the overall age uncertainty would be reduced to ~ ±60–65% of the sample age for Asp D/L ratios ranging from 0.1 to 0.6, respectively. Unlike the error in the D/L ratio, which compounds into increasing larger age uncertainties with increasing D/L, the effect of the uncertainty in temperature on the estimated age error is more uniform over the range of sample D/L ratios. If, for example, the uncertainty in the temperature were reduced from ±3 to ±1°C, then the age uncertainty would be reduced by about one-half (25–40% of the sample age for Asp D/L ranging from 0.1 to 0.6, respectively, assuming an uncertainty in D/L of 5%). The assumption that the average postdepositional temperature is known to within ±2°C (2 S.D.) is reasonable for Holocene samples, and for older samples from stratified lakes in temperature and high latitudes, or from marine settings. Given the relatively high precision of the Arrhenius parameters, they contribute little to the overall age uncertainty. Reducing the error in both parameters by 50% results in a <1% improvement in the age uncertainty. These estimated age errors assume no independent age control in the field area. The errors would be significantly reduced if independent numerical ages were available to calibrate the rate of racemization under local conditions (see below).

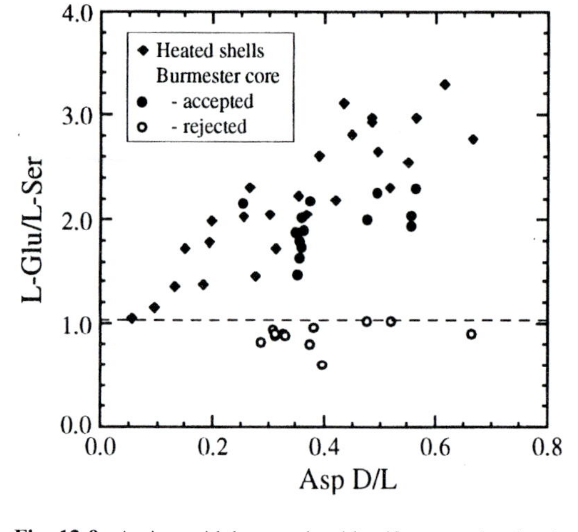

Fig. 12-9. Amino acid data used to identify contamination in ostracodes from the Burmester core, Utah. Samples that contained more L-serine than L-glutamic (L-Glu/L-Ser < 1.0) were rejected. These samples plot separately from the trend of increasing Glu/Ser with increasing age (Asp D/L) defined by the heated ostracodes and by other samples from the Burmester core that showed coherent downcore trends.

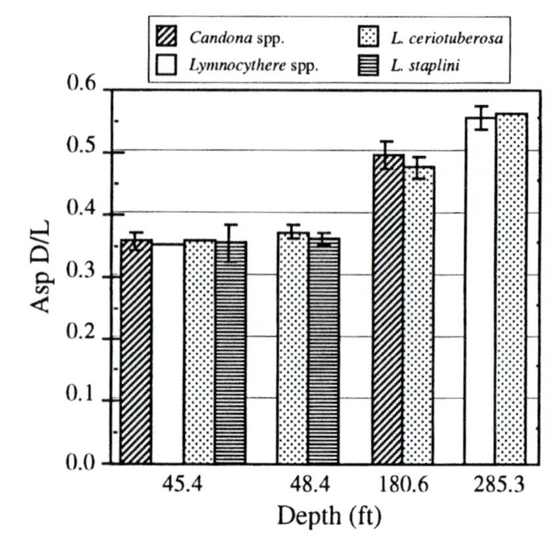

Fig. 12-10. Comparison of aspartic acid D/L ratios measured in different taxa from the same stratigraphic level of the Burmester core, Utah. Error bars are ±1 S.D.; "spp." indicates mixed species.

Burmester Core

The well-dated, ostracode-rich Burmester core provides an opportunity to evaluate the integrity of amino acid geochronology applied to ostracodes from a controlled stratigraphic setting. Because D/L ratios should increase downcore, these samples provide an independent test of the reliability of the method. Twenty-six samples of *Candona* or *Limnocythere* were analyzed from nine of the deep-lake or mud-flat units of the core. Seventeen of the ostracode samples were initially prepared by the U.S. Geological Survey then subsequently cleaned in the Amino Acid Geochronology Laboratory at Northern Arizona University; the other nine samples were prepared entirely by the Amino Acid Geochronology Laboratory, using unprocessed sediment directly from the core.

Not all of the samples yielded satisfactory results. Most (12 of 17) of the samples that were prepared by the Survey's laboratory yielded anomalous D/L ratios and amino acid concentrations. They contained high concentrations of L-Ser, which is indicative of modern contamination, and, in some cases, exhibited downcore reversals in D/L ratios. These samples were identified quantitatively and screened on the basis of the abundance of L-Ser relative to the more stable amino acid L-Glu. Samples in which the majority of subsamples contained a greater concentration of L-Ser than L-Glu were rejected. All of the heated samples and all of the samples from the Bonneville core that were processed from raw sediment in the Amino Acid Geochronology Laboratory showed L-Glu to L-Ser ratios greater than 1.0 (figure 12-9). The increase in the Glu/Ser ratio with age (as monitored by Asp D/L ratio) reflects the difference in the relative stability of the two molecules. This criterion appears to be a simple and reliable indicator of contamination for both ostracode genera analyzed in this study. Seven of the 14 samples that were accepted each included one subsample that was excluded on the basis of Glu to Ser ratios of less than 1.0.

Because most samples contained multiple taxa, the relative differences in the rate of racemization among taxa can be assessed. Several samples contained enough ostracodes so that monospecific samples could be analyzed separately (all subsamples from Burmester were made up of multiple shells). Although the data are too few for a confident determination of taxonomic differences, there is no significant difference in the mean D/L ratios measured among the taxa from the same stratigraphic level at either the genus or species level (figure 12-10, table 12-4). More recent analyses have demonstrated systematic differences in racemization rates among taxa, however.

The results of amino acid analyses show the expected curvilinear increase in D/L ratio with increasing age for all samples (figure 12-11A). D/L ratios in Asp and Glu increase systematically with age, where the age of each sample was calculated by linear interpolation between tephrochronologic and paleomagnetic data reported by Williams (1994). The D/L ratios for Asp and Glu were linearized according to the transformations derived from the laboratory-heated samples (figure 12-11B). Least-

TABLE 12-4 Aspartic acid (Asp) and glutamic acid (Glu) D/L ratios in ostracodes from the Burmester core, Utah. Only results from samples with L-Glu to L-Ser ratios > 1.0 are shown

Lab. no. (UAL)	Depth (ft)	Age[a] (ka)	Genus/species	Asp D/L[b] mean	Asp D/L[b] 1 S.D.	Glu D/L mean	Glu D/L 1 S.D.	n
2100	6.3	12	*Candona* spp.	0.265	0.009	0.079	0.003	6
2379	44.4	89	*L. ceriotuberosa*	0.374	0.002	0.118	0.008	5
2421	44.4	89	*L. staplini*	0.369	0.029	0.112	0.017	3
2175	44.5	89	*Candona* spp.	0.372	0.011	0.139	0.011	6
2176	44.5	89	*Limnocythere* spp.	0.367	0.006	0.116	0.005	4
2380	46.5	91	*L. ceriotuberosa*	0.369	0.005	0.121	0.003	4
2410	47.8	94	*L. ceriotuberosa*	0.378	0.009	0.125	0.006	4
2381	48.4	95	*L. ceriotuberosa*	0.389	0.011	0.131	0.007	5
2424	48.8	97	*L. staplini*	0.383	0.010	0.159	0.010	2
2419	180.6	392	*Candona* spp.	0.520	0.017	0.280	0.018	5
2407	180.6	393	*L. ceriotuberosa*	0.498	0.017	0.249	0.016	6
2144	285.2	621	*Limnocythere* spp.	0.581	0.019	0.326	0.023	6
2411	285.3	621	*L. ceriotuberosa*	0.579	0.019	0.306	0.024	6
2138	482.0	1119	*Candona* spp.	0.582	0.039	0.395	0.046	4

[a] Approximate ages based on independent tephra and paleomagnetic age controls from Williams (1994).
[b] Mean and standard deviation of "n" subsamples; each subsample was composed of multiple shells.

squares regression shows that the transformed D/L data conform to a linear function of age ($r^2 = 0.97$ for Asp and 0.98 for Glu, after omission of the oldest samples from this analysis). In an earlier analysis (Oviatt et al., 1999), prior to the development of the kinetic equation derived here, Asp D/L ratios from a partial data set were fitted with a parabolic function with equally good success. Ostracodes as old as ~1 Ma at this middle latitude site (current mean annual temperature = ~10°C) have sufficiently retained amino acids for accurate analyses.

DISCUSSION

The strong linear relationship between the transformed D/L ratios and sample ages in the Burmester core suggests that the effective temperature experienced by these samples was relatively constant. This is difficult to reconcile with independent evidence of dramatic climatic changes and fluctuating lake levels that must have alternately emerged and submerged the core site. Perhaps the interglacial-glacial temperature fluctuations are recorded by the second-order variability around the linear (isothermal) trend (figure 12-11). On the other hand, once the shells are buried to a depth of many tens of meters (depending on thermal conductivity), millennial-scale temperature changes are significantly attenuated and the principal control on temperature is the geothermal flux (e.g., Katz et al., 1979). Borehole temperatures measured in the northern Great Basin typically show geothermal gradients between 3 and 4°C per km (Borehole Temperatures and Climate Reconstruction; www.geo.lsa.umich.edu/~climate/). With increasing time, the decrease in the rate of racemization that is expected as the reaction approaches equilibrium is somewhat offset by the increased temperature as the shells become more deeply buried. Nonetheless, within the top 100 m of the Burmester core, analysis of the extent of racemization in ostracodes from additional levels, and reprocessing of samples from levels that were rejected because of contamination, should provide information on the paleotemperature history of the Salt Lake basin over the last ~1 Ma.

The accuracy of paleotemperature or age estimates derived from amino acid data depends upon an array of interdependent variables (McCoy, 1987). It also depends upon the specific application of the technique. In aminostratigraphy (e.g., Miller and Hare, 1980), differences in D/L ratios are used to differentiate stratigraphic units that might otherwise be lithologically identical, to assess the relative ages of deposits, or to correlate units across disjunct exposures. In this approach, the capacity of D/L ratios to resolve units of distinct ages is proportional to the rate of racemization, which, in turn, is a function of site temperature, fossil age, and the particular amino acid that is analyzed. Young fossils under high ambient temperatures using Asp have the highest potential for discerning small age differences.

The capacity of amino acids to resolve differences in sample age and paleotemperature also depends upon the accuracy of the D/L ratios. Although the analytical precision of the technique is high (better than 3%), the analysis is sensitive to contamination by modern amino acids at the sample and the subsample levels. The potential for contamination is

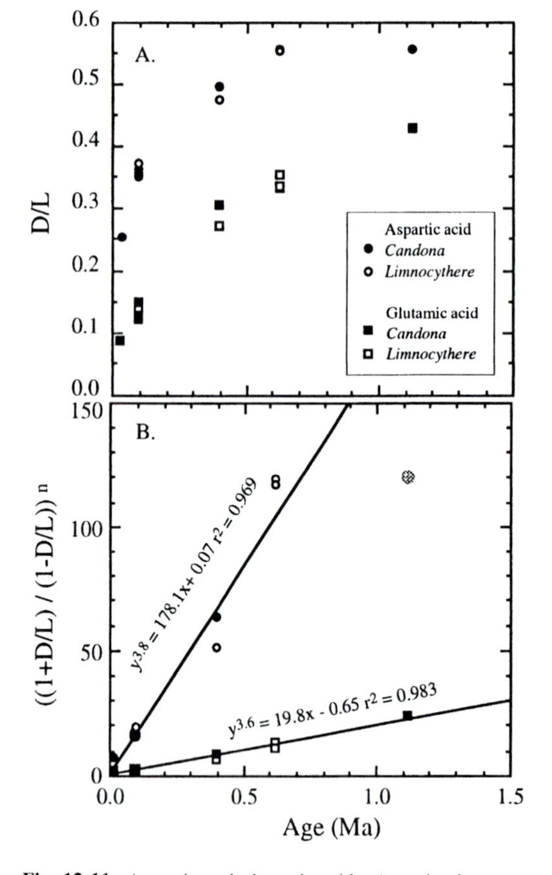

Fig. 12-11. Aspartic and glutamic acid D/L ratios in ostracodes of two genera from the Burmester core, Utah. (A) Untransformed D/L ratios. (B) D/L ratios transformed using equation 12-1, where exponents (3.6 for Asp and 3.8 for Glu) were chosen according to transformations that best fit the data from the laboratory-heated samples. Sample ages are based on independent tephra and paleomagnetic age control (Williams, 1994). The data point represented by the gray circle with a cross was excluded from the regression analysis. Data listed in table 12-4.

greater for microfossils than it is for larger mollusks, which are usually easier to clean and have a low surface area/mass ratio. The presence of contamination from unidentified source(s) in samples from the Burmester core underscores the sensitivity of the procedure. Using the Glu to Ser concentration ratio to recognize aberrant results objectively, and analyzing at least 10 subsamples, helps to circumvent the problem.

The most accurate numerical ages, and all paleotemperature estimates, are derived from D/L ratios that are tied to independently dated material from the same site. In this "calibrated" approach, the rate of racemization is determined for the site; this rate is then used to calculate the ages of samples of unknown age. The accuracy of the age estimate in this situation is directly proportional to the accuracy of the independent age control, and to the difference between the age of the undated sample and the control samples. It also depends upon the accuracy of the D/L ratios for both the calibrated and undated samples. In general, however, the overall age uncertainty is reduced considerably for calibrated ages.

In the absence of independent age control to calibrate the local rate of racemization, the precision of numerical ages depends upon, among other variables, the accuracy of the estimated effective temperature experienced by the sample since deposition. For most terrestrial settings, past temperatures are known within relatively broad ranges (e.g., the average Pleistocene temperature was 5–15°C lower than the average Holocene temperature). Because the rate of racemization is strongly dependent on temperature, an age estimate based upon such a wide range of temperature possibilities will be correspondingly broad. For some purposes, the ability to discern a Holocene shell from a late or middle Pleistocene shell is useful; for others, it is redundant.

Unlike materials collected from emerged deposits in terrestrial settings, the temperature history of ostracodes recovered in sediment cores from lacustrine and marine settings can often be estimated confidently within narrow limits. This is true for thermally stratified deep lakes from middle latitudes where bottom temperatures hover within 1° of the 4°C temperature of maximum-density fresh water. In these situations, the extent of racemization can be used to derive ages with a realistic uncertainty of 25–40% of the sample age (based on the error analysis presented above), without any independent age control. The accuracy will be improved, however, as independently dated samples from lake and marine cores are analyzed and used to develop a calibrated age equation.

Using reverse-phase HPLC, D/L ratios are measured in more than one amino acid. Theoretically, this should enable age estimates to be derived from several simultaneous equations. Where the data converge on a single age, the confidence in the results would be increased. On the other hand, the D/L data are strongly correlated (e.g., Goodfriend, 1991), and the errors associated with the age estimates for a given sample are controlled by the same underlying uncertainties. Nonetheless, in the future, a procedure is needed to maximize the geochronological information from diagenetic reactions involving different fossil biogenic molecules.

CONCLUSION

Ostracodes have outstanding potential for amino acid geochronology. Their shells retain high concentrations of amino acids and exhibit expected diagenetic trends under elevated temperatures and with increasing stratigraphic age. Routine analyses are now practical using new techniques that enable precise D/L separations from samples as small as a single shell (~0.02 mg). Deep-lake and marine environments under relatively stable temperatures and sampled in sediment cores are especially well suited to such analysis.

Acknowledgments: Jordon Bright and Rick Forester identified the ostracodes, and Jordon Bright prepared the samples for amino acid analysis and assisted in the laboratory. William Manley provided input for modeling reaction rates, and he and an anonymous reviewer made helpful comments on an earlier draft. The University of Utah (Department of Geology and Geophysics) allowed access to the Burmester core, and Bob Thompson and Jack Oviatt supplied the Burmester samples. Chris Luecke collected the dredge sample from Bear Lake. Funding was provided through NSF Grant EAR-9896251 and the U.S. Geological Survey's Western Lakes and Catchment Systems Studies (LACS).

NOTES

1. During laboratory hydrolysis under ~6 M HCl, any asparagine in fossil proteins undergoes deamidation to Asp (Zhao et al., 1989), and glutamine is transformed to Glu. The concentrations and D/L ratios of Asp and Glu may therefore include a small component of asparagine and glutamine, respectively, which were converted during sample preparation.
2. In figure 12-8, note that the solution to the age equation (curved line) falls slightly below the midpont of the error envelope. This accurately reflects the exponential relationship between temperature and the rate of racemization. In the Monte Carlo simulation, which uses 5000 iterations that include a normally distributed temperature error, the mean simulated age is slightly older than the age calculated from equation 12-3. The difference between the mean simulated age and the age calculated from equation 12-3 increases with increasing temperature uncertainty, and therefore a greater likelihood of a temperature significantly higher than the mean.

REFERENCES

Bate R. H. and East B. A. (1972) The structure of the ostracode carapace. *Lethaia* **5**, 177–194.

Eardley A. J., Shuey R. T., Gvosdetsky,V., Nash W. P., Picard M. D., Grey D. C., and Kukla G. J. (1973) Lake cycles in the Bonneville Basin, Utah. *Geol. Soc. Amer. Bull.* **84**, 211–216.

Forester R. M., Colman S. M., Reynolds R. L., and Keigwin L. D. (1994) Lake Michigan's Late Quaternary limnological and climate history from ostracode, oxygen isotope, and magnetic susceptibility. *J. Great Lakes Res.* **20**, 93–107.

Goodfriend G. A. (1991) Patterns of racemization and epimerization of amino acid in land snail shells over the course of the Holocene. *Geochim. Cosmochim. Acta* **55**, 293–302.

Goodfriend G. A. (1992) Rapid racemization of aspartic acid in mollusc shells and potential for dating over recent centuries. *Nature* **357**, 399–401.

Goodfriend G. A. and Meyer V. R. (1991) A comparative study of the kinetics of amino acid racemization/epimerization in fossil and modern mollusk shells. *Geochem. Cosmochim. Acta* **55**, 3355–3367.

Goodfriend G. A., Brigham-Grette J., and Miller G. H. (1996) Enhanced age resolution of the marine Quaternary record in the Arctic using aspartic acid racemization dating of bivalve shells. *Quat. Res.* **45**, 176–187.

Katz B. J., Harrison C. G. A., and Man E. H. (1979) The effects of the geothermal gradient on amino acid racemization. *Earth Planet. Sci. Lett.* **44**, 279–286.

Kaufman D. S. and Manley W. F. (1998) A new procedure for determining enantiomeric (D/L) amino acid ratios in fossils using reverse phase liquid chromatography. *Quat. Sci. Rev. (Quat. Geochron.)* **17**, 987–1000.

Light A. (1996) Amino acid paleotemperature reconstruction and radiocarbon shoreline chronology of the Lake Bonneville basin, USA. MA Thesis, University of Colorado, Boulder.

McCoy W. D. (1987) The precision of amino acid geochronology and paleothermometry. *Quat. Sci. Rev.* **6**, 43–54.

McCoy W. D. (1988) Amino acid racemization in fossil non-marine ostracod shells: A potential tool for the study of Quaternary stratigraphy, chronology, and palaeotemperature. In *Ostracodes in the Earth Sciences* (eds P. De Deckker, J.-P. Colin, and J.-P. Peypouquet), pp. 219–229. Elsevier.

Manley W. F., Miller G. H., and Czywczynski J. (2000) Kinetics of aspartic acid racemization in *Mya* and *Hiatella*: modeling age and paleotemperature of high-latitude Quaternary mollusks. In *Perspectives in Amino Acid and Protein Geochemistry* (eds G. A. Goodfriend, M. J. Collins, M. L. Fogel, S. A. Macko, and J. F. Wehmiller), pp. 202–218. Oxford University Press.

Miller G. H. (1985) Aminostratigraphy of Baffin Island shell-bearing deposits. In *Quaternary Environments: Baffin Island, Baffin Bay, and West Greenland* (ed. J. T. Andrews), pp. 394–427. Allen & Unwin.

Miller G. H. and Hare P. E. (1980) Amino acid geochronology: integrity of the carbonate matrix and potential of molluscan fossils. In *Biogeochemistry of Amino Acids* (eds P. E.

Hare, T. C. Hoering, and K. King, Jr.), pp. 415–444. Wiley.

Mitterer R. M. and Kriausakul N. (1989) Calculation of amino acid racemization ages based on apparent parabolic kinetics. *Quat. Sci. Rev.* **8**, 353–358.

Müller P. J. (1984) Isoleucine epimerization in Quaternary planktonic foraminifera: effects of diagenetic hydrolysis and leaching, and Atlantic–Pacific intercore correlations. *Meteor Forsch.-Ergenbnisse*, Reihe Co., No. 38, 25–47.

Oviatt C. G., Thompson R. S., Kaufman D. S., Bright J., and Forester R. M. (1999) Reinterpretation of the Burmester core, Bonneville Basin, Utah. *Quat. Res.*, **52**, 180–184.

Roof S. (1997) Comparison of isoleucine epimerization and leaching potential in the molluskan genera *Astarte, Macoma,* and *Mya. Geochim. Cosmochim. Acta* **61**, 5325–5333.

Stuiver M. and Reimer P. J. (1993) Extended ^{14}C data base and revised CALIB 3.0 ^{14}C age calibration program. *Radiocarbon* **35,** 215–230.

Thompson R. S., Whitlock C., Bartlein P. J., Harrison S. P., and Spaulding W. G. (1993) Climatic changes in the western United States since 18,000 yr B. P. In *Global Climates Since the Last Glacial Maximum* (eds H. E. Wright, Jr, J. E. Kutzbach, T. Webb, III, W. F. Ruddiman, F. A. Street-Perrott, and P. J. Bartlein), pp. 468–513. University of Minnesota Press.

Wehmiller J. F. (1984) Interlaboratory comparison of amino acid enantiomeric ratios in fossil Pleistocene mollusks. *Quat. Res.* **22,** 109–120.

Williams S. K. (1994) Late Cenozoic tephrostratigraphy of deep sediment cores from the Bonneville Basin, northwest Utah. *Geol. Soc. Amer. Bull.* **105**, 1517–1530.

Wurtsbaugh W. and Luecke C. (1995) Examination of the abundance and spatial distribution of forage fishes in Bear Lake (Utah/Idaho). Report, Utah Division of Wildlife Resources, Project F-47-R Study 5, 69 pp.

Zhao M., Bada J. L., and Ahern T. J. (1989) Racemization rates of asparagine–aspartic acid residues in lysozyme at 100°C as a function of pH. *Bioorgan. Chem.* **17**, 36–40.

13. Isoleucine epimerization in eggshells of the flightless Australian birds *Genyornis* and *Dromaius*

Gifford H. Miller, Charles P. Hart, E. Brendan Roark, and Beverly J. Johnson

The calcite eggshells of large flightless birds are often the most common vertebrate remains in the semiarid continental interior of Australia. Eggshells are predominantly of the extant emu, *Dromaius novaehollandiae*, and the ostrich-sized *Genyornis newtoni*, which became extinct about 50 ka ago (Miller et al., 1999). The eggshells of these two birds and their occurrence in Pleistocene sediment was first described by Williams (1981). Eggshells of the two taxa are similar in thickness (1.1 ± 0.1 mm), but *Genyornis* eggshells were larger, averaging 155 mm in length and 125 mm in width, as opposed to 135 mm and 90 mm, respectively, for *Dromaius* eggshells (Williams, 1981). Eggshells of large aquatic birds are found occasionally in eolian and littoral deposits surrounding modern salt lakes that contained permanent water in the Late Quaternary. Bird eggshells typically contain approx. 3% organic matter in a complex organic web (figure 13-1), about 50 times the concentrations in mollusks. The ability of the ratite eggshell to retain indigenous proteinaceous residues over geological time and the close approximation of epimerization kinetics to a first-order reversible reaction was originally documented by Brooks et al. (1990) using eggshells of the african ostrich, *Struthio camelus*. A similar integrity for eggshells of *D. novaehollandiae* is suggested by Miller et al. (1997), but supporting evidence for this contention has not been presented.

In view of the documented exceptional integrity of avian eggshells, systematic changes in protein residues retained within the eggshell may be used to reconstruct environmental temperatures or geologic time if reaction kinetics associated with protein degradation can be described. This is accomplished empirically by heating modern biominerals isothermally at temperatures between 80 and 160°C, temperatures at which the reactions may be monitored in real time. Reaction kinetics derived from the high-temperature simulations are then extrapolated to ambient temperatures to quantify changes observed over geologic time (cf. Bada and Schroeder, 1972; Mitterer, 1975; Miller, 1985; Brooks et al., 1990). This approach is predicated on the assumption that laboratory heating experiments reliably simulate reactions that occur at ambient temperatures. A lingering concern has been whether there may be one or more rate-limiting reactions (e.g., hydrolysis or decomposition reactions, such as decarboxylation, deamination, or deamidation), with sufficiently different activation energy from isoleucine epimerization that they would influence the apparent epimerization rate differently at ambient temperatures than they do at the elevated temperatures at which laboratory simulations are undertaken. These complications are largely untested, although the similarity in activation energies for most of the primary diagenetic reactions (29 ± 2 kcal/mol) suggests that the assumption might be valid. A further concern regards an initial phase of nonlinear kinetics observed in elevated-temperature simulations of Australian eggshells, and ostrich eggshells (P. E. Hare, personal communication, 1990; Brooks et al., 1990); it has always been assumed that this interval is negligible at ambient temperatures, but this assumption has not been rigorously evaluated.

In this Chapter, we present basic data on isoleucine epimerization kinetics in *Dromaius* eggshell based on isothermal experiments conducted on modern eggshell at 85, 110, 126, 143, and 160°C. These results are coupled with an extensive series of radiocarbon-dated Holocene eggshell samples from a variety of settings across Australia, for which reasonable assumptions can be made regarding their mean annual temperatures. Together, these two data sets allow assessment of Arrhenius parameters for isoleucine epimerization and various hydrolysis reactions in *Dromaius* eggshell. To test whether these elevated-temperature experiments reliably simulate digenetic reactions that occur at ambient temperatures, we conducted additional experiments at 143°C using Pleistocene eggshells of both *Dromaius* and the extinct *Genyornis*. We demonstrate that the rate of change in their A/I values at 143°C is the same for Pleistocene and modern eggshells despite vastly different thermal histories. An earlier attempt to address this assumption was conducted by Goodfriend and Meyer (1991) using land snails, although the study was restricted to the early stages of the isoleucine-epimerization reaction. To test whether hydrolysis can limit racemization at ambient temperatures, we compare the extent of isoleucine epimerization in different molecular-weight fractions for eggshells from drastically different thermal environments. We also describe

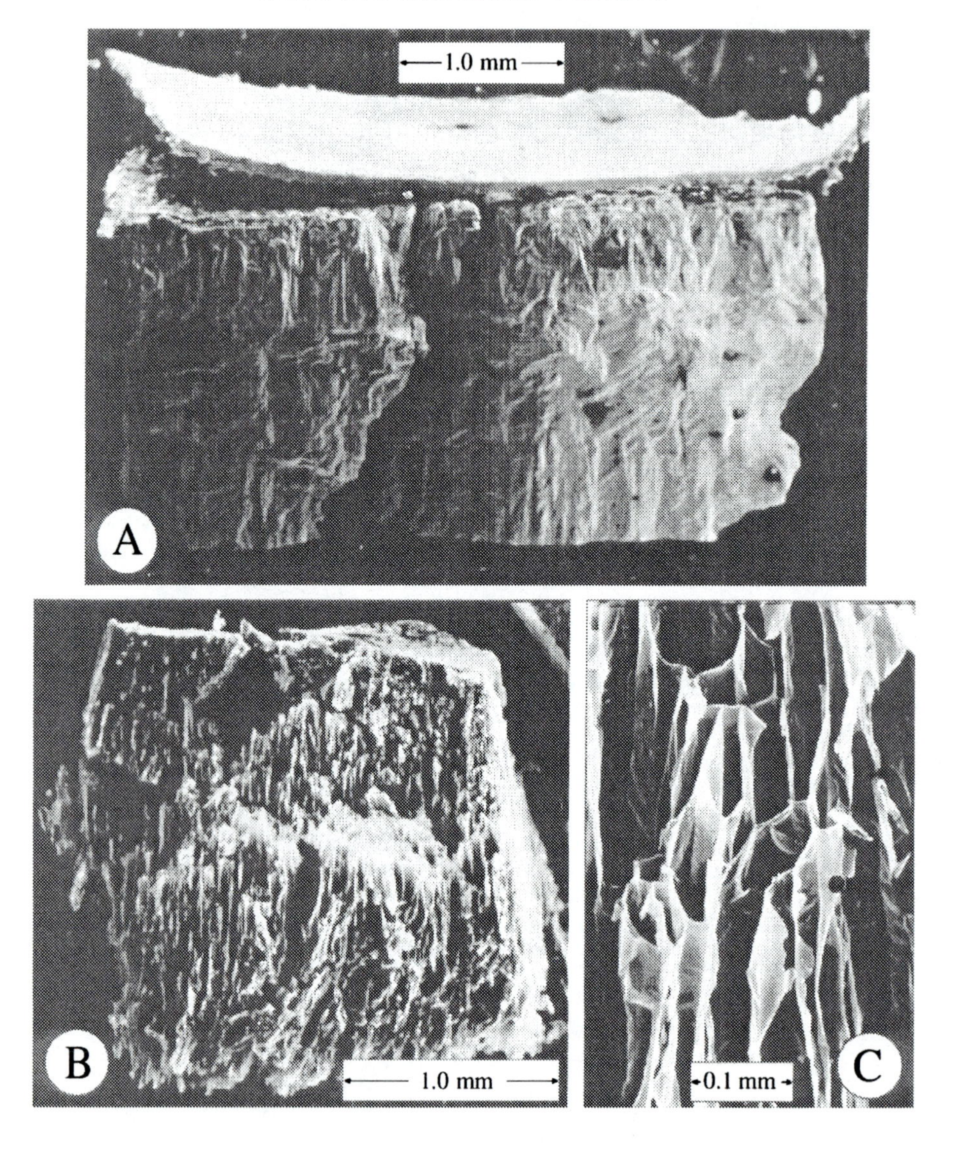

Fig. 13-1. SEM photomicrographs of modern ostrich eggshell demonstrate the relationship between organic and mineral phases. Single calcite crystals that extend through the eggshell are apparent in (A). The inner organic membrane is at the top of the image, beneath which the nucleation sites for calcification are visible. The calcite crystals grew across the eggshell to the outer membrane (not present). Following gentle decalcification of a similar-sized fragment (B), a "protein ghost" remains. At higher magnification (C), the highly structured protein sheaths are apparent, suggesting that the organic matter is contained within, rather than between, the calcite crystals. Scale bars indicate actual sizes. Modern *Dromaius* eggshells have a similar highly structured organic web, as would, presumably, *Genyornis* eggshell. Within a few decades to centuries under typical Australian environmental conditions, the bonds required to maintain the organic web are broken down.

experiments in which Pleistocene samples of *Dromaius* and *Genyornis* are heated at the same time/temperature pairs to derive Arrhenius parameters for the extinct *Genyornis*, despite the obvious lack of modern material. Finally, the potential and limitations of geological applications of the isoleucine-epimerization reaction in eggshell of these two taxa are assessed.

METHODS

Of the 20 or so common protein amino acids, only a few can be viewed as candidates for amino acid geochronology/paleothermometry, because of variations in their molecular stability (e.g., threonine, serine, and methionine are too unstable), their creation by the decomposition of more complex amino

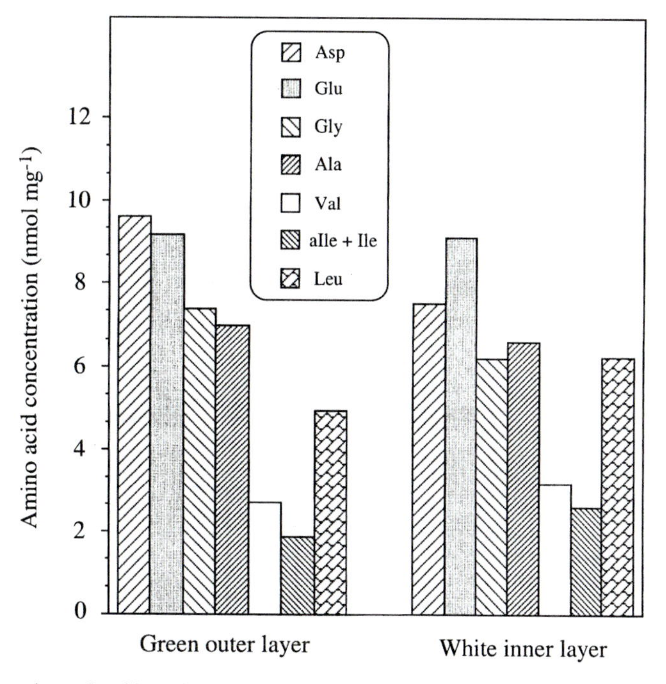

Fig. 13-2. The concentrations of stable amino acids from the outer, dark green layer of modern *Dromaius* eggshell and the inner, white layer of the same fragment. Analyses of two different eggshells are averaged. Amino acid concentrations in the inner white layer of these two samples are similar to the average of the larger fossil data set (figure 13-3). Abbreviations: Asp, aspartic acid; Glu, glutamic acid; Gly, glycine; Ala, alanine; Val, valine; Aile + Ile, the sum of D-alloisoleucine and L-isoleucine; Leu, leucine.

acids (e.g., alanine, glycine), or other competing reactions (e.g., glutamic acid). Of the potentially suitable molecules, isoleucine offers the advantages of a low epimerization rate, thermodynamic stability, and simple and reproducible separation and detection of its isomers. Furthermore, it cannot be readily created by the decomposition of more complex compounds. Unlike most other amino acids, isoleucine has two central carbon atoms, and over time the protein amino acid L-isoleucine racemizes primarily around the alpha-carbon atom (epimerization), creating the non-protein diastereomer D-alloisoleucine. The extent of isoleucine epimerization is expressed as A/I, the ratio of the D-isomer to its L-isomer parent. A/I values increase from approx. 0.02 in a modern eggshell to an equilibrium ratio of about 1.30; at 20°C the reaction requires approx. 250 ka to reach equilibrium. A number of recent reviews outline the principles and limitations of the method (e.g., McCoy, 1987; Miller and Brigham-Grette, 1989; Mitterer, 1993; Wehmiller, 1993; Bada et al., 1999), which need not be repeated here. Of particular relevance are recent studies defining the potential of isoleucine epimerization in ostrich eggshell (Brooks et al., 1990; Miller et al., 1992) and in Australian eggshells (Magee et al., 1995; Magee and Miller, 1998; Miller et al., 1997, 1999).

Cleaning Procedures

Prior to analysis, eggshells are mechanically cleaned by grinding with a Dremel tool to remove surface materials and an outer portion of the eggshell. In the case of *Dromaius* eggshell, which has a tripartite structure (Williams and Rich, 1991), the outer two layers are mechanically removed. The outermost layer of modern *Dromaius* eggshell is dark green and carries a different relative abundance of amino acids than does the denser interior layer (figure 13-2). Analyses of the outer and inner layers in fossil samples and in modern eggshell heated in the laboratory demonstrates that isoleucine epimerization proceeds approx. 10% faster, and A/I values are more variable in the outer layer than in the inner layer. Because the proportions of inner and outer layers are not constant among samples, and because of the greater variability found in the outer layer, we have adopted a preparation strategy that removes everything except the inner layer. *Genyornis* eggshell has a simpler structure, and mechanical cleaning is primarily designed to remove adhering clay and any secondary carbonate.

Once the eggshells are mechanically clean, an additional 33% of the eggshell is removed by the stoichiometric addition of 2 N HCl. The samples are placed in a vacuum centrifuge to ensure that the reaction proceeds to completion. This is especially important for very young samples, in which the organic web is so tightly cross-linked that it protects the carbonate matrix from reacting with the surrounding acid solution. Once the reaction is complete, the sample is considered clean. The clean fragment is broken with a pestle on weigh paper; clean tweezers are used to weigh fragment(s) totaling approx. 15 mg on an analytical balance and the samples are then transferred to sterile vials for preparation of the Free (those amino acids released from their original polypeptide chains by natural hydrolysis) and Total (Free plus peptide-bound amino acids) fractions.

Preparation of Free and Total Amino Acid Fractions

Subsamples of the clean eggshell are decalcified by dissolution in cold 7 N HCl (0.02 ml/mg eggshell). Once the sample has been decalcified, the glass walls behave as an ion-exchange surface, selectively sorbing high-molecular-weight residues. Any transfer of liquid will fractionate the sample on molecular weight size, altering the apparent A/I value (Miller, 1985; Mirecki, 1990). Consequently, Free and Total fractions are prepared from fragments of the same eggshell dissolved in separate vials. The 7 N HCl solution is spiked with the non-protein amino acid norleucine, at a concentration of 6.25×10^{-5} mol/L, to enable absolute concentrations for each amino acid to be determined. This is ten times the concentration we normally use for mollusk preparations, because of the higher concentration of amino acids in eggshell. For the Free fraction, the solution is simply desiccated at 80°C under a stream of N_2 to remove excess HCl; prior to analysis, each sample is rehydrated in 0.1 N HCl. For the Total fraction, the solution is sealed under N_2 in the same vial in which dissolution was accomplished, and individual amino acids are released by hydrolysis at 110°C for 22 hr. The sample is subsequently desiccated as for the Free fraction. Amino acids are separated by automated high-performance liquid chromatography (HPLC) utilizing ion-exchange resin and post-column derivitization with *o*-phthalaldehyde (OPA). Amino acids are identified by excitation with a long-wave UV source, with the concentration directly proportional to the intensity of fluorescence. No adjustments are made for slight differences in sensitivity to OPA for the different amino acids. Run parameters were optimized to provide baseline separation of D-alloisoleucine from L-isoleucine. This typically requires a slightly higher temperature (62°C) than in a standard amino acid analysis. A/I values are determined from peak height ratios; concentrations are derived by comparing peak areas of individual amino acids with the area of the norleucine spike. All samples are analyzed at least twice, with additional analyses if there is significant disagreement between the two. Synthetic standards containing a known amount of D-alloisoleucine are analyzed daily (about every 10th sample) to monitor instrumental precision, and interlaboratory standards (Wehmiller, 1984) are processed annually.

Heating Experiments

Isothermal elevated temperature experiments on modern and fossil eggshell were conducted by embedding one or more fragments of eggshell (50–200 mg) in a quartz sand matrix in a 18 × 250 mm sterile test tube. The sand was collected from an Australian sand dune typical of those containing eggshell and had been previously sterilized by pyrolysis at 550°C for 4 hr. A small amount of purified water (0.5 ml) was added to the sand to ensure adequate moisture for hydrolysis, and the tubes were heat-sealed under normal atmosphere to simulate natural conditions. Modern eggshells were obtained from freshly hatched nests from zoos or from the wild. For most experiments, two parallel series were processed using fragments from two different eggshells to test whether intereggshell variation in isoleucine epimerization kinetics occurs. The samples were placed in controlled-temperature ovens (Blue-M Stabil-Therm) that had been allowed to stabilize for at least one week at the chosen temperature. Temperatures were monitored throughout the experiment with a digital thermometer (Omega DP 460); temperature variations were rarely more than ±1°C from the chosen value. The test tubes containing eggshell embedded in moist sand were placed in the oven in preheated aluminum blocks to minimize the time required for the eggshell to reach oven temperature. Samples were removed at specific times, allowed to cool, then the eggshell extracted and prepared as described above.

RESULTS AND DISCUSSION

Amino Acid Composition in *Dromaius* and *Genyornis* Eggshell

Eggshells of both genera contain abundant amino acid residues, typically 50 times the concentration of common bivalve mollusks. For the most abundant amino acids (Asp, Glu, Gly, Ala), modern *Dromaius* eggshells ($n = 25$) contain about 10 nmol of each per mg eggshell, and about half that concentration for most other common amino acids (Thr, Ser, Val, Ile, Leu) (figure 13-2). Fossil *Dromaius* eggshells, even of late Holocene age, typically contain 20–30% less of the stable amino acids than is observed in freshly hatched eggshell.

In many biologically precipitated carbonates, the relative proportions of amino acids in the organic molecules contained within the biomineral can be used to aid in taxonomic identification (e.g., Kaufman et al., 1992). That is, they are conservative, exhibiting little variation across the range of

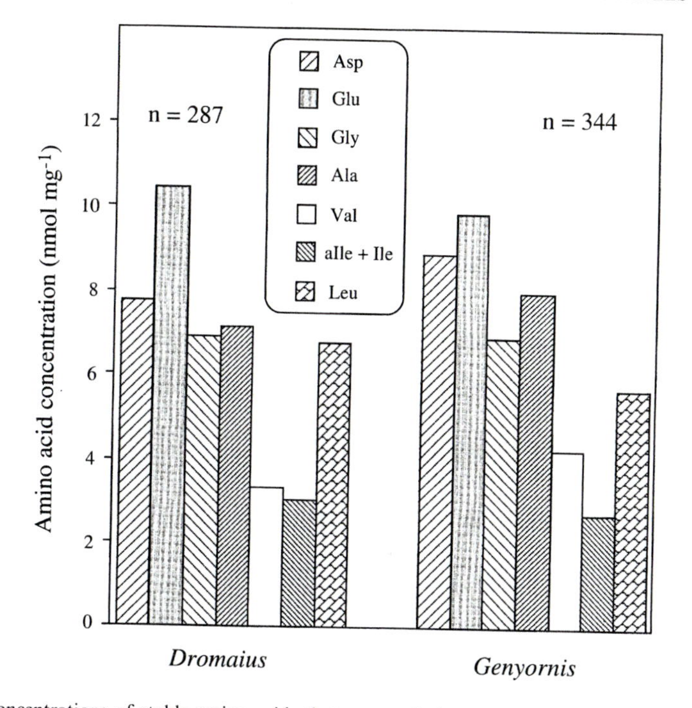

Fig. 13-3. Average concentrations of stable amino acids that are routinely quantified from fossil eggshells of *Dromaius* and *Genyornis*. *Dromaius* analyses are restricted to the inner layer of the eggshell. These samples are largely between 1 ka and 100 ka old. Diagnostic differences in the abundance patterns between the two taxa allow secure species differentiation, even if morphological differences are obscured by severe abrasion. Modern *Dromaius* eggshells are similar in relative composition, but tend to have slightly greater concentrations of each amino acid. Abbreviations are as in figure 13-2.

environments in which the organisms lived. Our analyses suggest that eggshells exhibit a similar taxonomic dependence. Because we have no modern eggshells of the extinct *Genyornis*, we can only compare Pleistocene samples to test whether there is a consistent difference in the amino acid concentrations in *Genyornis* and *Dromaius* eggshells. To make this comparison effectively, it is necessary to correct for the observed decrease in most amino acids with increasing sample age. We have made this compensation by plotting the concentration of each amino acid against the A/I measured in the same sample (a function of sample age), and fitting the data with a simple second-order polynomial. We have then used this expression to "reconstitute" the lost amino acids, and in so doing eliminate the differences related solely to sample age. We compared the reconstituted average concentrations of seven common, relatively stable amino acids in fossil *Dromaius* and *Genyornis* based on analyses of 287 different fossil *Dromaius* eggshell fragments and 344 eggshells of *Genyornis* (figure 13-3). The coefficient of variation in both data sets is approx. 13% for the absolute concentration of each amino acid, but the variations are systematic, such that the ratio of any one amino acid to another amino acid shows considerably less variation, typically approx. 5%. The relative abundances of these amino acids show little variation between eggshells from different regions. Although the concentrations and general pattern of relative abundances of amino acids in the two taxa are broadly similar, differences exist that are suitable for species identification in those instances in which eggshell morphology has been severely degraded; particularly diagnostic are differences in the relative abundances of Val, Ile, and Leu in eggshells of the two taxa.

Derivation of Arrhenius Parameters for Isoleucine Epimerization in *Dromaius* Eggshells

Modern *Dromaius* eggshells from zoo hatches were heated at temperatures of 85, 110, 126, 143, and 160°C for up to 12 time intervals, ranging from 1 hr to more than a year. The general pattern of protein diagenesis revealed through these simulations is summarized in figure 13-4, in which modern *Dromaius* eggshell was heated at 143°C in 12 steps to 500 hr, by which time the A/I had reached epimeric equilibrium. The most stable amino acids (represented by Glu in figure 13-4A) undergo an initial loss, then a very slow subsequent loss, remaining at nearly 90% of their initial concentrations at the end of the simulation. The least stable amino acids (represented by Ser in figure 13-4A) undergo rapid decomposition, dropping below

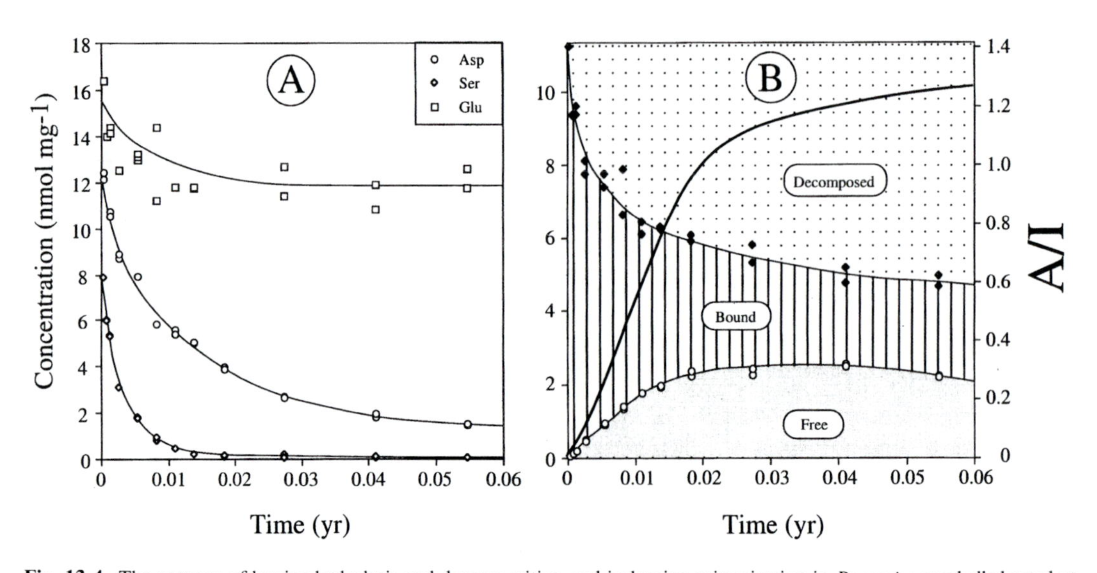

Fig. 13-4. The patterns of leucine hydrolysis and decomposition and isoleucine epimerization in *Dromaius* eggshells heated at 143°C. Eggshells from two different birds (from Denver Zoo) obtained shortly after hatching were used. A The concentration of total leucine (Leu; ◆) and free leucine (that fraction released by natural hydrolysis; ○) plotted against heating time. On the other y-axis, the extent of isoleucine epimerization (based on data in figure 13-5D) measured in the same samples is plotted. The lightly stippled region represents Leu lost to decomposition, or perhaps initially some diffusion. The vertically striped region represent the amount of Leu still peptide-bound, and the densely stippled region represents the amount of Free Leu. B. The concentrations of three other amino acids in the same experiment are plotted against time. The differences in loss reflect differences in the stability of each amino acid, which depends, in part, on interactions with the mineral phase.

10% of their initial concentrations within 100 hr. Amino acids of intermediate stability (represented by Asp in figure 13-4A) follow an intermediate decomposition path. The extent of hydrolysis is represented by changes in the concentration of Free and Total leucine (figure 13-4B). Initially, all of the individual amino acids are bound in polypeptide residues. The concentration of Free leucine increases from an initial value of near zero to a maximum after approx. 250 hr (by which time A/I is above 1.0), after which the rate of leucine hydrolysis is exceeded by the rate of Free leucine decomposition and the concentration of Free leucine declines. The proportion of leucine in the free state rises steadily to about 40% after 250 hr, then increases slowly to approx. 50% by 350 hr but changes little after that, suggesting that a balance is reached between hydrolysis and decomposition reactions.

The heating simulations provide the basis on which Arrhenius parameters for isoleucine epimerization are derived. For all temperatures, A/I values display a regular increase with increasing heating time (figure 13-5). Parallel simulations with fragments from different eggshells yielded statistically indistinguishable results. When plotted in such a way that A/I should be linearly related to time if epimerization were to obey first-order reversible kinetics as defined in equation 13-1, the slope of the linear segment is directly proportional to the forward rate constant, k_1.

$$\ln\left(\frac{1 + A/I}{1 - 0.77\,A/I}\right) = 1.77k_1t, \qquad (13\text{-}1)$$

where k_1 is the forward rate constant for isoleucine epimerization and t is time (in years).

In all five simulations, there is an initial period during which the rate of isoleucine epimerization is nonlinear (discussed below), after which the reaction follows a linear relationship as defined by equation 13-1. Once the linear portion is reached, the reaction continues to follow a linear relationship to an A/I value of approx. 1.0 in experiments at 143 and 160°C (y-axis value of approx. 2 in figure 13-5D and 5E); the lower temperature simulations are linear to the end of the experiment, at which point A/I < 1.

To constrain the rate of isoleucine epimerization at lower temperatures requires capitalizing on the natural experiments of the Holocene, for which radiocarbon provides the time constraint and the temperature of the historical period is assumed to be a reasonable approximation for the long-term Holocene average. We collected *Dromaius* eggshell from Holocene eolian sands in five separate regions of Australia: Lake Eyre's Madigan Gulf, northern Lake Frome, Port Augusta, Lake Victoria and

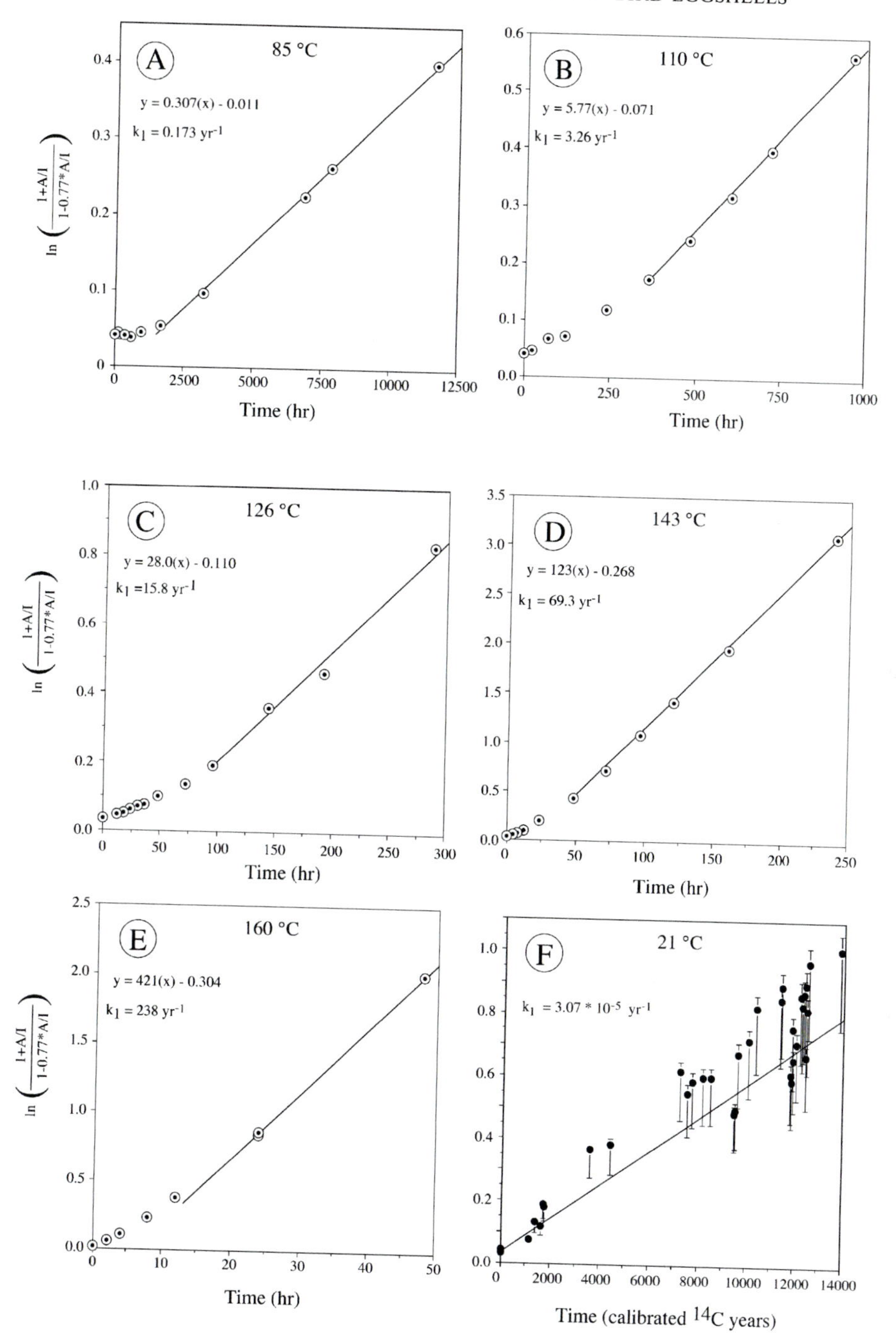

Fig. 13-5. The extent of isoleucine epimerization in modern *Dromaius* eggshell heated isothermally in the laboratory. The forward rate constant, k_1, has been calculated for each temperature from the slope of a linear regression fit to the data after the initial nonlinear phase. Following the initial interval of nonlinear kinetics, the reaction approximates linear kinetics to an A/I of at least 1.0 (2.0 on the x-axis as shown). The rate constants form the basis of the Arrhenius relationship (figure 13-7). Note scale change on the y-axes. A. Isothermal experiment at 85°C. B. Isothermal experiment at 110°C. C. Isothermal experiment at 126°C. D. Isothermal experiment at 143°C. E. Isothermal experiment at 160°C. F. Natural isothermal experiment at about 21°C, the current mean annual temperature around Madigan Gulf, Lake Eyre, from which we have an extensive series of radiocarbon-dated samples. The asymmetric error bars reflect uncertainties regarding burial histories (shallowly buried samples may reflect higher effective temperatures than more deeply buried samples at the same locality due to the greater amplitude of the annual temperature wave (see Wehmiller et al., Chapter 17).

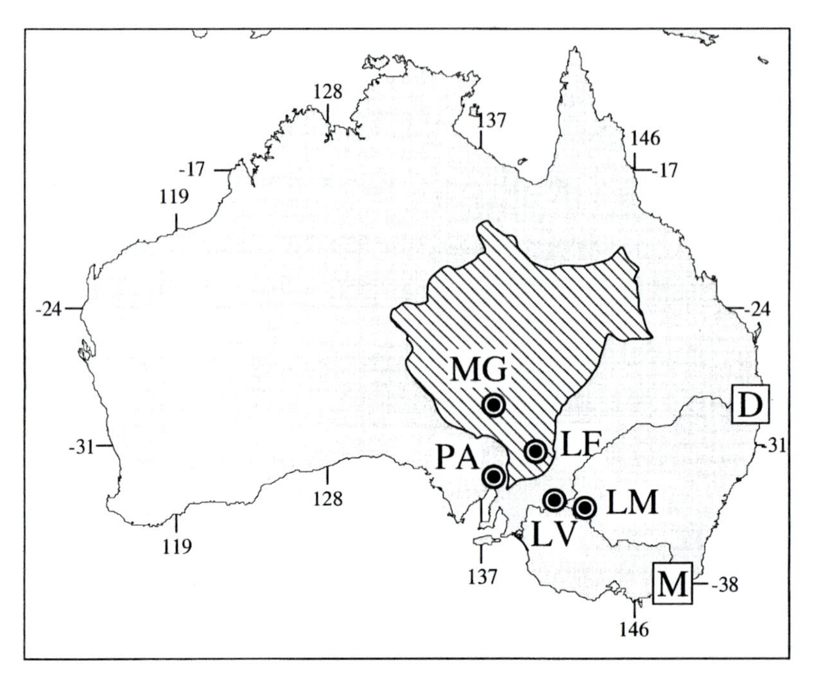

Fig. 13-6. Australia, showing the Lake Eyre Basin (diagonal hatching), the Darling (D) and Murray (M) rivers, and collection sites at Madigan Gulf, Lake Eyre (MG), Lake Frome (LF), Port Augusta (PA), Lake Victoria (LV) and Lake Mungo (LM).

Lake Mungo (figure 13-6). The mean annual temperature in these regions ranges from 21 to 17°C. A/I values in radiocarbon-dated eggshells are listed in table 13-1, along with the forward rate constant, k_1, calculated from a linear regression through the data for each region. The Madigan Gulf data set is shown in figure 13-5F.

An empirical derivation of Arrhenius parameters for isoleucine epimerization in *Dromaius* eggshell is made possible by coupling results from the five high-temperature experiments with the five radiocarbon-dated Holocene time series (figure 13-7). The activation energy is calculated to be 29.2 kcal/mol (equation 13-4),which is slightly higher than reported earlier on a less complete data set (29.1 kcal/mol; Miller et al., 1997). It is similar to the activation energy in *Struthio* eggshell (30.1 kcal/mol; Brooks et al., 1990; Miller et al., 1992), although isoleucine epimerization proceeds about twice as rapidly at ambient temperatures in *Dromaius* eggshell as it does in *Struthio* eggshell.

TABLE 13-1 Forward rate constants (k_1) for isoleucine epimerization based on high-temperature simulations (figure 13-5A–E) and a series of calibrated ^{14}C-dated samples < 15,000 years old from five different regions across Australia (figure 13-6). $Ln(k_1)$ is plotted against the reciprocal of temperature (K) in figure 13-7 to derive Arrhenius parameters

Region/Type	A/I	Calibrated age (yr)	Temperature (°C)	k_1 (yr^{-1})	$ln(k_1)$
"Fresh modern"			160	238	5.47
heated			143	69.3	4.24
eggshell			126	15.8	2.76
			110	3.26	1.18
			85	0.173	−1.75
Madigan Gulf	0.43	13,800	21.3	2.46×10^{-5}	−10.61
Lake Frome	0.41	14,545	20.6	2.50×10^{-5}	−10.59
Port Augusta	0.178	7,920	18.4	1.99×10^{-5}	−10.82
Lake Victoria	0.232	12,977	17.5	1.62×10^{-5}	−11.03
Lake Mungo	0.288	13,045	17.3	1.62×10^{-5}	−11.03

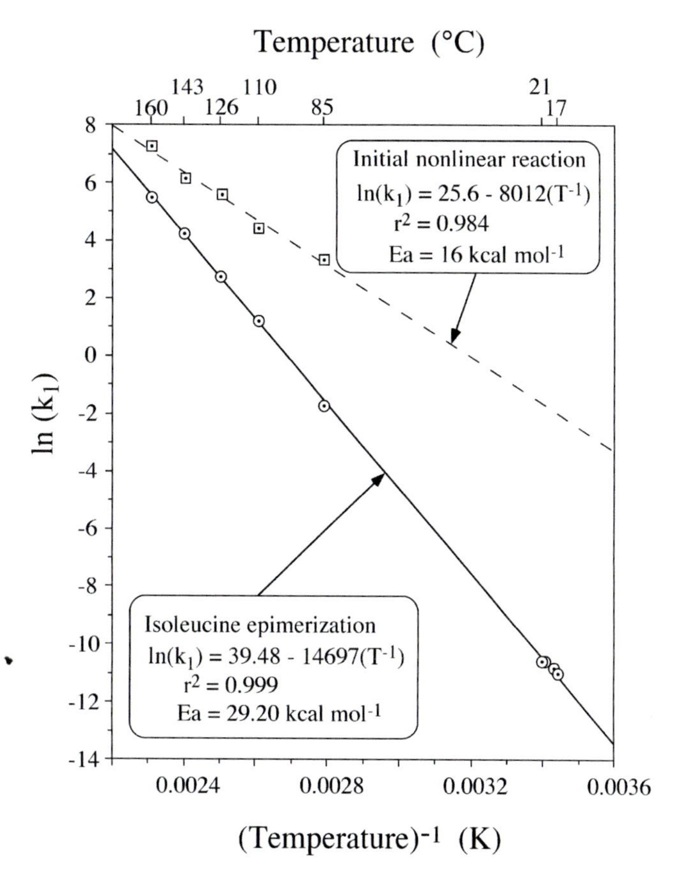

Fig. 13-7. Arrhenius plots of isoleucine epimerization and the initial nonlinear reaction in *Dromaius* eggshell. Isoleucine epimerization (solid line) is based on the five high-temperature simulations coupled with radiocarbon-dated Holocene time-series from five sites in Australia (figure 13-5A–F). Nonlinear reaction kinetics are derived solely from the high-temperature simulations, although the extrapolated rate for ambient temperatures is confirmed by empirical results from Lake Eyre (see Initial Nonlinear Kinetics section in the text).

Do High-temperature Experiments Reliably Simulate Reactions that Occur at Ambient Temperatures?

High-temperature simulations of modern specimens are the basis for deriving Arrhenius parameters for carbonate fossils, including eggshell (e.g., Brooks et al., 1990). To test whether these simulations reproduce reactions at ambient temperatures, but at an accelerated rate whereby they can be monitored in real time, we subjected Pleistocene *Dromaius* eggshell of two very different ages to incremental heating at 143°C, the same temperature at which we heated a modern series of *Dromaius* eggshell samples. The purpose of this experiment was to test whether the patterns of protein diagenesis were independent of the sample's thermal history, a prediction that must be met if the high-temperature simulations are reliable. Two parallel experiments were carried out for each fossil collection using two different fragments, with subsamples of each fragment heated for the same time intervals.

The younger Pleistocene sample (M92-A86) was a nearly complete eggshell found in eolian sand adjacent to Lake Eyre. It has a calibrated ^{14}C age of 13.8 ka, and analyses of nine separate fragments yield an average A/I of 0.41 ± 0.01; the two fragments used in this experiment had A/I values of 0.40 and 0.41. The older sample (M88-A21) was collected by D. Williams in 1983 from a fresh exposure into eolian sands at Lake Palankarinna (28°46.2′ S; 138°23.7′ E). Seven fragments of *Dromaius* eggshell analyzed from this collection yield an average A/I of 0.92 ± 0.03; the two fragments we used both had A/I values of 0.91. On the basis of a correlation to a nearby site where *Dromaius* eggshell samples have slightly higher A/I values, and from which two *Genyornis* eggshell samples have been dated by TIMS U-series (Miller et al., 1999), the Palankarinna samples are about 80 ka old. The younger sample experienced a

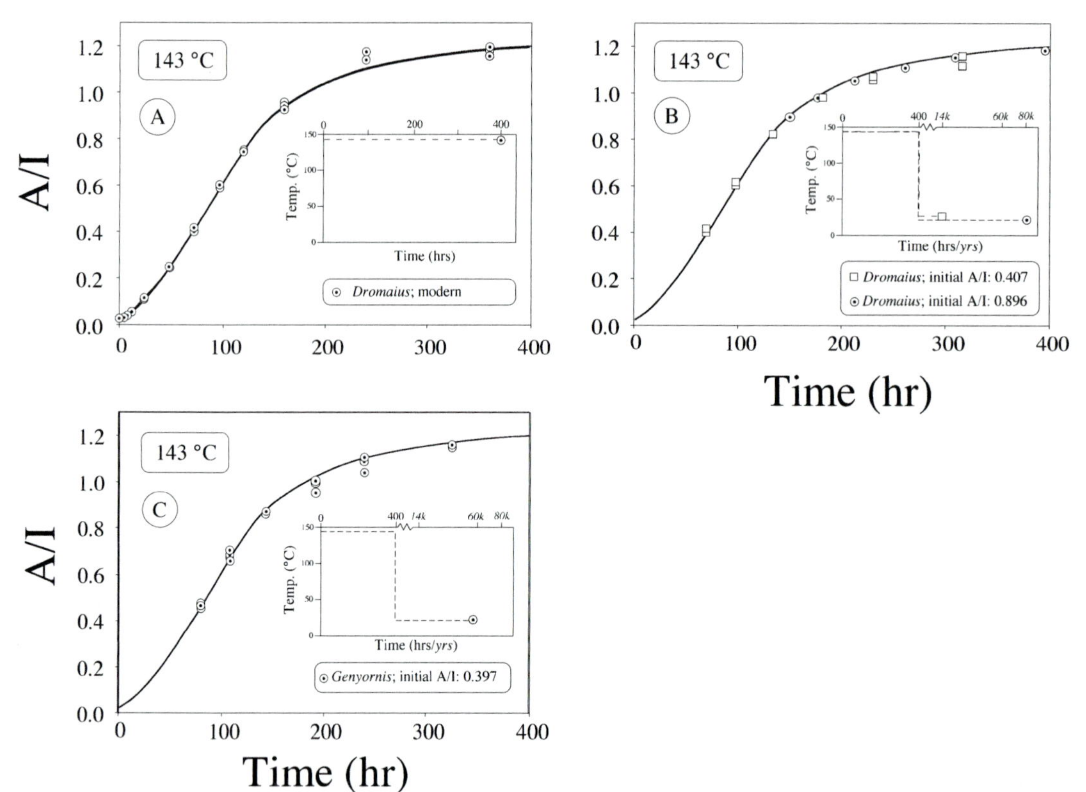

Fig. 13-8. Patterns of A/I increases in modern and Pleistocene eggshell heated at 143°C. The inset in each graph is a summary of the thermal history experienced by each sample. Note that there are scale changes on the time axes of the lower two inset graphs. In all cases, isoleucine epimerization in the Pleistocene samples follows the same pattern as that in the modern eggshell, despite dramatically different thermal histories. A. A/I increase in modern *Dromaius* eggshell heated at 143°C can be almost perfectly approximated by a third-order polynomial expression. This curve is reproduced in figures 13-8B, C. B. The A/I of two fossil *Dromaius* eggshells of very different initial A/I that were heated in the laboratory at 143°C are superimposed on the curve of figure 13-8A after adjustment of their heating times so that their initial A/I coincides with the curve. C. The A/I in three Pleistocene *Genyornis* eggshells with the same initial A/I heated in the laboratory at 143°C are superimposed on the curve of figure 13-8A after conversion of their measured A/I values to equivalent *Dromaius* values, using equation 13-2, and adjustment of their heating times so that the initial A/I value falls on the solid curve.

predicted mean temperature of 21°C (equations 13-1 and 13-4), whereas the older sample has been cooler, on average (16.5°C), because of the lower air temperature during the last glaciation. Thus, these samples have dramatically different thermal histories compared to the samples used to derive the rate constant at 143°C (isothermal at 143°C; figure 13-5D). The samples used in this test case spent 14 ka and 80 ka, respectively, at 10–21°C (accounting for Pleistocene temperature fluctuations), then from several hours to several days at 143°C.

To compare isoleucine epimerization in the heated fossil and the modern eggshell we derived a generic curve for A/I values in modern *Dromaius* eggshell heated at 143°C (figure 13-8A). This was accomplished by fitting a simple curve to the data in figure 13-5D. The curve, which passes through all data points except an anomalously high point at 240 hr, is used to represent the general pattern of A/I increase over time in modern *Dromaius* eggshell at 143°C.

In figure 13-8B, the observed increase in A/I values for the two fossil data sets of heated *Dromaius* eggshell are superimposed on the generalized curve from figure 13-8A by adding a sufficient constant to all of the heating times such that the initial A/I value for each of the fossil time series coincides with the generalized curve of figure 13-8A (74 hr added to M92-A86 and 149 hr added to M88-A21). Within the uncertainties of the data, both fossil samples coincide with the modern *Dromaius* pattern throughout the full length of the experiment.

This result demonstrates that isoleucine epimerization in *Dromaius* eggshell proceeds in a similar manner regardless of its thermal history. To achieve this condition, any complicating reactions, such as deamination, decarboxylation, or hydrolysis, must

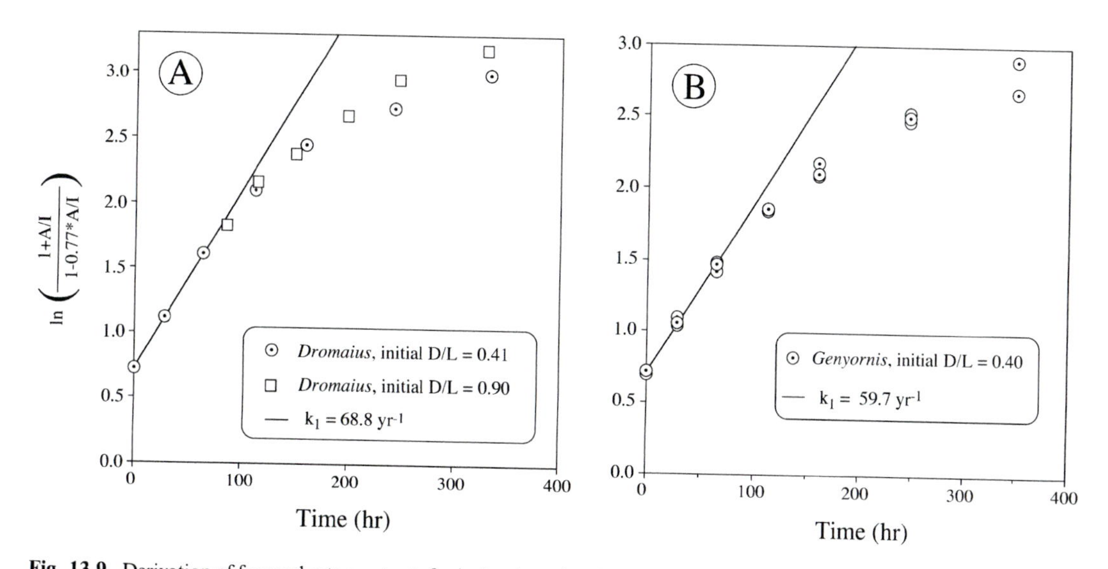

Fig. 13-9. Derivation of forward rate constants for isoleucine epimerization in Pleistocene eggshells that were then heated in the laboratory at 143°C. The reaction follows linear kinetics only to an A/I of approx. 1 (y-axis = 2). Rate constants are derived from linear regressions through the linear portion of the data only. The higher rate constant for *Dromaius* is consistent with the differences in A/I values observed where the two species are analyzed from the same deposit. A. *Dromaius* eggshell of two different ages with very different initial A/I. B. *Genyornis* eggshell in three different fragments with similar initial A/I.

have effectively the same activation energy as epimerization or must play an insignificant role in determining the epimerization rate. The latter is unlikely in light of experimental work demonstrating the dependency of epimerization rate on adjacent amino acids (Kriausakul and Mitterer, 1980). Consequently, we conclude that the activation energies of key rate-limiting reactions are close to 29 kcal/mol.

To test this conclusion further, the fossil-heating results are plotted in the same format as the modern heating simulations (figure 13-5A–E) in which linear segments correspond to first-order kinetics (figure 13-9A). Only the first 100 hr of the fossil simulation are linear in this plot, with the reaction deviating from linear kinetics in the later stages. The departure from linear kinetics occurs at an A/I value of approx. 1.0, and the forward rate constant calculated from the initial linear segment is 68.8 yr, which is indistinguishable from the rate constant derived from the modern *Dromaius* eggshell experiment at the same temperature ($69.3\,\text{yr}^{-1}$; figure 13-5D).

As a final test, we measured A/I values in the Free and Total amino acid fractions in many of the modern eggshells from the high-temperature simulations. Because isoleucine epimerization is most likely to occur when Ile is at a terminal position (Mitterer and Kriausakul, 1984), or during the actual hydrolysis step, A/I values in the Free fraction are always greater than those in the Total fraction. The difference between A/I values in the two fractions is a function of hydrolysis. Consequently, these data may be used to evaluate whether isoleucine hydrolysis proceeds differently relative to isoleucine epimerization at different temperatures (i.e., whether it has a different activation energy). A plot of Free vs. Total A/I values for all of the high-temperature data are described by a third-order polynomial regression ($r^2 = 0.998$; figure 13-10A), with no systematic variation for different temperatures. For a subset of the fossil data we also measured A/I values in the Free and Total fractions ($n = 48$). In figure 13-10B, these are superimposed on the polynomial regression from figure 13-10A. Although they exhibit more scatter about the line than do the heated modern samples, they fall close to the predicted line, without any systematic offset. These data support our contention above, that the activation energy for isoleucine hydrolysis is nearly identical to the activation energy for isoleucine epimerization. The data also imply that there is no preferential loss of low-molecular-weight free amino acids in the fossil samples, demonstrating the integrity of the eggshell and its approximation to closed-system behavior.

Collectively, these results indicate that the laboratory simulations at elevated temperatures provide a reliable means for monitoring reaction kinetics that proceed slowly over geological time at ambient temperatures.

Initial Nonlinear Kinetics

In the five high-temperature simulations (figure 13-5A–E), the proportion of time required before lin-

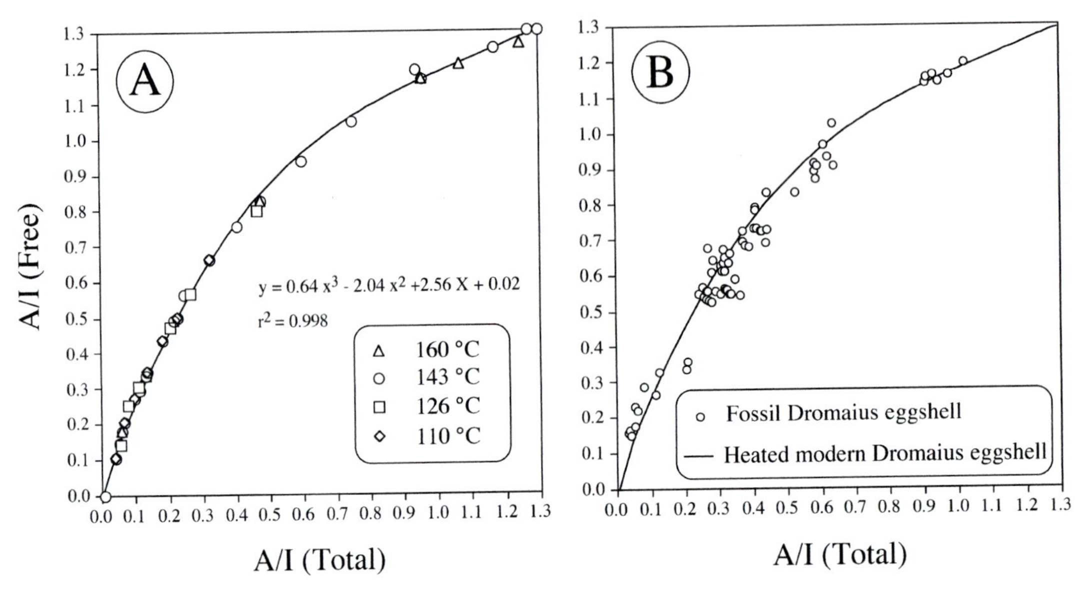

Fig. 13-10. A/I values in the Free fraction compared to the Total fraction in *Dromaius* eggshell. A. A/I in heated modern *Dromaius* eggshells are closely approximated by a third order polynomial expression. There is no apparent difference in the relationship between Free and Total A/I values between eggshells heated at 110, 126, 143 and 160°C. B. A/I in the Free and Total fractions of 48 fossil *Dromaius* eggshell samples are superimposed on the polynomial regression derived in figure 13-10A. The lack of any systematic differences related to temperature suggests that the isoleucine hydrolysis and epimerization reactions have similar activation energies.

ear kinetics are followed increases with higher temperatures, suggesting that the reaction governing the nonlinear portion has a very different activation energy than that for isoleucine epimerization. Although an initial lag in epimerization rate is predicted by the diagenetic model of Kriausakul and Mitterer (1980), it has not been reported in experimental studies. If a significant passage of time is required at ambient temperatures before the interval of nonlinear behavior is complete, the accuracy of quantitative estimates of time or temperature from epimerization data will be reduced.

We have evaluated the duration of the initial nonlinear interval in two ways: (1) by calculating the apparent Arrhenius parameters for the nonlinear reaction in the elevated-temperature experiments and extrapolating to ambient temperatures, and (2) by comparing experimentally the initial phases of isoleucine epimerization at 143°C in eggshell collected shortly after hatching ("fresh modern") and eggshell that remained in the natural environment for several decades after hatching ("aged modern") with our "modern zoo" simulations that were completed within a year of the eggs having been laid.

For the reaction that controls the initial nonlinear segment of epimerization, we define the apparent forward rate constant at any given temperature as the reciprocal of the x-axis (time) intercept of the linear portion of the reaction on isothermal plots of epimerization against time. Although not a true rate constant for the reaction, it is an objective measure that is proportional to the duration of the nonlinear interval, and thus is a suitable measure from which an activation energy may be derived. When the apparent rate constants derived from the high-temperature simulations are plotted on an Arrhenius diagram, they indicate a much lower activation energy than that for epimerization (~16 vs ~29 kcal/mol; figure 13-7), such that the duration of the nonlinear interval becomes increasingly less significant at lower temperatures. Most of our samples have mean annual temperatures of 17–22°C, for which the initial nonlinear segment is predicted from the Arrhenius parameters derived in figure 13-7 to last about 10 years, and therefore to have an inconsequential impact on quantitative geochronology or paleothermometry.

To test whether a few years at ambient temperatures is adequate to complete the nonlinear segment apparent in figure 13-5A–E, we heated multiple fragments of three additional samples, all collected from the Lake Eyre Basin, at 143°C for comparison with the "modern zoo" eggshell heated at the same temperature (figure 13-5D). One of the samples is an "aged modern" eggshell collected in 1994 AD from a cluster of fresh, but bleached white *Dromaius* eggshells that have a ^{14}C activity of 115.98% modern (AA-20890). Eggshell carbon comes from the food eaten by the bird and is equilibrated with atmospheric ^{14}C; for example, an emu eggshell collected live from south-central Australia in 1897 AD has a ^{14}C age of 105 ± 75 yr BP (AA-

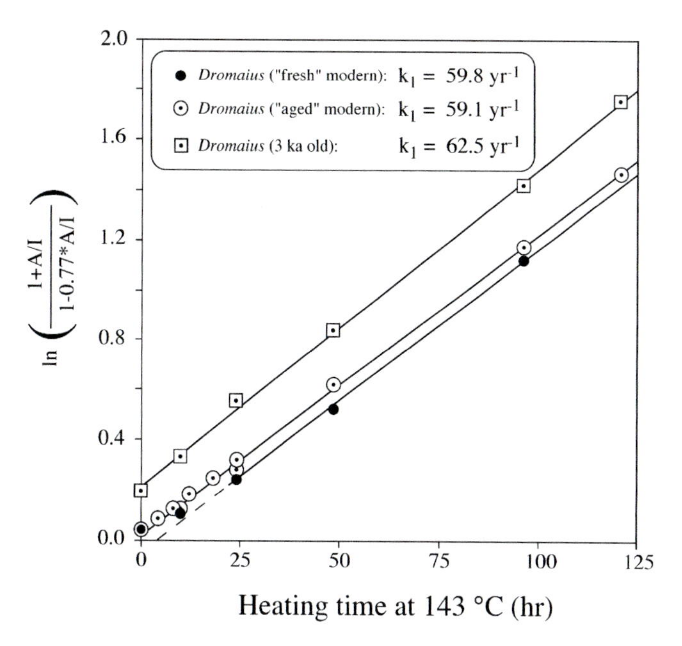

Fig. 13-11. Evaluation of the time required to pass through the initial nonlinear kinetic interval utilizing two "modern" *Dromaius* eggshells and one late Holocene *Dromaius* eggshell, all heated isothermally at 143°C. Only the "fresh modern" eggshell exhibits a nonlinear phase, although all three have similar k1 values.

16395). On the basis of the pattern of Southern Hemisphere atmospheric ^{14}C concentrations during the post-bomb era (J. C. Vogel, Personal communication, 1995), the atmosphere was at 115.98% modern in 1959 AD and again in approx. 1989 AD. Because the eggshell collection was extensively bleached, a process likely to require more than five years, we presume they are from 1959 AD, hence were approx. 35 years old when we undertook the heating simulation. The second sample, a "fresh modern" *Dromaius* eggshell, was collected in 1994 AD from a freshly hatched egg within 5 km of the "aged modern" sample. The third sample was a late Holocene eggshell that was about 3 ka old based on its A/I value (0.12) and mean annual temperature (21°C). The "modern zoo" eggshell used for the initial experiments were provided from a current year's hatch and the simulations occurred within a year of the egg being laid.

To test whether an initial nonlinear behavior in the isoleucine epimerization reaction occurred, subsamples were removed at frequent intervals during the early period of the of the experiment for all three samples. The results, plotted in figure 13-11, show that both the "aged modern" and late Holocene samples follow reversible first-order kinetics over the full length of the experiment with no indication of an initial nonlinear segment. In contrast, the "fresh modern" sample exhibits an initial nonlinear segment, although the duration is less than for the "modern zoo" sample (x-axis intercept for the "fresh modern" is 3 hr, whereas it is 19 hr for the "modern zoo" sample; figures 13-11 and 13-5D, respectively). These differences suggest that at ambient temperature around Lake Eyre (~21°C) less than 30 years is required before the isoleucine epimerization reaction passes through the nonlinear stage.

Despite the differences in their histories since being laid, the forward rate constants derived from the linear segments of the simulations for the three samples are almost identical (59.8 yr^{-1} for the "fresh modern", 59.1 yr^{-1} for the "aged modern", and 62.5 yr^{-1} for the 3-ka-old samples; figure 13-11), and they are similar to the "modern zoo" eggshell (69.3 yr^{-1}). This agreement is striking, particularly since the simulations in figure 13-11 were undertaken at a different time relative to the "modern zoo" simulation, and a difference of only 1°C in oven temperature will produce a 10% difference in the rate constant.

As a final check, the rate constants in the youngest radiocarbon-dated late Holocene samples from Madigan Gulf (1100 and 1600 yr BP) average 2.80×10^{-5} yr^{-1}, indistinguishable from the rate constant calculated from the entire set of radiocarbon-dated samples from Madigan Gulf (3.09×10^{-5} yr^{-1}; figure 13-5F). Thus, latest Holocene samples exhibit no detectable nonlinear segment.

Collectively, these data sets demonstrate that the initial nonlinear reaction, though accounting for a distinctly measurable portion of the time required to attain epimeric equilibrium at temperatures above 80°C, requires no more than a decade at ambient temperatures experienced by typical Australian samples. A decade at 20°C corresponds to a change in A/I value of only 0.0003, and hence does not compromise the application of racemization studies to quantitative geochronology or paleothermometry.

Loss of Amino Acids Over Time: Decomposition or Diffusion?

In the heating of modern *Dromaius* eggshell we noted earlier an apparent initial loss of amino acids, after which loss of the most stable amino acids decreases dramatically. Here we compare the concentrations of amino acids in *Dromaius* eggshells with a wide range of initial state, all heated at 143°C, to evaluate the cause of amino acid loss.

The change in concentration of three amino acids (Glu, Asp, Ser) with increasing time at 143°C in Pleistocene and modern eggshells is shown in figure 13-12A–D, with the time axis for the Pleistocene samples adjusted by adding the amount of time required for a modern sample to reach an equivalent A/I value when heated at the same temperature (figure 13-5D). The pattern for the "fresh modern" eggshell from Lake Eyre is superimposed on the data from the "modern zoo" specimen, demonstrating that despite their very different sources, both eggshells have similar concentrations of amino acids initially, and that their loss with increasing time at 143°C proceeds in a similar manner. A similar, but more pronounced, loss of amino acids has been observed in isothermally heated mollusks (e.g., Kimber and Griffen, 1987) and in time-series of fossil foraminifera (King, 1980; Müller, 1984).

Although the general pattern of changing amino acid concentrations is similar for the "aged modern" sample, initial concentrations are considerably lower for all amino acids, and the proportionate loss of amino acids for equal heating times is substantially less. The Pleistocene samples exhibit similar behavior to the "aged modern" sample when adjusted for their A/I value. Note, however, that glutamic acid (Glu), which degrades most slowly of all amino acids in eggshell, shows no trend in any of the "aged" samples. In contrast, Glu undergoes an initial decline in concentration in the heated "fresh" eggshell, then remains constant, but at a higher concentration than in the "aged" eggshell, even after long heating times.

In an earlier study in which ostrich eggshells were heated in excess aqueous solution incrementally to epimeric equilibrium, the concentration of amino acids in solution exhibited an initial slight rise during the earliest stages of diagenesis, then remained constant throughout the experiment, in contrast to mollusks under the same experimental conditions where the concentration of amino acids in solution increased steadily throughout the experiment (G. H. Miller and R. D. Ernst, unpubished data). It may be possible that a small subset of the organic matter in eggshell is vulnerable to rapid diffusional loss, but that most organic matter, including low-molecular-weight free amino acids, resists diffusional loss. Uncharacteristically, the rate of loss does not appear to be especially sensitive to temperature, given the high concentrations of Glu in the extensively heated modern samples (figure 13-12A).

Whatever the mechanism, the modest initial loss of amino acid that apparently occurs in eggshell is completed within a few decades at ambient temperatures, as stable amino acid concentrations in the "aged modern" (35-year-old) sample are similar to those in Pleistocene samples. To illustrate this point further, the mean concentration of Glu in over 400 late Quaternary *Dromaius* eggshells collected around Lake Eyre is 7.7 ± 1.4 nmol/mg, which is similar to the "aged modern" sample and, even at the 2σ uncertainty, well below that of "fresh modern" eggshells.

The available data suggest that there is an initial loss of approx. 20–30% of the total amino acid pool in *Dromaius* eggshells within the first few decades in the natural environment. A smaller loss is observed during the initial phases of the laboratory simulations. The mechanism that controls this loss remains unclear, but the patterns of epimerization, hydrolysis, and decomposition do not appear to be affected by it.

Derivation of Arrhenius Parameters for Isoleucine Epimerization in *Genyornis* Eggshells

Lacking modern eggshells, we have evaluated the Arrhenius parameters that describe isoleucine epimerization in eggshells of the extinct *Genyornis* by heating Pleistocene eggshell of *Genyornis* and *Dromaius* with similar initial A/I values at 143°C for up to 350 hr. Three different fragments from an extensive collection of fresh *Genyornis* eggshell collected at Coopers Dune just outside Port Augusta (figure 13-6), were used for this experiment. The initial A/I values are 0.390, 0.396 and 0.404, and the samples are estimated to be about 60 ka old. We compared isoleucine epimerization in these eggshells with the pattern observed in M92-A86 described above, a 14-ka-old *Dromaius* eggshell with an initial A/I of 0.41 ± 0.01; subsamples of both collections were placed in the same oven for the same time intervals (figure 13-8B, C). Replicate fragments analyzed at each time step yield similar A/I values throughout the experiment (e.g., figure 13-8C).

For *Genyornis* eggshell, the increase in A/I values deviates from first-order kinetics after 100 hours of heating, steadily decreasing in rate to the end of the experiment (figure 13-9B), a pattern similar to that observed in the comparable Pleistocene *Dromaius* eggshell (figure 13-9A). The forward rate constant calculated from the first three data points is $59.7\,yr^{-1}$, which compares with $68.8\,yr^{-1}$ in

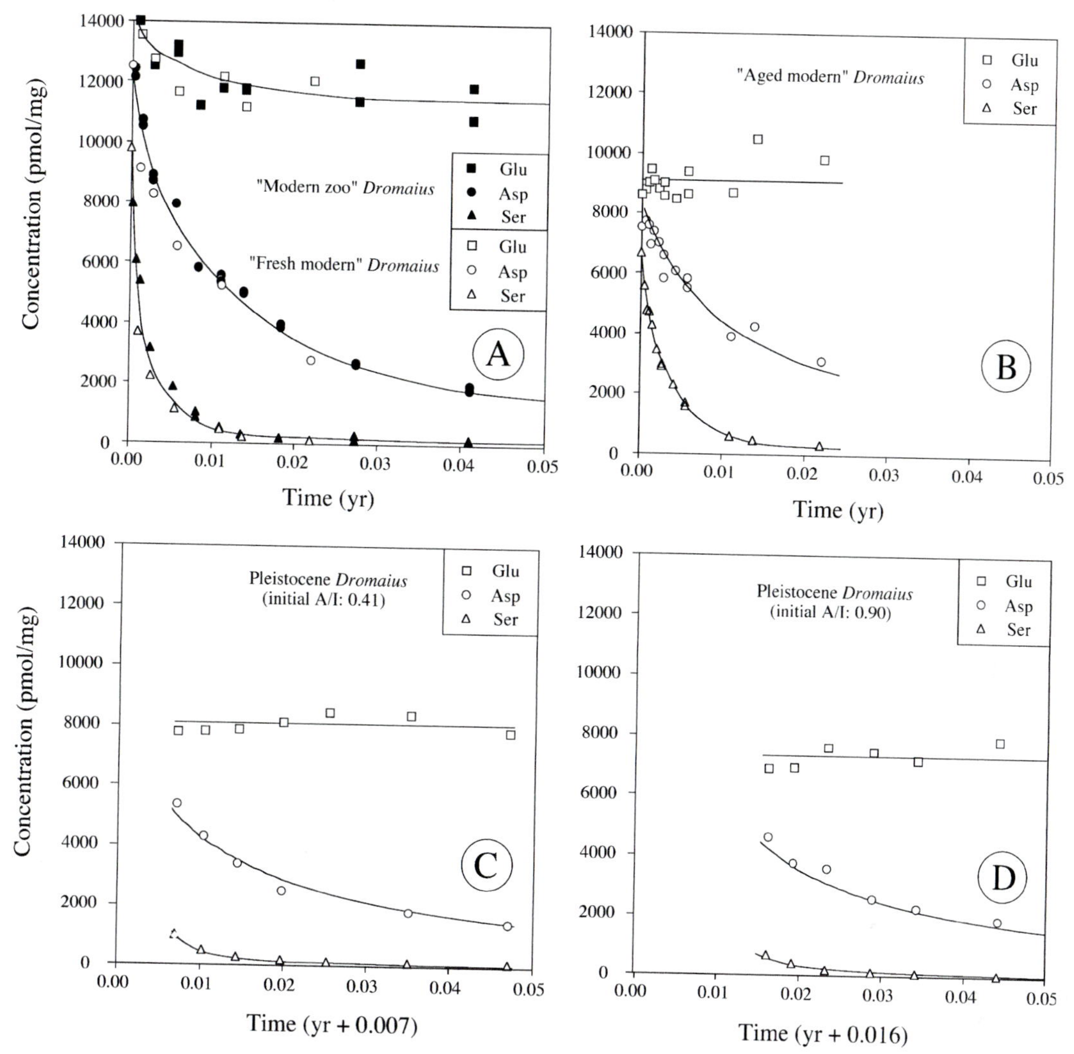

Fig. 13-12. Changing concentrations of three amino acids representing very stable (Glu), metastable (Asp) and least stable (Ser) amino acids in modern and fossil *Dromaius* eggshells heated at 143°C. The similarities between the four panels, despite dramatically different thermal histories, suggest that activation energies for the controlling reactions must be similar. The higher initial concentrations in the "fresh modern" eggshell relative to the other samples suggest that a subset of the amino acid pool may be subject to initial loss, but this appears to occur within a few decades at ambient temperatures. A. Amino acid concentrations in heated modern *Dromaius* eggshell derived from a zoo (solid symbols) and freshly hatched modern eggshells found in the wild in Australia (open symbols). There are no consistent differences between these two sources of eggshells. B. Amino acid concentrations in heated modern *Dromaius* eggshell that had been exposed at the surface for a few decades ("aged modern"). The initial concentrations are lower, and the loss with time, particularly for the stable amino acid Glu, is significantly lower than in the "fresh modern" samples. C, D. Amino acid concentrations in heated Pleistocene *Dromaius* eggshell; heating time is adjusted by the amount required for the "fresh modern" eggshell heated at 143°C to reach the same A/I value as the initial value in the fossils (figure 13-5).

Pleistocene *Dromaius* heated at the same temperature. Because the Pleistocene *Genyornis* and *Dromaius* simulations were conducted at the same time, and hence had identical temperatures, subtle differences in rate constants are significant. The ratio of the two rate constants is 1.15. To confirm this difference in epimerization rate, A/I values for the two taxa from this heating experiment are plotted in the same form as equation 13-1 (figure 13-13A), where they follow a simple linear relationship with a slope of 1.19 (similar to the ratio of their forward rate constants). Using the slope derived

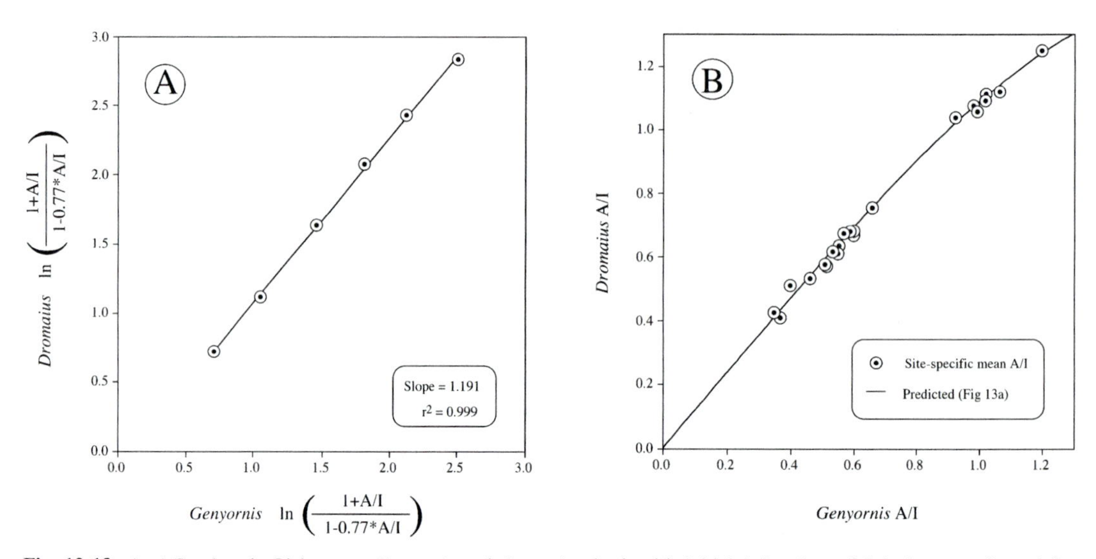

Fig. 13-13. A. A/I values in Pleistocene *Genyornis* and *Dromaius*, both with initial A/I values of 0.4, that were heated for identical time intervals at 143°C for up to 250 hr plotted in the form of equation 13-1. The slope of the regression line (1.19) defines the ratio of the forward rate constants for the two taxa. **B**. The relationship derived in figure 13-13A is used to derive an empirical prediction for the ratio of A/I in these taxa (equation 13-2 and solid line in figure 13-13B). The mean A/I values for each taxon in 22 Pleistocene sites where they occur together are superimposed on the predicted relationship (solid dots, figure 13-13B). The close correspondence of observed A/I in the two taxa in Pleistocene sites at ambient temperature with the ratio predicted from the simulation at 143°C suggests that the activation energy for isoleucine epimerization in *Genyornis* eggshells is not significantly different from that in *Dromaius* eggshells.

from figure 13-13A rather than the ratio of rate constants, because the slope is constrained by a larger data set, we calculate an empirical relationship to convert A/I values in *Genyornis* eggshell to equivalent A/I values expected in *Dromaius* eggshell from the same deposit (equation 13-2; solid curve in figure 13-13B).

$$y = -0.166x^3 + 0.092x^2 + 1.162x + 0.003 \quad r^2 = 0.999 \qquad (13\text{-}2)$$

where y = *Dromaius* A/I and x = *Genyornis* A/I.

In practical terms, for most of the samples encountered in central Australia (*Genyornis* A/I between 0.3 and 0.9), an equivalent *Dromaius* A/I value may be calculated by simply multiplying the *Genyornis* A/I value by 1.16.

Superimposed on the theoretical curve of figure 13-13B are all of the empirical data available to us where we have measured multiple individual samples of *Dromaius* and *Genyornis* eggshell from the same stratigraphic horizon (table 13-2). Fitting the empirical data with a third-order polynomial equation results in essentially the same expression as equation 13-2, with an r^2 of 0.998. The close correspondence of our expression derived by heating Pleistocene *Dromaius* and *Genyornis* eggshells and the ratio of the two A/I values where the eggshells of the two taxa are found together in a natural setting indicates that the activation energy for isoleucine epimerization in these two taxa is indistinguishable, and allows us to extend our applications in paleothermometry and geochronology into areas in which either of the two taxa occur.

The Arrhenius equations describing the temperature dependency of the rate of isoleucine epimerization in eggshells of *Genyornis* and *Dromaius* are:

$$\textit{Genyornis}: \quad \ln(k_{IG}) = 39.31 - 14697 \qquad (13\text{-}3)$$

$$\textit{Dromaius}: \quad \ln(k_{ID}) = 39.48 - 14697 \qquad (13\text{-}4)$$

where k_{IG} and k_{ID} are the forward rate constants in *Genyornis* and *Dromaius*, respectively, and T is temperature (K). The difference in intercepts is calculated from the ratio of the two rate constants determined at 143°C (figure 13-13A). Activation energies are demonstrated empirically to be identical (figure 13-13B).

These equations, when combined with equation 13-1, provide the general solutions relating time, temperature and A/I value for each taxon, whereby any of the three variables may be predicted if the other two are known.

Closed-System Behavior of Avian Eggshells

Indigenous protein residues retained by fossil eggshells are less susceptible to diffusional loss compared to other biominerals, even over long periods

TABLE 13-2 Australian sites from which multiple eggshell fragments of both *Genyornis* and *Dromaius* were collected from the same stratigraphic level. Average A/I values for each taxon are listed and plotted in figure 13-13B. The relationship between the two values is given by equation 13-2. General site locations are given in figure 13-6

Site	Figure 13-6 ref.	*Genyornis* mean A/I	*n*	*Dromaius* mean A/I	*n*	A/I *Dromaius*/ A/I *Genyornis*
Perry Sand Hills	LV	0.357	5	0.426	3	1.23
Wood Point	PA	0.365	8	0.411	5	1.13
Coopers Dune	PA	0.397	3	0.513	4	1.29
GN site, Hunt Pen.	MG	0.460	4	0.535	1	1.16
Mappa Callina Channel	MG	0.506	3	0.577	3	1.14
Cooper Mouth	MG	0.515	5	0.572	4	1.11
Williams Point	MG	0.533	50	0.616	11	1.16
Eric Is.	MG	0.549	5	0.610	3	1.11
East coast	MG	0.550	2	0.637	2	1.16
Cooper Mouth North	MG	0.550	5	0.630	3	1.15
KW Bluff	MG	0.567	4	0.673	2	1.19
Tick Cove	MG	0.588	4	0.679	4	1.15
KW Bluff	MG	0.597	6	0.667	6	1.12
Big John Ck	LF	0.598	5	0.682	2	1.14
Bully Sand Hill	MG	0.659	9	0.756	5	1.15
Putnamutanna	LF	0.922	5	1.038	2	1.13
Putnamutanna	LF	0.978	2	1.075	3	1.10
Putnamutanna	LF	0.989	3	1.055	2	1.07
Putnamutanna	LF	1.016	2	1.090	3	1.07
Putnamutanna	LF	1.017	3	1.112	3	1.09
Putnamutanna	LF	1.062	3	1.118	2	1.05
Putnamutanna	LF	1.198	2	1.249	3	1.04

of geological time (e.g., Brooks et al., 1990). Although the reason for the greater integrity of eggshell relative to mollusk shell has not been evaluated systematically, the results of our high-temperature simulations using modern and fossil eggshells helps to document the predictable behavior of protein diagenesis in these remarkable biominerals.

On the basis of the available evidence, we suggest that differences in the positions of organic material relative to the mineral phase in different carbonate fossils may explain the greater integrity observed in eggshells. Unlike most of the organic matter in mollusk shells, the function of which is primarily to nucleate calcite or aragonite crystals precipitated by the organism to enlarge its exoskeleton, organic matter in eggshell is apparently added to the calcite crystals to provide additional strength to the eggshell, and, in the process, to protect the developing embryo better. Nucleation of the calcite crystals in eggshells occurs on organic membranes on the inside of the eggshell (figure 13-1A), and the crystals then grow to an external membrane, outside the eggshell. Both membranes are rapidly degraded after hatching. The organic matter within the eggshell's mineral phase appears to be largely intracrystalline. We base this contention on the highly structured organic sheaths observed in decalcified modern eggshell (figure 13-1C), which are spaced at intervals of about 50 μm (the calcite crystals are approx. 200–500 μm wide). In contrast, organic matter in molluscan fossils is dominantly intercrystalline, although there are also minor amounts of intracrystalline organic material in these and most other biominerals (e.g., Berman et al., 1988). Additional studies are required to fully evaluate the relationship between the organic material and the mineral phase, but, as Collins and Riley (Chapter 11) note, targeting the intracrystalline organic matter may improve the precision of amino acid geochronology in molluscan samples. Intercrystalline organic matter may diffuse along crystal boundaries, and be lost from the biomineral, as suggested for mollusks by Sejrup and Haugen (1994), whereas intracrystalline residues in eggshells would lack such a pathway and consequently would diffuse far more slowly – an explanation first proposed informally by P. E. Hare (personal communication, 1990).

Limitations To Absolute Dating and Quantitative Paleothermometry

The predictable and reproducible behavior of protein diagenesis in general, and isoleucine epimerization specifically, in eggshells of the extant emu,

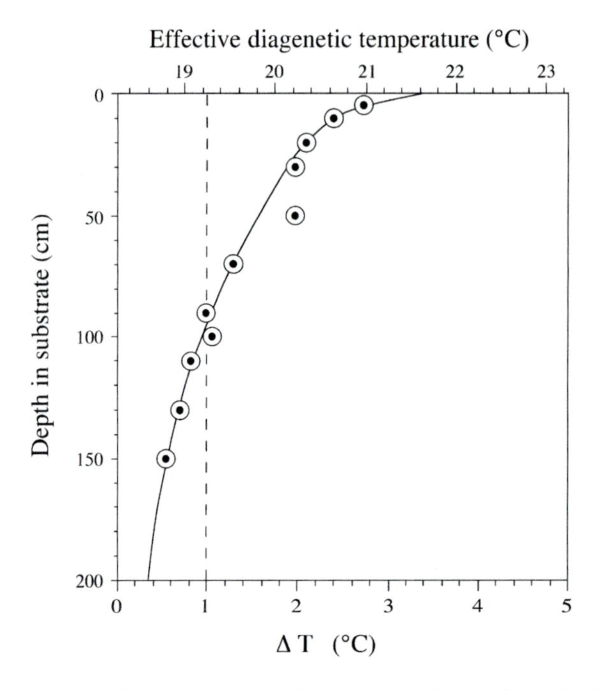

Fig. 13-14. Observed subsurface seasonal temperature fluctuations from Lake Mungo (figure 13-6) reconstructed from data in Ambrose (1984). The impact of seasonal heating is shown both as a change in the effective diagenetic temperature (top scale) and ΔT (lower scale), the difference between the arithmetic mean annual site temperature and the effective diagenetic temperature felt by the isoleucine epimerization reaction using equations 13-3 or 13-4. Because they experience seasonal thermal extremes, samples continuously buried in the upper 30 cm of the subsurface will epimerize much more rapidly than will deeply buried samples, despite all samples having the same arithmetic mean temperature. A minimum long-term burial depth > 1 m is preferred to minimize this effect.

D. novaehollandiae, and the extinct *G. newtoni*, suggests that A/I ratios may be used to provide secure dating control and reconstructions of past temperatures across much of the semiarid zone of Australia. The simplest applications, relative dating and correlation (aminostratigraphy) require no additional information, and may be applied to any region over which the temperature histories are likely to be similar (within 1°C). In such regions, sites with similar A/I values are correlative, and those with higher or lower A/I values are older or younger respectively.

Using a single equation with two unknowns to provide a quantitative estimate of age or temperature from A/I values measured in eggshell of either genus requires an independent assessment of either time or temperature. Most frequently, an independent age is sought. Once an independent assessment of sample age is obtained, an average rate constant can be computed (equation 13-1), and this can be applied to obtain dates from other sites with similar A/I values. However, Quaternary temperatures fluctuated widely, and a temperature history must be developed to define the rate constant adequately and to date sites reliably across the full range of A/I values. It is in the derivation of paleotemperatures that isoleucine epimerization provides the most precise constraints. If a time-series of independently dated sites can be developed, then A/I values in eggshells from those sites can be used to derive the temperature history of the region (e.g., Miller et al., 1997). Dating precision of $\pm 5\%$ is adequate for thermal history reconstructions, and even dating uncertainties of $\pm 10\%$ will allow temperature reconstructions of better than ± 1°C.

The primary limitations in the geological applications of A/I values in eggshells are due to thermal history uncertainties related to unconstrained changes in burial depth, and nonlinear kinetics for A/I > 1. Shallowly buried samples experience a greater amplitude of the annual temperature cycle than do more deeply buried samples. Because of the exponential dependency of epimerization rate on temperature (e.g., figure 13-7), the shallower samples will have higher A/I values than will more deeply buried samples of the same age. The importance of burial depth can be illustrated from the detailed temperature measurements of Ambrose (1984) at Lake Mungo (figure 13-6), where conditions are similar to those of many collecting sites in the interior. Closely spaced temperature recorders in the upper 1.5 m of the sediment column provide the basis upon which the difference between the arithmetic mean temperature and the "effective" temperature recognized by the epimerization reaction can be calculated (ΔT; figure 13-14). For samples buried less than 50 cm deep, the departure between the two is significant and theoretically would lead to higher A/I values than would be predicted from the mean annual site temperature, a problem originally referred to by Wehmiller (1977) as the "kinetically active" zone. However, practical examples testing this prediction have been difficult to find. Goodfriend (1991) reports no trend in the rate of racemization with burial depth in land snails of the same age from the Negev Desert, and a similar observation is reported by Ellis et al. (1996) for land snails from Texas.

The most extreme actual example we have documented that shows this effect is a nest of two whole (but crushed) *Dromaius* eggshells discovered by us emerging from the current land surface adjacent to Lake Eyre. The nest was in a thin (< 1 m thick) eolian unit of last glacial maximum age, common around Madigan Gulf (figure 13-6; Magee and Miller, 1998), and has almost certainly been very close to the surface since burial. The radiocarbon age of the eggshell is $27{,}345 \pm 530$ yr BP (AA-19014), and the A/I value averages 0.739. This A/I value contrasts with A/I values which range from 0.494 to 0.518 in five more deeply buried *Dromaius* eggshell of about the same age (^{14}C dates between 25,000 and 28,000 yr BP),. The difference in A/I values corresponds to a long-term average temperature 2.4°C higher for the shallowly buried nest than for the deeply buried samples. From figure 13-14, this temperature difference implies an average burial depth of only 10 cm over the past 27,000 yr. The nest was probably buried deeper than 10 cm for some part of its history, and subsequent deflation has brought it closer to the surface. Eggshell has a very short survival rate when exposed at the surface (a few decades), because the calcite eggshell is rapidly abraded by wind-blown quartz grains that dominate the semiarid zone.

This example demonstrates the need to evaluate each site in terms of its probable burial history. To minimize the effect of the annual temperature wave, sites at which eggshell has remained buried by more than 1 m are preferred. To illustrate the point with the 27 ka-yr-old sample discussed above, if the A/I value of a deeply buried eggshell is 0.49, then an eggshell from a ~1 m long-term depth, where the effective temperature is about 1°C higher than that experienced by a deeply buried sample (figure 13-14), is predicted to have a value of 0.57. If it were erroneously assumed that the temperature of the sample at a depth of 1 m was the same as for the deeply buried sample, the predicted age of the shallow sample would be 32 ka, 5 ka (20%) older than the actual age. With careful site observation, this is a reasonable approximation of the maximum likely error. In many collection sites, eggshell is concentrated as a lag in recent deflation hollows; consequently, burial histories may be difficult to assess.

The deviation of isoleucine epimerization from linear kinetics above an A/I value of approx. 1.0 precludes quantitative time/temperature calculations from the most epimerized samples. An A/I value of 1.0 corresponds to approx. 120,000 yr at Madigan Gulf, longer at cooler sites and less at warmer sites. Qualitative assessment of sample age is still possible, as A/I values exhibit a steady increase to equilibrium. It may be possible to model the nonlinear late stages of the reaction, and in so doing derive a quantitative prediction of time/temperature for the later stages of the reaction as well, although the uncertainties will be necessarily greater.

CONCLUSIONS

Isothermal simulations of time at five temperatures between 85 and 160°C conducted on freshly laid modern *Dromaius* eggshells, modern eggshells that had been aged for several years, and Pleistocene eggshells of *Dromaius* and the extinct *Genyornis*, coupled with a large data set of ^{14}C-dated Holocene *Dromaius* eggshell, provide the basis for evaluation of protein diagenesis in both taxa. The experimental results are used to derive Arrhenius parameters, and to evaluate whether high-temperature simulations reliably emulate the various diagenetic reactions that occur in eggshell at ambient temperatures. The primary findings of this study are listed below.

- Activation energy for isoleucine epimerization in both *Dromaius* and *Genyornis* is 29.2 kcal/mol, although the intercept is slightly different, with epimerization occurring in *Dromaius* about 1.16 times faster than in *Genyornis*. The reaction follows reversible first-order kinetics to an A/I of approx. 1.0; at higher A/I values epimerization proceeds regularly, although more slowly than is predicted by first-order kinetics.
- An initial nonlinear phase clearly evident at elevated temperatures, and predicted from theoretical modeling, is insignificant at ambient temperatures, requiring no more than a decade before linear kinetics are followed at 20°C.
- Several empirically derived lines of evidence indicate that the activation energies of hydrolysis and most decomposition reactions are similar to the activation energy for epimerization. Consequently, equations derived from the high-temperature simulations of time may be reliably extrapolated to derive rates for these reactions at ambient temperatures. These simulations allow reactions that occur slowly over geological time to be monitored reliably in real time in the laboratory.
- Avian eggshells more closely approximate a closed system than any other biomineral yet eval-

uated. Although the reasons for this have not been investigated explicitly, the available evidence suggests that most of the organic material in eggshells is intracrystalline, and hence resistant to diffusional loss. In contrast, organic matter is largely intercrystalline in molluscan shells, and low-molecular-weight organic residues may diffuse out growth bands, complicating the reaction kinetics.

- Because of the systematic behavior of protein diagenesis in avian eggshell, it is possible to derive more accurate predictions of sample age and integrated thermal history from A/I values in eggshells than for most other carbonate systems. The most serious limitations are the uncertain burial histories and the nonlinear behavior of reaction kinetics as epimerization approaches equilibrium. Careful evaluation of the former at each site can minimize the associated errors caused by long-term shallow burial, whereas additional modeling efforts may allow quantification of the reaction in the later, nonlinear stages.

Acknowledgments: P. E. Hare was the first to recognize the potential of ratite eggshells for amino acid racemization studies, and his advice and early encouragement are greatly appreciated. This research has been supported by the Paleoclimate Program of the Division of Atmospheric Sciences, National Science Foundation, under grants ATM-9311303 and ATM-9709806. Additional assistance has been provided by the University of Colorado Faculty Fellowship Program and the Research School of Earth Sciences, Australian National University. J. W. Magee of the Australian National University has been instrumental in the collection and interpretation of most of the eggshell samples used in this study, and his collaboration is gratefully acknowledged. N. Tuross, Smithsonian Institution, assisted with figure 13-1. The U.S. National Museum provided the emu eggshell collected live in 1897 AD from the Riverina region, Australia (USNM sample number 30377). We thank R. Ernst and A. Rosewater for assistance in the laboratory, and S. Webb, G. Atkin, O. Miller, and P. Clark for assistance and companionship in the field. We appreciate the logistical advice and access to Australian field sites provided by station owners and Aboriginal leaders.

REFERENCES

Ambrose W. R. (1984) Soil temperature monitoring at Lake Mungo: implications for racemization dating. *Australian Archaeology* **19**, 64–74.

Bada J. L. and Shroeder R. A. (1972) Racemization of isoleucine in calcareous marine sediments: kinetics and mechanism. *Earth Planet. Sci. Lett.* **15**, 1–11.

Bada J. L., Wang X. S., and Hamilton H. (1999) Preservation of key biomolecules in the fossil record: current knowledge and future challenges. *Phil. Trans. Roy. Soc. Ser. B.* **354**, 77–87.

Berman A., Addadi L., and Weiner S. (1988) Interactions of sea urchin skeleton macromolecules with growing calcite crystals: a study of intracrystalline proteins. *Nature* **331**, 546–548.

Brooks A., Kokis J. E., Hare P. E., Miller G. H., Ernst R. E., and Wendorf F. (1990) Chronometric dating of Pleistocene sites: protein diagenesis in ostrich eggshell. *Science* **248**, 60–64.

Collins, M. J. and Riley M. S. (2000) Amino acid racemization in biominerals: the impact of protein degrdation and loss. In *Perspectives in Amino Acid and Protein Geochemistry* (eds G. A. Goodfriend, M. J. Collins, M. L. Fogel, S. A. Macko, and J. F. Wehmiller), pp. 120–142. Oxford University Press.

Ellis G. L., Goodfriend G. A., Abbott J. T., Hare P. E., and von Endt D. W. (1996) Assessment of integrity and geochronology of archaeological sites using amino acid racemization in land snail shells: examples from central Texas. *Geoarchaeology* **11**, 189–213.

Goodfriend G. A. (1991) Holocene trends in ^{18}O in land snail shells from the Negev Desert and their implications for changes in rainfall source areas. *Quat. Res.* **35**, 417–426.

Goodfriend G. A. and Meyer V. R. (1991) A comparative study of the kinetics of amino acid racemization/epimerization in fossil and modern mollusk shells. *Geochim. Cosmochim. Acta* **55**, 3355–3367.

Kaufman D. S., Miller G. H., and Andrews J. T. (1992) Amino acid composition as a taxonomic tool for molluscan fossils: an example from Pliocene–Pleistocene Arctic marine deposits. *Geochim. Cosmochim. Acta* **56**, 2445–2453.

Kimber R. W. L. and Griffin C. V. (1987) Further evidence of the complexity of the racemization process in fossil shells with implications for amino acid racemization dating. *Geochim. Cosmochim. Acta* **51**, 839–846.

King K. J. (1980) Applications of amino acid biogeochemistry for marine sediments. In *Biogeochemistry of Amino Acids* (eds P. E. Hare, T. C. Hoering, and K. King, Jr.), pp. 377–391. Wiley.

Kriausakul N. and Mitterer R. M. (1980) Some factors affecting the epimerization of isoleucine in peptides and proteins. In *Biogeochemistry of Amino Acids* (ed P. E. Hare, T. C. Hoering, and K. King, Jr.), pp. 283–296. Wiley.

McCoy W. D. (1987) The precision of amino acid geochronology and paleothermometry. *Quat. Sci. Rev.* **6**, 43–54.

Magee J. W. and Miller G. H. (1998) Lake Eyre palaeohydrology from 60 ka to the present: beach ridges and glacial maximum aridity. *Palaeogeog. Palaeoclimatol. Palaeoecol.* **144**, 307–329.

Magee J. W., Bowler J. M., Miller G. H., and Williams D. L. G. (1995) Stratigraphy, sedimentology, chronology and paleohydrology of Quaternary lacustrine deposits at Madigan Gulf, Lake Eyre, South Australia. *Palaeogeog. Palaeoclimatol. Palaeoecol.* **113**, 3–42.
Miller G. H. (1985) Aminostratigraphy of Baffin Island shell-bearing deposits. In *Quaternary Environments: Eastern Canadian Arctic, Baffin Bay and West Greenland* (ed. J. T. Andrews), pp. 394–427. Allen and Unwin.
Miller G. H. and Brigham-Grette J. (1989) Amino acid geochronology: resolution and precision in carbonate fossils. *Quat Intl.* **1**, 111–128.
Miller G. H., Beaumont B., Jull A. J. T., and Johnson B. J. (1992) Pleistocene geochronology and palaeothermometry from protein diagenesis in ostrich eggshells: Implications for the evolution of modern humans. *Phil. Trans. Roy. Soc. Lond.* **337B**, 149–157.
Miller G. H., Magee J. W., and Jull A. J. T. (1997) Low-latitude glacial cooling in the Southern Hemisphere from amino acids in emu eggshells. *Nature* **385**, 241–244.
Miller G. H., Magee J. W., Johnson B. J., Fogel M. L., Spooner N. A., McCulloch M. T., and Ayliffe L. K. (1999) Pleistocene extinction of *Genyornis newtoni*: human impact on Australian megafauna. *Science* **283**, 205–208.
Mirecki J. E. (1990) Aminostratigraphy, geochronology, and geochemistry of fossils from late Cenozoic marine units in southeastern Virginia. Ph.D. Thesis, University of Delaware, Newark.
Mitterer R. M. (1975) Ages and diagenetic temperatures of Pleistocene deposits of Florida based on isoleucine epimerization in *Mercenaria. Earth Planet. Sci. Lett.* **28**, 275–282.
Mitterer R. M. (1993) The diagenesis of proteins and amino acids in fossil shells. In *Organic Geochemistry* (eds M. H. Engel and S. A. Macko), pp. 739–752. Plenum.
Mitterer R. M. and Kriausakul N. (1984) Comparison of rates and degrees of isoleucine epimerization in dipeptides and tripeptides. *Organic Geochem.* **7**, 91–98.
Müller P. J. (1984) Isoleucine epimerization in Quaternary planktonic foraminifera: effects of diagenetic hydrolysis and leaching and Atlantic–Pacific inter-core correlations. *"Meteor" Forschungsergerebnisse* Reihe C, **38**, 25–47.
Sejrup H. P. and Haugen J.-E. (1994) Amino acid diagenesis in the marine bivalve *Arctica islandica Linne* from NW-European sites: only time and temperature? *J. Quat. Sci.* **9**, 301–310.
Wehmiller J. F. (1977) Amino acid studies of the Del Mar, California, midden site: apparent rate constants, ground temperature models, and chronological implications. *Earth Planet. Sci. Lett.* **37**, 184–196.
Wehmiller J. F. (1984) Interlaboratory comparison of amino acid enantiomeric ratios in fossil Pleistocene mollusks. *Quat. Res.* **22**, 109–120.
Wehmiller J. F. (1993) Applications of organic geochemistry for Quaternary research: aminostratigraphy and aminochronology. In *Organic Geochemistry* (eds M. H. Engel and S. A. Macko), pp. 755–783. Plenum.
Wehmiller J. F., Stecher, III, H. A., York, L. L., and Friedman, I. (2000) The thermal environment of fossils: ground temperatures at aminostratigraphic sites on the U.S. Atlantic Coastal Plain. In *Perspectives in Amino Acid and Protein Geochemistry* (eds G. A. Goodfriend, M. J. Collins, M. L. Fogel, S. A. Macko, and J. F. Wehmiller), pp. 219–250. Oxford University Press.
Williams, D. L. G. (1981). *Genyornis* eggshell (Dromornithidae; Aves) from the Late Pleistocene of South Australia. *Alcheringa* **5**, 133–140.
Williams D. L. G. and Rich P. V. (1991) Fossil eggs from the Tertiary and Quaternary of Australia. In *Vertebrate Palaeontology of Australasia.* (eds P. V. Rich, J. M. Monaghan, R. F. Baird, and T. H. Rich), p. 1468. Monash University Publications Committee and Pioneer Design Studio.

14. The interpretation of aspartic acid racemization of dentine proteins

Emma R. Waite and Matthew J. Collins

Aspartic acid racemization has been used to determine age-at-death in forensic (e.g., Ohtani and Yamamoto, 1991a; Ritz et al., 1993; Mornstad et al., 1994) and archaeological science (Gillard et al., 1990, 1991; Carolan et al., 1997a). Results obtained from these studies have been variable. Age has been estimated to within ±0–3 years for modern freshly extracted teeth (e.g., Fu et al., 1995; Ohtani, 1995a), but for real cadavers in forensic and archaeological settings errors can be as much as ±13 years (for a skeletonized cadaver, Masters, 1986), ±15 years (Gillard et al., 1990; 18th century Spitalfields crypt) and ±24 years for Victorian skeletons (Carolan et al., 1997a). For modern freshly extracted teeth it is clear that the method works well, but the "black box" approach used ignores the underlying biochemical processes that lead to racemization. An improved understanding of the mechanisms of aspartic acid racemization in dentine is required if the difficulties with post-mortem cases are to be resolved.

Aspartic acid (Asp) is used because it racemizes more rapidly than other amino acids and therefore can provide the required resolution over forensic and recent archaeological time-scales. This fast rate of racemization arises because the mechanism is unusual. The complicated nature of the reactions of "aspartic acid racemization" means that there have been several oversimplifications made in the interpretation of *in vivo* racemization in human dentine. This chapter will discuss how these simplifications of the process may affect the interpretations of results and how they can be used as working models which can then be improved and refined.

OVERSIMPLIFICATIONS OF THE RACEMIZATION REACTION

Aspartic Acid Racemization Represents a Combined Signal

Measured "aspartic acid" (Asp) in dentine represents a combined signal from aspartic acid and its amidated form, asparagine (Asn), as well as stereoisomers produced during protein degradation. These compounds are related to each other through a series of decomposition and interconversion reactions illustrated in figure 14-1. However, acid hydrolysis, a prerequisite of amino acid analysis, converts all of these products to Asp. The combined signal from these compounds is therefore termed Asx, to distinguish it from the observed racemization of free Asp in solution (e.g., Bada, 1971).

Asparagine residues rapidly deamidate to form aspartic acid. The mechanism of deamidation is of profound importance in the racemization rate of Asx. Deamidation of asparaginyl residues at neutral and basic pH may involve intramolecular nucleophilic attack from the peptide amide nitrogen on the carbonyl of the side-chain amide forming a cyclic succinimide intermediate (Asu; Geiger and Clarke, 1987). The succinimide undergoes rapid epimerization, and upon decomposition it results in a mixture of D-Asp and L-Asp. Racemization of bound Asx is therefore more complex than for the free amino acid (e.g., Taft, 1956; Brinton and Bada, 1995) and is strongly influenced by local structure.

Deamidation of asparaginyl and isomerization of aspartyl residues may itself influence local structure via the formation of isoaspartate (iAsp), in which the side-chain carboxyl group becomes involved in peptide-bond formation (Geiger and Clarke, 1987, and see the reaction scheme in figure 14-1). Succinimides formed from synthetic peptides yielded aspartyl and isoaspartyl peptides in the ratio 1:3 after hydrolysis (Stephenson and Clarke, 1989). There is evidence for isoaspartyl residues derived from aspartyl residues in calmodulin (Ota and Clarke, 1989), glucagon (Ota et al., 1987), and β-linked peptides have been observed in the telopeptide regions of collagen (Garnero et al., 1997).

Asparaginyl residues racemize faster than their aspartyl counterparts (Stephenson and Clarke, 1989; Zhao et al., 1989). Isoaspartyl residues ought to form Asu at rates similar to those for aspartyl. However, the distortion of the backbone caused by the presence of an iAsp residue will alter conformation and potentially the ease of succinimide formation, e.g., Johnson et al. (1987) demonstrated increased enzyme activity when iAsp was replaced with Asp, and DiDonato et al. (1993) showed decreased renaturation when Asp was replaced by iAsp. A protein may contain aspartyl, asparaginyl, isoaspartyl, and succinimidyl residues,

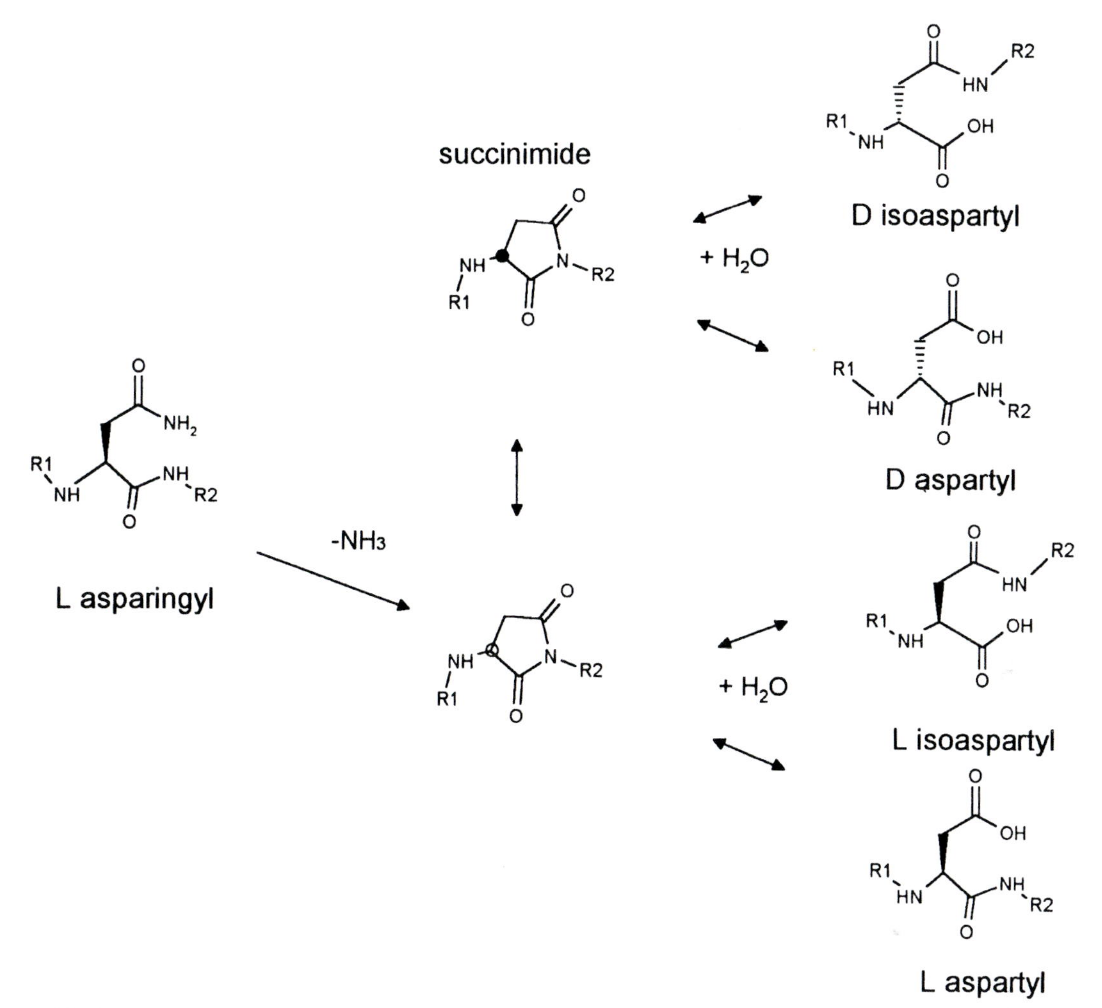

Fig. 14-1. Reaction scheme showing deamidation, racemization and isomerization via the formation of a succinimide.

all of which will have different rates of racemization, but which together will be reported as a single rate of "aspartic acid" racemization.

The relationship between the observed rate of Asx racemization in proteins and the underlying mechanisms is potentially complex. Despite this, interpretation of the rates of racemization of Asx in dentine has usually been derived from early work on the racemization of free amino acids. In solution, free Asp racemization follows a reversible first-order rate law, as follows:

$$\text{L-aspartic acid} \overset{k}{\leftrightarrow} \text{D-aspartic acid}$$

The rate of "racemization" (k) is estimated from the slope of the following logarithmic transformation of the D/L value:

$$\ln[(1 + \text{D/L})/(1 - \text{D/L})]_t - \ln[(1 + \text{D/L})/(1 - \text{D/L})]_{t=0} = 2k \cdot t$$

The transformation yields a straight line with the slope $2kt$ (where t is time) and has been widely adopted (see figure 14-2). The approach is relevant for free Asp, but not for free Asn (Brinton and Bada, 1995). Despite the potential complexity of kinetics in dentine proteins, only one group (Gillard et al., 1990, 1991) has not derived a rate constant using this model. Plotting the residuals from a linear regression of this transformation reveals a reasonable correspondence (at least for adults). However, the distribution of residuals is usually concave about the mean (figure 14-3), implying a deceleration in the overall rate with age. In studies of the "racemization" of Asp in non-mineralized collagen, this deceleration has been explained by differences in the rates of "racemization" at terminal, internal, and free amino acids (Smith and Evans, 1980). In more extensively racemized Asx, the pattern becomes more apparent (e.g., Goodfriend, 1991;

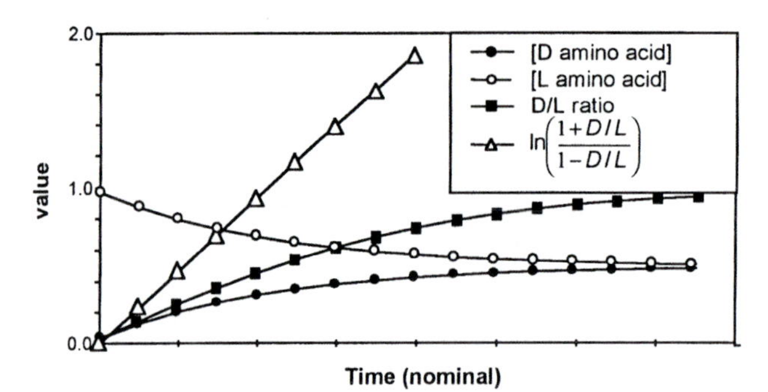

Fig. 14-2. Relative proportion of D- and L-amino acids, D/L value and effect of logarithmic transformation for free amino acids.

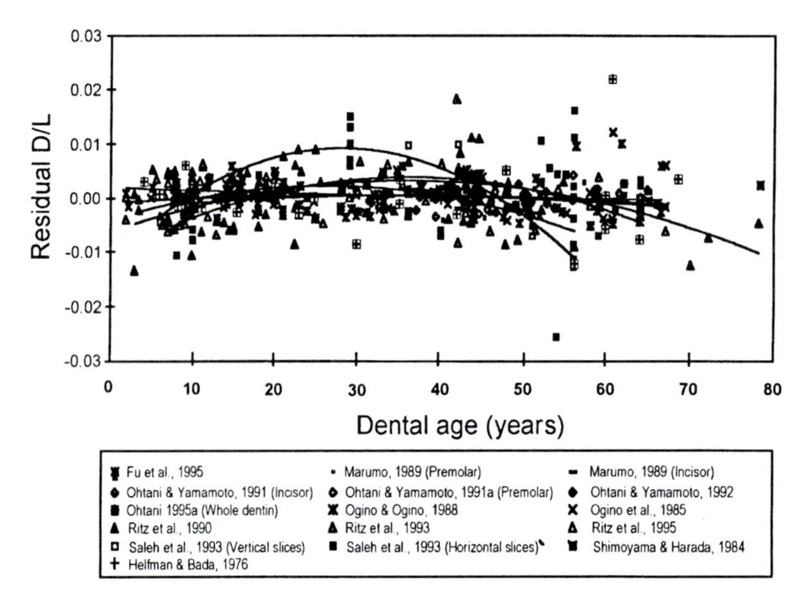

Fig. 14-3. Plot of residuals of linear regressions of D/L values from several data sets (total fraction of dentine).

Brinton and Bada, 1995; Goodfriend et al., 1996).

One partial explanation of the failure of Asx to follow reversible first-order kinetics is the decomposition of free Asn to slower racemizing Asp (Brinton and Bada, 1995). Deamidation of asparaginyl residues can also proceed by direct hydrolysis (Patel and Borchardt, 1990) but at neutral and alkali pH deamidation will normally occur via succinimide formation (with concomitant racemization). This process has been demonstrated in several studies on erythrocyte membrane (McFadden and Clarke, 1982; Lowenson and Clarke, 1991) and eye lens proteins (van Kleef et al., 1975), and may act as a control on the rate of protein turnover (Robinson et al., 1970; Robinson and Rudd, 1974). Deamidation of dentine proteins has not been investigated, but much of the Asx in reported sequences from non-collagenous proteins is initially present as Asn (e.g., 26% in human DMP1 (dentine matrix protein) gene; Hirst et al., 1997). Many of these Asn residues, lie amino- to residues which favor succinimide formation (e.g., Gly and Ser). Collins et al. (1999) have modeled nonlinear kinetics of Asx in terms of different rates for Asn and Asp residues, further subdivided into fast and slow pools based upon the residue which lies carboxyl- to Asx.

Soluble and Insoluble Components of Dentine

It has been widely observed that the soluble fraction isolated from dentine racemizes more rapidly than the total dentine extract (Ohtani and Yamamoto, 1991b; Ritz et al., 1993). A commonly made assumption is that upon extraction, the acid-soluble fraction of the dentine organic matrix consists of non-collagenous proteins (NCPs) and the insoluble fraction consists of collagen. Although this is broadly true, composition will vary depending upon the method of extraction employed and also the age of the dentine. The protein composition of the analyzed fraction is crucial in determining the rate of racemization. The factors which influence the type of proteins extracted in the soluble and insoluble fractions of dentine are discussed in the following sections.

Factors Affecting Increases in Collagen Solubility

Peptide-bond Hydrolysis

It is likely that peptide-bond hydrolysis of dentine collagen will occur, and since the tissue is metabolically isolated (Borggreven et al., 1979), damage to the collagen triple-helix will accumulate with age. We have argued that Asx residues in the collagen triple-helix cannot readily undergo succinimide formation (van Duin and Collins 1998), a view borne out by experimental observation (Sinex, 1960; Julg et al., 1987; Cloos, 1995). Decomposition of the triple-helix will reduce the energetic barrier to succinimide formation and thus increase the rate of observed racemization. In addition, the mechanism of succinimide formation from an asparagine residue has been shown to result not only in deamidation, but also in peptide-chain cleavage carboxyl- to the Asn residue (Blodgett et al., 1985; Geiger and Clarke, 1987; Voorter et al., 1988).

On the basis of experiments on the rate of degradation of heated *bone* collagen (Nielsen-Marsh et al., 2000) we estimate that, at 37°C, 6% of the insoluble collagen will be denatured to soluble gelatin over a period of 15 years. As gelatin is enriched in Gly, Pro, and Hyp relative to NCPs, deterioration of collagen should be expressed as an increase in these amino acids in the soluble protein fraction with age. Masters (1985) observed a 10% increase in Gly over 42 years, while Cloos (1995) reported a 6% increase over 25 years—rates somewhat lower than our prediction.

Pulverization

Pulverization of dentine before demineralization has been used in many racemization studies (e.g., Ritz et al., 1993; Mornstad et al., 1994), either to produce more homogeneous subsamples or to accelerate extraction. However, it has been suggested that the use of intense mechanical disintegration of calcified tissues could lead to degradation of the proteins (Linde, et al., 1980; Cloos, 1995). Cloos (1995) and Collins and Galley (1998) observed that pulverization of dentine and bone, respectively, increased the amount of collagen extracted in the soluble fraction. A higher collagen content in the soluble fraction will decrease the apparent rate of racemization (Collins and Galley, 1998). Ritz-Timme et al. (2000) argue that this process is responsible for the observation, made by Ohtani et al. (1998), of an apparent decrease in D/L value in soluble extracts of progressively finer (i.e., more mechanically damaged) size fractions of the same bone.

Collagen Decreases in Solubility as a Function of Cross-linking

Dentine collagen is poorly soluble (Veis and Scheuleter, 1964). With increasing age there is a progressive increase in collagen-fibre stability (e.g., Kohn and Rollerson, 1959; Walters and Eyre, 1983) which arises from changes in the number and type of covalent intermolecular cross-link. However, these changes are complex and poorly understood so it is not known precisely how much this will influence the pool of extracted proteins and thus the observed rate of racemization.

The Insoluble Component of Dentine Is Not Pure Collagen

Although the bulk of the insoluble fraction is collagen, Asx-rich NCPs may also be isolated in this fraction (Stetler-Stevenson and Veis, 1983; Fujisawa et al., 1994). Phosphophoryn (PP), a dentine phosphoprotein, binds strongly to collagen (Dickson et al., 1975; Carmichael et al., 1977), as does osteocalcin (Prigodich and Vesely, 1997).

The precise nature of the binding between PP and collagen is unknown. In competition adsorption studies, PP was noted to have a high affinity for collagen (Stetler-Stevenson and Veis, 1986), binding specifically to the "e" bands of collagen, which are regions of net positive charge at neutral pH (Traub et al., 1992). This evidence suggests an initial electrostatic attraction (e.g. Linde et al., 1981), but it may be that some PP does become covalently attached (Curley-Joseph and Veis, 1979; Lee and Veis, 1980; Fujisawa et al., 1984). Any Asx-rich NCPs contributing to the insoluble fraction would be anticipated to increase the observed rate of racemization.

Deterioration of NCPs May Accelerate Racemization

McCurdy et al. (1992) have demonstrated that the PP isolated from immature human tooth roots is very different from that isolated from mature roots, both quantitatively and qualitatively. The protein isolated from immature roots comprised more phosphoserine, serine and aspartic acid than the mature roots, and appeared as a distinct band after SDS-PAGE rather than a diffuse smear. The *in vivo* decomposition of PP may occur non-enzymatically. Tyler-Cross and Schirch (1991) observed cleavage of Asp–Ser, a common combination in dentine proteins, whereas phosphoserine residues may be cleaved *in vivo* by a β-elimination mechanism (Fujisawa et al., 1985; Masters, 1985).

Aspartic Acid Bonds are Particularly Sensitive to Acid Hydrolysis

Partridge and Davis (1950) and Schultz et al. (1962) observed preferential release of aspartic acid by partial hydrolysis with acetic acid and a weak mineral acid (0.03 N HCl), respectively. Substantial and specific cleavage at the *C*-terminal side of Asp residues was also demonstrated after hydrolysis in dilute HCl and formic acid, both at pH 2 (Inglis, 1983).

Takagi and Veis (1984) noted that aspartyl bonds are particularly sensitive to acid hydrolysis, and that, during acid demineralization, there may be partial degradation of human dentine PP. Differences in preparation method may lead to variation in the extent of induced hydrolysis. The rate of peptide-bond hydrolysis is sensitive to temperature, and therefore damage during acid demineralization can be minimized if procedures are conducted at low temperatures (Collins and Galley, 1998).

OVERLOOKING THE UNDERLYING COMPLEXITY

The reasonable adherence of Asx racemization in dentine to a first-order reversible rate law has meant that the underlying complexity has largely been ignored. However, the peculiarities of Asx racemization in dentine can only be understood in terms of a more complex model.

The Influence of pH

Ohtani (1995b) observed that the rate of racemization in dentine increased as a function of pH (figure 14-4) in contrast to the findings of studies in bone (Bada and Shou, 1980) which displayed little effect, presumably due to buffering effects of the apatite. Ohtani suggests that "this phenomenon is probably due to the difference in collagen between tooth and bone or the difference in method employed" (p. 185). An explanation of the contrast in the rate profile observed by Ohtani (1995b) for Asx in dentine (which increases with increasing pH) and that of free Asp (Bada, 1972; figure 14-4) was not offered. The differences between the latter two studies probably arises because the initial rate-controlling step in the racemization of dentine "Asx" (unlike that of the free amino acid) is via Asn decomposition to Asu, a mechanism which is favored by increasing pH (Capasso et al., 1993).

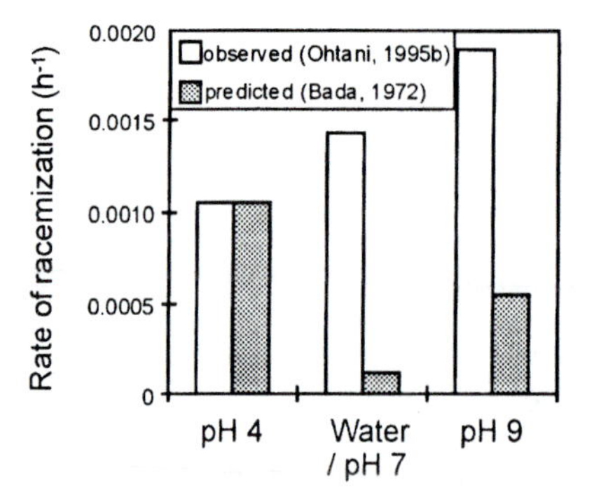

Fig. 14-4. Comparative rates of racemization of Asx in dentine and free Asp as a function of pH (corrected to pH 4).

The Influence of Initial Hydrolysis

If racemization is plotted against age at death for control teeth of known age a regression line is obtained, which, when extrapolated back to the y-axis, gives an intercept value which is greater than zero. This racemization at time zero has been attributed to racemization induced by acid hydrolysis during sample preparation (e.g., Gillard et al., 1991). However, the early parts of calibration curves are ill-defined as there are usually no data points for very young individuals.

Several groups have recognized hydrolysis as a source of racemization and have attempted to measure "typical" amounts of induced racemization in their samples. Figure 14-5 shows a comparison between various studies for the amount of hydrolysis-induced racemization. Many of the rates are very similar but the percent D-Asx values are shifted depending on the type of sample used and the initial amount of racemization in the sample. However, for this type of analysis to be useful, the value obtained must be relevant for the substrate being measured. A value of induced racemization for free Asp has no relevance to the amount induced from a mixture of acid-insoluble dentine proteins (Carolan et al., 1997b), as free and protein-bound amino acids will display different rates of racemization. Ohtani and Yamamoto (1991b) observed the progressive racemization of Asx induced by hydrolysis in 6 N HCl over a period of 72 hr (see figure 14-6). Using these data, it is possible to calculate the amount of induced racemization occurring during a standard (6 hr) hydrolysis. However, this value is five times lower than the intercept obtained from the regression of adult teeth from the same study[1], but similar to the value obtained for 50–250-day-old rat teeth (Ohtani et al., 1995; see figure 14-7).

In all studies of Asx racemization in human dentine, the intercept is higher than is estimated from the extent of induced racemization (for 6 hr, 100°C, 6 M HCl hydrolysis). The implication is that the initial rate of racemization is much faster than reported values for adult teeth (which are typically in the range $5–8 \times 10^{-4}$ yr^{-1}).

The rate of racemization of dentine in very young rat molars is very fast (Ohtani et al., 1995); the rate over the first 250 d of life was $k = 54 \times 10^{-4}$ yr^{-1}, an order of magnitude faster than that in adult human dentine[2] (Ohtani and Yamamoto, 1992).

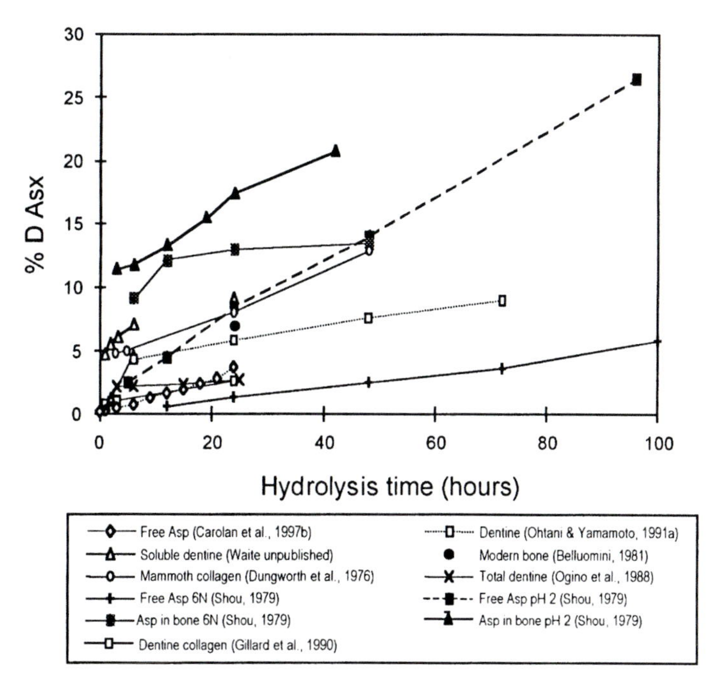

Fig. 14-5. Acid-induced hydrolysis from various studies.

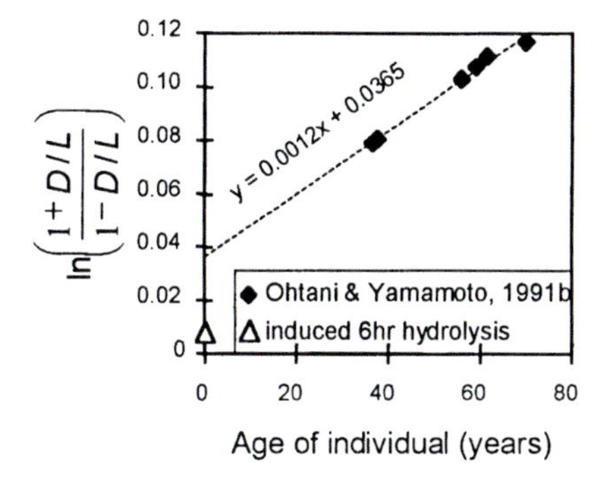

Fig. 14-6. Variation in the amount of induced hydrolysis determined by a single study.

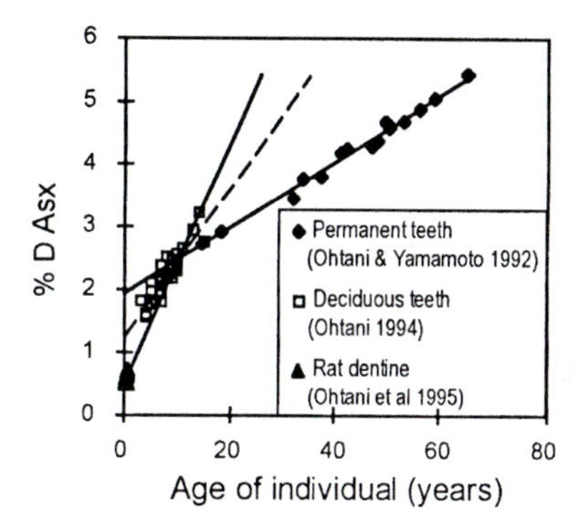

Fig. 14-7. Comparison of rates of racemization in human deciduous, permanent and rat teeth.

Ohtani et al. (1995) offer two explanations for the higher rate in rat teeth: (1) that the body temperature is higher in rats, (2) that the salivary pH is higher and that racemization is faster at alkali pH. However, the 1–2°C higher temperature should only make the rate 25% faster (not the 1000% observed), and buffering effects of tooth mineral should cancel out the higher pH of the saliva, (cf. Bada and Shou, 1980).

This example of a very fast initial racemization early in life may demonstrate what is happening in human teeth prior to eruption. The closest approximation to this is a study on deciduous teeth (Ohtani, 1994), in which the rate is intermediate between the rate in rat teeth and that in permanent human teeth (figure 14-5). The intercept obtained from the deciduous teeth is lower than that for the permanent teeth, but is still higher than the predicted intercept induced by hydrolysis. Integration of all four studies may give the most complete picture of the overall pattern of Asx racemization and the prediction of the intercept (Fig 14-7). If the initial rate of racemization in dentine is indeed as high as these combined data suggest, then the analysis of residuals from linear regressions would reveal a more curvilinear relationship if younger individuals were used in the regression analysis; this seems to be partly borne out by the data shown in figure 14-3.

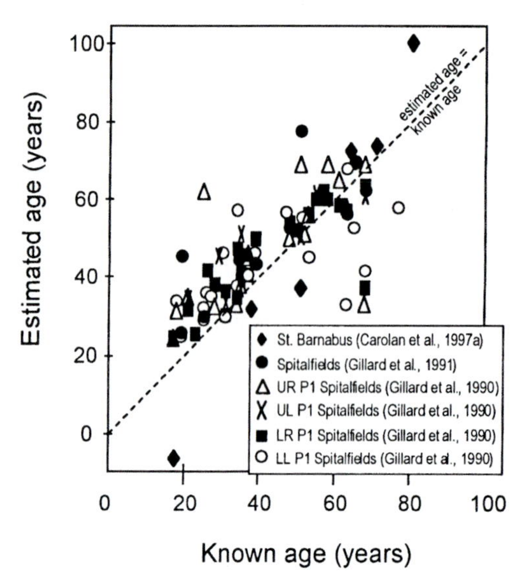

Fig. 14-8. Estimated age vs. known age for archaeological age-determination studies

High Temperature Experiments

Many authors have used *in vitro* high-temperature studies to investigate the kinetics of racemization (e.g., Hare and Mitterer, 1969; Bada and Schroeder, 1972; Shou, 1979; Bada and Shou, 1980; Mitterer and Kriausakul, 1989; Goodfriend and Meyer, 1991). This approach is complicated if reaction mechanisms at high and low temperatures are not comparable. Ohtani and Yamamoto (1992) investigated the relative rate of racemization in dentine and enamel at elevated temperatures (130–160°C) for periods of up to 1 hr. The samples were heated in open vessels, driving off most, but not all, of the water during the experiment (see Pineri et al., 1978). The results that they obtained were used to derive rate constants for *dry* (*sic*) dentine and enamel at low temperatures (15°C). In another study, the remarkable similarity in the rates of racemization of the soluble fractions extracted from teeth before and after heating at 150°C was used to advocate the use of burned bodies for racemization ageing (Ohtani et al., 1997; but see Waite and Collins, 1998). The extrapolation of high-temperature data to rates of Asx racemization observed *in vivo* is difficult because protein conformation (especially that of collagen) may change abruptly with temperature and therefore alter the ease of succinimide formation (Collins et al., 1999).

A Good Technique for Estimating Age at Death?

Many studies performed on freshly extracted modern teeth of known age have produced excellent calibration curves that can allow estimation of age at death to within ±3 years. In real forensic cases, it may not always be so simple. There have been some studies considering the effect of post-mortem conditions on age-at-death determinations (e.g., Masters, 1986; Ohtani et al., 1988; Ohtani, 1995c) which illustrate the application of the method, but these represent a limited data set. There is much to learn about the effects of different post-mortem conditions upon the accuracy of data obtained. In the case of fire victims (Ohtani et al., 1989), ages derived from burned dentine are highly overestimated.

Studies on historical samples have suggested that in older material the technique may become increasingly inaccurate. Gillard et al., (1990, 1991) reported age-at-death determinations for acid-insoluble dentine extracts from the largely 18th century Spitalfields crypt where coffin plaques gave the ages at death. Similarly, Carolan et al. (1997a) produced age estimates for similar extracts from Victorian samples of known age. In both cases, there were large errors in the ages estimated by aspartic acid racemization when compared to the known ages (see figure 14-8); the young produce age estimates that are always too old, and the old produce ages that are too young. It is difficult to explain why this might be so. Racemization of peptide-bound Asx in non-helical regions (telopeptides) of collagen will occur even if the remaining residues in the triple-helix are impeded. Racemization of all the telopeptide aspartyl residues will yield a D/L value ratio for total collagen of 0.09, a value within the range reported by Gillard et al. (1991) and Carolan et al. (1997a). We calculate (Collins et al., 1999) that this would be achieved in about 500 years at an effective burial temperature of 11°C, but, by the same argument, the telopeptide Asp should already be at equilibrium in adult dentine. The pattern of racemization observed in these samples appears to defy a simple kinetic explanation. However, using the Spitalfields samples, Child et al. (1993) were able to demonstrate the presence of microbes (and their associated enzymes) capable of promoting racemization. It should also be noted that similar patterns of error in age estimation were observed using traditional morphological techniques (Molleson et al., 1993), suggesting that the type of regression analysis used to calculate age may be partly responsible (see Aykroyd et al., 1997, 1999).

These problems with 100–200-year-old material may serve to highlight potential problems with younger forensic material (i.e., up to 50 years since death). We understand very little about diagenetic alteration of protein and the effects of different burial conditions on rates of racemization, aspects that are relevant to forensic, as well as archaeological, age determinations.

TABLE 14-1 An audit of the distribution of aspartyl and asparaginyl residues in dentine proteins. The Asx compositions were deduced from the gene sequences or protein was isolated from human dentine, except where indicated

Protein	Asn	Asx	Percentage of organic fraction of dentine	Percentage of total Asx residues in dentine			Source reference(s)
				Asn	Asp	Asx	
collagen triple-helix	1.8	4.8	90	82.1	78.9	53.2	Ayad et al. (1994)
telopeptides	0.0	8.0	0.1	0.0	0.3	0.1	Ayad et al. (1994)
phosphophoryn (DMP1 gene)	3.5	13.6	5–6	10.0	16.1	9.2	Hirst et al. (1997)
phosphoprotein		36.0	5–6	–	–	24.4	Takagi and Veis (1984)
biglycan	7.8	13.5	unknown	–	–	–	Fisher et al. (1989)[a]
decorin	5.0	12.5	unknown	–	–	–	Krusius and Ruoslahti (1986); Fisher et al. (1987)[a]; Vetter et al. (1993); Danielson et al. (1993)
matrix gla	6.8	8.7	0.5	1.8	0.3	0.5	Cancela et al. (1990)
osteocalcin	2.0	10.2	1–2	1.6	1.2	1.9	Poser et al. (1980)[a]; Celeste et al. (1986); Kiefer et al. (1990)
osteopontin	3.8	19.1	unknown	–	–	–	Young et al. (1990)
SPARC	4.6	14.6	0.4–0.6 (bovine dentine)	1.2	1.4	0.9	Maurer et al. (1995)
bone sialoprotein	8.0	13.6	0.5–0.8	2.7	1.1	9.2	Kim et al. (1994)
α2-HS-glycoprotein	3.6	9.5	0.2	0.4	0.3	0.2	Yoshioka et al. (1986); Lee et al. (1987); Kellerman et al. (1989)
serum albumin	2.9	9.1	0.2	0.3	0.4	0.2	Meloun et al. (1975)[b]; Lawn et al. (1981); Dugaiczyk et al. (1982)

[a] Isolated from human bone.
[b] Sequenced from blood protein.

FUTURE DIRECTIONS

In the forensic community, the construction of many good calibration curves is evidence that the aspartic acid racemization technique works in living dentine. The problems appear to arise when applying the method to archaeological material and, by extension, to exposed remains.

One improvement designed to aid interpretation when using a fraction of the total dentine proteins is to record the amino acid composition of the extract. As an example the "collagen" extracted by different methods (see Waite et al., 1999) will contain differing proportions of NCPs and this will lead to differences in the apparent rates of Asx racemization. Amino acid compositional data may help to identify this variation, but only rarely (e.g., Masters, 1985; Cloos, 1995) has it been reported. Table 14-1 summarizes the numbers of Asx residues in the various dentine matrix proteins and the contribution of each to the total Asx signal.

Interpretation of the rate of Asx racemization in dentine protein extracts would be improved if more were known about (1) the fraction of soluble collagen due to changes in (i) dentine age and (ii) pulverization; (2) the proportion of NCP isolated in the insoluble fraction; and (3) the rates and mechanisms of deamidation. We may then be in a better position to explore how post-mortem degradation processes can affect the results obtained from cadavers and human remains. In addition, knowledge of the reactions in dentine will help to define the best methods for analysis. The ultimate aim for forensic science would be to implement a standardized optimum extraction technique using simplified methods, so that the technique can be used with confidence in a legal framework.

NOTES

1. A further complication is that a typographical error has been made in the intercept reported on the graph of acid hydrolysis over time (Fig. 3, p. 125, of Ohtani and Yamamoto, 1991b).

2. Rat PPs differ from those of human dentine (Butler et al., 1983), being more acidic and perhaps more prone to degradation than those in human teeth.

REFERENCES

Ayad S., Boot-Handford R. P., Humphries M. J., Kadler K. E., and Shuttleworth C. A. (1994) *The*

Extracellular Matrix Facts Book. Academic Press, Harcourt Brace and Co.

Aykroyd R. G., Lucy D., Pollard A. M., and Solheim T. (1997) Technical note: Regression analysis in adult age estimation. *Amer. J. Phys. Anthropol.* **104**, 259–265.

Aykroyd R. G., Lucy D., Pollard A. M., and Roberts C. A. (1999) Nasty, brutish but not necessarily short: a reconsideration of the statistical methods used to calculate age at death from adult skeletal and dental age indicators. *Amer. Antiquity* **64**, 55–70.

Bada J. L. (1971) Kinetics of the non-biological decomposition and racemization of amino acids in natural waters. *Adv. Chem. Ser.* **106**, 309–331.

Bada J. L. (1972) Kinetics of racemization of amino acids as a function of pH. *J. Amer. Chem. Soc.* **94**, 1371–1373.

Bada J. L. and Schroeder R. A. (1972) Racemization of isoleucine in calcareous marine sediments: Kinetics and mechanism. *Earth Planet. Sci. Lett.* **15**, 1–11.

Bada J. L. and Shou P. M.-Y. (1980) Kinetics and mechanism of amino acid racemization in aqueous solution and in bones. In *Biogeochemistry of Amino Acids* (eds P. E. Hare, T. C. Hoering, and K. King, Jr.), pp. 235–255. Wiley.

Belloumini G. (1981) Direct aspartic acid racemization dating of human bones from archaeological sites of central and southern Italy. *Archaeometry* **23**, 125–137.

Blodgett J. K., Loudon G. M., and Collins K. D. (1985) Specific cleavage of peptides containing an aspartic acid (β-Hydroxyamic acid) residue. *J. Amer. Chem. Soc.* **107**, 4305–4313.

Borggreven J. M. P. M., Hoppenbrouwers P. M. M., and Gorissen R. (1979) Radiochemical determination of the metabolic activity of collagen in mature dentin. *J. Dent. Res.* **58**, 2120–2124.

Brinton K. L. and Bada J. L. (1995) Aspartic acid racemization and protein diagenesis in corals over the last 350 years—Comment. *Geochim. Cosmochim. Acta* **59**, 415–416.

Butler W. T., Bhown M., DiMuzio M. T., Cothran W. C., and Linde, A. (1983) Multiple forms of rat dentin phosphoproteins. *Arch. Biochem. Biophys.* **225**, 178–186.

Cancela L., Hsieh C. L., Francke U., and Price P. A. (1990) Molecular structure, chromosome assignment, and promoter organization of the human matrix Gla protein gene. *J. Biol. Chem.* **265**, 15040–15048.

Capasso S., Mazzarella L., Sica F., Zagari A., and Salvadori S. (1993) Kinetics and mechanism of succinimide ring formation in the deamidation process of asparagine residues. *J. Chem. Soc.-Perkin Trans.* **Ii**, 679–682.

Carmichael D. J., Dodd C. M., and Veis A. (1977) The solubilization of bone and dentin collagens by pepsin. Effect of cross-linkages and non-collagen components. *Biochim. Biophys. Acta* **491**, 177–192.

Carolan V. A., Gardner M. L. G., Lucy D., and Pollard A. M. (1997a) Some considerations regarding the use of amino acid racemization in human dentine as an indicator of age at death. *J. Forensic Sci.* **42**, 10–16.

Carolan V. A., Gardner M. L. G., and Pollard A. M. (1997b) Age determination via measurement of amino acid racemization in dental tissue. In *Archaeological Sciences 1995, Proceedings of a Conference on the Application of Scientific Techniques to the Study of Archaeology*. (*Oxbow Monograph 64*). (eds A. Sinclair, E. Slater, and J. Gowlett), Oxbow Books, pp. 349–357.

Celeste A. J., Buecker J. L., Kriz R., Wang E. A., and Wozney J. M. (1986) Isolation of the human gene for bone Gla protein utilizing mouse and rat cDNA clones. *EMBO J.* **5**, 1885–1890.

Child A. M., Gillard R. D., and Pollard A. M. (1993) Microbially induced promotion of amino acid racemization in bone: isolation of the microorganisms and detection of their enzymes. *J. Archaeol. Sci.* **20**, 159–168.

Cloos P. A. C. (1995) *Age-related Changes in Proteins in Human Dentine* M.Phil. Thesis, Det Kongelige Danske Kunstakademi.

Collins M. J. and Galley P. (1998) Towards an optimal method of archaeological collagen extraction; the influence of pH and grinding. *Ancient Biomolecules* **2**, 209–222.

Collins M. J., Waite, E. R., and van Duin A. C. T. (1999) Predicting protein decomposition: the case of aspartic acid racemization kinetics. *Phil. Trans. Roy. Soc.* **352**, 51–64.

Curley-Joseph J. and Veis A. (1979) The nature of covalent complexes of phosphoproteins with collagen in the bovine dentin matrix. *J. Dent. Res.* **58**, 1625–1633.

Danielson K. G., Fazzio A., Cohen I. R., Cannizzaro L., and Iozzo R. V. (1993) The human decorin gene—intron exon organization, discovery of 2 alternatively spliced exons in the 5′ untranslated region, and mapping of the gene to chromosome-12q23. *Genomics* **15**, 146–160.

Dickson I. R., Dimuzio D., Volpin D., Ananthanarayanan S., and Veis A. (1975) The extraction of phosphoproteins from bovine dentine. *Calcif. Tissue Res.* **19**, 51–61.

DiDonato A., Ciardiello M. A., Denigris M., Piccoli R., Mazzarella L., and Dalessio G. (1993) Selective deamidation of ribonuclease-A—isolation and characterization of the resulting isoaspartyl and aspartyl derivatives. *J. Biol. Chem.* **268**, 4745–4751.

Dugaiczyk A., Law S. W., and Dennison O. E. (1982) Nucleotide sequence and the encoded amino acids of human serum albumin mRNA. *Proc. Natl. Acad. Sci. U.S.A.* **79**, 71–75.

van Duin A. C. T. and Collins M. J. (1998) The effects of conformational constraints on aspartic acid racemization. *Org. Geochem.* **29**, 1227–1232.

Dungworth G., Schwartz A. W., and van de Leemput L. (1976) Composition and racemization of amino acids in mammoth collagen determined by gas and liquid chromatography. *Comp. Biochem. Physiol.* **53B**, 472–480.

Fisher L. W., Hawkins G. R., Tuross N., and Termine J. D. (1987) Purification and partial characterization of small proteoglycan-I and proteoglycan-II, bone sialoprotein-I and sialoprotein-II and osteonectin from the mineral. *J. Biol. Chem.* **262**, 9702–9708.

Fisher L. W., Termine J. D., and Young M. F. (1989) Deduced protein sequence of bone small proteoglycan I (biglycan) shows homology with proteoglycan II (decorin) and several nonconnective tissue proteins in a variety of species. *J. Biol. Chem.* **264**, 4571–4576.

Fu S.-J., Fan C.-C., Song H.-W., and Wei F.-Q. (1995) Age estimation using a modified HPLC determination of ratio of aspartic acid in dentin. *Forensic Sci. Intl.* **73**, 35–40.

Fujisawa R., Takagi T., Kuboki Y., and Sasaki S. (1984) Systematic purification of free and matrix-bound phosphophoryns of bovine dentin: presence of matrix-bound phosphophoryn as a distinct molecular entity. *Calcif. Tissue. Res.* **36**, 239–242.

Fujisawa R., Kuboki Y., and Sasaki S. (1985) *In vivo* cleavage of dentin phosphophoryn following β-elimination of its phosphoserine residues. *Arch. Biochem. Biophys.* **243**, 619–623.

Fujisawa R., Zhou H.-Y., and Kuboki Y. (1994) *In vitro* and *in vivo* association of dentin phosphophoryn with α1CB6 peptide of Type I collagen. *Connective Tissue Res.* **31**, 1–10.

Garnero P., Fledelius C., Ginetys E., Serre C.-M., Vignot E., and Delmas P. D. (1997) Decreased β-Isomerization of the C-terminal telopeptide of Type I collagen α-1 chain in Paget's disease of bone. *J. Bone Mineral Res.* **12**, 1407–1415.

Geiger T. and Clarke S. (1987) Deamidation, isomerization, and racemization at asparaginyl and aspartyl residues in peptides. Succinimide-linked reactions that contribute to protein degradation. *J. Biol. Chem.* **262**, 785–794.

Gillard R. D., Pollard A. M., Sutton P. A., and Whittaker D. K. (1990) An improved method for age at death determination from the measurement of D-aspartic acid in dental collagen. *Archaeometry* **32**, 61–70.

Gillard R. D., Hardman S. M., Pollard A. M., Sutton P. A., and Whittaker D. K. (1991) Determinations of age at death in archaeological populations using the D/L ratio of aspartic acid in dental collagen. In *Archaeometry '90* (eds E. Pernicka and G. A. Wagner), pp. 637–644, Birkhauser Verlag.

Goodfriend G. A. (1991) Patterns of racemization and epimerization of amino acids in land snail shells over the course of the Holocene. *Geochim. Cosmochim. Acta* **55**, 293–302

Goodfriend G. A. and Meyer V. R. (1991) A comparative study of the kinetics of amino-acid racemization and epimerization in fossil and modern mollusc shells. *Geochim. Cosmochim. Acta* **55**, 3355–3367.

Goodfriend G. A., Brigham-Grette J., and Miller G. H. (1996) Enhanced age resolution of the marine quaternary record in the arctic using aspartic acid racemization dating of bivalve shells. *Quat. Res.* **45**, 176–187.

Hare P. E. and Mitterer R. M. (1969) Laboratory simulation of amino acid diagenesis in fossils. *Carnegie Instn Wash. Yrbk* **67**, 205–208.

Helfman P. M. and Bada J. L. (1976) Aspartic acid racemization in dentine as a measure of aging. *Nature* **262**, 279–281.

Hirst K. L., Ibaraki-O'Connor I., Young M. F., and Dixon M. J. (1997) Cloning and expression analysis of the bovine dentin matrix acidic phosphoprotein gene. *J. Dental Res.* **76**, 754–760.

Inglis A. S. (1983) Cleavage at aspartic acid. *Meth. Enzymol.* **28**, 324–332.

Johnson B. A., Langmack E. L., and Aswad D. W. (1987) Partial repair of deamidation-damaged calmodulin by protein carboxyl methyltranferase. *J. Biol. Chem.* **262**, 12283–12287.

Julg A., Lafont R., and Perinett G. (1987) Mechanisms of collagen racemization in fossil bones: application to absolute dating. *Quat. Sci. Rev.* **6**, 25–28.

Kellerman J., Haupt H., Auerswald E.-A., and Mueller-Ester L. W. (1989) The arrangement of disulphide loops in human α2-HS glycoprotein. Similarity to the disulphide bridge structures of cystanins and kininogens. *J. Biol. Chem.* **264**, 14121–14128.

Kiefer M. C., Saphire A. C. S., Bauer D. M., and Barr P. J. (1990) The cDNA and derived amino-acid sequences of human and bovine bone Gla protein. *Nucleic Acids Res.* **18**, 1909–1909.

Kim R. H., Shapiro H. S., Li J. J., Wrana J. L., and Sodek J. (1994) Characterization of the human bone sialoprotein (BSP) gene and its promoter sequence *Matrix Biol.* **14**, 31–40.

Kohn R. R. and Rollerson E. (1959) Effect of age and heat on human collagenous tissue. *Arch. Pathol.* **68**, 316–321.

Krusius T. and Ruoslahti E. (1986) Primary structure of an extracellular matrix proteoglycan core protein deduced from cloned cDNA. *Proc. Natl. Acad. Sci. U.S.A.* **83**, 7683–7687.

Lawn R. M., Adelman J., Bock S. C., Franke A. E., Houck C. M., Najarian R. C., Seeburg P. H., and Wion K. L. (1981) The sequence of human serum albumin cDNA and its expression in *Escherichia coli*. *Nucleic Acids Res.* **9**, 6103–6114.

Lee S. L. and Veis A. (1980) Studies on the structure and chemistry of dentin collagen-phosphophoryn covalent complexes. *Calcif. Tissue Intl.* **31**, 123–134.

Lee C.-C., Bowman B. H., and Yang F. (1987) Human α-2-HS-glycoprotein—the α-chain and β-chain with a connecting sequence are encoded by a single mRNA transcript. *Proc. Natl. Acad. Sci. U.S.A.* **84**, 4403–4407.

Linde A., Bhown M., and Butler W. T. (1980) Non-collagenous proteins of dentin. A re-examination of proteins from rat incisor dentin utilizing techniques to avoid artifacts. *J. Biol. Chem.* **255**, 5931–5942.

Linde A., Bhown M., and Butler W. T. (1981) Non-collagenous proteins of rat dentin. Evidence that phosphoprotein is not covalently bound to collagen. *Biochim. Biophys. Acta* **704**, 235–239.

Lowenson J. D. and Clarke S. (1991) Spontaneous degradation and enzymatic repair of aspartyl and asparaginyl residues in aging red-cell proteins analyzed by computer-simulation. *Gerontology* **37**, 128–151.

McCurdy S. P., Clarkson B. H., and Feagin F. F. (1992) Comparison of phosphoprotein isolated from mature and immature human tooth roots. *Arch. Oral Biol.* **37**, 1057–1065.

McFadden P. N. and Clarke S. (1982) Methylation at D-aspartyl residues in erythrocytes: Possible step in the repair of aged membrane proteins. *Proc. Natl. Acad. Sci, U.S.A.* **79**, 2460–2464.

Marumo T. (1989) Age estimation by amino acid racemization in dentin—application of fractionation and extraction. *Kanagawashigaku* **24**, 290–300.

Masters P. M. (1985) *In vivo* decomposition of phosphoserine and serine in noncollagenous protein from human dentine. *Calcif. Tissue Intl.* **37**, 236–241.

Masters P. M. (1986) Age at death determinations for autopsied remains based on aspartic-acid racemization in tooth dentin—importance of postmortem conditions. *Forensic Sci. Intl.* **32**, 179–184.

Maurer P., Hohenadl C., Hohenester E., Gohring W., Timpl R., and Engel J. (1995) The C-terminal portion of BM-40 (sparc/osteonectin) is an autonomously folding and crystallisable domain that binds calcium and collagen-IV. *J. Mol. Biol.* **253**, 347–357.

Meloun B., Moravek L., and Kostka V. (1975) Complete amino acid sequence of human serum albumin. *FEBS Lett.* **58**, 134–137.

Mitterer R. M. and Kriausakul N. (1989) Calculation of amino acid racemization ages based on apparent parabolic kinetics. *Quat. Sci. Rev.* **8**, 353–357.

Molleson T., Cox M., Waldron A. H., and Whittaker D. K. (1993) The middling sort. In *Research Report 86. The Spitalfields Project, Volume II—The Anthropology*. Council for British Archaeology.

Mornstad H., Pfeiffer H., and Teivens A. (1994) Estimation of dental age using HPLC-technique to determine the degree of aspartic acid racemization. *J. Forensic Sci.* **39**, 1425–1431.

Nielsen-Marsh C. N., Gernaey A. M., Turner-Walker G., Hedges R. E. M., Pike A. G., and Collins M. J. (2000) Chemical diagenesis of the protein and mineral fractions of bone. In *Osteology Current Practice and Future Prospects* (eds M. Cox and S. Mays), pp. 439–454. Greenwich Medical Media.

Ogino T. and Ogino H. (1988) Application to forensic odontology of aspartic acid racemization in unerupted and supernumerary teeth. *J. Dental Res.* **67**, 1319–1322.

Ogino T., Ogino H., and Nagy B. (1985) Application of aspartic acid racemization to forensic odontology: post mortem designation of age at death. *Forensic Sci. Intl.* **29**, 259–267.

Ogino T., Ogino H., Hobo T., and Ogino H. (1988) Effect of acid hydrolysis on the racemization method using amino acid racemization in teeth. *Bunseki Kagaku* **37**, 476–481.

Ohtani S. (1994) Age estimation by aspartic acid racemization in dentin of deciduous teeth. *Forensic Sci Intl.* **68**, 77–82.

Ohtani S. (1995a) Estimation of age from dentin using the racemization reaction of aspartic acid. *Amer. J. Forensic Med. Pathol.* **16**, 158–161.

Ohtani S. (1995b) Estimation of age from dentin by utilizing the racemization of aspartic acid: Influence of pH. *Forensic Sci. Intl.* **75**, 181–187.

Ohtani S. (1995c) Estimation of age from the teeth of unidentified corpses using amino acid racemization with reference to actual cases. *Amer. J. Forensic Med. Pathol.* **16**, 238–242.

Ohtani S. and Yamamoto K. (1991a) Estimation of ages from racemization of an amino acid in teeth—assessment of errors under various experimental conditions. *Japan. J. Legal. Med.* **45**, 124–147.

Ohtani S. and Yamamoto K. (1991b) Age estimation using the racemization of amino acid in human dentin. *J. Forensic Sci.* **36**, 792–800.

Ohtani S. and Yamamoto K. (1992) Estimation of age from a tooth by means of racemization of an amino-acid, especially aspartic acid—comparison of enamel and dentin. *J. Forensic Sci.* **37**, 1061–1067.

Ohtani S., Kato S., Sugeno H., Sugimoto H., Marumo T., Yamazaki M., and Yamamoto K. (1988) A study on the use of the amino acid racemization method to estimate the ages of unidentified cadavers from their teeth. *Bull. Kanagawa Dental College* **16**, 11–21.

Ohtani S., Sugeno H., Marumo T., and Yamamoto K. (1989) Two cases of age estimation from teeth of a burned body using amino acid racemization. *Japan. J. Legal Med.* **43**, 191–197.

Ohtani S., Matsushima Y., Ohhira H. and Watanabe A. (1995) Age-related changes in D-aspartic acid of rat teeth. *Growth Devel. Ageing* **59**, 55–61.

Ohtani S., Matsushima Y., Kobayashi Y., and Kishi K. (1998) Evaluation of aspartic acid racemization ratios in the human femur for age estimation. *J. Forensic Sci.* **43**, 949–953.

Ohtani S., Yamada Y., and Yamamoto I. (1997) Age estimation from racemization rate using heated teeth. *J. Forensic Odonto-Stomatol.* **15**, 9–12.

Ota I. M. and Clarke S. (1989) Enzymatic methylation of L-isoaspartyl residues derived from aspartyl residues in affinity-purified calmodulin. The role of conformational flexibility in spontaneous isoaspartyl formation. *J.Biol.Chem.* **264**, 54–60.

Ota I. M., Ding L., and Clarke S. (1987) Methylation at specific altered aspartyl and asparaginyl residues in glucagon by the erythrocyte protein carboxyl methyltransferase. *J.Biol.Chem.* **262**, 8522–8531.

Partridge S. M. and Davis H. F. (1950) Preferential release of aspartic acid during the hydrolysis of proteins. *Nature* **165**, 62–63.

Patel K. and Borchardt R. T. (1990) Chemical pathways of peptide degradation 2. Kinetics of deamidation of an asparaginyl residue in a model hexapeptide. *Pharmaceut. Res.* **7**, 703–711.

Pineri M. H., Escoubes M., and Roche G. (1978) Water–collagen interactions: calorimetric and mechanical experiments. *Biopolymers* **17**, 2799–2815.

Poser J. W., Esch F. S., Ling N. C., and Price P. A. (1980) Isolation and sequence of the vitamin K-dependent protein from human bone. *J. Biol. Chem.* **255**, 8685–8691.

Prigodich R. V. and Vesely M. R. (1997) Characterization of the complex between bovine osteocalcin and type I collagen. *Arch. Biochem. Biophys.* **345**, 339–341.

Ritz-Timme S., Schutz H.-W., and Collins M. J. (2000) Commentary on Ohtani S., Matsushima Y., Kobayashi Y., Kishi K. (2000) Evaluation of aspartic acid racemization ratios in the human femur for age estimation. J. Forensic Sci. 1998; 43: 949–953. *J. Forensic Sci.* **44**, 874–875.

Ritz S., Schutz H.-W., and Schwarzer B. (1990) The extent of aspartic acid racemization in dentin: a possible method for a more accurate determination of age at death? *Zeitschrift Rechtsmedzin.* **103**, 457–462.

Ritz S., Stock R., Schutz H.-W., and Kaatsch H.-J. (1995) Age estimation in biopsy specimens of dentin. *Intl. J. Leg. Med.* **108**, 135–139.

Ritz S., Schutz H.-W., and Peper C. (1993) Postmortem estimation of age at death based on aspartic-acid racemization in dentin—its applicability for root dentin. *Intl. J. Legal Med.* **105**, 289–293.

Robinson A. B. and Rudd C. J. (1974) Deamidation of glutaminyl and asparaginyl residues in peptides and proteins. *Curr. Top. Cell. Regul.* **8**, 247–294.

Robinson A. B., McKerrow J. H., and Cary P. (1970) Controlled deamidation of peptides and proteins: An experimental hazard and a possible biological timer. *Proc. Natl. Acad. Sci. U.S.A.* **66**, 753–757.

Saleh N., Deutsch D., and Gil-Av E. (1993) Racemization of aspartic acid in the extracellular matrix proteins of primary and secondary dentin. *Calcif. Tissue Intl.* **53**, 103–110.

Schultz J., Allison H., and Grice M. (1962) Specificity of the cleavage of proteins by dilute acid. I. Release of aspartic acid from insulin, ribonuclease, and glucagon. *Biochemistry* **1**, 694–698.

Shimoyama A. and Harada K. (1984) An age determination of an ancient burial mound man by apparent racemization reaction of aspartic acid in tooth dentine. *Chem. Lett.* **10**, 1661–1664.

Shou P. M. Y. (1979) *Kinetics and Mechanisms of Several Amino Acid Diagenetic Reactions in Aqueous Solutions and in Fossils*. Ph.D. Thesis, University of California, San Diego.

Sinex F. M. (1960) Ageing and the lability of irreplaceable molecules—II. The amide groups of collagen. *J. Gerontol.* **15**, 15–18.

Smith G. G. and Evans R. C. (1980) The effect of structure and conditions on the rate of racemization of free and bound amino acids. In *Biogeochemistry of Amino Acids* (eds P. E. Hare, T. C. Hoering, and K. King, Jr.), pp. 257–282. Wiley.

Stephenson R. C. and Clarke S. (1989) Succinimide formation from aspartyl and asparaginyl peptides as a model for the spontaneous degradation of proteins. *J. Biol. Chem.* **264**, 6164–6170.

Stetler-Stevenson W. G. and Veis A. (1983) Bovine dentin phosphophoryn: composition and molecular weight. *Biochemistry* **22**, 4326–4335.

Stetler-Stevenson W. G. and Veis A. (1986) Type I collagen shows a specific binding affinity for bovine dentin phosphoryn. *Calcif. Tissue Intl.* **38**, 135–141.

Taft R. V. (1956) *Steric Effects in Organic Chemistry*. Wiley.

Takagi Y. and Veis A. (1984) Isolation of phosphophoryn from human dentin organic matrix. *Calcif. Tissue Intl.* **36**, 259–265.

Traub W., Jodaikin A., Arad T., Veis A., and Sabsay B. (1992) Dentin phosphophoryn binding to collagen fibrils. *Matrix* **12**, 97–201.

Tyler-Cross R. and Schirch V. (1991) Effects of amino acid sequence, buffers, and ionic strength on the rate and mechanism of deamidation of asparagine residues in small peptides. *J. Biol. Chem.* **266**, 22549–22556.

van Kleef F. S. M., DeJong W. W., and Hoenders H. J. (1975) Stepwise degradation and deamidations of the eye lens protein α-crystallin in ageing. *Nature* **258**, 264–266.

Veis A. and Scheuleter R. J. (1964) The macromolecular organization of dentine matrix collagen. I: Characterization of dentine collagen. *Biochemistry* **3**, 1650–1657.

Vetter U., Vogel W., Just W., Young M. F., and Fisher L. W. (1993) Human decorin gene—intron exon junctions and chromosomal localization. *Genomics* **15**, 161–168.

Voorter C. E. M., de Haard-Hoekman W. A., van den Oetelaar P. J. M., Bloemendal H., and de Jong W. W. (1988) Spontaneous peptide bone cleavage in aging alpha-crystallin through a succinimide intermediate. *J. Biol. Chem.* **263**, 19020–19023.

Waite E. R. and Collins M. J. (1998) Response to paper by Ohtani S., Yamada Y., and Yamamoto I. (1997) Age estimation from racemization rate using heated teeth. *J. Forensic Odonto-Stomatol.* **15**, 9–12. *J. Forensic Odonto-Stomatol.* **16**, 20–21.

Waite E. R. Collins M. J., Ritz S. Schutz H.-W., Cattaneo C, and Boorman, H. (1999) A review of the methodological aspects of aspartic acid racemization analysis for use in forensic science. *Forensic Sci. Int.* **103**, 113–124.

Walters C. and Eyre D. R. (1983) Collagen cross-links in human dentin: increasing content of hydroxypyridinium residues with age. *Calcif. Tissue Intl.* **35**, 401–405.

Yoshioka Y., Gejyo F., Marti T., Rickli E. E., Burgi W., Offner G. D., Troxler R. F., and Schmid K. (1986) The complete amino-acid-sequence of the α-chain of human-plasma α-2HS-glycoprotein. *J. Biol. Chem.* **261**, 1665–1676.

Young M. F., Kerr J. M., Termine J. D., Wewer U. M., Wang M. G., McBride O. W., and Fisher L. W. (1990) cDNA cloning, mRNA distribution and hereogeneity, chromosomal location and RFLP analysis of human osteopontin (OPN). *Genomics* **7**, 491–502.

Zhao M. X., Bada J. L., and Ahern T. J. (1989) Racemization rates of asparagine and aspartic-acid residues in lysozyme at 100°C as a function of pH. *Bioorganic Chem.* **17**, 36–40.

15. Interpretation of D/L-amino acid data from agricultural soils

Ronald W. L. Kimber and Charles V. Griffin

It is recognized that the D/L values used in the amino acid racemization (AAR) dating technique are calculated from the abundances of D- and L-enantiomers of amino acids derived from mixtures of many types of peptides of differing chain lengths (Kimber and Hare, 1991) together with any free amino acids present, usually known as "total amino acids". Variations in these parameters, as well as age, determine the degree of racemization measured. Interpretation of the data is, therefore, complex even with single entities such as shell, coral or bone. In soils, the complexity and, therefore, the potential usefulness of the technique for dating purposes would seem to be reduced further because of the presence of both living and dead microorganisms, as well as plant material. Many living microorganisms contain D-enantiomers of Glu, Ala and, to a lesser extent Asp, in their cell walls that tend to confound interpretation of D/L values. In addition, in most soils organic matter is continually being added, ensuring that, at least in the topsoil, much of the residual organic material is "modern". Some useful information on the age of soils and sediments has, nevertheless, been reported by us (Kimber and Griffin, 1988; Griffin, 1990; Kimber et al., 1994) as well as by other workers (Mahaney, et al., 1986; Mahaney and Rutter, 1989) using the AAR technique. Most of these reports have been largely in support of other evidence of age and in specific locations where some of the potential contribution of variables was reduced in effect. For example, confirmation of ages originally determined by tephrochronology and radioisotope dating in tephra–loess–paleosol sequences in New Zealand (Kimber et al., 1994).

To a great extent, effective age assessments of discrete materials, such as shells, using the AAR technique are possible only if they are largely isolated from external effects. Age assessments are best made where protein/peptide hydrolysis is minimal so that the breakdown of protein material, to peptides of various chain lengths, occurs at a relatively consistent and steady rate. Practically speaking, this usually means that assessments should be made either early or late in the racemization process. Most research has shown that a transitional period of variable rate occurs between two relative linear rates—a rapid rate early on, and a much slower one late in the racemization process. It is also important for protein material to be in an environment where little contamination from newer or older amino acid material has occurred. Clearly, using these criteria to assess the usefulness of the technique in agricultural soils suggests that it is unlikely to be useful for dating in most situations.

In order to attempt to overcome some of the problems caused by such an open system we developed further the technique devised by Cheng et al. (1975), and expanded by Limmer and Wilson (1980), whereby a "protected" fraction of peptide material entrapped in the soil matrix is released by treatment of the soil with HF. This fraction seems to be less affected by environmental influences, particularly the rate of natural peptide hydrolysis, than the material not entrapped (Kimber et al., 1994). Our modified procedure produced amino acids derived from normal HCl peptide hydrolysis of the soils whose D/L values were usually lower than those produced from HCl hydrolysis of peptide material released from the soil matrix by HF treatment. The proposal was that fractions released by HCl, which produced low amino acid D/L ratios, contained more recently added material from plant residues and bacteria, whereas the fractions released by HF were less accessible to contamination from these sources. However, the results were variable and inconsistent (Griffin and Kimber, 1988). This could be be partly due to the contribution of D-enantiomers naturally occurring in bacterial cell walls, in particular from Glu and Ala. Overall, the results were different from those derived from tephra–loess–paleosol sequences in which D/L values for the "HCl" fractions were much greater than for the "HF" fractions and increased more rapidly going down the profile (Kimber et al., 1994). It seemed that an understanding of the kinetics of L- to D-amino acid conversion would enhance our knowledge of the system and enable us to evaluate more definitively the usefulness of the application of AAR dating to soils and perhaps to suggest other possible outcomes from the data produced.

In this regard, many soils in Australia are overutilized, particularly in marginal lands, but also in better soils (because of the use of inappropriate farming techniques). The result is that many of these soils become degraded in terms of nutrient content, physical structure, and biological diversity. There is consequential enhancement of erosion by wind and water and general landscape degradation. Over-use has resulted in poor crop yields, with large tracts of land being removed from cropping rotations and requiring many years and remedial action to return to anything approaching its original state.

Any technique that could be developed to give a general measure of the extent of soil degradation could be useful for assisting environmentalists to prescribe the best balance of soil usage to achieve sustainable farming. We report, here, a potential use for the kinetic information on amino acid D/L value changes derived from the heating of soils at various temperatures.

EXPERIMENTAL METHODS

Soils

Soil samples (Urrbrae fine sandy loam; Piper, 1938: Rhodoxeralf, Great Soil Group) collected from two plots of a series of long-term rotation trials at the Waite Agricultural Research Institute, Adelaide, South Australia (Griffin and Kimber, 1988) were used for the kinetics experiments. The first plot had been under continuous wheat/fallow (WF) cropping for 60 years and the second had been subject to 40 years of permanent pasture (PP). Although the two soils were pedogenetically identical, the WF soil had become highly degraded because of the highly exploitative nature of the rotation treatment used compared to the PP soil. Total hydrolysable N was 0.3 mg/g compared to 0.8 mg/g for the non-degraded soil. (Ladd and Russell, 1983). Typical general soil analytical data for the WF (cf. PP) soil included: clay content, 13% (17); pH 5.9 (5.5); organic C, 1.01% (2.80); total N, 0.11% (0.27); C/N, 10 (10.4) (Golchin et al., 1995).

Heating Experiments

One gram each of representative samples from the rotation trial soils was heated at 140°C in an oven with 5 ml of double-distilled water in hydrolysis tubes sealed with Teflon-lined screw caps. Heating was continued for up to 32 days, and duplicate tubes removed for analysis at various times during this period. Supplementary experiments were carried out on the PP soil at 105 and 160°C.

Analysis Procedure

The procedure used was our previously developed method (Griffin and Kimber, 1988; Kimber et al., 1994). Each sample (both solids and aqueous material) from the heating experiments was heated directly with HCl (adjusted to 6 M) at 110°C for 16 hr in Teflon-sealed screw-capped tubes to release the majority of the amino acids. The insoluble material remaining was sedimented by centrifugation in polypropylene tubes, washed with double-distilled water, and the combined supernatants evaporated to dryness at 40–50°C on a rotary film evaporator ("HCl" fraction). The insoluble material remaining was shaken with 5.0 M HF/0.1 M HCl at 25°C for 24 hr to release strongly bound and entrapped peptide material, and the liquor was removed by freeze-drying. This residue was hydrolyzed with 6 M HCl, as previously, to produce more amino acids, and the solution was then reduced to dryness by rotary film evaporation (RFE) ("HF" fraction). Both fractions ("HCl" and "HF", respectively) were fluoridated by suspension in 5 M HF, which was then evaporated by freeze drying and subjected to clean-up (using ion-exchange chromatography in polypropylene columns) and elution with 0.5 M HF. After further freeze-drying the anion was reconverted to the Cl^- ion by the addition of 5 M HCl followed by removal by RFE. This was followed by cation exchange, elution of amino acids with ammonium hydroxide, and derivatization prior to gas chromatographic analysis for D- and L-enantiomers of the amino acids, as previously described (Kimber and Griffin, 1987).

Preliminary analysis of the data was achieved by describing the change in D/L values of the amino acids from the two soil treatments, two extraction procedures, and three hydrolysis temperatures by using mathematically (where possible) constructed trend lines. These indicated the trends in the changes of D/L values over time during the hydrolysis process, rather than providing precise values. They are particularly inaccurate early in the heating period.

RESULTS AND DISCUSSION

Examples of the results are given below for changes in the D/L ratios of several amino acids derived from the two soils (PP and WF) heated at 140°C (figures 15-1 and 15-2). In general, the "HCl" fraction of the WF samples showed distinctly greater racemization rates than the "HF" fraction, and both rates were much greater than those from samples from the PP soils (Griffin, 1990). This is evident for valine (Val), allo-isoleucine/isoleucine (Aile/Ile) and alanine (Ala), but is more complex for aspartic acid (Asp) (figures 15-1 and 15-2). The effect of varying the temperature, as demonstrated by experiments with the PP soil, was to increase the racemization rate when carried out at 160°C and to reduce it at 105°C, but overall the general trends were the same. The result shown in figure 15-3 for Val is a typical example. In general, during heating, concentrations of the L-enantiomers declined rapidly while those of the D-enantiomers increased slightly; the degree to which this occurred was dependent upon temperature and time.

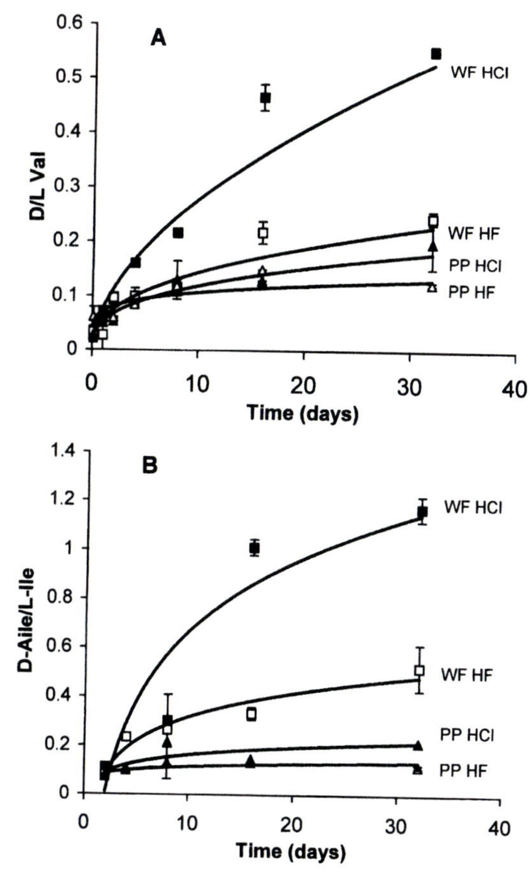

Fig. 15-1. Changes in D/L values of (A) valine and A/I values of (B) D-alloisoleucine/L-isoleucine isolated from soils (WF and PP), sampled from the Waite Campus permanent rotation field trials and heated in water at 140°C. These changes are indicated by trend lines (power for Val and logarithmic for Aile/Ile) for the "HCl" and "HF" fractions, and by bars to indicate standard deviations for measurements from each sampling period.

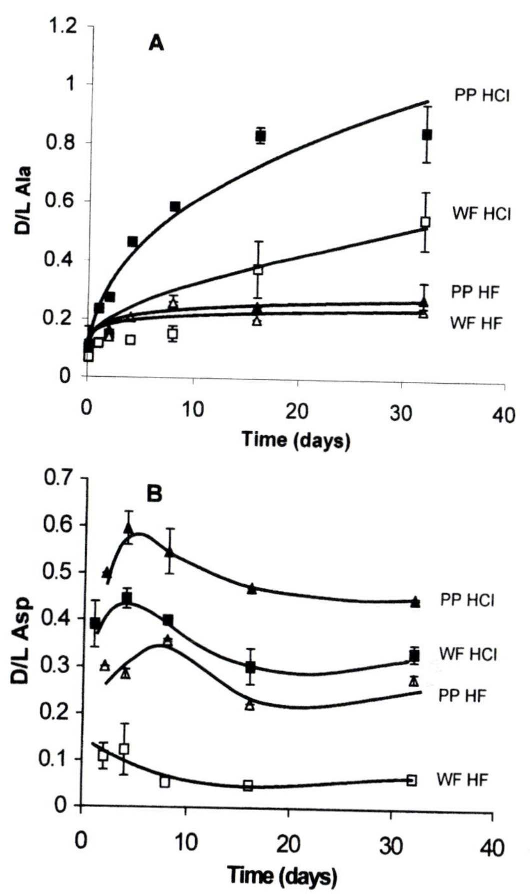

Fig. 15-2. Changes in D/L values of (A) alanine and (B) aspartic acid isolated from soils (WF and PP), sampled from the Waite Campus permanent rotation field trials and heated in water at 140°C. These changes are indicated by trend lines (power for Ala, WF HCl, logarithmic for PP HCl and PP HF, and free trace for WF HF; free trace for Asp PP HCl, WF HCl and PP HF, and polynomial for Asp WF HF) and by bars to indicate standard deviations for measurements from each sampling period.

Little is known about the racemization of amino acids in soils, but heating experiments have the potential to provide very useful kinetics data about the racemization process in this environment, as they have for aging of other fossil materials (Hare and Mitterer, 1969; Vallentyne, 1969; Williams and Smith, 1977; Kriausakul and Mitterer, 1980; Kimber et al., 1986; Kimber and Griffin, 1987). These heating experiments enable a more coherent comparison of the racemization process in soils to be made in terms of rate and complexity. Although the rates of the natural chemical processes are enhanced by the heating process, and there may be some doubt as to whether the heating conditions reflect what has happened during the sample's previous history, there can be no doubt that the resulting data reflect the conditions at the start of heating.

The findings of this investigation provide a good explanation for apparent contradictory results from our agricultural soils compared to previous work with buried palaeosols. The observed D/L values of amino acids from natural (unheated) agricultural soil (Griffin and Kimber, 1988) from the "HF" fractions were generally greater than those from the "HCl" fractions (table 15-1). This may be due to an age effect whereby the organic matter in the "HCl" fraction is diluted with modern material from the dynamic agricultural soil environment, but the material in the "HF" fraction is more pro-

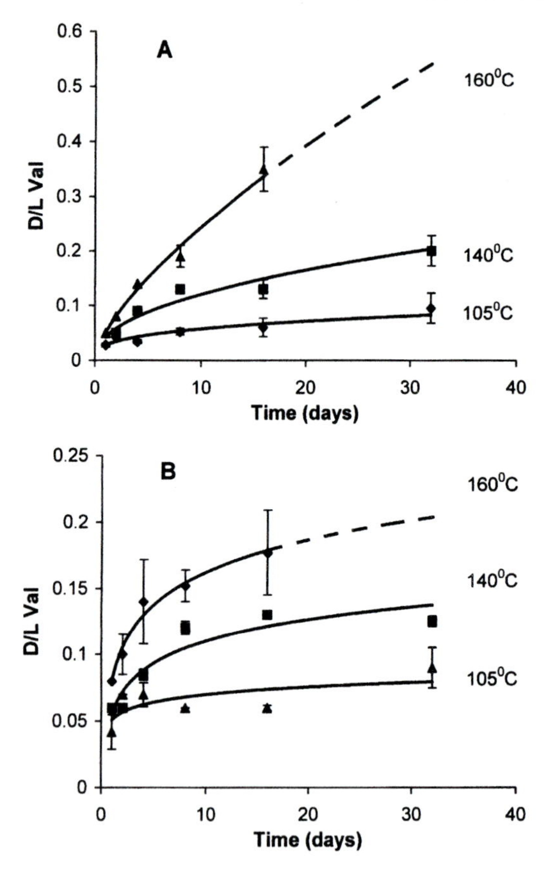

Fig. 15-3. Effect of temperature on the rate of racemization of valine isolated from the PP soil sampled from the Waite Campus permanent rotation field trials (heated in water), indicated by trend lines (power) for the "HCl" fraction (A), and (logarithmic) for the 'HF' fraction (B). Bars indicate standard deviations for measurements from each sampling period.

TABLE 15-1 D/L values of four amino acids from "HCl" and "HF" fractions from wheat/fallow rotation and permanent pasture soils (from Griffin and Kimber, 1988)

	Amino acid			
Fraction	Val	Leu	Ala	Asp
WF "HCl"	0.026	0.048	0.101	0.100
	(0.010)[a]	(0.012)	(0.009)	(0.026)
PP "HCl"	0.027	0.056	0.119	0.115
	(0.011)	(0.012)	(0.010)	(0.009)
WF "HF"	0.033	0.079	0.057	0.067
	(0.020)	(0.062)	(0.017)	(0.054)
PP "HF"	0.062	0.171	0.126	0.130
	(0.018)	(0.046)	(0.046)	(0.062)

[a] Standard deviation.

tected from contamination. This interpretation, however, may be confounded in the cases of Ala and Asp, where the L-enantiomers would be expected to be utilized more readily than the naturally present D-enantiomers, resulting in higher D/L values. With our buried paleosols (Kimber et al., 1994), D/L values from "HCl" fractions increased down the profile (reflecting an age increase, as determined by tephrochronology) more rapidly than those from the "HF" fraction (0.2–0.9 and 0.17–0.57, respectively, from a 0.5 m to a 4 m depth). These observations are consistent with the results of the heating studies (a simulated aging process) described here, whereby the rates of racemization of the "HCl" fractions were greater than those of the "HF" fraction (despite the fact that their initial D/L values were often greater, e.g., for Val and Leu in both soils, table 15-1). In addition, each fraction had a distinctive rate of racemization for each amino acid. Thus, in a normal "modern" agricultural soil, D/L values of the "HF" fraction may be greater than D/L values of the "HCl" fraction but have a slower rate of racemization than the latter upon heating. However, in a buried paleosol, the D/L values in the "HCl" fraction are greater than the D/L values of the "HF" fraction because of the greater racemization rates of the former and the comparatively reduced contamination and interaction with modern material owing to its buried location.

The examples given in figures 15-1 and 15-2A for Val, Aile/Ile and Ala showed greater racemization rates for the WF soil than for the PP. The difference between the three amino acids was only in their rates, with the more stable Val only being 60% racemized in the 32 d heating period, whereas both Aile/Ile and Ala were almost fully racemized (i.e., a 50:50 mixture of D- and L-enantiomers). The D/L values of the PP fractions increased only slightly during the period to approximately 0.2 with little difference between the "HF" and "HCl" fractions. The observation that Val racemized more slowly than Aile/Ile in the WF soil would be surprising if made with fossil shells or other discrete entities containing protein material. However, one cannot take it to be so in such a complex environment as soil, where many possible factors may cause different relative rates of racemization between the two amino acids, e.g., ease of release from the soil matrix, differential catalytic effects, etc.

On the other hand, Asp under the same conditions behaved quite differently. For each fraction (except WF "HF") and soil, a rapid increase in D/L value was followed by a fall (figure 15-2B). We have reported this phenomenon previously for heated fossil shells (Kimber and Griffin, 1987) and for a natural tephra–loess–paleosol sequence in

New Zealand (Kimber et al., 1994). A further complication with this amino acid is that the overall extent of racemization of each fraction is different, with the PP "HCl" fraction being the most racemized and the WF "HF" fraction the least racemized. It is likely that the reversal effect resulted from the lower stability of Asp in the various peptide species present as precursors similar to those discussed elsewhere (Kimber and Griffin, 1987; Kimber and Hare, 1992). This result illustrates, again, the difficulties that can occur when using Asp for AAR dating if this possibility is overlooked. Some other amino acids were also less than satisfactory for various reasons, including stability problems and analytical difficulties in separating them effectively from other soil organic substances. In any event, it has been shown by many workers (Kriausakul and Mitterer, 1980; Hearty et al., 1986; Kimber and Griffin, 1987) that Val and Aile/Ile are two of the amino acids that produce consistently reliable AAR data from fossiliferous material.

Our earlier work on the unheated soils (Griffin and Kimber, 1988) showed no difference between D/L values of the "HCl" fractions of the WF and PP soils, whereas the D/L values from the "HF" fractions of the WF soil were significantly lower than those from the PP soil (table 15-1). A naive interpretation might be that the organic matter in the WF soil was younger than that from the PP soil, although this would be at variance with other results (Ladd and Butler, 1966; Kimber and Searle, 1970). These authors based their conclusions that the WF soil contained older organic matter, on spectroscopic and pyrolysis gas–liquid chromatographic evidence, respectively, as well as nutritional evidence that the WF soil was very degraded (hydrolyzable N content, 0.3 mg/g). The WF soil had been degraded by many years of "mining" of nutrients for plant growth, because of the nature of the rotation. During this process, easily available nutrients, in particular nitrogen, would be used first; thereafter, microbiological exploitative "mining" processes would remove the more hydrophobic and other recalcitrant material, allowing the now more accessible peptides to be hydrolyzed more readily during our heating experiments. We believe that the formation of a greater proportion of smaller peptides (more end-groups) during this process has resulted in a greater degree of racemization. Within this context, the HCl hydrolysis process would release the more available material and the following HF process the less available material. This is particularly evident in the "HCl" fraction into which has been extracted the more easily accessible amino acids. This scenario is consistent with our results, in which the rate of racemization of amino acids in the WF "HCl" fraction was greater than in the "HF" fraction. In the latter, which was composed of peptide material extracted from the more protected sites in the soil matrix, the entrapped peptide material was less accessible to water for hydrolysis and more of the larger peptides remained; consequently, less racemization occurred. It is reasonable also to suggest that this fraction was less contaminated by modern material, which, on the other hand, was relatively easily accessible in the soil for extraction into the "HCl" fraction. This accounts for the observed differences in rates of racemization of the amino acids derived from these two sources.

Although it may seem unlikely that this modified AAR method could produce data that were more useful than that produced by other methods, e.g., ^{14}C dating, for estimating the age of agricultural soils, nevertheless, one of us (Griffin, 1990) has devised a method with potential for assessing the "apparent" age of these soils, using kinetics information from the experiments described above. In that work, the changes in rates of racemization were used to determine relative ages of the "HCl" and "HF" fractions by the use of Arrhenius equations. Extrapolation of the Arrhenius plots to ambient temperature provided the rates of racemization for uncalibrated age assessment. This exercise clearly demonstrated that the "HF" fraction in the PP soil contained older material, the resulting apparent mean residence time based on Val being an order of magnitude greater for the "HF" fraction. It is interesting that E_a values for Val, as determined from heating experiments (figure 15-3), are similar to those of the >1000/molecular-weight fraction of some mollusk species and much lower than those for their total hydrolysates (Kimber and Griffin, 1987). This is consistent with a complex and "protected" environment in which the parent medium-length peptides reside.

Usefulness As a Measure of the Extent of Degradation of the Soil

The results presented here show clear differences in the D/L values for a degraded soil (WF) and a non-degraded (PP) soil. The scheme shown in figure 15-4 gives a representation of the types of processes that may have resulted in the trends in D/L values observed in the experiments. Figure 15-4 depicts a typical aggregate from the PP soil where the normal soil structure would include numerous micropores that contain organic matter, including protein material and peptides. Many of these aggregates would also have layers of polyphenolics and other hydrophobic materials ("a *mélange* of biological chemicals" according to Oades, et al., 1988) adhering to surfaces, blocking the pores, and restricting chemical and microbiological degradation. On the other hand, aggregates from the WF soil have lost much of this protection and the less readily available nutrients can be scavenged from the soil. During this process, more intractable material is metabolized by microorganisms and used as a "subsistence diet". The concept of "protection" is very

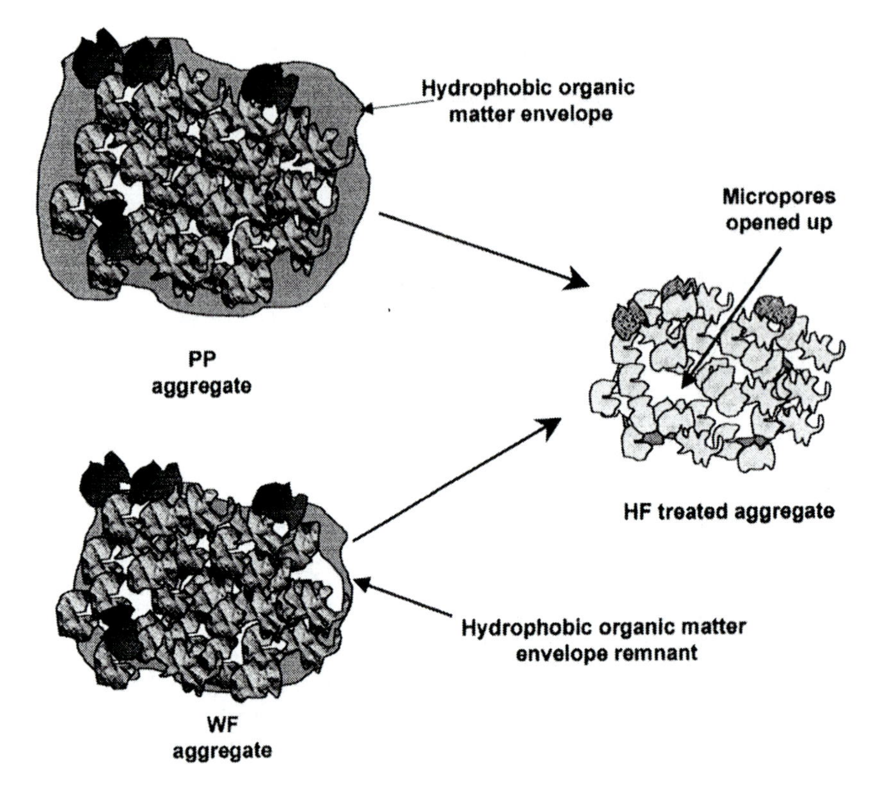

Fig. 15-4. Schematic diagram of the types of changes that occur in a soil aggregate when a soil is degraded, typified by conversion from a PP-type soil to a WF-type soil, and following treatment with HF.

important in this respect, as is that of "mining" of nutrients. Clearly, the most available nutrients are used first by plants and microorganisms in the soil. If some of the potential nutrients, e.g., peptide material, are in "protected" locations, either strongly sorbed to surfaces or within the interstices of the soil matrix, surrounded by hydrophobic moieties (cf. Knicker and Hatcher, 1997), etc., not only does "mining" not occur so readily but the material is also likely to contain older amino acids.

Much new plant material and microbial biomass in soil is relatively hydrophobic and so is less accessible to hydrolysis, even during our heating experiments, than soil organic matter that has aged and degraded. This seems to be the situation with the PP and WF soils. The PP soil contains much new organic matter (Oades et al., 1988) that is highly hydrophobic and not easily hydrolyzed during the pyrolysis period, resulting in low D/L values. On the other hand, the WF soil has been depleted of nutrients as a result of the exploitative microbial subsistence phase breaking down many of the hydrophobic molecules and other more recalcitrant materials exposing peptides to easy hydrolysis and racemization during the period of heating.

Collection of more data from soils with differing extents of degradation should enable a model to be constructed to measure the extent of degradation of soils with unknown histories. Currently we can say that after 32 days of heating in water at 140°C, a D/L value for Val of ~0.2 indicates a non-degraded soil and ~0.6 indicates a highly degraded soil under the conditions used and for the soil type examined. The same can be said for Aile/Ile (0.2 and 1.2, respectively) and for Ala (0.25 and 0.9, respectively). Clearly, we cannot say at this stage how widely applicable these observations may prove to be; many more soil types with differing extents of degradation need to be assessed before the universality of the method can be confirmed.

CONCLUSIONS

AAR techniques provide a unique approach to the examination of the behavior of amino acid material in the soil system. The evidence presented above indicates that AAR data derived from heating studies provide a basis for the development of a useful tool for the assessment of the extent of degradation of soils. If these data can be factored into models used to describe the extent of degradation of soils, the modified AAR process could prove useful for assessing the suitability of agricultural soils for continued cropping within a sustainable agricultural system and could assist in the handling of environmental concerns associated with soil and land conservation.

Acknowledgments: Much of the work described here resulted from the research of C.V.G. for a Ph.D. at the University of Adelaide. Facilities were provided by CSIRO Land and Water and financial support by G. H. Chenery. We very much appreciate the travel grant awarded to R.W.L.K. by the Geophysical Laboratory of the Carnegie Institution of Washington to enable the presentation at the Washington conference on Perspectives in Amino acid and Protein Geochemistry, and for the opportunity to honour the life and science of Dr P. E. Hare.

REFERENCES

Cheng C.-N., Schufeldt R. C., and Stevenson F. J. (1975) Amino acid analysis of soils and sediments: extraction and desalting. *Soil Biol. Biochem.* **7**, 143–151.

Golchin A., Clarke P., Oades J. M., and Skjemsted J. O. (1995) The effects of cultivation on the composition of organic matter and structural stability of soils. *Aust. J. Soil Res.* **33**, 975–993.

Griffin C. V. (1990) *Amino Acid Racemization in Agricultural Soils*. Ph.D Thesis, University of Adelaide.

Griffin C. V. and Kimber R. W. L. (1988) Racemization of amino acids in agricultural soils: an age effect? *Aust. J. Soil Res.* **26**, 531–536.

Hare P. E. and Mitterer R. M. (1969) Laboratory simulation of amino acid diagenesis in fossils. *Carnegie Instn Wash. Yrbk.* **67**, 205–208.

Hearty P. J., Miller G. H., Stearns C. E., and Szabo B. J. (1986) Aminostratigraphy of Quaternary shorelines in the Mediterranean basin. *Bull. Geol. Soc. Am.* **97**, 850–858.

Kimber R. W. L. and Griffin C. V. (1987) Further evidence of the complexity of the racemization process in fossil shells with implications for amino acid racemization dating. *Geochim. Cosmochim. Acta* **51**, 839–846.

Kimber R. W. L. and Griffin C. V. (1988) Amino acid racemization dating of soils and sediments. In *Archaeometry: Australasian Studies 1988*, (ed. J. R. Prescott), pp. 22–29. University of Adelaide Printing Service.

Kimber R. W. L. and Hare P. E. (1992) Wide range racemization of amino acids in peptides from human fossil bone. *Geochim. Cosmochim. Acta* **56**, 739–743.

Kimber R. W. L., Griffin C. V., and Milnes A. R. (1986) Amino acid racemization dating: evidence of apparent reversal in aspartic acid racemization with time in shells of *Ostrea*. *Geochim. Cosmochim. Acta* **50**, 1159–1161.

Kimber R. W. L. and Searle L. E. (1970) Pyrolysis GC of soil organic matter. 2. The effect of extractant and soil history on the yields of products from pyrolysis of humic acids. *Geoderma* **4**, 57–71.

Kimber R. W. L., Milnes A. R., and Kennedy N. (1994) Amino acid racemization dating of soils: evidence of good correlation of D/L values of aspartic acid with depth and age in an acid soil sequence. *Aust. J. Earth Sci.* **41**, 19–26.

Knicker H. and Hatcher P. G. (1997) Survival of protein in an organic-rich sediment: Possible protection by encapsulation in organic matter. *Naturwissenschaften* **84**, 231–234.

Kriausakul N. and Mitterer R. M. (1980) Comparison of isoleucine epimerization in a model dipeptide and fossil protein. *Geochim. Cosmochim. Acta.* **44**, 753–758.

Ladd J. N. and Butler J. H. A. (1966) Comparison of some properties of soil humic acids and synthetic phenolic polymers incorporating amino derivatives. *Aust. J. Soil Res.* **4**, 41–54.

Ladd J. N. and Russell J. S. (1983) Soil Nitrogen. In *Soils: An Australian Viewpoint*, pp. 569–607. CSIRO: Melbourne/Academic Press London.

Limmer A. W. and Wilson A. T. (1980) Amino acids in buried paleosols, *J. Soil Sci.* **31**, 147–153.

Mahaney W. C. and Rutter N. W. (1989) Amino acid D/L ratio distributions in two late Quaternary soils in the Afroalpine zone on Mt. Kenya, East Africa. *Catena* **16**, 205–214.

Mahaney W. C., Boyer M. G., and Rutter N. W. (1986) Evaluation of amino acid composition as a geochronometer in buried soils on Mt. Kenya, East Africa. *Geol. Phys. Quat.* **40**, 171–183.

Oades J. M., Waters A. G., Vassallo A. M., Wilson M. A., and Jones G. P. (1988) Influence of management on the composition of organic matter in a red-brown earth as shown by ^{13}C nuclear magnetic resonance. *Aust. J. Soil Res.* **26**, 289–299.

Piper C. S. (1938) The red–brown earths of South Australia. *Trans. Roy. Soc. S. Aust.* **62**, 53–100.

Rutter N. W., Crawford R. J., and Hamilton R. D. (1985) Amino acid racemization dating. In *Dating Methods of Pleistocene Deposits and their Problems. Geoscience Canada (Reprint Series)*, vol. 2 (ed. N. V. Rutter), pp. 23–30.

Vallentyne J. R. (1969) Pyrolysis of amino acids in Pleistocene *Mercenaria* shells. *Geochim. Cosmochim. Acta* **33**, 1453–1458.

Williams K. M. and Smith G. G. (1977) A critical evaluation of the application of amino acid racemization geochronology and geothermometry. *Orig. Life* **8**, 91–144.

16. Kinetics of aspartic acid racemization in *Mya* and *Hiatella*: modeling age and paleotemperature of high-latitude quaternary mollusks

William F. Manley, Gifford H. Miller, and Julie Czywczynski

The extent of amino acid racemization in protein residues of Quaternary mollusks has been utilized for decades to estimate the age of fossil collections or the temperature history of fossil localities (see reviews by Miller and Brigham-Grette, 1989; Mitterer, 1993; Wehmiller, 1993). Studies of amino acid geochronology and paleothermometry have commonly been based on the epimerization reaction of isoleucine (the conversion of L-isoleucine to D-alloisoleucine). Such studies have been important for research on Arctic paleoclimate (e.g., Miller and Hare, 1980; Miller et al., 1989; Mangerud and Svendsen, 1992). More recently, Goodfriend et al. (1996) demonstrated that greater resolution for high-latitude sites is possible through analysis of a faster racemizing amino acid, aspartic acid (Asp). The new approach was coupled with preliminary kinetic models of Asp racemization in the two genera of bivalve mollusks common to high-latitude settings—*Mya* and *Hiatella* (Goodfriend et al., 1996). However, the available kinetics data are limited; heating experiments, for example, were conducted at only one temperature, and only five fossil collections were included to determine rates of racemization at ambient temperatures. Also, our knowledge of Asp kinetics in a variety of carbonate fossils has been hindered in the past by the inability analytically to resolve a related amino acid, asparagine (Asn), which converts to Asp during laboratory-induced hydrolysis and possibly under natural conditions (cf. Brinton and Bada, 1995; Goodfriend and Hare, 1995; cf. Collins et al., 1999). The potential of Asp racemization to constrain the timing and magnitude of high-latitude paleoenvironmental change would benefit from a more thorough understanding of Asp kinetics and the interaction between Asp and Asn.

In this chapter, we develop new kinetic models relating Asp racemization to time and temperature for *Mya* and *Hiatella*. The models are based on Hyd (hydrolyzed, or total) D/L Asp results from 3 high-temperature simulation experiments and for 12 radiocarbon-dated fossil collections. Hyd D/L Asp ratios were quantified with reverse-phase high-performance liquid chromatography (RPHPLC; Kaufman and Manley, 1998). We utilized a new class of transformation that linearizes data approaching an asymptote of 1.0, and rates of Asp racemization thus calculated were used to determine Arrhenius parameters. The kinetic models are illustrated via calculation of the effective diagenetic temperature (EDT) for eight fossil collections from the Eastern Canadian Arctic, Iceland, and Svalbard. We also measured the concentration and D/L ratios of Free Asp and Free Asn in one of the heating experiments for *Mya*. These data indicate that Asn is present in *Mya*, racemizes faster than does Asp, and appears to deamidate to Asp prior to laboratory preparation of samples. The empirical kinetic models yield equations that can be utilized by aminostratigraphers to quantify climate change with more resolution than was previously available with kinetic models of isoleucine epimerization.

Hyd D/L Asp as reported here is the racemization ratio for aspartic acid as measured on sample hydrolysates. Any asparagine present within samples is converted through deamidation to aspartic acid during laboratory acid hydrolysis (cf. Brinton and Bada, 1995; Goodfriend and Hare, 1995). Thus, Hyd D/L Asp ratios reflect the combined racemization of both aspartic acid and asparagine. Unless otherwise noted, we use the term "Asp racemization" to refer generally to racemization apparent from Hyd D/L Asp ratios.

Many studies have taken advantage of the relatively fast racemization of Asp in a variety of carbonate fossils. Heating experiments have been conducted on ostrich eggshell (Goodfriend and Hare, 1995), ostracodes (Kaufman, 2000), gastropods (Goodfriend, 1992; Goodfriend and Meyer, 1991; Goodfriend et al., 1995), bivalves (Goodfriend and Stanley, 1996; Goodfriend et al., 1996), and other mollusks (Kimber and Griffen, 1987). In addition, several studies have examined the kinetics of Asp racemization in time-series of fossils under nearly isothermal conditions (e.g., gastropods, Goodfriend, 1991; corals, Goodfriend et al., 1992; foraminifera, Harada and Handa, 1995). Asp racemization in carbonate fossils from high-latitude settings has received relatively limited attention (Rutter et al., 1980; Goodfriend et al.,

1996; Manley et al., 1998). Goodfriend et al. (1996) conducted a heating experiment at 100°C for *Mya arenaria* and *Hiatella arctica*, incorporated the Hyd D/L Asp results for five fossil collections into their kinetic models, and utilized the preliminary models to estimate ages and temperature depressions for glacial periods in Alaska, the Eastern Canadian Arctic, and Svalbard. This study expands upon that of Goodfriend et al. (1996) with more data, a slightly different approach, presentation of the kinetic equations, and quantification of both Asp and Asn.

MATERIALS AND METHODS

The heating experiments were conducted on modern paired valves of *Mya truncata* and *H. arctica* collected from lower tidal flats near Pangnirtung, Baffin Island, Eastern Canadian Arctic (66°08′N, 65°44′W). The specimens were cleaned with brush and water to remove preserved soft-body parts and periostracum, then broken into small pieces (130–250 mg, avoiding hinge and edges). Two fragments from separate individuals were placed with 4.5 g of sterile quartz sand and 0.7 ml of deionized water in each of three series of glass tubes, which were flame-sealed, placed in stable ovens at 85, 110, and 135°C, and withdrawn at prescribed intervals. Preheated aluminum blocks were used for all test tubes at 85 and 135°C, and the first 8 of 20 tubes withdrawn for each species from 110°C. Temperature fluctuations during the experiments were less than ±0.3°C.

After laboratory heating, shell fragments were removed from the glass tubes and prepared according to standard procedures. Fossil samples (table 16-1) were also prepared with the following steps.

TABLE 16-1 Radiocarbon ages, mean annual temperatures, and Hyd D/L Asp ratios for fossil collections

Site	Genus	^{14}C Lab. no.	Reported age (^{14}C yr BP)	ΔR[a]	Calibrated age[b] (cal yr BP)	Ref.[c]	Temp (°C)[d]	Hyd D/L Asp[e]
Samples utilized in kinetic model								
Kapp Ekholm, W. Svalbard	*Mya*	two[f]	8675 ±110	70	9660	10	−6	0.134 ±0.006 (5)
Kapp Ekholm, W. Svalbard	*Mya*	T-8534	9080 ±110	70	10,060	10	−6	0.124 ±0.006 (5)
Brøggerhalvøya, W. Svalbard	*Mya*	GX-8590	10,390 ±315	70	11,360	2, 7	−6	0.199 ±0.012 (2)
Loks Island, SE Baffin Is.	*Mya*	AA-5835	10,555 ±75	140	11,530	11, 12	−6	0.161 ±0.003 (3)
Brøggerhalvøya, W. Svalbard	*Mya*	DIC-3122	10,850 ±90	70	12,270	7, 8, 12	−6	0.130 ±0.020 (5)
Tromsø, Norway	*Mya*	T-2378	11,990 ±130	60	13,450	12	3	0.285 (1)
Aalesund, Norway	*Mya*	T-3705	11,090 ±140	5	12,620	3	7	0.304 ±0.005 (2)
Bangor, Maine	*Mya*	Y-2207	12,440 ±120	−95	14,710	1	7	0.293 ±0.017 (3)
Noble Inlet, SE Baffin Is.	*Hiatella*	GSC-3404	8240 ±90	140	9090	5	−6	0.140 ±0.012 (6)
Brøggerhalvøya, W. Svalbard	*Hiatella*	DIC-3122	10,850 ±90	70	12,270	7, 8, 12	−6	0.128 ±0.019 (4)
Enfield, Maine	*Hiatella*	Y-2205	12,210 ±100	−95	14,390	1, 12	7	0.267 ±0.016 (3)
SW British Columbia	*Hiatella*	I(GSC)-6	12,625 ±450	390	14,300	3	10	0.328 ±0.008 (3)
Samples utilized to illustrate the kinetics model								
ILC-Z, SE Baffin Is.	*Mya*	several[g]	11,000 ±125	140	12,370	13	−6	0.150
Poole Point, W. Svalbard	*Hiatella*	AA-19035	11,340 ±150	70	13,250	14	−6	0.147 ±0.013 (5)
Lake Linnévatnet, W. Svalbard	*Hiatella*	two[h]	10,505 ±105	70	11,570	6, 9	−6	0.196 (1)
Borgarfjördur, W. Iceland	*Mya*	Lu-2193	12,465 ±110	140	14,360	4	5	0.458 ±0.010 (3)
Brøggerhalvøya, W. Svalbard	*Mya*				~125,000	2, 7	−6	0.214 ±0.008 (3)
Brøggerhalvøya, W. Svalbard	*Hiatella*				~125,000	2, 7	−6	0.213 ±0.015 (6)

[a] ΔR values, local deviations from a globally averaged marine-reservoir correction, estimated from Stuiver and Braiunas (1993).

[b] Reported ^{14}C ages were brought to a common format (^{13}C-corrected, normalized to $\delta^{13}C = -25‰$, with error terms of ±1 S.D.), then calibrated using the CALIB 3.03c program (Stuiver and Reimer, 1993) with the marine sample data sets of Stuiver and Braziunas (1993) and Bard et al. (1993).

[c] References for stratigraphic context and radiocarbon ages: 1, Stuiver and Borns (1975); 2, Miller (1982); 3, Miller (1985); 4, Ingólfsson (1988); 5, Miller et al. (1988); 6, Werner (1988); 7, Miller et al. (1989); 8, Forman (1990); 9, Mangerud and Svendsen (1990); 10, Mangerud and Svendsen (1992); 11, Kaufman et al. (1993); 12, Goodfriend et al. (1996); 13, Kaufmann and Manley (1998); 14, Andersson et al. (1999).

[d] Present mean annual air temperature, used as an estimate of effective diagenetic temperature.

[e] Values are mean ± S.D. (number of shells analyzed per collection). For ILC-Z, tens of shells were ground to a powder before analysis (Kaufman and Manley, 1998).

[f] The age shown here is an average of two dates from nearby, physically correlative collections (T-8323 and T-8333).

[g] The age shown here is an average of six dates on this collection (AA-0655, GSC-5036, GSC-5037, QC-480A, QC-480C, QL-1173).

[h] The age shown here, for a core depth of 511 cm, is estimated by assuming a linear sedimentation rate between dates AA-3186 at 490 cm (10,250 ± 80) and Ua-1350 at 642 cm (11,640 ± 130).

Fragments were mechanically cleaned using a rotary grinder to remove outer carbonate layers. An additional 10–33% of shell material by weight was removed by stoicheometric addition of 2 M HCl. Fragment subsamples were then dissolved in 7 M HCl using a ratio of 0.02 ml/mg shell and hydrolyzed under N_2 for 6 hr at 110°C in the resulting solution of 5.9 M HCl (see the results of the hydrolysis test in Kaufman and Manley, 1998). These hydrolysis conditions differ from those used by Goodfriend et al. (1996; 20 hr at 100°C). Hydrolysates were evaporated to dryness under N_2 at approx. 80°C for 20 min, and reconstituted with 0.01 M HCl with 0.77 mM sodium azide (a bacterial inhibitor).

Amino acid concentrations and ratios were quantified using RPHPLC, in a manner similar to that of Kaufman and Manley (1998; see also Fitznar et al., 1999). The liquid chromatograph consists of a Shimadzu LC-10AD quaternary pump with an He degasser, an HP1100 autosampler, a Hypersil BDS C18 column (5 µm, 4.0 mm × 250 mm), a Hitachi F1000 fluorescence detector (with excitation and emission wavelengths of 335 nm and 445 nm, respectively), and an HP3395 integrator. Eluent A consists of 23 mM sodium acetate with 1.5 mM sodium azide, adjusted to pH 6.00 with 10% acetic acid. Eluent B is 93% methanol and 7% acetonitrile. The pre-column derivatizing reagent [260 mM isobutyryl-L-cysteine (IBLC) and 170 mM *o*-phthaldialdehyde (OPA) in a 1.0 M potassium borate buffer solution of pH 10.4] is automatically mixed with each sample prior to injection. A linear gradient analysis was performed at 1.0 ml/min and approx. 25°C, from 5% eluent B upon injection to 23–27% eluent B at 45 min. Although nine or more stereoisomer pairs—including D-alloisoleucine/L-isoleucine—can be quantified with longer analytical runs (Kaufman and Manley, 1998), we elected to limit our analyses to L- and D-Asp, L- and D-Glu (glutamic acid), L- and D-Ser (serine), and L- and D-Asn, when present. The minimum detection threshold is approximately 1 pmol enantiomer.

D/L Asp ratios were determined by division of peak areas, and calibrated to correct for differential fluorescence detection. Thus, the raw D/L ratios were divided by the values measured for a synthetic racemic mixture (Sigma No. DLAA-24; D/L Asp = 0.957 ± 0.031, $n = 79$; mean ± 1 S.D.). Usually, two or three analytical runs were made for each sample solution. Over the three months of sample analyses, results for interlaboratory comparison powders were comparable to values obtained by gas chromatography (Wehmiller, 1984) and by another RPHPLC analysis (Kaufman and Manley, 1998). Included in this assessment is a new powder designated "ILC-Z" (see table 16-1 and Kaufman and Manley, 1998). Our Hyd D/L Asp results are as follows [mean ± 1 S.D. (number of analyses)]: ILC-Z, 0.150 ± 0.004 (46); ILC-A, 0.363 ± 0.007 (20); ILC-B, 0.690 ± 0.007 (15); and ILC-C, 0.865 ± 0.016 (12). Analytical errors for replicate runs of heating-experiment sample solutions, analyzed over days or a few weeks, average 2% (coefficient of variation). Variation is higher between the two fragments in each test tube for any given temperature and species, with average intershell variability of 5%.

Separate subsamples were prepared and analyzed for Free Asp and Free Asn. The hydrolysis step (above) was omitted. Raw D/L Asn values were calibrated to correct for differential fluorescence (the racemic value is 1.078 ± 0.031, $n = 52$). Free D/L Asp results for the interlaboratory standards are: ILC-Z, 0.241 ± 0.059 (7); ILC-A, 0.611 ± 0.004 (4); ILC-B, 0.864 ± 0.008 (4); and ILC-C, 0.929 ± 0.010 (3). Neither enantiomer of Asn was detected in the ILC samples, except for ILC-Z, which yielded a mean Free D/L Asn value of 0.263 ± 0.066 (12). Analytical errors for Free D/L measurements in the *Mya* samples heated at 110°C are higher than for corresponding Hyd D/L measurements, averaging 7% for Free D/L Asp and 16% for Free D/L Asn. Intershell variability is 4% for Free D/L Asp and 7% for Free D/L Asn. Relative concentrations of Free Asn and Free Asx (Asp + Asn) were calculated directly from peak areas. Relative concentrations of Free Asx and Hyd Asp were calculated by normalizing to concurrently analyzed standards of known concentration.

RACEMIZATION RESULTS

Modern, unheated specimens of *Mya* and *Hiatella* show small but measurable Hyd D/L Asp ratios. For *Mya*, Hyd D/L Asp is 0.040 ± 0.003 (4 individuals), and for *Hiatella*, Hyd D/L Asp is 0.033 ± 0.002 (4 individuals). These values reflect racemization induced during sample preparation, probably during hydrolysis.

The isothermal heating experiments show that racemization of aspartic acid increases systematically with time in a strongly nonlinear fashion (figure 16-1, table 16-2). Initial rates of racemization are very high, slowing over time and with increasing D/L ratio to yield convex-upward curves for both species. Racemization of aspartic acid is in general much faster than epimerization of D-alloisoleucine and L-isoleucine (A/I, as estimated with the kinetics model of Miller, 1985). For example, Asp racemization is initially about 100 times faster than A/I epimerization. By the time Hyd D/L Asp reaches 0.200, Asp racemization is still approx. 30 times faster than A/I epimerization. Rates of racemization and epimerization converge in the heating experiments after Hyd D/L Asp reaches approx. 0.580 and A/I reaches approx. 0.130. At each temperature, Asp racemization is slightly faster in *Mya* than in *Hiatella* (figure 1). In none of the experiments does Hyd D/L Asp reach equilibrium. The

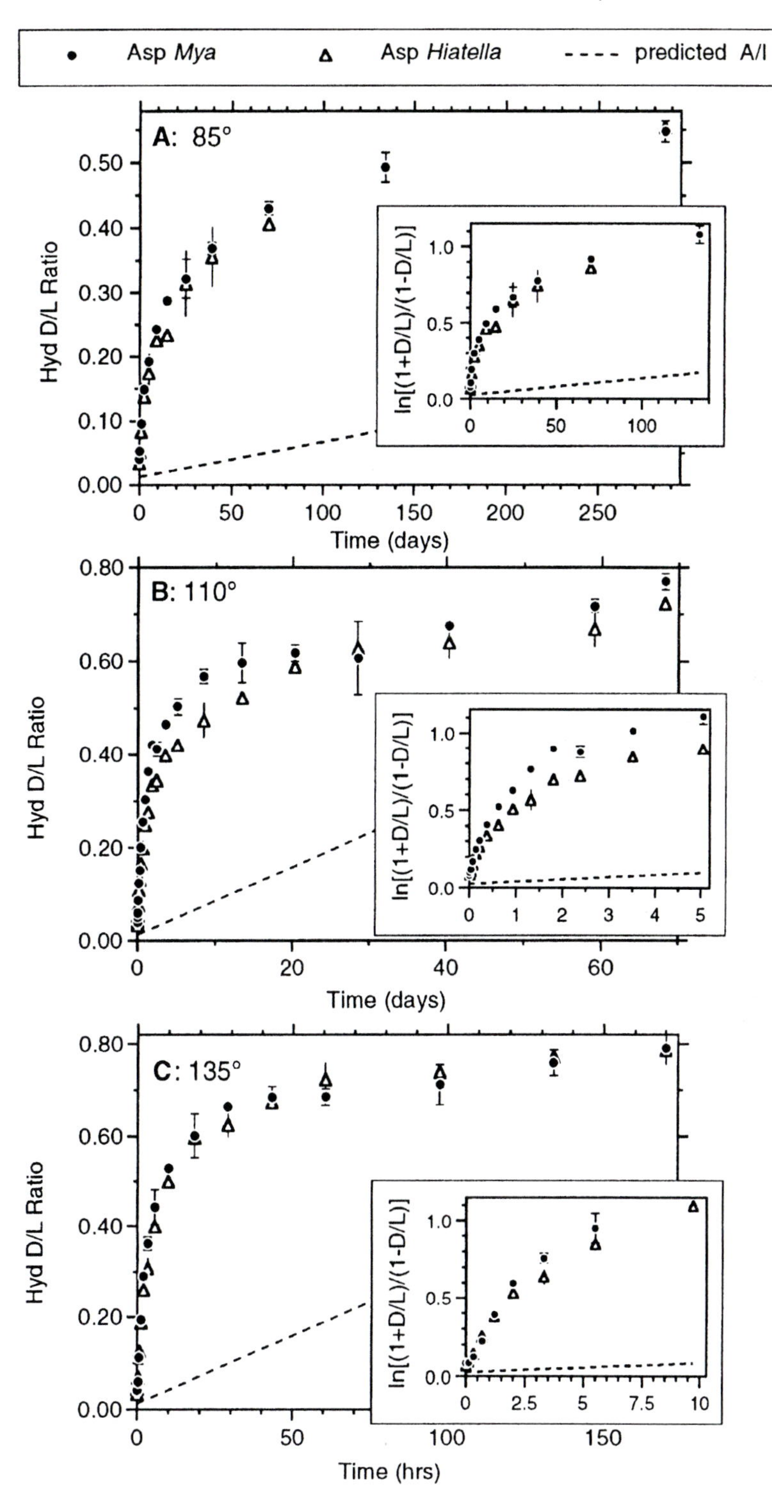

Fig. 16-1. Curvilinear increase in Hyd D/L Asp ratios for *Mya* and *Hiatella* heated at 85, 110, and 135°C (cf. Table 16-2). Inset plots show log normal transformations for Hyd D/L Asp ratios <0. 5. Broken lines show the trends expected for isoleucine epimerization (based on the kinetic model of Miller, 1985). The curves illustrate that rates of racemization are initially high, and decrease with time and increasing D/L ratio. Rates of Asp racemization are slightly higher, on average, for *Mya* than for *Hiatella*. Error bars show intershell variability (±1 S.D.).

TABLE 16-2 Hyd D/L Asp results for the heating experiments. D/L values are shown with mean ratio for two heated specimens, ± 1 S.D. Each specimen solution was quantified with an average of 2.2 analytical runs

Time (d)	*Mya* D/L Asp	CV[a]	*Hiatella* D/L Asp	CV[a]
At 85°C				
0.16	0.053 ±0.001	2.3	0.051 ±0.000	0.1
0.83	0.096 ±0.000	0.2	0.083 ±0.008	9.9
2.33	0.148 ±0.009	6.2	0.137 ±0.009	6.5
4.96	0.192 ±0.004	2.0	0.174 ±0.029	16.6
9.06	0.243 ±0.007	3.0	0.225 ±0.009	4.2
14.79	0.287 ±0.007	2.6	0.233 ±0.007	3.1
24.71	0.322 ±0.030	9.3	0.314 ±0.050	15.9
38.79	0.369 ±0.009	2.4	0.355 ±0.046	12.9
70.08	0.430 ±0.010	2.3	0.406 ±0.002	0.6
133.96	0.493 ±0.022	4.4	0.572[b] ±0.003	0.6
286.04	0.548 ±0.016	2.9	0.554 ±0.005	0.9
At 110°C				
0.01	0.051 ±0.002	4.8	0.041 ±0.003	7.9
0.03	0.059 ±0.005	9.2	0.053 ±0.004	7.3
0.07	0.086 ±0.001	0.8	0.075 ±0.011	14.4
0.14	0.124 ±0.005	3.7	0.108 ±0.001	1.0
0.22	0.152 ±0.004	2.9	0.131 ±0.011	8.4
0.37	0.201 ±0.008	3.9	0.166 ±0.008	5.1
0.62	0.255 ±0.008	3.0	0.199 ±0.014	6.8
0.93	0.304 ±0.009	3.1	0.248 ±0.000	0.2
1.31	0.363 ±0.001	0.3	0.276 ±0.030	10.9
1.80	0.421 ±0.007	1.8	0.334 ±0.006	1.7
2.38	0.412 ±0.016	4.0	0.344 ±0.008	2.3
3.50	0.466 ±0.011	2.4	0.398 ±0.013	3.1
5.04	0.503 ±0.018	3.7	0.420 ±0.006	1.5
8.41	0.568 ±0.015	2.6	0.474 ±0.036	7.6
13.40	0.596 ±0.042	7.1	0.522 ±0.001	0.1
20.28	0.617 ±0.017	2.8	0.588 ±0.012	2.0
28.52	0.606 ±0.078	12.8	0.627 ±0.008	1.2
40.36	0.675 ±0.011	1.6	0.639 ±0.033	5.1
59.13	0.718 ±0.014	1.9	0.668 ±0.037	5.6
68.19	0.770 ± 0.017	2.2	0.723 ±0.010	1.4
At 135°C				
0.005	0.041 ±0.001	1.8	0.044 ±0.005	10.9
0.013	0.060 ±0.011	18.3	0.069 ±0.005	6.7
0.028	0.111 ±0.014	12.5	0.124 ±0.013	10.5
0.050	0.193 ±0.002	0.9	0.187 ±0.008	4.4
0.083	0.290 ±0.009	3.2	0.260 ±0.007	2.7
0.138	0.362 ±0.014	3.8	0.308 ±0.021	6.8
0.229	0.443 ±0.039	8.9	0.399 ±0.003	0.9
0.405	0.529 ±0.012	2.2	0.499 ±0.011	2.2
0.755	0.601 ±0.048	8.0	0.597 ±0.006	1.0
1.206	0.664 ±0.006	0.9	0.624 ±0.025	4.0
1.796	0.684 ±0.024	3.4	0.674 ±0.008	1.2
2.520	0.685 ±0.018	2.6	0.723 ±0.035	4.8
4.062	0.712 ±0.044	6.1	0.739 ±0.006	0.8
5.580	0.759 ±0.028	3.6	0.771 ±0.015	1.9
7.084	0.790 ±0.007	0.9	0.786 ±0.029	3.7

[a] Coefficient of variation (S.D. dividied by the mean, expressed as a percentage).
[b] Single outlier omitted from Figure 16-1.

form of the racemization curves is similar to that shown by Goodfriend et al. (1996) for *H. arctica* and *M. arenaria* at 100°C. Also, the strongly convex-upward curves are similar to plots of Hyd D/L Asp in heating experiments and fossil time-series for gastropods (e.g., Goodfriend and Meyer, 1991; Goodfriend, 1992), coral (Goodfriend et al., 1992), ostrich eggshell (Goodfriend and Hare, 1995), and ostracodes (Kaufman, 2000).

To improve estimates of the temperature dependence of the racemization rate, we have expanded our kinetic modeling to include Hyd D/L Asp results for 12 radiocarbon-dated fossil collections (table 16-1). These fossils range in age from 8240 ^{14}C yr BP to 12,625 ^{14}C yr BP, and were collected from latitudes spanning 45–79° N. The present mean annual temperature at each collection site approximates the average temperature experienced by the fossils after death (cf. Miller, 1985; Goodfriend et al., 1996). Six other fossil collections (table 16-1) were analyzed to illustrate application of the kinetics models (see below). Five of the collections in table 16-1 were also analyzed by Goodfriend et al. (1996), who reported Hyd D/L Asp values on average 17% higher than those determined here; this discrepancy is at least partly due to differences in hydrolysis procedures (cf. Kaufman and Manley, 1998).

RACEMIZATION KINETICS

To model the racemization of aspartic acid with respect to time and temperature, two relationships must be quantified. First, the shape or form of the racemization curve at high temperatures (in heating experiments over short periods of time) must be modeled to simulate the increase in Hyd D/L Asp over time at low temperatures (in natural environments over centuries or millennia). In other words, a transformation of Hyd D/L Asp must be found that linearizes the data with respect to time, irrespective of temperature. Second, rates of increase in the linearized data are calculated for various temperatures to quantify the temperature dependence of the racemization reaction. If these two relationships can be adequately resolved with a quantitative model, then time, temperature, and D/L ratio define a three-dimensional surface, and each of the variables can be estimated if the other two are known. This approach, presented below, follows closely those taken previously (e.g., Schroeder and Bada, 1976; Miller, 1985; McCoy, 1987; Goodfriend et al., 1996; see also Collins et al., 1999).

TRANSFORMATION FUNCTIONS

We chose to limit our kinetic modeling of Hyd D/L Asp to values <0.5, for three reasons. First, an initial test of numerous transformation functions (see below) found that the functions fit the data best when D/L ratios >0.5 were omitted. This result is not surprising; other studies have shown that the interplay of amino acid and polypeptide decomposition, hydrolysis, and racemization complicates the racemization reaction beyond an initial predictable phase (e.g., Kriausakul and Mitterer, 1978; Wehmiller, 1993). Second, after Hyd D/L Asp of approx. 0.5 is reached, the rate of Asp racemization slows to approximate the rate of isoleucine epimerization, which can then be used to further gauge the age or paleothermometry of molluscan samples. Finally, a range in Hyd D/L Asp of 0.0–0.5 is sufficient to encompass the range of values expected for mid- to late Quaternary *Mya* and *Hiatella* at high latitudes.

A test of the $\ln[(1 + D/L)/(1 - D/L)]$ transformation indicates that Asp racemization in *Mya* and *Hiatella* does not follow first-order linear kinetics, as does isoleucine epimerization in many mollusks up to A/I values of approx. 0.35 (e.g., Miller, 1985; Goodfriend and Meyer, 1991). Plots of the transformed Hyd D/L Asp values are strongly convex-upward (figure 16-1). Poor fits by linear regression are expressed by average r^2 values of 0.83 for *Mya* and 0.84 for *Hiatella*. This deviation from first-order kinetics is typical for Asp (e.g., Goodfriend and Meyer, 1991; Goodfriend et al., 1996), and suggests that aspartic acid racemization is complicated—beyond the interplay of decomposition, hydrolysis, and racemization—by the conversion of asparagine to aspartic acid (cf. Goodfriend, 1991; Brinton and Bada, 1995; Goodfriend and Hare, 1995; Collins et al., 1999).

Power-law functions previously utilized in Asp kinetic studies provide nearly linear transformations, but fail to transform low D/L Asp values adequately. Goodfriend et al. (1995) and Goodfriend and Stanley (1996) found that D/L Asp raised to the power of two or three formed a linear relationship with time for bivalves and gastropods. Furthermore, Goodfriend et al. (1996) found best fits with power exponents of 3.6 overall for *M. arenaria*, 1.7 specifically for *M. arenaria* D/L Asp values <0.25, and 2.8 for *Hiatella*. With our heating-experiment results, the best power-law fits were obtained with exponents of 3.6 for *Mya* and 3.4 for *Hiatella* (r^2 = 0.977 and 0.982, respectively). However, these transformations do not adequately linearize D/L values less than approx. 0.17, which, after transformation, display concave-upward trends. Parabolic transformations (e.g., Mitterer and Kriausakul, 1989; Wehmiller, 1993) were not tested because raising time to an exponent < 1 is equivalent to raising D/L to an exponent > 1.

This finding led us to examine a mathematical class of functions that linearizes curves approaching an asymptote. The new transformation, $[(1 + D/L)/(1 - D/L)]^n$, is a logical choice for modeling enantiomeric ratios approaching an equilibrium value of 1.00, and provides a better linear fit to the Asp data than does a simple power-law function. Best-fit exponents for this transformation, as determined by the highest average r^2 value for the three heating experiments, are 3.4 for *Mya* and 3.3 for *Hiatella* ($r^2 = 0.988$ and 0.981, respectively). The $[(1 + D/L)/(1 - D/L)]^n$ transformation reduces scatter about the regression line throughout the range of D/L Asp values for each of the three temperatures.

However, the $[(1 + D/L)/(1 - D/L)]^{3.4}$ and $[(1 + D/L)/(1 - D/L)]^{3.3}$ transformations do not linearize the D/L Asp results equally well for each of the three temperatures. The transformed data are slightly convex-upward for the 85°C series, approximately linear for the 110°C series, and strongly concave-upward for the 135°C series (figure 16-2). The effect is pronounced for Hyd D/L Asp values less than approx. 0.12, corresponding to fragments that had been heated at 135°C for less than 40 min. Thus, rates of Asp racemization in the highest temperature experiment are initially slower than expected from the form of the curves at 85 and 110°C. The difference in the pattern of Asp racemization at 135°C, compared with the pattern closely matched at 85 and 110°C, may be the result of the following two factors. (1) At 135°C, the interplay of decomposition, hydrolysis, racemization, and deamidation of asparagine to aspartic acid may result in a kinetic pathway of racemization unlike that followed at lower temperatures (cf. Collins et al., 1999). (2) The glass tubes used in the heating experiments (filled with 4.5 g sand, 0.7 ml water, and two mollusk fragments each) experienced thermal inertia, such that the mollusk fragments required a significant period of time to attain isothermal conditions—an effect probably experienced during the relatively short periods of heating in the 135°C experiment. In either case, it is apparent that the 85 and 110°C racemization curves are similar in form, and that the 135°C racemization curve is unlikely to represent the pattern of racemization at temperatures less than 135°C.

Thus, to linearize the D/L Asp ratios with respect to time, we relied on the transformation that best fit the results of the 85 and 110°C simulations (figure 16-3). As determined by the highest average r^2 value for these two series, exponents for the $[(1 + D/L)/(1 - D/L)]^n$ function are 3.9 for *Mya* ($r^2 = 0.994$) and 4.2 for *Hiatella* ($r^2 = 0.995$). As shown in figure 16-3, this approach linearizes the data well. What scatter does exist is due to intershell variability rather than analytical error. For comparison, the best-fit exponent for Hyd D/L Asp in ostracodes is 3.6 (Kaufman, 2000), indicative of slightly lower curvature. We assume that the form of the racemization reaction for relatively short periods of time at 85 and 110°C adequately represents the shape of the racemization reaction for centuries or millennia at ambient temperatures of approx. −15–10°C. This assumption is common to kinetic modeling, and is partly validated by the findings of Goodfriend and Meyer (1991), who showed that Asp racemization in gastropods at 106.5°C

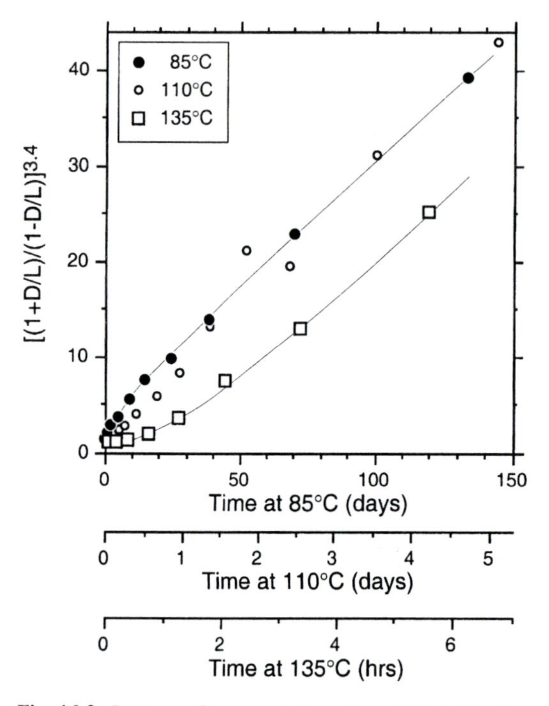

Fig. 16-2. Patterns of increase in transformed Hyd D/L Asp values with time for *Mya* at 85, 110, and 135°C, using the transformation function that provides the highest average r^2 value. Transformed values depict a slightly convex-upward trend for the 85°C series, a linear relationship for 110°C, and a strongly concave-upward bend for 135°C. The transformation function overcorrects for a less convex-upward relationship between Hyd D/L Asp and time for the highest temperature experiment. Lines are visual best fits for the 85 and 135°C data.

followed the same pattern of increase with time as did Asp racemization in fossil gastropods at nearly isothermal conditions of approx. 19°C.

Calculation of Arrhenius Parameters

Although Hyd D/L Asp doesn't follow first-order linear kinetics, the heating experiments indicate that racemization progresses in a systematic fashion, and the data provide an empirical relationship from which a kinetic model can be constructed. Using an approach similar to that taken previously with Hyd D/L Asp in mollusks (Goodfriend et al., 1995, 1996), we have modified kinetic equations (Schroeder and Bada, 1976; Miller et al., 1983; McCoy, 1987) to incorporate the best-fit transformations.

The equation that relates the transformed D/L values to the rate of racemization at a given temperature is:

$$\left(\frac{1 + \mathrm{D/L}}{1 - K'\mathrm{D/L}}\right)^n = (1 + K')kt + C \qquad (16\text{-}1)$$

where D/L is the Hyd D/L Asp ratio, K′ is the reciprocal of the equilibrium constant, k is the forward rate constant for a given temperature (yr^{-1}), t is the age of the sample (yr), and C is an analytical constant equivalent to left side of the equation at time zero. For enantiomers such as Asp, we can assume that K′ is 1.00. As described above, n = 3.9 for *Mya* and 4.2 for *Hiatella*. Values of C are 1.37 for modern, unheated *Mya* with a Hyd D/L Asp value of 0.040, and 1.32 for modern, unheated *Hiatella* with a Hyd D/L Asp value of 0.033. Thus, the rate constant is given by 0.5 times the slope of each line shown in figure 16-3. For *Mya* at 85, 110, and 135°C, the rate constants are $9.01 \times 10^1\ yr^{-1}$, $2.61 \times 10^3\ yr^{-1}$, and $2.77 \times 10^4\ yr^{-1}$, respectively. For *Hiatella*, the rate constants are $9.54 \times 10^1\ yr^{-1}$, $1.59 \times 10^3\ yr^{-1}$, and $3.74 \times 10^4\ yr^{-1}$, respectively. Note that the magnitude of the rate constant is dependent on the exponent used in the transformation, but the higher rates for *Mya* in part reflect a slightly higher rate of racemization compared to *Hiatella* (figure 16-1).

Rate constants were determined for each fossil collection (table 16-1) by subtracting C from the transformed D/L value, and dividing the result by twice the calibrated age. The EDT experienced by the fossils, all post-glacial in age, can be approximated by the mean annual air temperature (MAT) at each site (the same approach taken by, e.g., Miller, 1985, and Goodfriend et al., 1996; cf. McCoy, 1987; see below).

The Arrhenius equation, which relates the rate constant to temperature, is:

$$k = Ae^{(-E_a/RT)} \qquad (16\text{-}2)$$

where A is the frequency factor (yr^{-1}), E_a is the activation energy (in kcal/mol or kJ/mol), R is the gas constant (0.001987 kcal/K·mol or 0.008314 kJ/mol), and T is the absolute temperature (in K). Concordant with the equation, the Arrhenius plots (figure 16-4) show the natural log of the rate constant versus the inverse of absolute temperature, defining the temperature dependence of the racemization reaction. Least-squares linear regression lines were calculated for data from the 85, 110°C, and fossil series, and provide good fits ($r^2 = 0.992$ for *Mya* and 0.993 for *Hiatella*). The ln(k) values for the fossil collections fall into the same trend indicated by the 85 and 110°C values, indicating that the temperature dependence of the racemization reaction does not appear to vary with temperature (cf. Goodfriend, 1997; Collins et al., 1999). The slopes of the lines, after multiplication with the gas constant, yield E_a values of 30.08 kcal/mol (125.9 kJ/mol) for *Mya* and 30.16 kcal/mol (126.2 kJ/mol) for *Hiatella*. The intercepts of the regression lines yield lnA values of 46.94 yr^{-1} for *Mya* and 46.83 yr^{-1} for *Hiatella*.

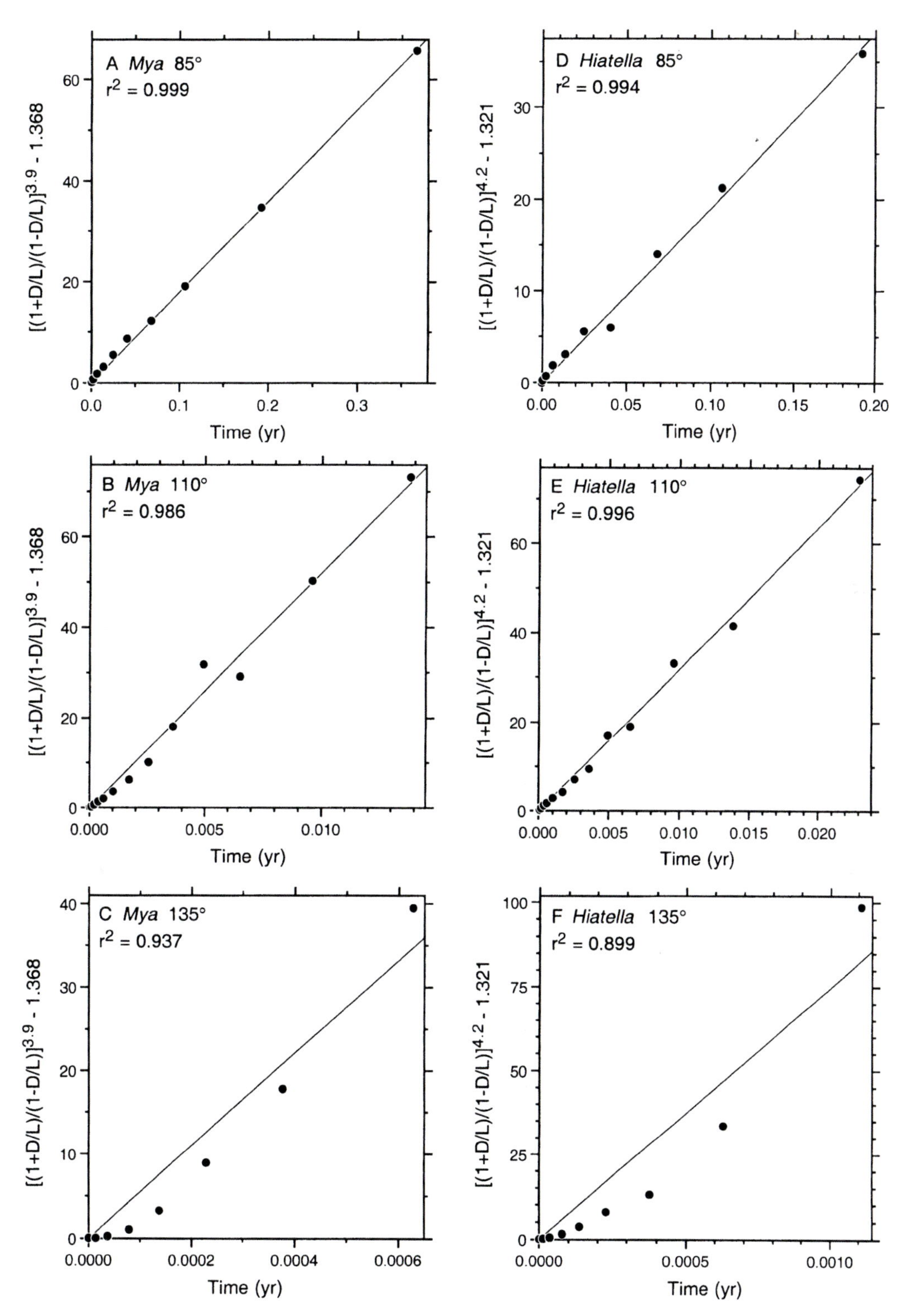

Fig. 16-3. Plots of transformed Hyd D/L Asp values versus time for *Mya* and *Hiatella* in the three heating experiments. The transformed values have been corrected for racemization induced during laboratory hydrolysis by subtraction of the value C, which corresponds to the Hyd D/L Asp ratio measured in modern, unheated specimens. Rates of racemization were calculated from the slopes of the linear regression lines, which were forced through the origin.

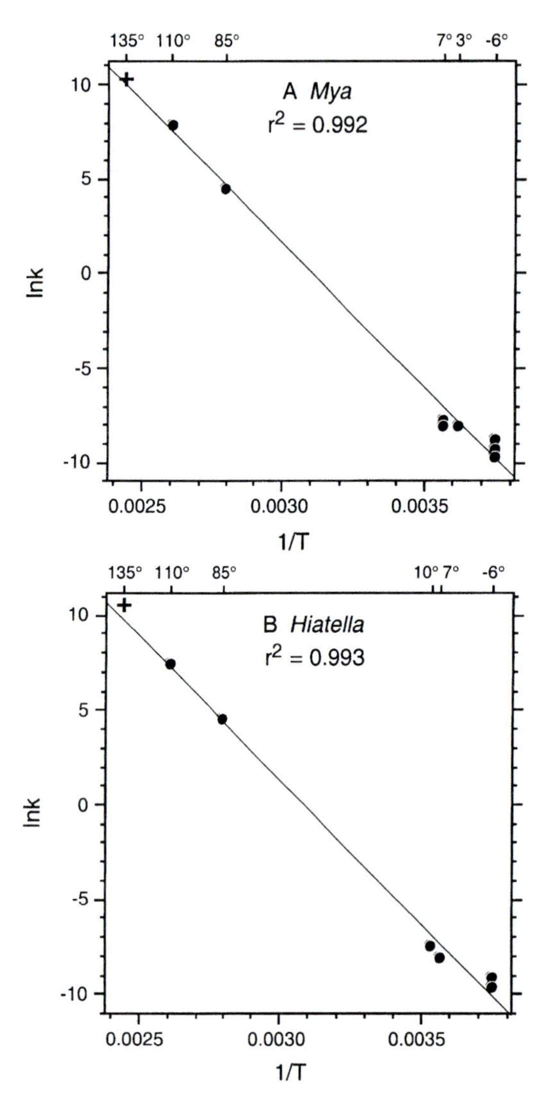

Fig. 16-4. Arrhenius plots for *Mya* and *Hiatella*, relating the natural log of the rate constant to the inverse of temperature (in Kelvin). Linear regressions describe the temperature dependency of the racemization reaction. Crosses show the results of the 135°C heating experiment, which were excluded from the regression and from the determination of the Arrhenius parameters. Corresponding temperatures (in °C) are indicated on the upper axis of each plot.

Data from the 135°C simulation were omitted from calculation of the Arrhenius parameters, because of the poor fit of the transformed D/L values versus time (figure 16-3C, F). The corresponding data points, shown for comparison on the Arrhenius plots (figure 16-4), lie close to the regression lines, but overestimate the rate constants expected for that temperature.

Although strict comparisons are limited by the dependence of Arrhenius parameters on the transformation function used, the E_a values above indicate that the temperature dependence of Asp racemization in *Mya* is similar to that in *Hiatella*. The E_a values are similar to those of 30.6 kcal/mol (128.0 kJ/mol) and 30.0 kcal/mol (125.5 kJ/mol) calculated by Goodfriend et al. (1996) for Hyd D/L Asp in *Mya* and *Hiatella*.

Temperature, Time, and D/L Equations

Equations 16-1 and 16-2 above combine to yield a three-dimensional surface relating the variables of D/L, time, and temperature (cf. Miller et al., 1983; Miller, 1985; McCoy, 1987). Substitution yields the following equation to estimate temperature:

$$T = \frac{-E_a}{R \ln\left[\dfrac{\left(\dfrac{1+D/L}{1-D/L}\right)^n - C}{2At}\right]} \tag{16-3}$$

In this case, T (in Kelvin) is the effective diagenetic temperature (EDT), the temperature experienced by a fossil integrated over time, incorporating the effects of exponentially higher racemization rates during periods of above-average temperature (Wehmiller, 1977). The EDT may in some cases be higher than a corresponding MAT. For Holocene or postglacial fossils that have been buried beneath ≥ 2 m of overlying sediment for much of their history, seasonal oscillations are dampened, and the EDT is close to the MAT integrated through time (Miller and Brigham-Grette, 1989). For fossils that have experienced glacial/interglacial oscillations in climate, marine submergence, or extended cover under a warm-based ice sheet, the EDT may be higher than the MAT averaged through time. McCoy (1987) propagated errors through the Arrhenius equations and concluded that EDT estimates calculated in this fashion carry a precision of about 3°C.

Equations 16-1 and 16-2 can also be combined to calculate the EDT for an interval of time bracketed by samples of known age and Hyd D/L Asp. The resulting equation, from McCoy (1987), is:

$$T(t_2 - t_1) = \frac{E_a}{R \ln\left(\dfrac{A(t_2 - t_1)}{k_2 t_2 - k_1 t_1}\right)} \tag{16-4}$$

where $T_{(t2-t1)}$ is the EDT (in Kelvin) for the interval of time between t_2 (the age of the older sample) and t_1 (the age of the younger sample). Variables k_2 and k_1 are the rate constants for the older and younger samples, respectively, calculated with equation 16-1. Potential error is approx. 3°C, but can be greater if the differences in age or D/L Asp are close in magnitude to the uncertainties associated with their measurement.

More precise is the measurement of the *difference* in temperature between an interval of time (bracketed by two samples of ages t_2 and t_1) and the subsequent period of time (between t_1 and the present). The corresponding calculation, indicated in Miller et al. (1987) and McCoy (1987), is simply the difference between two EDT values:

$$\Delta T = T_1 - T_{(t_2 - t_1)} \tag{16-5}$$

where ΔT is the difference in EDT as described above, T_1 is the EDT for the younger sample, and $T_{(t_2-t_1)}$ is calculated from equation 16-4. The resultant uncertainty for ΔT is commonly 1–2°C (McCoy, 1987).

Substitution of equations 1 and 2 also yields the following equation to estimate the age, t, of a sample in years:

$$t = \frac{\left(\frac{1 + \mathrm{D/L}}{1 - \mathrm{D/L}}\right)^n - C}{2Ae^{(-E_a/RT)}} \tag{16-6}$$

Fossil ages can be calculated in this fashion with a precision generally no better than ±40–50% (McCoy, 1987; cf. Kaufman, 2000). Much of this error is a result of uncertainty in the estimate of the EDT. If an EDT can be estimated from an independently dated calibration sample, and if the EDT and resulting rate constant are assumed to apply to other samples from the locality, a precision of approx. 15% is possible (McCoy, 1987). In unusually favorable settings (warm, nearly isothermal conditions during the late Holocene), the age-resolving power of D/L Asp is of the order of decades (Goodfriend, 1992; Goodfriend et al., 1992, 1995; Goodfriend and Rollins, 1998). However, given common uncertainties in EDT for high-latitude Quaternary sequences, it seems "unwise to attempt the conversion of D/L ratios to absolute age" (Miller and Brigham-Grette, 1989), and the application of amino acid geochronology is best utilized in the relative-age sense of "aminostratigraphy" (Miller and Hare, 1980) or with an appropriate expression of uncertainty.

For the sake of comparison, it is sometimes instructive to estimate a Hyd D/L Asp value for a given pair of age and paleotemperature estimates. The equation is:

$$\mathrm{D/L} = \frac{\left[2Ate^{(-E_a/RT)} + C\right]^{1/n} - 1}{\left[2Ate^{(-E_a/RT)} + C\right]^{1/n} + 1} \tag{16-7}$$

where D/L is the predicted Hyd D/L Asp ratio, and the other variables are as defined for previous equations. The precision of D/L prediction is strongly weighted by uncertainties in estimating EDT. On the basis of the approach taken by McCoy (1987), we estimate that the resulting uncertainty is approx. 0.07.

The data from the heating experiments (figure 16-1) and most of the fossils incorporated into the kinetics model (table 16-1) indicate that Asp racemizes faster in *Mya* than in *Hiatella* (consistent with the results of Goodfriend et al., 1996). In plots of predicted D/L versus age (figure 16-5), this is illustrated by steeper curves for *Mya* than for *Hiatella*. For example, 12,000-year-old mollusks with EDT values of −5°C are expected to yield Hyd D/L Asp ratios of 0.15 for *Mya* and 0.12 for *Hiatella*. Similarly, 125,000-year old mollusks with the same EDT values should yield ratios of about 0.37 for *Mya* and 0.31 for *Hiatella*. The taxonomic effect is reversed for A/I ratios: isoleucine epimerizes *more slowly* in *Mya* than in *Hiatella* under ambient conditions (Goodfriend et al., 1996).

APPLICATION OF THE KINETIC MODEL

Results for fossil mollusks from the Eastern Canadian Arctic, Svalbard, and Iceland (table 16-1) illustrate the ability of the kinetic model to estimate paleotemperature. For example, a collection

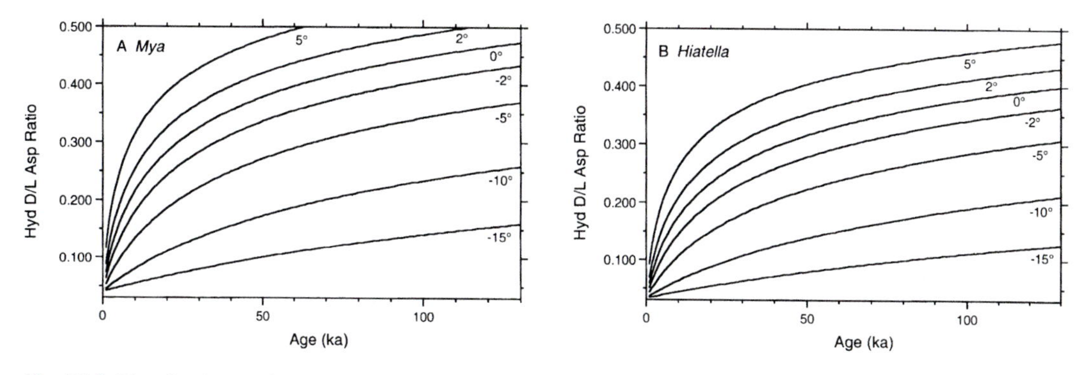

Fig. 16-5. Plots for *Mya* and *Hiatella* of Hyd D/L Asp ratios predicted by the kinetic models from combinations of age and EDT. Each curve represents the increase in racemization expected under isothermal conditions (°C).

from southeastern Baffin Island (interlaboratory comparison sample "ILC-Z") was dated to 12.4 ka (thousands of cal yr BP), and gave a Hyd D/L Asp ratio of 0.150. For postglacial samples such as this, the MAT integrated through time is expected to be close to the current MAT (slightly warmer during the middle Holocene, and slightly cooler during the early and late Holocene). For deeply buried collections such as this, the EDT should be close to the integrated MAT. The model estimates an EDT of −5°C, which is close to the current MAT of −6°C, and well within the uncertainty of 3°C given above.

Two collections from western Svalbard illustrate a contrast in thermal history related to terrestrial versus subaqueous conditions. Five *Hiatella* samples from shallow marine sediments at Poole Point were uplifted into a terrestrial environment shortly after their death at 13.3 ka (Andersson et al., 1999). The average Hyd D/L Asp of 0.147 requires an EDT of −4°C, which is similar to the current MAT of −6°C. In contrast, a single *Hiatella* sample from marine sediments below nearby Lake Linnévatnet experienced warmer, subaqueous temperatures after its death at 11.6 ka (table 16-1; Mangerud and Svendsen, 1990). The Hyd D/L Asp ratio of 0.196 requires an EDT of 0°C. This result is tentative, given a D/L analysis of only one, rather than several, mollusks per collection. Nonetheless, the EDT agrees (within the expected error) with the current lake-bottom temperature of approx. +2°C (Miller and Brigham-Grette, 1989). The difference in EDT between the two collections, 4°C, illustrates the ability of the kinetic model to differentiate thermal histories.

Another example depicts the influence of geothermal activity on ground temperature. A collection of 14.4 ka *Mya* samples from western Iceland, 10 m below the top of a coastal exposure (Ingólfsson, 1988), yielded an average Hyd D/L Asp ratio of 0.458. The site lies 5 km from an active geothermal field, where boiling water reaches the surface. Geothermal gradients in the region are locally variable, but can be as high as 1°C/m. The EDT estimated for the collection is 10°C, which is significantly above the current MAT of 5°C. The difference between the two appears to reflect a local geothermal gradient of about 0.5°C/m.

Another comparison estimates the temperature depression associated with glaciation of western Svalbard. A collection of 12.3 ka *Hiatella* samples from raised marine sediments at Brøggerhalvøya (Episode A, Miller, 1982; Miller et al., 1989) yielded a Hyd D/L Asp ratio of 0.128 and an estimated EDT of −5°C. A nearby collection of *Hiatella* samples from raised marine sediments thought to date from the last interglacial period (125 ka, isotope stage 5e, Episode C) gave a Hyd D/L Asp ratio of 0.213. Using equation 16-4, we calculate an EDT of −12°C for the time period 125–12.3 ka, bracketing the Weichselian glaciation (encompassing isotope stages 5d–2). Thus, the kinetic model for *Hiatella* estimates a reduction in the EDT of 7°C for the Weichselian, relative to the postglacial period. *Mya* samples from the same 12.3 ka collection gave a Hyd D/L Asp of 0.130 and an estimated EDT of −6°C. *Mya* from the same 125 ka collection gave a Hyd D/L Asp of 0.214. The EDT estimated with these data for the Weichselian is −13°C, with an EDT depression of 7°C, in agreement with the value calculated from the *Hiatella* results. However, ground temperatures during the two main Weichselian glacial stades were probably lower, because the ΔEDT result includes integration of the slightly warmer temperatures expected for an interstadial (isotope stage 3), and significantly warmer temperatures (at approx. 0°C) during several millennia of marine submergence as well as cover under a warm-based ice sheet. Thus, we estimate a depression in MAT of ≥ 7°C for the early and late Weichselian glacial periods, relative to the present interglacial period. This finding compares favorably with minimum temperature depressions, calculated in similar fashion, of 4°C for the Weichselian glaciation on Norway, 9°C for the late Devensian glacial period in Scotland (Miller et al., 1987), and 9°C for last-glacial cooling in the southern hemisphere (Miller et al., 1997).

ASPARTIC ACID AND ASPARAGINE DYNAMICS

The strongly nonlinear racemization kinetics for aspartic acid, as measured in hydrolysates, may be a result of the interaction between asparagine and aspartic acid. The former converts to the latter through the process of deamidation—during the laboratory-induced, acid hydrolysis step and/or during fossil diagenesis under ambient conditions. Thus, the Hyd D/L Asp values we have measured may represent the combined racemization of Asp and Asn.

Goodfriend and Hare (1995) outlined three related hypotheses to explain the sharp, convex-upward bend apparent in Hyd D/L Asp kinetics. First, the initial period of rapid increase may be dominated by racemization of Asn, with the later, slower increase in D/L ratio dominated by racemization of Asp. Indeed, Asn should racemize faster than Asp at neutral and basic pH, based on mechanistic considerations (Bada, 1985; Collins et al., 1999), and Asn appears to racemize faster than Asp in aqueous Free solutions and for some enzymes (Zhao et al., 1989; Brinton and Bada, 1995). "Hypothesis 1" can be further categorized by considering racemization and deamidation rates (Goodfriend and Hare, 1995). "Hypothesis 1A" posits that Asn racemization is initially very fast, but slows as the D/L Asn ratio approaches 1.0. In this case, the bend in Hyd D/L Asp kinetics is due to the slowing of Asn racemization, and samples experience little or no deamidation prior to laboratory-induced acid hydrolysis. In contrast, "Hypothesis 1B" is that racemization of Asn is initially fast, but that the bend coincides with progressive deamidation of Asn to the slower racemizing Asp. This line of reasoning is supported by studies that show that the rate of deamidation is comparable to the rate of racemization in aqueous Free solutions and in some polypeptides (Geiger and Clarke, 1987; Brinton and Bada, 1995). Furthermore, the processes are interlinked; in some peptides the deamidation of Asn produces a succinimide intermediate, for which rates of racemization and hydrolysis are particularly high (Geiger and Clarke, 1987; Collins et al., 1999). Finally, "Hypothesis 2" outlined by Goodfriend and Hare (1995) is that the observed nonlinear pattern is due to the inherent kinetics of Asp racemization, with different rates of racemization for different moieties of Asp (Free, terminal-position, or internally bound in polypeptides), and with little or no contribution to Hyd D/L Asp kinetics by Asn.

To clarify the interplay of hydrolysis, deamidation, and racemization of Asp and Asn in mollusks, we examined the D/L ratios and concentrations of Free Asp and Free Asn in *Mya* samples heated at 110°C (figure 16-6; table 16-3). Unlike with previous analytical techniques, both Asp and Asn are quantifiable with our reverse-phase liquid chromatography. Asparagine is present in the Free fraction, and the increase in Free D/L Asn is initially faster than that in Free D/L Asp or Hyd D/L Asp (figure 16-6A). For example, by the time Hyd D/L Asp reaches 0.20, the Free D/L Asn ratio is approx. 0.41, but the Free D/L Asp ratio is only approx. 0.23. Thereafter, Free D/L Asn increases more slowly whereas Free D/L Asp continues to increase at a moderate pace.

The concentration data show that Free amino acids are liberated by hydrolysis during the heating experiment, as expected (figure 16-6B, right axis). Free Asn and Free Asp (= Free Asx) initially account for less than 1% of the total Asx (measured as Hyd D/L Asp), consistent with the limited extent of hydrolysis during the earliest phase of the experiment. The percent Free rises to 3% of the Total Asx by the time Hyd D/L Asp reaches 0.50.

The data also show that Asn makes up a decreasing proportion of Free amino acids over time (figure 16-6B, left axis). During the earliest phase of the experiment, Free Asn is a significant fraction of the Free Asx pool, accounting for 15–18% of Free Asx. However, by the time Hyd D/L Asp has reached 0.34, the percentage of Asn in the Free pool has dropped to 9%. This decrease in the proportion of Free Asn is largely due to an increase in the concentration of Free Asp. After Hyd D/L Asp reaches a value of 0.57, Free Asn is no longer detectable, and the concentration of Free Asp continues to climb.

The results shown in figure 16-6 allow us to test among the hypotheses above to clarify the

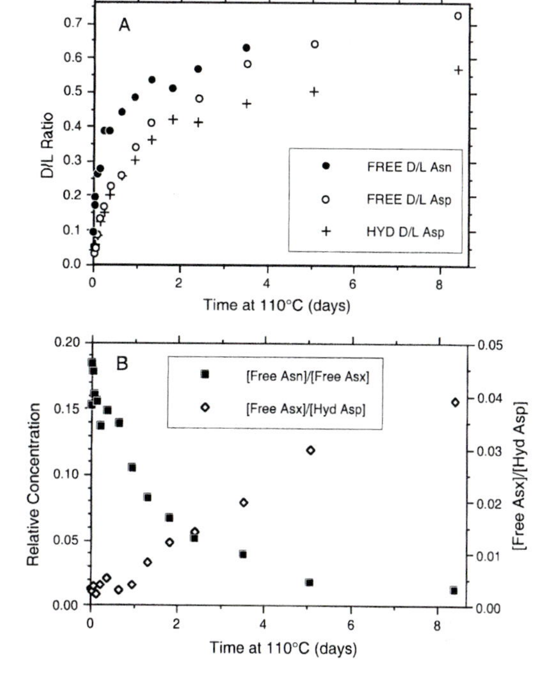

Fig. 16-6. Changes in D/L ratios and concentration for aspartic acid and asparagine in *Mya* heated at 110°C (cf. table 16-3). A. Comparison of Free and Hyd ratios. B. Increase in the percentage of Free amino acids (proportion of Free Asx to Hyd Asp, right axis), and concurrent decrease in the proportion of Free Asn to Free Asx (left axis).

dynamics of aspartic and asparagine racemization in *Mya*. First, the presence of Asn in the Free fraction indicates that it is present (to an uncertain extent) in the polypeptide-bound fraction of amino acids (figure 16-7). This finding is consistent with the observations of Weiner (1979), who found that Asn and/or glutamine (Gln) comprises approx. 10–20% of combined pool of Asp + Asn + Glu + Gln in mollusks. Second, in our heating experiment, higher D/L ratios in Free Asn than in Free Asp demonstrate that racemization is inherently faster in Asn than in Asp (without necessarily passing through a succinimide intermediate; cf. Collins et al., 1999). Thus, Asn racemization appears to be a significant contributor to apparent Hyd D/L Asp kinetics, and we can discount "Hypothesis 2" above. Differing rates of racemization for different moieties of Asp may still be an important factor, but this is not the only factor responsible for the nonlinear kinetics of Hyd D/L Asp.

The data also suggest that Asn is undergoing deamidation to Asp. Asparagine in the polypeptide-bound fraction should continue to liberate Free Asn through hydrolysis over the course of the experiment, as it did during the earliest phase. Indeed, the rise in Free D/L Asn (figure 16-6A)

TABLE 16-3 Free D/L Asp and D/L Asn results for *Mya* in the 110°C heating experiment. D/L values are shown with the mean ratio for "*n*" heated specimens, ±1 S.D

Time (days)	Free D/L Asp	CV[b]	*n*	Free D/L Asn[c]	CV	*n*
0.000[a]	0.050 ±0.013	25.4	4	0.093 ±0.031	33.2	3
0.010	0.032 ±0.004	14.2	2	0.194 ±0.037	19.0	2
0.031	0.046 ±0.002	3.6	2	0.170		1
0.073	0.085 ±0.005	5.8	2	0.259		1
0.138	0.134 ±0.011	8.5	2	0.276 ±0.019	7.0	2
0.222	0.168 ±0.002	1.4	2	0.385 ±0.004	1.0	2
0.372	0.226 ±0.002	0.8	2	0.387 ±0.010	2.5	2
0.624	0.259 ±0.009	3.3	2	0.440 ±0.046	10.4	2
0.935	0.340 ±0.013	3.7	2	0.485 ± 0.029	6.0	2
1.313	0.410 ±0.014	3.5	2	0.533 ±0.061	11.5	2
1.801	0.509 ±0.014	2.8	2	0.512		1
2.381	0.482 ±0.046	9.6	2	0.566 ±0.011	2.0	2
3.500	0.582 ±0.003	0.6	2	0.627 ±0.009	1.5	2
5.043	0.643 ±0.016	2.5	2	N.D.		
8.413	0.726 ±0.035	4.8	2	N.D.		
13.398	0.751 ±0.041	5.5	2	N.D.		
20.281	0.788 ±0.025	3.1	2	N.D.		
28.517	0.801 ±0.006	0.7	2	N.D.		
40.358	0.839 ±0.004	0.5	2	N.D.		
59.128	0.876 ±0.026	2.9	2	N.D.		
68.191	0.881 ±0.006	0.6	2	N.D.		

[a] Surprisingly, unheated specimens of modern *Mya* yielded measurable Free D-ASP AND D-Asn, above background analytical levels. Reacemization in these samples may have occurred during the life of the mollusks, during a year of sample storage, or during routine-sample preparation for free analyses (no laboratory heating, but minor grinding and acid dissolution).
[b] CV, As in table 16-2.
[c] N.D. D/L Asn was non-detectable for some analyses.

suggests that Free Asn is being produced after passage through terminal peptide positions, where rates of racemization are expected to be high (by analogy with A/I; Kriausakul and Mitterer, 1978). However, concentrations of Free Asn remain low, and concentrations of Free Asp climb rapidly with increasing heating time. These observations imply that Free Asn is converting to Free Asp shortly after hydrolysis (figure 16-7). Incorporation of the higher racemized Free Asn into the Free Asp population would help to explain the continued, moderate rise in Free D/L Asp after Free D/L Asn begins to rise more slowly (figure 16-6A). Furthermore, a portion of the Asn in the peptide-bound fraction may be undergoing deamidation to Asp before (or during) hydrolysis (figure 16-7). In any case, Free Asn concentrations fail to rise during the experiment, and Asn appears to undergo significant deamidation before reaching equilibrium D/L ratios of 1.0. Thus, "Hypothesis 1A" above can be discounted; the rate of deamidation to Asp is significant, and is comparable to the rates of Asn and Asp racemization.

The results of the heating experiment additionally characterize the style and rates of racemization within the polypeptide-bound population of amino acids. Specifically, Free Asn and Asp concentrations are too low ($<3\%$ of total) to account independently for the nonlinear rise in Hyd D/L Asp (figure 16-7). The D/L ratios for peptide-bound Asx, back-calculated by weighting D/L values by concentration, are $<1\%$ lower than Hyd D/L Asp values (for Hyd D/L Asp <0.5; figure 16-7). In other words, Hyd D/L Asp ratios are primarily a measure of racemization within peptide-bound Asp and Asn. Thus, peptide-bound Asx follows nearly the same nonlinear, convex-upward pattern of racemization exhibited by Hyd D/L Asp. In turn, this suggests that the peptide-bound fraction of Asx includes a faster racemizing Asn that is concurrently deamidating to the slower racemizer Asp.

Complete characterization of Asn racemization, deamidation, and hydrolysis is not possible because it is analytically difficult to measure the concentration or D/L ratio of peptide-bound asparagine. Furthermore, the high-temperature experiment may not fully characterize Asp/Asn dynamics if rates of hydrolysis, deamidation, and racemization do not share the same temperature dependency. The balance among these processes may be different for ambient temperatures (Goodfriend and Hare, 1995; Collins et al., 1999). However, the available information suggests that the nonlinear

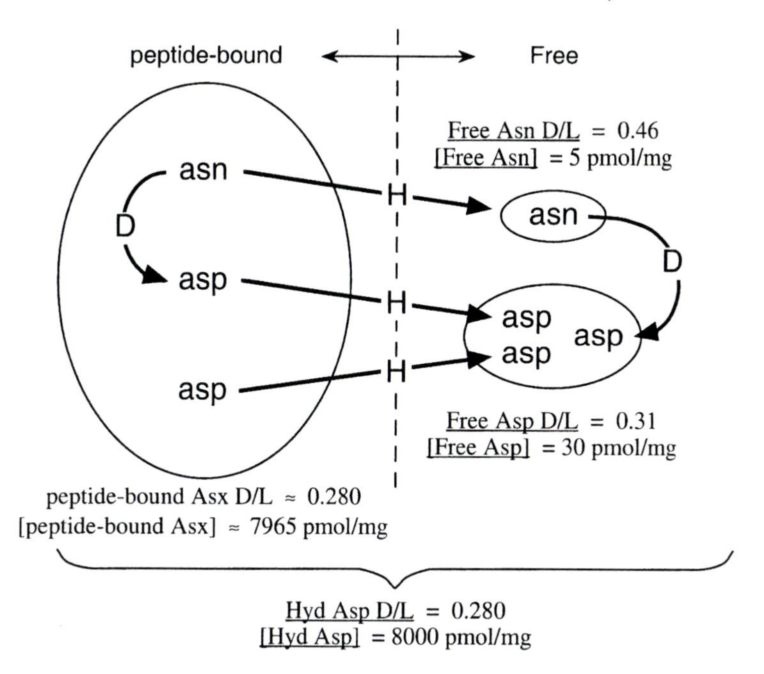

Fig. 16-7. Schematic representation of the interplay among deamidation (D), hydrolysis (H), and racemization, as determined or suggested by the data from *Mya* samples heated at 110°C. Underlined quantities are those that have been measured directly by chromatography, and correspond with values reached for a Hyd D/L Asp ratio of 0.280. The original peptide-bound population of Asx (Asp + Asn) has undergone partial hydrolysis to produce measurable Free Asn and Free Asp. A portion of the Asp bound in polypeptides may have been produced from deamidation of Asn. Similarly, a portion of the Free Asp appears to have been produced from deamidation of Free Asn. By the time Hyd D/L Asp has reached 0.280, Free Asp and Free Asn exhibit relatively high D/L ratios, but comprise only 0.4% of the total concentration of Asp and Asn (pmol per mg shell). The D/L ratio calculated for peptide-bound Asx is only 0.1% lower than the Hyd D/L Asp ratio.

kinetics of Hyd D/L Asp in *Mya* is due to a combination of a higher rate of racemization in Asn than in Asp and loss of the faster racemizer, over through time, via deamidation (i.e., "Hypothesis 1B"). This finding may be broadly applicable to the nonlinear kinetics of Hyd D/L Asp in other types of biominerals, including *Hiatella*.

CONCLUSIONS

Mya and *Hiatella* samples heated in the laboratory at 85, 110, and 135°C, combined with radiocarbon-dated fossils, allow us to quantify the kinetics of Hyd D/L Asp with respect to time and temperature. This study expands upon a preliminary analysis of Asp kinetics in *Mya* and *Hiatella* by Goodfriend et al. (1996). Racemization of aspartic acid, as measured in hydrolysates, does not follow reversible, first-order linear kinetics, partly, at least, because of the conversion of asparagine to aspartic acid. Nonetheless, racemization proceeds in a systematic fashion, and the empirical model derived here provides reasonable estimates of EDT for a set of fossil samples. The specific conclusions are outlined below.

- Rates of racemization apparent for aspartic acid in heated *Mya* and *Hiatella* samples are initially very high, decrease as D/L values increase, and describe strong convex-upward curves of Hyd D/L Asp versus time. The racemization curves are similar in shape to those measured for gastropods, corals, eggshell, and ostracodes.
- Asp racemization is slightly faster in *Mya* than in *Hiatella*, as observed in the heating experiments and in a limited set of fossil collections. Up to a Hyd D/L Asp of approx. 0.5, Asp racemization is much faster than is isoleucine epimerization.
- A new class of mathematical transformations provides the best approach for linearizing Hyd D/L Asp with respect to time. The functions—$[(1+D/L)/(1-D/L)]^{3.9}$ for *Mya* and $[(1+D/L)/(1-D/L)]^{4.2}$ for *Hiatella*—quantify the observed shape of the racemization curves better than do simple power-law transformations.
- Asp racemization rates calculated from the 85, 110°C, and fossil series provide good fits on Arrhenius plots, characterizing well the temperature dependence of the racemization reaction. Activation energies and intercept values thus derived are 30.1 kcal/mol (125.9 kJ/mol) and

46.9 yr^{-1} for *Mya* and 30.2 kcal/mol (126.2 kJ/mol) and 46.8 yr^{-1} for *Hiatella.*

- Kinetic equations incorporating the Arrhenius parameters provide a means of estimating the age or temperature history of fossil collections. Estimates of EDT are inherently more precise than are estimates of age.
- The kinetic models for *Mya* and *Hiatella* yield: (1) a reasonable EDT estimate of −5°C for a 12 ka sample from southeastern Baffin Island, where the current MAT is −6°C; (2) a difference in EDT of 4°C between two collections from western Svalbard, reflecting the influence of subaerial versus subaqueous thermal conditions over the last 12 ka; (3) an EDT of 10°C for 14 ka fossils from western Iceland, consistent with geothermal heating above the current MAT of 5°C; and (4) a minimum depression of 7°C in MAT for the Weichselian glacial period on western Svalbard, relative to the postglacial period.
- The ability of reverse-phase liquid chromatography to measure concentrations and D/L ratios of Free aspartic acid and Free asparagine allows us to clarify the mechanisms of Hyd D/L Asp kinetics. The nonlinearity of aspartic acid racemization observed in hydrolysate samples is due, at least in part, to deamidation of fast-racemizing asparagine to slower racemizing aspartic acid in the Free amino acid pool and particularly in the peptide-bound population of amino acids.

Acknowledgments: We thank Darrell Kaufman (Northern Arizona University) for numerous discussions and valued input. Mike Kaplan (University of Colorado) collected modern specimens from Pangnirtung Fiord. Clifford Woodward (Hewlett Packard) helped us to improve the chromatography. Fossil collections were kindly provided by Jan Mangerud (University of Bergen), Minze Stuiver (University of Washington) and Ólafur Ingólfsson and Torbjörn Andersson (University of Göteborg). Comments from Steve Roof (Hampshire College) and an anonymous reviewer helped to improve an earlier draft. This study was funded by National Science Foundation grant OPP-9529350.

REFERENCES

Andersson T., Forman S., Ingólfsson Ó., and Manley W. F. (1999) Late Quaternary environmental history of central Prins Karls Forland, western Svalbard. *Boreas* **8**, 292–307.

Bada J. L. (1985) Racemization of amino acids. In *Chemistry and Biochemistry of the Amino Acids* (ed. G. C. Barrett), pp. 399–414. Chapman and Hall.

Bard E., Arnold M., Fairbanks R. G., and Hamelin B. (1993) ^{230}Th–^{234}U and ^{14}C ages obtained by mass spectrometry on corals. *Radiocarbon* **35**, 191–200.

Brinton K. L. F. and Bada J. L. (1995) Comment on "Aspartic acid racemization and protein diagenesis in corals over the last 350 years" by G. A. Goodfriend, P. E. Hare, and E. R. M. Druffel. *Geochim. Cosmochim. Acta* **59**, 415–416.

Collins M. J., Waite E. R., and van Duin A. C. T. (1999) Predicting protein decomposition: the case of aspartic-acid racemization kinetics. *Phil. Trans. Roy. Soc. Lond. Ser. B* **354**, 51–64.

Fitznar H. P., Lobbes J. M., and Kattner G. (1999) Determination of enantiomeric amino acids with high-performance liquid chromatography and pre-column derivatisation with *o*-phthaldialdehyde and *N*-isobutyrylcesteine in seawater and fossil samples (mollusks). *J. Chromatogr.* **A832**, 123–132.

Forman S. L. (1990) *Svalbard Radiocarbon Date List I.* Institute of Arctic and Alpine Research Occasional Paper No. 47.

Geiger T. and Clarke S. (1987) Deamidation, isomerization, and racemization of asparaginyl and aspartyl residues in peptides. *J. Biol. Chem.* **262**, 785–794.

Goodfriend G. A. (1991) Patterns of racemization and epimerization of amino acids in land snail shells over the course of the Holocene. *Geochim. Cosmochim. Acta* **55**, 293–302.

Goodfriend G. A. (1992) Rapid racemization of aspartic acid in mollusc shells and potential for dating over recent centuries. *Nature* **357**, 399–401.

Goodfriend G. A. (1997) Aspartic acid racemization and amino acid composition of the organic endoskeleton of the deep-water colonial anemone *Gerardia*: Determination of longevity from kinetic experiments. *Geochim. Cosmochim. Acta* **61**, 1931–1939.

Goodfriend G. A. and Hare P. E. (1995) Reply to the Comment by K. L. F. Brinton and J. L. Bada on "Aspartic acid racemization and protein diagenesis in corals over the last 350 years". *Geochim. Cosmochim. Acta* **59**, 417–418.

Goodfriend G. A. and Meyer V. R. (1991) A comparative study of the kinetics of amino acid racemization/epimerization in fossil and modern mollusk shells. *Geochim. Cosmochim. Acta* **55**, 3355–3367.

Goodfriend G. A. and Rollins H. B. (1998) Recent barrier beach retreat in Georgia: dating exhumed salt marshes by aspartic acid racemization and post-bomb radiocarbon. *J. Coastal Res.* **14**, 960–969.

Goodfriend G. A. and Stanley D. J. (1996) Reworking and discontinuities in Holocene sedimentation in the Nile Delta: documentation from amino acid racemization and stable isotopes in mollusk shells. *Mar. Geol.* **129**, 271–283.

Goodfriend G. A., Hare P. E., and Druffel E. R. M. (1992) Aspartic acid racemization and protein diagenesis in corals over the last 350 years. *Geochim. Cosmochim. Acta* **56**, 3847–3850.

Goodfriend G. A., Kashgarian M., and Harasewych M. G. (1995) Use of aspartic acid racemization and post-bomb ^{14}C to reconstruct growth rate and longevity of the deep-water slit shell *Entemnotrochus adasonianus*. *Geochim. Cosmochim. Acta* **59**, 1125–1129.

Goodfriend G. A., Brigham-Grette J., and Miller G. H. (1996) Enhanced age resolution of the marine Quaternary record in the Arctic using aspartic acid racemization dating of bivalve shells. *Quat. Res.* **45**, 176–187.

Harada N. and Handa N. (1995) Amino acid chronology in the fossil planktonic foraminifera, *Pulleniatina obliquiloculata* from Pacific Ocean. *Geophys. Res. Lett.* **22**, 2353–2356.

Ingólfsson Ó. (1988) Glacial history of the lower Borgarfjordur area, western Iceland. *Geol. Föreningen* **110**, 293–309.

Kaufman D. S. (2000) Amino acid racemization in ostracodes. In *Perspectives in Amino Acid and Protein Geochemistry* (eds G. A. Goodfriend, M. J. Collins, M. L. Fogel, S. A. Macko, and J. F. Wehmiller), pp. 145–160. Oxford University Press.

Kaufman D. S. and Manley W. F. (1998) A new procedure for determining DL amino acid ratios in fossils using Reverse Phase Liquid Chromatography. *Quat. Sci. Rev.* **17**, 987–1000.

Kaufman D. S., Miller G. H., Stravers J. A., and Andrews J. T. (1993) An abrupt early Holocene (9.9–9.6 kyr BP) ice stream advance at the mouth of Hudson Strait, Arctic Canada. *Geology* **21**, 1063–1066.

Kimber R. W. L. and Griffen C. V. (1987) Further evidence of the complexity of the racemization process in fossil shells with implications for amino acid racemization dating. *Geochim. Cosmochim. Acta* **51**, 839–846.

Kriausakul N. and Mitterer R. M. (1978) Isoleucine epimerization in peptides and proteins: kinetic factors and application to fossil proteins. *Science* **201**, 1011–1014.

McCoy W. D. (1987) The precision of amino acid geochronology and paleothermometry. *Quat. Sci. Rev.* **6**, 43–54.

Mangerud J. and Svendsen J. I. (1990) Deglaciation chronology inferred from marine sediments in a proglacial lake basin, western Spitsbergen, Svalbard. *Boreas* **19**, 249–272.

Mangerud J. and Svendsen J. I. (1992) The last interglacial–glacial period on Spitsbergen, Svalbard. *Quat. Sci. Rev.* **11**, 633–664.

Manley W. F., Miller G. H., Mangerud J., and Svendsen J. I. (1998) Aspartic acid racemization in arctic marine bivalves: improved age resolution for Quaternary glacial and sea-level histories at Brøggerhalvøya and Kapp Ekholm, western Svalbard. *Program and Abstracts, Perspectives in Amino Acid and Protein Geochemistry*, p. 55 (abstr.). Geophysical Lab, Carnegie Institution of Washington, Washington DC.

Miller G. H. (1982) Quaternary depositional episodes, western Spitsbergen, Norway: Aminostratigraphy and glacial history. *Arct. Alp. Res.* **14**, 321–340.

Miller G. H. (1985) Aminostratigraphy of Baffin Island shell-bearing deposits. In *Quaternary Environments: Baffin Island, Baffin Bay, and West Greenland* (ed. J. T. Andrews), pp. 394–427. Allen and Unwin.

Miller G. H. and Brigham-Grette J. (1989) Amino acid geochronology: resolution and precision in carbonate fossils. *Quat. Intl.* **1**, 111–128.

Miller G. H. and Hare P. E. (1980) Amino acid geochronology: integrity of the carbonate matrix and potential of molluscan fossils. In *Biogeochemistry of Amino Acids* (eds P. E. Hare, T. C. Hoering, and K. King, Jr), pp. 415–444. Wiley.

Miller G. H., Sejrup H. P., Mangerud J., and Andersen B. G. (1983) Amino acid ratios in Quaternary molluscs and foraminifera from western Norway: correlation, geochronology and paleotemperature estimates. *Boreas* **12**, 107–124.

Miller G. H., Jull A. J. T., Linick T., Sutherland D., Sejrup H. P., Brigham J. K., Bowen D. Q., and Mangerud J. (1987) Racemization-derived late Devensian temperature reduction in Scotland. *Nature* **326**, 593–595.

Miller G. H., Hearty J., and Stravers J. A. (1988) Ice-sheet dynamics and glacial history of southeasternmost Baffin Island and outermost Hudson Strait. *Quat. Res.* **30**, 116–136.

Miller G. H., Sejrup H. P., Lehman S. J., and Forman S. L. (1989) Glacial history and marine environmental change during the last interglacial–glacial cycle, western Spitsbergen, Svalbard. *Boreas* **18**, 273–296.

Miller G. H., Magee J. W., and Jull A. J. T. (1997) Low-latitude glacial cooling in the southern Hemisphere from amino-acid racemization in emu eggshells. *Nature* **385**, 241–244.

Mitterer R. M. (1993) The diagenesis of proteins and amino acids in fossil shells. In *Organic Geochemistry* (eds M. H. Engel and S. A. Macko), pp. 739–753. Plenum Press.

Mitterer R. M. and Kriausakul N. (1989) Calculation of amino acid racemization ages based on apparent parabolic kinetics. *Quat. Sci. Rev.* **8**, 353–357.

Rutter N. W., Crawford R. J., and Hamilton R. (1980) Correlation and relative age dating of Quaternary strata in the continuous permafrost zone of northern Yukon with D/L ratios of aspartic acid of wood, freshwater molluscs, and bone. In *Biogeochemistry of Amino Acids* (eds P. E. Hare, T. C. Hoering, and K. King, Jr), pp. 463–475. John Wiley and Sons.

Schroeder R. A. and Bada J. L. (1976) A review of the geochemical applications of the amino acid racemization reaction. *Earth-Sci. Rev.* **12**, 347–391.

Stuiver M. and Borns H. W. (1975) Late Quaternary marine invasion in Maine: its chronology and associated crustal movement. *Geol. Soc. Amer. Bull.* **86**, 99–104.

Stuiver M. and Braziunas T. F. (1993) Modeling atmospheric ^{14}C influences and ^{14}C ages of marine samples to 10,000 BC. *Radiocarbon* **35**, 137–189.

Stuiver M. and Reimer P. J. (1993) Extended ^{14}C data base and revised CALIB 3.0 ^{14}C age calibration program. *Radiocarbon* **35**, 215–230.

Wehmiller J. F. (1977) Amino acid studies of the Del Mar, California, midden site: apparent rate constants, ground temperature models, and chronological implications. *Earth Planet Sci. Lett.* **37**, 184–196.

Wehmiller J. F. (1984) Interlaboratory comparison of amino acid enantiomeric ratios in fossil Pleistocene mollusks. *Quat. Res.* **22**, 109–120.

Wehmiller J. F. (1993) Applications of organic geochemistry for Quaternary research: aminostratigraphy and aminochronology. In *Organic Geochemistry* (eds M. H. Engel and S. A. Macko), pp. 755–783. Plenum Press.

Weiner S. (1979) Aspartic acid-rich proteins: major components of the soluble organic matrix of mollusk shells. *Calcif. Tissue Intl.* **29**, 163–167.

Werner A. (1988) Holocene glaciation and climatic change, Spitsbergen, Svalbard. Ph. D. Thesis, University of Colorado, U.S.A.

Zhao M., Bada J. L., and Ahern T. J. (1989) Racemization rates of asparagine–aspartic acid residues in lysozyme at 100°C as a function of pH. *Bioorg. Chem.* **17**(1), 36–40.

17. The thermal environment of fossils: effective ground temperatures at aminostratigraphic sites on the U.S. Atlantic Coastal Plain

John F. Wehmiller, Hilmar A. Stecher III, Linda L. York, and Irving Friedman

A variety of geochemical, hydrological, and geomorphic studies require an understanding of the thermal environment of the shallow subsurface (< 10 m below land surface). Several recent studies demonstrate a growing interest in the use of ground temperature data to document environmental change (Chapman et al., 1992; Pollack et al., 1998) and the role of shallow ground temperature changes in biogeochemical cycles (Goulden et al., 1998). Amino acid racemization (AAR) geochronology, or aminostratigraphy, requires information on the thermal history, including the modern thermal environment of fossil samples analyzed. Aminostratigraphic studies usually assume a simple relationship between air and ground temperatures, but this assumption has rarely been evaluated with site-specific studies. Discrepancies in aminostratigraphic correlation of sites along the Atlantic Coastal Plain (ACP) might be explained by large variations in late Quaternary temperature histories along a north–south gradient through the region (Corrado et al., 1986; Hollin and Hearty, 1990; Wehmiller et al., 1988, 1992). To understand the potential for these variations and the magnitude of their impact on AAR results, it is necessary to examine the relationship between current air and ground temperatures throughout this same region.

Most studies of ground temperatures emphasize data appropriate for either geothermal or agricultural research, hence they cover depth ranges that are either deeper or shallower, respectively, than those of interest here. Because most of the ACP Quaternary mollusks used for aminostratigraphic purposes have been buried at depths ranging from 1–5 m below land surface, we focus here on the recording of temperatures within this same depth range. Although these molluscan samples originally accumulated in shallow marine sands at water depths ranging to approx. 5 m, their confining units have been *emergent* for most of their history because of tectonic and eustatic sea-level factors. Other studies that include ground temperature data from 0.5–5 m include those of Chang (1958), Toy et al. (1978), Norton et al. (1981), Kawanishi (1983), Environment Canada (1984), Tezcan (1992), Lewis (1998), and Hwang (1995). Buntebarth (1984) presents summaries of the patterns of diurnal and seasonal temperature change as functions of depth below land surface. Specific discussions of the relationship between shallow thermal environments and AAR results are found in Wehmiller (1977), Goodfriend (1987), Miller and Brigham-Grette (1989), Hollin et al. (1993), Wehmiller et al. (1995), Ellis et al. (1996), Oches et al. (1996), and Sinibaldi et al. (1999), but relatively few aminostratigraphic studies have actually incorporated multi-year ground temperature records for the regions of study.

In this chapter, we present data for ground temperatures measured during 1994–1999 at depths ranging to 3 m at sites between Maine and Florida along the U.S. ACP and we relate these observations to the basic principles and assumptions of AAR. The primary purpose of this chapter is to document the relationship between mean or effective ground temperatures at selected depths and mean or effective air temperatures derived from climate data. Long-term climatological observations are much more easily obtained than the logistically difficult ground temperature data reported here. We also include comparisons with limited ground temperature data from other sites in North America in order to extend our observed air–ground temperature relationship.

METHODS

Methods of Temperature Recording

Two instrumental approaches were used for obtaining ground temperatures. These methods were chosen for their economy and ability to function without servicing or exposure of any above-ground equipment for periods of up to one year. The recording intervals, and the sites at which these records were obtained, are summarized in figure 17-1 and table 17-1. Detailed discussions of the methods involved in this work are presented below.

During the interval from mid-1994 through late 1996 (especially 1994–1995), we used a system of Ambrose diffusion cells to acquire data for effective ground temperatures, specifically defined here as

TABLE 17-1 Temperature-probe locations and summary of sites used in this study for ground temperature measurements, using either electronic (direct or remote) or Ambrose cell devices. Latitude, longitude, site description, and method of augering are included

State/site	Abbreviation[a]	N Lat.	W Long.	Probe type[b]	Auger method[c]	Observation interval	Site characteristics
Maine							
Walpole: Darling	DMC-shd[a]	43 56.4	69 34.4	H	2, 3	11/96–11/98	shaded and exposed sites; glacial gravel; affected by frost-heaving
Marine Center	DMC-sun[a]	43 56.4	69 34.4	H	2, 3	11/96–11/98	
Massachusetts							
Woods Hole	WH Sun[a]	41 38.8	70 38.7	H	2, 3	11/96–11/98	unshaded; well drained
	WH Shade[a]	41 36.6	70 38.2	H	2, 3	11/96–11/98	partial shade; well drained
Maryland							
Chestertown	Chtwn(sun)[a]	39 17.2	75 55.3	H	2	5/98–5/99	unshaded; well drained
	Chtwn(shd)[a]	39 17.5	75 55.9	H	2	5/98–5/99	partial shade; well drained
Delaware							
Newark	Nrk	39 40.6	75 44.9	(A)	1	8/94–8/95	sandy; partial shade; well drained; disturbed site (underground utilities)
Wilmington	Wilm	39 47.1	75 30.9	A, H, D	A1 H2 D1	D: 11/94–6/99; A 11/94–11/95; H variable	moist; partial shade; tight clay soil
Dover	Dvr	39 08.1	75 29.6	H	2, 3	1/96–11/96	medium sand; partial shade; well drained
Milton	Milton	38 46.4	75 18.6	A, H	A1, 2 H2, 3	H: 3/95–11/96; A: 11/94–11/95	medium sand; partial shade; well drained
Lewes Univ. Campus	LC[a]	38 46.8	75 09.8	H	2, 3	1/96–6/99	medium sand; no shade; well drained
Cape Henlopen Park	CHSP[a]	38 46.4	75 05.7	(A), H	A2 H2, 3	H: 2/97–1/99; A: 9/94–9/95	dune sand; no shade; water table at ca. 1 m (Ambrose site) dune/maritime forest shade; water table at ca. 3 m (Hobo site)

Location	Site	Lat.	Long.	Type	Auger	Period	Description
Virginia							
Nassawadox	Custis	37 29.2	75 50.1	H	2	12/96–11/98	partial shade; well drained
Wallops	Wallops	37 56.2	75 29.0	A	2	8/94–8/95	partial shade; well drained
Oyster	LTER	37 17.6	75 55.9	A	1, 2	8/94–8/96	unshaded grass cover; well drained
Virginia Beach	Gomez	36 46.9	76 12.1	A, H	A1 H2	A: 8/95–8/96; H: 8/95–1/97	partial shade; disturbed site (commercial excavation activity)
North Carolina							
Duck	Duck	36 10.9	76 45.0	A	2	9/94–9/95	open coastal dune; no shade; well drained
Greenville	GNC[a]	35 35.44	77 16.9	A, H	A 2 H 2, 3	A: 8/94–8/95; H: 1/96–1/99	partial shade; well drained
Cape Hatteras	CHNS[a]	35 15.2	75 32.0	A, H	A 2 H 2, 3	A: 8/94–8/95; H: 1/96–1/99	partial shade; well drained sand
Beaufort	Duke	34 42.9	76 40.4	A	1	8/94–8/95	open to partial shade; well drained sand
Wilmington	UNCW	34 13.5	77 52.7	A	1	8/94–8/95	shaded; well drained sand
South Carolina							
Conway	Conway	33 50.2	79 02.9	A, H	A1, 2 H3	A: 8/94–8/96; H: 9/95–7/96	partial shade; sand; well drained
Waties Island	Waties	33 50.9	78 34.6	A	1	8/94–8/95	shaded; well drained sand
Georgetown	Brch	33 20.1	79 12.2	(A)	1	8/94–8/95	open and shaded; water table at 1 m
Charleston (Ft Johnson)	FTJ[a]	32 45.2	79 53.9	A, H	A1, H2	A: 8/94–8/95; H: 2/96–12/98	Ambrose sites exposed and possibly disturbed; Hobo sites partially shaded, well-drained
Charleston (Forest Service)	FES[a]	32 48.0	80 04.2	H	2	2/96–12/98	shaded; water table at ca. 2 m
Georgia							
Skidaway	Skid	31 59.7	81 00.9	A	1	8/94–8/95	partial shade; water table at ca. 1 m
Florida							
Tampa	Tam-shd[a]	28 04.2	82 23.4	H	2	11/97–11/98	shaded; sandy, well drained; water table at ca. 2 m during wet season
	Tam-sun[a]	28 04.6	82 23.4	H	2	11/97–11/98	exposed; sandy, well drained

[a] Sites for which recording continues into 2000; these sites are the focus of the data analysis summarized in table 17-3.
[b] A = Ambrose; H = Hobo; D = Direct; (A) Ambrose data not plotted in figure 17-7 because of limited results.
[c] 1 = 1″ soil auger; 2 = 3″ bucket auger, 3 = 6″ bucket auger.

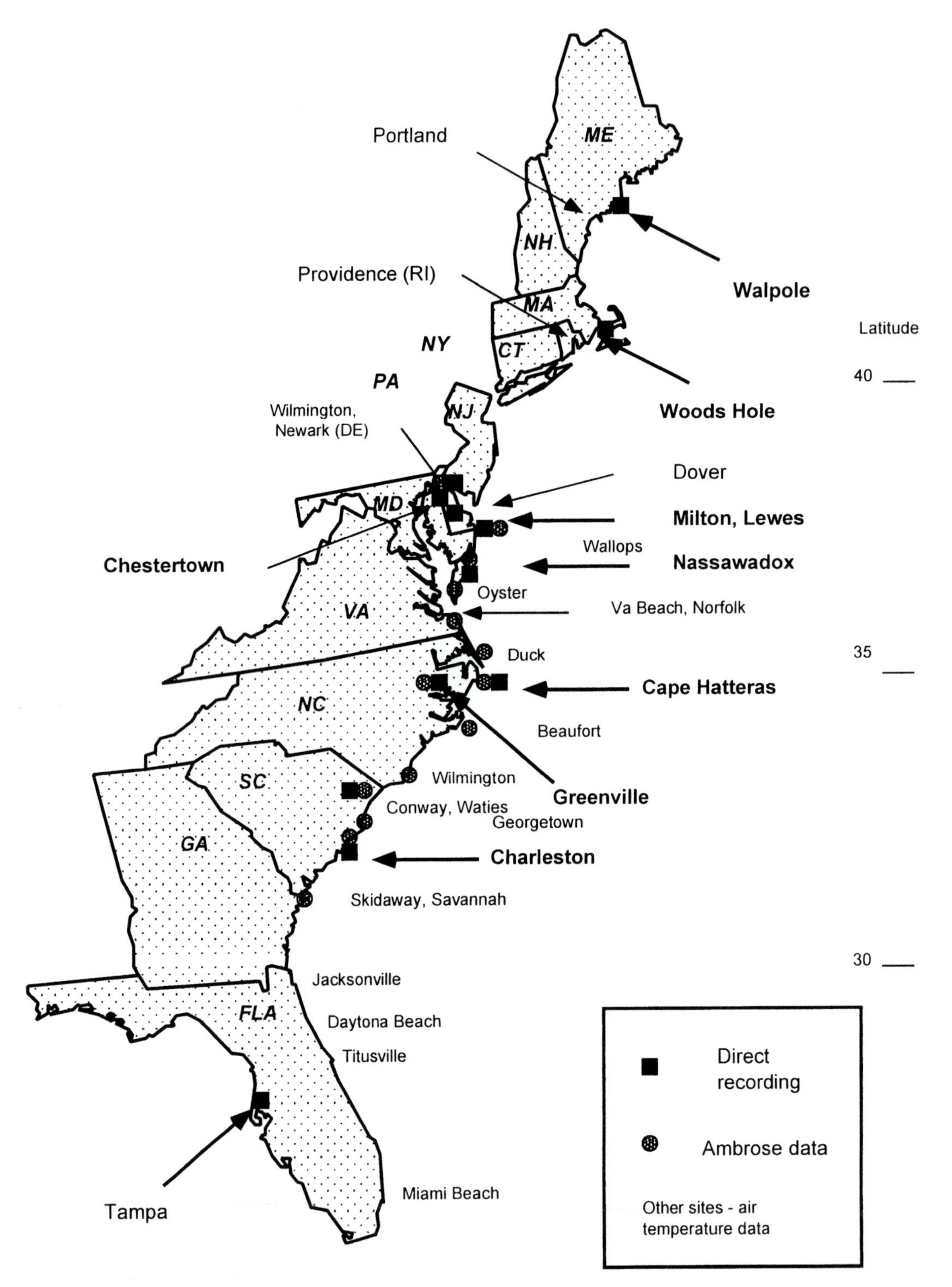

Fig. 17-1. Map of the U.S. Atlantic coast from Maine to Florida, showing sites at which either direct recording or Ambrose cell effective temperature measurements have been made. See tables 17-1 and 17-2 for additional locality information. "Direct recording" identifies sites where daily ground temperatures have been obtained; "Ambrose data" identifies sites where Ambrose cells were deployed during all or part of the 1994–1996 interval. Other sites are used for climatological records.

Ambrose effective temperatures. Effective temperatures are those temperatures that represent the integrated kinetic effect of all temperatures to which an Ambrose cell has been exposed. The Ambrose effective temperature is also important for AAR studies, so the Ambrose cell approach has yielded data that are highly relevant to the goals of this research. The details of Ambrose cells and other chemical methods of measuring effective ground temperatures are described in Norton et al. (1981) and Trembour et al. (1986, 1988). Ambrose cells consist of a sealed polycarbonate tube filled with desiccant and sealed within a glass tube containing water. These nested pairs of tubes are placed inside individual brass capsules that are positioned at different depths in a 1″ (2.5 cm) ID PVC tube (usually 2 m long) that is filled with insulation to prevent convection within the tube. The rate of diffusion of water through the polycarbonate tube wall is a function of temperature, and the total amount of water diffused is a function of the integrated temperature history of the cell. The tube design allows the brass capsules to provide thermal contact with the walls of the hole in which the PVC tube is placed. Most of the PVC tubes were configured so that Ambrose cells were located at depths of 0.25, 0.5, 1.0, and 2.0 m in a single PVC pipe assembly. Other length/depth configurations are possible because of the interchangeable nature of the threaded brass and PVC components. The top end of the PVC tube is fitted with a metal cap that permits relocation of the device with a metal detector. At most sites, two such probes were installed in separate holes, usually within 10 m of each other. Occasionally larger distances (~100 m) were involved to test for the effects of different exposure conditions. Because the Ambrose effective temperature is calculated from the weight gain of the polycarbonate diffusion cell over the entire interval (including transport and return to the laboratory), care must be taken to keep the cells dry and sealed when they are not actually deployed.

Electronic Data-Loggers

Although the Ambrose method requires no electronic instrumentation(which might be vulnerable to field conditions), the obvious disadvantage of the Ambrose method is the lack of actual temperature readings during the period of interest. As the present study progressed, we employed small electronic data-loggers in place of the Ambrose cells, using methods of installation that are modeled on the Ambrose cell/PVC-pipe design.

The data-loggers used for this work were manufactured by the Onset Computer Corp. under the model name "HOBO." The devices used had a temperature range of $-5°$ to $+37°C$, with 255 digital bins, resulting in a resolution of 0.16–0.20°C under the conditions used. The manufacturer's stated temperature accuracy is 0.2°C under these conditions. In practice, the loggers were compared in batches in the laboratory between field deployments at multiple temperatures (including 0°C). In all but one instance, the loggers met the stated specifications. In most cases, the data-loggers were programmed to function for 360 d, and in this mode the loggers' capacity of 1800 data points permits a temperature reading every 4.8 hr.

Two types of waterproof housing were used to deploy the loggers. Free-buried capsules were used for depths up to 1 m, and capsules attached to insulation-filled PVC pipes were used for depths of 1 m or more. The free-burial capsules were fabricated either from galvanized iron-pipe nipple and caps, or from PVC pipe with a galvanized iron plug. In these cases, all threaded joints were coated with a Teflon pipe-sealing compound. Capsules for the deeper probes were made from 1.5″ copper pipe with copper caps at either end. One cap was soldered, after which it was pressure-fit into 1.5″ Schedule-30 thin-walled PVC pipe. The data-logger was wrapped in plastic or aluminum foil and placed in the copper capsule along with a bag containing approximately 3 g desiccant. The second copper cap was then installed and sealed using electrical splice sealant and rubber electrical splice tape. The rubber splice tape was covered with standard vinyl electrical tape (for mechanical strength and abrasion resistance). Approximately 50% of each copper capsule was left uncovered by tape to ensure adequate thermal contact. The probes were deployed so that the capsule was centered at the desired depth. Consequently, the Hobo recorders represent the average temperature for a depth interval of approximately 5 cm around the mid-point depth of each capsule. As with the Ambrose cell installations, the PVC pipe was filled with insulation (vermiculite, crushed Styrofoam, or compressed plastic bags), and the top was sealed with a galvanized iron plug to allow recovery using a metal detector. All of the Hobo recorders were installed in separate holes, rather than being arranged in a vertical series like the Ambrose cells. Pipe ends were usually buried approximately 5 cm below land surface. All of the necessary components for the installation are generally available from commercial plumbing or electrical suppliers. Additional information about the design, construction, and maintenance of these devices is available from the authors.

Direct Recording

At one site (Wilmington, DE; J.F.W.'s back yard), we employed direct electronic measurement of ground temperature, using a battery-operated thermometer with a remote thermistor buried at a depth of 1 m. The above-ground thermometer was enclosed in a weather-proof metal box, and air and ground temperatures were recorded daily. This method has been used in conjunction with the installation of both Ambrose and Hobo devices

(Ambrose for one year, Hobo for part of a year) at the same site so that direct comparisons of the different methods can be made.

Methods of Installation of Equipment

During the course of this study, three methods of preparation of holes for probe installation were used, as summarized in table 17-1. Standard soil augers (1″) were used for most Ambrose probe installations in August, 1994, but all later Ambrose installations and all data-logger probe installations were conducted using larger diameter (3″ or 6″) bucket augers. The larger diameters were required for the Hobo probes because the pipes have larger diameters than the Ambrose probes. Although a disadvantage of the larger diameter augers is the greater disturbance of the hole (thus inhibiting restoration of thermal equilibrium), the advantage of these larger holes is their ease of refilling in a manner that ensures tight repacking of the hole, which was achieved using either $\frac{1}{4}$″ metal rods (for the small holes) or 1″-diameter wooden rods (for the larger holes). Small-diameter holes are difficult to refill uniformly in a manner that eliminates all air pockets, especially in dry, coarse soil. Problems related to air pockets are suspected of causing anomalous Ambrose results, particularly for some of the earliest probe installations. Because of this potential problem, we suggest that the most reliable Ambrose results are those for the cells at the very bottom of the hole (usually 2 m), a position at which the hole-refilling methods are most likely to have eliminated air pockets. Additionally, all those sites where probes penetrated the water table should give reliable results. Holes in very dry soil can also be repacked using water if it is available.

Site Selection

Sites for the installation of temperature-recording equipment were selected using a number of criteria, primarily the security of the equipment (which would remain buried without the need for further servicing) for at least one year. Although no equipment is exposed above ground with the devices used, it was necessary to have local support or interest from personnel at each site. Therefore, almost all of the selected sites are on land related to federal, state, or private scientific or educational activity. Information about each site is found in table 17-1, and figure 17-1 shows the locations of these sites. Most are in areas of late Quaternary coastal deposits, usually < 10 m elevation, with sandy, well-drained soil. The Greenville, NC site is at a higher elevation in more compact pre-Quaternary marine strata. The Wilmington, DE site is in thick (> 3 m) regolith formed on Piedmont crystalline rocks. The Walpole, ME and Woods Hole, MA sites are north of the Quaternary glacial limit in the eastern U.S., hence the deposits at these these sites are generally sand and gravel.

During the course of this work, several sites were used for experimental purposes for short intervals. Operation at these sites was later discontinued to permit extension of the data recording effort to a larger region and to eliminate repetition of results in smaller areas. Additionally, after the data for the first year were obtained, we focused our efforts on obtaining data from sites that were not fully exposed to the sun, although we continue with a few exposed vs. shaded comparisons where appropriate. Lewes, Dover and Milton, DE and Conway, SC were used for initial experimentation with the Hobo devices or for comparison of the Hobo and Ambrose devices. Both the Lewes and the Dover sites were used for testing different PVC pipe designs for the Hobo recorders, and the results confirmed the reliability of our design methods. In particular, data for both free-burial and pipe-mounted recorders buried at 0.5 and 1.0 m were identical, indicating that the vertical pipe did not affect heat transmission between the recorder and the land surface.

Three sites used for this study are considered "disturbed" by human activities and the results are not given great emphasis. The Newark, DE site is on the campus of the University of Delaware and is affected by underground utilities. The Gomez Pit site (Virginia Beach, VA) is a commercial excavation that has been the focus of AAR studies for nearly 15 years (Mirecki et al., 1995). Both Ambrose and electronic devices were installed at Gomez Pit, but the data reveal unusual depth distributions of effective temperature that may be the result of the excavation activities—the probes were installed approximately 50 m from any excavated pit wall, but the land surface at the site of the probes had been disturbed. The Fort Johnson, SC site for the Ambrose cell installation (1994–1995) was within approximately 20 m of an underground sewer line, revealed only after the initial installation. Subsequent years of recording at Fort Johnson have been conducted at a more protected and unaltered site.

Regional Climate Data (Air Temperatures)

Air temperature records for the ACP have been obtained from two primary sources—the U.S. Northeast and Southeast Regional Climate Data Centers in Ithaca, NY and Columbia, SC, respectively. Because our focus is on the relationship of latitudinal trends in ground temperature to latitudinal trends in long-term air temperature, we emphasize the 30-year (1961–1990) mean air temperature data in this report. Detailed analysis of the specific relationship between ground and air temperatures for particular time intervals is deferred to a later publication. The sites for which climate data have been summarized are listed in table 17-2, along with relevant 30-year mean annual air temperatures.

Notation

For the following discussions, the following notation is used for citation of specific temperature values presented in either graphical or tabular form. Equation 17-1 (below) is the basis for the calculation of many of these values.

T_a = 1961–1990 mean annual air temperature.

$T_{a/aar}$ = 1961–1990 effective annual air temperature, for racemization, assuming 27 kcal/mol (113 kJ/mol), derived from mean monthly air temperatures and sine-modeling (see equation 17-1, below).

$T_{a/amb}$ = 1961–1990 effective annual air temperature, for ambrose cells, assuming 10 kcal/mol (41.8 kJ/mol), derived from mean monthly air temperatures.

$T_{g/m}$ = Mean ground temperature from sine-modeling.

$T_{g/amb}$ = Measured Ambrose cell effective ground temperature for specific one-year interval noted in table 17-1, as described in the *Methods* section.

$T_{g/aar}$ = Effective ground temperature for racemization calculated by sine-modeling from electronic or direct measurements (daily data), for a specific measurement interval, assuming an activation energy of 27 kcal/mol (113 kJ/mol).

$T_{g/amb'}$ = Effective ground temperature for the Ambrose reaction, calculated as for $T_{g/aar}$ but assuming 10 kcal/mol activation energy (41.8 kJ/mol).

Data Reduction

Ambrose effective temperatures are based on cell weights obtained before and after deployment, according to the procedures presented in Trembour et al. (1986). All data are presented in graphical form (figure 17-7, later) as the derived effective temperature (based on the activation energy for diffusion of water through the polycarbonate cell wall), rather than the raw weight differences.

Upon retrieval of the Hobo data-loggers, data were downloaded and processed through the use of a Microsoft Excel spreadsheet. In order to standardize the frequency of data from experiments of different duration and to provide a synchronized time-frame, daily averages (five data points/day were obtained in most cases) were calculated so that all data for each day are reduced to a single data point. This procedure allows greatly reduced file sizes whilst maintaining the necessary time-resolution. The reference point for all these data is 1 January, 1995, and all time data points are numbered from this date. Graphs of temperature data are plotted with axes labeled as "95-day" to portray these time-series.

Temperature maxima and minima were estimated by modeling the daily average data to a sine function using KALEIDAGRAPH 3.0.5 (Abelbeck Software). Data were fitted to the formula:

$$y = A + B\{\sin[(x - C)/D]\} \qquad (17\text{-}1)$$

where y = daily average temperature and x = "95-day". The four parameters, A–D, characterize the curve: A is the mean annual temperature; B is the amplitude; C is the "lag", the Julian day on which the temperature rises to the annual mean; and D is the wavelength/2Π. Figure 17-2 illustrates the application of this model to data from two sites. For results presented herein, D was fixed to a constant value of 58.131 d; model fits generated allowing D to float were virtually identical to those derived from fixed D values in all cases for which at least 12 months of data were modeled. This approach to sine-curve fitting follows similar methods presented by Norton et al. (1981) and Miller and Brigham-Grette (1989).

Air temperature data, as well as daily averages from the electronic recorders, are converted into effective temperatures using methods similar to those summarized in Norton et al. (1981: Appendix XVII-I, XVII-II). These methods include the use of equations employing activation energies appropriate for either Ambrose cells or AAR, as outlined below.

Ambrose cell polycarbonate diffusion: approx. 10 kcal/mol (41.8 kJ/mol); Trembour et al. (1986, 1988):

$$\ln k(\text{yr}^{-1}) = 18.00 - 5405/T(T = \text{temperature}, ^\circ\text{K}) \qquad (17\text{-}2)$$

AAR: approx. 27 kcal/mol (113 kJ/mol); (Bada and Schroeder, 1972; Mitterer, 1975; Kaufman and Brigham-Grette, 1993; Wehmiller, 1993):

$$\log k(\text{yr}^{-1}) = 15.77 - 5939/T(T = \text{temperature}, ^\circ\text{K}) \qquad (17\text{-}3)$$

Equations 17-2 and 17-3 are given in the forms originally published in the cited references.

Effective temperatures are calculated using sinusoidal modeling based on mean values and amplitudes (the "A" and "B" parameters from equation 17-1) by calculating the model-predicted temperature and rate constant in regular increments through an annual temperature cycle. The annually averaged rate constant derived from these calculations can then be converted into the effective annual temperature for that particular temperature record. This procedure can be used for any activation

TABLE 17-2 Summary of U.S. Atlantic coast air temperature data and results of modeling for calculation of effective air temperatures in °C. Mean air temperature data (1961–1990) for selected coastal sites from Maine to Florida. Some of these air temperature recording sites are 50 km or more from the ground temperature sites used in the present study. The "30-yr air T_{eff} AAR" value ($T_{a/aar}$) is derived from the 30-year monthly averages by calculating a rate constant (k) for each monthly average (see equation 17-3) and then claculating the effective annual temperature from the average of the 12 monthly k values. Sine-modeling (figure 17-2) is used for models 1, 2, and 3: Model 1 effective temperatures are derived from the averages for the maximum and minimum months only; Model 2 is based on fitting a sine curve to the monthly averages, and Model 3 is also based on this fitting but is corrected to better approximate T_{eff} when model parameters are derived from daily data. The details of this comparison are reviewed in appendix 3. Model comparisons have not been conducted for all of the sites listed; $T_{a/aar}$ values listed in column 4 are considered most useful for the discussions in this chapter

State/site	Latitude (°N)	**T_a 1961–1990 30-year mean air (°C)**	**$T_{a/aar}$ 30 yr air T_{eff} AAR (°C)**	$T_{a/aar}$ Model 1 (°C)	$T_{a/aar}$ Model 2 (°C)	$T_{a/aar}$ Model 3 (°C)	Uncorrected mean A[a] (°C)	Uncorrected amplitude B[a] (°C)	Delay C[a]	R	Corrected mean (°C)	Corrected amplitude (°C)
ME												
Eastport	44.9	**6.28**	**10.68**	10.34	10.73	11.10	6.281	11.377	4.356	0.997	6.527	11.582
Portland	43.6	**7.44**	**12.82**	12.78	12.85	13.24	7.442	12.838	4.266	0.998	7.669	13.089
MA/RI												
Boston	42.3	**10.71**	**15.61**	15.66								
Providence	41.75	**10.24**	**15.19**	15.23	15.16	15.49	10.236	12.234	4.313	0.999	10.417	12.466
NJ/DE												
Atlantic	39.3	**11.66**	**16.39**	16.40								
Wilmington PR	39.76	**12.1**	**16.23**	16.57	16.87	17.16	12.093	12.074	4.171	0.996	12.243	12.301
Dover	39.15	**13.49**	**18.03**	17.75								
Lewes	38.75	**13.29**	**17.53**	17.27	17.56	17.82	13.289	11.346	4.282	0.999	13.420	11.550

MD/VA												
Salisbury Aprt	38.3	**13.28**	**17.66**	17.62								
Painter	37.5	**14.31**	**18.33**	18.20								
Norfolk	36.8	**15.37**	**19.26**	19.03	19.28	19.49	15.373	10.838	4.247	0.999	15.470	11.026
NC												
Manteo	35.95	**16.81**	**20.13**	18.80								
Greenville	35.58	**15.99**	**19.63**	19.29	19.67	19.86	15.994	10.484	4.118	0.998	16.081	10.661
Cape Hatteras	35.25	**16.78**	**19.84**	19.53	19.83	19.99	16.777	9.469	4.395	0.999	16.851	9.615
Wilmington	34.2	**17.59**	**20.77**	20.51								
SC/GA												
Myrtle Beach	33.8	**18.12**	**21.17**	21.44								
Conway	33.85	**17.66**	**20.87**	20.47								
Charleston	32.78	**18.57**	**21.52**	21.10	21.56	21.68	18.572	9.391	4.134	0.997	18.617	9.534
Savannah	32.05	**19.2**	**21.96**	21.57								
FLA												
Jacksonville	30.31	**20.4**	**22.62**	22.15	22.65	22.72	20.398	8.093	4.150	0.996	20.413	8.196
Titusville	28.55	**22.05**	**23.44**	22.76	23.46	23.47	22.045	6.332	4.239	0.992	22.033	6.380
Tampa	27.95	**22.46**	**23.93**	23.19	23.96	23.98	22.457	6.538	4.231	0.993	22.438	6.592
Miami Beach	25.7	**24.55**	**25.16**	24.85	25.17	25.11	24.551	4.142	4.461	0.994	24.498	4.122
			avg k from monthly mean k values	sine fit to max and min monthly k	sine fit to 12 monthly means	sine fit to 12 monthly means (corrected)						

[a] Coefficients in equation 17-1.

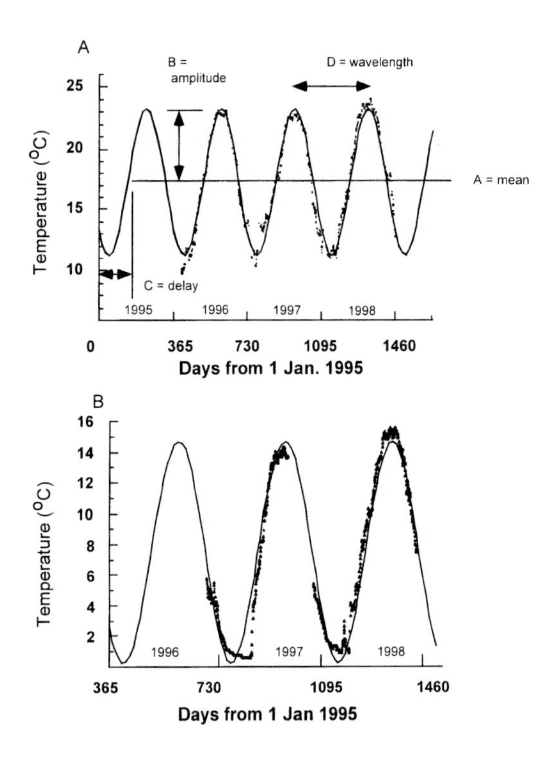

Fig. 17-2. Examples of curve-fitting to multi-year temperature records. A. Three years of temperature data, one datum per day, from a 1 m recorder depth at Cape Hatteras, NC, fitted to the sine-model in equation 17-1. The x-axis presents days from 1 January, 1995. The four curve-fitting parameters (A, B, C, and D) are defined in this figure and summarized in the text. For all the results presented herein, the parameter related to the year length (D) was set to the theoretical value of 58.131 d ($365.25/2\pi$). The curve shown here is for this three-parameter fit; the other parameters are as follows: $A = 17.253$; $B = 5.590$; $C = 142.37$; $R = 0.9845$. A four-parameter fit to these same data (one that does not fix D) yields the following: $A = 17.238$; $B = 5.996$; $C = 140.27$; $D = 58.279$; $R = 0.9825$. Note that the two parameters A and B (mean and amplitude) are very similar for the two fitting approaches. These are the critical values for derivation of effective temperatures. B. Similar example showing the deviation from the sine model for the highest latitude site studied in this work (DMC, Walpole, ME), for late 1996 to late 1998 (95 d axis starts at 1 January, 1995). The major deviation from the simple sine curve (seen especially in the spring of 1997) is interpreted to be the effect of snow cover, with snow insulation slowing the winter cooling and then producing a rapid spring temperature rise during snow-melt. The three- and four-parameter curve fits (see figure 17-2A for definitions) are as follows: $A = 7.467$; $B = 7.191$; $C = 140.21$; $R = 0.9782$ (3-parameter); $A = 7.488$; $B = 7.362$; $C = 165.18$; $D = 56.56$; $R = 0.9838$ (four-parameter). Although the differences between the three-parameter and four-parameter constants are slightly greater than for curve 17-2A, the effective temperatures (for racemization) calculated from the different pairs of constants are within 0.11°C.

energy that might be assumed or modeled for AAR or other geochemical reactions. The electronic ground temperature data were modeled at daily intervals. Monthly mean air temperature data were modeled using a similar equations except that D was set to 0.5236 (i.e., 12/2П). For some sites, two years of daily mean air temperatures were modeled, as well as the monthly means for the same period, in order to compare the parameters derived from these two approaches. As expected, the amplitude (B in equation 17-1) for monthly air temperature means is smaller than that for daily air temperature means, but there is strong linear correlation of A and B between the two methods (see appendix 3). This relationship was used to "correct" effective temperatures derived from mean monthly air data to what they would be if mean daily air data were used. The effective air temperatures derived from monthly or daily data are presented in table 17-2.

The variability of air temperature means or amplitudes derived from analysis of daily or monthly data, or from simple interpolation from maxima and minima, is usually within 0.5°C, depending on both the length and interannual variability of the record. Although the use of daily mean air temperature data is preferred for the most reliable calculation of effective air temperatures, mean monthly air temperature data are much simpler to use because of the smaller number of data points and their smoothing effect on the natural variability of the daily data.

RESULTS

Air Temperature Data

Figure 17-3 presents data for mean and effective air temperatures (T_a and $T_{a/aar}$) for sites from Maine to Florida, as listed in table 17-2. For discussion purposes, the mean and effective air temperatures presented in figure 17-3 serve as the reference temperatures for all ground temperature data presented here, even though the actual air temperatures during the periods of ground temperature measurements were not equal to these 30-year averages. The deviations of mean monthly temperatures during our study interval from the 1961–1990 averages are shown in figure 17-4. This figure shows that most of the 1994–1995 interval during which Ambrose cells were installed was above the 30-year average by approximately 1–3°C. Most of 1996–1997 was near to, or slightly below, this average, with the exception of early 1997. Most of 1998 was above the 30-year average. The only exceptions to these generalizations, which are based on data from the region from Delaware to South Carolina, were noted for 1994–1995 data from Maine and Rhode Island, but Ambrose cells were not deployed at these higher latitude sites. Interannual differences are apparent in all the

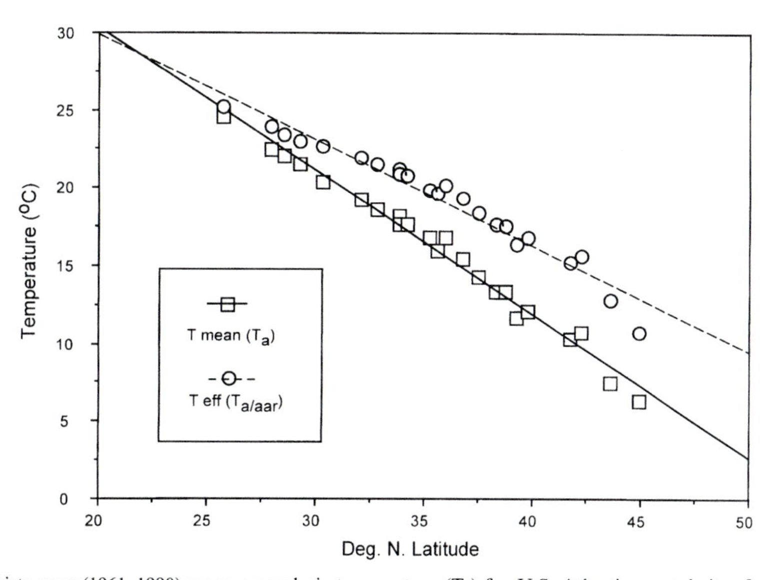

Fig. 17-3. Thirty-year (1961–1990) mean annual air temperature (T_a) for U.S. Atlantic coastal sites from Maine to Florida plotted vs. latitude. Data from Northeast and Southeast regional climate data centers. Also plotted is the racemization effective annual air temperature ($T_{a/aar}$), derived from the mean monthly data for the same 1961–1990 interval. The effective temperature is calculated for AAR, assuming an activation energy of 27 kcal/mole (113 kJ/mole). See text for notations. The regressions for the lines shown are as follows: mean temperature: $T_a = 48.74 - 0.922(\text{lat.})$, $R^2 = 0.985$; effective temperature: $T_{a/aar} = 43.58 - 0.682(\text{lat})$, $R^2 = 0.956$.

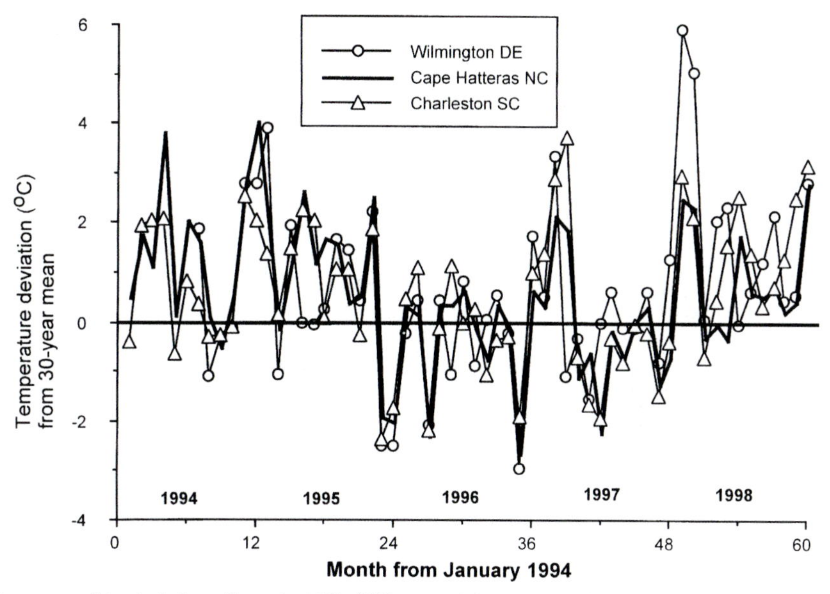

Fig. 17-4. Summary of the deviations (from the 1961–1990 average) in mean monthly air temperature for three climate record sites between Delaware and South Carolina (see figure 17-1) for the interval during which ground temperature data are reported here. Data from Southeastern and Northeastern regional climate data centers, Columbia, SC, and Ithaca, NY, respectively. For this entire region, with some exceptions, much of 1994–1995 was warmer than the long-term average, 1996 was equal or slightly cooler than this average, 1997 was mixed, and 1998 was generally warmer than the 30-year average. Data for other sites in the region (not plotted, for the sake of clarity) are similar to the trends depicted for these three sites, which span the latitude range for all the Ambrose cells that were deployed early in this study. In order to simplify the discussion in this Chapter, all ground temperature data for the study period are compared with the 1961–1990 average.

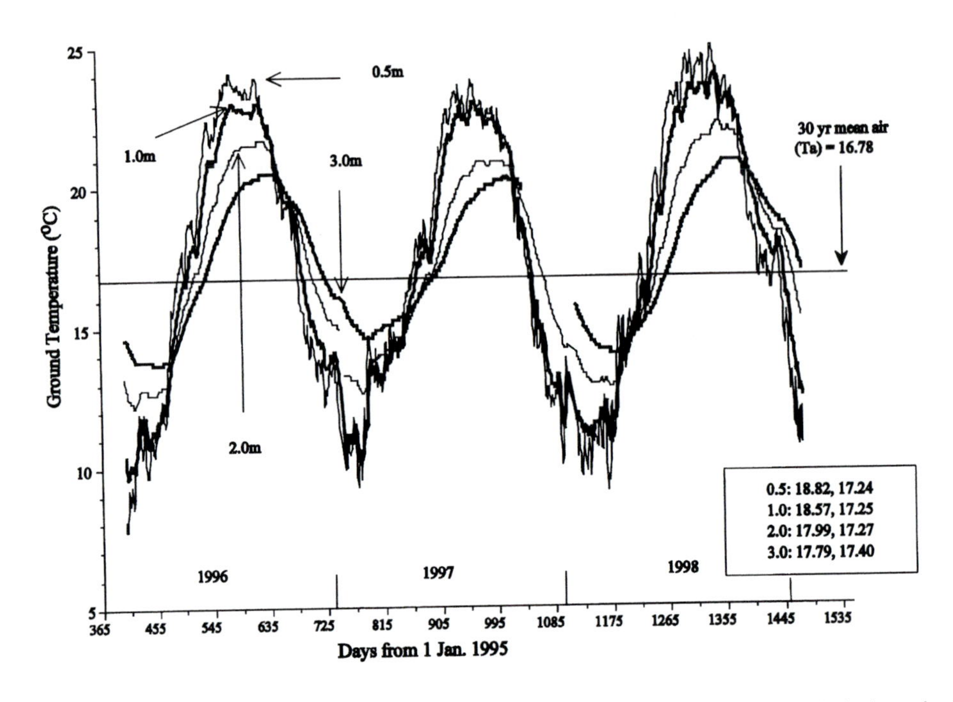

Fig. 17-5. Three years of electronic (direct-recording) data from Cape Hatteras, NC, for four depths. The horizontal axis shows days from 1 January, 1995. The 30-year mean air temperature is 16.78°C. Inset values show effective ground temperatures ($T_{g/aar}$) and mean ground temperatures for the four depths indicated. Data are summarized in table 17-3.

ground temperature records for which many years of data are available. Because ground temperature recording devices were not installed for simultaneous intervals, and because of lags between air and ground temperature, it is more convenient to compare all ground temperature data to the long-term mean air temperatures. Nevertheless, it is important to note that the mean of June 1994 to September 1995 (when most of the Ambrose cells were deployed) air temperatures shown in figure 17-4 was between 2.2 and 3.9°C greater than the mean for the same sites between January 1996 and December 1998 (when most of the Hobo devices were deployed). The maximum differences between these mean values are observed for the Norfolk and Cape Hatteras recording stations.

Ground-temperature Data

Representative electronic data are shown in figures 17-5 and 17-6 in two forms: as a time-series for a single site (Cape Hatteras, NC) at different depths (with calculated AAR effective temperatures from equation 17-3) and for a single depth (2 m) at different sites over the latitude range of this study. Table 17-3 presents a summary of the sine-modeling constants for all of the ground-temperature data, showing the sites, depths, and intervals modeled. Data for some sites are not included (Wilmington, Dover, Milton, Conway, Gomez) because of disturbed ground effects or duplication of record, but these data are all presented in appendix 1.

Ambrose Results

The Ambrose cell effective temperature data are presented in figures 17-7A–E. These figures show the actual Ambrose effective temperatures (plotted vs. depth) derived from the cell weights, for the measurement intervals indicated in table 17-1. The data are grouped by region (with some overlap), from Delaware (figure 17-7A), Virginia (figure 17-7B), central North Carolina (figures 17-7C, D), southeastern North Carolina and northeastern South Carolina (figure 17-7E), and central South Carolina and Georgia (figure 17-7F). Almost all the data presented in these figures were obtained during the interval between August 1994 and September 1995, a period of higher-than-average air temperatures for the U.S. Atlantic coast (figure 17-4). For most of Ambrose probe data, the difference between results for different probes is a result of exposure differences between the two probes, with the higher Ambrose effective temperatures associated with the more exposed sites. Some of the Ambrose results clearly deviate from the expected trend of decreasing effective temperature with depth. There is no definite explanation for

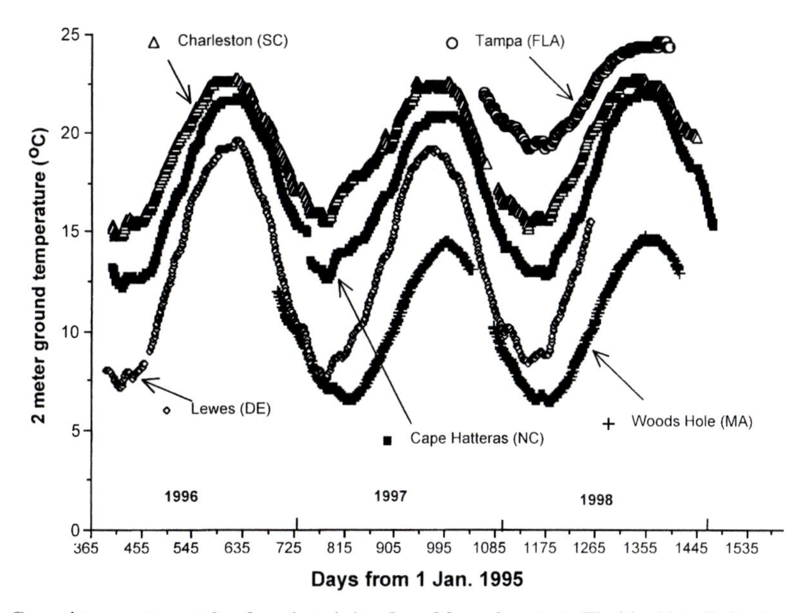

Fig. 17-6. Ground temperature at 2 m for selected sites from Massachusetts to Florida. Not all sites have been monitored for same interval, but recording continues at all sites shown. Gaps in the data represent intervals when recording ceased at the end of a 360 d record, prior to replacement with new devices. The Charleston and Lewes sites are more exposed than are the other three sites; all are in well-drained material. Data for the 2 m depth are not available for the site in Maine. Effective temperatures for these depths/sites ($T_{g/aar}$; see table 17-3) are as follows: Woods Hole (shd), 10.46°C; Lewes (LC), 13.79°C; Cape Hatteras (CHNS), 17.27°C; Charleston (FTJ), 19.16°C; Tampa (sun), 23.85°C.

these deviations, although either cell leakage or air pockets within the probe holes are the most probable causes.

Depth Relationships of Ground Temperature/ Electronic Data

The data summarized in table 17-3 and figure 17-5 conform to expectations and demonstrate several principles that must be met if the results are to be considered valid. The mean temperatures for all depths are generally within 0.5°C, particularly for shaded sites. The amplitude of annual variation decreases with depth, hence calculated effective temperatures also decrease with depth. The seasonal delay in minima or maxima also increases with depth, demonstrating the heat-transmission properties of the soil. Finally, there is less sensitivity to abrupt climate changes with increasing depth. Interestingly, however, several records (particularly those for the Fort Johnson, SC site; see figure 17-15, later) demonstrate "heat transmission events" at all depths, during which large rainfall events (hurricanes or tropical storms noted in the daily climate records from the SE Region Climate Data Center) have rapidly raised or lowered temperatures at depths of even 2 m, depending upon the actual soil temperature gradient at the time of the rainfall event.

Relationship Between Effective Ground Temperature and 30-year Mean Annual Air Temperature

Figure 17-8 shows $T_{g/aar}$, the effective ground temperature (calculated from equations 17-1 and 17-3) for each site, plotted vs. depth. The data point for 4 m is *not* an effective ground temperature measurement, but rather the 30-year mean air temperature (T_a)—this point is shown for comparison with the $T_{g/aar}$ values. Exponential regressions to these data are summarized in table 17-4. Also shown in table 4 is the regression for the Ambrose cell data of Miller and Brigham-Grette (1989) for a desert site with a mean annual air temperature of approximately 17°C. Figure 17-8 and table 17-4 show that effective ground temperatures ($T_{g/aar}$) decrease with depth at every site (contrary to some of the Ambrose cell data) but that the differences between effective ground temperatures ($T_{g/aar}$) and mean air temperature (T_a) are not uniform throughout the region of study. Exposure effects are the primary cause of these differences, as demonstrated by the results for shaded vs. exposed sites (DMC, WH, LC/CHSP, and TAM), seen also in the shape coefficients for the curves (table 17-4). The regressions summarized in table 17-4 can be evaluated for their reliability by calculating a predicted temperature for a depth of 20 m, a depth that is likely to be unaffected by either

TABLE 17-3 Comparison of 30-year mean air temperature with effective ground temperature (°C). Daily average ground temperature data for each depth at each site were fitted to equation 17-1 to generate modeled mean (A), amplitude (B), and delay (C) values. Effective temperatures derived from thse data are also shown, for ambrose ($T_{g/amb'}$) and racemization ($T_{g/aar}$) reactions. See equations 17-2 and 17-3, respectively. See appendices 2 and 3 for additional information. Results are shown to nearest 0.1°C for simplicity. Reported recording intervals for all depths at a single site may not be identical because of missing data or equipment failure

Record location	Observation interval	T_a 30-year mean air[a]	$T_{g/m}$ modeled mean (A)	Modeled amplitude (B)	Modeled delay (C)	$T_{a/aar}$ (air)[b]	$T_{g/aar}$ (ground)	$T_{a/amb}$ (air)[c]	$T_{a/amb'}$ (ground)	Sine fit R value
Portland (ME) air	1961–1990	7.44				12.82		9.75		
DMC sun 1 m	11/23/96–11/11/98		10.2	7.3	140		12.2		11.0	0.988
DMC sun 1.5 m	11/23/96–11/8/97		9.8	6.0	155		11.2		10.4	0.994
DMC shd 0.5 m	11/23/96–11/11/98		7.5	7.2	140		9.4		8.3	0.972
DMC shd 1 m	11/23/96–11/11/98		7.5	6.0	148		8.9		8.0	0.979
Providence (RI) air	1961–1990	10.24				15.19		12.36		
WH sun 0.5 m	12/16/97–11/12/98		13.0	9.7	122		16.3		14.4	0.990
WH sun 1 m	11/25/96–11/12/98		12.4	8.0	133		14.7		13.3	0.991
WH sun 2 m	11/25/96–11/12/98		12.1	5.6	152		13.3		12.6	0.991
WH shd 0.5 m	12/16/97–11/12/98		10.8	6.4	140		12.3		11.4	0.994
WH shd 1 m	11/25/96–11/12/98		10.6	5.2	154		11.6		11.0	0.994
WH shd 2 m	11/25/96–11/12/98		10.5	4.0	171		11.1		10.7	0.996
Dover (DE) air	1961–1990	13.49				18.03		15.44		
Chtwn sun 0.5 m	5/26/98–5/19/99		13.7	10.0	127		17.1		15.1	0.990
Chtwn sun 1 m	5/26/98–5/19/99		13.8	9.0	136		16.6		15.0	0.994
Chtwn sun 2 m	5/26/98–5/19/99		13.9	6.2	157		15.3		14.4	0.997
Chtwn shd 0.5 m	5/26/98–5/19/99		13.2	8.2	130		15.6		14.1	0.986
Chtwn shd 1 m	5/26/98–5/19/99		12.8	6.5	143		14.4		13.4	0.993
Chtwn shd 2 m	5/26/98–5/19/99		13.1	5.1	157		14.1		13.5	0.997
Chtwn shd 3 m	5/26/98–5/19/99		13.0	3.6	175		13.5		13.3	0.998
Chtwn shd 6 m (in water table)	5/26/98–5/19/99		12.6	1.2	254		12.6		12.6	0.996
Lewes (DE) air	1961–1990	13.29				17.53		15.12		
LC 0.5	1/28/96–6/10/99		14.4	9.5	123		17.5		15.7	0.978
LC 1	1/28/96–6/10/99		14.1	7.9	132		16.3		15.0	0.987
LC2	1/28/96–6/10/99		13.8	5.6	151		15.0		14.3	0.990
CHSP 0.25	2/8/97–1/2/98		13.0	8.8	120		15.8		14.1	0.971
CHSP 0.5	2/3/97–1/24/99		13.8	7.7	126		15.9		14.6	0.975
CHSP 1	2/3/97–1/24/99		13.8	7.0	130		15.6		14.5	0.980
CHSP 2	2/3/97–1/24/99		13.9	4.5	148		14.7		14.2	0.989
CHSP 3	2/3/97–1/24/99		14.2	2.7	163		14.5		14.3	0.992

Painter (VA) air	1961–1990	14.32				18.34		16.03		
Custis 0.5	12/22/96–12/18/97		13.7	8.4	127		16.2		14.7	0.971
Custis 1.0	12/20/96–11/6/98		14.0	6.8	135		15.7		14.7	0.981
Custis 2.0	12/20/96–11/16/98		14.2	4.4	158		14.9		14.5	0.989
(Norfolk, Va records also used for comparison with Custis site)										
Hatteras (NC) air	1961–1990	16.78				19.84		18.04		
CHNS 0.1	2/6/96–1/13/99		17.0	7.5	127		19.0		17.8	0.956
CHNS 0.25	1/22/97–1/13/99		17.4	6.9	130		19.1		18.0	0.963
CHNS 0.5	2/6/96–1/13/99		17.2	6.6	135		18.8		17.9	0.978
CHNS 1	2/6/96–1/13/99		17.3	6.0	142		18.6		17.8	0.982
CHNS 2	2/6/96–1/13/99		17.3	4.4	159		18.0		17.5	0.986
CHNS 3	2/6/96–11/3/99		17.4	3.2	176		17.8		17.6	0.988
GNC (NC) air	1961–1990	15.99				19.63		17.54		
GNC 0.5	2/7/96–1/13/99		16.4	7.0	126		18.2		17.1	0.972
GNC 1	2/7/96–1/13/99		16.3	5.8	137		17.6		16.8	0.982
GNC 2	2/7/96–1/13/99		16.4	4.2	156		17.1		16.7	0.986
Charleston (SC) air	1961–1990	18.57				21.52		19.9		
FT J 0.1	2/7/96–11/26/98		18.9	7.7	113		20.9		19.7	0.950
FT J 0.5	12/17/96–11/26/98		19.2	6.2	102		20.6		19.7	0.973
FT J 1	2/7/96–11/26/98		19.1	5.3	133		20.2		19.5	0.982
FT J 2	2/7/96–11/26/98		19.2	3.5	151		19.6		19.4	0.989
FES 0.5	2/7/96–11/26/98		17.9	5.4	129		19.0		18.3	0.972
FES 1	2/7/96–11/26/98		17.9	5.5	142		18.7		18.2	0.980
FES 2	2/7/96–11/26/98		17.8	3.3	162		18.2		18.0	0.986
Tampa (FLA) air	1961–1990	22.46				23.93		23.03		
Tam shd 0.5 m	11/28/97–10/24/98		22.1	4.6	128		22.9		22.4	0.973
Tam shd 1 m	11/28/97–10/24/98		22.0	3.6	140		22.5		22.2	0.987
Tam shd 2 m	11/28/97–10/24/98		22.0	2.6	156		22.2		22.1	0.993
Tam sun 0.5 m	11/28/97–10/24/98		24.8	6.4	111		26.2		25.3	0.954
Tam sun 1 m	11/28/97–10/24/98		24.4	5.0	122		25.3		24.7	0.979
Tam sun 2 m	11/28/97–10/24/98		23.9	3.3	142		24.3		24.0	0.993

[a] See table 17-2.
[b] See table 17-2: derived from mean monthly air temperature data.

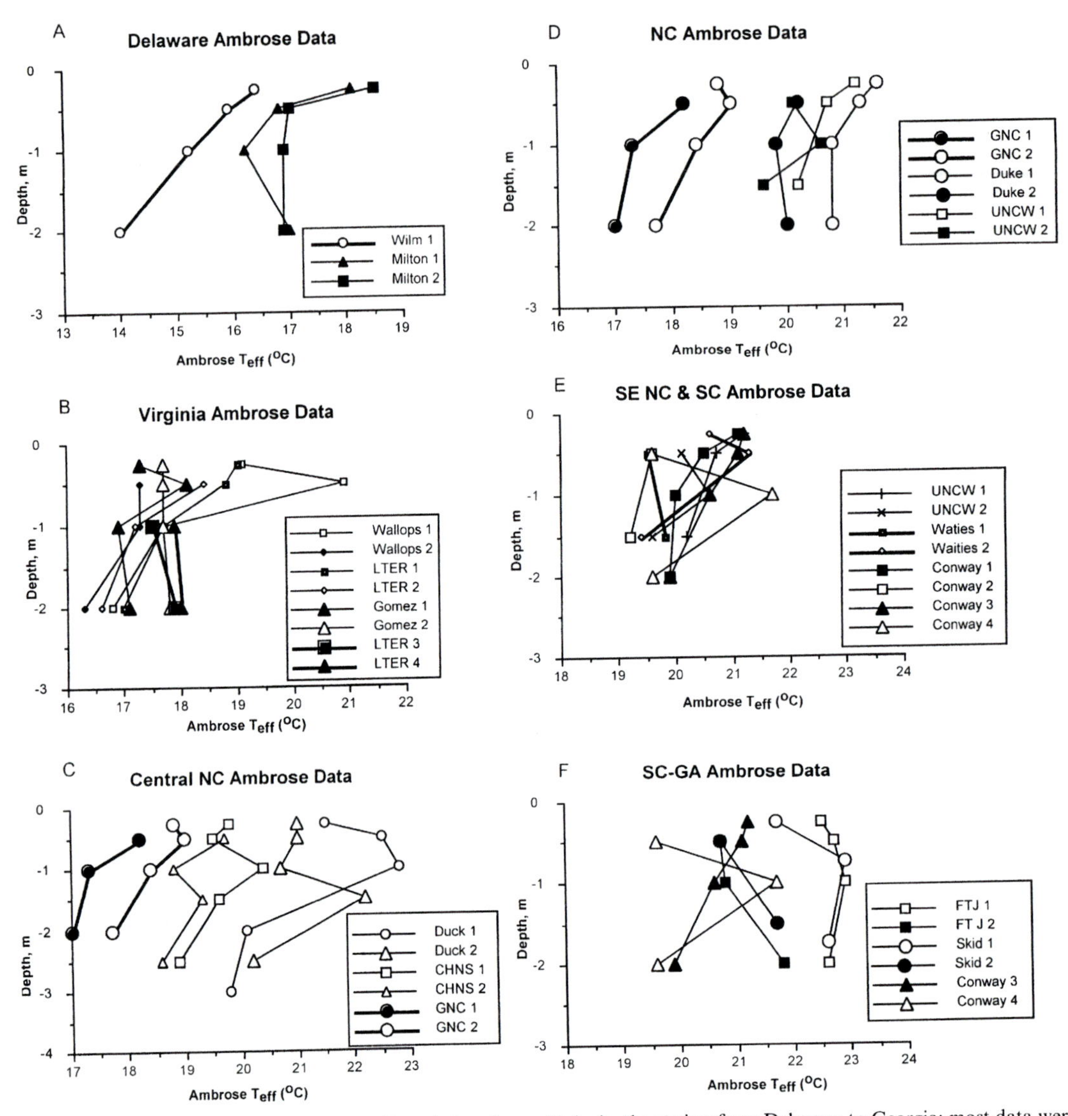

Fig. 17-7. Ambrose effective temperature ($T_{g/amb}$) data for multiple depths at sites from Delaware to Georgia; most data were obtained between August 1994 and October 1995. Some data are plotted in two consecutive figures for comparison. A. Wilmington and Milton, DE (Milton 1 and 2, adjacent probes, identical exposure). B. Wallops Island, Oyster (LTER 1–4), and Gomez Pit, VA (Wallops 1 and 2, identical unshaded exposure; LTER 1 and 2, 3 and 4, consecutive years; identical unshaded exposure; Gomez 1 and 2, identical partial shade exposure). C. Central and NE NC: Duck, Cape Hatteras National Seashore, and Greenville (Duck 1 and 2, identical unshaded exposure; CHNS 1 and 2, identical partial shade; GNC 1 shaded; GNC 2 unshaded). D. Central and SE NC: Greenville, Duke University Marine Laboratory (Beaufort) and University of North Carolina at Wilmington, NC (GNC 1 shaded; GNC 2 unshaded; Duke 1 unshaded, Duke 2 shaded; UNCW 1 and 2, identical partial shade). E. Southeastern NC and northeastern SC: University of North Carolina at Wilmington, Waities Island and Conway, SC (UNCW 1 and 2, identical partial shade; Waties 1 and 2, identical partial shade; Conway 1 and 2, 3 and 4 consecutive years, identical partial shade). F. SC and GA: Conway and Charleston, SC; Skidaway Island (Savannah), GA (FTJ 1, unshaded; FTJ 2, shaded; Skid 1 unshaded; Skid 2 shaded; Conway 3 and 4 identical partial shade).

seasonal changes or geothermal effects. In most cases, the regressions do predict temperatures within 1°C of the mean air temperature. In a few cases (LC, Custis, FES) the differences between these predictions and mean air temperatures are nearly 2°C; the value of the regression coefficient is not always an accurate measure of the reliability of the equation.

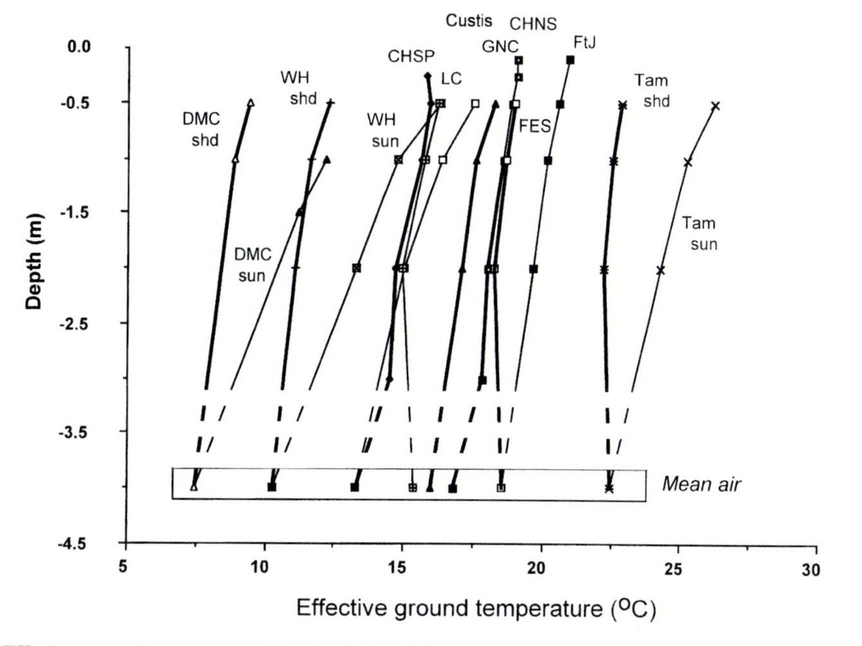

Fig. 17-8. Effective ground temperatures ($T_{g/aar}$) calculated from electronic data for different depths at sites from Maine to Florida. Calculations are as discussed in text. Heavier lines indicate trends for shaded sites. Points plotted at 4 m depth actually represent the regional 30-year mean annual air temperature (table 17-3) for comparison with effective ground temperatures. Effective ground temperatures ($T_{g/aar}$) at depths to at least 3 m are always greater than, or equal to, mean air temperatures; effective air temperatures ($T_{a/aar}$; see table 17-2) are usually greater than the effective ground temperatures portrayed here. See table 17-4 for a summary of reqressions describing these data.

TABLE 17-4 Best-fit logarithmic regressions of effective ground temperature ($T_{g/aar}$) vs. depth, as plotted in figure 17-8. Regressions for data from the Lehner site (Arizona) (Miller and Brigham-Grette, 1989) are shown for the effective temperature for amino acid racemization. Equations 17-2 and 17-3 can be used to convert from the original Ambrose cell data to AAR effective temperature, and examples of this relationship are found in table 17-3. The maximum depth of a recorder is listed, and the T(20) value is the temperature predicted by the regression for a depth of 20 m. General equation: Temperature = M + Q(log [depth]), where depth is in meters

Site	M	Q	R^2	Max. depth	T(20) (°C)
WH sun	14.76	−5.01	0.999	2.0	8.2
WH shd	11.66	−2.06	0.999	2.0	9.0
Chtwn shd	14.68	−2.55	0.966	6.0	11.4
LC	16.28	−4.20	0.999	2.0	10.8
CHSP	15.26	−1.34	0.819	3.0	13.5
Custis	15.63	−2.13	0.977	2.0	12.9
CHNS	18.37	−0.90	0.878	3.0	17.2
GNC	17.62	−1.88	0.997	2.0	15.2
FT J	20.08	−0.97	0.911	2.0	18.8
FES	18.63	−1.25	0.997	2.0	17.0
Tam shd	22.53	−1.06	0.997	2.0	21.2
Tam sun	25.23	−3.24	0.999	2.0	21.0
Lehner[a]	19.95	−3.71	0.992	1.5	15.1

[a] Original data from Miller and Brigham-Grette (1989).

Relationship Between AAR Effective Ground Temperature and Latitude and Mean Annual Air Temperature

The effective ground temperature for all sites at 2 m ($T_{g/aar}$) is plotted against latitude in figure 17-9, which also includes the 30-year mean annual air temperature for the same region. The effective ground temperature regressions for shaded and exposed sites are as follows:

2 m shaded:

$$T_{g/aar}(°C) = 44.58 - 0.785(°N.\,lat.)R^2 = 0.961; \qquad (17\text{-}4)$$

2 m exposed:

$$T_{g/aar}(°C) = 46.23 - 0.796(°N.\,lat.)R^2 = 0.991. \qquad (17\text{-}5)$$

There is good agreement in the slopes of these two regressions, and there is an offset of approximately 1.6°C. Note that the air temperature regression shown in the caption for figure 17-9 is for the same latitude range as the actual ground temperature sites, to avoid the effects on this regression

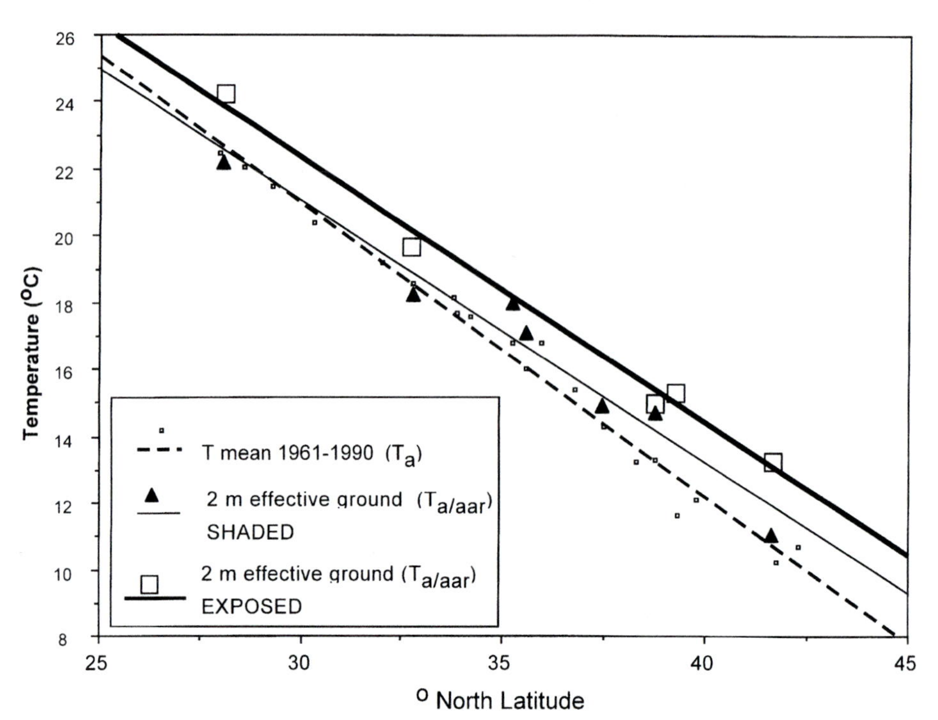

Fig. 17-9. Effective ground temperatures ($T_{g/aar}$; table 17-3) and mean air temperature plotted vs. latitude for the study region. The mean air temperature trend is from figure 17-3, limited to data from within the same latitude range as the recording sites. $T_{g/aar}$ values for 2 m depth at shaded and exposed sites are plotted separately to note possible differences in either regression slopes or intercepts. The regressions on the three data sets shown here are as follows: mean air 1961–1990: $T_a = 47.34 - 0.879(\text{lat})$, $R^2 = 0.985$ (note that this regression differs from that in figure 17-3 because of a narrower latitude range of data); 2 m shaded effective temperature: $T_{g/aar} = 44.58 - 0.785(\text{lat.})$, $R^2 = 0.961$; 2 m unshaded effective temperature: $T_{g/aar} = 46.23 - 0.796(\text{lat.})$, $R^2 = 0.991$. The regressions for *all* of the 2 m data points (shaded and unshaded) are as follows: 2 m effective ground temperature: $T_{g/aar} = 45.13 - 0.786(\text{lat.})$, $R^2 = 0.950$; 2 m mean ground temperature: $T_{g/m} = 46.54 - 0.846(\text{lat.})$, $R^2 = 0.970$. These and other regressions are discussed in the text.

from sites outside the specific region of ground temperature data (as would be the case if the regression from figure 17-3 were used). Because of the small number of data points (and the variable intervals over which the data were obtained), the parallel regressions for shaded and exposed sites are not significantly different statistically (the 95% confidence interval is approximately 5°C). Identification of exposed sites is relatively simple, but there is greater ambiguity regarding the thermal environment of "shaded" sites because the shaded areas surrounding many of these sites are, in some cases, limited to as little as 800 m^2. In spite of these uncertainties, the trends seen in figure 17-9 make physical sense, and the trends for the three sites where there is a clear contrast in exposure between ground temperature recording stations (Woods Hole, Charleston, and Tampa) support the conclusion implied by the offset between the regressions for the shaded and exposed sites.

The regressions of $T_{g/aar}$ for four depths, vs. latitude, for shaded and exposed sites combined, are as follows:

0.5 m:

$$T_{g/aar}(°C) = 47.31 - 0.821(°\text{N. lat.})\ R^2 = 0.802 \tag{17-6}$$

1 m:

$$T_{g/aar}(°C) = 46.42 - 0.810(°\text{N. lat.})\ R^2 = 0.860 \tag{17-7}$$

2 m:

$$T_{g/aar}(°C) = 45.13 - 0.786(°\text{N. lat.})\ R^2 = 0.950 \tag{17-8}$$

3 m:

$$T_{g/aar}(°C) = 51.04 - 0.947(°\text{N. lat.}) \quad \text{(only two data points for 3 m)} \tag{17-9}$$

Equations 17-4–17-6 indicate a steady progression to lower intercept values and more gradual slopes,

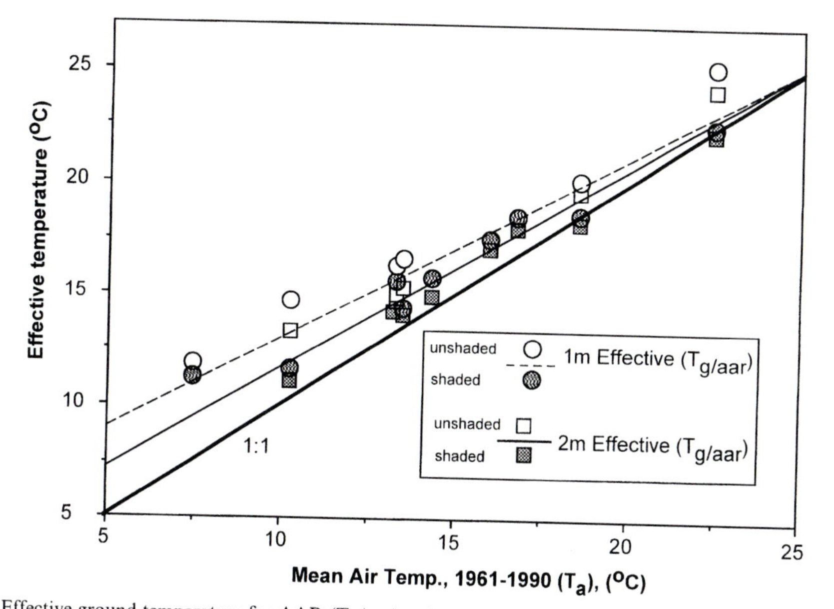

Fig. 17-10. Effective ground temperature for AAR (Tg/aar) at 1 and 2 m depths (see table 17-3) vs. 30-year mean air temperature (T_a). Data for all sites (shaded and unshaded) are plotted. Open symbols represent unshaded sites. The heavy line is the 1:1 trend. Regressions are presented in the text as equations 17-10–17-15.

consistent with the trend seen in figures 17-5 and 17-8 for decreasing effective temperature with depth.

The relationships shown in figure 17-9 are presented in a slightly different manner in figure 17-10, where $T_{g/aar}$ values for 1 and 2 m are plotted against 30-year mean annual temperatures (T_a). The regressions for these relationships are as follows:

For unshaded sites:

1 m:

$$T_{g/aar} = 5.47 + 0.844(T_a),\ R^2 = 0.980 \quad (17\text{-}10)$$

2 m:

$$T_{g/aar} = 3.31 + 0.908(T_a),\ R^2 = 0.984 \quad (17\text{-}11)$$

For shaded sites:

1 m:

$$T_{g/aar} = 4.71 + 0.788(T_a),\ R^2 = 0.970 \quad (17\text{-}12)$$

2 m:

$$T_{g/aar} = 2.51 + 0.881(T_a),\ R^2 = 0.978 \quad (17\text{-}13)$$

For shaded and unshaded sites combined:

1 m:

$$T_{g/aar} = 4.95 + 0.809(T_a),\ R^2 = 0.925 \quad (17\text{-}14)$$

2 m:

$$T_{g/aar} = 2.72 + 0.898(T_a),\ R^2 = 0.953 \quad (17\text{-}15)$$

As in the discussion of figure 17-9, there can be some uncertainty about the characteristics of "shaded" sites, so the significance of the statistics in this data set may be somewhat limited. Nevertheless, the relationships between the slopes or intercepts in equations 17-4–17-15 are all internally consistent with the physical characteristics of the recording sites and with theoretical trends of effective temperature vs. latitude.

DISCUSSION

Figures 17-9 and 17-10 represent the primary goal of this paper: to quantify the relationship between ground and air temperatures over a broad latitude range of the U.S. ACP. Figures 17-7–17-10 all demonstrate that effective ground temperatures at the study depths are generally between 0.5 and 3°C greater than long-term mean annual air temperatures at the sites. Care must be taken not to overinterpret the regressions presented with these figures, as the ground temperature records are limited in both number of sites and length of record. The following sections focus on specific aspects of the results presented here.

The Relationship Between Ambrose Cell Data at 2 m and 30-year Mean Air Temperature

Figure 17-11 presents the Ambrose cell data ($T_{g/amb}$) at a 2 m depth (the depth for which we have greatest confidence in the results, as reviewed in the *Methods* section) vs. the 30-year mean annual

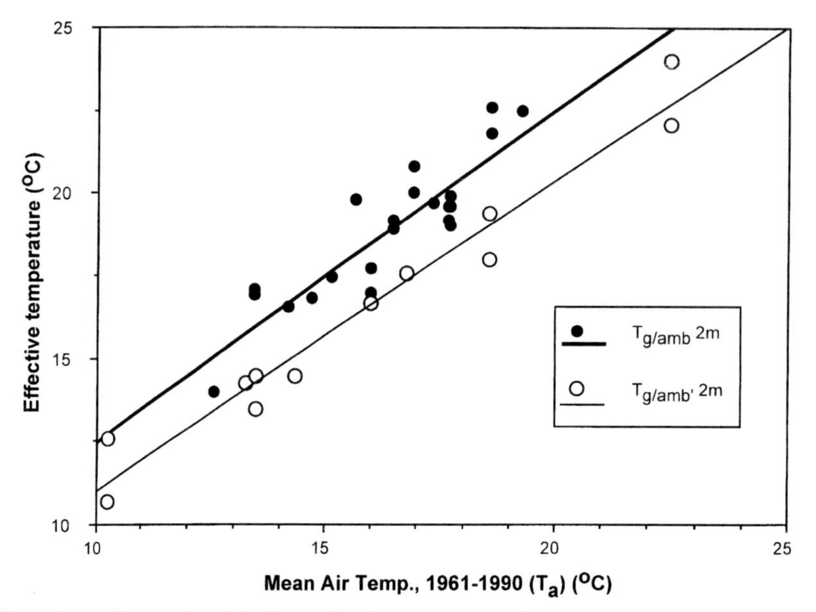

Fig. 17-11. Comparison of mean air and Ambrose effective temperatures. Thirty-year mean air temperatures (T_a) as in figures 17-3 and 17-9; Ambrose cell data ($T_{g/amb}$) for 2 m depth (interval 1994–1995), plotted as solid circles. Ambrose effective temperatures ($T_{g/amb'}$) derived from direct recording data (table 17-3), shown as open circles. The systematic difference between the $T_{g/amb}$ and $T_{g/amb'}$ results is a consequence of the general difference in air temperatures for the two measurement intervals (figure 17-4). Data for all sites (exposed and shaded) are shown. Regressions for combined data (shaded and exposed) are as follows: 2 m Ambrose (obs) : $T_{g/amb} = 2.42 + 1.0\,(T_a)$, $R^2 = 0.79$; 2 m Ambrose (calc): $T_{g/amb'} = 1.64 + 0.94\,(T_a)$, $R^2 = 0.96$. The regression shown for $T_{g/amb'}$ is for a wider range of mean air temperatures (i.e., latitude) than that for $T_{g/amb}$; the regression of $T_{g/amb'}$ vs. T_a for the actual narrower range of latitude and mean air temperatures of the Ambrose cell sites is: 2 m Ambrose (calc): $T_{g/amb'} = 1.83 + 0.91\,(T_a)$, $R^2 = 0.94$.

air temperature (T_a). Also plotted are the Ambrose effective temperatures derived from electronic observations beginning in 1996 ($T_{g/amb'}$; see table 17-3). The Ambrose 2 m effective temperatures ($T_{g/amb}$) measured for the 1994–1995 interval consistently plot above the values calculated from electronic data obtained in subsequent years. This difference (reflected in the intercept values of the regressions shown in the caption for figure 17-11) represents the difference in air temperatures (2–3°C; see above) during the different intervals of measurement (figure 17-4). Early in this study, the relationship between the Ambrose data and mean air temperature, combined with the depth profiles of Ambrose temperatures ($T_{g/amb}$) at Fort Johnson and Skidaway (figure 17-7) initially led to the conclusion that there might be a significant difference in gradient between ground and air temperatures between the mid-Atlantic and southeastern regions of the U.S. (Wehmiller et al., 1996). More recent electronic data, and reconsideration of the quality of some of our initial measurements, indicate that this conclusion is probably not correct. Further data (both longer time-series and a greater density of recording sites) should resolve this question.

Direct Comparisons of Ambrose Cell Data with Electronic Data; Implications for Measurement Methods

During the course of this study, there were a few intervals during which Ambrose cells and electronic or direct recording devices were installed at a single site simultaneously. Only at the Wilmington, DE site was there continuous direct recording of ground temperature during the entire interval of installation of an Ambrose cell at 1 m. The data for this comparison are shown in figure 17-12. In this case, the Ambrose effective temperature calculated for 1 m from the direct measurements ($T_{g/amb'}$) is slightly (approx. 0.2°C) above the Ambrose cell determination ($T_{g/amb}$) for this depth.

Other comparisons of Ambrose cell data with effective temperatures from electronic recording are shown in figure 17-12 (for Greenville and Cape Hatteras, NC), but these comparisons are less conclusive because there was no overlap in the time intervals of the recorders and the Ambrose cells. In these cases, the difference between electronic recorders and Ambrose cells is probably the result of the general difference in air temperatures for the Ambrose interval (1994–1995) compared with the electronic interval (1996–1998),

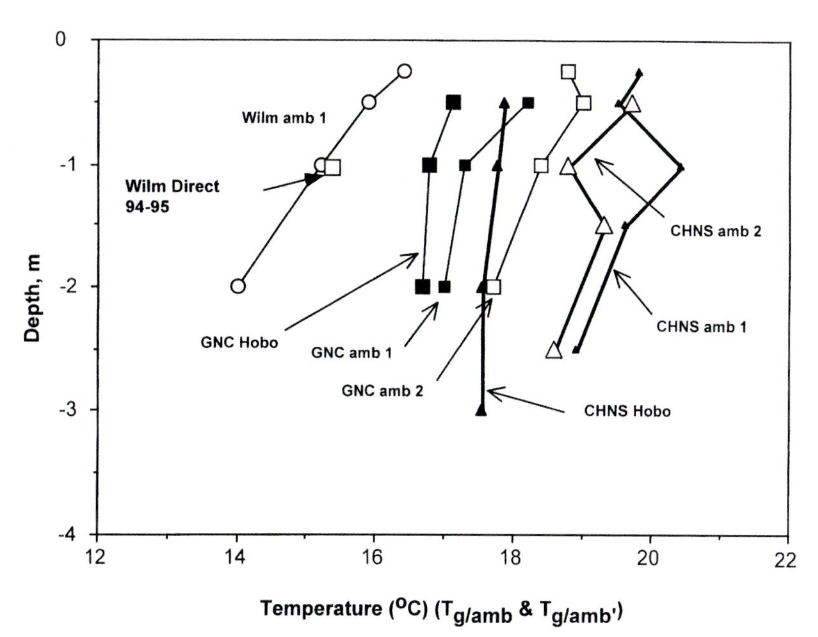

Fig. 17-12. Example of a comparison of Ambrose cell data with direct recording data the same site. Ambrose data for Wilmington, DE, and Greenville and Cape Hatteras, NC, are compared with direct records at these same sites. Only the Wilmington data are for identical measurement intervals, and the agreement between the two methods for the 1 m depth (the only depth at which direct data were obtained) is 0.2°C. See figure 17-7 for Ambrose data; GNC and CHNS Hobo data are from table 17-3. Wilmington "Direct" represents daily data, recorded via a buried electric thermometer, for the same measurement interval as the Ambrose data. Hobo recorders were installed at the Wilmington site for comparison with the buried electric thermometer, but Hobo recorders did not operate for the entire interval for which Ambrose cells were installed at this site.

as summarized in figure 17-4. Similar conclusions can be derived from comparisons of Ambrose and electronic data for Lewes/Milton, DE, Gomez Pit, VA, and Conway, SC, the three other areas where both electronic recorders and Ambrose cells have functioned during intervals of at least partial overlap.

Interannual Differences in Ground Temperature

Because the observation intervals summarized in table 17-1 span several years, it is possible to compare the relationships of interannual ground temperatures to interannual air temperature records. Figure 17-13 presents the 1996, 1997, and 1998 2 m depth temperatures for the Cape Hatteras site. The data in figure 17-13 can be compared with the mean monthly air temperature data for these three years (figure 17-4). The contrast in spring air temperatures was also recorded in the ground temperatures, even at 2 m. Using the maxima and minima from annual records as reference points for sinusoidal modeling of effective temperature, the 1997 $T_{g/aar}$ is approximately 0.8–0.9°C cooler than either the 1996 or 1998 effective temperatures. Similar effects are noted in the results for Charleston, SC, Greenville, NC, and Lewes, DE, the other sites for which two years of data are available.

Exposure and/or Distance-from-Coast Effects

Figures 17-14, 17-15, and 17-16 demonstrate the magnitude of local variability that can be observed for measuring sites that are at the same latitude or otherwise close enough to be considered to have "identical" temperature histories for any aminostratigraphic application. Figure 17-14 shows the differences in the annual 2 m ground temperature cycle for Greenville and Cape Hatteras, NC. All depths at these two sites show that the coastal site (Cape Hatteras) has warmer effective temperatures than the inland site, even though both are at essentially the same latitude. The difference in effective ground temperatures ($T_{g/aar}$) between Greenville and Cape Hatteras (0.9°C) is larger than the differences in both mean (approx. 0.5°C) and effective air temperatures between the two sites (see table 17-2).

Figure 17-15 presents the three-year 2 m ground temperature record for two sites, in the Charleston, SC area, with different drainage and exposure conditions. The Fort Johnson (Ft J) site for electronic recording is in well-drained Pleistocene coastal sands with partial shade (the Ambrose site at Fort Johnson was in the same sediments, in a more

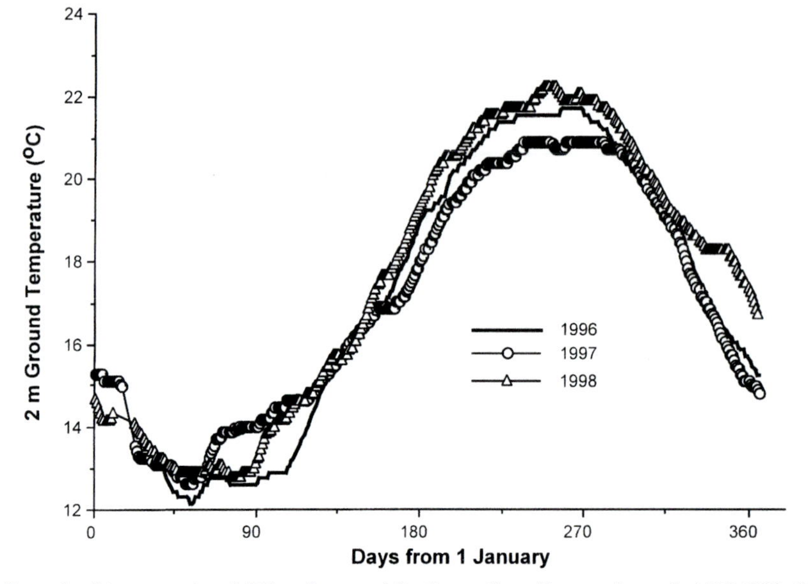

Fig. 17-13. Example of interannual variability: the record for 2 m at Cape Hatteras for early 1996–1998. See figure 17-4 for comparison of mean monthly air temperature differences for this interval, particularly for the early months of each year.

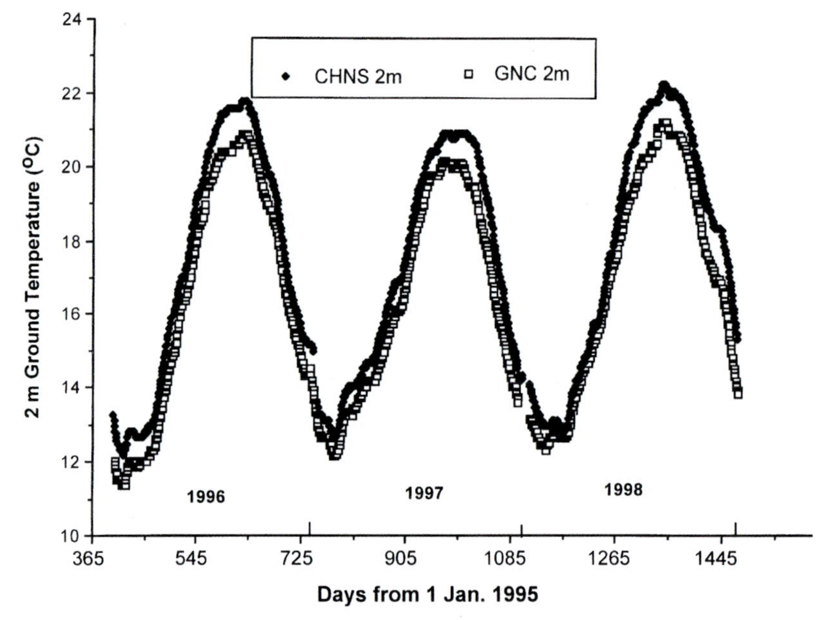

Fig. 17-14. Three years of direct recording at 2 m for coastal (CHNS) and inland (GNC) sites in central North Carolina (Cape Hatteras and Greenville, respectively). The major difference between the records is in the summer maximum; the difference in effective temperatures ($T_{g/aar}$) for the two sites is consistent with the difference in mean air temperatures for the two sites (table 17-3).

exposed site probably affected by underground utilities). The Forestry Experiment Station (FES) is also in Pleistocene coastal sands, but in a poorly drained and well-shaded location. There is a difference of more than 1.5°C in the AAR effective temperature ($T_{g/aar}$) for these two sites (table 17-3), demonstrating the effect of the major difference in summer maxima for the two sites.

Ambrose effective temperatures ($T_{g/amb}$) for three central North Carolina sites (all located within 1° latitude) are shown in figure 17-7C. Although there is some scatter for each Ambrose probe, the major conclusion to be drawn from these results is that exposure effects can be very significant. The difference between the Greenville and Duck, NC 2 m Ambrose cell effective temperatures ($T_{g/amb}$) is

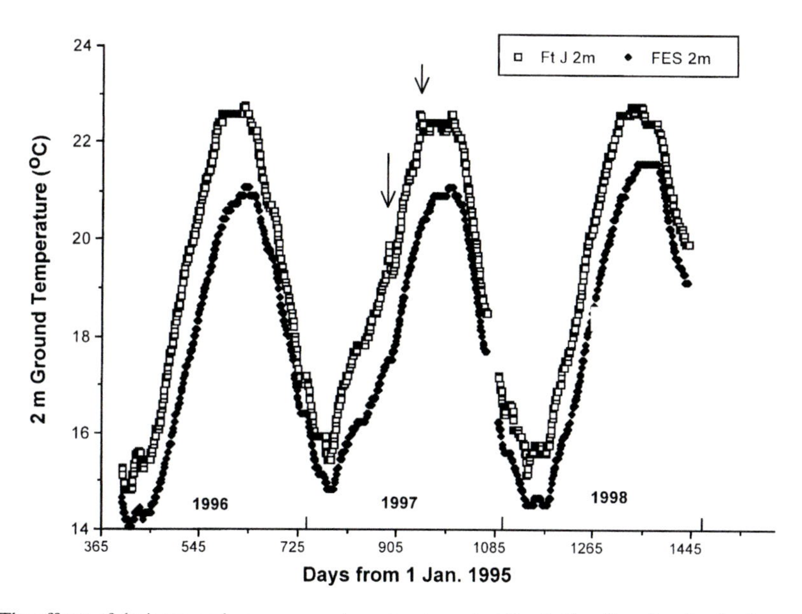

Fig. 17-15. The effects of drainage and exposure on two sites in central South Carolina: 2 m depth, three years of recording. The Fort Johnson site (Ft J) is in well-drained sandy soil, with partial shade. The Forest Service site (FES) is shaded with poor to moderate drainage. The difference in effective temperatures ($T_{g/aar}$) for these two records is approximately 1.4° (table 17-3). Arrows indicate intense rainfall events that were recorded by abrupt temperature changes at all depths at Ft J; the permeability at this site allows these events to be recognized throughout the depth range recorded, if there is sufficient contrast between rainwater temperature and ground temperature at the time of the event.

approximately 3°C. Given the differences in activation energies for Ambrose cell hydration and AAR (equations 17-2 and 17-3), the AAR effective temperature ($T_{g/aar}$) difference between these sites is somewhat greater than 4°C. The difference between the two Ambrose probes at Greenville is due to contrasting exposure conditions; the two Cape Hatteras probes were in identical exposure conditions (well-drained, partial shade), as were the Duck probes (exposed coastal dune).

Other sites also show the effects of exposure differences, as demonstrated in table 17-3 and discussed in association with the regressions presented in figure 17-9. The WH and DMC shaded and exposed ground temperature data indicate that these effects can be quite significant in higher latitude sites. The LC and CHSP results (Lewes, DE) also represent exposed and shaded sites, respectively, and the shaded Custis, VA site has slightly lower effective temperatures than the exposed LC site, even though LC is north of Custis. The recently obtained Tampa (FLA) data are important because they extend the latitude range of our observations of the exposure effect. Other examples of shaded vs. exposed contrasts in ground temperature are found in Norton et al. (1981); differences of 2–3°C were often observed at a 2 m depth for sites with extreme differences in exposure.

The Ground Temperature Record in Higher Latitude Sites

Figures 17-2A, B demonstrate the differences in curve shapes for the winter ground temperature minima for sites with contrasting winter climatology. Figure 17-2B, for the DMC site in coastal Maine, shows the blanketing effect of snow cover in early 1997 (cf. Goodrich, 1982; Goulden et al., 1998), permitting the shallow subsurface to remain warmer than in the absence of snow cover. Similar effects were noted in a short-term (6-week) deployment of recorders (sampling at 1 hr intervals) for test purposes at multiple depths (between 0.1 and 0.5 m) at Lewes, DE during the winter of 1995–1996, when snow cover was observed to maintain ground temperatures above those expected from the downward trends prior to the snowfall (H. A. Stecher, III, unpublished). Although our results are quite limited, it appears from the trends in figures 17-9, 17-10, and 17-17 (below), that there is greater variability in $T_{g/aar}$ at the higher latitude sites, presumably because of the effect of contrasting exposures on snow cover, which can lead to significant differences in winter ground temperature minima. Although the temperatures shown in figure 17-2B never actually reached the freezing point, it is quite likely that the temperature record shown in figure 17-2B was also affected by the thermal buffering effect of freezing at shallower depths (the "zero curtain" effect, discussed by Outcalt et al.,

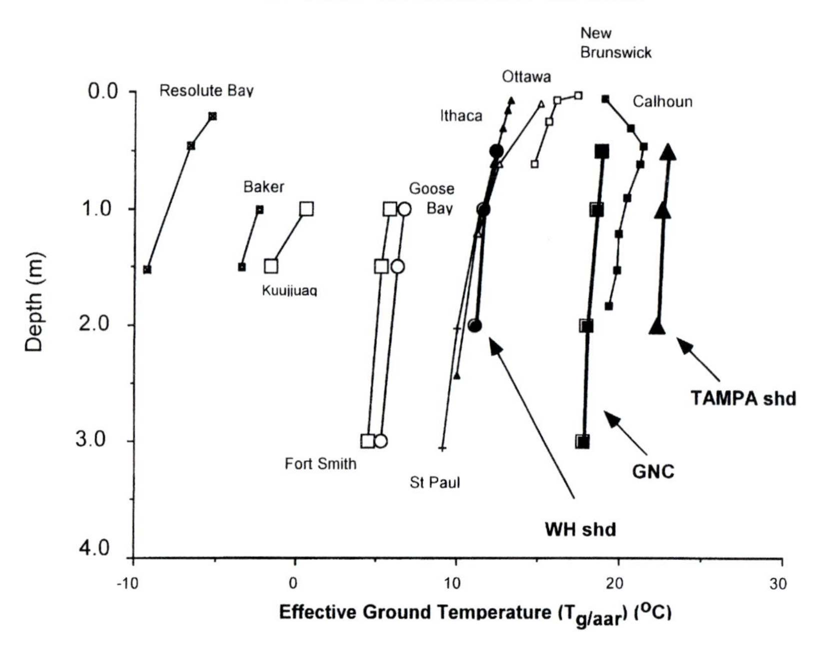

Fig. 17-16. Comparison with data from other sites, eastern or central North America. Data are from Chang (1958), Environment Canada (1984), Baker and Ruschy (1993), and this work. Effective ground temperatures ($T_{g/aar}$) are derived from published mean monthly ground temperatures or from table 17-3 for Woods Hole, Greenville, and Tampa. All sites except Calhoun, SC show steady decreases in effective temperature with depth, and the results obtained in this work are quite consistent with those obtained by earlier workers. The mean annual air temperatures for the regions represented by these sites are as follows (Willmott et al., 1981): Resolute Bay, NWT, −16.1°C; Baker, NWT, −12.2°C; Kuujjuaq, QUE, −5.6°C; Fort Smith, NWT, −3.6°C; Goose Bay, LAB, −0.2°C; Ottawa, Ont, 5.4°C; St. Paul, MN, 7.4°C; Ithaca, NY, 9°C; New Brunswick, NJ, 13°C; Woods Hole, MA, 10.2°C; Greenville, NC (GNC), 16°C; Calhoun, SC, 18.5°C; Tampa, FLA, 22.4°C.

1990). The combination of our results and especially those from other sources (e.g., Chang, 1958; Environment Canada, 1984; Goulden et al., 1998) indicates that the difference between effective ground temperature ($T_{g/aar}$) and mean air temperature (T_a) at high latitude is greater than at low latitudes because winter air temperatures fall far below freezing, whereas winter ground temperatures remain relatively high because of snow cover and/or latent heat effects that delay or mitigate cooling.

Comparison with Other Ground Temperature Data for Eastern North America

Effective ground temperature ($T_{g/aar}$) data for other sites in eastern North America are plotted in figure 17-16. Data for three sites (Woods Hole, Greenville, and Tampa) are also shown for comparison. Effective temperatures were calculated for mean monthly temperatures reported in the references cited in the figure 17-16 caption. With the exception of the Calhoun, SC site, all data show the expected trend of decreasing effective temperature with depth. Figure 17-17 presents the data shown in figure 17-16 in comparison to air and ground temperatures, including air and effective ground temperature data and regressions from figures 17-3 and 17-9. Although the number of data points for the higher latitude range (north of 43°N.) shown in figure 17-17 is rather small, it appears that there is greater variability in $T_{g/aar}$ (both coastal and inland sites are included in the plot, the latter apparently being more affected by ground freezing) and perhaps a lower slope of $T_{g/aar}$ vs. latitude compared with the lower latitude trend seen in both figures 17-9 and 17-16. The air and ground temperature data presented in figure 17-17 can also be represented by single linear regressions as follows:

Air: $$T_a = 47.88–0.906(\text{lat.}\ ^\circ\text{N}) \quad R^2 = 0.985 \qquad (17\text{-}16)$$

Ground:

$$T_{g/aar} = 41.86–0.712(\text{lat.}\ ^\circ\text{N}) \quad R^2 = 0.921. \qquad (17\text{-}17)$$

The contrast between equations 17-16 and 17-17 and those given for the regressions in the two separate latitude regions plotted in figure 17-17 is not large, hence there may not be any statistically sig-

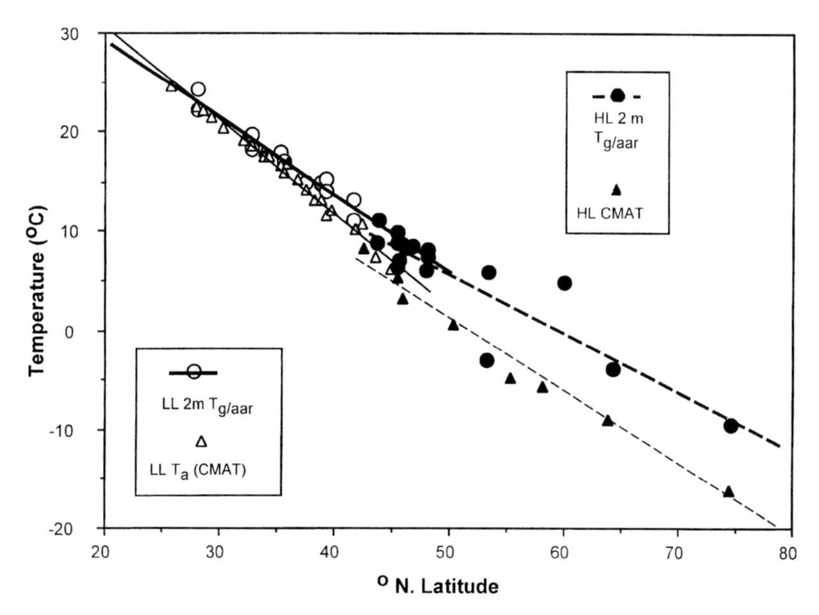

Fig. 17-17. Mean air and effective ground temperatures for low-latitude (LL) and high-latitude (HL) sites along the U.S. Atlantic coast (from figure 17-3) and eastern and central Canada (from figure 17-16 and associated references). The high-latitude effective ground temperatures at 2 m depth are interpolated or extrapolated from the trends seen in figure 17-16. The regressions on the four lines shown in this figure are as follows. LL T_a (low-latitude current mean annual air temperature; from figure 17-3): $T_a = 48.74 - 0.922(\text{lat.})$, $R^2 = 0.985$ [latitudes $< 45°$ N]. LL 2 m $T_{g/aar}$ (2 m effective temperature, shaded and unshaded combined, from figure 17-9): $T_{g/aar} = 45.13 - 0.786(\text{lat.})$, $R^2 = 0.950$ [latitudes $< 45°$ N]. HLT_a (high-latitude current mean annual air temperature; from figure 17-3): $T_a = 38.42 - 0.747(\text{lat.})$, $R^2 = 0.979$ [latitudes $> 43°$ N]. HL 2 m $T_{g/aar}$ (2 m effective temperature, from figure 17-16): $T_{g/aar} = 35.67 - 0.598(\text{lat})$, $R^2 = 0.813$ [latitudes $> 43°$ N].

nificant difference in the air/ground temperature relationships for these two regions. Nevertheless, the variability in the effective ground temperature at higher latitude sites is important for modeling the thermal history of Quaternary samples (see below).

Implications of Ground Temperature Differences for Sites with Similar or Identical Current Mean Annual Air Temperatures

The results presented here indicate that differences in effective ground temperature of 1°C or more can be encountered for sites that would usually be interpreted to have "similar or identical" mean air temperatures (i.e., sites at the same latitude). Examples of these contrasts can be seen in the Greenville/Cape Hatteras/Duck data (figures 17-7C, 17-8, and 17-14) and the Ft J and FES data (figures 17-8 and 17-15). The extreme difference of 3–4°C. between Duck and Greenville can be explained by the contrasting environments of these two sites, but such a difference should serve as a reminder of the potential significance of these contrasts.

It is useful to consider the potential magnitude of local variability in ground temperatures on numerical age estimates derived from AAR data. These calculations always require selection of a particular kinetic model and associated assumptions about the temperature dependence of the racemization rate constants. An example calculation is depicted in figure 17-18, using a parabolic kinetic model (Mitterer and Kriausakul, 1989) and the assumption of an 8%/deg. variation in the parabolic rate constant (Kaufman and Brigham-Grette, 1993). This calculation is focused on *only* the question of local variability in ground temperatures, and not on the more difficult problem of contrasts in climate histories for samples with different ages but similar modern temperatures.

Assuming that a given region can be characterized by a "single" mean annual air temperature, figure 17-18 shows how a ±8% variation in a parabolic rate constant can affect age estimates. The sample calculation is as follows: a sample with a D/L value of 0.25 and an assigned (calibration) age of 100,000 years is used to estimate the age of sample with a D/L value of 0.50. If the two samples are assumed to have identical effective temperatures, the more racemized sample will have an age estimate of 400,000 years. For 1.0°C cooler or warmer effective temperatures, the older sample would be estimated to be either 342,000 or 467,000 years old, respectively. These calculations demonstrate that a ±1.0°C uncertainty in modern effective ground temperatures could explain most or all of the variation (usually of the order of ±10%) in a

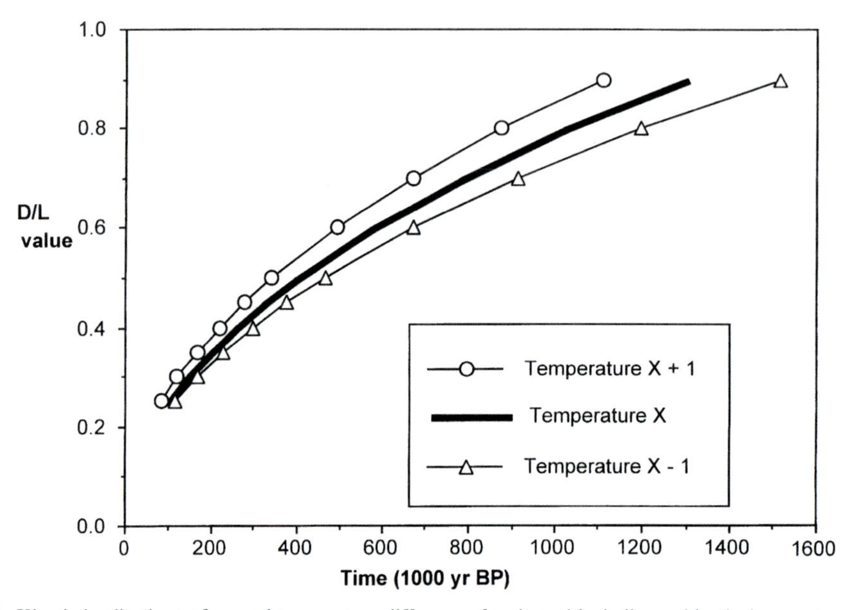

Fig. 17-18. Kinetic implications of ground temperature differences for sites with similar or identical current mean annual air temperatures. Parabolic kinetics are assumed, with a temperature dependence of 8%/°C in the parabolic rate constant. The two curves above and below the central curve represent the envelope of uncertainty produced by a ±1°C range in effective diagenetic temperature. See text for discussion.

specific aminozone for an area of similar modern mean air temperatures (e.g., Wehmiller et al., 1988, 1992), without invoking diagenetic effects on D/L variability. Greater aminozone variability can be explained by the larger differences in effective temperature that can be encountered in comparisons of results for samples with substantially different burial depths, especially when contrasting exposure also affects the sites (compare Greenville and Duck, NC). These differences can be particularly important for comparisons of results for Holocene samples, because of the potentially large proportion of their history that could have been at elevated temperatures associated with shallow burial (Wehmiller, 1977; Miller and Brigham-Grette, 1989). Additionally, there is often a certain amount of variability in an aminozone that can be "inherited" from the early phases of accumulation of the analyzed samples, simply because of the finite range of ages of samples in a "single unit" and the variability in early thermal history for these samples because of different burial depths (see figure 17-6 of Wehmiller et al., 1995).

Implications of Modern Ground Temperatures for Modeling of the Temperature Histories of Pleistocene Fossils

Figures 17-3, 17-9, and 17-17 indicate that the modern latitudinal gradient of mean annual air temperature is steeper than that of effective air or 2 m ground temperature, so the difference between effective ground temperature and mean air temperature increases with latitude. The greater seasonal variation of solar insolation at higher latitudes causes a greater difference between effective and mean air temperatures at these latitudes, and high-latitude effective ground temperatures remain substantially higher than mean air temperatures because of the thermal "buffering" provided by the latent heat of ground ice formation (see, for example, figure 1 of Goulden et al., 1998). Because a *steepened* latitudinal gradient of effective temperatures between approx. 28 and 36°N must be invoked to reconcile the correlation of independently dated aminozones along the US Atlantic coast (Wehmiller et al., 1988; Hollin and Hearty, 1990; Wehmiller, 1993, 1997), it is important to speculate on the possible causes of such steepening, which has not been required to explain the gradients of D/L values observed in two other north–south transects in the U.S. (Kennedy et al., 1982; Oches et al., 1996).

Because the difference between effective and mean temperatures is directly related to seasonal amplitude, it is theoretically possible to reduce effective temperatures simply by reducing the seasonal amplitude, without changing the mean temperature. A simple model of the Pleistocene climate history for coastal sites along the U.S. ACP (approx. latitude 40–30° N) would suggest that northern climate zones (those characterized by the higher-latitude trends seen in figure 17-17) shifted southward to this region during full glacial intervals (see, for example, Bartlein et al., 1998), while the coastal sites themselves shifted inland as sea level fell. During these colder intervals, the mean temperatures of the ACP sites would have decreased (relative to today) but the decrease in effective temperature, or the change in latitudinal gradient of effective temperature, would have depended upon the relative changes in seasonal insolation at different latitudes and on possible shifts to a different relationship between air and ground temperatures (like that represented for the higher latitude region of figure 17-17).

A detailed model of possible late Quaternary variation in seasonal insolation along the U.S. ACP is beyond the scope of this chapter, but some estimate of the potential magnitude of variation in seasonality can be made using the data from Berger and Loutre (1991). Data for late Quaternary solar insolation at latitudes 30 and 60° N (Berger and Loutre, 1991) indicate that seasonality differences in radiation (expressed as June–December differences in Wm^{-2}) were in phase over the past 130 kyr, but that the magnitudes of these differences were not always identical. For most of this time interval, the June–December difference at both 30 and 60° N was greater than the modern difference (approx. 200 Wm^{-2}). Furthermore, the June–December difference was usually about 10 $\pm 10\,Wm^{-2}$ greater at 30° N than at 60° N, potentially yielding somewhat steeper (< 10%?) latitudinal gradients of effective temperature if no other factors (such as atmospheric circulation, snow cover, or formation of ground ice) acted to modify the impact of these insolation variations. On the basis of the estimates of actual temperature reductions along the ACP from Bartlein et al. (1998), the seasonality contrast and resultant change in the late Pleistocene effective temperature gradient do not appear to be large enough to reconcile the steeper-than-present (25% or greater?) latitudinal correlation of ACP-calibrated aminozones (Wehmiller et al., 1988; Hollin and Hearty, 1990; Wehmiller et al., 1992; Wehmiller, 1993, 1997).

The observations presented here focus on the thermal environment of samples along the U.S. ACP, but these discussions have broad implications for any model of the Quaternary history of samples from other environments. A variety of thermal histories can be envisioned for the Quaternary fossils studied by aminostratigraphers. Even for the group of marine mollusks, all of which were initially deposited under water, sample histories ranging from burial under ice (glaciers) to full exposure in cemented reef or beach-rock can be envisioned. These factors probably contribute to uncertainties in regional aminozones that are greater than would be predicted from analytical uncertainties alone. The wide range of D/L values observed for several Eemian sites in northwestern Europe (Miller and Mangerud, 1985; Sejrup and Haugen, 1994) may be a consequence of widely contrasting thermal histories in an environment where ice cover and dramatic seasonality contrasts (compared with today) have been recorded. In spite of the fact that differences between modern mean air temperature for sites within a study region are the most convenient proxy for differences in temperature histories of samples from that region, the factors outlined here can "decouple" this relationship. Extrapolation of theoretical aminostratigraphic correlation trends over wide latitude or thermal ranges must be done with an appreciation of the potential uncertainties inherent in these extrapolations.

CONCLUSIONS

The major focus of this Chapter has been to present methods for recording ground temperatures and establishing the relationship between effective ground temperatures and long-term mean air temperatures for the U.S. ACP. The aminostratigraphic data for this area are extensive but problematic, because latitudinal correlations of calibrated enantiomeric ratios are not wholly consistent with modern latitudinal gradients of mean air temperature. Although the ground temperature data reported here do not permit detailed modeling of the entire late Quaternary temperature history for the region, the following conclusions seem appropriate.

1. Modern effective ground temperatures for the region can be predicted by a simple linear relationship to modern air temperatures, and mean monthly air temperature data (rather than mean annual or mean daily temperatures) are appropriate for these comparisons.
2. Effective ground temperatures are greater than long-term mean air temperatures and decrease with depth (figure 17-8).
3. The difference between effective ground temperatures and mean air temperatures increases with latitude; this relationship can be perturbed by ground-freezing effects and/or ice cover (figures 17-9, 17-10, and 17-17).
4. Local influences can introduce significant differences in the profile of effective ground temperature vs. depth (figures 17-8 and 17-9); depths above the water table are particularly vulnerable to these factors. Our results do not include observations from below the water table (with the exception of one site), although it is likely that many of the ACP mollusk samples used for AAR studies have been below the water table for much of their history.
5. Proximal sites (those that would be assumed to have "similar or identical temperature histories" in aminostratigraphic correlation) with similar modern exposure conditions are likely to have modern effective ground temperatures within 1°C (for similar depths); extreme contrasts in

modern effective ground temperatures for proximal sites might be as large as 4°C.

6. For regional aminostratigraphic studies, it is strongly recommended that a suite of sites be used for temperature recording, representing a broad range of modern air temperatures, depths and exposure conditions. Ideally, records of at least one full year should be obtained, although longer records will yield more reliable averages for a given site or region; these longer records, if continuously maintained, may also provide insights regarding future environmental change on both local and regional scales (Berger, 1998), as interannual variabilty in air temperatures is also reflected in ground temperatures.
7. Although histories of late Quaternary vegetation cover or soil moisture may be as difficult to quantify as those for late Quaternary temperature change at a specific site, it is important that aminostratigraphic data for sites with similar modern air temperatures but potentially different ground temperature histories be compared within the context of these possible differences. Depths >2 m are ideal for sampling, >1 m are acceptable. If shallowly buried samples must be analyzed, comparisons of AAR data from these samples with data from other more deeply buried samples of the same age should be made if at all possible. The impact of shallow burial on the AAR results is a function of the relative proportion of a sample's age spent at elevated ground temperatures; for Pleistocene samples this interval may be less than 5%, but for Holocene samples it might be 50% or more.
8. Correlations of aminostratigraphic data for regions of contrasting ground temperature history (involving burial depth, exposure, ground moisture, or ground-freezing effects) can have inherent uncertainties that could be quite large. Isochrons that predict D/L as a function of latitude or temperature usually invoke modern air temperatures as the controlling variable (e.g., Hollin and Hearty, 1990; Wehmiller, 1992; Murray-Wallace, 2000); the discussion related to figure 17-18 implies that an envelope of uncertainty of *at least* 10% of the isochron D/L value should be included in these broad correlation models.

Although this study was initiated using the Ambrose cell techniques, it has become clear that a more informative tool for ground temperature studies is the remote electronic data-logger that we began to incorporate into our study in 1995. Most of the conclusions outlined above are based on the electronic records only, although many could also be derived from Ambrose data alone. As the major costs of regional ground temperature surveys such as that presented here are in the logistics of travel and instrument deployment, the benefits of the electronic recorders far outweigh the relatively small cost difference between the electronic and Ambrose equipment. Consequently, it is recommended that future studies of ground temperature for aminostratigraphic purposes employ electronic recorders wherever possible, although some combination of Ambrose and electronic recorders could be used if necessary.

Acknowledgments: Partial support for this work from National Science Foundation Grant No. EAR9315052 to the University of Delaware is gratefully acknowledged, as is support from the College of Arts and Science, University of Delaware. Special thanks go to all those who have assisted in the extensive field efforts involved in this project: Eric Oches, University of South Florida; Daniel Belknap, University of Maine; Scott Harris, Coastal Carolina University; June Mirecki, College of Charleston; James Black, Todd Keyser, and Andrew Urie, University of Delaware; Steve Colman, Robert Oldale, and Rob and Kama Thieler, U.S. Geological Survey, Woods Hole, MA; Marcia Lyons, Cape Hatteras National Seashore; Paula Wehmiller. Additionally, the following individuals have assisted with site selection and the long-term observation of recorder installations: Stanley Riggs, East Carolina University; William Birkemeier, U.S. ACE, Field Research Facility, Duck, NC; Cliff Davis, Duke University Marine Laboratory; William Cleary, UNC, Wilmington; Paul Gayes, Coastal Carolina University; Dennis Allen and Joe Schubauer-Berigan, Baruch Institute; Craig Onqué and Steve Temmermand, Wallops Island Marine Consortium; Terry Thompson and Randy Carlson, Nature Conservancy/LTER; Foster Fulsom, Fort Johnson Marine Laboratory; Carl Trettin, U.S. Forest Service; Lee Knight, Skidaway Institute of Oceanography; Patrick Cooper, Cape Henlopen State Park; Mark Noll, Dover Air Force Base. Special thanks go to Gifford Miller and Eric Oches for many helpful comments during the review of the manuscript.

REFERENCES

Bada J. L. and Schroeder R. A. (1972) Racemization of amino acids in calcareous marine sediments: kinetics and mechanism. *Earth Planet. Sci. Lett.* **15**, 1–11.

Baker D. G. and Ruschy D. L. (1993) The recent warming in eastern Minnesota shown by ground temperatures. *Geophys. Res. Lett.* **20**, 371–374.

Bartlein P. J., Anderson K. H., Anderson P. M., Edwards M. E., Mock C. J., Thompson R. S., Webb R. S., Webb T., III, and Whitlock C. (1998) Paleoclimate simulations for North America over the past 21,000 years: features of the simulated climate and comparisons with paleoenvironmental data. *Quat. Sci. Rev.* **17**, 549–585.

Berger A. R. (1998) Environmental change, geoindicators, and the autonomy of Nature. *GSA Today* **8(1)**: 3–8.

Berger A. and Loutre M. F. (1991) Insolation values for the climate of the last 100 million years. *Quat. Sci. Rev.* **10**, 297–317.

Buntebarth G. (1984) *Geothermics*. Springer-Verlag.

Chang J.-H. (1958) *Ground Temperatures*, vols I and II. Blue Hill Meteorological Observatory, Harvard University, Massachusetts, U.S.A.

Chapman D. S., Chisholm T. J., and Harris R. N. (1992) Combining borehole temperature and meteorological data to constrain past climate change. *Palaeogeog. Palaeoclimat. Palaeoecol.* **98**, 269–281.

Corrado J. C., Weems R. E., Hare P. E., and Bambach R. K. (1986) Capabilities and limitations of applied aminostratigraphy as illustrated by analyses of *Mulinia lateralis* from the late Cenozoic marine beds near Charleston, South Carolina. *S. Carolina Geol.* **30**, 19–46.

Ellis G. L., Goodfriend G. A., Abbott J. T., Hare P. E., and von Endt, D. W. (1996) Assessment of integrity and geochronology of archeological sites using amino acid racemization in land snail shells: examples from central Texas. *Geoarchaeology* **11**, 189–213.

Environment Canada (1984) *Canadian Climate Normals*, vol. 9. 1951–1980. Ottawa, Canada. Canadian Government Publishing Center.

Goodfriend G. A. (1987) Chronostratigraphic studies of sediments in the Negev Desert using amino acid epimerization analysis of land snail shells. *Quat. Res.* **28**, 374–392.

Goodrich L. E. (1982) The influence of snow cover on the ground thermal regime. *Canadian Geotechnical J.* **19**, 421–432.

Goulden M. L., Wofsy S. C., Harden J. W., Trumbore S. E., Crill P. M., Gower S. T., Fries T., Daube B. C., Fan S.-M., Sutton D. J., Bazzaz A., and Munger J. W. (1998) Sensitivy of boreal forest carbon balance to soil thaw. *Science* **279**, 214–217.

Hollin J. T. and Hearty P. J. (1990) South Carolina interglacial sites and Stage 5 sea levels. *Quat. Res.* **33**, 1–17.

Hollin J. T., Smith F. L., Enourf J. T., and Jenkins D. G. (1993) Sea-cave temperature measurements and amino acid geochronology of British late Pleistocene sea stands. *J. Quat. Sci.* **8**, 359–364.

Hwang S.-J. (1995) The effects of soil moisture on the energy balance at the bare soil surface. *Environ. Res. Cent. Res. Pap.* **17**, 1–51.

Kaufman D. L. and Brigham-Grette J. (1993) Aminostratigraphic correlations and paleotemperature implications, Pliocene–Pleistocene high sea-level deposits, northwestern Alaska. *Quat. Sci. Rev.* **12**, 21–34.

Kawanishi H. (1983) Annual variations of ground and streamflow temperatures in small forested valleys. *J. Hydrol.* **66**, 253–267.

Kennedy G. L., Lajoie K. R., and Wehmiller J. F. (1982) Aminostratigraphy and faunal correlations of late Quaternary marine terraces, Pacific coast U.S.A. *Nature* **299**, 545–547.

Lewis T. (1998) The effect of deforestation on ground surface temperatures. *Global Planet. Change* **18**, 1–13.

Miller G. H. and Brigham-Grette J. (1989) Amino acid geochronology: resolution and precision in carbonate fossils. *Quat. Intl.* **1**, 111–128.

Miller G. H. and Mangerud J. (1985) Aminostratigraphy of European marine interglacial deposits: *Quat. Sci. Rev.* **4**, 215–278.

Mirecki J. E., Wehmiller J. F., and Skinner A. (1995) Geochronology of Quaternary Coastal Plain deposits, southeastern Virginia, USA. *J. Coastal Res.* **11**, 1135–1144.

Mitterer R. M. (1975) Ages and diagenetic temperatures of Pleistocene deposits of Florida based upon isoleucine epimerization in *Mercenaria. Earth Planet. Sci. Lett.* **28**, 275–282.

Mitterer R. M. and Kriausakul N. (1989) Calculation of amino acid racemization ages based on apparent parabolic kinetics. *Quat. Sci. Rev.* **8**, 353–357.

Murray-Wallace C. V. (2000) Quaternary coastal aminostratigraphy—Australian data in a global context. In *Perspectives in Amino Acid and Protein Geochemistry* (eds G. A. Goodfriend, M. J. Collins, M. L. Fogel, S. A. Macko, and J. F. Wehmiller), pp. 279–300. Oxford University Press.

Norton D. R., Friedman I., and Olmsted F. H. (1981) *Ground Temperature Measurements, Professional Paper 1203*. U.S. Geological Survey.

Oches E. A., McCoy W. D., and Clark P. U. (1996) Amino acid estimates of latitudinal temperature gradients and geochronology of loess deposition during the last glaciation, Mississippi Valley, United States. *Bull. Geol. Soc. Amer.* **108**, 892–903.

Outcalt S. I., Nelson F. E., and Hinkel K. M. (1990) The zero-curtain effect: heat and mass transfer across an isothermal regime in freezing soil. *Water Resources Res.* **26**, 1509–1516.

Pollack H. N., Huang S., and Shen P.-Y. (1998) Climate change record in subsurface temperatures: a global perspective. *Science* **282**, 279–281.

Sejrup H. P. and Haugen J.-E. (1994) Amino acid diagenesis in the marine bivalve *Arctica islandica* from northwest European sites: only time and temperature? *J. Quat. Sci.* **9**, 301–309.

Sinibaldi M., Goodfriend G. A., Barkay G., Marchese S., and Gil-Av E. (1999) Aspartic acid racemization in teeth from Ketef Hinnom, a late Holocene burial cave in Jerusalem: evidence for reuse. *Geoarchaeology* **14**, 441–454.

Tezcan A. K. (1992) Formula for the calculation of temperature at 1 m depth in Turkey. *Geothermics* **21**, 415–417.

Toy T. J., Kuhaida A. J., and Munson B. E. (1978) The prediction of mean monthly soil temperature from mean monthly air temperature. *Soil Sci.* **126**, 181–189.

Trembour F. W., Friedman I., Jurceka F. J., and Smith F. L. (1986) A simple device for integrating temperature, relative humidity, or salinity over time. *J. Atmos. Oceanic Tech.* **3**, 186–190.

Trembour F. W., Smith F. L., and Friedman I. (1988) Diffusion cells for integrating temperature and humidity over long periods of time. *Proceedings, Materials Research Society Symposium*, vol. 125. *Materials Stability and Environmental Degradation* (eds A. Barkatt, E. D. Verink, Jr, and L. R. Smith), pp. 375–381. Materials Research Society.

Wehmiller J. F. (1977) Amino acid studies of the Del Mar, California midden site: apparent rate constants, ground temperature models, and chronological implications. *Earth Planet. Sci. Lett.*, **37**, 184–196.

Wehmiller J. F. (1992) Aminostratigraphy of southern California Quaternary marine terraces. In *Quaternary Coasts of the United States: Marine and Lacustrine Systems, Special Publication No. 48* (eds C. H. Fletcher, III and J. F. Wehmiller), pp. 317–321. Society for Sedimentary Geology.

Wehmiller J. F. (1993) Applications of organic geochemistry for Quaternary research: aminostratigraphy and aminochronology, In *Organic Geochemistry* (eds M. Engel and S. Macko), pp. 755–783. Plenum.

Wehmiller J. F. (1997) Aminostratigraphy and aminochronology: attempts to resolve issues identified by comparison of U-series and racemization data for Quaternary fossils, US Atlantic coastal plain. *EOS, Trans. Amer. Geophys. Un.* 78(46), F789.

Wehmiller J. F., Belknap D. F., Boutin B. S., Mirecki J. E., Rahaim S. D., and York L. L. (1988) A review of the aminostratigraphy of Quaternary mollusks from United States Atlantic Coastal Plain sites. In *Dating Quaternary Sediments, Special Paper 227* (ed. D. L. Easterbrook), pp. 69–110. Geological Society of America.

Wehmiller J. F., York, L. L., Belknap D. F., and Snyder S. W. (1992) Theoretical correlations and lateral discontinuities in the Quaternary aminostratigraphic record of the U.S. Atlantic Coastal Plain. *Quat. Res.* **38**, 275–291.

Wehmiller J. F., York L. L., and Bart M. L. (1995) Amino acid racemization geochronology of reworked Quaternary mollusks on US Atlantic coast beaches: implications for chronostratigraphy, taphonomy, and coastal sediment transport. *Mar. Geol.* **124**, 303–337.

Wehmiller J. F., Friedman I., and Stecher H. A., III (1996) Latitudinal gradients of effective ground temperature, US Atlantic coastal plain: geochronological implications. *Geol. Soc. Amer. Abst. Progs.* **28**, A117.

Willmott C. J., Mather J. R., and Rowe C. M. (1981) *Average Monthly and Annual Surface Air Temperature and Precipitation Data for the World, Parts I and II.* Publications in Climatology, University of Delaware Center for Climatic Research, Newark, DE, U.S.A.

APPENDIX 1

Results of sine-modeling of ground temperatures for sites that have been used for method development, or where results are from sites that are no longer in use. For comparison, see table 17-3. Dover, DE results include identical depth probes with slightly different experimental designs that were being tested during method-development stages. Additional data were obtained from LC site (Lewes, DE) for periods that were too short (3 months or less) for sine-modeling. (See table over page.)

APPENDIX 2. COMPARISON OF EFFECTIVE TEMPERATURE CALCULATIONS USING SINE-MODELING AND DAILY TEMPERATURES

The plots of daily mean temperature vs. date (figure 17-2) indicate that there can be substantial, short-term departures from the sine model. This is true particularly for the air temperature, in which deviations of ±10°C are common, and for the shallower ground temperature records. Because the effective temperature calculated herein for a given record was based on the mean and the amplitude taken from the sine model (see *Methods* section and figure 17-2), the actual effective temperature will be slightly different. In order to test the effect of these deviations from ideal sinusoidal behavior on the calculated effective temperatures, we performed the following sensitivity analysis on several one-year records.

First, the data were plotted and curve-fitted in KALEIDAGRAPH, as described in the *Methods* section and summarized in tables 17-2 and 17-3. These model parameters were then used to recreate "expected" temperatures for each day of the year. Each of these expected daily temperatures was used to calculate a rate constant for that day, using equation 17-3 (see text). All 365 daily rate constants are then averaged to yield a mean annual rate constant, which is then converted (via equation 17-3) to an effective annual temperature. This *modeled* effective annual temperature can be compared with the effective temperature derived directly from daily average temperature measurements, which can be processed to yield daily rate constants that are then averaged

TABLE 17-A2-1

Site	Modeled interval	Modeled mean (°C)	Modeled amplitude (°C)	Modeled delay (d)	$T_{g/aar}$ AAR T_{eff} (°C)	Sine fit R value
Wilm 1m Therm.	11/13/94–6/9/99	12.7	7.29	141.1	14.66	0.9918
Wilm 1m	1/15/97–1/5/98	12.47	7.06	139.32	14.31	0.9971
Dover 0.5m	1/24/96–11/7/97	12.77	10.33	123.37	16.41	0.9864
Dover 1m-a[a]	1/24/96–11/7/97	12.89	8.47	132.75	15.45	0.9938
Dover 1m-b[a]	1/24/96–11/7/97	12.78	9.24	130.28	15.77	0.9928
Dover 2m-a[a]	1/24/96–11/7/97	12.72	6.54	150.27	14.31	0.9979
Dover 2m-b[a]	1/24/96–11/7/97	12.94	7.06	146.64	14.77	0.9969
Dover 3m	1/24/96–11/7/97	13.07	4.81	168.28	13.96	0.9988
Milton 1.3	3/9/95–11/7/96	14.19	8.43	132.13	16.71	0.9918
Milton 2.25	11/25/95–11/7/96	13.49	5.87	152.22	14.78	0.9972
Gomez 1.3	9/23/95–8/19/96	14.95	6.77	148.72	16.63	0.9931
Gomez 2	2/5/96–1/9/97	15.15	6.27	153.63	16.6	0.9971
Conway 1.3	9/22/95–7/20/96	18.3	4.58	150.11	19.08	0.9902

[a] Adjacent probes, separate holes approx. 5 m apart, used for tests of reproducibility and alternate equipment designs. See table 17-3 for comparison.

over the same 365-day interval to yield the "true" effective temperature.

The results for five records are summarized in table 17-A2-1. We chose two representative air temperature records because they should give the greatest deviation between methods. Two ground temperature records were chosen which were expected to give the greatest deviations between the two methods: one is particularly shallow (0.1 m below ground) and the other was from the ME site that had substantial snow cover and showed a systematic deviation from sinusoidal behavior (see fig. 17-2B). Finally, a 2 m record is included to show the effect of this comparison for deeper records.

The effective air temperatures calculated from the measured data are higher than those calculated from the sine-model parameters. The air temperature records for Charleston, SC, and Portland, ME, show significant differences (0.53°–0.75°C) between the effective temperatures calculated from the model parameters and from the actual measurements.

For the ground temperatures, however, even the cases where the greatest effect is expected show less than a 0.05°C difference between the two ways of calculating the effective temperature. Indeed, for the 2 m record presented, there is only a 1/100th of a degree of difference. This indicates that the sine model is an accurate way to calculate effective ground temperatures.

APPENDIX 3. COMPARISON OF EFFECTIVE TEMPERATURE CALCULATIONS USING SINE-MODELED DAILY OR MONTHLY AIR TEMPERATURES

As explained in the *Methods* section, the sine-model parameters derived from plots of temperature vs.

TABLE 17-A2-2

Site	Actual mean[a]	Modeled mean[b]	Modeled amplitude[b]	T_{eff} model[b,c] parameters	T_{eff} using actual[c] measurements	T_{eff} difference
Portland ME air	7.924	7.923	11.802	12.625	13.382	0.757
Charleston SC air	19.258	19.259	8.494	21.747	22.282	0.535
DMC 1 m shade	6.832	6.899	5.990	8.305	8.351	0.046
Ft Johnson 0.1 m	19.278	19.242	7.478	21.212	21.257	0.045
Ft Johnson 2 m	19.280	19.275	3.323	19.692	19.703	0.011

[a] The "actual mean" temperature for each site was calculated as the mean of the average daily values reported for these sites.
[b] These model parameters differ from those in table 17-3 because different measurement intervals are modeled.
[c] T_{eff} is for racemization [27 kcal/mol (113 kJ/mol)] as discussed in text.

TABLE 17-A3-1 Annual mean air temperatures, amplitudes, and delay for eight sites derived from both daily and monthly records for the period 1 Jan. 96 to 31 Dec. 97

Site[a]	Actual mean[b]	Mean (daily)[c]	Mean (monthly)[d]	Amplitude (daily)[c]	Amplitude (monthly)[c]	Delay (daily)[c]	Delay (monthly)[c]
Portland, ME	7.630	7.634	7.435	12.449	12.235	115	4.30
Providence, RI	10.450	10.504	10.307	11.944	11.723	114	4.26
Wilmington, DE	12.130	12.147	11.971	12.133	11.905	113	4.22
Lewes, DE	15.050	15.049	14.899	11.236	10.997	111	4.18
Norfolk, VA	15.210	15.230	15.133	10.506	10.352	114	4.23
Cape Hatteras, NC	16.650	16.620	16.574	9.285	9.159	118	4.36
Greenvile, NC	15.910	15.915	15.864	10.362	10.222	109	4.10
Charleston, SC	18.550	18.558	18.504	8.772	8.633	108	4.04

[a] Air temperature recording sites from NE and SE regional climate data centers.
[b] The "actual mean" temperature for each site for the period 1 Jan. 96 to 31 Dec. 97 was calculated as the mean of the average daily values reported for these sites. Note that these values differ from the 1961–1990 data shown in tables 17-2 and 17-3.
[c] "Mean (daily)" temperatures, "Amplitude (daily)", and "Delay (daily)" were calculated by applying the sine model to the plot of mean daily air temperatures from NE and SE regional climate data centers. Note the difference between these values and those for 1961–1990 in tables 17-2 and 17-3.
[d] "Mean (monthly)" temperatures, "Amplitude (monthly)", and "Delay (monthly)" were calculated by applying the sine model to the plot of mean monthly air temperatures taken from the NE and SE regional climate data centers.

time are slightly different depending on whether daily or monthly temperatures are used. Specifically, monthly data tend to smooth out the highs and lows resulting in smaller amplitude (parameter B; see figure 17-2) compared to daily data. While the daily data give a more accurate estimate of the true annual temperature, monthly mean air temperature data are much more convenient for use in the calculations presented here. Table 17-A3-1 presents the comparison of effective temperatures derived by sine-modeling of daily and monthly means for 1996–1997. The differences between the daily and monthly model parameters can be used to "correct" monthly model data to the more accurate daily parameters.

The means and amplitudes calculated using the two different methods exhibit a strikingly linear relationship. The mean and amplitude modeled from monthly data are related to the mean and amplitude modeled from daily data as follows:

$$\text{monthly mean} = -0.3536 + 1.0166(\text{daily mean}); \quad R^2 = 1.0 \quad (17\text{-}3\text{-}1)$$

$$\text{monthly amplitude} = 0.145 + 0.9698(\text{daily amplitude}); \quad R^2 = 1.0 \quad (17\text{-}3\text{-}2)$$

As expected, the effect on the means is much less than that on the amplitudes, but in both cases the relationship is so tight that parameters derived from the monthly data can reasonably be used to estimate those that would come from the daily data. The $T_{a/aar}$ Model 2 and Model 3 values presented in table 17-2 represent the results of these uncorrected and corrected means and amplitudes, respectively.

V. AMINO ACID RACEMIZATION DATING APPLICATIONS

18. Revised aminostratigraphy for land–sea correlations from the northeastern North Atlantic margin

D. Q. Bowen

The difficulty in establishing land–sea correlations for the glacial–interglacial Pleistocene record of mid-latitudes over the last million years is well known (Kukla, 1977). It arises largely from the fragmented record of terrestrial deposits that cannot be addressed by traditional means of subdivision and correlation based on paleontology. There are insufficient paleontologic first-appearance and extinction events because of the relatively short time duration, and the "biochronology" scale based on small mammalian fossils in Europe is not underpinned by a definitive geochronology. Such difficulties are compounded by the slow development of geochronologic methods as a means of resolving problems of subdivision and correlation.

A provisional aminostratigraphy based on D-Aile/L-Ile measurements on terrestrial gastropods and bivalves has provided a subdivision and correlation of the British Pleistocene glacial–interglacial sequence. But converting that relative amino acid time-scale into a geochronologic one in calendar years relies on the availability and applicability of new and applicable geochronologic methods. Many of these are still in their early stages of development. The proposed strategy, therefore, is to organize and define formally defined lithostratigraphic units, or parts of such units, as aminozones. Conversion of that aminostratigraphy into an amino acid geochronologic scale is the means of land–sea correlations with oxygen isotope stratigraphy. Such correlations place local and regional terrestrial events within the global marine oxygen-isotope stratigraphic framework, from which inferences of global ice volume and the orbital forcing of the climate system are drawn. Thus, aminostratigraphy and amino acid geochronology make a powerful contribution to greater understanding of the evolution of the Quaternary climate system.

DEFINITION

It follows that it is desirable to revise earlier definitions of aminostratigraphy (Miller and Hare, 1980) so as to be consistent with the style of other definitions in the *International Stratigraphic Guide* (Salvador, 1994). *Aminostratigraphy* is the element of stratigraphy that deals with the distribution of fossils with similar amino acid D/L values in the stratigraphic record, and the organization of strata into units on the basis of the fossils they contain. It may be used over any limited geographic region that is characterized by similar mean annual and seasonal temperature regimes and which was uniformly affected by paleoclimatic changes. Amino acid racemization (AAR) ratios in monogeneric samples, or different genera that racemize at similar rates, may be used directly to subdivide rock sequences and to correlate stratigraphic units.

Aminostratigraphic units are bodies of rock strata that are defined or characterized on the basis of the racemization ratios of particular amino acids in the fossils they contain. An *aminozone* is a regional aminostratigraphic unit used for relative age dating and correlation. It is based on the AAR ratios of fossils contained in lithostratigraphic units. (This definition of "aminozone" is based largely on that of Nelson 1982). *Amino acid geochronology* is a geochronologic framework of aminozones with numerical ages provided by independent geochronologic dating methods.

STANDARD CLASSIFICATION AND AMINOSTRATIGRAPHY

The subdivision and classification of glacial–interglacial Quaternary deposits in the British Isles has a long history, but the forerunner of modern classification was the recognition of Lowestoft and Gipping glaciations in Eastern England by Baden-Powell (1948). Subsequently, on the basis of vegetational history inferred from pollen data, it was proposed that the glacial–interglacial sequence, from oldest to youngest, was as follows: Lowestoft glacial (Anglian), Hoxnian interglacial, Gipping glacial (Wolstonian), and Ipswichian interglacial (West, 1963). Defined as chronostratigraphic stages, they became the basis of a formal standard classification (table 18-1) (Mitchell et al., 1973).

Two concomitant developments of some importance occurred in the same year. First, officers of the British Geological Survey showed that there was no lithostratigraphic evidence for the Gipping glaciation: that is, for glacial deposits intermediate in age between the Hoxnian and Ipswichian (Bristow and Cox, 1973). The problem of this missing, post-Hoxnian but pre-Ipswichian, glaciation (Gipping) had, however, been anticipated by Mitchell et al., (1973), who thought it could be recognized at Nechells and Quinton near Birmingham, U.K., where Hoxnian interglacial deposits separated lower and upper glacial deposits. This region lay outside the limit of the Late Devensian glaciation. Curiously, they located their type locality not there, but farther east at Wolston in Warwickshire. However, the Wolstonian type locality was unrelated to any interglacial marker horizon—the fundamental basis of the classification (Bowen, 1978). The second major development was the greater complexity of climate system variability inferred from oxygen isotopic analysis of core V23-283 by Shackleton and Opdyke (1973). By implication, terrestrial sequences would be more complex than hitherto believed (Bowen, 1978).

It was also unsatisfactory that, except for the Devensian, all of the glacial–interglacial stages are defined in eastern England and are incapable of recognition further afield, other than by fitting local evidence, sometimes arbitrarily, into the standard classification. The recognition of a number of inferred biostratigraphically fixed points in time when pollen analysis indicated the development of a mixed oak–deciduous forest in eastern England failed to take into account wider regional variability in the flora. Subsequently, additional events were proposed on the basis of pollen, faunal, geomorphic, and lithostratigraphic evidence (as well as geochronologic ages). These included evidence from Marsworth (Shotton et al., 1983), Stanton Harcourt (Briggs et al., 1985), and Ilford and Aveley (Sutcliffe, 1964; Sutcliffe and Bowen, 1973). Some of the first measurements of D-Aile/L-Ile ratios at British sites, measured on *Corbicula fluminalis* from Ilford (Miller et al., 1978).

The first systematic attempt to propose an aminostratigraphy for all of southern Britain was based on D-Aile/L-Ile measurements on terrestrial gastropods and bivalves from the type-localities of Mitchell et al. (1973) as well as other sites (Bowen et al., 1989). The aminostratigraphic region lies between 51 and 53° N, and between 2° E and 4° W. One site lies outside this at Grace near Amiens in the Somme Valley, Northern France at 50° N. The mean annual temperature (MAT) of this region, including the Somme Valley, is $9.95° \pm 0.45°$. It is based on six stations: Kew, London (10.6°); Cromer, eastern England (9.6°), Rugby (9.3°) and Oxford (10.1°), English Midlands, Shanklin (10.1°), Isle of Wight, and Amiens (10°). The Pleistocene record suggests that all these sites experienced similar variations in paleotemperature (Sibrava et al., 1986).

It is of prime importance that the integrity of each aminozone is validated by lithostratigraphic succession and, when appropriate, by geomorphic separation. This ensures that the analytical strategy is based on samples with well-controlled geologic age differences. All of the genera used epimerize at approximately the same rate, as is indicated by the similarity of ratios from different species collected from the same lithostratigraphic unit. This allows a wider application than would have otherwise have been possible had the investigation been generically restricted. The D-Aile/L-Ile ratios from the type localities of the standard glacial–interglacial classification at West Runton (Cromerian), Hoxne (Hoxnian), and Bobbitshole (Ipswichian) showed clear aminostratigraphic separation. In addition, the separate status of an aminozone, intermediate in age between Hoxnian and Ipswichian, was confirmed at Stanton Harcourt, Aveley, Ilford, and elsewhere. A major surprise was the discovery that interglacial deposits at Swanscombe near London, previously considered to be Hoxnian, represented an additional interglacial event intermediate in age between the Cromerian and Hoxnian (Bowen et al., 1989; Bowen 1992). A further, pre-Cromerian, aminozone was named after organic deposits at Waverley Wood, Warwickshire (Shotton et al., 1993). Previously, these were thought to be post-Hoxnian in age (Mitchell et al., 1973).

Land–sea correlation of the aminozones with oxygen isotope stratigraphy was effected by a limited number of independent geochronologic ages and interpolation between these ages (table 18-1; figure 18-1). The geochronology was provided at: Halling (radiocarbon: 13 ka); Bacon Hole, Gower (uranium (U)-series: 124 ± 5.4), which has the same D-Aile/L-Ile ratios as Bobbitshole (Ipswichian); and Hoxne (electron-spin resonance: 319 ± 38 ka). In addition, D-Aile/L-Ile ratios of 0.43 ± 0.02 were measured on shells collected from magnetically reversed (Matuyama Chron) high terrace deposits in the Somme Valley at Grace near Amiens (Bates, 1993). The three geochronologic ages, together with the Somme Valley D-Aile/L-Ile ratios that are thought to lie close to the Matuyama–Brunhes reversal (780 ka), lent confidence to the correlation of the other intervening aminozones of Waverley Wood with stage 15, Cromerian with Stage 13, and Swanscombe with stage 11 (Bowen et al., 1989) (table 18-1). A further factor in the correlation of Swanscombe with oxygen isotope stage 11 is that it is younger than till of the Anglian glaciation at Hornchurch (Figure 18-2), a glaciation correlated with the stage 12 Elster glaciation of Germany (Sarnthein et al., 1986; Bowen and Sykes, 1988).

CRITICISMS OF THE AMINOSTRATIGRAPHY

Criticisms of the aminostratigraphy arose largely from a reluctance to recognize that the standard classification of the British glacial–interglacial

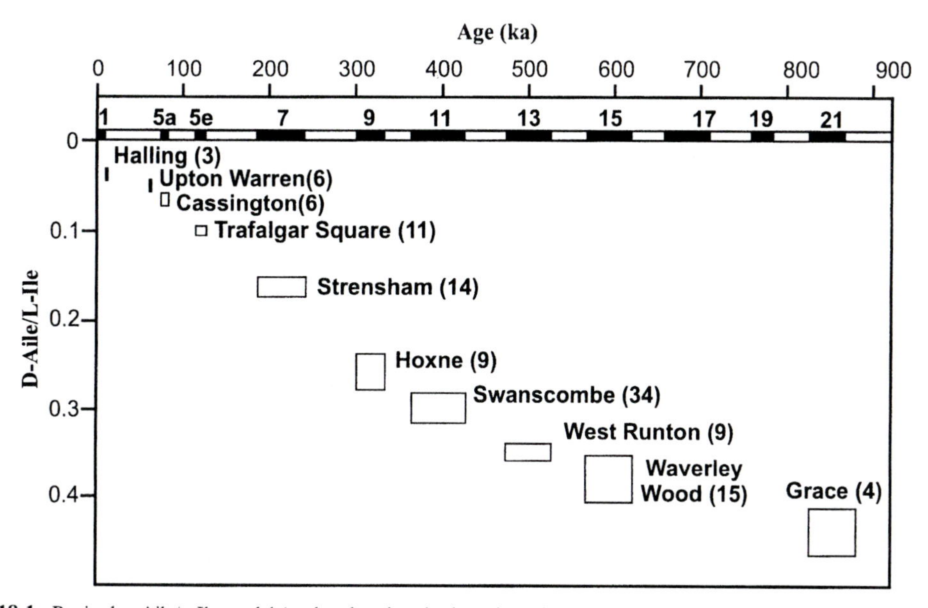

Fig. 18-1. Revised D-Aile/L-Ile model (updated and revised version of Bowen et al., 1989). The vertical scale is an amino acid (D-Aile/L-Ile) relative time scale from youngest (top) to the oldest (bottom). The horizontal scales show a time-scale (in thousands of years) and the duration (bars) of numbered oxygen isotope stages. High ice-volume events (ice ages) are indicated by even numbers. Low-ice-volume events (interglacials) are indicated by odd numbers. The vertical side of each rectangle represents one standard deviation of D-Aile/L-Ile measurements. The horizontal side of each rectangle represents the duration of the odd-numbered oxygen isotope stage with which the aminozone is correlated through an independent geochronology (table 18-1). All ratios refer to peak height measurements and methods of analysis outlined by Bowen et al. (1985). The number of samples measured to calculate the mean D-Aile/L-Ile ratio and one standard deviation for each aminozone is shown in parentheses.

TABLE 18-1 Standard classification (1973), aminozones (1989 and 1999), lithostratigraphy, geochronology, and land–sea correlations with oxygen isotope stratigraphy (δ^{18}O). The italicized lithostratigraphic names refer to units in the Severn Valley (see figures 18-3 and 4)

1973 Classification	1989 Aminozone	D-Aile/L-Ile (1989)	D-Aile/L-Ile (1999)	1999 Aminozone	1999 Lithostratigraphy	Geochronology (ka)[b]	δ^{18}O	Polarity[d]
Devensian	Halling	0.04 ± 0.01 (3)	0.04 ± 0.01 (3)	Halling	Halling Member	13.1 ± 0.2 (^{14}C)[11]	2	BC
					Holt Heath Member[a]	< 30(^{14}C)[10]		
			0.07 ± 0.01 (5)	Upton Warren	*Upton Warren Member*	> 42(^{14}C)[9]	3	BC
	Upton Warren	0.07 ± 0.01 (5)	0.08 ± 0.01 (6)	Cassington	Cassington Member	~ 80(OSL)[8]	5a	BC
Ipswichian	Ipswichian	0.09 ± 0.01 (6)	0.1 ± 0.01 (11)	Trafalgar Square	Trafalgar Square Member	124 ± 5.4 (U)[c7]	5e	BC
					Ridgacre Formation[a]	159.5 ± 13 (^{36}Cl)[6]	6	BC
Wolstonian	Stanton Harcourt	0.16 ± 0.02 (26)	0.17 ± 0.01 (14)	Strensham	*Strensham Member*	~ 2000 (OSL)[c5]	7	BC
					Bushey Green Member		8	BC
Hoxnian	Hoxnian	0.26 ± 0.01 (4)	0.26 ± 0.02 (9)	Hoxne	Hoxne Formation	319 ± 38 (ESR)[4]	9	BC
					Spring Hill Member		10	BC
	Swanscombe	0.3 ± 0.02 (14)	0.3 ± 0.02 (34)	Swanscombe	Swanscombe Member	471 ± 15 (TL)[c3]	11	BC
						~ 400 (U)[c3]	11	BC
Anglian					*Woolridge Member*[a]		12	BC
Cromerian	Cromerian	0.35 ± 0.01 (9)	0.35 ± 0.01 (9)	West Runton	West Runton Member	~ 500 (ESR)[2]	13	BC
	Waverley Wood	0.38 ± 0.02 (3)	0.38 ± 0.03 (15)	Waverley Wood	*Waverley Wood Member*		15	BC
					Kenn Formation[a]		16	BC
		0.43 ± 0.02 (4)	0.43 ± 0.02 (4)	Grace	Grace Formation	810 ± 140 (ESR)[1]	21	M

[a] Till or outwash deposits. [b] References: [1]Laurent et al. (1998); [2]Rink et al. (1996); [3]Lewis in Bowen et al. (1999); [4]Gladfelter et al. (1993); [5]Scott and Buckingham (1997); [6]Phillips et al. (1994); [7]Bowen et al. (1989); [8]Maddy et al. (1998); [9]Bowen (1999); [10]Shotton (1967); [11]Kerney (1971). [c] Age comes from a different site in the same aminozone. All ratios refer to peak height measurements and methods of analysis outlined in Bowen et al. (1985). [d] Polarity: BC, Brunhes Chron; M, Matuyama Chron.

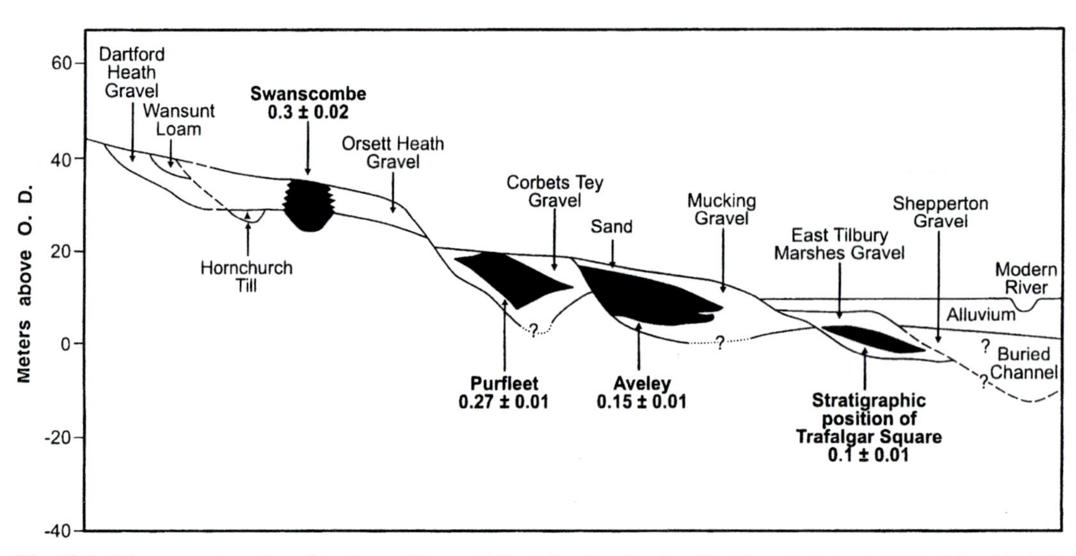

Fig. 18-2. Diagrammatic section of the Lower Thames Valley below London (modified from Bridgland, 1994) with D-Aile/L-Ile ratios for interglacial deposits (black). Compare with tables 18-1 and 18-2. Fluvial deposits are unshaded. Note the Hornchurch Till correlated with stage 12. Interglacial deposits are correlated as: Swanscombe with stage 11; Purfleet with stage 9; Aveley with stage 7; and Trafalgar Square with substage 5e.

sequence was redundant. Particular criticism stemmed from an unwillingness to accept that the Swanscombe and Hoxne aminozones represented separate interglacials after the Anglian glaciation. Because Hoxnian interglacial deposits are found in depressions on the surface of Anglian glacial deposits, it was thought that they represented the interglacial period immediately following that glaciation; thus, the case for recognizing an earlier post-Anglian Swanscombe interglacial period would be untenable. The fallacy of that argument, however, is evident when it is appreciated that Holocene deposits are also found in depressions on the surface of Anglian glacial deposits. Further attempts to show that the Swanscombe and Hoxne interglacials are the same age, by using either pollen or mammalian paleontology, are also flawed because they rely on facies–flora or facies–fauna controlled by climate and not time. Neither is based on the first appearance or extinction of fossil taxa (Bowen, 1978). Aminostratigraphic correlations between Swanscombe and other interglacial deposits were subject to further criticism. That made between Barnham and Swanscombe by Ashton et al. (1994) was criticized by West and Gibbard (1985), who broadened their criticism to object to inappropriate land–sea correlations as well as the role of orbital forcing in providing explanations for terrestrial events. In reply, Ashton, Bowen, and Lewis systematically addressed and rejected all their criticisms (in West and Gibbard, 1995).

Others, whilst accepting the correlation of the Hoxne aminozone with oxygen isotope stage 9, argued that deposits of the Anglian glaciation should be correlated with stage 10 (Sumbler, 1995). The fallacy of that argument has already been addressed, but, more powerfully, IGCP Project 24 (*Glaciations in the Northern Hemisphere*) concluded that there was no evidence in Europe for glaciation of that age (Sibrava et al., 1986). Raymo (1997) also suggested that because stage 10 followed closely on the peak interglacial conditions of stage 11, there might not have been enough time to develop a large ice-sheet. It is, perhaps, significant that the relatively brief duration of oxygen isotope stage 10 (52 ka) may be reflected in the relatively small difference in D-Aile/L-Ile ratios between the Swanscombe stage 11 aminozone (0.3 ± 0.017) and the Hoxne stage 9 aminozone (0.26 ± 0.02). It is only possible for aminostratigraphy to support a stage 10 glaciation if substages of stage 7 are recognized from the D-Aile/L-Ile data. But the combined evidence of aminostratigraphy, lithostratigraphy, geomorphology, and further geochronologic ages (below) militate against such a possibility.

Various criticisms of anomalous D-Aile/L-Ile ratios show scant appreciation of the sources of analytical error common to all geochronologic methods. Miller and Brigham-Grette (1989), Wehmiller (1990), and Kaufman and Miller (1992) have reviewed the problems of sampling strategy and laboratory procedures for amino acid analysis. In our laboratory, chromatograms were classified on a scale of 1 to 5. Data from classes 1 and 2 were rejected. The classification criteria are as follows: (5) Thr and Ser separated, all other amino acids resolved to within 5% of the peak height; (4) Thr and Ser separated, Aile, Ile, and Leu sepa-

TABLE 18-2 Aminostratigraphy of valley systems in southern Britain. **Bold**: aminozone type localities and lithostratigraphic units with D-Aile/L-Ile ratios characteristic of those aminozones (table 18-1 and figure 18-1). Lithostratigraphic names are from Bowen et al. (1999)

$\delta^{18}O$ Stage	$\delta^{18}O$ Stage age (ka)	D-Aile/L-Ile (1999)	Aminozone	Severn Valley & Bristol area	Avon Valley	Lower Thames	Eastern England valleys
5e	114–127	0.1 ± 0.01 (11)	**Trafalgar Square**		**New Inn Bed**	**Trafalgar Square Member**	**Tattershall Bed** **Ravenstone Member** **Bobbitshole Formation**
6	127–186			Ridgacre Formation[a] Kidderminster Member	Cropthorne Member	Mucking & Taplow Gravel	Stoke Goldington Member Thorpe Member
7	186–242	0.17 ± 0.01 (14)	**Strensham**	**Strensham Member**	**Ailstone Bed**	**Aveley, Crayford, Ilford: Uphall**	**Stutton Formation** **Kirkby Bed** **Hartigan's Pit Member**
8	242–301			Bushley Green Member	Pershore Member	Corbets Tey & Basal Mucking Gravel	
9	301–312	0.26 ± 0.02 (9)	**Hoxne**	**Hill House Bed**	**Frog Member**	**Ilford: Cauliflower** **Purfleet** **Belhus Park**	**Hoxne Formation** **Woodston Beds** **March Formation**
10	312–364			Spring Hill Member		Orsett Heath & basal Corbets Tey gravel	
11	364–426	0.3 ± 0.02 (34)	**Swanscombe**			**Swanscombe, Ingress Vale & Little Thurrock & Sugworth**	**Barnham & Elveden Formations** **West Stow Formation** **Nar Member**
12	427–474			Woolridge Member[a]	Wolston Formation[a]	Orsett Heath basal gravel & Hornchurch till[a]	
13	474–528	0.35 ± 0.01 (9)	**West Runton**				
15	568–621	0.38 ± 0.03 (15)	**Waverley Wood**	**Yew Tree Formation**	**Waverley Wood Member**		
16	621–659			Kenn Formation[a]		"Northern Drift"?[a]	

[a] Till or outwash deposits.

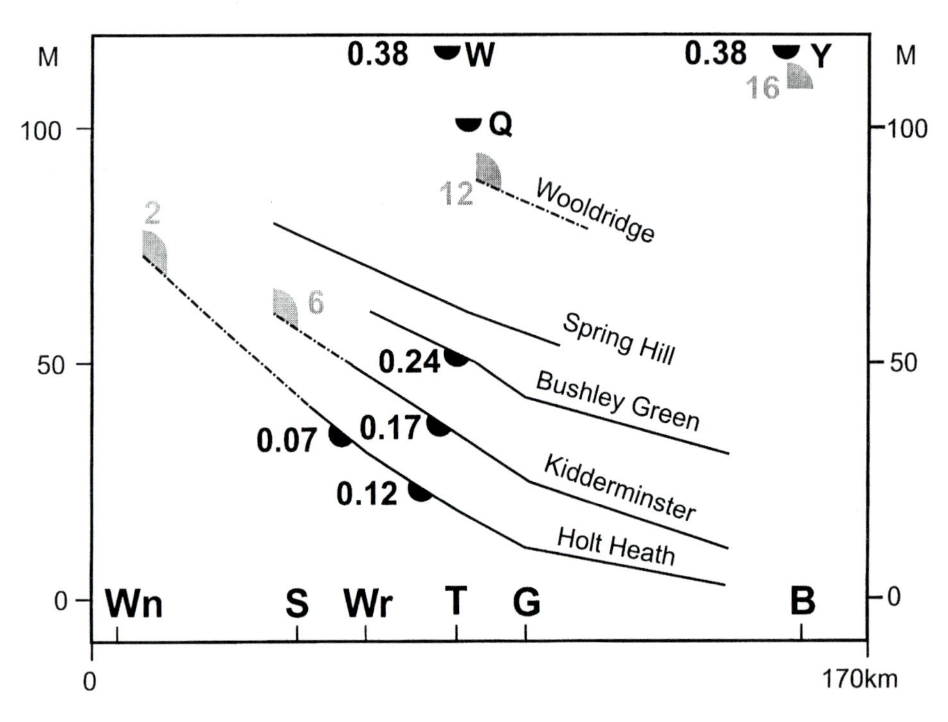

Fig. 18-3. Diagrammatic representation of lithostratigraphic units and aminostratigraphy in the Lower Severn and Avon valleys. Half-circles: interglacial deposits with mean D-Aile/L-Ile ratios (compare with table 18-1 and figure 18-1). Quarter circles: ice-margins and oxygen-isotope stage number with which the glaciation is correlated. Continuous lines: upper terraced surface of named lithostratigraphic units. Broken lines: terraced surface of outwash terraces. W, Waverley Wood; Y, Yew Tree Farm; Wn, Wolverhampton; S, Stourport; Wr, Worcester; T, Tewkesbury: G, Gloucester; B, Upper Bristol Channel (based on Maddy et al., 1995; Bowen, 1999).

rated to within 5% of peak height, and other amino acids resolved to within 10% of peak height; (3) Thr and Ser poorly or not resolved, and Aile, Ile and Leu separated to within 15% of peak height ratio of D-Aile/L-Ile approximately equal to peak area ratio, or short run (i.e., amino acids up to Val eluted quickly); (2) peak height ratio greater than peak area ratio, or Aile, Ile or Leu peaks misshapen, or Aile, Ile or Leu not separated to within 5% of peak height, or baseline noise very high, or amino acid composition or concentration unusual for the species analyzed; and (1) very poor separation of amino acids.

Anomalously low D-Aile/L-Ile ratios from West Runton (Comerian) were resolved by additional and more satisfactory analyses, consistent with regional lithostratigraphy, by two independent laboratories (University of Colorado and London University). Gibbard (1994) and Scourse (1997) maintained that anomalous D-Aile/L-Ile ratios at Purfleet in the Lower Thames Valley were incorrect. Instructively, these were on *C. fluminalis*, measured some ten years apart by two independent laboratories (Miller et al., 1978; Bowen et al., 1989). Yet, both sets of measurements produced anomalous high ratios given the stratigraphic and geomorphic context of that area. Subsequent measurements on different species at Purfleet, however, are consistent with local lithostratigraphy and geomorphology (Bowen et al., 1996). It appears that *C. fluminalis* sometimes yields anomalous high D-Aile/L-Ile ratios. But at other sites, such as Strensham (stage 7) (de Rouffignac et al., 1994) and Waverely Wood (stage 15) (Shotton et al., 1993), the D-Aile/L-Ile ratios measured on *C. fluminalis* are the same as those measured for other species. Given such uncertainty, correlation using D-Aile/L-Ile ratios only for *C. fluminalis* is provisional, as is the case between the Waverley Wood aminozone and Yew Tree Formation near Bristol. But to suggest that the age of glacial deposits (Kenn Formation) underlying the Yew Tree Formation has been made by "pigeon-holing" events into an oxygen-isotope stratigraphy (Scourse, 1997) is gratuitous and misleading (see *Extent and timing of glaciation*). The integrity of the Waverley Wood aminozone, based on the lithostratigraphic Waverley Wood Member, is secure because it lies in a lithostratigraphic and aminostratigraphic sequence that ranges in age from Early to Late Pleistocene (Maddy in Bowen et al., 1999).

NEW EVIDENCE FOR TESTING THE AMINOSTRATIGRAPHY

The original aminostratigraphy, amino acid geochronology, and land–sea correlation proposed by Bowen et al. (1989) is tested in two ways: first, by further validating the integrity and separation of the aminozones by additional D-Aile/L-Ile measurement at the original sites as well as from new sites where local lithostratigraphy and geomorphology confirm separation in time; and second, by using new, independent geochronologic age estimates. As a result, some of the original aminozones have been redefined at new type localities where lithostratigraphic and geomorphic relationships are clearer.

Lithostratigraphic and geomorphic separation of the aminozones is demonstrable in the valley of the Warwickshire Avon upstream of Stratford-upon-Avon (Maddy et al., 1991). But the outstanding validation of the aminostratigraphy is in the Lower Thames Valley, where interglacial deposits occur in a lithostratigraphic sequence of mainly cold-climate fluvial gravels (figure 18-2, tables 18-1 and 18-2) (Bridgland, 1994). The superposition of the lithostratigraphic units and aminozones in the Lower Thames Valley, where average long-term uplift rates are relatively slow, is replaced by their geomorphic separation further upstream in the Thames (figure 18-3) where relatively higher long-term uplift rates have been calculated (Maddy, 1997). The successful extension of the aminostratigraphy to the principal valley systems of lowland England is further confirmation of its validity (table 18-2).

New geochronologic ages obtained by different methods are shown in table 18-1. The age of the Grace aminozone, now refined by ESR dating, allows correlation with Stage 21 (Laurent et al., 1998). An ESR age at West Runton (Rink et al., 1996) shows that the Cromerian sediments at that site correspond to the youngest event of the Dutch Cromerian complex. Thermoluminescence (TL) and U-series ages for the Swanscombe aminozone were obtained from West Stow in Norfolk (table 18-1). A revised ESR age at Hoxne confirms the original correlation with stage 9 (table 18-1). Although Strensham now replaces Stanton Harcourt as a type locality (de Rouffignac et al., 1994) the OSL (optically stimulated luminescence) age for that aminozone is from the original site (Scott and Buckingham, 1997). Fluvial gravels of the Kidderminster Station Member of the Severn Valley Formation overlie interglacial deposits at Strensham. Traced upstream, they merge with outwash related to the Ridgacre glacial deposits. Cosmogenic ^{36}Cl rock exposure ages from erratic boulders on the surface of the Ridgacre Formation of 159.5 ± 13 ka show that the glaciation occurred during the oxygen isotope stage (Phillips et al., 1994). This provides a further, although indirect, geochronologic age control for the Strensham aminozone (Stage 7).

Correlation of the Upton Warren aminozone with substage 5a (Bowen et al., 1989) has been revised because of the discovery of a new site at Cassington, near Oxford (Maddy et al., 1998). The Cassington aminozone has slightly higher D-Aile/L-Ile ratios (0.08 ± 0.01) than Upton Warren (0.07 ± 0.01) and has an OSL age of 80 ka that places it in substage 5a (Maddy et al., 1998). It seems that Upton Warren is somewhat younger, although its radiocarbon dates (between 40 and 50 ka) are still believed to be minimum ages.

The revised aminostratigraphy is based primarily on new geochronologic ages for the aminozones (table 18-1). New D-Aile/L-Ile data from lithostratigraphic and geomorphic settings has extended the aminostratigraphy throughout southern England (Green et al., 1996; Keen et al., 1997; Bowen et al., 1999) (table 18-2). It also provides a framework for evaluating the number, extent, and timing of lowland glaciations during the Brunhes Chron (figure 18-3).

EXTENT AND TIMING OF GLACIATION

Possibly the oldest known glacial deposits are those of the Kenn Formation near Bristol (Gilbertson & Hawkins 1978), where they are overlain by freshwater interglacial deposits at Yew Tree Farm (Andrews et al., 1984) (table 18-1). These, provisionally correlated with the Waverley Wood aminozone, suggest an oxygen isotope stage 15 age (Bowen et al., 1989). Thus, they provide a minimum age for the underlying Kenn Formation, which may be oxygen isotope Stage 16 in age (Bowen 1994, Bowen et al., 1999). The Anglian (Stage 12) glaciation is represented by the glacial deposits of the Woolridge Member (table 18-1, figure 18-3). The timing of the Ridgacre glaciation is consistent with the aminostratigraphy and confirmed by ^{36}Cl rock exposure ages to have occurred during oxygen isotope Stage 6. Late Devensian glaciation (Stage 2) is confirmed by radiocarbon ages.

Thus, it is possible to establish the number, extent and timing of four major glaciations, correlated with oxygen isotope stages 16, 12, 6, and 2, from the aminostratigraphy (figure 18-4). Glaciations during stages 18, 14, 10, and 8 failed to extend as far south as the Late Devensian ice-margin (Bowen, 1999). This is consistent with proposals for only four extensive glaciations during the last 900 ka, based on a global stratigraphic synthesis (GSS97) from 11 new δ^{18}O records (Raymo, 1997). Raymo suggested that these could only have occurred when July insolation at 65° N had been unusually low for more than a full precessional (>21 ka) cycle brought about by eccentricity (100 ka) modulation of the precession component of northern hemisphere summer insolation. In Ireland, however,

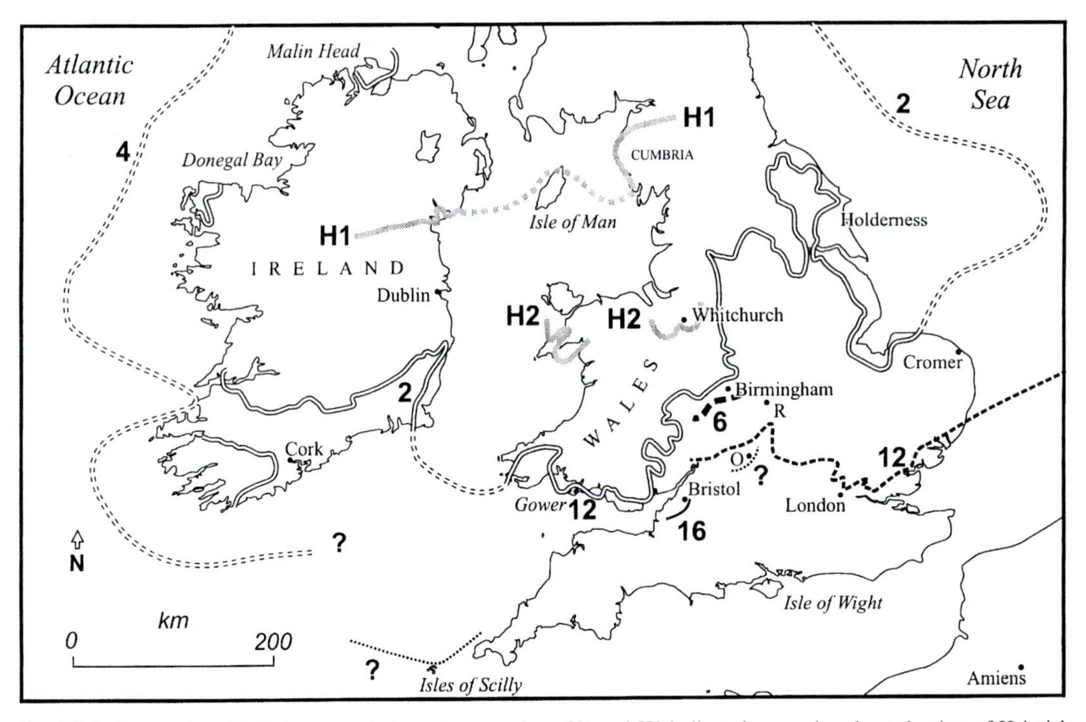

Fig. 18-4. Ice margins with their oxygen isotope stage numbers. H1 and H2 indicate ice margins about the time of Heinrich Events (Bowen 1999). Question mark indicates uncertainty.

the greatest extent of ice during the last glacial cycles occurred not during oxygen isotope Stage 2, but during stages 4 and 3. This may be a function of enhanced precipitation owing to the westerly position of Ireland (Bowen, 1999).

Acknowledgments: The valuable comments of the reviewers Daniel Belknap, Gifford Miller, and John Wehmiller are acknowledged with many thanks. I also thank Alun Rogers (Cardiff University), who prepared the diagrams.

REFERENCES

Andrews J. T., Gilbertson D. D., and Hawkins A. B. (1984) The Pleistocene succession of the Severn Estuary: a revised model based on amino-acid racemization studies. *J. Geol. Soc. Lond.* **141**, 967–974.

Ashton N., Bowen D. Q., Holman A., Hunt C., Irving B., Kemp R. A., Lewis S. G., McNabb J., Parfitt S., and Seddon M. B. (1994) Excavations at the Lower Palaeolithic site at East Farm, Barnham, Suffolk, 1989–1992. *J. Geol. Soc. Lond.* **151**, 599–607.

Baden-Powell D. F. W. (1948) The chalky boulder clays of Norfolk and Suffolk. *Geol. Mag.* **85**, 279–96.

Bates M. R. (1993) Quaternary Aminostratigraphy in Northwestern France. *Quat. Sci. Rev.* **12**, 793–810.

Bowen D. Q. (1978) *Quaternary Geology: a Stratigraphic Framework for Multidisciplinary Work*. Pergamon.

Bowen D. Q. (1992) Aminostratigraphy of non-marine mollusca in Southern Britain. *Sveriges Geol. Under.* **81**, 65–67.

Bowen D. Q. (1994) The Pleistocene history of North West Europe. *Sci. Progr.* **76**, 209–223.

Bowen D. Q. (1999) Only four major '100 ka' glaciations during the Brunhes Chron? *Intl. J. Earth Sci. (Geol. Rundsc.)*, **88**, 276–284.

Bowen D. Q., and Sykes G. A. (1988) Correlation of marine events and glaciations on the northeast Atlantic margin. *Phil. Trans. Roy. Soc.* **B318**, 619–635.

Bowen D. Q., Sykes G. A., Reeves A., Miller G. H., Andrews J. T., Brew J. S., and Hare P. E. (1985) Amino acid geochronology of raised beaches in south-west Britain. *Quat. Sci. Rev.* **4**, 279–318.

Bowen D. Q., Hughes S., Sykes G. A., and Miller G. H. (1989) Land–sea correlations in the Pleistocene based on isoleucine epimerization in non-marine molluscs. *Nature* **340**, 49–51.

Bowen D. Q, Sykes G. A., Maddy D., Bridgland D. R., and Lewis S. G. (1996) Aminostratigraphy and amino acid geochronology of English lowland valleys: the Lower Thames in context. In

The Quaternary of the Lower Reaches of the Thames (eds Bridgland D. R., Allen P., and Haggart B. A.), pp. 61–63. Quaternary Research Association, Durham.

Bowen D. Q., Cameron D., Campbell S., Gibbard P. L., Holmes J., Lewis S., Maddy D., McCabe A. M., Preece R., Sutherland D. G., and Thomas G. S. P. (1999) *Revised Correlation of Quaternary Deposits in the British Isles. Special Report No. 23.* Geological Society of London.

Bridgland D. R. (1994) *Quaternary of the Thames.* Chapman & Hall.

Briggs D. J., Coope G. R., and Gilbertson D. D. (1985) The chronology and environmental framework of early man in the Upper Thames Valley: a new model. *Brit. Arch. Rep.* **137**, 1–176.

Bristow C. R., and Cox F. C. (1973) The Gipping Till: a reappraisal of East Anglian glacial stratigraphy. *J. Geol. Soc. Lond.* **129**, 1–37.

Gibbard P. L. (1994) *Pleistocene History of the Lower Thames Valley.* Cambridge University Press.

Gilbertson D. D., and Hawkins A. B. (1978) The col-gully and glacial deposits at Court Hill, Clevedon, near Bristol, England. *J. Glac.* **20**, 173–188.

Gladfelter B. G., Wymer J. J., and Singer R. (1993) *The Lower Palaeolithic Site at Hoxne, England.* University of Chicago Press.

Green C. P., Coope G. R., Jones R. L., Keen D. H., Bowen D. Q., Currant A. P., Holyoak D. T., Ivanovich M., Robinson J. E., Rogerson R. J., and Young R. C. (1996) Pleistocene deposits at Stoke Goldington, in the valley of the Great Ouse, UK. *J. Quat. Sci.* **11**, 59–87.

Kaufman D. S., and Miller G. H. (1992) Overview of amino acid geochronology. *Comp. Biochem. Physiol.* **102B**, 199–204.

Keen D. H., Coope G. R., Jones R. L., Field M. H., Griffiths H. I., Lewis S. G., and Bowen D. Q. (1997) Middle Pleistocene deposits at Frog Hall Pit, Stretton-on-Dunsmore, Warwickshire, English Midlands, and their implication for the age of the type Wolstonian. *J. Quat.Sci.* **12**, 183–208.

Kerney M. P. (1971) A Middle Weichselian deposit at Halling, Kent. *Proc. Geol. Assoc.* **82**, 1–11.

Kukla G. J. (1977) Pleistocene land–sea correlations. *Earth Sci. Rev.* **13**, 307–374.

Laurent M., Falgueres J. J., Bahain Rousseau L., and Van Vliet Lanoe B. (1998) ESR dating of Quartz extracted from Quaternary and Neogene sediments: method, potential and actual limits. *Quat. Intl.* **17**, 1057–1062.

Maddy D. (1997) Uplift driven valley-incision and river terrace formation in southern England. *J. Quat. Sci.* **12**, 539–545.

Maddy D., Keen D. H., Bridgland D. R., and Green C. P. (1991) A revised model for the Pleistocene development of the River Avon, Warwickshire. *J. Geol. Soc. Lond.* **148**, 473–484.

Maddy D., Green C. P., Lewis S. G., and Bowen D. Q. (1995) Pleistocene geology of the Lower Severn Valley, U. K. *Quat. Sci. Rev.* **14**, 209–222.

Maddy D., Lewis S. G., Scaife R. G., Bowen D. Q., Coope G. R., Green C. P., Hardaker T., Keen D. H., Rees-Jones J., Parfitt S., and Scott K. (1998) The Upper Pleistocene Deposits at Cassington, near Oxford, UK. *J. Quat. Sci.* **13**, 205–231.

Miller G. H., and Brigham-Grette J. (1989) Amino acid geochronology: resolution and precision in carbonate fossils. *Quat. Intl.* **1**, 111–128.

Miller G. H., and Hare P. E. (1980) Amino acid geochronology: integrity of the carbonate matrix and potential of molluscan fossils. In *Biogeochemistry of Amino Acids* (eds P. E. Hare, T. C. Hoering, and K. King, Jr), pp. 415–444. New York.

Miller G. H., Hollin J. T., and Andrews J. T. (1978) Aminostratigraphy of U.K. Pleistocene deposits. *Nature* **281**, 215–278.

Mitchell G. F., Penny L. F., Shotton F. W., and West R. G. (1973) A correlation of Quaternary Deposits in the British Isles. *Special Report No. 4.* Geological Society of London.

Nelson A. R. (1982) Quaternary glacial and marine stratigraphy of the Qivitu Peninsula northern Cumberland Peninsula, Baffin Island. *Geol. Soc. Amer. Bull.* **92**, 512–518.

Phillips F. M., Bowen D. Q., and Elmore D. (1994) Surface exposure dating of glacial features in Britain using cosmogenic chlorine-36: preliminary results. *Min. Mag.* **58A**, 722–723.

Raymo M. E. (1997) The timing of major climate terminations. *Paleoceanography* **12**, 577–585.

Rink W. J., Schwarcz H. P., Stuart A. J., Lister A. E., Marseglea E., and Brennan B. J. (1996) ESR dating of the type Cromerian Freshwater Bed at West Runton, UK. *Quat. Sci. Rev.* **15**, 727–738.

de Rouffignac C., Bowen D. Q., Coope G. R., Keen D. H., Lister A. M., Maddy D., Robinson, E., Sykes G. A., and Walker M. J. C. (1994) Late Middle Pleistocene deposits at Upper Strensham, Worcestershire, England. *J. Quat. Sci.* **10**, 15–31.

Salvador A. (ed.) (1994) *International Stratigraphic Guide*, 2nd edn. Geological Society of America.

Sarnthein M., Stremme H. E., and Mangini A. (1986) The Holstein interglaciation: time stratigraphic position and correlation to stable isotope stratigraphy of deep sea sediments. *Quat. Res.* **26**, 283–298.

Scott K., and Buckingham C. (1997) Quaternary fluvial deposits and palaeontology at Stanton Harcourt, Oxfordshire. In *The Quaternary of the South Midlands and the Welsh Marches* (eds S. G. Lewis and D. Maddy), pp. 115–126. Quaternary Research Association.

Scourse J. D. (1997) Transport of the Stonehenge Bluestones: testing the glacial hypothesis. In *Science and Stonehenge* (eds B. Cunliffe, and C. Renfrew), pp. 271–314. Oxford University Press.

Shackleton N. J., and Opdyke N. D. (1973) Oxygen isotope and palaeomagnetic stratigraphy of Equatorial Pacific core V28-238: oxygen isotope temperatures and ice volumes on a 10^5 year and a 10^6 year scale. *Quat. Res.* **3**, 39–55.

Shotton F. W. (1967) Age of the Irish Sea Glaciation of the Midlands. *Nature* **215**, 136–137.

Shotton F. W., Sutcliffe A. J., Bowen D. Q., Currant A. P., Coope G. R., Harmon R. S., Shackleton N. J., Stringer C. B., Turner C., West R. G., and Wymer J. J. (1983) Interglacials after the Hoxnian in Britain. *Quat. News.* **39**, 19–25.

Shotton F. W., Keen D. H., Coope G. R., Currant A. P., Gibbard, P. L., Alto M., Peglar S. M., and Robinson J. E. (1993) The Pleistocene deposits at Waverley Wood Pit, Warwickshire, England. *J. Quat. Sci.* **8**, 293–325.

Sibrava V., Bowen D. Q., and Richmond G. M. (1986) Quaternary glaciations in the Northern Hemisphere. *Quat. Sci. Rev.* **5**, 1–510.

Sumbler M. G. (1995) The terraces of the rivers Thame and Thames and their bearing on the chronology of glaciation in central and eastern England. *Proc. Geol. Assoc. Lond.* **106**, 93–106.

Sutcliffe A. J. (1964) The mammalian fauna. In *The Swanscombe Skull. Occasional Paper 20.* (ed. C. D. Ovey), Royal Anthropological Institute Occasional Paper **20**, 85–111.

Sutcliffe A. J., and Bowen D. Q. (1973) Preliminary report on excavations in Minchin Hole, April–May, 1973. *News. Wm Pengelly Cave Stud. Trust* **21**, 12–25.

Wehmiller J. F. (1990) Amino acid racemization: applications in chemical taxonomy and chronostratigraphy of Quaternary fossils. In *Skeletal Biomineralization: Patterns. Processes and Evolutionary Trends* (ed. J. G. Carter), pp. 287–318. New York.

West R. G. (1963) Problems of the British Quaternary. *Proc. Geol. Assoc. Lond.* **74**, 147–186.

West R. G. and Gibbard P. L. (1995) Discussions on excavations at the Lower Palaeolithic site at East Farm, Barnham, Suffolk, 1989–1992. Reply by: Ashton N. M., Bowen D. Q., and Lewis S. G. *J. Geol. Soc. Lond.* **152**, 570–574.

19. Aminostratigraphy of two Pleistocene marine sequences from the Mediterranean coast of Spain: Cabo de Huertas (Alicante) and Garrucha (Almería)

Trinidad Torres, Juan F. Llamas, Laureano Canoira, F. Javier Coello, Pilar García-Alonso, and José E. Ortiz

Sea-level variations and the subsequent changes in coastline morphology have been subjects of interest for some time. The record of Quaternary coastal evolution is usually quite fragmentary because the emergent sedimentary record almost always corresponds with highstand or interglacial periods, whereas cold periods are reflected in erosive platforms, scarps, or the accumulation of eolian deposits (dunes). In the area of the present study (and elsewhere) along the southeastern coast of Spain, morphological criteria have served as the primary basis for the relative dating of marine terraces (Goy and Zazo, 1982, 1988; Lario et al., 1993; Fumanal et al., 1997). Basin evolution criteria have recently been used to analyze sea level variations (Bardají et al., 1995, 1997; Zazo et al., 1997), as has palaeomagnetism (Bardají et al., 1995), despite the usually unfavorable sediment grain size (sand and gravel).

Corals found in Quaternary coastal units have been successfully employed for uranium (U)-series dating (e.g., Stein et al., 1991, 1993). Because corals are very scarce in the Mediterranean environment, it has been necessary to work with mollusk shells for U-series chronology (Dumas, 1981; Goy and Zazo, 1988; Causse et al., 1993; Goy et al., 1993; Hillaire-Marcel et al., 1996). Because, during their growth, members of the Mollusca incorporate very little uranium into their shells, it is necessary to assume that these samples incorporated uranium shortly after death and remained closed systems during most of their subsequent time as fossils. U-series dating can also employ the isochronal diagram method (e.g., Causse et al., 1993).

The Mediterranean environment may be described as an area of alpine influence in which there is still neotectonic activity. In Upper Pliocene and Quaternary times, the neotectonic process was common and almost continuous, and Mediterranean marine terraces were uplifted, sunken, faulted and tilted (Dumas et al., 1988; Goy and Zazo, 1989; Rey and Fumanal, 1996).

Geochronology based on geochemical methods may be affected by a number of factors. Amino acid racemization (AAR) has been employed in a number of studies of coastal terrace sequences (e.g., Wehmiller and Belknap, 1982; Wehmiller, 1990; Kauffman, 1992; Wehmiller et al., 1992, 1995; Murray-Wallace, 1995) and is applied in the present study. A large number of samples must be analyzed because of several problems encountered in the study area:

1. Contamination by organic matter may be a problem because marine terraces constitute an excellent place for seabirds to land, and for plants to grow, even in the arid environment of the Mediterranean. Algae are frequent also, and the terraces, mostly the lowermost ones, suffered (and continue to suffer) wave spray with recent organic matter arrival.
2. Marine terrace deposits are not usually particularly thick (< 1 m) and do not efficiently insulate sediments and fossils from solar heating.
3. No single genus of mollusks is found at all terrace sites of interest, so a large number of genera must be analyzed to permit correlation between all the study sites.
4. Reworking of older shells into younger deposits is also encountered.

Related AAR applications to Mediterranean coastal terraces include those described by Hearty et al. (1986), Belluomini et al. (1986, 1993), Hearty (1987), Dumas et al. (1988), Ochietti et al. (1993), and Hillaire-Marcel et al. (1996)

The Tyrrhenian Cycle started during or before the Last Interglacial (oxygen isotope stage 5). This cycle reflected a general warming of the water in the Mediterranean owing to the inflow of warm seas through the Straits of Gibraltar, and is described—as regards the Mediterranean area—as a complex of coastal deposits with typical faunal remains: *Strombus bubonius*, *Patella ferruginea*, and *Conus mediterraneus*. On the Atlantic Coast of the Iberian Peninsula, deposits of similar age occur, but without warm-water marker mollusk species (cf. Hearty, 1987; Hillaire-Marcel et al., 1996).

According to the aforementioned authors, Tyrrhenian coastal evolution might have occurred in three different ways: (1) from a single, long, still-standing sea level; (2) from a slow, but continuous, sea-level change; or (3) from two or three periods of

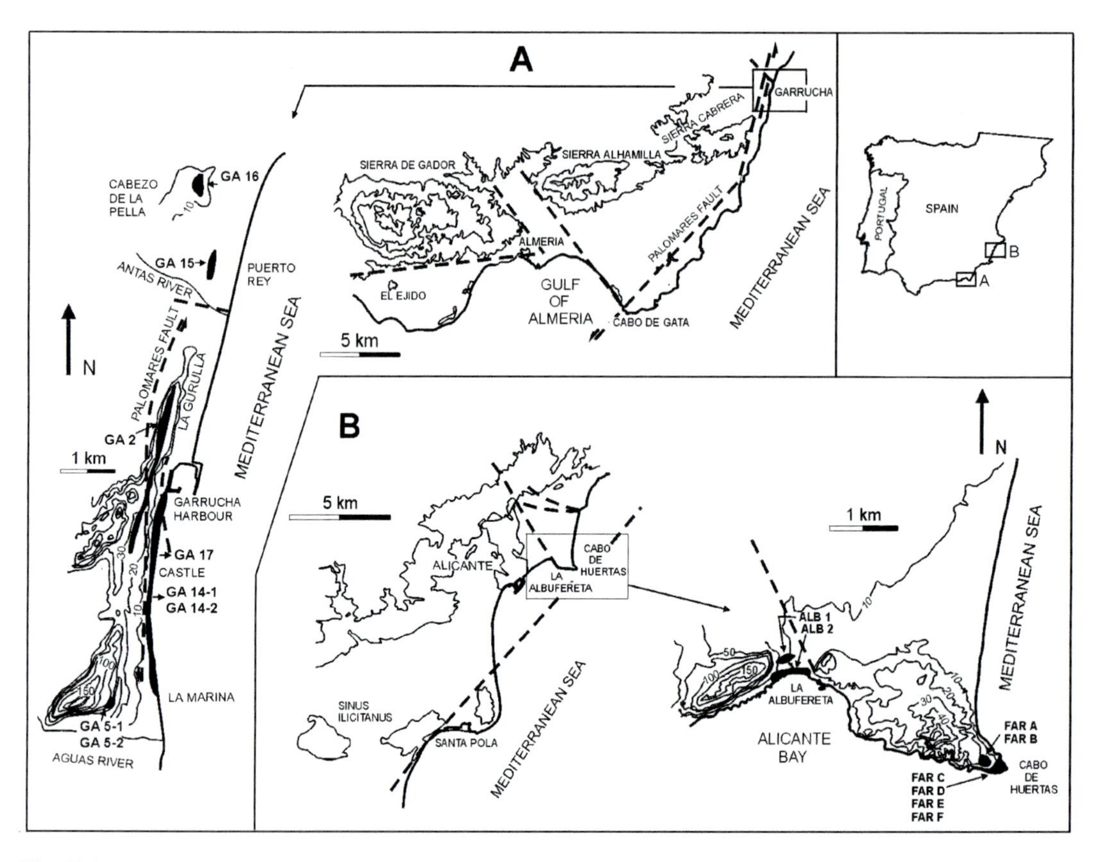

Fig. 19-1. Geographical and geological situation of areas studied. A. Cabo de Huertas zone. B. Garrucha zone. Faults controlling coastal and inland morphology are indicated.

highstand sea levels. We have selected the Cabo de Huertas site because of the diversity and abundance of fossil specimens of the Mollusca. The Garrucha-area site was chosen for purposes of comparison, since it has a similar number of marine terraces with an almost identical arrangement.

REGIONAL GEOLOGY AND GEOMORPHOLOGY

Cabo de Huertas (FAR and ALB Zones)

The Cabo de Huertas area is on the Mediterranean coast of the Iberian Peninsula (figure 19-1) north of the city of Alicante, and marks the abrupt end of the sinus of Elche (*Sinus Ilicitanus*), a recent flat depression filled with Upper Pleistocene–Holocene deposits. Two different zones comprise the area: Albufereta (ALB) and Cabo de Huertas s.s. (FAR).

The substratum of the Cabo de Huertas section (FAR) consists of thin-bedded, green–yellow marine sandstones and siltstones of Upper Tortonian age, dipping to the west. The modern Cabo de Huertas morphology reflects its Pleistocene evolution: the flat top was a Middle Pleistocene marine abrasion platform at 29 m a.s.l. (above sea level). The abrupt front reflects Pre-Tyrrhenian/Tyrrhenian cliff building. Minor forms at its base are two beach deposits (at 2 and 0.2 m a.s.l.) from the Tyrrhenian cycle, the higher one being capped by two dune units with intervening colluvial sediment. To the west, near the Serra Grossa highs on the Albufereta beach (ALB), are two small Pleistocene marine terraces at 10 m (ALB1) and 0.2 m (ALB2) a.s.l.

The Cabo de Huertas (FAR and ALB) area is in the Betic structural dominion, where strong Alpine tectonics have occurred. Neotectonics and seismicity are factors controlling recent morphological evolution in the zone. According to Fumanal et al. (1997), there are two fault systems in the area: SW–NE and SE–NW, the former controlling coastline morphology.

Garrucha (GA)

The Garrucha area (GA) is in the province of Almería, 200 km to the SW of Alicante (figure 19-1B). In this area, a complex marine terrace series has been identified at six different points: La Marina (GA5-1,2), La Gurulla (GA2), Castle

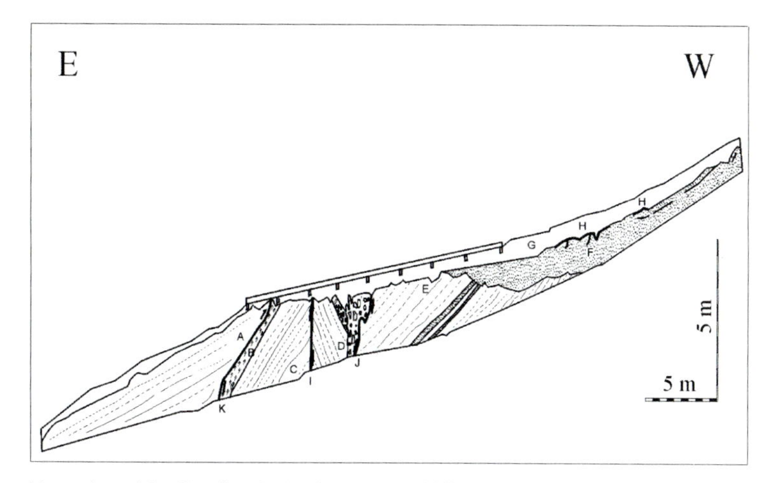

Fig. 19-2. Stratigraphic section of La Gurulla (GA2) plus a 25 m middle terrace. A, C, D, and E, Pliocene–Lower Pleistocene silt and fine sand; E, with two quartzous gravel intercalations; F, tilted gravelly Pleistocene terrace with marine mollusks; G, sandy and gravelly red soil; H, 10-cm-thick laminated caliche laminae broken and dipping eastwards; D′, fault breccia blocks; I and J, faults linked to the strike–slip fault complex of the Palomares fault (the latter showing sand injection along the fault surface); K, fault associated with gravitational processes.

(GA14-1,2; GA17), Puerto Rey (GA15), and Cabezo de la Pella (GA16).

The highest levels (La Marina GA5-1,2) are near the northern margin of the Aguas river. The uppermost stratigraphic record is represented by a wave-abrasion platform carved on Triassic dolostones, with patches of gravel strongly cemented by iron oxides and carbonate. No organic remains have been found here. This surface is 45 m a.s.l. The GA5-1 (40 m a.s.l.) and GA5-2 (38 m. a. s.l) marine terrace levels lie on Triassic slates and dolostones.

The intermediate terrace level, La Gurulla (GA2), consists of a long, narrow strip of hills running along the present shoreline, 28 m a.s.l., the upper houses of the village of Garrucha having been built on their surface. Near the northern limit of the studied area, there is a mesa-like surface, the Cabezo de la Pella (GA16), located at 25 m a.s.l. The lack of connection between the La Gurulla (GA2) deposits and Cabezo de la Pella (GA16) may be explained by the erosion of the Antas River, which had previously (between the intermediate marine terrace and Tyrrhenian cycle terrace deposition times) built a fan delta system. An example of the geological framework of the Garrucha (GA) area appears in the La Gurulla (GA2) intermediate terrace level, along a road cut inside the village (figure 19-2). At this point, the intermediate terrace (GA2) marine deposits are composed of strongly cemented quartz, carbonate, and slate (mostly micaschist) gravels, which are strongly carbonate-cemented. They are capped by a 5–10-cm-thick caliche and a 30–40-cm-thick sandy (gravelly) red soil that is missing in other La Gurulla hill points (because of erosion). The terrace lies unconformably on "Pliocene s.l." carbonate (nodulous) yellow silts that show alternating beds of gravel and sand along the road-cutting. The "Pliocene s.l." complex dips strongly seawards, and is affected by two faults related to the Palomares fault (a strike–slip fault), which remained active for a long time and were responsible for the tilting of the intermediate marine-terrace level, as seen in the caliche beds.

The Tyrrhenian cycle terraces appear along the beach at Garrucha Castle (GA14-1,2; GA17). At the northern limit of the village, they disappear as a result of neotectonic and Antas river erosive action. The GA17 terrace reappears at Puerto Rey (GA15), a tourist resort.

The Garrucha zone is in the internal Betic realm, an area that was, and still is, under strong neotectonic influence. The most prominent phenomenon is the Palomares fault. Since the end of the Pliocene, the Sierra Cabrera has risen 300 m. The effect of this fault on marine terrace distribution will be evident in further discussion.

STRATIGRAPHY

Cabo de Huertas Area (Cabo de Huertas s.s. (FAR) and Albufereta (ALB))

The first description of Cabo de Huertas Pleistocene stratigraphy appears in Dumas (1981), a paleontological study of Tyrrhenian deposits being included in Sillero et al. (1993).

Cabo de Huertas s.s. (FAR)

In the Cabo de Huertas s.s. Section (FAR), the highest and oldest, Pleistocene marine-terrace deposits are at 29 m a.s.l. (FAR-A in figure 19-3). They consist of highly porous, sandy, biomicritic, algal limestone, 1.5 m in thickness, with a "trivial"

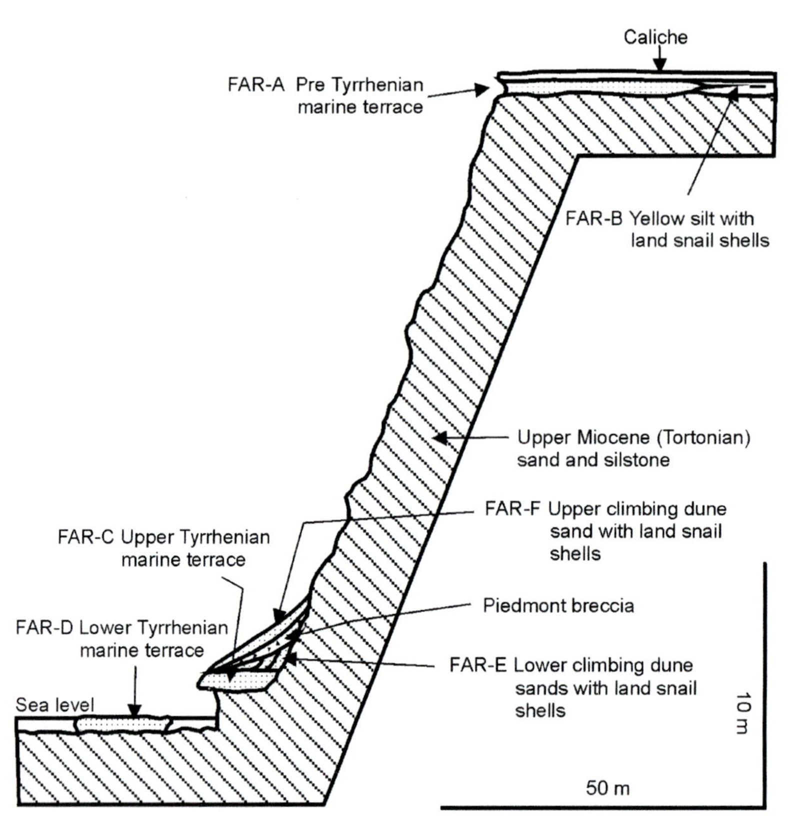

Fig. 19-3. Cabo de Huertas s.s. stratigraphic section. FAR-A, uppermost marine terrace; FAR-B, eolian silts; FAR-C, Tyrrhenian upper terrace; FAR-D, Tyrrhenian lower terrace; FAR-E, lower eolian sands; FAR-F, upper eolian sands.

(as a water temperature indicator) fossil assemblage. Landward, this marine deposit "interfingers" with yellow calcareous silts, of unknown thickness, which contain abundant land snail (Helicacea) shells (FAR-B in figure 19-3). According to Dumas (1981), uranium/thorium (U/Th) dating of a *Glycymeris* shell was beyond the range of the method (>350 ky).

The oldest (highest) Tyrrhenian level (FAR-C in figure 19-3) consists of carbonate-cemented bioclastic sand and gravel, measuring 2 m in thickness. The bed is very rich in fossils, with abundant "Senegalian complex" representatives. *Glycymeris* sp. shells are not frequent. Land-snail shells are not rare at the top of the strata. A date of 91,000 ± 1,000 yr BP was obtained using ^{234}U/^{230}Th dating of a *Strombus* sp. shell (Dumas, 1981).

The latest Tyrrhenian deposits are at the present sea level. They consist of bioclastic sandstone and gravel measuring 35 cm in thickness (FAR-D in figure 19-3). Their faunal content is similar to that of the upper Tyrrhenian level, although the density of fossils is lower. *Glycymeris* sp. shells are not frequent.

The uppermost Tyrrhenian marine terrace is covered with a complex of continental deposits, and the complex bottom is composed of eolian sands (climbing dunes) (FAR-E in figure 19-3) with abundant land-snail shells. These eolian sands are covered with unconsolidated piedmont breccia, of 0.5 m average thickness, and at their top there is a 2–3-m-thick eolian sand and silt-bed alternation with very abundant land snail shells (FAR-F in figure 19-3). In the uppermost part, some land snail shells were found within infilled burrows (of rodents?). According to Dumas (1981), the lower eolian beds are beyond the range of the ^{14}C dating method (on Helicacea) and the uppermost ones are ^{14}C dated at 22,800 ± 1000 yr BP.

In the Albufereta zone (ALB), the uppermost marine terrace (ALB-1) at 10 m a.s.l., consists of a 2-m-thick medium- to coarse-grained brown sand with medium-scale trough cross-bedding. The thickness is 0.5 m, and the faunal content is very sparse.

The lowermost marine terrace (ALB-2) in the Albufereta zone is 20–30 cm a.s.l. and consists of bioclastic very coarse sand/very fine gravel. Immature individual bivalve shells were recovered and analyzed.

Garrucha Area (La Marina GA5-1,2; La Gurulla GA2; Castle GA14-1,2 and GA17; Puerto Rey GA-15; Cabezo de la Pella GA16)

La Marina (GA5-1,2)

The highest marine deposits appear in La Marina (GA5-1,2) at 35 and 40 m a.s.l. The highest (GA5-1) consists of 1.2 m of medium/fine-sized gravel in 20–30-cm-thick sets with low-angle planar cross-bedding. These are strongly carbonate cemented and are capped by a 10-cm-thick caliche. The fossil remains consist of strongly eroded and well-rounded *Glycymeris* sp. fragments. The lower deposit (GA5-2) shows similar characteristics to that described above (GA5-1). The GA5-1 and GA5-2 outcrops have been destroyed by building since our 1997 field campaign.

La Gurulla (GA2)

The intermediate terrace level, La Gurulla hill (GA2) at 28 m a.s.l., consists of 1.4–2.1 m of strongly cemented quartzous, calcareous and slaty conglomerates, usually in disorganized 0.3–0.5-m-thick sets. The faunal content is poor. This level lies on strongly fault-affected Pliocene–Lower Pleistocene materials. The anomalous tilting of the terrace may be observed (figure 19-2).

Cabezo de la Pella (GA16)

At the northern end of the studied zone is the Cabezo de la Pella (GA-16), a small mesa-like hill (25 m a. s.l.) capped by 1-m-thick coarse conglomerates. The faunal content is poor. During the summer of 1999 it was destroyed because of building.

Castle (GA14-1,2; GA17)

The lower (Tyrrhenian) terraces 3.0 and 0.5 m a.s.l. (GA14-1 and GA14-2) outcrop at the southern limit of the Garrucha village, below the Castle (the lowermost one makes up the current beach-rock). The uppermost level (GA14-2) is 2.6 m thick and consists of strongly carbonate-cemented calcareous, slaty and quartzous horizontally bedded gravel. The lowermost level (GA14-1) is also conglomeratic. A minor fault linked to the Palomares fault affected the lower terrace. GA17 is a sampling station located in a building-foundation trench; on the basis of geomorphology, this may be correlated with the Tyrrhenian (GA14-2) upper terrace at the Castle.

Puerto Rey (GA15)

The only Tyrrhenian terrace at this site (GA15) consists of sediments that are finer grained than those at the Castle (GA14-1,2 and GA17). They consist of very fine gravels, which are strongly carbonate-cemented and cross-bedded, with large-sized gravel interbeds.

The faunal content of the Castle (GA14-1,2 and GA17) and Puerto Rey (GA15) Tyrrhenian deposits is virtually monospecific, consisting of 10-cm-thick lenses of well-preserved *Glycymeris* valves.

SAMPLE PREPARATION

Shells were carefully washed with running water and cleaned by extensive dissolution in 2 N HCl. In bivalve shells of sufficient size, a discoid sample was removed from an area as close as possible to the beak, using a hollow diamond drill. For the Gastropoda, samples were drilled from the last whorl.

For the chemical preparation we used the method described by G. A. Goodfriend and V. Meyer (personal communication, 1992) (see also Goodfriend and Mitterer, 1988; Goodfriend and Meyer, 1991; and Goodfriend, 1991). In short, this method is as follows: hydrolysis of approx. 80 mg shell was performed using a mixture of 12 N hydrochloric acid (2.9 µl/mg) and 6 N hydrochloric acid (100 µl) in sealed test tubes at 100°C for 20 hr under an inert (N_2) atmosphere. After sample desalting with HF and vacuum-drying, derivatization was carried out by (1) esterification with thionyl chloride in isopropanol, and (2) reaction with 150 µl trifluoroacetic acid anhydride (TFAA) (25% in dichloromethane).

Later, the samples were dissolved in *n*-hexane and injected into a Hewlett-Packard 5890 Series II gas chromatograph with an HP 6850 automatic sample injector. We used helium as the carrier gas and a Chirasil-Val column (25 m; Alltech). Integration of the peak areas was carried out using the PEAK96 integration program from Hewlett-Packard. To control our gas chromatography (GC) analysis results we have analyzed Interlaboratory Comparison (ILC) samples (Wehmiller, 1984) (table 19-1).

RESULTS AND DISCUSSION

For the purposes of regional comparison, we employ here the aminozones defined by Hearty (1987) for Tyrrhenian deposits from the Mediterranean island of Mallorca. According to data from the National Meteorological Institute, the current mean annual temperatures (CMATs) for both Mallorca and Cabo de Huertas range between 15.0 and 17.5°C, whereas in Garrucha, the CMAT is 17.5–20.0°C. In any case, a major factor might be the very shallow burial depth (tables 19-2 and 19-3) for almost all the samples, because of the thinness of the marine-terrace deposits.

Cabo de Huertas (FAR and ALB)

Cabo de Huertas s.s. (FAR)

In the Cabo de Huertas s.s. section (FAR), 181 samples have been analyzed (table 19-2). Because of the difficulty of obtaining samples of a single

TABLE 19-1 Comparison of results of analysis of Interlaboratory Comparison exercise standards (Wehmiller, 1984) in the Biomolecular Stratigraphy Laboratory of the Madrid School of Mines (Torres et al., 1997) and previously reported values (Wehmiller, 1984)

Amino acid[a]	ILC samples	ILC-A		ILC-B		ILC-C	
D-Aile/L-Ile	ILR	0.212	±0.072	0.540	±0.162	1.250	±0.030
	BMSL	0.180		0.650		1.245	
Pro	ILR	0.278	±0.068	0.595	±0.210	0.810	±0.260
	BMSL	0.195		0.507		0.786	
Leu	ILR	0.196	±0.042	0.497	±0.098	0.833	±0.086
	BMSL	0.182		0.444		0.849	
Asp	ILR	0.378	±0.056	0.705	±0.056	0.894	±0.158
	BMSL	0.373		0.728		0.914	
Phe	ILR	0.239	±0.040	0.583	±0.108	0.873	±0.178
	BMSL	0.220		0.608		0.885	
Glu	ILR	0.203	±0.022	0.432	±0.034	0.849	±0.070
	BMSL	0.185		0.426		0.832	

[a] D-Aile/L-Ile = D-alloisoleucine/L-isoleucine, Pro = D/L proline, Leu = D/L leucine, Asp = D/L aspartic acid, Phe = D/L phenylalanine, Glu = D/L glutamic acid.

genus from all sites, samples of a wide variety of genera, or even of unidentified fragments, have been sampled and analyzed to obtain reference values for neighboring uplifted or tilted areas. We present the results as a descriptive statistical table (table 19-4) of D-alloisoleucine/L-isoleucine (D-Aile/L-Ile), D/L leucine (Leu), D/L glutamic acid (Glu) and D/L aspartic acid (Asp) values and in plots of mean D-Aile/L-Ile values (and ranges) for all genera (figure 19-4).

The average D/L values of Ile, Leu, Glu, and Asp for samples from the Cabo de Huertas area (FAR) (table 19-4 and figure 19-4), imply that there is good chronostratigraphic agreement between FAR-A (uppermost marine terrace) and FAR-B (yellow eolian silts). In spite of the similarity of racemiza-

TABLE 19-2 Analyzed samples from the Cabo de Huertas zone, showing stratigraphic levels and genera. Burial conditions

Localities[a]	Material	Distance (m) from: front	Distance (m) from: top
Cabo de Huertas Section (FAR)			
FAR-A: Pre-Tyrrhenian marine terrace	Echinodermata (4), Fragmenta indet. (9), *Glycymeris* sp. (11), *Patella* sp. (7), *Spondylus* sp. (4)	0.40	0.50
FAR-B: Pre-Tyrrhenian aeolian silts	Helicacea (6)	0.40	0.50
FAR-C: Tyrrhenian upper marine terrace	*Arca* sp. (7), *Monodonta* sp. (10), Helicacea (10), *Osilinus* sp. (5), *Glycymeris* sp. (15), Ostreiodae (5), *Lima* sp. (10), *Patella* sp. (10), *Spondylus* sp. (10), *Venus* sp. (6), *Lucinella* sp. (4)	0.50	1.50
FAR-D: Tyrrhenian lower marine terrace	*Monodonta* sp. (6), Ostreidae (6), *Glycymeris* sp. (11), *Pecten* sp. (2), *Venus* sp. (6)	0.40	0.10
FAR-E: Lower aerolian sands	Helicacea (10)	0.30	1.20
FAR-F: Upper aerolian sands	Helicacea (5)	0.40	2.0
Albufereta zone (ALB)			
ALB-1: upper level	*Glycymeris* sp. (12)	0.40	0.50
ALB-2: lower level	*Arca* sp. (3), *Cardium* sp. (10), *Glycymeris* sp. (2), *Lucinella* sp. (2), *Tellina* sp. (2), *Venus* sp. (3)	0.10	0.00

[a] FAR = Cabo de Huertas s.s., ALB = Albufereta.

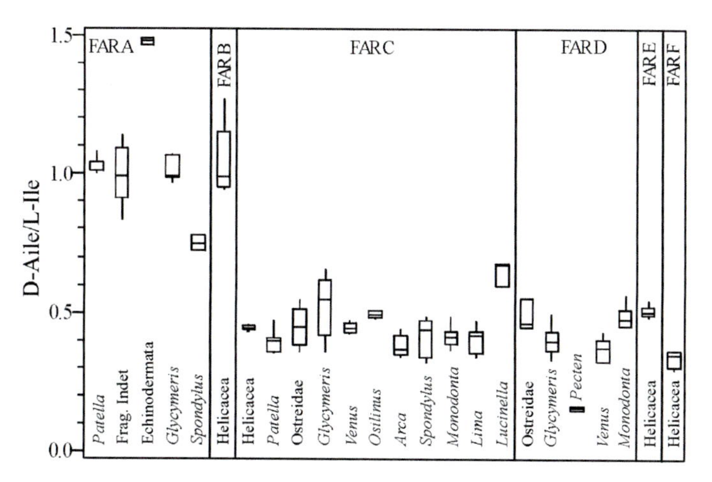

Fig. 19-4. Box-and-whisker graphs (median, Q_1, Q_3, min. and max.) of D-Aile/L-Ile ratios in samples from the Cabo de Huertas s.s. (FAR) section.

tion results from these two sites, we have not established relationships between racemization rates in the Helicacea and marine bivalves. According to our experience, gastropods racemize faster than bivalves, but at high D-Aile/L-Ile values, differences among genera are reduced.

Levels FAR-C (upper Tyrrhenian level) and FAR-D (lower Tyrrhenian level) show slight differences in their racemization ratios (figure 19-4), with the racemization ratios being slightly lower in the lower Tyrrhenian terrace (FAR-D). In general, members of the Gastropoda show higher racemization values than similar-age bivalves; the lowest racemization values always appear in bivalve samples, and members of the Echinodermata racemize faster than members of the Mollusca. This has also been observed in Holocene marine samples from the Gulf of Cadiz (cf. Torres et al., 1996). In the FAR-C and FAR-D stratigraphic levels, the racemization ratios of Gastropoda samples seemed to be more homogeneous than those in Pelecypoda samples.

TABLE 19-3 Analyzed samples from the Garrucha zone, showing stratigraphic levels and genera. Burial conditions

Localities[a]	Material	Distance (m) from: front	Distance (m) from: top
Garrucha zone (GA)			
GA5-1: La Marina highermost terrace (sublevel 1)	*Glycymeris* sp. (8)	0.30	0.30
GA5-2: La Marina highermost terrace (sublevel 2)	*Glycymeris* sp. (15)	0.30	1.00
GA2: La Gurulla intgermediate terrace level	*Glycymeris* sp. (11), *Cardium* sp. (8), *Venus* sp. (7)	0.30	0.50
GA16: Cabezo de La Pella intermediate terrace level	*Glycymeris* sp. (8)	0.30	0.50
GA14-2: The Castle upper Tyrrhenian level	*Glycymeris* sp. (10)	0.20	0.50
GA17: The Castle upper Tyrrhenian level	*Glycymeris* sp. (4)	>0.50	1.20
GA14-1: The Castle lower Tyrrhenian level	*Glycymeris* sp. (9), *Cardium* sp. (4), *Conus* sp. (4), *Strombus* sp. (1)	0.10	0.10
GA15: Puerto Rey Tyrrhenian level	*Glycymeris* sp. (16), *Cardium* sp. (16) *Conus* sp. (4), *Strombus* sp. (1)	>2.00	0.50

[a] GA = Garrucha.

TABLE 19-4 Descriptive statistics of amino acid epimerization/racemization of samples from Cabo de Huertas s.s. (FAR)

Location/Genus	D/L	*n*	min	max	mean	S.D.[a]
FAR-A						
Patella	D-Aile/L-Ile	7	1.00	1.08	1.03	0.027
	Leu	7	0.74	0.76	0.75	0.008
	Asp	7	0.71	0.74	0.73	0.009
	Glu	7	0.58	0.60	0.59	0.006
Frag. indet.	D-Aile/L-Ile	9	0.83	1.14	1.00	0.104
	Leu	9	0.55	0.72	0.65	0.068
	Asp	9	0.68	0.72	0.70	0.011
	Glu	9	0.59	0.72	0.65	0.041
Echinoder.	D-Aile/L-Ile	4	1.48	1.51	1.49	0.013
	Leu	4	0.76	0.83	0.80	0.031
	Asp	4	0.78	0.79	0.79	0.005
	Glu	4	0.90	0.92	0.90	0.010
Glycymeris	D-Aile/L-Ile	11	0.97	1.07	0.01	0.038
	Leu	11	0.75	0.88	0.82	0.036
	Asp	11	0.67	0.69	0.68	0.007
	Glu	11	0.63	0.71	0.65	0.031
Spondylus	D-Aile/L-Ile	4	0.72	0.78	0.75	0.029
	Leu	4	0.48	0.55	0.53	0.031
	Asp	4	0.70	0.73	0.71	0.011
	Glu	4	0.44	0.47	0.46	0.010
FAR-B						
Helicacea	D-Aile/L-Ile	5	0.94	1.27	1.04	0.133
	Leu	6	0.74	0.98	0.83	0.086
	Asp	6	0.76	0.84	0.80	0.034
	Glu	6	0.70	0.78	0.73	0.025
Patella	D-Aile/L-Ile	10	0.35	0.47	0.39	0.037
	Leu	10	0.31	0.36	0.33	0.016
	Asp	10	0.55	0.57	0.56	0.008
	Glu	10	0.25	0.28	0.26	0.009
Glycymeris	D-Aile/L-Ile	13	0.36	0.66	0.53	0.102
	Leu	14	0.24	0.53	0.36	0.098
	Asp	15	0.36	0.80	0.57	0.015
	Glu	15	0.30	0.57	0.45	0.095
FAR-C						
Helicacea	D-Aile/L-Ile	10	0.43	0.46	0.44	0.010
	Leu	10	0.45	0.49	0.48	0.012
	Asp	10	0.55	0.60	0.57	0.015
	Glu	10	0.38	0.40	0.39	0.006
Ostreidae	D-Aile/L-Ile	5	0.36	0.55	0.45	0.072
	Leu	5	0.31	0.47	0.41	0.066
	Asp	5	0.51	0.72	0.60	0.098
	Glu	5	0.28	0.50	0.40	0.080
Venus	D-Aile/L-Ile	6	0.42	0.47	0.44	0.018
	Leu	6	0.28	0.48	0.34	0.074
	Asp	6	0.61	0.64	0.62	0.011
	Glu	6	0.35	0.45	0.39	0.035
Osilinus	D-Aile/L-Ile	5	0.48	0.51	0.50	0.014
	Leu	5	0.41	0.46	0.43	0.022
	Asp	5	0.55	0.56	0.56	0.003
	Glu	5	0.41	0.41	0.41	0.003
FAR-C						
Arca	D-Aile/L-Ile	9	0.34	0.44	0.38	0.039
	Leu	9	0.34	0.43	0.39	0.029
	Asp	9	0.55	0.61	0.59	0.021
	Glu	9	0.32	0.41	0.37	0.024
Spondylus	D-Aile/L-Ile	9	0.32	0.49	0.41	0.067
	Leu	9	0.30	0.51	0.40	0.095
	Asp	10	0.49	0.80	0.62	0.123
	Glu	9	0.26	0.49	0.38	0.073
Monodonta	D-Aile/L-Ile	10	0.36	0.49	0.42	0.036
	Leu	8	0.38	0.46	0.41	0.026
	Asp	10	0.58	0.61	0.60	0.007
	Glu	9	0.33	0.44	0.38	0.032
Lima	D-Aile/L-Ile	10	0.34	0.47	0.40	0.045
	Leu	10	0.18	0.32	0.24	0.058
	Asp	10	0.50	0.54	0.52	0.015
	Glu	10	0.24	0.27	0.25	0.012
Lucinella	D-Aile/L-Ile	3	0.60	0.68	0.65	0.042
	Leu	3	0.51	0.63	0.59	0.067
	Asp	3	0.69	0.76	0.73	0.039
	Glu	3	0.46	0.51	0.49	0.029
Monodonta	D-Aile/L-Ile	6	0.45	0.57	0.49	0.042
	Leu	6	0.42	0.52	0.49	0.033
	Asp	6	0.60	0.62	0.61	0.011
	Glu	6	0.41	0.46	0.44	0.018
Ostreidae	D-Aile/L-Ile	3	0.45	0.55	0.49	0.057
	Leu	3	0.30	0.46	0.39	0.081
	Asp	6	0.54	0.57	0.56	0.009
	Glu	5	0.31	0.41	0.36	0.041
FAR-D						
Glycymeris	D-Aile/L-Ile	9	0.33	0.50	0.40	0.050
	Leu	9	0.27	0.41	0.35	0.044
	Asp	10	0.40	0.59	0.50	0.058
	Glu	10	0.29	0.43	0.37	0.041
Pecten	D-Aile/L-Ile	2	0.11	0.13	0.12	0.011
	Leu	2	0.08	0.09	0.09	0.011
	Asp	2	0.03	0.39	0.21	0.253
	Glu	2	0.15	0.17	0.16	0.011
Venus	D-Aile/L-Ile	6	0.32	0.43	0.37	0.042
	Leu	6	0.27	0.40	0.33	0.057
	Asp	6	0.52	0.60	0.57	0.031
	Glu	6	0.30	0.38	0.34	0.029
FAR-E						
Helicacea	D-Aile/L-Ile	9	0.48	0.55	0.51	0.020
	Leu	10	0.56	0.64	0.59	0.025
	Asp	10	0.67	0.71	0.69	0.012
	Glu	10	0.44	0.46	0.45	0.008
Helicacea	D-Aile/L-Ile	4	0.29	0.36	0.34	0.033
	Leu	5	0.48	0.60	0.54	0.043
	Asp	5	0.62	0.68	0.66	0.027
	Glu	5	0.38	0.45	0.43	0.027

[a] S.D., Standard deviation.

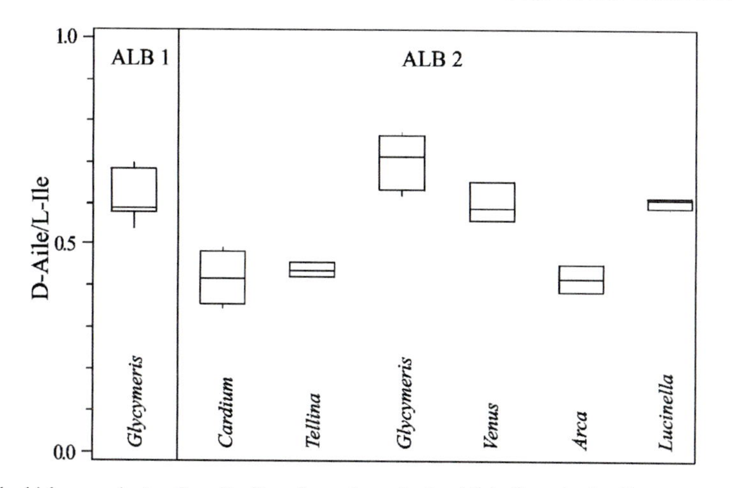

Fig. 19-5. Box-and-whisker graphs (median, Q_1, Q_3, min. and max.) of D-Aile/L-Ile ratios in all genera from Albufereta (ALB).

AAR ratio in land snail shells from FAR-E and FAR-F (eolian sands) levels are markedly consistent with the stratigraphy of these two units. However, an isolated land snail shell recovered from the topmost part of the upper Tyrrhenian terrace (FAR-C) had lower D/L values than Helicacea samples from the overlying eolian bed (FAR-E). This anomaly might be explained by a more rapid burial of the FAR-C shell from climbing dunes, thus reducing initial racemization due to heating by insolation.

Comparison of the D-Aile/L-Ile values (figure 19-4) confirms that there is an important gap between Pre-Tyrrhenian marine terraces (FAR-A) (beyond the range of the U/Th dating) and Tyrrhenian marine terraces (FAR-C and FAR-D). However, the epimerization ratio differences are seen to be far more marked in *Glycymeris* sp. and *Patella* sp. than in *Spondylus* sp. Racemization ratios in land snail shells also reflect the aforementioned stratigraphic gap. Average D-Aile/L-Ile ratios from samples from units FAR-C and FAR-D confirm their chronostratigraphic differentiation.

According to Hearty (1987), a complex Tyrrhenian aminozone (zones E and F–G of Hearty) is recognized on the Mediterranean island of Mallorca.

The mean D-Aile/L-Ile values from the F–G aminozone on Mallorca range between 0.51 and 0.57 in *Glycymeris* sp. and 0.40 and 0.45 in *Arca* sp.; in the Tyrrhenian upper deposits of the Cabo de Huertas (FAR-C), the values are 0.53 and 0.38, respectively.

Aminozone E was U-series-dated (*Cladocora caespitosa*, a coral) at 129 ± 7 ky. The mean D-Aile/L-Ile value was 0.40 ± 0.03 in *Glycymeris* sp. and 0.27 in *Arca* sp. For all samples from the Tyrrhenian lower terraces of Mallorca, the D-Aile/L-Ile values range between 0.38 and 0.46 in *Glycymeris* sp. and between 0.27 and 0.34 in *Arca*. In *Glycymeris* sp. from the Tyrrhenian lower deposits of the Cabo de Huertas (FAR-D), the D-Aile/L-Ile averages 0.40.

La Albufereta (ALB)

In the Albufereta zone (ALB), 34 samples have been analyzed. The results (table 19-5 and figure 19-5) do not reveal marked chronological differences between the two terraces. In fact, racemization data for Ile, Glu, and Asp from the only common species (*Glycymeris* sp.) show lower values in the higher terrace (ALB1) than in the lower one (ALB2). Although there might be a tectonic explanation for this apparent age relationship (as stated by Dumas, 1981), it is also possible that this inversion of racemization results is due to thermal and diagenetic factors.

Garrucha (GA)

In the Garrucha zone, 109 samples were analyzed (tables 19-3 and 19-6) and D-Aile/L-Ile values are plotted in figure 19-6 for each site in the region. The analytical results from the uppermost marine terraces in the La Marina area (GA5-1 and GA5-2) samples are not good: some D-Aile/L-Ile ratios are erratic and D/L leucine peaks have not always been identified. Nevertheless, D-Aile/L-Ile values are slightly higher for the higher of these two units (GA5-1). D/L Asp and Glu ratios (table 19-6) from the lower stratigraphic level (GA5-2) are higher than those in the uppermost level (GA5-1). These data imply that both terraces were deposited within a relatively brief period of time, and that it is impossible to distinguish between them using AAR.

The *Glycymeris* sp. D-Aile/L-Ile values seen in figure 19-6 indicate that the upper terraces in the La Marina area (GA5-1 and GA5-2) and the higher terrace in the Garrucha village zone (GA2) are similar in age.

TABLE 19-5 Descriptive statistics of amino acid of epimerization/racemization of samples from Albufereta (ALB)

Locality/Genus	D/L	*n*	min	max	mean	S.D.
ALB1						
Glycymeris	D-Aile/L-Ile	7	0.54	0.70	0.61	0.058
	Leu	8	0.32	0.55	0.48	0.072
	Asp	8	0.45	0.63	0.51	0.059
	Glu	8	0.44	0.54	0.48	0.036
ALB2						
Glycymeris	D-Aile/L-Ile	4	0.62	0.77	0.70	0.067
	Leu	4	0.48	0.58	0.52	0.046
	Asp	4	0.68	0.81	0.75	0.054
	Glu	4	0.47	0.56	0.51	0.039
Cardium	D-Aile/L-Ile	8	0.35	0.49	0.42	0.059
	Leu	8	0.29	0.46	0.37	0.067
	Asp	10	0.24	0.57	0.49	0.096
	Glu	7	0.26	0.36	0.32	0.035
Tellina	D-Aile/L-Ile	2	0.42	0.46	0.44	0.023
	Leu	2	0.39	0.43	0.41	0.030
	Asp	2	0.51	0.56	0.54	0.033
	Glu	2	0.37	0.40	0.39	0.022
Venus	D-Aile/L-Ile	3	0.56	0.65	0.60	0.046
	Leu	3	0.41	0.48	0.44	0.037
	Asp	3	0.70	0.72	0.71	0.008
	Glu	3	0.46	0.47	0.46	0.007
Arca	D-Aile/L-Ile	3	0.39	0.45	0.42	0.034
	Leu	3	0.35	0.37	0.36	0.007
	Asp	3	0.50	0.55	0.52	0.028
	Glu	3	0.37	0.41	0.39	0.021
Lucinella	D-Aile/L-Ile	4	0.58	0.61	0.60	0.013
	Leu	4	0.52	0.57	0.54	0.018
	Asp	4	0.64	0.71	0.69	0.033
	Glu	4	0.46	0.47	0.47	0.004

TABLE 19-6 Descriptive statistics of amino acid of epimerization/racemization ratios of samples from Garrucha (GA)

Locality/Genus	D/L	*n*	min	max	mean	S.D.
ALB1						
Glycymeris	D-Aile/L-Ile	3	0.77	0.99	0.90	0.114
	Leu	0	—	—	—	—
	Asp	6	0.67	0.83	0.75	0.070
	Glu	6	0.97	0.77	0.71	0.038
GA5-2						
Glycymeris	D-Aile/L-Ile	11	0.73	1.18	0.93	0.144
	Leu	3	0.83	0.01	0.90	0.093
	Asp	12	0.64	0.98	0.81	0.079
	Glu	11	0.64	0.83	0.76	0.060
GA2						
Glycymeris	D-Aile/L-Ile	11	0.67	1.06	0.90	0.143
	Leu	8	0.73	0.88	0.82	0.063
	Asp	10	0.60	0.85	0.77	0.070
	Glu	9	0.62	0.76	0.69	0.038
Cardium	D-Aile/L-Ile	8	0.40	0.70	0.56	0.094
	Leu	8	0.28	0.52	0.44	0.075
	Asp	8	0.64	0.77	0.70	0.040
	Glu	8	0.40	0.60	0.53	0.058
Venus	D-Aile/L-Ile	7	0.56	0.71	0.64	0.056
	Leu	7	0.44	0.57	0.52	0.040
	Asp	7	0.73	0.78	0.76	0.016
	Glu	7	0.50	0.56	0.54	0.022
GA16						
Glycymeris	D-Aile/L-Ile	6	0.72	0.79	0.75	0.023
	Leu	6	0.49	0.57	0.52	0.027
	Asp	6	0.59	0.80	0.64	0.083
	Glu	6	0.54	0.63	0.59	0.033
GA14-2						
Glycymeris	D-Ale/L-Iso	9	0.48	0.76	0.59	0.086
	Leu	5	0.44	0.56	0.51	0.050
	Asp	10	0.65	0.86	0.78	0.064
	Glu	10	0.45	0.70	0.63	0.075
GA17						
Glycymeris	D-Aile/L-Ile	4	0.47	0.53	0.49	0.024
	Leu	4	0.53	0.72	0.59	0.091
	Asp	4	0.72	0.78	0.75	0.027
	Glu	4	0.47	0.54	0.50	0.033
GA-14-1						
Glycymeris	D-Aile/L-Ile	4	0.39	0.57	0.49	0.086
	Leu	4	0.39	0.59	0.47	0.092
	Asp	9	0.70	0.84	0.76	0.041
	Glu	8	0.48	0.59	0.55	0.037
Strombus	D-Aile/L-Ile	2	0.51	0.57	0.54	0.042
	Leu	2	0.47	0.54	0.50	0.050
	Asp	4	0.63	0.79	0.74	0.052
	Glu	4	0.55	0.69	0.64	0.061
GA15						
Glycymeris	D-Aile/L-Ile	12	0.49	0.60	0.52	0.027
	Leu	14	0.46	0.67	0.52	0.063
	Asp	15	0.64	0.79	0.71	0.052
	Glu	14	0.45	0.58	0.50	0.043
Cardium	D-Aile/L-Ile	4	0.40	0.48	0.43	0.033
	Leu	4	0.41	0.48	0.44	0.029
	Asp	4	0.55	0.60	0.58	0.018
	Glu	4	0.35	0.40	0.37	0.023
Conus	D-Aile/L-Ile	4	0.49	0.67	0.58	0.100
	Leu	4	0.34	0.46	0.41	0.050
	Asp	4	0.58	0.70	0.64	0.052
	Glu	4	0.38	0.50	0.46	0.058

The D-Aile/L-Ile averages in *Glycymeris* sp. samples from the Cabezo de la Pella (GA16) terrace are lower than at La Gurulla (GA2). We conclude that, in spite of their similar elevations, the Cabezo de la Pella (GA16) terrace is more recent than the La Gurulla terrace (GA2). We explain this as being a case of local uplift related to the Palomares fault. The *Glycymeris* sp. samples from the Cabezo de la Pella terrace (GA16) show an average D-Aile/L-Ile value of 0.75, thus differing markedly from the higher terraces at La Marina (GA5) (0.90) and La Gurulla (GA2) (0.90).

There is an important chronostratigraphic gap between the intermediate terrace level of La Gurulla (GA2) hill and the Castle Tyrrhenian levels (GA14-1,2 and GA17).

In the Tyrrhenian upper level from the Castle (GA14-2), *Glycymeris* sp. D/L values are higher than those from the lower level (GA14-1). The average D-Aile/L-Ile value in *Glycymeris* sp. from GA17—a deeply buried sampling point—is 0.49,

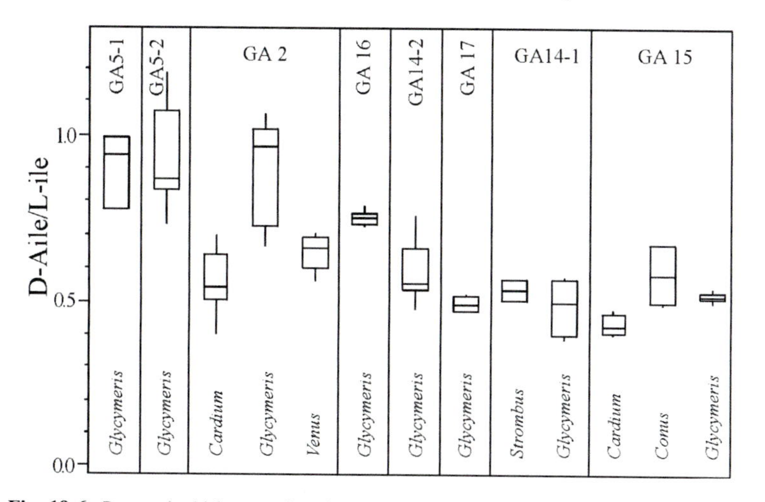

Fig. 19-6. Box-and-whisker graphs of D-Aile/L-Ile ratios in all genera from Garrucha (GA).

which is much lower than at the Castle (GA14-2) (0.59), perhaps because of weathering and heating (from shallow burial) of the latter samples. We consider that the GA-17 D-Aile/L-Ile value (0.49) represents the "true" average epimerization value of this terrace level when deeply sampled. We correlate the Tyrrhenian level from Puerto Rey (GA-15) with the Tyrrhenian upper terrace at the Castle (GA14-2). In the Tyrrhenian upper terrace at the Castle (GA14-2), the average epimerization ratio in *Glycymeris* sp. is 0.59. Taking into account the CMAT difference between Mallorca and Cabo de Huertas and the coast of Almería, as well as the influence of shallow sample burial, this value agrees well with its Mallorca equivalent of 0.51–0.57 (F–G aminozone, Hearty, 1987) and with the Cabo de Huertas (FAR-C) value of 0.53. If we consider that the mean D-Aile/L-Ile value in *Glycymeris* sp. from deeply buried (>1.2 m) strata is 0.49 (GA17), and 0.52 at Puerto Rey (GA15), we may conclude that there is a very good fit between Hearty's (1987) Mallorca values and those from the Garrucha area. The higher mean D-Aile/L-Ile and the abnormally high variability of D-Aile/L-Ile values in shallower samples may be explained by differential diagenesis or heating in the shallow subsurface.

Glycymeris sp. samples from the Tyrrhenian lower terrace at the Castle (GA14-1) have an average D-Aile/L-Ile value of 0.49; it seems that samples have been influenced by the shallow burial condition. In Mallorca, Hearty (1987) found an average value range of 0.34–0.46, and at Cabo de Huertas (FAR-D) the average value is 0.40.

The average D-Aile/L-Ile ratio in *Glycymeris* sp. from the Puerto Rey (GA15) Tyrrhenian terrace is 0.52, which is very similar to the value for the Tyrrhenian upper terrace at GA17. Taking into account the fact that these samples were deeply buried, and the fact that their epimerization ratios are lower than those from the upper Tyrrhenian terrace of Garrucha Castle (GA14-2), where sampling was performed in shallower beds, the Puerto Rey (GA15) deposits may be correlated with the Tyrrhenian higher terraces at El Castillo (GA14-2) and Cabo de Huertas (FAR-C), and with Hearty's (1987) E aminozone from Mallorca.

On the basis of mean D-Aile/L-Ile values for *Glycymeris* sp. (figure 19-7) five different aminozones can be distinguished in the studied area: (1) Uppermost Cabo de Huertas (FAR-A); (2) La Marina (GA5-1,2) and La Gurulla (GA-2); (3) Cabezo de la Pella (GA-16); (4) Upper Tyrrhenian deposits (FAR-C, GA14-1, and GA17); and (5) Lower Tyrrhenian deposits (FAR-D and GA14-2).

According to the maximum D-Aile/L-Ile values found, it seems evident that our aminostratigraphy is incomplete and does not cover the whole Pleistocene record, whose lower part could be in the Cuesta Colorada section (Torres et al., 1997) 10 km north of Almería city. The gap found between D-Aile/L-Ile values from the studied terraces suggests that an unknown number of intermediate terraces are lacking. Uppermost Pleistocene and Holocene deposits do not outcrop as terraces in the studied area, but occur as bay-infills.

Evaluation of the Reliability of D/L Amino Acid Values

In view of the large number of D/L values obtained in this study, it is important to develop criteria for the reliability of different amino acids (D-Aile/L-Ile, Leu, Glu, and Asp) in the most common genus used for aminostratigraphy in the region, i.e., *Glycymeris*.

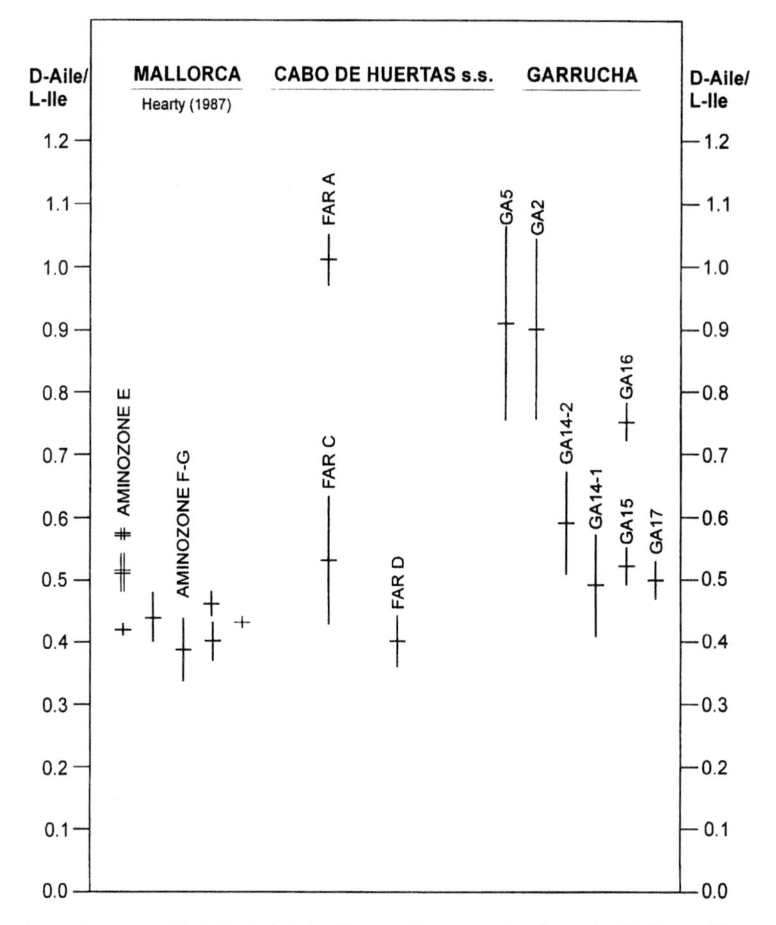

Fig. 19-7. Comparison of mean D-Aile/L-Ile ($\pm 1\sigma$) in *Glycymeris* sp. samples from the Mallorca (Hearty, 1987), Cabo de Huertas (FAR), and Garrucha (GA) areas. D-Aile/L-Ile values from F–G aminozone localities from Mallorca (solid lines). D-Aile/L-Ile values from the E aminozone are shown as double lines. GA14-1,2 epimerization values are higher for Mallorca because of the shallow burial of the samples.

Geomorphic characteristics are the first criteria to be applied to such a test, because in most terrace systems age increases with height above sea level (table 19-7). We selected the Garrucha (GA) area and Tyrrhenian terraces (FAR-C and FAR-D) from the Faro de Huertas zone. For further statistical calculations, we added a coefficient based on height a.s.l., i.e., 1 for the highest and 5 for the lowest.

TABLE 19-7 Stratigraphic classification of Garrucha and Cabo de Huertas marine terraces, according to their current topographical position above sea level

Topographical position	Localities
Highest	1 GA5
↓	2 GA2
	3 GA16
	4 GA14-2, GA15, GA17, FAR C
Lowest	5 GA14-1, FAR D

To evaluate possible effects of weathering or surficial heating, we classified the sampling points into three groups: shallow (GA2, GA5, GA14-1, GA14-2), intermediate (GA16), and deep-buried (GA15, GA17, FAR-C). Samples from (FAR-D) at the surf zone are included in the deep-burial group.

Table 19-8 shows the results of calculating the Spearman's rank correlation coefficient for the four amino acids, relative to the rank position of the sampled terraces. The results obtained demonstrate that the order of reliability is: D-Aile/L-Ile > Leu > Glu > Asp.

In order to test these results, we later calculated the Cronbach α-coefficient, using all *Glycymeris* sp. samples (FAR-A ones included), as a way of evaluating the overall reliability of the amino acids used. This method makes it possible to determine

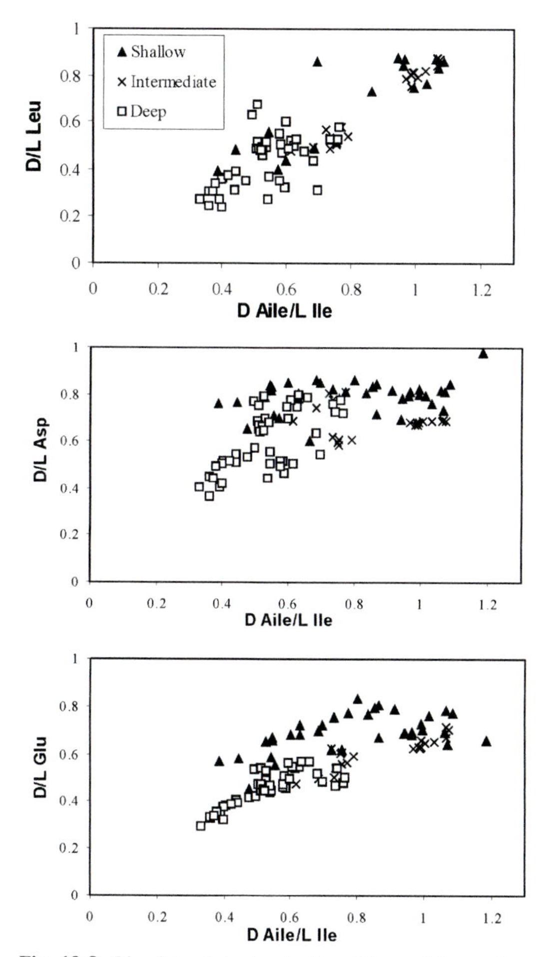

Fig. 19-8. Bivariate plots of D/L Asp, Glu, and Leu values vs. D-Aile/L-Ile values for *Glycymeris* sp. as a function of burial conditions (shallow, intermediate, and deep). The main localities sampled are indicated (ALB and GA-15 excepted).

the contribution made by the reliability of each item (variable) used, a comparison then being performed between overall reliability (after eliminating each item) and the reliability obtained using all the items. The α-value decreases as the different variables are removed stepwise, one by one, this decrease being much higher as the contribution of this variable to the final result becomes more important. The new reliability scale is: Leu > Glu > D-Aile/L-Ile > Asp, although the differences between the first three amino acids are very small.

To analyze this pattern further, D/L values of Leu, Glu, and Asp were plotted in relation to D-Aile/L-Ile, as a function of depth (figure 19-8). D/L Leu values were found to correlate well with D-Aile/L-Ile values for all depths. On the other hand, for both Glu and Asp, a good correlation with D-Aile/

TABLE 19-8 Spearman's rank correlation coefficient for the amino acids compared with the rank positions of the sampled terraces

	Rank correlation coefficient			
	D-Aile/L-Ile	Leu	Asp	Glu
Leu	0.782			
Asp	0.656	0.749		
Glu	0.809	0.789	0.865	
Relative position	0.843	0.711	0.493	0.710

L-Ile values was found only for deeply buried samples. This suggest that samples of shallow or intermediate depth may be diagenetically altered with respect to Asp and Glu.

The above discussion is summarized in figure 19-9A, which shows the results obtained using the mean D/L values in *Glycymeris* sp. in relation to the five age groups defined using elevation (table 19-7). D-Aile/L-Ile allow the sites to be correctly classified; Leu maintains the order, but does not allow a distinction to be made between two of them; Glu behaves in a similar way, but the result obtained reverses the observed field order between the two sites; and, lastly, Asp is not valid, in this case, for the desired objective.

The relationships seen for *Glycymeris* sp. samples in figure 19-9A can also be seen using the data for all the genera studied (figure 19-9B), although the relationships are less evident because data from different genera were combined. In spite of this, D-Aile/L-Ile and Leu still define the same aminozones as seen in figure 19-9A. The reliability of Asp and Glu acid varies considerably; this is probably related to the different kinetic behavior of these amino acids observed for different genera (Lajoie et al., 1980; Torres et al., 1997).

The lower reliability of Asp in old samples, as in this case, can be explained because of the extensive racemization reached (Asp is the "fastest" amino acid studied) in most of the terraces sampled but the influence of the "apparent reversal kinetics" detected in leached mollusk shells (Kimber et al., 1986; Kimber and Griffin, 1987) and in dentine (Torres et al., 1999) must be taken into account. In contrast, in well-preserved Plio–Pleistocene bivalves in the Arctic, Asp appears to give highly reliable results, showing less among-individual variation than D-Aile/L-Ile values (Goodfriend et al., 1996). In the context of the present study, Asp and Glu show a greater sensitivity to the varying preservation conditions of the different sites. In fact, the more superficial, less protected sites show significantly faster kinetics than the deeper, more protected, ones.

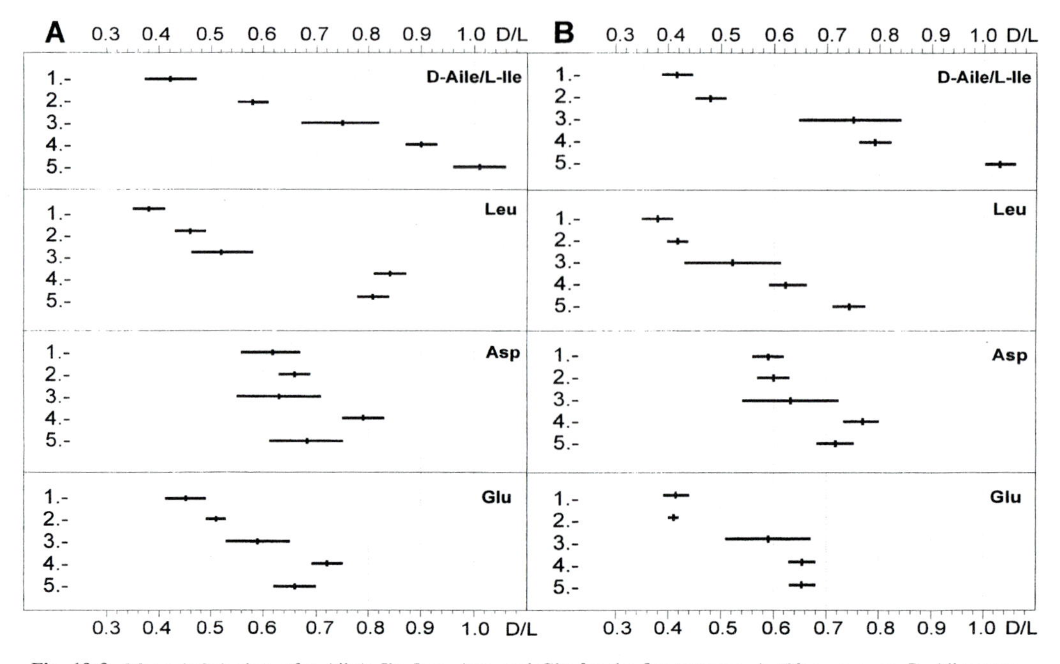

Fig. 19-9. Mean (±2σ) plots of D-Aile/L-Ile, Leu, Asp, and Glu for the five terraces. A. *Glycymeris* sp. B. All genera.

CONCLUSIONS

At Cabo de Huertas (FAR and ALB) there is a system of well-resolved marine terraces. Even in areas of tectonics, it appears that terraces have been deformed in regional units, so clear geomorphological relationships have been preserved. AAR analysis resolves the two Tyrrenian terraces (FAR-C and FAR-D) and permits correlation with the aminozones of the island of Mallorca (Hearty, 1987). The younger terrace (FAR-D) is dated at 129 ky (±7 ky). The U-series date of 91 ky (±1 ky) for a *Strombus* sp. sample from the Cabo de Huertas Upper Tyrrhenian terrace (FAR-C) must be rejected because of the low reliability of mollusk shells for U-series dating. A wide chronostratigraphical gap (220 ky minimum) exists between the older Pleistocene marine deposits (FAR-A) and the others (FAR-C and FAR-D). The confused aminostratigraphic situation of the Albufereta (ALB-1 and ALB-2) may be explained in terms of neotectonics, as stated by Dumas (1981), or, more probably, because of differential diagenesis and solar heating linked to their different burial conditions.

An extensive series of different molluscan genera has been analyzed, making it possible to establish an amino AAR for future comparisons with other Mediterranean marine deposits. In general, members of the Gastropoda show higher racemization ratios than do bivalves (Torres et al., 1996). In the Tyrrhenian levels (FAR-C and FAR-D), the racemization ratios of Gastropoda samples seemed to be more uniform than in that in Pelecypoda samples. We can conclude that in the stratigraphical record of Cabo de Huertas (FAR-C and FAR-D) at least two Tyrrhenian age sea levels are recorded.

In the Garrucha (GA) area, two Tyrrhenian-age (GA14-1, GA17, GA15, and GA14-2) sea levels are also recorded. The distribution of the older marine terraces is tectonically controlled and linked to the Palomares fault, a complex strike–slip fault system that is still active. In fact, according to D-Aile/L-Ile ratios for *Glycymeris* sp., the uppermost terrace levels (GA5-1 and GA5-2), now at 40 m a.s.l., were coeval with the intermediate terrace (GA2) level, the current height of which is at 28 m a.s.l. because of sinking and tilting linked to the Palomares fault. We explain the high D-Aile/L-Ile epimerization ratio variability as a diagenesis/thermal history-linked effect.

Neotectonics allows us to explain why the Cabezo de la Pella (GA16) terrace (25 m a.s.l.) cannot be correlated with La Gurulla (GA2) at 28 m a.s.l., the latter being younger according to their amino acid D/L ratios. Neotectonics probably acted as a preservational factor for some intermediate old marine deposits.

On the basis of the D-Aile/L-Ile ratios of *Glycymeris* sp., five high-sea-level events (aminozones) can be distinguished (figure 19-7): (1) Uppermost Cabo de Huertas (FAR-A); (2) La Marina (GA5-1,2) and La Gurulla (GA2); (3) Cabezo de la Pella (GA16); (4) Tyrrhenian upper terraces: Garrucha Castle (GA14-2, GA17), Puerto Rey (GA15), and upper Cabo de Huertas (FAR-C);

and (5) Tyrrhenian lower terraces: Garrucha Castle (GA14-1) and lowermost Cabo de Huertas (FAR-D).

We observed that the reliability of the different amino acids for the correct classification of the sites studied varies in the order D-Aile/L-Ile > Leu > Glu > Asp. The reliability of Glu and Asp is not sufficient to guarantee correct classification of all the sites. In particular, the reliability of Glu decreases significantly when samples belonging to different species are used, whereas D-Aile/L-Ile and Leu allow for correct site classification using different species.

Acknowledgments: We are deeply indebted to Dr V. Meyer of the University of Bern (now at the EMPA St. Gallen Organization) for the considerable help she provided in the setting up of our laboratory. Dr G. Goodfriend of the Carnegie Institution at Washington (now at the George Washington University, Washington DC) sent us the analysis protocol and GC program. Dr J. F. Wehmiller sent us standards that were very helpful in checking our analytical results. We would also like to thank the editors (Dr J. F. Wehmiller and Dr G. Goodfriend) and the referees (Dr W. McCoy and one anonymous referee) for their critical reading of the manuscript and their useful suggestions The Biomolecular Stratigraphy Laboratory has been partially funded by ENRESA (Spanish National Company for Radioactive Waste Management).

REFERENCES

Bardají T., Goy J. L., Mörner N. A., Zazo C., Silva P. G., Somoza L., Dabrio C., and Baena J. (1995) Towards a Plio-Pleistocene chronostratigraphy in Eastern Betic Basins (S.E.) Spain, *Geodin. Acta* **8-2**, 112–126.

Bardají T., Goy J. L., Silva P. G., Zazo C., Mörner N. A., Somoza L., Dabrio C. J., and Baena J. (1997) The Plio Pleistocene boundary in south east Spain: a review. *Quat. Intl.* **40**, 27–32.

Belluomini G., Branca M., Delitalia L., Pecorini G., and Spano C. (1986) Isoleucine epimerization dating of quaternary marine deposits in Sardinia, Italy. Z. *Geomorph. N.F.* Bd **62**, 109–117.

Belluomini G., Manfra L., and Proposito A. (1993) Una recente aminocronologia dei depositi marini Pleistocenici dell area di Montalto di Castro e Tarquinia (Viterbo). *Il Quaternario* **6**, 241–248.

Causse C., Goy J. L., Zazo C., and Hillaire-Marcel C. L. (1993) Potential chronologique (Th/U) des faunes Pléistocenes méditerranéennes: example des terrasses marines des regions de Murcie et Alicante (Sud-Est de l'Espagne). *Geodin. Acta* **6**, 121–134.

Dumas B. (1981) La Région d'Alicante (abstr.). In *Libret Guide Table Ronde sur le Thyrrenien d'Espagne*. INQUA. Comm. Lignes de Rivages (coord. E. Aguirre), 45–75.

Dumas B., Gueremy P., Hearty P. J., Lehnaff R., and Raffy J. (1988) Morphometric analysis and amino acid geochronology of uplifted shorelines in a tectonic region near Reggio Calabria, South Italy. *Palaeogeo. Palaeoclim. Palaeoecol.* **68,** 273–289.

Fumanal M. P., Rey J., Usera J., Martínez J., Mateu G., Blázquez A. M., and Ferrer C. (1997) El Proyecto La Nao: evolución cuaternaria del litoral meridional valenciano. *Cuaternario Ibérico*, 98–112.

Goodfriend G. A. (1991) Patterns of racemization and epimerization of amino acids in land snail shells over the course of the Holocene. *Geochim. Cosmochim. Acta* **55**, 293–302.

Goodfriend G. A. and Meyer V. (1991) A comparative study of the kinetics of amino acid racemization/epimerization in fossil and modern mollusk shells. *Geochim. Cosmochim. Acta* **55**, 293–302.

Goodfriend G. A., and Mitterer R. M. (1988) Late Quaternary land snails from the North coast of Jamaica: local extinctions and climate change. *Palaeogeo. Palaeoclim. Palaeoecol.* **63**, 293–311.

Goodfriend G. A., Brigham-Grette J., and Miller G. H. (1996) Enhanced age resolution of the marine Quaternary record in the Artic using aspartic acid dating of bivalve shells. *Quat. Res.* **45**, 176–187.

Goy J. L. and Zazo C. (1982) Niveles marinos cuaternarios y su relación con la neotectónica en el litoral de Almería (España). *Bol. Roy. Soc. Española Hist. Natl. (Geol.)* **80**, 171–184.

Goy J. L. and Zazo C. (1988) Sequences of Quaternary marine levels in Elche Basin (Eastern Betic Cordillera, Spain). *Palaeogeo. Palaeoclim., Palaeoecol.* **68**, 301–310.

Goy J. L. and Zazo C. (1989) The role of neotectonics in the morphologic distribution of the Quaternary marine and continental deposits of the Elche Basin, southeast Spain. *Tectonophysics* **163**, 219–225.

Goy J. L., Zazo C., Bardají T., Somoza L., Causse C., and Hillaire-Marcel C. (1993) Eléments d'une chronostratigraphie du Tyrrhénien des régions d'Alicante-Murcie, Sud-Est de l'Espagne. *Geodin. Acta* **6**, 103–119.

Hearty P. J., Miller G. H., Stearns C. E., and Szabo B. J. (1986) Aminostratigraphy of Quaternary shorelines around the Mediterranean basin. *Geol. Soc. Amer. Bull.* **97**, 850–858.

Hearty P. J. (1987) New data on the Pleistocene of Mallorca. *Quat. Sci. Rev.* **6**, 245–257.

Hillaire-Marcel C., Gariépy C., Ghaleb B., Goy J. L., Zazo C., and Cuerda J. (1996) U-series measurement in Tyrrhenian deposits from Mallorca. Further evidence for two last-interglaciar high sea levels in the Balearic islands. *Quat. Sci. Rev.* **15**, 53–62.

Kauffman D. (1992) Aminostratigraphy of Pliocene–Pleistocene high sea level deposits, Nome coastal plain and adjacent nearshore area, Alaska. *Geol. Soc. Amer. Bull.* **104**, 40–52.

Kimber R. W. L. and Griffin C. V. (1987) Further evidence of the complexity of the racemization process in fossil shells with implications for amino acid dating. *Geochim. Cosmochim. Acta* **16**, 839–846.

Kimber R. W. L., Griffin C. V., and Milnes A. R. (1986) Amino acid racemization dating: Evidence of apparent reversal in aspartic acid racemization with time in shells of Ostrea. *Geochim. Cosmochim. Acta* **56**, 1159–1161.

Lajoie K. R., Wehmiller J. F., and Kennedy G. I. (1980) Inter- and intrageneric trends in apparent racemization kinetics of amino acids in Quaternary mollusks. In *Biogeochemistry of Amino Acids* (eds P. E. Hare, T. C. Hoering, and K. King, Jr), pp. 305–340. Wiley.

Lario J., Zazo C., Somoza L., Goy J. L., Hoyos M., Silva P. G., and Hernández-Molina F. J. (1993) Los episodios marinos cuaternarios de la costa de Málaga (España). *Rev. Soc. Geol. España* **6**, 41–46.

Murray-Wallace C. V. (1995) Aminostratigraphy of Quaternary coastal sequences in Southern Australia—an overview. *Quat. Intl.* **26**, 69–86.

Ochietti S., Raynal J. P., Pichet P., and Texier P. (1993) Aminostratigraphie du dernier cycle climatique au Maroc atlantique, de Casablanca à Tanger. *C.R. Acad. Sci. Paris* **317**, Ser. II, 1625–1632.

Rey J. and Fumanal M. P. (1996) The Valencian coast (Western Mediterranean): neotectonics and geomorphology. *Quat. Sci. Rev.* **15**, 789–802.

Sillero C., Vives F., Martín J. M., and Ródenas A. (1993) Un pequeño testimonio tirreniense en el cabo de la Huerta. *Cidaris* **2**, 25–29.

Stein M., Wassenbourg G. J., and Lajoie K. R. (1991) U-series ages of solitary corals from the California coast by mass spectrometry. *Geochim. Cosmochim. Acta* **57**, 3709–3722.

Stein M., Wassenbourg G. J., Chen J. H., and Zhu Z. R. (1993) TIMS U-series dating and stable isotopes of the last interglacial events in Papua New Guinea. *Geochim. Cosmochim. Acta* **57**, 2541–2554.

Torres T., Canoira L., Coello F. J., García-Alonso P., García-Cortés A., Llamas J., Mansilla H., Nestares T., Peláez A., and Somoza, L (1996) Caracterización geoquímica orgánica de los moluscos holocenos del Golfo de Cádiz (Andalucía, España). *Geogaceta* **19**, 150–153.

Torres T., García-Alonso P., Canoira L., Coello F. J., García-González L., Nestares T., Peláez A., and Rodríguez-Alto N. (1997) Racemización de aminoácidos de braquiópodos y pelecípodos de la sección de Cuesta Colorada (Almería, SE de España) *Geogaceta* **21**, 207–210.

Torres T., Llamas J. F., Canoira L., and García-Alonso P. (1999) Aspartic acid racemization in the dentine of bears. Tooth dentine amino acids versus mollusca amino acids. In *Advances in Biochirality* (eds G. Palyi, C. Zucchi, and L. Caglioti), pp. 247–256. Elsevier Science.

Wehmiller J. F. and Belknap D. F. (1982) Amino acid estimates, Quaternary Atlantic Coastal Plain: comparison with U-series dates, biostratigraphy and palaeomagnetic control. *Quat. Res.* **18**, 311–336.

Wehmiller J. F. (1984) Interlaboratory comparison of amino acid enantiomeric ratios in fossil Pleistocene mollusks. *Quat. Res.* **22**, 109–120.

Wehmiller J. F. (1990) Amino acid racemization: applications in chemical taxonomy and chronostratigraphy of Quaternary fossils. In *Skeletal Biomineralization* (ed. J. G. Carter), pp. 536–608. Van Nostrand.

Wehmiller J. F., York L. L., Belknap D. F., and Snider S. W. (1992) Theoretical correlations and lateral discontinuities in the Quaternary aminostratigraphical record of the U.S. Atlantic Coastal Plain. *Quat. Res.* **38**, 275–291.

Wehmiller J. F., York L. L., and Bart M. L. (1995) Amino acid racemization geochronology of reworked Quaternary mollusks on U.S. Atlantic coast beaches: implications for chronostratigraphy, taphonomy and coastal sediment transport. *Marine Geol.* **124**, 303–337.

Zazo C., Goy J. L., Hillaire-Marcel C., Hoyos M., Cuerda J., Ghaleb B., Bardají T., Dabrio C. J., Lario J., Silva P. G., González A., González F., and Soler V. (1997) El nivel del mar y los interglaciales cuaternarios. Su registro en las costas peninsulares e insulares españolas. Cuaternario Ibérico Congress Abstracts 23–32.

20. Quaternary coastal aminostratigraphy: Australian data in a global context

Colin V. Murray-Wallace

Historically, literature on the Late Quaternary (last 140 ka) coastal evolution of Australia has been dominated by case studies that have relied upon radiocarbon dating, this method having been widely applied to the dating of carbon-bearing materials from a range of stratigraphical contexts (e.g., case studies in Thom, 1984). Studies of older Pleistocene (> 50 ka) coastal successions and landforms have been frustrated, however, by the limited number of techniques appropriate to the highly specific character of Australian coastal environments. For instance, the general paucity of corals in the temperate carbonate coastal successions of southern Australia has significantly restricted the application of the uranium-series disequilibrium method of dating in this region. Other characteristics of Quaternary coastal successions in Australia that impinge on the application of dating methods include the widely separated occurrences of shelly facies and the consequent problems of stratigraphical correlation with potential for homotaxial correlations. The general absence of emergent coastal deposits and well-defined sequences of marine terraces, conferred by the continent's intra-plate tectonic setting and widespread occurrence of stable Precambrian cratons, has also frustrated attempts to establish the relative ages of some coastal successions. With some notable exceptions (e.g. Huntley et al., 1993, 1994; Sherwood et al., 1994), few researchers have applied luminescence or electron spin resonance methods to assess the age of Quaternary coastal successions in Australia. In view of these considerations, amino acid racemization (AAR) reactions have been extensively applied to the dating of Quaternary coastal sedimentary successions in Australia since the early 1980s (Murray-Wallace, 1995).

The first studies to apply AAR reactions in Australian coastal stratigraphical investigations involved site-specific case studies (e.g. von der Borch et al., 1980; Belperio et al., 1984; Kimber and Milnes, 1984). In a study of the Late Quaternary lithostratigraphy of part of the Coorong coastal plain, near Robe, South Australia, von der Borch et al. (1980) used AAR to verify last interglacial uranium-series ages previously determined in aragonitic fossil mollusks and mud, because of concerns about the dating of these materials by the uranium-series method (figure 20-1). Their amino acid results were consistent with the uranium-series results, and a similarity was noted in the extent of racemization in the genus *Katelysia* with *Chione* and *Protothaca* from a marine terrace of equivalent age in San Diego, California.

In a subsequent study, Belperio et al. (1984) reported the extent of leucine racemization in specimens of the arcoid bivalve *Anadara trapezia*, to verify their lithostratigraphical correlation of the last interglacial Glanville Formation between Gulf St. Vincent and northern Spencer Gulf in South Australia (figure 20-1). In a pilot study, Kimber and Milnes (1984) reported results for a wider range of amino acids for several marine-mollusk genera from Holocene and Late Pleistocene coastal deposits south of Adelaide.

Hearty and Aharon (1988) reported on the extent of isoleucine epimerization in the giant clam *Tridacna gigas* from the Huon Peninsula, Papua New Guinea, and from a range of sites in the Great Barrier Reef, northern Queensland. Their results are particularly important for modeling the regional (latitudinal) differences in the degree of isoleucine epimerization, associated with contrasting diagenetic temperatures. In particular, they noted that specimens of *T. gigas*, dated by the uranium-series method at 127 ± 2.7 ka, yielded a mean isoleucine-epimerization value of 1.08 ± 0.08. The data of Hearty and Aharon (1988) suggest that equilibrium values for isoleucine are attained within approximately 200 ka from the onset of epimerization. The mean sea-surface temperature around the Huon Peninsula is approximately 27°C.

Since these early works, a systematic attempt has been undertaken to examine specific time-slices of the Quaternary coastal record of Australia. Identification of glacial-age coastal deposits on the outer continental shelf of New South Wales, confirmation of the extensive distribution of coastal successions of last interglacial age (oxygen isotope substage 5e) and their neotectonic significance, and the dating of biostratigraphically important taxa from older Pleistocene successions, represent some of the achievements in the application of AAR reactions to the stratigraphical analysis of the Australian Quaternary marginal marine record.

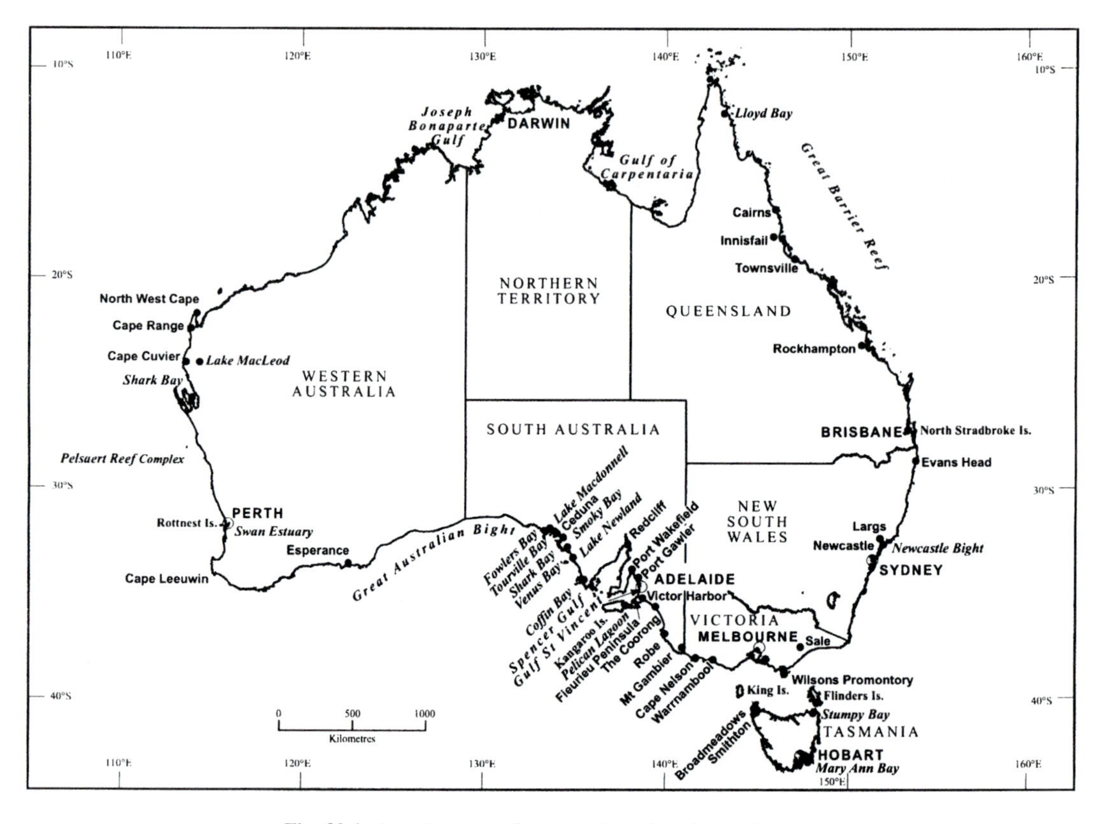

Fig. 20-1. Location map of sample sites referred to in the text.

The following review provides a synthesis of aspects of these studies and evaluates Australian aminostratigraphic data in a global context. The work draws from previously published studies as well as from hitherto unpublished data.

ANALYTICAL METHODS

In this work, the quantification of amino acid residues in Holocene and Pleistocene fossil marine mollusks was determined using the established gas chromatographic procedures formalized by Frank et al. (1977). Full details of the analytical methods are provided elsewhere (Murray-Wallace, 1993), and a condensed summary is presented here.

Surface encrustations of soil or sediment on fossil mollusks were removed using dental tools; this followed by cleaning in distilled water in an ultrasonic bath. The outer surfaces of the shells were then removed in dilute HCl. Wherever possible, the hinge region was selected for analysis. Samples were hydrolyzed for 16 hr at 110°C in 8 M HCl. Following cation-exchange isolation of the amino acid residues, samples were freeze-dried and derivatized. Chromatography of the *N*-pentafluoropropionyl D,L-amino acid 2-propyl esters was undertaken using a Hewlett-Packard 5890 Series II gas chromatograph with a 25 m, coiled, fused, silica capillary column with the stationary phase Chirasil L-Val, a flame ionization detector and helium carrier gas. D/L ratios were determined on the basis of peak-area calculations.

Samples of fossil mollusks for AAR analysis have been obtained from a range of coastal and marine facies in Australia. Shell samples have been collected from deeply buried situations (i.e., > 1 m) to minimize the effects of diurnal temperature changes, such that temperatures associated with longer-term climatic change represent the dominant influence on diagenetic racemization. Fossil mollusks have been collected from natural outcrop exposures, recent excavations, drill holes, and vibracores.

Amino acid analyses have been restricted to specimens that show minimal evidence of diagenetic modification (e.g., minimal inversion of metastable aragonite to calcite), and shells with chalky surfaces have been excluded. Large stout shells such as *A. trapezia*, *Katelysia* sp., *Glycymeris* sp. and *Pecten fumatus* have been favored, as they provide the opportunity to undertake more stringent sample pretreatments (i.e., removal of weathered surfaces of shells). X-ray diffraction has revealed the presence of minimal calcite in the specimens analyzed by AAR (< 2%). Surprisingly, aragonite was, over-

TABLE 20-1 Analytical results of the Interlaboratory Comparison samples of Wehmiller (1984b)

Amino acid	Samples ILC-A	ILC-B	ILC-C
Ala	0.429	0.785	0.976
Val	0.214	0.442	0.751
Leu	0.214	0.617	0.834
Asp	0.383	0.688	0.823
Phe	0.234	0.640	0.830
Glu	0.194	0.424	0.760
Pro	0.286	0.680	0.903
Aile/Ile	0.258	–	1.271

whelmingly, the primary mineral in specimens of the bivalve mollusk *Katelysia rhytiphora* from the Plio-Pleistocene Jandakot Member of the Yoganup Formation in the Perth Basin (Murray-Wallace and Kimber, 1989). Results of the analysis of Interlaboratory Comparison (ILC) samples of Wehmiller (1984b) are presented in table 20-1.

METHODOLOGICAL CONSIDERATIONS

The potential sources of uncertainty in the measurement of amino acid residues in fossils, calibration of AAR data by an independent dating method, and a detailed understanding of the morphostratigraphical context of the fossils intended for dating were all carefully considered in this investigation. These issues are briefly reviewed here as a prelude to discussion of specific AAR results from Australia.

Sources of Uncertainty

Several studies have attempted to quantify the different sources of uncertainty inherent in the use of amino acid enantiomeric ratios to assign relative and numeric ages to Quaternary deposits (Brigham, 1983; Murray-Wallace and Kimber, 1987; Miller and Brigham-Grette, 1989; Murray-Wallace, 1995). Much of the uncertainty in the measurement of amino acid D/L ratios in fossils involves both systematic and unsystematic errors which relate to sample preparation and analysis, differences between samples, and differences within and between deposits (figure 20-2). In particular,

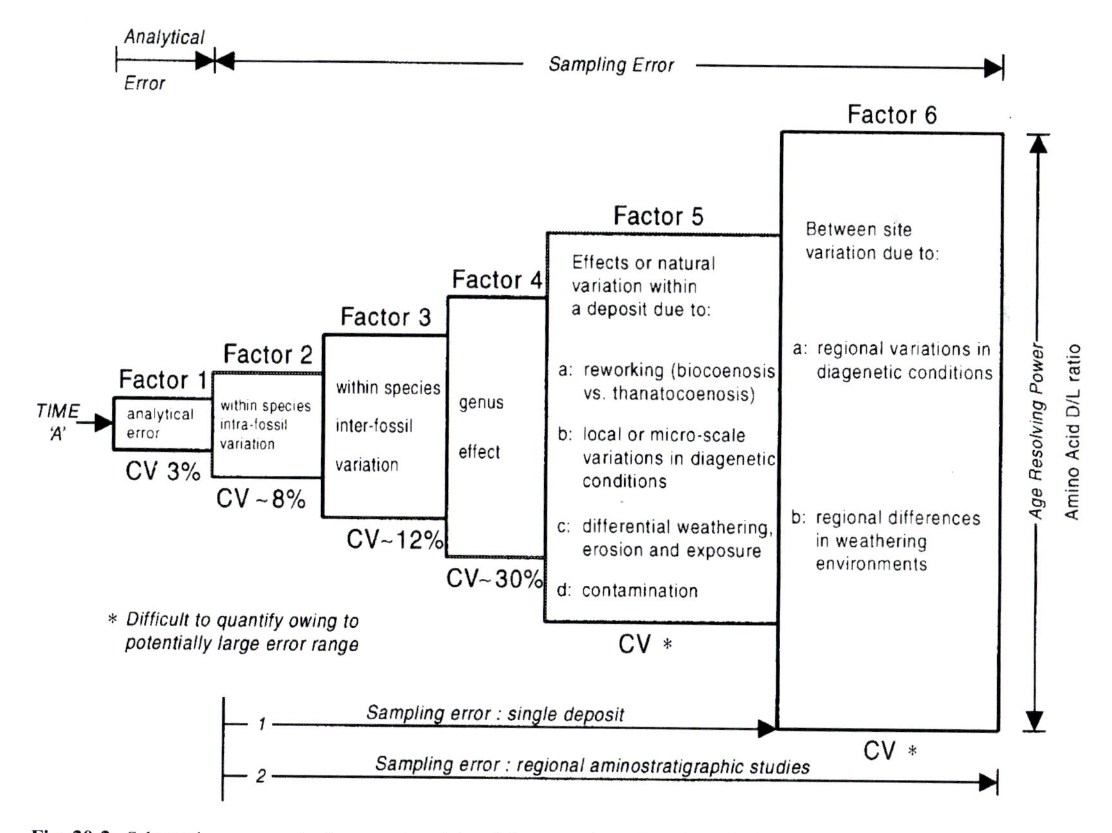

Fig. 20-2. Schematic assessment of parameters determining enantiomeric ratio variation in fossil mollusks from Quaternary coastal sedimentary successions. Note that the range of coefficients of variation is dependent upon the amino acid selected. Coefficients of variation reported above represent maximum values for several amino acids and are for the total acid hydrolysate.

sources of uncertainty include the following: analytical precision (e.g. sample pretreatment, preparation, and gas chromatography: $<3\%$); within-specimen intra-fossil variation ($<8\%$), within-genus, between-sample variation ($<12\%$); generic effects; natural variation within deposits; and regional variations in diagenetic conditions (Murray-Wallace, 1995). The latter three variables are more difficult to quantify and need to be evaluated on a site-specific basis. The ultimate age-resolving power of the AAR method is a function of the sum of all of the aforementioned variables.

Calibration of AAR Data

The complex nature, and the difficulty of modeling racemization kinetics in natural systems has required the use of independent methods of dating to calibrate the extent of racemization attained with respect to time. The controversy concerning the nature of racemization kinetics in fossil marine mollusks (Kvenvolden et al., 1979, 1981; Wehmiller, 1981, 1982, 1984a) reaffirmed earlier findings (Wehmiller and Hare, 1971; Bada and Schroeder, 1972; Mitterer, 1975) that racemization in these matrices does not conform to first-order, extended linear reaction kinetics. Collectively, these investigations illustrated the need for additional frameworks to calibrate AAR data. Further attempts at mathematical modeling of the nature of racemization kinetics observed in fossil mollusks have included natural logarithmic curve fits (Wehmiller et al., 1988) and a model based upon apparent parabolic kinetics (Mitterer and Kriausakul, 1989; Murray-Wallace and Kimber, 1993).

In Australia, several approaches have been adopted for the calibration of AAR data. Fossil mollusks younger than 25 ka (early last glacial maximum; oxygen isotope stage 2) and Holocene have been directly calibrated using radiocarbon dating (e.g., Murray-Wallace et al., 1996a: tables 20-2 and 20-3). Wherever possible, the extent of racemization has been determined in fossil mollusks also dated using the radiocarbon method. This has involved either analysis of a subsample of the same specimen dated by radiocarbon, or, preferably, the analysis of the HCl residue from the CO_2 evolution procedure of radiocarbon dating. The latter analytical protocol ensures that the radiocarbon analyzed is directly associated with the residual amino acids also analyzed by AAR (Murray-Wallace and Bourman, 1990). Radiocarbon ages reported in this work have been corrected for the marine-reservoir effect for southern Australian ocean surface waters, using the Gillespie and Polach (1979) value of -450 ± 35 years, and converted to sidereal years using the calibration program of Stuiver and Reimer (1993).

The extent of racemization of amino acids in fossil mollusks older than 25 ka has been calibrated using a variety of methods that include uranium-series disequilibrium (Sherwood et al., 1994), electron-spin resonance (Murray-Wallace and Goede, 1995) and luminescence (Murray-Wallace et al., 1996b) and representative examples are presented in table 20-2. In the case of uranium-series disequilibrium and electron-spin resonance, AAR analyses have been undertaken on fossil specimens from the same *in situ* death assemblages.

Caution was exercised in the use of luminescence dating to calibrate the results of AAR as these two methods normally date different events. In the case of AAR in fossil mollusks, the event dated is the formation of the shell layer sampled, which in short-lived mollusks cannot be statistically distinguished from the death of the organism (Wehmiller et al., 1995). In many geological contexts, this event will pre-date deposition. In contrast, the event generally quantified by luminescence methods is the time of deposition. An exception to this, however, can be found in the Pleistocene barrier-lagoon succession of the Coorong coastal plain in South Australia (Murray-Wallace et al., 1996b; figure 20-1). Here, deep cuttings through eolianite coastal barrier successions reveal that during deposition, migrating dunes cascaded into the back-barrier lagoon. Intertidal mollusks, such as *A. trapezia* and *Katelysia* sp., which formerly lived within the back-barrier lagoons moved up through the rapidly accumulating sediment to escape suffocation, but could not keep pace with the rate of sediment supplied by the migrating eolian dunes and were therefore asphyxiated. Thus, in this case, the depositional event dated by luminescence (dune formation) is coeval with the death of the lagoonal mollusks (dated by racemization), permitting the comparison of these two methods. The same processes of sedimentation can be observed today in the Coorong Lagoon, a modern analogue of the Pleistocene succession of barriers.

Australian Stratigraphical Framework: General Remarks

Pleistocene and Holocene coastal sediments in Australia occur in a range of morphostratigraphical contexts (Belperio et al., 1995). The sediments are typically unlithified or only partially lithified, are host to diverse molluscan fossil assemblages, and either crop out or occur in shallow subcrop. Coastal barriers and associated back-barrier lagoon and estuarine sedimentary environments represent a substantial portion of the Australian coastline, particularly in southern Australia, and equivalents are well represented in the Quaternary record of this continent. Palaeosols and a range of calcretes and associated hardpans, as well as other subaerial exposure features (e.g., rhizocretions, solution pipes, rillen) are well developed within the Pleistocene temperate carbonate successions of southern and western Australia. In large parts of

TABLE 20-2 Extent of AAR (total acid hydrolysate) in Pleistocene marine mollusks also calibrated by an independent dating method

Locality/Stratigraphic unit	Dating method and material dated[a]	Age (ka)	Reference	CMAT[a] (°C)	D/L Leu	D/L Val
Late Pleistocene						
Robe, SA, Woakwine Range, lagoon facies	U/Th; *Anadara trapezia*	125 ±10	Schwebel (1978)	14.7	0.35 ±0.01	0.22 ±0.03
Robe, SA, Woakwine Range, aeolian dune facies	TL; quartz sand	132 ±6 132 ±6 118 ±4	Huntley et al., (1993, 1994)			
Lake Hawdon, SA, Woakwine Range, lagoon facies	ESR; *Katelysia* sp.	126	R. Grün (pers. comm. 1998)			
Streaky Bay, SA, Glanville Formation	U/Th; *Goniopera* sp.	124 ±9	Daniel et al. (1998)	16.9	0.44	0.33 ±0.02
Outer continental shelf, NSW	^{14}C; *Pecten fumatus*	17,320 ±220 (SUA-3050)	Ferland et a. (1995)	15.0	0.19	0.20 ±0.02
Middle Pleistocene						
Mt Gambier, SA, Burleigh Range	TL; quartz sand	237 ±16	Murray-Wallace et al. (1996b)	13.2	–	0.39 ±0.07
Hunter Valley, NSW, Drill Hole 6	U/Th; *Anadara* sp.	181.7 ±10.5	This study	17.5	0.65 ±0.01	0.46 ±0.003
Hopkins River, VIC	U/Th; *Anadara* sp.	193 ±10	Sherwood et al. (1994)	13.7	0.50 ±0.05	0.34 ±0.03

[a] CMAT, current mean annual temperature.

TABLE 20-3 Extent of AAR (total acid hydrolyate) in radiocarbon-dated Holocene marine mollusks from southern Australia

Locality	Species	Radiocarbon laboratory code	^{14}C Calendric age (cal. BP)[a]	Amino acid D/L ratio[b]				Depth of burial (m)	CMAT (°C)
				Val	Leu	Phe	Glu		
Sir Richard Peninsula, SA	*Donax deltoides*	SUA-2885	410 ±120	0.04 ±0.001	–	0.09 ±0.001	0.08 ±0.001	near surface	15.7
King Island, Three Rivers Creek	*Patella laticostata*	SUA-2927	790 ±130	0.01	0.05 ±0.02	0.04 ±0.01	0.05 ±0.001	1	12.5
Sir Richard Peninsula, SA	*Donax deltoides*	SUA-2884	1700 ±170	0.06 ±0.005	–	0.11 ±0.01	0.11 ±0.01	near surface	15.7
Sir Richard Peninsula, SA	*Donax deltoides*	SUA-2881	2380 ±200	0.07 ±0.01	–	0.19 ±0.01	0.12 ±0.01	near surface	15.7
Sir Richard Peninsula, SA	*Donax deltoides*	SUA-2882	3180 ±190	0.06	–	0.14	0.14	near surface	15.7
Admiralty Bay, King Island	mixed-species assemblage	SUA-2926	4560 ±240	0.05 ±0.004	0.12 ±0.004	0.14 ±0.002	0.12 ±0.002	1	12.5
Smoky Bay, SA	*Fulvia tenuicostata*	CS-407	5580 ±320	0.08 ±0.01	0.15 ±0.002	0.28(?) ±0.002	0.13 ±0.06	3.38–3.48	16.9
Hervey Bay, QLD	*Anadara trapezia*	SUA-2515	6050 ±120	0.14 ±0.02	0.23 ±0.002	–	–	2.8–3.2	16.9
Shell Pits Point, TAS	*Glycymeris striatularis*	SUA-2928	7120 ±150	0.09 ±0.01	–	0.16 ±0.05	0.15 ±0.01	0.6	12.5
Smoky Bay, SA	*Katelysia scalarina*	CS-452	7300 ±610	0.07 ±0.005	0.24 ±0.01	–	0.14 ±0.001	4.88–4.96	16.9
Bass Basin, Bass Strait Core #23	*Fulvia tenuicostata*	SUA-2389	10,690 ±200 (uncalibrated)	0.08 ±0.001	0.19 ±0.01	0.21 ±0.01	0.15 ±0.02	0.34	~ 12.5

[a] Radiocarbon ages have been corrected for the marine reservoir effect for southern Australian ocean surface waters (-450 ±35 years; Gillespie and Polach, (1979) and converted to sidereal years BP using the calibration program of Stuiver and Reimer (1993). Radiocarbon error terms are reported at the 2σ level.
[b] Amino acids: Val, valine; Leu, leucine; Phe, phenylanine; Glu, glutamic acid.

southern Australia, such as the Coorong coastal plain, the presence of thick calcrete hardpans has preserved the original morphology of the coastal barrier eolianite successions (Murray-Wallace et al., 1998).

Holocene coastal sedimentary successions have been deposited in similar environments to their Pleistocene equivalents, but tend to be strongly reduced, in contrast to oxidized Pleistocene sediments, particularly for estuarine successions. Holocene coastal successions also occur seaward of their Pleistocene equivalents. Within the Holocene coastal successions, paleo-sea-level indicators such as intertidal shell and mangrove facies tend to occur within ± 1 m of the present-day mean sea level, and were deposited since the culmination of the post-glacial marine transgression, which in southern Australia appears to have occurred some 7 ka (calendar years) ago (Thom and Chappell, 1975).

Australian coastal environments have experienced minimal tectonic dislocation throughout the Quaternary, because of the intra-plate setting of the continent as well as the widespread distribution of stable Precambrian cratons on which large sectors of the coastline have developed. Australia is the flattest continent and its tectonic stability is also reflected by the very limited neotectonic uplift evident in coastal environments. Neotectonic-induced variation in the elevation of last interglacial facies is presented in figure 20-3. The plot has largely been established based on the aminostratigraphical dating of fossil mollusks such as *A. trapezia* and *Katelysia* sp. from these successions (Murray-Wallace and Belperio, 1991; Belperio et al., 1995; Bourman et al., 1999). The aminostratigraphical results reveal that, over distances of hundreds of kilometers, intertidal coastal facies of the last interglacial maximum (oxygen isotope Substage 5e, approx. 132–118 ka) occur at elevations less than 4 m above present sea level (APSL), and the most consistent shoreline datum is 2 m APSL on the Eyre Peninsula coast, developed on the Precambrian Gawler Craton (Murray-Wallace and Belperio, 1991; Daniel et al., 1998; figure 20-3). The highest occurrence of last interglacial coastal sediments of shallow water origin is at Mary Ann Bay in Tasmania, where the upper erosional surface of a shallow subtidal shell bed occurs at 24 m APSL (Murray-Wallace and Goede, 1995). On the Australian mainland, the highest recorded last interglacial shelly intertidal facies occurs at 18 m APSL near the Quaternary hot-spot volcanic province of Mt. Gambier in South Australia (Murray-Wallace et al., 1996b). Both these examples represent localised neotectonic anomalies to the general stability of the Australian continent and appear to relate to inferred incipient hot-spot activity preceding volcanism in the former (Bowden and Colhoun, 1984) and restricted intra-plate volcanism in the latter (Murray-Wallace et al., 1998).

Knowledge of the neotectonic behavior of the Australian continent is particularly important when evaluating AAR results for oxygen isotope stage 5 deposits from this region. Given the negligible tectonic uplift experienced in Australian coastal environments, intertidal facies within interstadial sedimentary successions, important indicators for the delineation of former sea levels, occur below present sea level. Thus, coastal sediments of oxygen isotope substages 5c (approx. 105 ka) and 5a (approx. 82 ka) consistently occur below present sea level around Australia (e.g., Spencer Gulf region; Hails et al., 1984). Exceptions include emergent eolianites such as the Robe Range on the Coorong coastal plain in South Australia (Murray-Wallace et al., 1998). However, shell beds associated with the Robe Range coastal eolianite barrier occur some 12–15 m below present sea level, and it is only the aeolian dune facies of this barrier which today occurs above present sea level and gives rise to a prominent cliffed coast. This is an important consideration for evaluating the stage 5 amino acid data from Australia, as it indicates that variations inherent in the last interglacial *sensu stricto* AAR data set, cannot be attributed to the inadvertent sampling of younger coastal deposits of stage 5 (i.e., substages 5c and 5a rather than substage 5e).

The preservation potential of older interglacial deposits in the coastal lithostratigraphical record represents a further issue relating to the neotectonic inheritance of the Australian region. In "stable" areas subject to multiple marine transgression and regression, there is a greater likelihood that substantial temporal gaps will characterize the lithostratigraphical record of these settings (e.g., erosion of older estuarine valley fill successions during glacial low sea-stand events). Thus, one of the principal aims of applying AAR reactions to age assessments of coastal sequences in Australia has been to assess the completeness of the stratigraphical record.

In the following discussion, Australian coastal deposits which have yielded a continental-scale aminostratigraphy will be briefly reviewed within the framework of the marine oxygen isotope record (Imbrie et al., 1984; Bassinot et al., 1994). The discussion is divided into two main sections that review coastal deposits associated with interglacial and interstadial highstand events and deposition during glacial lowstands.

SEA LEVEL HIGHSTANDS

Early Pleistocene

The oldest fossil mollusks analyzed in Australia by the AAR method are from three units provisionally assigned as Plio–Pleistocene and Early Pleistocene ages on the basis of biostratigraphical evidence. The stratigraphic units include the Jandakot Member of

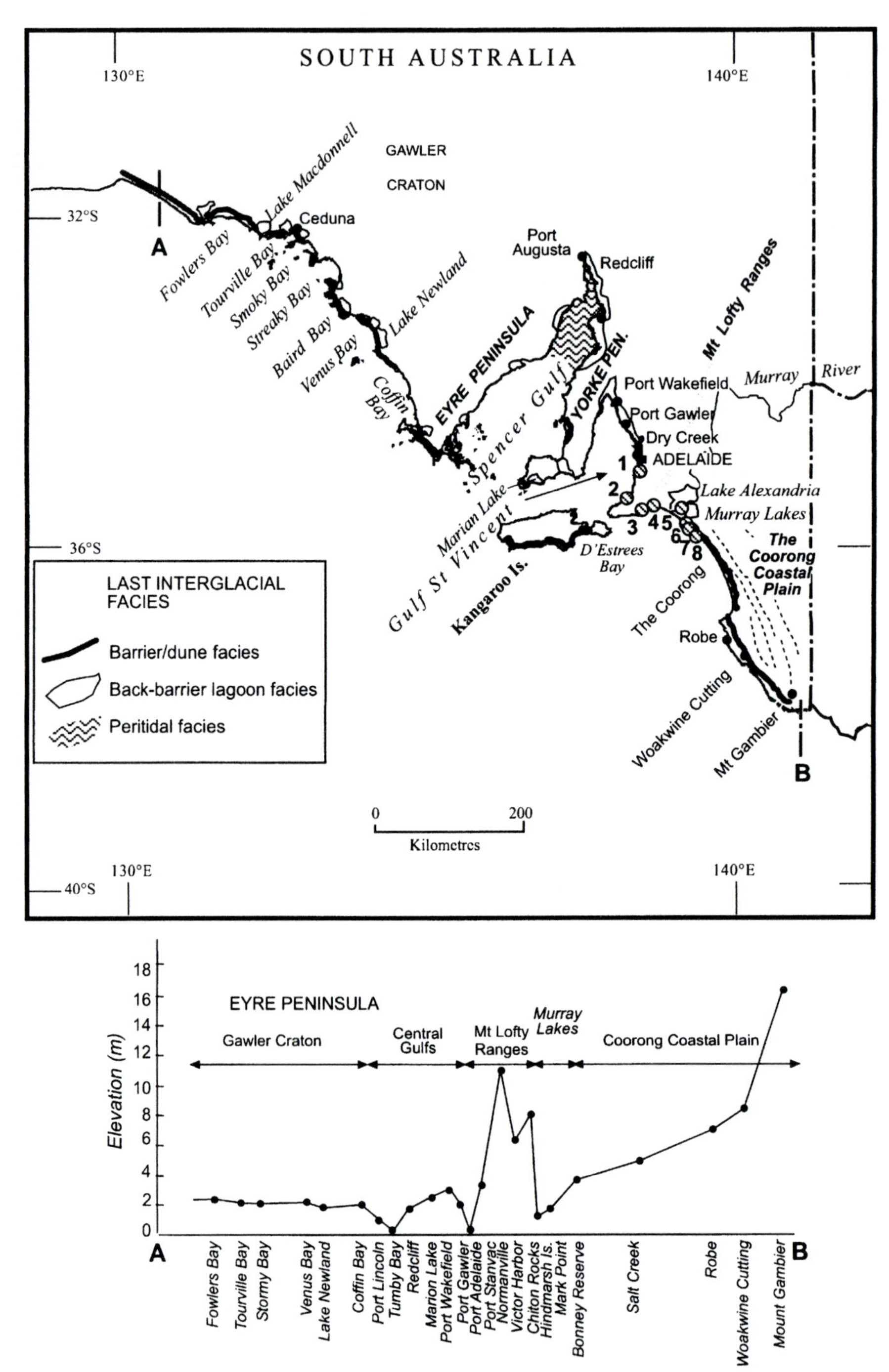

Fig. 20-3. Present-day elevation of intertidal facies of the last interglacial shoreline (oxygen isotope Substage 5e) between Fowlers Bay and Mount Gambier, South Australia (transect A-B in location map). The geochronological evidence for the age of these coastal successions is largely based on aminostratigraphy, calibrated by other geochronological methods (Murray-Wallace and Belperio, 1991; Belperio et al., 1995; Bourman et al., 1999). The most consistent substage 5e shoreline datum is from western Eyre Peninsula, the site of the stable Precambrian Gawler Craton, where a value of 2 m APSL is consistently registered. Greater variation is apparent for the central gulfs, Spencer and St. Vincent, and some of this may relate to regional hydro-isostatic feedback effects, as is evident for Holocene successions. The Mt Lofty Ranges reveal ongoing uplift and the Coorong coastal plain reflects spatially varying epeirogenic uplift associated with Late Quaternary volcanism. Additional sample localities: 1 – Port Stanvac; 2 – Normanville; 3 – Victor Harbor; 4 – Chiton Rocks; 5 – Hindmarsh Island; 6 – Mark Point; 7 – Bonney Reserve; and 8 – Salt Creek (refer to map).

the Yoganup Formation, an entirely subsurface marine unit of Plio–Pleistocene age in the Perth Basin, Western Australia, the Roe Calcarenite of the Eucla Platform in Western Australia, and the Memana Formation of Flinders Island; the latter two were provisionally assigned to the Early Pleistocene (Ludbrook, 1978; Sutherland and Kershaw, 1971). A summary of the foraminiferal and molluscan biostratigraphical evidence for the ages of these shell-rich coastal and shallow marine successions is presented in Murray-Wallace (1995). At present, the biostratigraphical evidence is not of sufficient resolution to correlate these successions to specific oxygen isotope stages. These fossils are important, however, for providing some indication of the extent of racemization of amino acids in fossils of the terminal Pliocene to the Early Pleistocene age.

Middle Pleistocene

Pre-Stage 7 Deposits

A unit of skeletal carbonate sands and muds of shallow marine origin occurs beneath a 16-m-thick sequence of limnic sands, organic muds, and marls at Egg Lagoon on King Island (figure 20-1). The marine sands were first described by Jennings (1959), who assigned a last interglacial age to the succession based on lithostratigraphical correlations with similar sediments elsewhere on King Island and in northwestern Tasmania. Specimens of the cockle *K. rhytiphora* from the skeletal carbonate sands, selected for AAR analysis, were obtained from an auger hole (AH8 of D'Costa et al., 1993) in Egg Lagoon at a depth of 10.3–10.5 m below the ground surface. The shells included moderately well-preserved disarticulated bivalves and large fragments of valves. A preliminary thermoluminescence age of > 78 ka (minimum age) on quartz sand from an adjacent auger hole (AH10, 8–8.2 m) was reported by D'Costa et al. (1993).

Penultimate Interglaciation (Oxygen Isotope Stage 7)

Several coastal deposits of oxygen isotope stage 7 age, containing assemblages of fossil mollusks, have been confidently identified on the basis of uranium-series (Sherwood et al., 1994), luminescence (Murray-Wallace et al., 1996b) and electron-spin resonance dating (Hewgill et al., 1983; Goede, 1989). Aminostratigraphy has yielded results consistent with these methods of age assessment. Coastal facies of stage 7 age have been identified on the basis of aminostratigraphy from Peppermint Grove in the Swan River estuary in Perth, Western Australia (Murray-Wallace and Kimber, 1989), Redcliff, northern Spencer Gulf (Murray-Wallace et al., 1988), and the southern Coorong coastal plain, South Australia (Murray-Wallace et al., 1996b), the Hopkins River estuary near Warrnambool, Victoria (Sherwood et al., 1994), and from the Hunter River valley fill near Newcastle, New South Wales (the latter being previously unreported).

Coastal deposits of penultimate interglacial age appear to be far less extensive than equivalent deposits of last interglacial or Holocene age; this distribution probably reflects their lower preservation potential in the stratigraphical record (i.e., erosion during subsequent lowstands as well as reworking during subsequent transgressions). The penultimate interglacial deposits occur in a range of lithological and morphostratigraphical contexts, and thus no single variable can be attributed to their preservation. In the case of Peppermint Grove, the Coorong coastal plain, and the Hopkins River, these deposits are mantled by thick calcrete profiles that have developed on calcarenites and, accordingly, have hindered erosional processes. Together with the deposit at Redcliff, all are above present sea level and indicate that the sea level was higher during oxygen isotope stage 7 than was otherwise indicated by the oxygen isotope record (cf. Imbrie et al., 1984).

At Peppermint Grove in the Swan River estuary near Perth, richly fossiliferous bioclastic carbonate estuarine sands mantled by a thick succession of eolianites crop out between 1 and 7.3 m APSL. The unit, termed the Peppermint Grove Member of the Tamala Formation (Murray-Wallace and Kimber, 1989), has been assigned to oxygen isotope stage 7 on the basis of electron-spin resonance (Hewgill et al., 1983) and aminostratigraphy (Murray-Wallace and Kimber, 1989).

Pedogenically modified and poorly sorted mottled muds containing *A. trapezia* and abraded foraminifera occur in shallow subcrop (< 2 m) and landward of last interglacial successions at Redcliff in northern Spencer Gulf, South Australia. These sediments were termed the "Older Pleistocene marine beds" by Hails et al. (1984), who identified equivalent peritidal facies (stratigraphically below sediments securely dated as last interglacial) in submarine vibracores from northern Spencer Gulf (Belperio et al., 1984; Hails et al., 1984). The extent of racemization of amino acids in specimens of *A. trapezia* from the Older Pleistocene marine beds is significantly higher than that for equivalent shells of last interglacial age (Murray-Wallace et al., 1988).

The geological attributes of the Coorong coastal plain have been extensively studied given that the region represents one of the few areas in Australia subject to demonstrable tectonism. A "staircase" of interglacial coastal barriers that have developed during each interglacial since at least the Brunhes–Matuyama geomagnetic polarity reversal occurs across a low-gradient, 100-km-wide coastal plain (Huntley et al., 1993, 1994; Murray-Wallace et al., 1998). Near Robe, an aeolian facies of the coastal barrier associated with oxygen isotope stage 7 has yielded a luminescence age of 230 ± 11 ka (Huntley et al., 1993). Large fragments of the bivalve mollusk *K. rhytiphora*, from a flint cobble beach facies from the same barrier structure (Burleigh Range) but to the south near Mount Gambier, have also revealed higher extents of AAR than is evident for fossil mollusks of last interglacial age (figure 20-4). A luminescence analysis on quartz sand from the same deposit yielded an age of 237 ± 16 ka (Murray-Wallace et al., 1996b).

A course-grained calcarenite deposit within the Hopkins River estuary (Site 4 of Sherwood et al., 1994) contains a transported assemblage of *A. trapezia* as well as other mollusk species. Uranium–thorium ages of 193 ± 10, 215 ± 15, and 253 ± 15 ka obtained for *A. trapezia* are suggestive of a stage 7 age for this deposit. AAR analyses have also been undertaken for *A. trapezia* from the same shell bed.

Recent drilling along the axis of the Hunter River valley fill near Newcastle, New South Wales has revealed the presence of a laterally persistent unit of fossiliferous estuarine muds (R. Boyd, personal communication, 1997). Specimens of disarticulated *A. trapezia* were sampled from drill hole DH6 (at a depth of 52.3 m), drilled through the core of the last interglacial inner barrier, a large barrier complex described in detail by Thom et al. (1992). Stratigraphic analysis revealed the presence of a 5-m-thick, laterally persistent muddy unit below the last interglacial barrier facies. A uranium-series age of 181,700 ± 10,500 yr was obtained on a specimen of *A. trapezia* from the same lithostratigraphical unit at a downcare depth of 51.2 m. The AAR results for the specimen of *A. trapezia* from drill hole DH6 are discussed later in this work.

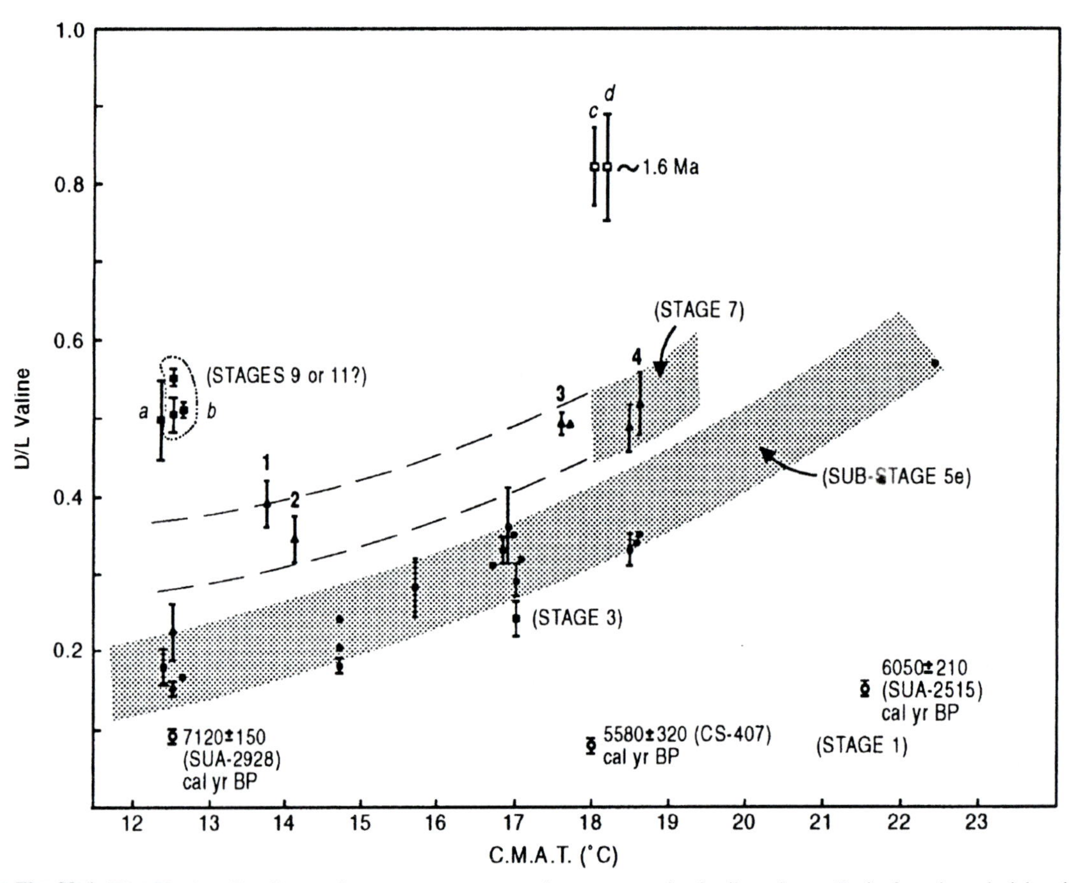

Fig. 20-4. Plot of D/L valine data against current mean annual temperature for fossil marine mollusks from interglacial and interstadial highstand deposits from southern Australia. Open circles represent results for radiocarbon dated Holocene fossil mollusks. The last interglacial (oxygen isotope substage 5e) data set (solid circles) is highlighted within the shaded envelope and illustrates the strong temperature dependence of racemization reactions. Results for the penultimate interglaciation (Stage 7) are represented by open triangles and display a similar trend to the Substage 5e data set. Stage 7 data include: (1) the Burleigh Range, Coorong coastal plain near Mount Gambier; (2) Hopkins River estuary, near Warrnambool; (3) Hunter Valley drill hole (DH6), near Newcastle; and (4) "Older Pleistocene marine beds", northern Spencer Gulf, South Australia. Additional samples include; (a) Memana Formation, Flinders Island; (b) Egg Lagoon, King Island; (c) Jandakot Member of the Yoganup Formation, Perth Basin; and (d) Roe Calcarenite, Eucla Platform, Western Australia.

Late Pleistocene

Last Interglaciation (Oxygen Isotope Substage 5e)

In contrast to the penultimate interglaciation, coastal deposits of the last interglaciation are preserved far more extensively around Australia, and the morphostratigraphical characteristics and significance of these successions have been extensively documented (Ludbrook, 1984; Murray-Wallace and Belperio, 1991; Belperio et al., 1995, Belperio, 1995). Estuarine facies of these successions contain some faunal elements that no longer live in the coastal waters of a large sector of southern Australia and which include the arcoid bivalve *A. trapezia* (Murray-Wallace et al., 2000) and the megascopic foraminifer *Marginopora vertebralis* (Cann and Clarke, 1993).

In view of the paucity of corals in southern Australia, the majority of uranium-series analyses reported from last interglacial coastal successions have been undertaken on corals from the northern parts of the western and eastern coasts of Australia. A list of uranium-series ages is provided by Murray-Wallace (1995), and more recent results have been reported by Stirling et al. (1995) and Zhu et al. (1993). The few results available from southern Australia, where the majority of AAR studies have been undertaken, have yielded ages consistent with oxygen isotope substage 5e. For example, Schwebel (1978) reported an age of 125 ± 10 ka for aragonitic mollusks from the back-barrier lagoon facies of the Woakwine Range, a major coastal barrier complex on the Coorong coastal plain in the southeast of South Australia. Luminescence ages of 132 ± 6, 132 ± 9 and 118 ± 4 ka have been reported for the aeolian dune facies of the Woakwine Range (Huntley et al., 1993, 1994). An electron-spin resonance age of 126 ka has also been obtained recently for the cockle *K. rhytiphora* from the back-barrier lagoon facies of the Woakwine Range at the southernmost point of Lake Hawdon South (R. Grün, personal communication, 1998).

In a heterogeneous deposit showing evidence of reworking, specimens of *Turbo undulatus* from shelly gravels 4 m APSL at Point Richie, near Warrnambool, have yielded uranium-series ages of 114 ± 10 ka and 140 ± 10 ka, which suggests that some of the fossil mollusks, at least, are of last interglacial age. Other specimens from the same deposit have yielded uranium-series ages consistent with oxygen isotope stage 7.

Daniel et al. (1998) reported a uranium-series age of 124 ± 9 ka for the solitary coral *Goniopora somaliensis* from Streaky Bay, South Australia (figure 20-3). *A. trapezia* from related deposits at Denial Bay and Tourville Bay, and *K. rhytiphora* from Lake Newland, all three sites being near Streaky Bay, have yielded amino acid D/L values consistent with the last interglacial age (oxygen isotope substage 5e) reported by Daniel et al. (1998) and with more extensive amino acid data from southern Australia.

Interstadial Stage 3 (Approx. 60–30 ka)

A sequence of carbonate muds with a diverse and well-preserved assemblage of fossil mollusks, dominated by *Ostrea* sp., *Katelysia* sp., *Fulvia* sp. and the gastropod *Batillaria (Zeacumantus) diemenensis*, occurs beneath a sheet drape of Holocene sediments in Gulf St. Vincent, South Australia (Cann et al., 1988; Murray-Wallace et al., 1993). Vibracores of these sediments were obtained in the central portion of the Gulf, in present water depths of less than 40 m (Cann et al., 1988). Extensive radiocarbon and AAR analyses indicated an interstadial age (Stage 3) for these strata. Using benthic foraminifera as surrogate paleo-water depth indicators, Cann et al. (1988, 1993) suggested that, during Stage 3, a maximum high sea-level stand of −22 m was attained in Gulf St. Vincent. AAR results for this sedimentary succession are presented in table 20-4 and figure 20-4.

Holocene

Fossil mollusks of Holocene age have been sampled from a wide range of settings in southern Australia, and the sampling has been undertaken primarily to assist in the calibration of AAR kinetics. Representative results for the extent of AAR evident in Holocene marine mollusks are presented in table 20-3.

SEA-LEVEL LOWSTANDS

Glacial Maxima

Few detailed geochronological studies have been undertaken on glacial age lowstand deposits from the continental shelf of Australia. Early studies concentrated only on the last glacial maximum (e.g., Phipps, 1970; Veeh and Veevers, 1970). Recently, however, Ferland and Roy (1997) reported the results of a detailed study of the lithostratigraphy of a richly fossiliferous, temperate carbonate succession from the New South Wales outer continental shelf. Their study revealed alternating successions of shell-rich skeletal carbonate sand interbedded with quartz-carbonate sand depleted in macrofossils. They concluded that carbonate deposition occurred during glacial lowstands, whereas siliciclastic sediments were deposited during highstands on the outer continental shelf.

Ferland and Roy (1997) reported numerous radiocarbon ages which equate with the last glacial maximum (oxygen isotope stage 2) for fossil *P. fumatus*, *Glycymeris* sp., and other shallow-water marine mollusks, as well as a suite of "background" results according to the radiocarbon method. They concluded that all the shell-rich units, by analogy

TABLE 20-4 Extent of AAR (total acid hydrolysate) in Pleistocene marine mollusks from southern Australia

Locality and stratigraphic unit	Species	Number of individuals analyzed	Amino acid D/L ratio[a]				CMAT (°C)	Geological age or $\delta^{18}O$ stage
			Val	Leu	Phe	Glu		
NSW continental shelf Core FT-BMR 19.3A	*Anadara trapezia*	1	0.11	–	0.26 ±0.01	0.17 ±0.01	17.0	Stage 2
Gulf St. Vincent, SA; Cores SV#4 & SV#5	*Katelysia rhytiphora* and *Fulvia tenuicostata*	12	0.24 ±0.03	0.29 ±0.03	–	–	17.0	Stage 3
Redcliff, SA, "Older Pleistocene marine beds"	*Anadara trapezia*	4	0.48 ±0.01	–	0.89 ±0.03	0.62 ±0.02	19.0	Stage 7
Hunter Valley Drill hole DH6	*Anadara trapezia*	2	0.46 ±0.01	0.65 ±0.01	–	–	17.5	Stage 7
Hopkins River, Vic.	*Anadara trapezia*	4	0.34 ±0.03	0.50 ± 0.11	0.55 ±0.11	0.47 ±0.06	13.7	Stage 7
Burleigh Range, near Mt Gambier, SA	*Katelysia rhytiphora*	5	0.39 ± 0.07	–	0.69 ±0.06	0.52 ±0.04	13.2	Stage 7
Swan Estuary, WA; Peppermint Grove Member, Tamala Formation	*Anadara trapezia*	3	0.48 ±0.04	0.52 ±0.02	0.82 ±0.02	0.62 ±0.02	18.5	Stage 7
	and *Katelysia rhytiphora*	4	0.57 ±0.04	0.68 ±0.05	0.89	0.63	18.5	Stage 7
Egg Lagoon, King Island; unnamed shell bed	*Katelysia scalarina*	3	0.48 ±0.09	0.74 ±0.13	0.73 ±0.02	0.54 ±0.07	12.5	Stages 9 or 11
	and *K. rhytiphora*	1	0.51 ±0.01	0.70 ±0.03	0.64 ±0.02	0.55 ±0.04	12.5	
Flinders Island Memana Formation	*Katelysia rhytiphora*	2	0.45	0.58 ±0.18	0.84 ±0.09	0.53 ±0.01	12.5	Early Pleistocene
	Glycymeris striatularis	1	–	0.69 ±0.05	0.67 ±0.02	0.64 ±0.03		
	Divalucina cumingi	1	–	–	0.93 ±0.01	0.07 ±0.01		
Perth Basin, WA; Jandakot Member, Tamala Formation	*Kaletlysia rhytiphora*	12	0.82 ±0.05	0.64 ±0.11[b]	0.95 ±0.02	0.78 ±0.01	18.5	Plio/Pleistocene
Eucla Platform, WA; Roe Calcarenite	*Katelysia* sp.	7	0.82 ±0.07	0.77 ±0.11[b]	–	–	18.2	Plio/Pleistocene

[a] Amino acids: Val, valine; Leu, leucine; Phe, phenylalanine; and Glu, glutamic acid (total acid hydrolyzate).
[b] D/L leucine value artificially low resulting from the coelution of β-Ala with L-Leu.

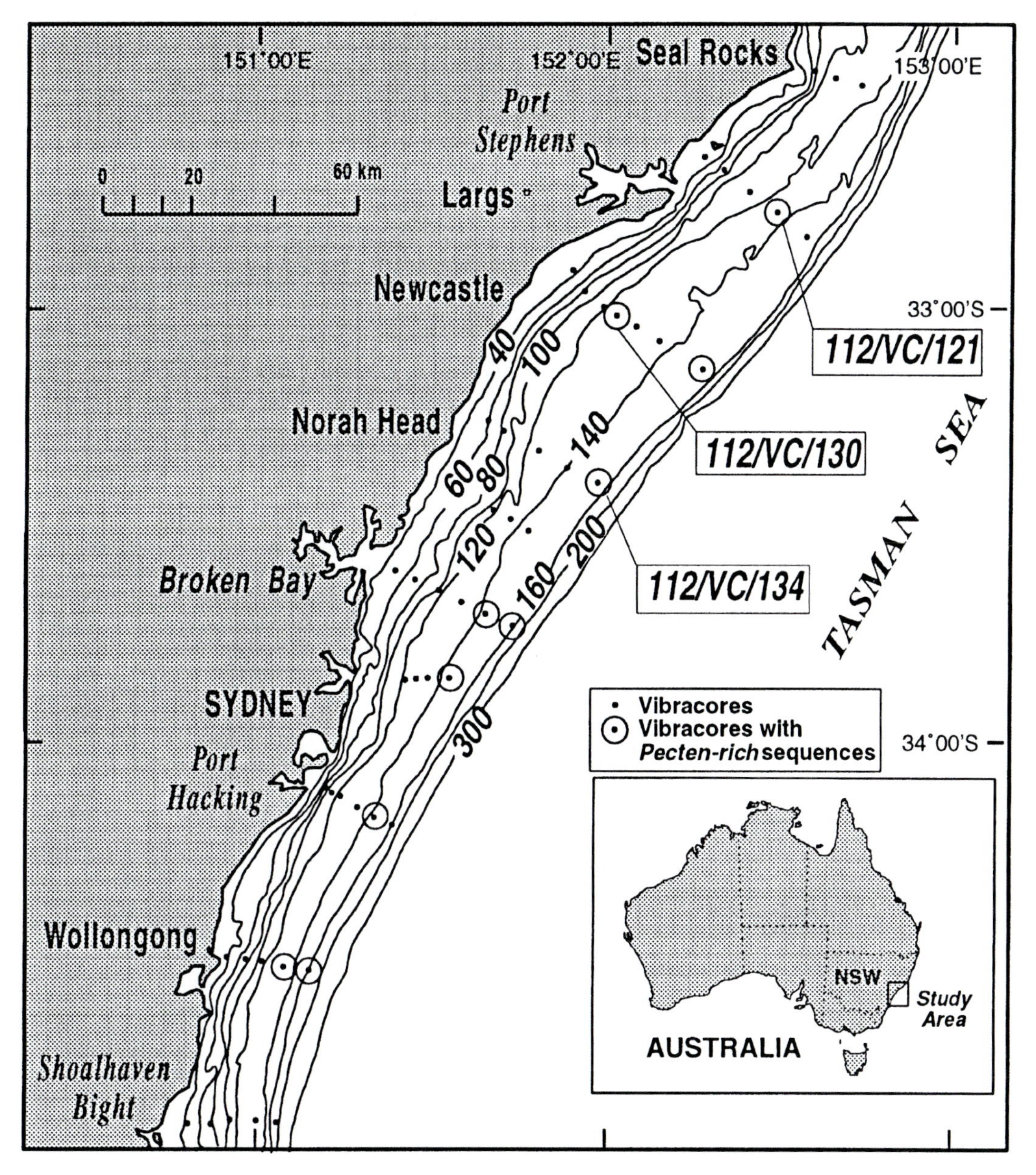

Fig. 20-5. Location of three marine vibracores, the subject of AAR analyses, collected from the New South Wales continental shelf. The cores studied include 112/VC/130, 112/VC/121 and 112/VC/134 collected in present water depths of 123, 139, and 150 m, respectively. Bathymetry is in meters.

with the uppermost shelly unit from each core dated as last glacial age, were deposited during glacial low-sea-level events. In contrast, the units devoid of macrofossils were considered to represent high-stand deposits.

The AAR results reported here are based on fossil mollusks collected from three vibracores in present water depths greater than 120 m (figure 20-5). AAR analyses on *P. fumatus* and *Glycymeris striatularis*, which have yielded "background" radiocarbon results, indicate the presence of sediments of oxygen isotope Stages 8 and 6, in addition to the last glacial maximum (Ferland et al., 1995; Murray-Wallace et al., 1996a). A summary of these data and a comparison with the onland record from the valley fill succession of Hunter River are presented later in this work.

AMINOSTRATIGRAPHY OF AUSTRALIAN QUATERNARY COASTAL SUCCESSIONS

Aminostratigraphy: Definition

Arguably the most widely used application of AAR reactions has been in the aminostratigraphical (chronostratigraphical) subdivision of sedimentary successions, particularly in coastal studies, and a large literature has emerged on this subject (e.g., Miller et al., 1979; Kennedy et al., 1982; Miller and Mangerud, 1985; Wehmiller et al., 1988; Hsu et al., 1989; Aguirre et al., 1995; Murray-Wallace, 1995; Bowen et al., 1998). *Aminostratigraphy* is defined here as a chemostratigraphical method, based on chronostratigraphical principles and involves the classification of strata into chronostratigraphical units or *aminozones* that correspond with intervals of time, termed *geochronological units* (Salvador, 1994). Litho-, bio-, and morphostratigraphical evidence as well as calibration by other geochronological methods, is generally used to support these chronostratigraphically defined units. The fundamental unit of aminostratigraphy, the *aminozone*, refers to a single time-stratigraphical unit representing a well-defined interval of time and is delineated by the clustering (equivalence) of amino acid enantiomeric ratios on replicate fossils from a single, mappable lithostratigraphical unit. According to Miller and Hare (1980), this approach involves the least ambiguous application of AAR data, as it largely overcomes the uncertainties associated with racemization kinetic models and diagenetic temperature histories. Traditional practice has been the application of this tool in restricted geographic settings to overcome regional uncertainties in diagenetic temperature histories.

Overview of Results

A summary of the principal results of AAR analyses on fossil mollusks from coastal interglacial highstand deposits is given in figure 20-4 for the amino acid valine. Results for other amino acids for mollusks from interglacial highstand deposits are presented in tables 20-3–20-5.

Representative results for the extent of AAR in fossil marine mollusks of last interglacial age from southern Australia are given in table 20-5 and figure 20-4. The extent of valine racemization in fossil *Katelysia* sp. and *A. trapezia*, two intermediate racemizing mollusk genera, conforms with an exponential trend in the extent of racemization with diagenetic temperature and, as a corollary, current mean annual air temperature (figure 20-4). Thus, for fossils from sites characterized by higher current mean annual air temperatures, a greater extent of racemization is noted, consistent with theoretical predictions. Accordingly, the extent of AAR essentially doubles for every 5°C increase in current mean annual temperature for last interglacial mollusks from southern Australia. A noteworthy point is that, apart from a restricted area in the interior of Tasmania and the Snowy Mountains in southeastern Australia (areas well removed from coastal settings), Australia was ice-free during the Quaternary. Latitudinal paleo-temperature gradients for Australian coastal localities are therefore likely to be defined by smooth functions, by analogy with present-day temperature gradients. This represents a relevant consideration for global aminostratigraphic correlations, particularly if data are included from northern hemisphere settings formerly close to ice margins during glacial maxima.

The results for the last interglacial fossil mollusks are based on samples collected from southern Australia between Shark Bay, Western Australia (24°50′S, 113°30′E), and Mary Ann Bay, Tasmania (42°53′S, 147°19′E; figures 20-1 and 20-4), span a current mean annual air temperature range from 22.4–12.5°C. All the results plotted in figure 20-4 are for fossil mollusks from deeply buried contexts (i.e., > 1 m). The envelope representing the last interglacial data set illustrates the degree of variation that may be expected for a single isotope stage, in this case Substage 5e (approx. 132–118 ka).

Less data are available for the penultimate interglaciation (oxygen isotope Stage 7); however, the available evidence indicates that a similar latitudinal gradient in extent of racemization with respect to diagenetic temperature histories is apparent (figure 20-4). The extent of racemization of amino acids in fossil mollusks assigned to oxygen isotope Stage 7 on the basis of independent dating methods is consistently higher, and is significantly different from the results for the last interglacial fossil mollusks (figure 20-4).

The extent of racemization of amino acids in a specimen of *A. trapezia* at 52.3 m in drill hole DH6 from Hunter Valley is significantly greater than that typically obtained for last interglacial *A. trapezia* from the same geographic region, such as the locally well-known last interglacial reference section at Largs (Thom and Murray-Wallace, 1988; Leary, 1992; Murray-Wallace et al., 1996c). By analogy with results from Peppermint Grove and Redcliff, two sites with similar current mean annual temperatures, a penultimate interglacial age is assigned to the *Anadara* specimen from the Hunter Valley fill (figure 20-4). The uranium-series age of 181 ± 10.5 ka for an *Anadara* specimen from the same stratigraphic unit within drill hole DH6 strengthens the validity of the AAR results.

A comparable extent of AAR is evident for fossil mollusks from the Memana Formation, Flinders Island, and marine sediments at Egg Lagoon (figure 20-4). The results for *Katelysia* from marine sedi-

TABLE 20-5 Extent of AAR (epimerization) in Last Interglacial (oxygen-isotope Substage 5e) marine mollusks from southern Australia

Species	Number of individuals analyzed	Locality	Stratigraphic unit	CMAT[a] (°C)	Amino acid D/L ratio[b]		
					Val	AIle/Ile	Leu
Anadara trapezia	1	Shark Bay	Dampier Limestone	22.4	0.57 ±0.03	–	0.57
Katelysia rhystiphora	4	Minim Cove, Swan River, WA	Minim Cove Member, Tamala Formation	18.5	0.33 ±0.02	0.35 ±0.02	0.48 ±0.03
Katelysia scalarina	8	Minim Cove, Swan River, WA	Minim Cove Member, Tamala Formation	18.5	0.34 ±0.03	0.39 ±0.04	0.49 ± 0.03
Anadara trapezia	1	Redcliffe, Swan River, WA	Tamala Formation	18.5	0.35	–	0.55 ±0.03
Anadara trapezia	4	Tourville Bay, SA	Glanville Formation	16.9	0.36 ±0.05	0.47 ±0.02	0.45 ±0.02
Anadara trapezia	3	Denial Bay, SA	Glanville Formation	16.9	0.33 ±0.02	0.44 ±0.03	0.44 ±0.01
Katelysia rhytiphora	1	Lake Newland, SA	Glanville Formation	16.9	0.35 ±0.01	0.45 ±0.01	0.47 ±0.01
Anadara trapezia	3	2 km north of Port Wakefield, SA	Glanville Formation	17.0	0.29 ±0.02	0.35 ±0.02	0.42 ±0.02
Anadara trapezia	12	Port Wakefield, SA	Glanville Formation	17.0	0.32 ±0.06	0.43 ±0.02	0.51 ±0.02
Katelysia rhytiphora	5	Port Wakefield, SA	Glanville Formation	17.0	0.32 ± 0.04	0.45 ±0.07	0.51 ±0.07
Mactra australis	6	Normanville, SA	Glanville Formation	16.0	0.28 ±0.01	0.36 ±0.02	0.28 ±0.02
Mactra australis	2	Hindmarsh Is., SA	Glanville Formation	16.0	0.26 ±0.03	0.39 ±0.01	0.37 ± 0.02
Anadara trapezia	2	Chiton Rocks, SA	Glanville Formation	16.0	0.29 ±0.01	0.49 ±0.01	0.41 ±0.01
Anadara trapezia	3	Dry Creek, Adelaide, SA	Glanville Formation	16.9	0.31 ±0.01	0.39 ±0.01	–
Anadara trapezia	3	Hindmarsh Valley, SA	Glanville Formation	15.7	0.28 ±0.04	0.35 ±0.05	0.45 ±0.05
Anadara trapezia	2	Largs, NSW	–	17.9	0.30 ±0.01	0.45 ±0.02	0.45 ±0.04
Anadara trapezia	2	Woakwine Range, Lagoon facies, SA	Glanville Formation	14.7	0.22 ±0.03	–	0.35 ±0.01
Katelysia rhytiphora	4	Woakwine Range, Lagoon facies, SA	Glanville Formation	14.7	0.20 ±0.02	–	0.34 ±0.01
Katelysia rhytiphora	2	East Dairy Range, Lagoon facies, SA	Glanville Formation	14.7	0.18 ±0.01	–	0.34 ±0.01
Glycymyeris (Tucetilla) striatularis	3	Montagu, TAS	Mella Sand	12.5	0.15 ±0.01	–	0.30 ±0.04
Glycymeris (Tucetilla) striatularis	6	Broadmeadows, TAS	Mella Sand	12.5	0.18 ±0.02	–	0.31 ±0.04
Glycymeris (Tucetilla) striatularis	4	Mowbray Swamp, TAS	Mella Sand	12.5	0.20 ±0.02	–	–
Fulvia tenuicostata	4	Mary Ann Bay, TAS	–	12.5	0.22 ±0.04	–	0.39 ±0.02
Katelysia scalarina	3	Yellow Rock River, King Island	–	12.5	0.15 ±0.01	–	–
Katelysia rhytiphora	6	North East River, Flinders Island	–	12.5	0.14 ±0.02	–	0.36 ±0.05

[a] CMAT, current mean annual temperature (°C).
[b] Amino acids: Val, valine; Aile/Ile, D-alloisoleucine/L-isoleucine; Leu, leucine (total acid hydrolyzate).

ments from Egg Lagoon, King Island clearly preclude the last interglacial age assigned to these strata by Jennings (1959). If the Early Pleistocene age assigned to the Memana Formation on the basis of biostratigraphical evidence remains valid, then the AAR results from this unit require a reappraisal. By analogy with the amino acid data from the Jandakot Member of the Yoganup Formation, Perth Basin, and the Roe Calcarenite from the Eucla Platform, Western Australia, the extent of AAR for the Memana Formation appears to be lower than the predicted value, even allowing for the lower diagenetic temperature history likely for King Island (figure 20-4). Further analyses are clearly required to assess the integrity of these fossils for the retention of indigenous amino acids, particularly lower molecular-weight peptide residues, as the lower-than-predicted extent of racemization evident in the total acid hydrolysate may reflect *in situ* leaching and hence the preferential loss of these more labile fractions from the fossils.

The amino acid results for Holocene fossils also highlight the influence of diagenetic temperatures on racemization rates (figure 20-4).

Land–Sea Correlations

A preliminary attempt at land–sea correlations using the AAR data for fossil mollusks from the New South Wales outer continental shelf, with results from the Hunter River valley fill, between Newcastle and Largs (figure 20-5) is presented in table 20-6. The data are grouped into six time-slices that equate with oxygen isotope stages 1, 2, 5e, 6, 7 and 8 of the deep-sea record (Imbrie et al., 1984; Bassinot et al., 1994). Oxygen isotope stages 1, 5e and 7 represent interglacial estuarine deposits from Hunter Valley, whereas glacial stages 2, 6 and 8 relate to the shallow-water shelly facies from the outer continental shelf of New South Wales (the latter being based on vibracores collected offshore from Newcastle) (figure 20-5). Amino acid analyses were undertaken on the fossil mollusks *A. trapezia* and *P. fumatus*, two genera present in both on land and offshore settings; the number of specimens analyzed (partly a function of their abundance within deposits) is given in table 20-6.

Independent evidence for the age of some of these deposits is available. The Holocene specimen is from a midden located within a golf course at Fullerton Cove, near Newcastle. A radiocarbon age of 3280 ± 90 yr BP (Beta-51088) was obtained for a specimen of *A. trapezia* from this deposit (Leary, 1992). Charcoal was not found within the midden, suggesting that the deposit was not subject to heating by camp-fires. A comparable extent of racemization has been noted for other marine mollusks of mid-Holocene age from natural shell beds in southern Australia (Murray-Wallace, 1995).

Mollusks assigned to Stage 2 have also been analyzed by the radiocarbon method and have yielded ages that range from early- to late-last glacial maximum (i.e., from 27 to 16 ka; Ferland and Roy, 1997). Subsamples of individual *P. fumatus* from this group have been analyzed both by accelerator mass spectrometry radiocarbon dating and by AAR (Murray-Wallace et al., 1996a). A valve of *A. trapezia* has also yielded an age of 18,400 ± 280 yr BP (Beta-45,438).

Six valves of *A. trapezia* collected from the Largs shell bed yielded an uncorrected radiocarbon age of 34,390 ± 370 yr BP (SUA-3008). This result is regarded as a minimum age, reflecting the incorporation, during late diagenesis, of approximately 2% radiocarbon with a modern activity, which could not be isolated during sample pretreatment. A lithostratigraphical correlation with a coralline facies at Grahamstown points to a last interglacial age for the Largs deposit (Thom and Murray-Wallace, 1988). Corals from beneath the Inner Barrier at Grahamstown yielded uranium-series ages of 142 ± 12 and 143 ± 13 ka (Marshall and Thom, 1976). For the older specimens of *A. trapezia*, analyzed by AAR from the Hunter Valley drill hole (DH6), a uranium-series age of 181.7 ± 10.5 ka has been obtained, thus indicating the presence of Stage 7 and Substage 5e deposits within the valley fill succession.

An increase in the extent of AAR with age is generally evident in this land–sea correlation. Although some overlap in the results is apparent for Substage 5e and Stage 6, several factors may be contributing to this. For example, difficulties arise in land–sea comparisons because of the contrasting diagenetic temperature histories likely to have been experienced by these settings, despite their geographic proximity. The fossil mollusks from the outer continental shelf are unlikely to have ever been subaerially exposed, particularly in the case of the mollusks in Core 112/VC/134 collected in a present water depth of 150 m. In contrast, the shell bed at Largs is overlain by a 4-m-thick unit of poorly sorted estuarine sands and clays and is likely to have experienced a higher diagenetic temperature history. Allowing for uncertainties in modern temperature records and their short duration, a difference of approx. 2°C is likely between the shelf core sites and the Largs deposit at any given time. As suggested for the U.S. Middle Atlantic continental shelf (Groot et al., 1995), Pleistocene temperature variations offshore in the New South Wales continental shelf are likely to have been smaller (i.e., for those areas not subaerially exposed) than those of terrestrial settings (i.e., Largs deposit). Therefore, the extent of racemization of amino acids in the younger (last interglacial) mollusks at Largs is likely to be slightly higher with respect to age when compared with the mollusks considered here to be of penultimate glacial age from the offshore setting (table 20-6).

A further consideration relates to the relative age of Substage 5e and Stage 6 deposits in this correla-

TABLE 20-6 Aminostratigraphical land–sea correlations of fossil marine mollusks of the New South Wales outer continental shelf with estuarine mollusks from the Hunter Valley, New South Wales. Amino acids: Leu, leucine; Val, valine; Glu, glutamic acid; Ala, alanine

Hunter Valley, New South Wales (this study)		New South Wales, outer continental shelf (Murray-Wallace et al., 1996a)	
Holocene (stage 1)			
Fullerton Cove Golf Course Midden			
3280 ±90 yr BP (Beta-51088)			
Anadara trapezia ($n = 2$)			
Leu 0.16			
Ala 0.13			
Glu 0.10			
Last glacial maximum (stage 2)		*Pecten fumatus*	
		Early	Late
		($n = 11$)	($n = 9$)
		Leu 0.30 ±0.04	0.20 ±0.04
		Val 0.27 ±0.05	0.19 ± 0.04
		Glu 0.27 ±0.03	0.19 ±0.03
		Ala 0.43 ±0.06	0.28 ±0.05
		Anadara trapezia ($n = 1$)	
		Leu 0.22	
		Glu 0.17 18,400 ±280 yr BP	
		Ala 0.43	
Last interglacial maximum (substage 5e)			
Largs			
Pecten fumatus ($n = 1$)	*Anadara trapezia* ($n = 4$)		
Leu 0.40 ±0.03	0.46 ±0.02		
Val 0.40 ±0.005	0.30 ±0.01		
Glu 0.28 ±0.005	0.36 ±0.02		
Penultimate glaciation (stage 6)		*Pecten fumatus* ($n = 21$)	
		Leu 0.37 ±0.02	
		Val 0.35 ±0.02	
		Glu 0.31 ±0.03	
		Ala 0.52 ±0.05	
Penultimate interglaciation (stage 7)			
Hunter Valley drill hole DH6@52.3 m			
Anadara trapezia ($n = 1$)			
Leu 0.65 ±0.005			
Val 0.46 ±0.003			
Ala 0.80 ±0.001			
Pre-penultimate glaciation (stage 8)		*Pecten fumatus* ($n = 8$)	
		Leu 0.51 ±0.02	
		Val 0.50 ±0.02	
		Glu 0.42 ±0.02	
		Ala 0.72 ±0.03	

tion. For example, the estuarine facies at Largs may relate to the early part of the last interglacial maximum, and therefore may potentially be only a few thousand years younger than the fossils interpreted here as correlating with Stage 6. At present, the AAR method does not have sufficient resolution to distinguish subtle differences in age for materials that are potentially hundreds of thousands of years old.

QUATERNARY AMINOSTRATIGRAPHY: AUSTRALIAN DATA IN A GLOBAL CONTEXT

Sufficiently large data sets now exist for the extent of AAR in Quaternary marine mollusks to permit an assessment of the potential of aminostratigraphical correlation of coastal sequences on a global scale. Global aminostratigraphical correlations were originally undertaken on an incidental basis as part of larger investigations (e.g., von der Borch et al., 1980). Hearty and Miller (1987), however, first formalized global aminostratigraphical correlation based on an extensive compilation (3000 shells) of the extent of isoleucine epimerization in last interglacial *sensu lato* mollusks, from a range of locations around the Circum-Atlantic, Mediterranean and South Pacific Oceans. Since the study of Hearty and Miller (1987), others have attempted similar correlations (e.g., Hollin and Hearty, 1990; Hearty and Kindler, 1995; Murray-Wallace, 1995).

Global aminostratigraphical correlations require considerable caution as they rest heavily on the following: (1) rigorous interlaboratory comparisons of amino acid standards (e.g., Wehmiller, 1984b); (2) due acknowledgment of the parameters that may influence diagenetic racemization, especially the diagenetic temperature history; (3) reliable calibration of amino acid data by independent geochronometrical methods, in conjunction with rigorous geomorphological, lithostratigraphical and paleontological evidence of age; and (4) knowledge of the nature of racemization kinetics in different genera of fossil mollusks.

Marine mollusks, independently established by other dating methods as being of last interglacial age (oxygen isotope Substage 5e), from northern hemisphere and southern Australian sites with comparable current mean annual temperatures and extents of racemization (epimerization) point to the possibility of global correlation (figure 20-6). Results for the amino acid isoleucine are depicted

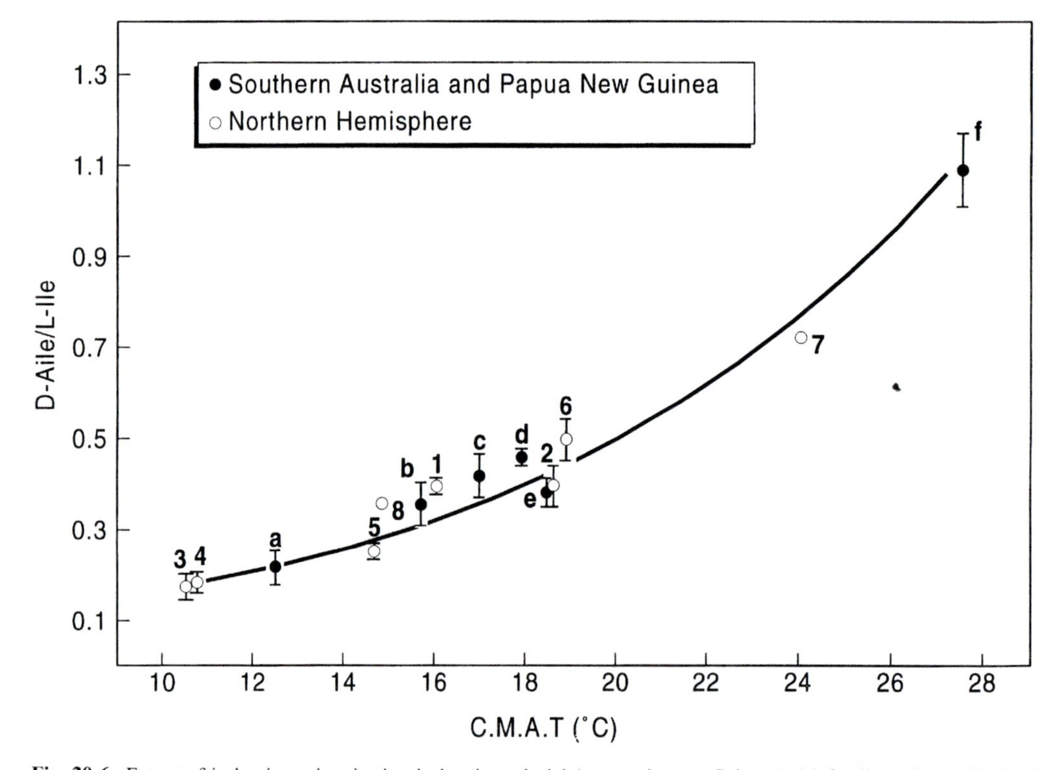

Fig. 20-6. Extent of isoleucine epimerization in last interglacial (oxygen isotope Substage 5e) fossil marine mollusks plotted against the current mean annual temperature (CMAT°C) for coastal successions from northern and southern hemisphere sites. Amino acid data from southern Australia and Huon Peninsula, Papua New Guinea are represented by solid circles. Results from northern hemisphere localities are indicated by open circles. Sample localities: (1) Latina, Italy (Hearty and Dai Pra, 1986); (2) Tunisia (Miller et al., 1986); (3) and (4) The Netherlands (Miller and Mangerud, 1985); (5) Corsica (Hearty et al., 1986); (6) Tunisia (Hearty et al., 1986); (7) Florida (Mitterer, 1975); and (8) San Diego, California (Karrow and Bada, 1980). Southern Australia and Papua New Guinea sites include: (a) Mary Ann Bay, Tasmania; (b) Victor Harbor, South Australia; (c) Port Wakefield, South Australia; (d) Largs, New South Wales; and (e) Minim Cove, Western Australia (Murray-Wallace et al., 1991). Isoleucine data for *Tridacna gigas* are from the Late Pleistocene marine terraces of the Huon Peninsula, Papua New Guinea (Hearty and Aharon, 1988).

in figure 20-6. These data show an exponential trend in the extent of isoleucine epimerization for isochronous fossils from sites of contrasting present-day temperatures (and as a corollary, contrasting diagenetic temperatures) that is similar to the trend apparent in the results for valine from southern Australia (cf. figure 20-4).

Fossil mollusks from well-buried contexts and with reliable independent evidence of age were selected for this comparison. Mollusk genera from northern hemisphere sites include *Glycymeris* sp., *Arca* sp. from Italy and Tunisia (Miller et al., 1986; Hearty and Dai Pra, 1986), *Mercenaria* samples from Florida (Mitterer, 1975), *Chione* from San Diego County, California (Karrow and Bada, 1980), and *Bittium reticulatum* and *Cardium edule* from The Netherlands (Miller and Mangerud, 1985). Species included in the compilation from southern Australia include *Katelysia* sp., *A. trapezia*, and *G. striatularis* and represent species that undergo moderate and comparable rates of racemization.

It is particularly noteworthy that last interglacial fossils from different regions but with equivalent current mean annual temperatures (e.g., Minim Cove, Western Australia and Tunisia) are characterized by similar extents of isoleucine epimerization. This suggests that these fossils have experienced equivalent *effective diagenetic temperature* histories (i.e., the integrated effect of all of the temperatures experienced by these fossils during diagenesis is comparable).

Future challenges in global aminostratigraphical correlation include the following: (1) extending the geographical and temporal coverage of AAR (epimerization) data; (2) improving the age-resolving power of the method, especially for oxygen isotope Stage 5; and (3) applying these data sets to regional paleoclimatic reconstructions.

SUMMARY

1. The extent of AAR in fossil marine mollusks from coastal and shallow marine successions has provided a valuable supplementary framework for subdivision of the Quaternary marginal marine record of southern Australia.
2. Confirmation of the extensive distribution of coastal successions of the last interglacial maximum (oxygen isotope Substage 5e) has been established using AAR, calibrated by independent dating methods, from key reference sections where more than one method of dating could be undertaken. These studies have helped to resolve variations in the elevation of intertidal facies of last interglacial coastal successions and, accordingly, have quantified subtle neotectonic variations in an otherwise tectonically stable continent.
3. Directly calibrated by radiocarbon dating, AAR has identified coastal facies of last glacial age (oxygen isotope Stage 2), as well as stages 6 and 8 on the outer continental shelf of New South Wales in a zone where present water depths exceed 120 m.
4. AAR has also assisted in the assessment of the completeness of the Australian marginal marine lithostratigraphical record. In particular, successions of penultimate interglacial age (oxygen isotope Stage 7) have been identified by aminostratigraphy, calibrated by independent methods of dating. The results indicate that sea levels during Stage 7 were higher than those otherwise predicted on the basis of the oxygen isotope record.

Acknowledgments: This research has been supported by the Australian Research Council and the Quaternary Environments Research Centre at the University of Wollongong. Rick Oches, John Wehmiller, and an anonymous reviewer are thanked for their constructive reviews of this work. This paper is a contribution to IGCP Project 437 (*Coastal Environmental Change During Sea-level Highstands: a Global Synthesis with Implications for Management of Future Coastal Change*).

REFERENCES

Aguirre M. L., Bowen D. Q., Sykes G. A., and Whatley R. C. (1995) A provisional aminostratigraphical framework for late Quaternary marine deposits in Buenos Aires province, Argentina. *Marine Geol.* **128**, 85–104.

Bada J. L. and Schroeder R. A. (1972) Racemization of isoleucine in calcareous marine sediments: kinetics and mechanism. *Earth Planet. Sci. Lett.* **15**, 1–11.

Bassinot F. C., Labeyrie L. D., Vincent E., Quidelleur X., Shackleton N. J., and Lancelot Y. (1994) The astronomical theory of climate and the age of the Brunhes–Matuyama magnetic reversal. *Earth Planet. Sci. Lett.* **126**, 91–108.

Belperio A. P. (1995) Quaternary. In *The Geology of South Australia*, vol. 2. *The Phanerozoic* (eds J. F. Drexel and W. V. Preiss), pp. 219–280. South Australia, Geological Survey, Bulletin 54.

Belperio A. P., Murray-Wallace C. V., and Cann J. H. (1995) The last interglacial shoreline in southern Australia: morphostratigraphic variations in a temperate carbonate setting. *Quat. Intl.* **26**, 7–19.

Belperio A. P., Smith B. W., Polach H. A., Nittourer C. A., De Master D. J., Prescott J. R., Hails J. R., and Gostin V. A. (1984) Chronological studies of the Quaternary marine sediments of northern Spencer Gulf, South Australia. *Marine Geol.* **61**, 265–296.

Bourman R. P., Belperio A. P., Murray-Wallace C. V., and Cann J. H. (1999) A last interglacial embayment fill at Normanville, South Australia and its neotectonic implications. *Roy Soc. South Aust. Trans.* **123**, 1–15.

Bowden A. R. and Colhoun E. A. (1984) Quaternary emergent shorelines of Tasmania. In *Coastal Geomorphology in Australia* (ed. B. G. Thom), pp. 313–342. Academic Press.

Bowen D. Q., Pillans B., Sykes G. A., Beu A. G., Edwards A. R., Kamp P. J. J., and Hull A. G. (1998) Amino acid geochronology of Pleistocene marine sediments in the Wanganui Basin: a New Zealand framework for correlation and dating. *J. Geol. Soc. Lond.* **155**, 439–446.

Brigham J. K. (1983) Intrashell variations in amino acid concentrations and isoleucine epimerization ratios in fossil *Hiatella arctica*. *Geology* **11**, 509–513.

Cann J. H. and Clarke J. (1993) The significance of *Marginopora vertebralis* (foraminifer) in surficial sediments at Esperance, Western Australia and in last interglacial sediments in northern Spencer Gulf, South Australia. *Marine Geol.* **111**, 171–187.

Cann J. H., Belperio A. P., Gostin V. A., and Murray-Wallace C. V. (1988) Sea-level history 45,000 to 30,000 yr BP, inferred from benthic foraminifera, Gulf St Vincent, South Australia. *Quat. Res.* **29**, 153–175.

Cann J. H., Belperio A. P., Gostin V. A., and Rice R. L. (1993) Contemporary benthic foraminifera in Gulf St Vincent, South Australia, and a refined late Pleistocene sea level history. *Aust. J. Earth Sci.* **40**, 197–211.

Daniel R., von der Borch C., Veeh H., James N., and Bone Y. (1998) Late Quaternary evolution of Streaky Bay, South Australia: relative sea levels and timing. In *Australian and New Zealand Geomorphology Group, Eighth Biennial Conference, Abstract Booklet* (ed. R. P. Bourman), Goolwa, South Australia. p. 15.

D'Costa D. M., Grindrod J., and Ogden R. (1993) Preliminary environmental reconstructions from late Quaternary pollen and mollusc assemblages at Egg Lagoon, King Island, Bass Strait. *Aust. J. Ecol.* **18**, 351–366.

Ferland M. A. and Roy P. S. (1997) Southeastern Australia: a sea-level dependent, cool-water carbonate margin. In *Cool-water Carbonates, SEPM Special Publication* vol. 56 (eds N. P. James and J. D. A. Clarke), pp. 37–52.

Ferland M. A., Roy P. S., and Murray-Wallace C. V. (1995) Glacial lowstand deposits on the outer continental shelf of southeastern Australia. *Quat. Res.* **44**, 294–299.

Frank H., Nicholson G. J., and Bayer E. (1977) Rapid gas chromatographic separation of amino acid enantiomers with a novel chiral stationary phase. *J. Chromat. Sci.* **15**, 174–176.

Gillespie R. and Polach H. A. (1979) The suitability of marine shells for radiocarbon dating of Australian Prehistory. In *Proceedings of the Ninth International Conference on Radiocarbon Dating* (eds R. Berger and H. Suess), pp. 404–421. University of California Press.

Goede A. (1989) Electron spin resonance—a relative dating technique for Quaternary sediments near Warrnambool, Victoria. *Aust. Geograph. Std.* **27**, 14–30.

Groot J. J., Benson R. N., and Wehmiller J. F. (1995) Palynological, foraminiferal and aminostratigraphic studies of Quaternary sediments from the U. S. Middle Atlantic upper continental slope, continental shelf and coastal plain. *Quat. Sci. Rev.* **14**, 17–49.

Hails J. R., Belperio A. P., Gostin V. A., and Sargent G. E. G. (1984) The submarine Quaternary stratigraphy of northern Spencer Gulf, South Australia. *Marine Geol.*, **61**, 345–372.

Hearty P. J. and Aharon P. (1988) Amino acid chronostratigraphy of late Quaternary coral reefs: Huon Peninsula, New Guinea, and the Great Barrier Reef, Australia. *Geology* **16**, 579–583.

Hearty P. J. and Dai Pra G. (1986) Aminostratigraphy of Quaternary marine deposits in the Lazio Region of Central Italy. *Zeit. Geomorph.*, Suppl.-Bd., 62, 131–140.

Hearty P. J. and Kindler P. (1995) Sea-level highstand chronology from the stable carbonate platforms (Bermuda and The Bahamas). *J. Coastal Res.* **11**, 675–689.

Hearty P. J. and Miller G. H. (1987) Global trends in isoleucine epimerization: data from the Circum-Atlantic, the Mediterranean and the South Pacific. *Geol. Soc. Amer. Abstracts* p. 698.

Hearty P. J., Miller G. H., Stearns C. E., and Szabo B. J. (1986) Aminostratigraphy of Quaternary shorelines in the Mediterranean Basin. *Geol. Soc. Amer., Bull.* **97**, 850–858.

Hewgill F. R., Kendrick G. W., Webb R. J., and Wyrwoll K.-H. (1983) Routine ESR dating of emergent Pleistocene marine units in Western Australia. *Search* **14**, 215–217.

Hollin J. T. and Hearty P. J. (1990) South Carolina interglacial sites and stage 5 sea levels. *Quat. Res.* **33**, 1–17.

Hsu J. T., Leonard E. M., and Wehmiller J. F. (1989) Aminostratigraphy of Peruvian and Chilean Quaternary marine terraces. *Quat. Sci. Rev.* **8**, 255–262.

Huntley D. J., Hutton J. T., and Prescott J. R. (1993) The stranded beach-dune sequence of south-east South Australia: a test of thermoluminescence dating, 0–800 ka. *Quat. Sci. Rev.* **12**, 1–20.

Huntley D. J., Hutton J. T., and Prescott J. R. (1994) Further thermoluminescence dates from the dune sequence in the southeast of South Australia. *Quat. Sci. Rev.* **13**, 201–207.

Imbrie J., Hays J. D., Martinson D. G., McIntyre A., Mix A. C., Morley J. J., Pisias N. G., Prell W. L., and Shackleton N. J. (1984) The orbital theory of Pleistocene climate: support from a revised chronology of the marine $\delta^{18}O$ record. In *Milankovitch and Climate*, Part 1, NATO ASI Series C, vol. 126, (eds A. Berger, J. Imbrie, J. Hays, G. Kukla, and B. Saltzman), pp. 269–305. Reidel.

Jennings J. N. (1959). The coastal geomorphology of King Island, Bass Strait in relation to changes in the relative level of land and sea. *Rec. Queen Vic. Mus., Launceston* **11**, 1–39.

Karrow, P. F. and Bada J. L. (1980) Amino acid racemization dating of Quaternary raised marine terraces in San Diego County, California. *Geology* **8**, 200–204.

Kennedy G. L., Lajoie K. R., and Wehmiller J. F. (1982) Aminostratigraphy and faunal correlations of late Quaternary marine terraces, Pacific Coast, USA. *Nature* **299**, 545–547.

Kimber R. W. L. and Milnes A. R. (1984) The extent of racemization of amino acids in Holocene and Pleistocene marine mollusks in southern South Australia: preliminary data on a time-framework for calcrete formation. *Aust. J. Earth Sci.* **31**, 279–286.

Kvenvolden K. A., Blunt D. J., and Clifton H. E. (1979) Amino acid racemization in Quaternary shell deposits at Willapa Bay, Washington. *Geochim. Cosmochim. Acta.* **43**, 1505–1520.

Kvenvolden K. A., Blunt D. J., and Clifton H. E. (1981) Age estimations based on amino acid racemization: reply to comments by J. F. Wehmiller. *Geochim. Cosmochim. Acta.* **45**, 265–267.

Leary S. (1992) Aminostratigraphy of the Newcastle Bight embayment, with particular reference to a Late Pleistocene deposit at Largs. Bsc(Hons) Thesis, University of Newcastle, Australia, 87 pp.

Ludbrook N. H. (1978) Quaternary molluscs of the western part of the Eucla Basin. *Bull., Geol. Surv. Western Australia*, vol. 125.

Ludbrook N. H. (1984). *Quaternary Molluscs of South Australia.* Handbook No. 9. Department of Mines and Energy, South Australia.

Marshall J. F. and Thom B. G. (1976) The sea level in the last interglacial. *Nature* **263**, 120–121.

Miller G. H. and Brigham-Grette J. (1989) Amino acid geochronology: resolution and precision in carbonate fossils. *Quat. Intl.* **1**, 111–128.

Miller G. H. and Hare P. E. (1980) Amino acid geochronology: integrity of the carbonate matrix and potential of molluscan fossils. In *Biogeochemistry of Amino Acids* (eds P. E. Hare, T. C. Hoering, and K. King, Jr), pp. 415–451. Wiley.

Miller G. H. and Mangerud J. (1985) Aminostratigraphy of European marine interglacial deposits. *Quat. Sci. Rev.* **4**, 215–278.

Miller G. H., Hollin J. T., and Andrews J. T. (1979) Aminostratigraphy of UK Pleistocene deposits. *Nature* **281**, 530–543.

Miller G. H., Paskoff R., and Stearns C. E. (1986) Amino acid geochronology of Pleistocene littoral deposits in Tunisia. *Zeit. Geomorph.* Suppl.-Bd., **62**, 197–207.

Mitterer R. M. (1975) Age and diagenetic temperatures of Pleistocene deposits of Florida based on isoleucine epimerization in *Mercenaria. Earth Planet. Sci. Lett.* **28**, 275–282.

Mitterer R. M. and Kriausakul N. (1989) Calculation of amino acid racemization ages based on apparent parabolic kinetics. *Quat. Sci. Rev.* **8**, 353–357.

Murray-Wallace C. V. (1993) A review of the application of the AAR reaction to archaeological dating. *The Artefact* **16**, 19–26.

Murray-Wallace C. V. (1995) Aminostratigraphy of Quaternary coastal sequences in southern Australia—an overview. *Quat. Intl.* **26**, 69–86.

Murray-Wallace C. V. and Belperio A. P. (1991) The last interglacial shoreline in Australia—a review. *Quat. Sci. Rev.* **10**, 441–461.

Murray-Wallace C. V. and Bourman R. P. (1990) Direct radiocarbon calibration for amino acid racemization dating. *Aust. J. Earth Sci.* **37**, 365–367.

Murray-Wallace C. V. and Goede A. (1995) Aminostratigraphy and electron spin resonance dating to Quaternary coastal neotectonism in Tasmania and the Bass Strait islands. *Aust. J. Earth Sci.* **42**, 51–67.

Murray-Wallace C. V. and Kimber R. W. L. (1987) Evaluation of the amino acid racemization reaction in studies of Quaternary marine sediments in South Australia. *Aust. J. Earth Sci.* **34**, 279–292.

Murray-Wallace C. V. and Kimber R. W. L. (1989) Quaternary marine aminostratigraphy: Perth Basin, Western Australia. *Aust. J. Earth Sci.* **36**, 553–568.

Murray-Wallace C. V. and Kimber R. W. L. (1993) Further evidence for apparent 'parabolic' racemization kinetics in Quaternary molluscs. *Aust. J. Earth Sci.* **40**, 313–317.

Murray-Wallace C. V., Kimber R. W. L., Gostin V. A., and Belperio A. P. (1988) Amino acid racemisation dating of the "Older Pleistocene Marine Beds", Redcliff, northern Spencer Gulf, South Australia. *Roy. Soc. South Aust., Trans.* **112**, 51–55.

Murray-Wallace C. V., Belperio A. P., Picker, K., and Kimber R. W. L. (1991) Coastal aminostratigraphy of the last interglaciation in southern Australia. *Quat. Res.* **35**, 63–71.

Murray-Wallace C. V., Belperio A. P., Gostin V. A., and Cann J. H. (1993) Amino acid racemization and radiocarbon dating of interstadial marine strata (oxygen isotope stage 3), Gulf St. Vincent, South Australia. *Marine Geol.* **110**, 83–92.

Murray-Wallace C. V., Ferland M. A., Roy P. S., and Sollar A. (1996a) Unravelling patterns of reworking in lowstand shelf deposits using amino acid racemisation and radiocarbon dating. *Quat. Sci. Rev. (Quat. Geochron.)*, **15**, 685–697.

Murray-Wallace C. V., Belperio A. P., Cann J. H., Huntley D. J., and Prescott J. R. (1996b) Late Quaternary uplift history, Mount Gambier region, South Australia. *Zeit. Geomorph.* Suppl.-Bd. 106, 41–56.

Murray-Wallace C. V., Leary S., and Kimber R. W. L. (1996c) Amino acid racemisation dating of a last interglacial estuarine deposit at Largs, New South Wales. *Linn. Soc., New South Wales, Proc.* **117**, 213–232.

Murray-Wallace C. V., Belperio A. P., and Cann J. H. (1998) Quaternary neotectonism and intraplate volcanism: the Coorong to Mount Gambier Coastal Plain, Southeastern Australia: a review. In *Coastal Tectonics*, Special Publications, vol. 146 (eds I. S. Stewart and C. Vita-Finzi), pp. 255–267. Geological Society, London.

Murray-Wallace C. V., Beu A. G., Kendrick G. W., Brown L. J., Belperio A. P., and Sherwood J. E. (2000) Palaeoclimatic implications of the occurrence of the arcoid bivalve *Anadara trapezia* (Deshayes) in the Quaternary of Australasia. *Quat. Sci. Rev.* **19**, 559–590.

Phipps C. V. G. (1970) Dating eustatic events from cores taken in the Gulf of Carpentaria and samples from the New South Wales continental shelf. *Aust. J. Sci.* **32**, 329–330.

Salvador A. (1994) (ed.) *International Stratigraphic Guide*, 2nd edn. Geological Society of America.

Schwebel D. A. (1978) *Quaternary Stratigraphy of the Southeast of South Australia.* Unpublished Ph. D. Thesis, The Flinders University of South Australia.

Sherwood J., Barbetti M., Ditchburn R., Kimber R. W. L., McCabe W., Murray-Wallace C. V., Prescott J. R., and Whitehead N. (1994) A comparison study of Quaternary dating techniques applied to sedimentary deposits in southwest Victoria, Australia. *Quat. Sci. Rev. (Quat. Geochron.)* **13**, 95–110.

Stirling C. E., Esat T. M., McCulloch M. T., and Lambeck K. (1995) High-precision U-series dating of corals from Western Australia and implications for the timing and duration of the last interglacial. *Earth Planet. Sci. Lett.* **135**, 115–130.

Stuiver M. and Reimer P. J. (1993) Extended ^{14}C database and revised CALIB radiocarbon calibration program. *Radiocarbon* **35**, 215–230.

Sutherland F. L. and Kershaw R. C. (1971) The Cainozoic geology of Flinders Island, Bass Strait. *Roy. Soc. Tasm., Pap. Proc.* **105**, 151–175.

Thom B. G. (ed.) (1984) *Coastal Geomorphology in Australia.* Academic Press.

Thom B. G. and Chappell J. (1975) Holocene sea levels relative to Australia. *Search* **6**, 90–93.

Thom B. G. and Murray-Wallace C. V. (1988) Last interglacial (Stage 5e) estuarine sediments at Largs, New South Wales. *Aust. J. Earth Sci.* **35**, 571–574.

Thom B. G., Shepherd M., Ly C. K., Roy P. S., Bowman G. M., and Hesp P. A. (1992) *Coastal Geomorphology and Quaternary Geology of the Port Stephens–Myall Lakes area.* Monograph No. 6. Department of Biogeography and Geomorphology, ANU.

Veeh H. H. and Veevers J. J. (1970) Sea level at −175 m off the Great Barrier Reef 13,600 to 17,000 years ago. *Nature* **226**, 536–537.

von der Borch C. C., Bada J. L., and Schwebel D. A. (1980) Amino acid racemization dating of late Quaternary strandline events of the coastal plain sequence, southern South Australia. *Roy. Soc. South Aust., Trans.* **104**, 167–170.

Wehmiller J. F. (1981) Kinetic model options for interpretation of amino acid enantiomeric ratios in Quaternary mollusks: comments on a paper by Kvenvolden et al. 1979. *Geochim. Cosmochim. Acta* **45**, 261–264.

Wehmiller J. F. (1982) A review of amino acid racemization studies in Quaternary mollusks: stratigraphic and chronologic applications in coastal and interglacial sites, Pacific and Atlantic Coasts, United States, United Kingdom, Baffin Island and tropical islands. *Quat. Sci. Rev.*, **1**, 83–120.

Wehmiller J. F. (1984a) Relative and absolute dating of Quaternary mollusks with amino acid racemization: evaluation, applications and questions. In *Quaternary Dating Methods* (ed. W. C. Mahaney), pp. 171–193. Elsevier Science.

Wehmiller J. F. (1984b) Interlaboratory comparison of amino acid enantiomeric ratios in fossil Pleistocene mollusks. *Quat. Res.*, **22**, 109–120.

Wehmiller J. F., Belknap D. F., Boutin B. S., Mirecki J. E., Rahaim S. D., and York L. L. (1988) A review of the aminostratigraphy of Quaternary mollusks from United States Atlantic Coastal Plain sites. In *Dating Quaternary Sediments* (ed. D. J. Easterbrook), Geol. Soc. Amer. Special Paper 227 pp. 69–110.

Wehmiller J. F. and Hare P. E. (1971). Racemization of amino acids in marine sediments, *Science* **173**, 907–911.

Wehmiller J. F., York L. L., and Bart M. L. (1995). Amino acid racemization geochronology of reworked Quaternary mollusks on the U.S. Atlantic coast beaches: implications for chronostratigraphy, taphonomy and coastal sediment transport. *Marine Geol.* **124**, 303–337.

Zhu Z. R, Wyrwoll K.-H., Collins L. B., Chen J. H., Wasserburg G. J., and Eisenhauer, A. (1993) High-precision U-series dating of last interglacial events by mass spectrometry: Houtman Abrolhos Islands, Western Australia. *Earth Planet. Sci. Lett.* **118**, 281–293.

21. Amino acid geochronology of Quaternary coastal terraces on the northern margin of Delaware Bay, southern New Jersey, U.S.A.

Michael L. O'Neal, John F. Wehmiller, and Wayne L. Newell

Emergent Quaternary coastal–marine terraces have long been interpreted as highstand deposits (Cooke, 1958; Oaks and Coch, 1963) resulting from fluctuations in climate-driven eustatic sea level (Flint, 1971; Bloom and Yonekura, 1985). Oaks and Coch (1963) recognized that terraces of the U.S. Atlantic coastal plain may be complex features formed from several deposits from many sea-level events. The estuarine terraces flanking the northern margin of Delaware Bay in southern New Jersey (figure 21-1) have been studied by numerous researchers during the past century. Comprising the Cape May Formation, these sediments have alternately been interpreted as fluvial, glacial outwash, estuarine and marine deposits of Sangamonian to Wisconsinan age (Lewis and Kummel, 1912; Salisbury and Knapp, 1917; MacClintock and Richards, 1936; MacClintock, 1944; Gill, 1962; Owens and Minard, 1979; Newell et al., 1988). Newell et al. (1995) redefined the Cape May Formation as three allostratigraphic units representing early and late phases of the Sangamonian interglacial and the middle Wisconsinan glacial.

In this report we present aminostratigraphic and geophysical evidence for the further subdivision of the Cape May Formation into six late Pleistocene allostratigraphic units. Initially recognized in ground-penetrating radar profiles (O'Neal, 1997), these sea-level highstand sequences form two composite terraces along the northern margin of Delaware Bay, separated by a multi-genetic escarpment (figure 21-2). The older upper terrace ranges in elevation from +18 m to +9 m above mean sea level (MSL) and is underlain by four estuarine–marine sequences inset into and/or overtopping one another. In turn, these units are inset into the fluvial Mio–Pliocene Bridgeton Formation and the marine Miocene Cohansey Formation. The younger lower terrace slopes bayward from +6 m to MSL and contains two vertically adjacent estuarine–marine highstand deposits and the modern salt marsh. Separating the two terraces is the variably sloping Cedarville Scarp (Newell et al., 1995) that was created by erosion during at least two sea-level highstand events, which also deposited the sequences underlying the lower terrace.

Amino acid analyses were conducted on fossil mollusk samples collected from the Morie Company sand and gravel pit near Mauricetown, New Jersey (figures 21-2 and 21-3). Three of the allostratigraphic units were recognized in Morie Pit by lithostratigraphic and ground-penetrating radar analyses. From these studies, we conclude that the upper two units in Morie Pit correlate with two phases of marine oxygen isotope stage 9, whereas the lower unit correlates with stage 11. Furthermore, this geochronology has allowed us to interpret the six units of the Cape May Formation identified nearby in ground-penetrating radar profiles (O'Neal, 1997) as a fairly complete record of late Pleistocene sea-level highstand events, including deposits of stages 5, 7, 9, 11, and probably 13. These age estimates fit well with other amino-acid-derived geochronologies for coastal stratigraphic units in the U.S. Mid-Atlantic coastal plain (Belknap, 1979; Wehmiller and Belknap, 1982; Wehmiller et al., 1988; Toscano et al., 1989; Groot et al., 1990; York, 1990; Toscano, 1992; Toscano and York, 1992; Wehmiller et al., 1992; Mirecki et al., 1995).

SITE DESCRIPTION

The Morie Company sand and gravel pit is located on the western flank of the Maurice River, landward of the Cedarville Scarp separating the upper and lower bay-margin terraces (figure 21-2). Vibracores DCV1 and DCV2 were collected in the floor of Morie Pit, at an elevation of approximately MSL (figure 21-3). The sand and gravel overburden up to land surface at +8.5 m MSL had been removed during excavation of the pit. DCV1 was taken close to the pit wall adjacent to the land surface ground-penetrating radar profile DC1. The pit wall was scraped to expose a fresh surface at the location marked "SC" in figure 21-3, and a lithologic profile was completed. The combined scraped section (MPSS) and subsurface section of DCV1 were combined as MPC1 (Morie Pit Composite), seen in figure 21-4. Since this composite section is fairly close to profile DC1, the lithologies observed were used to interpret and develop a profile to cor-

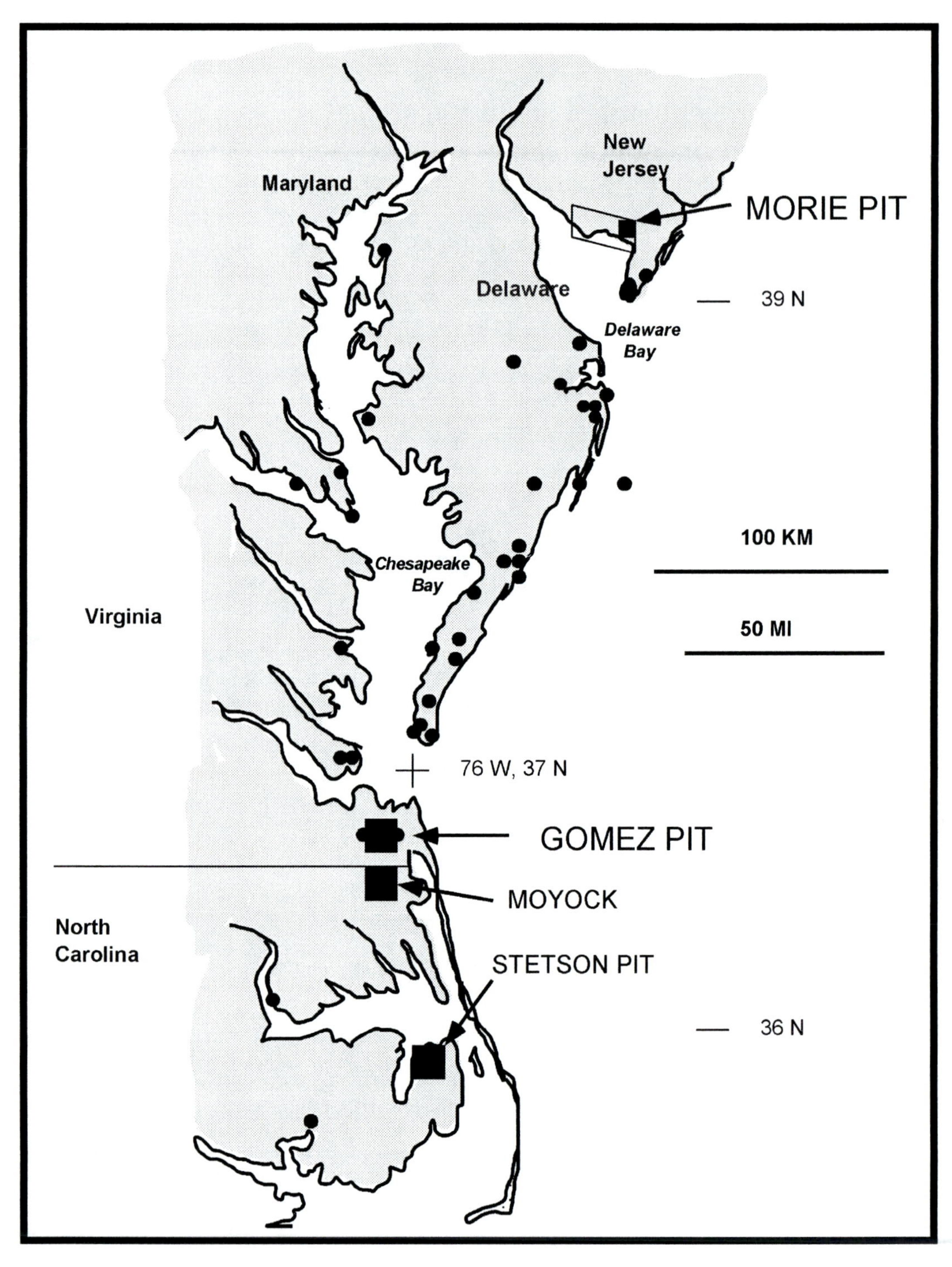

Fig. 21-1. Map of U.S. mid-Atlantic coastal plain, showing the study area (rectangle) of southern New Jersey and the Morie and Gomez Pits, southern NJ and southeastern VA, respectively. The aminostratigraphy of Gomez Pit, Stetson Pit, Moyock, and other unidentified sites (solid circles) in this map is summarized in figure 21-5 and in Groot et al., 1990; Genau et al., 1994; Mirecki et al., 1995; and Wehmiller, 1993, 1997. The Gomez, Stetson, and Moyock pits all have independent U-series geochronologic calibration (Groot et al., 1990).

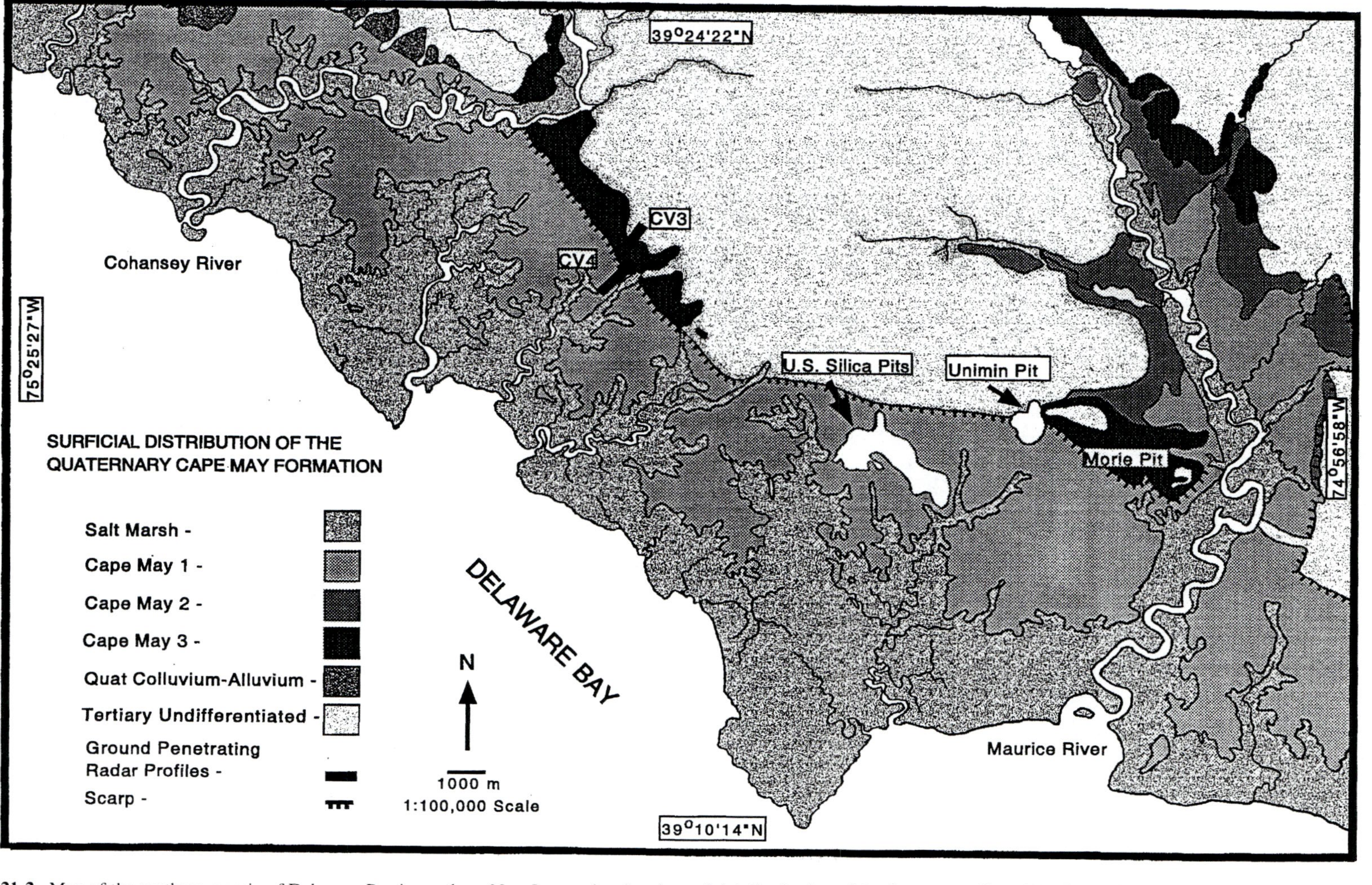

Fig. 21-2. Map of the northern margin of Delaware Bay in southern New Jersey, showing the surficial distribution of the Quaternary Cape May Formation in the study area (adapted from Newell et al., 1995). The locations of Morie Pit and ground-penetrating radar profiles CV3 and CV4 (O'Neal, 1997) are shown in relation to the Cedarville Scarp.

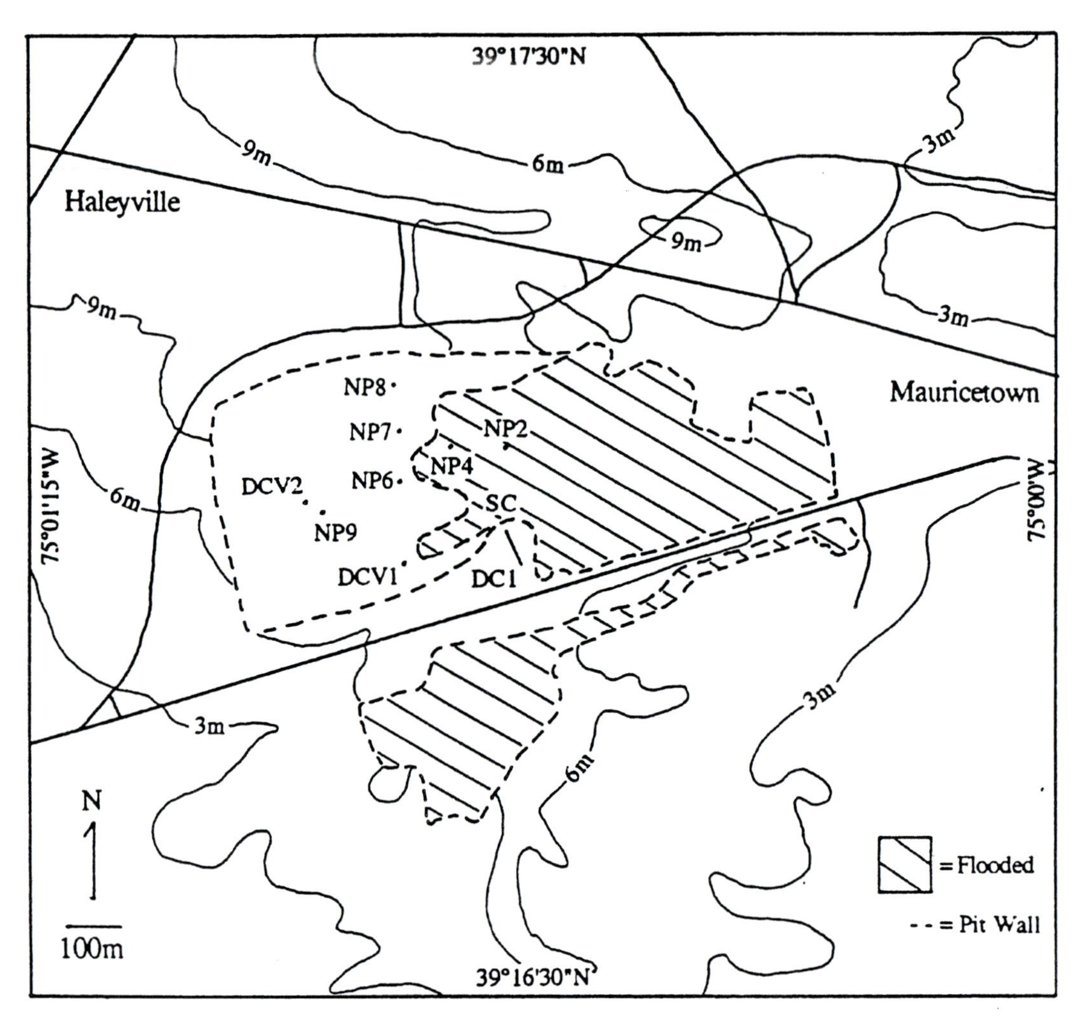

Fig. 21-3. Location map of the Morie Company sand pit west of Mauricetown, New Jersey, showing ground-penetrating radar profile DC1, vibracores DCV1 and DCV2, a scraped section (SC), and Morie Company cores NP2, 4, 6, 7, 8, and 9 (adapted from O'Neal, 1997).

relate with allostratigraphic units defined further to the west by O'Neal (1997).

As shown in figure 21-4, a 1.5-m-thick gravelly sand is present at the top of MPC1. This unit is interpreted as a regressive phase of allostratigraphic unit 4, and lies above reflector C in profile DC1. From +2 m to +6.5 m MSL, the section shows a gravelly sand unit, overlain by a thick, well-sorted, very fine sand. The underlying unconformity is believed to be correlative with reflector B in profile DC1, the sediments above being representative of a transgressive lag and sandy tidal-flat or barrier-bar deposit correlative with the lower section of unit 4 identified elsewhere.

The presence of north- and south-dipping, low-angle cross-beds, in 5–10 cm sets within the fine sand, supports the shallow, sandy tidal-flat depositional environment interpretation. At the base of this section, the lithology is variable. Tracing the unconformity laterally across the pit wall revealed relief of as much as 1.5 m, with the gravel lag deposit resting on sands, gravels, and silty sands with compacted mud balls. All of these features support the interpretation of this unconformity as a transgressive ravinement surface.

The 5 m of MPC1 below this unconformity, from +2 m to −3 m MSL, are interpreted to correlate with allostratigraphic unit 3, as defined by O'Neal (1997) in other bay-margin locations (figure 21-2). An unconformable sandy gravel lag deposit rests above a fossiliferous silty sand at −3 m MSL. This lower unconformity is interpreted to be coincident with reflector A in profile DC1, representing a transgressive ravinement surface.

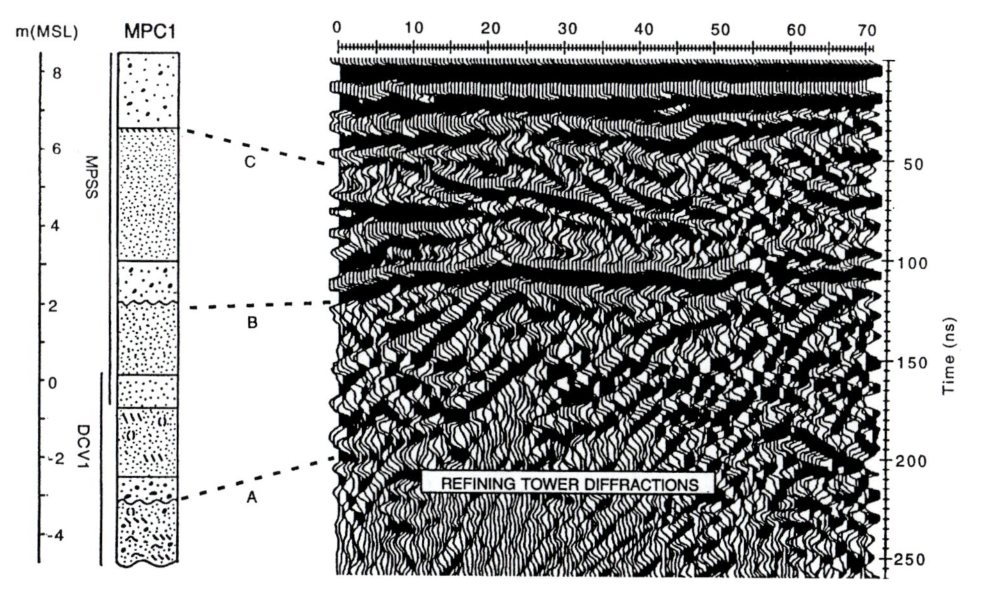

Fig. 21-4. Correlation of prominent reflectors A, B, and C in ground-penetrating radar line DC1 with unconformities identified in composite section MPC1 (MPSS = scraped section, SC, in pit wall and vibracore DCV1), as located in figure 21-3 (modified from O'Neal, 1997).

Above the gravel lag associated with the lower unconformity in MPC1 is a 1.5-m-thick layer of well-sorted, very fine silty sand containing abundant fossil mollusk (*Mercenaria*) shells. An overlying 0.75-m-thick layer of coarse sand separates this fossiliferous unit from a 2-m-thick layer of well-sorted, very fine/fine sand lying immediately below the unconformity between units 3 and 4 at +2 m MSL. The lower gravel lag, fossiliferous fine sand, and coarse sand sequence is interpreted to be the thin, landward extent of a transgressive system tract, with the overlying fine sand probably reflecting a highstand systems tract that has been incised by the transgressive ravinement of allostratigraphic unit 4 above.

The silty sand below the unconformity at the base of unit 3 contains abundant oyster (*Crassostrea*) shells in its lower section and a few *Mercenaria* shells in its upper section. This fossiliferous unit is interpreted to be correlative with allostratigraphic unit 2 of O'Neal (1997), on the basis of elevation and relative stratigraphic position. The vibracore DCV2 extends from MSL to −4.5 m MSL, and reveals a sequence similar to that of DCV1, over the same elevation interval, but without macro fossils.

AMINOSTRATIGRAPHY

Aminostratigraphic Background: Mid-Atlantic Coastal Plain

In this study, relative and numerical age estimates are derived from measurements of D/L values in fossil *Mercenaria* specimens collected from Morie Pit and other sites in the mid-Atlantic region of the eastern U.S. The post-depositional racemization of L-amino acids into an equilibrium mixture of D- and L-amino acids has been used extensively in this region as a chronostratigraphic tool (Wehmiller et al., 1988; Wehmiller, 1993).

Wehmiller et al. (1988) delineated an aminostratigraphic framework for the U. S. Atlantic coastal plain. Regions of similar (within 3°C) current mean annual air temperatures (CMATs) were defined (regions I–V, from north to south), and all amino acid data from these individual regions were then interpreted with the assumption that all samples from a particular aminostratigraphic region experienced similar effective temperatures. This assumption is fundamental to both relative and numerical applications of aminostratigraphy, although it is generally accepted that the assumption is not correct in detail. Contrasting burial conditions and steepened latitudinal temperature gradients during full-glacial climates are among the many factors that affect any estimate of effective temperatures for specific sites or samples in a given aminostratigraphic region. The present study lies at the northern edge of Aminostratigraphic Region II of Wehmiller et al. (1988). This region reaches from 36°N to 39.5°N (southeastern Virginia to southern New Jersey), where CMAT values range from approximately 16 to 13°C, respectively (see figure 21-1).

The aminostratigraphy of Region II has been summarized in several publications during the past two decades. Following the initial work of

Belknap and Wehmiller (1980), additional studies include those of Wehmiller et al. 1988; York et al., 1989; Groot et al., 1990; Wehmiller et al., 1992; Genau et al., 1994; and Mirecki et al., 1995. Mirecki et al. (1995), along with Wehmiller et al. (1998), discussed the lithostratigraphy and aminostratigraphy of a section at Gomez Pit, Virginia Beach, VA, an important calibration site (corresponding to marine isotope stage 5) near the southern edge of Aminostratigraphic Region II (figure 21-1). Published and unpublished data from Gomez Pit are cited in this Chapter for comparison with the Morie Pit results. Morie Pit *Mercenaria* samples are observed to be approximately 40% more racemized than those from Gomez Pit, hence the general conclusion presented here is that the Morie Pit section represents Pre-stage 5 deposition. This conclusion modifies previous age estimates (Newell et al., 1995), but is consistent with stratigraphy developed from geophysical, geomorphic, and lithological analysis of the region (O'Neal, 1997).

Morie Pit Aminostratigraphic Methods

Field Sampling

Fossil mollusks from Morie Pit first became available to researchers at the University of Delaware in 1988, when dredging operations began to encounter a richly fossiliferous mud unit below the water level in the pit. *Mercenaria* specimens were collected from the dredging spoil piles, but no *in situ* specimens were available. Amino acid enantiomeric ratio analyses of approximately one dozen of these dredge-spoil *Mercenaria* specimens were conducted between 1989 and 1993, and some results have been reported (Groot et al., 1990; Lacovara, 1990, 1997; Wehmiller, 1997). These results, and many others from Aminostratigraphic Region II, are summarized in figure 21-5. Analytical results reported in

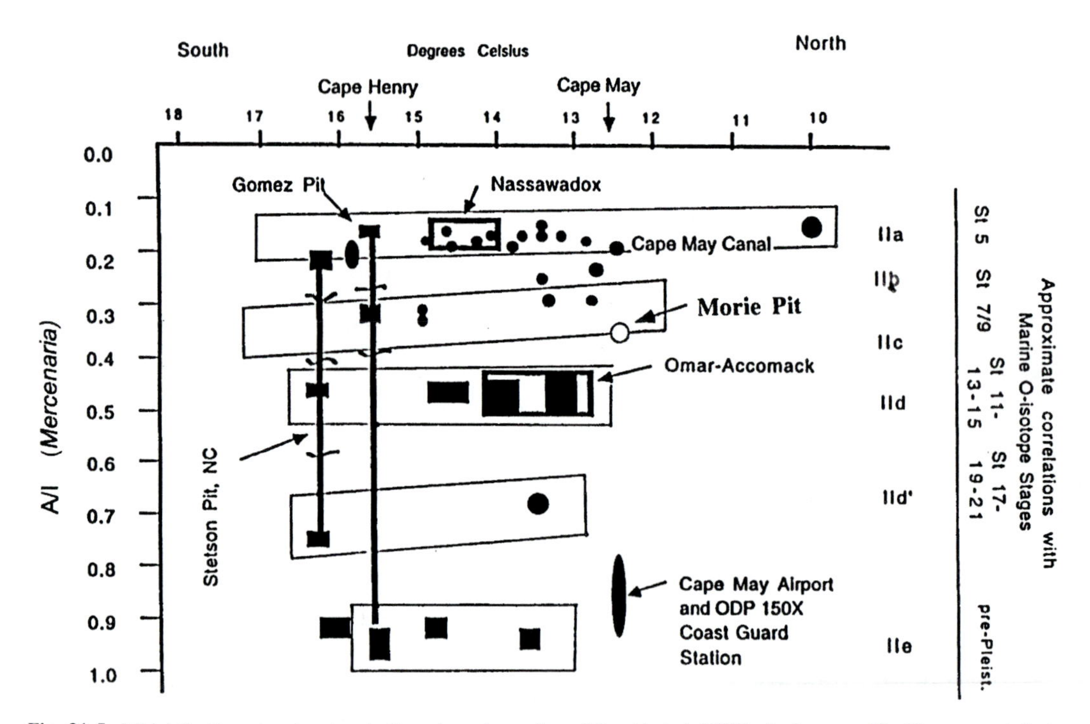

Fig. 21-5. Mid-Atlantic regional aminostratigraphy redrawn from Mirecki et al. (1995). Aminozones IIa–IIe represent clusters of A/I values in *Mercenaria* or its equivalents. See Groot et al. (1990) for AAR site (solid circle) information. Superposed aminostratigraphic sections at Stetson Pit, NC (York et al., 1989), and Gomez Pit, VA, are shown for reference. Aminostratigraphic data for sites within the Nassawadox and Omar–Accomack (Colman and Mixon, 1988) barrier complexes are shown within rectangles that include data from aminozones IIa and IId. AAR data from Cape May Airport, NJ (Lacovara, 1990), and ODP 150X Coast Guard Station, NJ (Wehmiller, 1997), boreholes are shown as ellipses. AAR data from Morie Pit, NJ (this study) are shown as an open circle. Because of differences between analytical methods used in this and previous studies, Morie Pit data are approx. 10% greater than most of the other aminozone values. Size of symbols represents range of measured A/I values and reflects both sample size and inter-shell variability.

this Chapter differ slightly in numerical value from those seen in figure 21-5 for Gomez and Morie pits because of different sampling procedures, discussed below.

Approximately 75 new specimens of *Mercenaria* were collected from Morie Pit in December, 1995. Although these specimens were also obtained by dredging and therefore were not seen in their natural depositional position, the stratigraphic context of the shells is clear, as they were collected immediately adjacent to the dredging operation. Additionally, the shells were collected within days of dredging; consequently, they had not been affected by the unknown exposure times of the 1988 spoil pile collection. The 1995 analyses are the focus of the present study, although a few of the original 1988 specimens were reanalyzed to compare new sampling and analytical strategies with those used in the original studies.

The 1995 dredging operation removed material from the floor of the pit near the side wall adjacent to the site of ground-penetrating radar profile DC1 (figure 21-4). Sediment was removed from 3 to 4 m below the pit floor, which is within 1 m of MSL. A slurry pipe had expelled coarse dredge and excavation material away from the site. Numerous *Mercenaria* and *Crassostrea* shells, as well as abundant green-stained, granular to coarse sand, were present in this material. Evidence of recent dredging included preserved flow structures in the sand, moist compacted host sediment in the dredged bivalves, and general contrast with previous dredge deposits that had become desiccated from exposure.

The *Mercenaria* specimens occurred in two distinct color populations. Although both groups were composed of a range of similar-sized specimens of comparable preservation, one population had a strong orange iron-oxide stain, in contrast to the more typical gray appearance of the other group. This color difference also suggests that the exposure time of these specimens was quite short, as other shells on older spoil piles do not demonstrate such a contrast. The possibility that these two color populations represent source units with different ages was explored by racemization analysis of six specimens of each color.

Intrashell Sampling

Over the course of our studies of *Mercenaria* aminostratigraphy, we have undertaken investigation of the intrashell variability of amino acid enantiomeric ratios. Intrashell variability of D/L values within *Mercenaria* has been studied previously (Hare and Mitterer, 1969; Belknap, 1979; McCartan et al., 1982; York, 1990; Mirecki, 1990) and has generally been shown to be less than 5% at a relatively coarse sampling scale (analyzed fragments generally weighing between 0.5 g to 1.0 g). The magnitude of this variability is comparable to that observed in other taxa (Brigham, 1983; Hearty et al., 1986; Miller and Brigham-Grette, 1989; Goodfriend et al., 1997), and most workers agree that such variability can be tolerated, especially if the same sample position is used in aminostratigraphic analysis of multiple specimens.

More recent intrashell studies of *Mercenaria* (Mirecki, 1990; Wehmiller et al., 1998; J. F. Wehmiller, L. L. York, and M. L. O'Neal, unpublished data) have demonstrated that diagenetic alteration may introduce substantial (up to 50%) variation in D/L values in some (but not all) shells from a single unit, particularly if the shells are sampled on a finer scale (analyzed fragments weighing less than 0.1 g). Visible alteration of the shells with these large variabilities may not always be obvious, so assessment of diagenetic alteration by other methods is recommended (Krantz et al., 1996). Most of the specimens studied here were examined by using either thin sections or Fiegl's solution (Friedman, 1959) staining of polished shell cross-sections.

Figure 21-6 shows the positions used for subsampling of *Mercenaria* for aminostratigraphic analysis. The positions routinely used are identified as X, Y, and Z: X and Y represent the inner homogeneous layer, although position X (usually characterized as the "hinge region") has a much more complex structure than that seen at position Y. Position X has been the position routinely sampled in almost all previous research at the University of Delaware—the selection of position X often results from the fact that the hinge area is often most robust and best preserved, particularly for fragmented samples. Growth banding in the inner layer builds inward during the lifetime of the mollusk. Position Z represents the middle layer, where growth banding builds radially and in which annual growth banding is much more clearly resolved than in the inner layer. Preserved growth bands in fossils may actually facilitate open-system behavior for the middle layer, particularly from position Z to the growth edge (Krantz et al., 1996). If intrashell variability greater than approximately 6% is observed, the D/L values for shells with this larger variability are generally highest for the inner layer at position Y (Wehmiller et al., 1998; J. F. Wehmiller, L. L. York, and M. L. O'Neal, unpublished data). Occasionally, the outer growth edge is also analyzed, as this position was usually sampled by Mitterer in previous work (Mitterer, 1975; McCartan et al., 1982).

Analytical results were obtained using both high-performance liquid chromatography (HPLC) and gas chromatography (GC) methods that are routinely employed at the University of Delaware Aminostratigraphy Laboratory (Belknap, 1979; McCartan et al., 1982; Wehmiller, 1984; York, 1990; Mirecki et al., 1995; Wehmiller and Miller, 2000). HPLC was used for the rapid processing of multiple samples, yielding data on the amino acid composition and ratio of D-alloisoleucine to L-isoleucine, the two enantiomers (diastereoisomers)

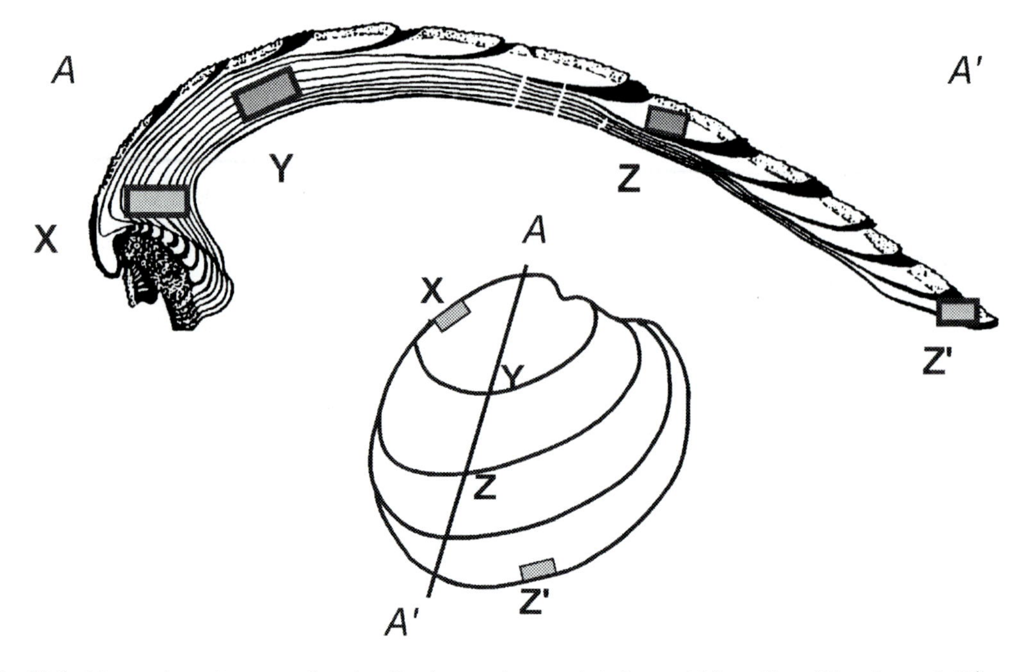

Fig. 21-6. *Mercenaria* specimen sampling sites. Previous analyses used shell material from hinge (X) and growth (Z′) areas. Sample sites Y and Z, taken along section line A (described in Pannella and MacClintock, 1968) were used for analysis in this study.

that are resolvable by this analytical method. Over the course of the past eight years, minor adjustments in hardware and reagents for the HPLC system have resulted in systematically higher (<5%) ratios of D-allo to L-iso (hereafter referred to as A/I) using present methods, compared with those produced at Delaware in the mid to late 1980s. Several previously studied shells from Gomez Pit were reanalyzed during this study to account for these higher values in comparisons made with Morie Pit shells. Ongoing comparison of new and previous analyses occurs regularly, including the analysis of Interlaboratory Comparison samples (Wehmiller, 1984).

GC was used to obtain D/L values for as many as seven amino acids (alanine, aspartic acid, glutamic acid, leucine, phenylalanine, valine, and occasionally isoleucine). Samples were analyzed using Chirasil-Val analytical procedures described elsewhere (Frank et al., 1977). Desalting methods evolved from conventional ion-exchange to the CaF_2 precipitation method used by Goodfriend and Meyer (1991). During the course of this study, GC analyses were conducted using either a 50 m × 0.25 mm ID Chirasil-Val column installed in a Perkin-Elmer 3920 gas chromatograph with a flame ionization detector, or a 25 m × 0.25 mm ID Chirasil-Val column installed in a Hewlett-Packard 6890 gas chromatograph with a flame-ionization detector. Additionally, the GC analyses were usually conducted on the same total hydrolysate prepared for HPLC analysis.

RESULTS

Table 21-1 presents the results of enantiomeric ratio analyses of Morie Pit samples, including three of the spoil-pile samples collected in 1988. Results for Gomez Pit samples (reanalyzed using the sampling techniques described above) are also included in table 21-1, because of the important calibration role that this site has for the mid-Atlantic region (Mirecki et al., 1995; Wehmiller et al., 1997, 1998). Data in table 21-1 are given as either A/I values or D/L leucine values at positions X, Y, and Z for different samples. Data for alanine D/L values at position Y are also given. Alanine, being more extensively racemized than either leucine or isoleucine, appears to have a more rapid relative rate that is useful for the resolution of relatively small age differences. Other amino acid D/L values are all internally consistent with the intrashell trends reported by other investigators (Lajoie et al., 1980; Goodfriend and Meyer, 1991). Representative GC data for the Gomez Pit calibration site and southern New Jersey are summarized in figure 21-7.

The A/I values for the two differently colored populations at Morie Pit do not indicate different ages for these samples. The orange-stained samples

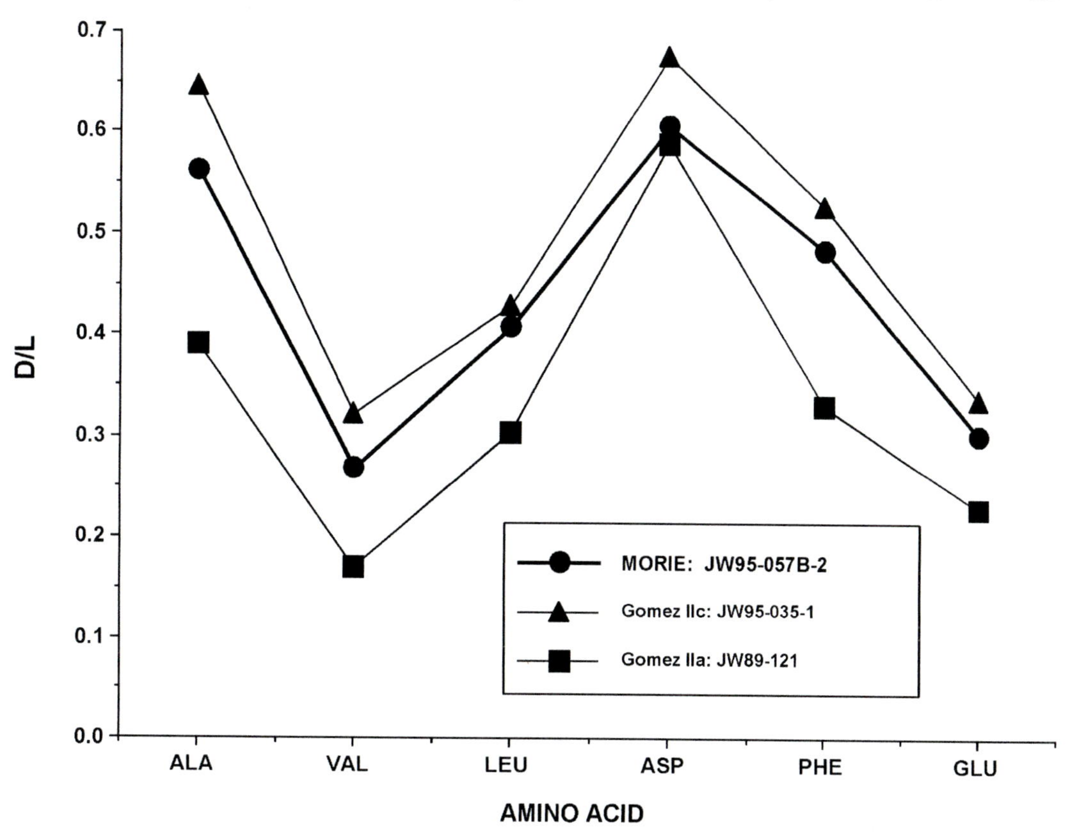

Fig. 21-7. D/L values for six amino acids in representative samples from Morie Pit and the two superposed aminozones at Gomez Pit (aminozones IIa and IIc). See table 21-1 for sample numbers and data for other samples from these same sites. Values reported are means from at least three chromatograms; standard deviation is $\leq 3\%$.

yield mean A/I values at positions Y and Z of 0.372 ± 0.013 ($n = 4$) and 0.377 ± 0.027 ($n = 5$), respectively, whereas the gray samples yield mean A/I values at these same positions of 0.366 ± 0.034 ($n = 3$) and 0.353 ± 0.007 ($n = 4$). The grand mean A/I values for seven position-Y analyses and nine position-Z analyses are 0.368 ± 0.023 and 0.366 ± 0.023, respectively, demonstrating little intrashell difference. Figure 21-8 demonstrates this intrashell relationship for Morie Pit and shells from two aminozones at Gomez Pit. This figure shows that the intrashell difference is usually, but not always, less than 6% of the grand mean values for a site. Some of the spoil-pile samples from Morie Pit (88-69-1 and 88-69-2) show particularly large *intrashell* variation that is perhaps indicative of their exposure on the spoil pile for an undetermined interval. The areas of figure 21-8 labeled Gomez IIa and Gomez IIc represent two superposed aminozones documented with data for samples collected between 1984 and 1989 at extensive excavated outcrops in Gomez Pit (Mirecki et al., 1995; Wehmiller et al., 1998). Sample 95-057C-11 from Morie Pit yielded an A/I value of 0.53, which is substantially above all other results from this collection. Without any particular evidence to the contrary, it is assumed that this shell was reworked from some older unit, although no units with this ratio have been sampled in the region. Leucine D/L values for three Morie Pit shells show excellent precision (0.40 ± 0.01) and an intrageneric relationship to the corresponding A/I values (mean 0.366 ± 0.002) that is consistent with previous observations (Wehmiller et al., 1988).

In view of the mean A/I values of 0.375 and 0.359 for the orange and gray groups of *Mercenaria*, respectively, the two populations were interpreted to be of equivalent age, some other factor being responsible for the color variation. For this reason, vibracore DCV1 was located in the floor of Morie Pit, in an undisturbed section immediately adjacent to the dredging-operation site.

Vibracore DCV1 (basal section of MPC1) recovered two silty sand units, separated by a green-stained granular to coarse sand interval overlying an eroded surface (figure 21-4). It is believed that this same sand was present at the surface-collection site that yielded the different color groups of

TABLE 21-1 Amino acid D/L values for samples from Morie Pit compared with data from Gomez Pit (VA) and Moyock (NC) (see figure 21-1). Values reported are means from multiple chromatograms; standard deviation is $\leq 3\%$

Lab. number[b]	Locality	Sample[a]	UDAMS[c]	Color	A/I pos. X[d]	A/I pos. Y	A/I pos. Z	Leu pos. X	Leu pos. Y	Leu pos. Z	Ala pos. Y
950383/384	Morie	88-69-1	05083	–		0.31	0.41				
950385/386	Morie	88-69-2	05083	–		0.33	0.42				
950387/388	Morie	88-69-3	05083	–		0.36	0.35				
960170/134	Morie	95-057B-2	05143	orange		0.368	0.379		0.41	0.41	0.57
960136	Morie	95-057B-4	05143	orange			0.359				
960169	Morie	95-075B-5	05143	orange		0.391					
960135	Morie	95-057C-1	05143	orange			0.417				
960168/133	Morie	95-057C-2	05143	orange		0.365	0.385		0.4		0.598
960167/137	Morie	95-057C-5	05143	orange		0.365	0.345		0.39		0.57
960127/128	Morie	95-057B-6	05143	gray		0.363	0.35				
960126	Morie	95-057B-7	05143	gray		0.345					
960130	Morie	95-057C-7	05143	gray		0.335					
960131	Morie	95-057C-8	05143	gray			0.36				
960166/132	Morie	95-057C-10	05143	gray		0.402	0.358				
960129	Morie	95-057C-11	05143	gray			0.53				
960195	Morie DCV1	96-139-1a	05144	(frag.)		0.332					
960196	Morie DCV1	96-139-1b	05144	(frag.)			0.407				
960076/78/79	Gomez IIa	gp278	06056	–	0.16	0.23	0.22	0.21	0.19	0.27	0.47
950337	Gomez IIa	89-139	06076a	–		0.2			0.25		0.43
950339/340	Gomez IIa	89-141	06076a	–		0.21	0.22				
960080	Gomez IIa	89-121	06076a	–		0.24			0.29		0.4
950393	Gomez IIc	88-46-1	06074	–		0.36	0.38		0.41		0.69
950439	Gomez IIc	95-35-1	06207	–		0.41			0.43		0.64
960082/83/85	Moyock	93-61	07251	–				0.22	0.31	0.26	0.43

[a] Sample numbers refer to specific shells.
[b] Laboratory numbers refer to separate fragments analyzed from a single shell.
[c] UDAMS refers to University of Delaware AMStratigraphy localities, as described in Wehmiller et al. (1988).
[d] A/I = D-alloisoleucine/L-isoleucine; Leu = D/L leucine; Ala = D/L alanine; shell sampling positions as in figure 21-6.

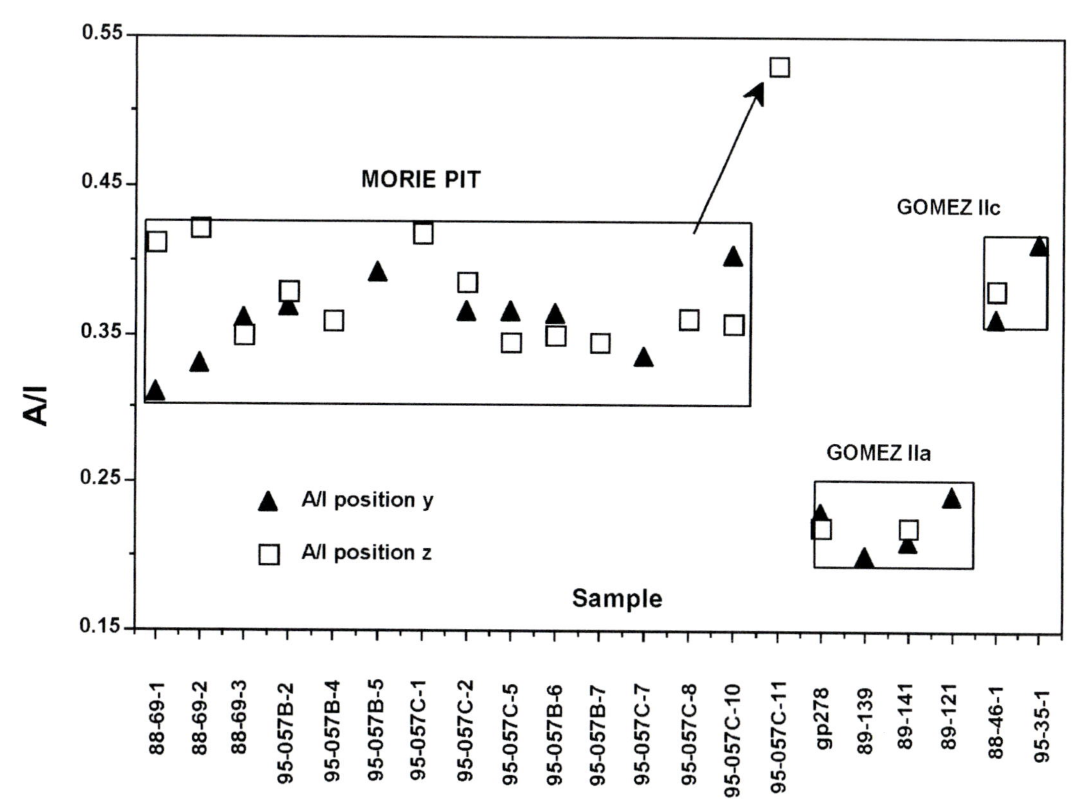

Fig. 21-8. Variability in intrashell A/I values is shown for sample positions Y and Z (see figure 21-6) of *Mercenaria* collected from Morie Pit, New Jersey, and Gomez Pit, Virginia. Differences in intrashell A/I values are less than 6% of the grand mean A/I for Morie Pit shells and Gomez Pit aminozone IIa and IIc shells, respectively. Morie Pit spoil-pile samples 88-69-1 and 88-69-2 show a large A/I variability, probably indicative of their prolonged exposure.

Mercenaria. This surface is interpreted to be the transgressive ravinement separating allostratigraphic units 2 and 3 of O'Neal (1997).

In DCV1, the silty sand above this green-sand unconformity was predominantly gray in color, with the upper 30 cm bearing an orange oxidation stain. Although no fossil material was recovered from this section of the core, it is believed that the two *Mercenaria* groups collected from the pit floor are from this same horizon.

Numerous *Crassostrea* shells were collected from the silty sand below the unconformity, explaining their presence at the surface collection site. A small number of *Mercenaria* shells and fragments were recovered from immediately below the unconformity.

The results of the amino acid analyses performed on core DCV1 shells are presented in table 21-1. The values of 0.322 and 0.407, taken from two locations on the same shell, yield a mean value of 0.365. This mean is almost identical to the mean of 0.367 for the *Mercenaria* from the dredge spoil, assumed to have come from the overlying unit.

The presence of an obvious lithologic unconformity between the upper and lower shell beds strongly suggests that the two units resulted from two separate depositional episodes. Stratigraphic evidence collected from the Morie Pit and nearby (O'Neal, 1997) supports this interpretation and places the deposits in two distinct sea-level highstand events. Although the age difference between these deposits is not resolvable geochemically using the amino acid racemization (AAR) data obtained in this study, the interpretation herein is that the *Mercenaria* specimens collected from above and below the unconformity at −3 m MSL in core DCV1 are from two separate Pleistocene estuarine–marine transgressions.

DISCUSSION

Age Estimates for Morie Pit Shell Samples

Figures 21-7 and 21-8 demonstrate the aminostratigraphic relationship between the 1995 Morie Pit samples to those from the two aminozones at Gomez Pit. Figure 21-8 shows that the mean A/I value of 0.367 from Morie Pit *Mercenaria* samples is slightly less than the mean value from Gomez Pit Aminozone IIc and is substantially greater than the

Gomez Pit Aminozone IIa A/I values. The Gomez section, described in detail elsewhere (Mirecki et al., 1995; Wehmiller et al., 1998) represents two superposed fossiliferous units, the upper one having U-series coral dates in the range 75–80 kyr (Szabo, 1985; Wehmiller et al., 1997). Although this date is controversial because the present elevation (approx. +1 m) of the dated corals is not consistent with global ice-volume models for late stage 5 (e.g., Cronin et al., 1981), there is little question that Gomez Pit Aminozone IIa must be assigned to isotope stage 5, with an age estimate of between 80 and 130 kyr. Figure 21-7 presents D/L values for other amino acids for representative samples from Morie Pit and Gomez IIa and IIc units. This figure also demonstrates that the Morie Pit samples have D/L values between those of the two Gomez aminozones, although the relative differences among these samples are not identical for all amino acids. Aspartic acid is particularly notable in this regard, but its relationship to the other amino acid D/L values is affected by its very rapid initial rate of racemization (Lajoie et al., 1980; Goodfriend and Meyer, 1991). If one assumes that there is no difference in effective temperatures between the Morie Pit and the Gomez Pit, then the qualitative conclusion that results from the comparisons in figures 21-7 and 21-8 is that the Morie Pit unit is older than Gomez Pit Aminozone IIa and younger than Gomez Pit Aminozone IIc. The assumption of equivalent effective temperatures is undoubtedly incorrect, as there is a difference today of 2.5°C, and the more northerly Morie Pit probably experienced greater full-glacial temperature reductions. Correcting for this plausible temperature difference increases the age difference between the Morie Pit and Gomez IIa samples.

Numerical ages for the Morie Pit samples and section can be calculated in several ways. The simplest of these is the use of parabolic kinetics [$A/I = k(t)^{0.5}$ (Mitterer and Kriausakul, 1989), where t = time or sample age, k = rate constant for specific effective temperature] to estimate the age difference between the Gomez Pit calibration samples and the Morie Pit samples, just as done by Mirecki et al. (1995) to estimate age differences for samples from the superposed section at Gomez Pit. In the Mirecki et al. (1995) study, a mean A/I value of 0.17 was used as a calibration value, and ages of 80, 100, and 120 kyr were assigned to these samples for calculation of parabolic rate constants. The youngest of these ages is the actual U-series age of corals found at the site, whereas the oldest of these ages is more consistent with the timing of the sea-level maximum of substage 5e and the actual elevation of the Gomez Pit marine unit from which the dated corals were sampled.

The calibration A/I values from Mirecki et al. (1995) are not used here because of differences in sampling and analytical methods. Reanalysis of selected Gomez Pit samples at position Y (figure 21-6, table 21-1) yield A/I values between 0.20 and 0.24. A mean A/I value of 0.22 for the Gomez IIa aminozone is used for parabolic kinetic model age estimation of the Morie Pit shells, and table 21-2 gives the summary of these calculations, which are given for three different calibration ages (representing early to late stage 5). Additionally, the calculations are made assuming no effective temperature difference between Gomez and Morie,

TABLE 21-2 Age calculations based on different kinetics model options and different assumptions regarding the effective temperature difference between the calibration site (Gomez Pit) and Morie Pit. Parabolic kinetics are described in the text. Results of calculations using the nonlinear model of Boutin (in Wehmiller et al., 1988) are also shown. The parabolic and nonlinear models yield different (but internally consistent) age estimates

	Morie Pit age estimates (kyr)		
Assigned age for Gomez calculation:	80	100	120
Parabolic age estimate for Morie Pit (kyr)			
Gomez Pit calibration A/I = 0.22			
Morie Pit measured A/I = 0.367 ± 0.023			
k Morie = k Gomez (no difference in T_{eff})	223	278	334
k Morie = 0.846*k Gomez	311	389	467
(Morie T_{eff} 2°C lower than Gomez)			
Boutin model based on leucine			
Calibration leucine D/L = 0.30			
Morie Pit leucine = 0.42			
No temperature difference	210	260	328
2°C temperature difference	274	343	429
Average of six amino acids (parabolic model)			
no temperature difference	182	228	273
2°C temperature difference	255	318	382

and for an assumed 2°C difference in effective temperature. For this latter calculation, an 8%/°C reduction in the parabolic rate constant is assumed on the basis of the observations of Kaufman and Brigham-Grette (1993). This temperature sensitivity is consistent with all previous equations for the temperature dependence of fossil racemization kinetics (summarized in Wehmiller, 1993; Wehmiller and Miller, 2000).

As shown in table 21-2, the range of parabolic model age estimates (based on A/I values) for Morie Pit is from approximately 220 kyr to 335 kyr for the assumption of no temperature difference between Gomez and Morie. This range shifts to older ages (approx. 310–470 kyr) using the assumption of a 2°C difference in effective temperature between the two sites. This latter range is considered more reasonable, given the present difference in temperatures between the sites (Wehmiller et al., 1996). Model ages derived from the leucine kinetic model presented in Wehmiller et al. (1988) yield a range from 274 kyr to 429 kyr for the assumed 2°C difference in effective temperature, and parabolic kinetics applied to A/I plus five amino acids (data from figure 21-7) yield a range (the average of six individual age estimates for all but aspartic acid) from 255 kyr to 382 kyr, for this same 2°C difference in effective temperature. All of the above age estimates for Morie Pit samples would increase if the effective temperature difference between Morie and Gomez were assumed to be greater than 2°C.

Figure 21-9 demonstrates the relationship between the parabolic model age estimates in table 21-2 and the marine isotope record. This graphical representation of the age uncertainties associated with calibrated numerical age estimates presents these calculations in the context of the natural sea-level fluctuations that were responsible for deposition of the units under study. This figure demonstrates the actual time intervals during which deposition would have been possible, given a particular range of kinetic model age estimates. The figure also demonstrates that the typical range of A/I values seen among multiple samples at a single site can easily translate into a range of age estimates of 50 kyr or more, particularly when the extent of racemization is large enough so that the rate of racemization (the slope of D/L vs. time) has slowed significantly from the initial phases. Intrasample variability can also introduce significant uncertainty in these kinetic model extrapolations.

Chronostratigraphic Correlations and Pleistocene Sea-Level Implications

The best age estimate for the Morie Pit shells (mean A/I value 0.367 ±0.023) corresponds to either stage 9 or stage 11 of the marine isotope record (table 21-2, figure 21-8). These age estimates for the Morie Pit site are greater than stage 5, the traditionally assumed age for the Cape May Formation ("last interglacial" or "Sangamonian"), but the existence of at least one southern New Jersey aminozone with lower D/L values than Morie Pit (Groot et al., 1990) supports this pre-last interglacial interpretation. A stage 7 age assignment cannot be eliminated on the basis of the numerical calculations alone, but geomorphic, stratigraphic, and relative-elevation arguments support the older age estimates. Because of the general lack of "calibration sites" along the U.S. Atlantic coastal plain (and associated uncertainties, see Wehmiller et al., 1992), it is important to demonstrate how the physical stratigraphy and geomorphology in the immediate area (figure 21-2) support this stage 9/11 age estimate.

This interpretation is strengthened by the unconformity between the two fossil beds identified in Morie Pit, which suggests separation by a regression and subsequent transgression rather than continuous deposition. The extent of the associated allostratigraphic unit 3, overlying the unconformity, seen in Morie Pit and throughout the northern Delaware Bay margin, suggests a sea-level highstand of at least 5 m above modern MSL for the deposition of the unit.

It therefore seems unlikely, when marine isotope stages are compared, that the upper sequence (units 3 and 4) in Morie Pit was deposited during stage 7. According to the oxygen isotope curve, the sea-level high during stage 7 was probably lower and of shorter duration than that of stage 9. If allostratigraphic unit 2 in Morie Pit is correlative with deposition during the high sea level of stage 11, the sea level during stage 7 would not have been high enough to overtop the unit 2 deposits to the extent observed in units 3 and 4 (stage 9 interpretation). By the same argument, if unit 2 is correlative with stage 9, then the entire sequence above the unconformity must also be of stage 9 age. This latter possibility would suggest a single highstand, with minor sea-level fluctuations, and an extremely high sediment supply.

In either scenario, the upper deposit in Morie Pit, allostratigraphic unit 4, is correlated here as probable stage 9 age, owing to its elevation and stratigraphic position between older deposits below, and younger deposits bayward of the pit. Allostratigraphic units 3 and 2 are correlated with isotope stages 9 and 11 in this study, respectively.

Although a great deal of uncertainty exists between the marine $\delta^{18}O$ curve and an inferred record of global sea level, the overall pattern of sea-level highs and ages fits reasonably well with the highstand stratigraphy and amino acid age estimates derived from Morie Pit.

During the exploration phase of excavation by the Morie Company, several split-spoon cores (NP = New Pit) were collected in the area of the present-day pit (figure 21-3). Lithologic samples from six of these cores were supplied by the Morie Company (J. Zadorozny, personal communication, 1995) and were used, in conjunction with

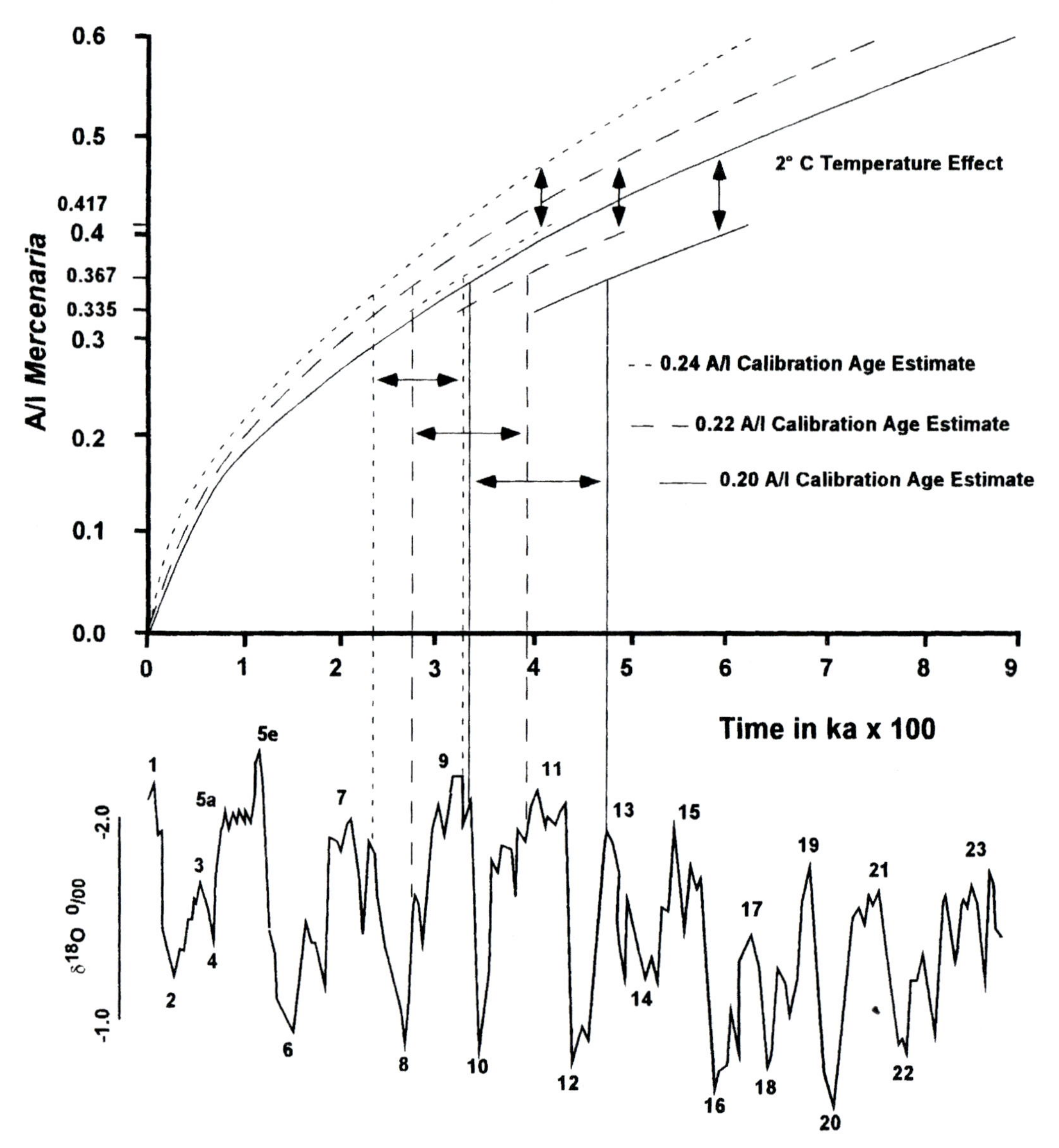

Fig. 21-9. Age estimates for the A/I mean of *Mercenaria* from Morie Pit are modeled with parabolic kinetic curves using 100 ka calibration A/I values of 0.24, 0.22, and 0.20 (adapted from O'Neal, 1997). Age-estimate ranges are generated from each curve by comparison with companion curves representing a 2°C CMAT difference between southern New Jersey and southeastern Virginia (Gomez Pit). Correlation of age ranges with the sea-level highstand record is made by comparison with the marine $\delta^{18}O$ curve (modified from Morrison, 1991).

vibracores and scraped sections obtained during this study, to construct a litho- and aminostratigraphic cross-section of the pit as it would have existed prior to excavation.

Figure 21-10, a NNW-to-SSE trending cross-section, was drawn using split spoon cores NP8, NP7, and NP6, and composite section MPC1. The transgressive ravinement surface of possible stage 9 age is clearly represented by the sloping unconformity seen above the base of the section. This surface is incised into sediments of probable stage 11 age or older. The silty to muddy sands of the lowermost unit contain *Crassostrea* beds overlain by *Mercenaria* horizons, indicative of a shift from quiescent waters to a more energetic environment and/or from lower to higher salinity levels. This sequence was probably deposited in or near the flooded mouth of an ancestral Maurice River.

Units representing two sea-level cycles, likely to be two phases of stage 9, are preserved above the lower unconformity. The lower unit, allostratigraphic unit 3, exhibits transgressive lag to nearshore to sandy tidal-flat lithologies. The upper unit, allostratigraphic unit 4, repeats this sequence

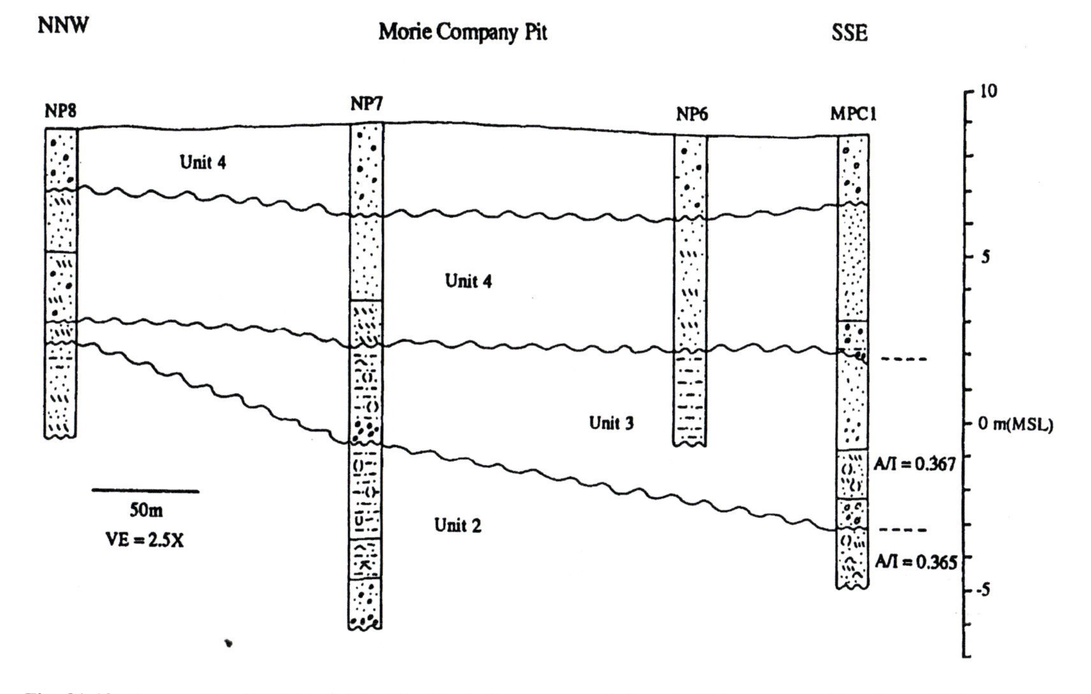

Fig. 21-10. Cross-section (NNW to SSE) of the Morie Company sand pit west of Mauricetown, New Jersey. Split-spoon cores NP8, -7, and -6 are used in conjunction with the combined section, MPC1 (see figure 21-3) to show the relationship between allostratigraphic units 2, 3, and 4 (O'Neal, 1997). Amino acid data are given for units 3 and 2, respectively, above and below the prominently sloping unconformity. VE = vertical exaggeration.

above an erosional surface thought to represent a minor fluctuation in sea level. The elevations and lithologies of units 3 and 4 suggest similar sediment sources and distribution patterns. A regressive phase of the upper unit is also present at the top of the section.

On the basis of the elevation of unconformities in the Morie Pit stratigraphic section and their correlative major reflectors on nearby ground-penetrating radar profiles (O'Neal, 1997), we have projected the aminostratigraphic age estimates obtained in the pit to corresponding allostratigraphic units underlying the two terraces along the northern Delaware Bay margin. Figure 21-11 shows the position of the Morie Pit stratigraphic sequence in relation to the six allostratigraphic units of O'Neal (1997) along a ground-penetrating radar-generated interpretive cross-section, obtained near Cedarville, New Jersey (GPR profiles CV3 and CV4 located in figure 21-2).

Morie Pit was excavated on the upper terrace, landward of the Cedarville Scarp, at a land-surface elevation of approximately 9 m above MSL. The slope and elevation of the bounding unconformities between allostratigraphic units 2, 3, and 4 in the pit correlate closely with the unconformities separating allostratigraphic units imaged in the composite cross-section shown in figure 21-11. The stratigraphic sequence and position beneath the upper terrace, in relation to the scarp, is the same for allostratigraphic units 2, 3, and 4 within the pit and in the radar profiles. Each unit exhibits lithologies indicating similar environments of deposition at each correlative location. Whilst these three units are herein correlated with stages 9 and 11, the next older unit, allostratigraphic unit 1, may be correlated with the preceding interglacial stage in the oxygen isotope record with an apparent sea level high enough to have deposited the unit in its position relative to the other units on the upper terrace. Although no direct evidence of the age of this unit was collected in the present study, oxygen isotope stage 13 is a likely candidate.

Separated from the upper terrace by the Cedarville Scarp, two allostratigraphic units underlie the lower terrace, which do not occur under the upper terrace (units 5 and 6). Figure 21-11 depicts a complex geometry of sediment deposits and infilled paleovalleys, interpreted from the cross-cutting relationships of major reflectors in the radar records obtained at this location (O'Neal, 1997), where units 5 and 6 are seen to overlie units 2 and 3. The transgressions associated with the deposition of the upper two units (5 and 6) beneath the lower terrace appear to have created the escarpment. Only two major interglacial episodes (stage 7 and sub-

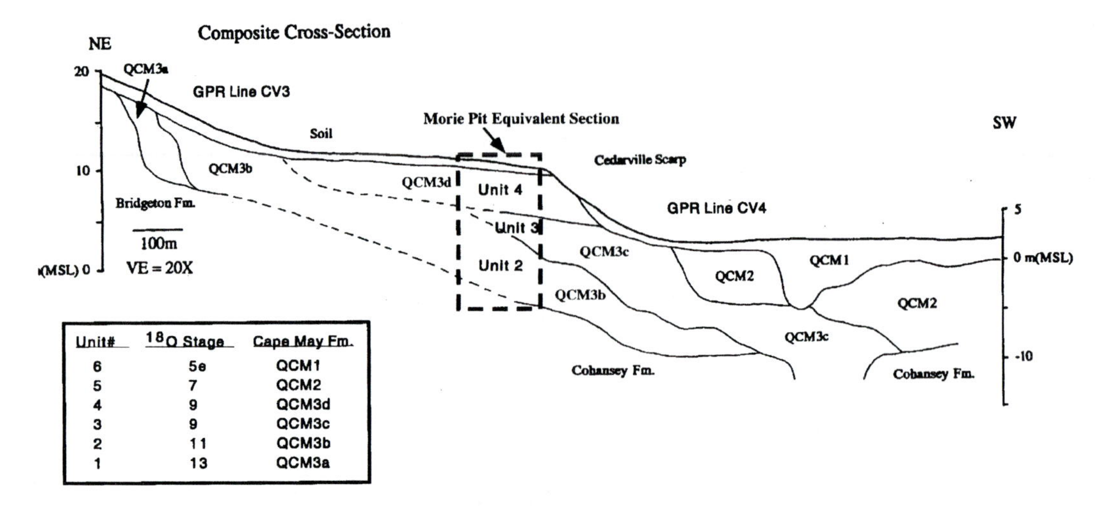

Fig. 21-11. Composite cross-section derived from ground-penetrating radar profiles CV3 and CV4 (see figure 21-2), showing the internal stratigraphy of the Cape May Formation on the northern margin of Delaware Bay. Members and submembers, identified as allostratigraphic units in radar sections and boreholes by O'Neal (1997), are shown within the framework of the Cape May Formation of Newell et al. (1995). A stratigraphic section equivalent to that of Morie Pit is shown in relation to the Cedarville Scarp. VE = vertical exaggeration.

stage 5e) appear in the marine $\delta^{18}O$ curve between stage 9 (the youngest deposit on the upper terrace) and the Holocene. On the basis of these stratigraphic and geomorphic relationships, it is reasonable to correlate allostratigraphic units 5 and 6 with oxygen isotope stage 7 and substage 5e, respectively.

Allostratigraphic units 1, 2, 3, and 4, underlying the upper terrace then, are correlated with stages 13, 11, and two phases of stage 9, in an offlapping, bayward sequence and subdivide the Cape May 3 unit of Newell et al. (1995) into four subunits, QCM3a, QCM3b, QCM3c, and QCM3d, respectively. The ages of these units are correlative with Pleistocene sea-level highstands at approximately 475, 400, and 320–310 ka, respectively.

Underlying the lower terrace are two highstand units (allostratigraphic units 5 and 6), which are correlated with stage 7 and substage 5e on the basis of stratigraphic and geomorphic position, and elevation. These two units correspond with Cape May 2 (QCM2) and Cape May 1 (QCM1) of Newell et al. (1995), respectively, and were deposited during sea-level highstands at approximately 210 and 125 ka. The erosion of the Cedarville Scarp separating the two terraces took place during both of these sea-level highs.

CONCLUSIONS

Throughout nearly a century of formal study of the terrace deposits on the northern Delaware Bay margin, the establishment of a detailed chronology has been problematic because of a lack of suitable materials and dating methods, with most investigators assigning a "Sangamonian or younger" age to the Cape May Formation. Although AAR techniques have been used to provide age estimates for Quaternary deposits in the Atlantic coastal plain for some 20 years, regional correlation of aminostratigraphic units has proved difficult. Few sites exist which clearly provide age calibrations for the AAR method or contain sufficiently long sections to allow for detailed stratigraphic interpretations. Comparison and correlation of aminostratigraphic data obtained by several workers over a long period has proved to be complicated by often-unpredictable intrasample variability in D/L values. The present work extends several studies of intrashell variation in AAR values and includes observations from the Morie (New Jersey) and Gomez (Virginia) pits (this study and Wehmiller et al., 1998). These investigations imply that natural diagenetic alteration can introduce significant variability in D/L values within a single shell, and suggest a more consistent sampling approach (as the common hinge position of *Mercenaria* may not be the best point for sampling, in spite of its convenience).

With data from the Gomez Pit, Virginia (Mirecki et al., 1995) as a calibration, the aminostratigraphy of the Morie Pit, New Jersey, combined with geophysical, lithostratigraphic and geomorphic investigations of estuarine-terrace deposits along the northern shore of Delaware Bay (O'Neal, 1997), sheds new light on the preserved record of Quaternary sea-level fluctuations in southern New Jersey. Mapped predominantly as the Cape May Formation, these deposits may contain one of the more complete sections of Quaternary highstand stratigraphy in this section of the Atlantic coastal plain. With six allostratigraphic units being identified as estuarine–marine highstand deposits in both ground-penetrating radar profiles and lithostratigraphic sections, it is clear that the deposition of the Cape May Formation spanned several transgressions during the Pleistocene. The AAR kinetic model age of the Morie Pit fossil mollusks supports the geomorphic interpretation that a large volume of the preserved section is of marine oxygen isotope stage 9 and/or 11 age. This conclusion is further supported by the presence of aminostratigraphic units of stages 9–11 age in coastal deposits on the southern side of the bay in Delaware (Groot et al., 1990; Ramsey, 1993; Andres and Ramsey, 1995). On a global perspective, the stage 9–11 age estimate for the bulk of the mid- to late Quaternary highstand record in this region fits well with the virtually uninterrupted high sea level during this time period (relatively short stage 10 regression) inferred from the marine oxygen isotope record (Shackleton, 1987).

Acknowledgments: This work was supported by NSF Grant EAR9315052 to the University of Delaware (Geology Department). The authors thank M. Bart, K. J. Lacovara, S. McGeary, and J. Zadorozny for their contributions to this project.

REFERENCES

Andres A. S., and Ramsey K. W. (1995) Geologic map of the Seaford area, Delaware. *Del. Geol. Sur. Geol. Map* **9**.

Belknap D. F. (1979) Application of amino acid geochronology to stratigraphy of Late Cenozoic marine units of the Atlantic Coastal Plain. Ph.D. dissertation, 552 p., University of Delaware, Newark.

Belknap D. F., and Wehmiller J. F. (1980) Amino acid racemization in Quaternary mollusks; examples from Delaware, Maryland, and Virginia. In *Biogeochemistry of Amino Acids* (eds P. E. Hare, T. C. Hoering, and K. King, Jr.), pp. 401–414. Wiley.

Bloom A. L., and Yonekura N. (1985) Coastal terraces generated by sea-level change and tectonic uplift. In *Models in Geomorphology* (ed. M. J. Woldenberg), 139–154. Allen and Unwin.

Brigham J. K. (1983) Intrashell variations in amino acid concentrations and isoleucine epimerization ratios in fossil *Hiatella arctica*. *Geology* **11**, 509–513.

Colman S. M., and Mixon R. B. (1988) The record of major Quaternary sea-level changes in a large coastal plain estuary, Chesapeake Bay, eastern United States. *Palaeogeog. Palaeoclim. Palaeoecol.* **68**, 99–116.

Cooke C. W. (1958) Pleistocene shore lines in Maryland. *Geol. Soc. Amer. Bull.* **69**, 1187–1190.

Cronin T. M., Szabo B. J., Ager T. A., Hazel J. E., and Owens J. P. (1981) Quaternary climates and sea levels of the U.S. Atlantic Coastal Plain. *Science* **211**, 233–240.

Flint R. F. (1971) *Glacial and Quaternary Geology*. Wiley.

Frank H., Nicholson G. J., and Bayer E. (1977) Rapid gas chromatographic separation of amino acid enantiomers with a novel chiral stationary phase. *J. Chromatog. Sci.* **15**, 174–176.

Friedman G. M. (1959) Identification of carbonate minerals by staining methods. *J. Sed. Pet.* **29**, 87–97.

Genau R. B., Madsen J. A., McGeary S., and Wehmiller J. F. (1994) Seismic-reflection identification of Susquehanna River paleochannels on the mid-Atlantic coastal plain. *Quat. Res.* **42**, 166–175.

Gill H. E. (1962) Ground-water resources of Cape May County, New Jersey—salt-water invasion of principal aquifers. *New Jersey Dept. Cons. Econ. Dev., Sp. Rep.* **18**, 1–171.

Goodfriend G. A., and Meyer V. R. (1991) A comparative study of the kinetics of amino acid racemization/epimerization in fossil and modern mollusk shells. *Geochim. Cosmochim. Acta* **55**, 3355–3367.

Goodfriend G. A., Flessa K. E., and Hare P. E. (1997) Variation in amino acid epimerization rates and amino acid composition among shell layers in the bivalve Chione from the Gulf of California. *Geochim. Cosmochim. Acta* **61**, 1487–1493.

Groot J. J., Ramsey K. W., and Wehmiller J. F. (1990) Ages of the Bethany, Beaverdam, and Omar formations of southern Delaware. *Delaware Geol. Surv. Rep. Inv.* **47**, 1–19.

Hare P. E., and Mitterer R. M. (1969) Laboratory simulation of amino acid diagenesis in fossils. *Carnegie Instn Wash. Yrbk* **67**, 205–208.

Hearty P. J., Miller G. H., Stearns C. E., and Szabo B. J. (1986) Aminostratigraphy of Quaternary shorelines in the Mediterranean Basin. *Geol. Soc. Amer. Bull.* **97**, 850–858.

Kaufman D. S., and Brigham-Grette J. (1993) Aminostratigraphic correlations and paleotemperature implications, Pliocene—Pleistocene high-sea-level deposits, northwestern Alaska. *Quat. Sci. Rev.* **12**, 21–33.

Krantz D. E., Stecher H. A., III, Wehmiller J. F., Kaufman A., Lord C. J., and Macko S. A. (1996) Diagenesis of Pleistocene mollusk shells from the US Atlantic coastal plain: processes and scales. *Proc. Geol. Soc. Amer. Ann. Mtg.*, p. A117 (abstr.).

Lacovara K. J. (1990) Reference section for the Cape May Formation: Lithology, palynology, and amino acid data. *Proc. Geol. Soc. Amer. Northeast. Sec. Mtg.*, p. A29 (abstr.).

Lacovara K. J. (1997) Definition and evolution of the Cape May and Fishing Creek Formations in the Middle Atlantic Coastal Plain of southern New Jersey. Ph.D. Thesis, University of Delaware, Newark.

Lajoie K. R., Wehmiller J. F., and Kennedy G. L. (1980) Inter- and intra-generic trends in apparent racemization kinetics of amino acids in Quaternary mollusks. In *Biogeochemistry of Amino Acids* (eds P. E. Hare, T. C. Hoering, and K. King, Jr.), pp. 305–340. Wiley.

Lewis J. V., and Kummel H. B. (1912) Geologic map of New Jersey, 1910–1912. *New Jersey Geol. Surv.*

McCartan L., Owens J. P., Blackwelder B. W., Szabo B. J., Belknap D. F., Kriausakul N., Mitterer R. M., and Wehmiller J. F. (1982) Comparison of amino acid racemization geochronometry with lithostratigraphy, biostratigraphy, uranium-series coral dating, and magnetostratigraphy in the Atlantic Coastal Plain of the Southeastern United States. *Quat. Res.* **18**, 337–359.

MacClintock P., and Richards H. G. (1936) Correlation of late Pleistocene marine and glacial deposits of New Jersey and New York. *Geol. Soc. Amer. Bull.* **47**, 289–338.

MacClintock P. (1944) Marine topography of the Cape May Formation. *J. Geol.* **51**, 458–472.

Miller G. H., and Brigham-Grette J. (1989) Amino acid geochronology: resolution and precision in carbonate fossils. *Quat. Intl.* **1**, 111–128.

Mirecki J. E. (1990) Aminostratigraphy, geochronology, and geochemistry of fossils from late Cenozoic marine units in southeastern Virginia. Ph.D. Thesis, University of Delaware, Newark.

Mirecki J. E., Wehmiller J. F., and Skinner A. F. (1995) Geochronology of Quaternary coastal plain deposits, southeastern Virginia, U.S.A. *J. Coast. Res.* **11**, 1135–1144.

Mitterer R. M. (1975) Ages and diagenetic temperatures of Pleistocene deposits of Florida based upon isoleucine epimerization in *Mercenaria*. *Earth Plan. Sci. Lett.* **28**, 275–282.

Mitterer R. M., and Kriausakul N. (1989) Calculation of amino acid racemization ages based on apparent parabolic kinetics. *Quat. Sci. Rev.* **8**, 353–357.

Morrison R. B. (1991) Introduction. In *Quaternary Nonglacial Geology; Conterminous U.S., The Geology of North America*, vol. K-2 (DNAG) (ed. R. B. Morrison), pp. 1–12. Geological Society of America.

Newell W. L., Powars D. S., Owens J. P., and Schindler J. S. (1995) Surficial Geological Map of New Jersey: Southeastern Sheet. *USGS Open File Report* 95–272.

Newell W. L., Wyckoff J. S., Owens J. P., and Farnsworth J. (1988) Cenozoic Geology and Geomorphology of Southern New Jersey Coastal Plain. *SEFOP Guidebook*.

Oaks R. Q., Jr, and Coch N. K. (1963) Pleistocene sea levels, Southeastern Virginia. *Science* **140**, 979–983.

O'Neal M. L. (1997) Ground penetrating radar analysis of composite Quaternary coastal terraces on the northern margin of the Delaware Bay. Ph.D. Thesis, University of Delaware, Newark.

Owens J. P., and Minard J. P. (1979) Upper Cenozoic sediments of the Lower Delaware Valley and the northern Delmarva Peninsula, New Jersey, Pennsylvania, Delaware and Maryland. *USGS Prof. Paper* 1067-D.

Pannella G., and MacClintock C. (1968) Biological and environmental rhythms reflected in molluscan shell growth: paleobiological aspects of growth and development: a symposium (ed. D. B. Macurda). *J. Paleont. Suppl.* **42**, 64–80.

Ramsey K. W. (1993) Geologic map of the Milford and Mispillion River quadrangles. *Del. Geol. Sur. Geol. Map* **8**.

Salisbury R. D., and Knapp G. N. (1917) The Quaternary formations of southern New Jersey. *New Jersey Geol. Sur. Final Rep.* **8**, 1–218.

Shackleton N. J. (1987) Oxygen isotopes, ice volume and sea level. *Quat. Sci. Rev.* **6**, 183–190.

Szabo B. J. (1985) Uranium-series dating of fossil corals from marine sediments of the United States Coastal Plain *Geol. Soc. Amer. Bull.* **96**, 398–406.

Toscano M. A. (1992) Record of oxygen isotope stage 5 on the Maryland Inner Shelf and Atlantic Coastal Plain—a post-transgressive-highstand regime. In *Quaternary Coasts of the United States, Marine and Lacustrine Systems SEPM Special Publication* (eds C. H. Fletcher, III and J. F. Wehmiller), pp. 89–99.

Toscano M. A., and York L. L. (1992) Quaternary stratigraphy and sea-level history of the U.S. Middle Atlantic Coastal Plain. *Quat. Sci. Rev.* **11**, 301–328.

Toscano M. A., Kerhin R. T., York L. L., Cronin T. M., and Williams S. J. (1989) Quaternary stratigraphy of the inner continental shelf of Maryland. *Maryland Geol. Sur. Rep. Inv.* **50**, 1–116.

Wehmiller J. F. (1984) Interlaboratory comparison of amino acid enantiomeric ratios in fossil mollusks. *Quat. Res.* **22**, 109–120.

Wehmiller J. F. (1993) Applications of organic geochemistry for Quaternary research: aminostratigraphy and aminochronology. In *Organic Geochemistry* (eds M. Engel and S. Macko), pp. 755–783. Plenum.

Wehmiller J. F. (1997) Data report: aminostratigraphic analysis of mollusk specimens: Cape May coast guard station borehole. In *Proceedings of the Ocean Drilling Program, Scientific Results, 150X* (eds K. G. Miller and S. W. Snyder), pp. 355–357. ODP.

Wehmiller J. F., and Belknap D. F. (1982) Amino acid age estimates, Quaternary Atlantic Coastal Plain: comparison with U-series dates, biostratigraphy, and paleomagnetic control. *Quat. Res.* **18**, 311–336.

Wehmiller J. F., and Miller G. H. (2000) Aminogstratigraphic dating methods in Quaternary geology. In *Quaternary Geochronology, Methods and Applications* (eds J. S. Noller, J. M. Sowers, and W. R. Lettis). American Geophysical Union, Reference Shelf v. 4, pp. 187–222.

Wehmiller J. F., Belknap D. F., Boutin B. S., Mirecki J. E., Rahaim S. D., and York L. L. (1988) A review of the aminostratigraphy of Quaternary mollusks from United States Coastal Plain sites. In *Dating Quaternary Sediments*, Geological Society of America Special Paper, vol. 227 (ed. D. L. Easterbrook), pp. 69–110. Geological Society of America.

Wehmiller J. F., York L. L., Belknap D. F., and Snyder S. W. (1992) Theoretical correlations and lateral discontinuities in the Quaternary aminostratigraphic record of the U.S. Atlantic Coastal Plain. *Quat. Res.* **38**, 275–291.

Wehmiller J. F., Krantz D. E., Simmons K, Ludwig K. R., Markewich H. W., Rich F., and Hubert R. C. Jr (1997) US Atlantic coastal plain late Quaternary geochronology: TIMS U-series dates continue to indicate 80 kyr sea level at or above present. *Progr. Abstr. Geol. Soc. Amer. Ann. Mtg.*, p. A346 (abstr.).

Wehmiller J. F., Lamothe M., and Noller J. S. (1998) Comparison of approaches to dating Atlantic coastal plain sediments, Virginia Beach, VA. In *Dating and Earthquakes: Review of Quaternary Geochronology and its Applications to Paleoseismology NUREG/CR 5562* (eds J. M. Sowers et al.), 4-3–4-31. U.S. Nuclear Reg. Comm.

York L. L. (1990) Aminostratigraphy of the U.S. Atlantic coast Pleistocene deposits: Maryland continental shelf and North and South Carolina coastal plain. Ph.D. Thesis, University of Delaware, Newark.

York L. L., Wehmiller J. F., Cronin T. M., and Ager T. A. (1989) Stetson Pit, Dare County, North Carolina: an integrated chronologic, faunal, and floral record of subsurface coastal Quaternary sediments. *Palaeogeog. Palaeoclim. Palaeoecol.* **72**, 115–132.

22. Chronostratigraphy of sediments in the southern Gulf of California, based on amino acid racemization analysis of mollusks and coralline algae

Glenn A. Goodfriend, Jochen Halfar, and Lucio Godinez-Orta

Although amino acid racemization (AAR) dating has been used primarily for correlation and dating of pre-radiocarbon (> 40,000 yr BP) deposits, it has also proved to be a useful method for analysis of stratigraphic integrity in younger deposits. In particular, it has been used to determine whether mollusk assemblages within a particular stratum are of mixed or uniform age, which is significant in determining the time of sediment accumulation or the ages of the fauna within the sediments (Goodfriend, 1987, 1989; Goodfriend and Stanley, 1996; Murray-Wallace et al., 1996; Kowalewski et al., 1998). AAR has the advantage over radiocarbon dating in this regard because the analyses are relatively easy to carry out, so large numbers of samples can be dated. In the present study, this method is applied to the evaluation of the chronostratigraphy of littoral zone sediments based on cores from La Paz Bay (Bahía de la Paz), in the southern Gulf of California. The protein amino acid L-isoleucine racemizes[1] to the nonprotein amino acid D-alloisoleucine over time, so the ratio of D-alloisoleucine to L-isoleucine (A/I) gives a measure of relative age.

The modern and Quaternary carbonate environments in the La Paz area in the southwestern Gulf of California (figure 22-1) are largely unstudied. They represent one of only a few carbonate settings along the entire eastern Pacific margin in which such carbonate-rich sediments are found. Carbonates in the tectonically young rift basin of the Gulf of California form in areas where the narrow continental shelf widens enough to provide sufficient space to accommodate carbonate formation. The most extensive shelf width is in the La Paz area, where carbonates form in small pocket bays and shallow underwater ridges to a depth of 40 m.

The oceanography of the surface-water masses in the southern Gulf of California is influenced by both the warm Eastern Tropical Pacific water and the cold California Current (Alvarez-Borrego and Lara-Lara, 1991). The interplay of both currents is reflected by the extreme temperature regime, with winter temperatures dropping to 18°C and summer temperatures reaching up to 31°C. This results in an unusual composition of the subtropical to warm-temperate La Paz carbonates, which are made up of equal percentages of coral and rhodolith (coralline red algal) material and smaller amounts of mollusk fragments and foraminifera.

Corals, present only as small patch reefs composed of the genera *Pocillopora* and *Porites*, are at their limit of their existence, because of low winter sea-surface temperatures. Living rhodoliths of the genus *Lithophyllum* (Riosmena-Rodriguez, 1997) therefore dominate the modern shallow-water environments. Rhodoliths (figure 22-2) are nodules or unattached branched growths with a nodular form, composed principally of coralline red algae (Bosence, 1983). They generally live in the shallow subtidal zone but can grow to a water depths of 100 m, depending on the amount of light available for photosynthesis. Their high-Mg calcite skeletons can exhibit different growth morphologies, ranging from spherical to ellipsoidal, and can be thickly or thinly branching, with individual branches reaching up to 1 cm in diameter. Rhodolith morphology is a result of frequent overturning due to water motion (during storms) or biogenic activity (bottom-feeding fish and infaunal or epifaunal organisms) (Bosellini and Ginsburg, 1971). Even though they cover large areas of the modern shallow-water carbonate environments, rhodoliths contribute only about the same amount of carbonate to modern sediments as do the much less abundant corals, which are confined to small patch reefs. This is because of the slow vertical growth rate of rhodoliths (250–400 μm/yr) as compared to corals (2.5–3 cm of vertical extension/yr). Carbonate environments which are co-dominated by corals, rhodoliths, and mollusks have been frequently described in the fossil record (e.g., Carannante et al., 1988; Bryan and Hugglestun, 1991; Fravega et al., 1993; Carannate and Simone, 1995) but are rare in modern settings. Hence, the La Paz carbonates provide a modern model for subtropical to warm-temperate carbonate formation and they can be compared with similar fossil settings.

The study of sediments in the La Paz Bay was undertaken in order to (1) determine if the modern Gulf of California environments are representative of the Holocene and Pleistocene, (2) calculate sedi-

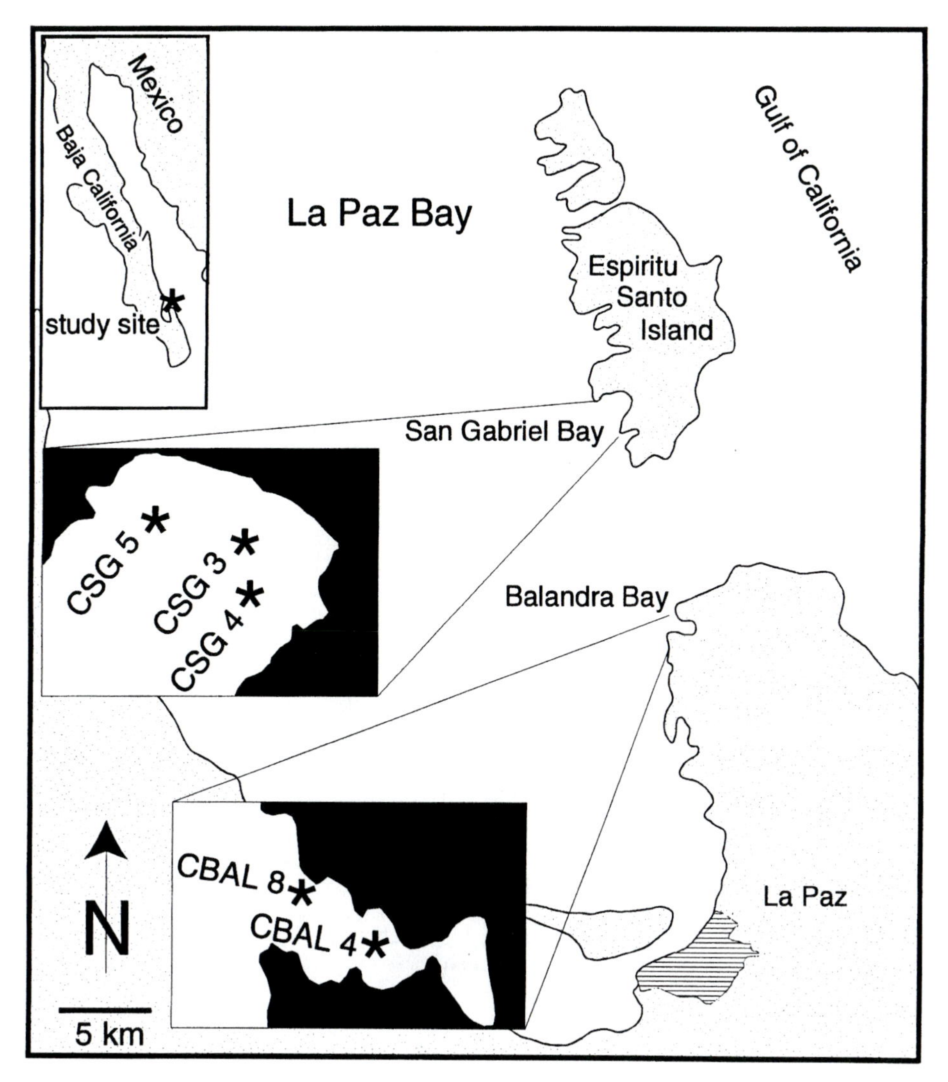

Fig. 22-1. Location map of Baja California and the La Paz area, showing locations of core sites.

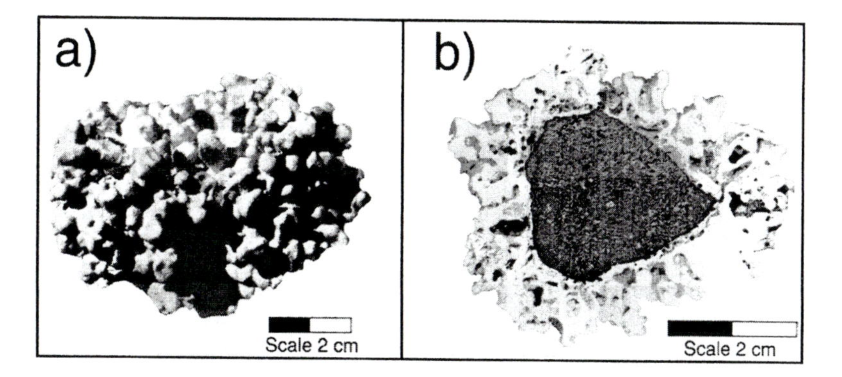

Fig. 22-2. Photograph of rhodolith. Whole (a) and section (b), showing the volcanic pebble nucleus around which rhodolith formed.

ment-accumulation rates in this non-tropical regime, and (3) determine whether changes of biogenic composition occurred over time. For this purpose, a series of sediment cores was obtained from the La Paz shallow-water settings. In this chapter, we present results concerning the chronology of these cores, emphasizing the use of AAR for chronostratigraphic downcore analysis, which preceded the sedimentological investigations.

MATERIALS AND METHODS

Vibracores were obtained from shallow water (< 2 m depth) in two pocket bays near La Paz during March, 1996. Cores CSG 3, CSG 4, and CSG 5 were collected along a NW–SE transect in San Gabriel Bay (figure 22-1, table 22-1), all approx. 800 m from the shoreline and 600 m apart. San Gabriel Bay is located 25 km north of the city of La Paz, on the uninhabited Isla Espiritu Santo. The shallow pocket bay faces westward to La Paz Bay and is protected from the open Gulf of California. Apart from a mangrove zone in the eastern part of the bay, sediments consist of almost pure carbonates to a water depth of 30 m. Small *Pocillopora* patch reefs, which represent the northernmost coral patch reefs in the Gulf of California, are prominent features of San Gabriel Bay. Another series of cores, CBal 4 and CBal 8, were collected in Balandra Bay on top of an intertidal sandbar and in a subtidal sandflat, respectively. Balandra Bay, located on the northern La Paz Peninsula (figure 22-1), is a narrow eastward-facing bay, that features extensive intertidal carbonates sand flats and a large mangrove lagoon. Living corals are almost completely absent from this bay.

Cores were obtained by drilling from a small boat in water depths ranging from 0.5 and 2 m. The vibracoring device was attached to a 2.5-inch (6.2-cm)-diameter 6-m-long aluminum irrigation pipe, which served as both the core barrel and the liner. The degree of compaction from coring was evaluated by measuring the depth to the sediment surface inside the core liner, while it was still in place in the sediment. On this basis, a uniform decompaction factor of 1.4 was applied to all core data. The core barrel was subsequently split with a circular saw, one half was preserved with epoxy resin, and the other half was sampled at 4 cm intervals.

For the AAR analyses, rhodolith (*Lithophyllum* sp.) fragments and bivalve (*Chione californiensis*) shells or fragments were picked from the cores. Samples sizes were 15–70 mg for *Chione* and 15–60 mg for the rhodoliths. Where possible, analyses of *Chione* were based on pieces from outside the pallial line, containing the middle and outer shell layers. However, this was not always possible, because of the fragmentary nature of many of the specimens. Previous work on variation in isoleucine epimerization rates among different shell layers of another species of *Chione* showed that rate differences of up to 15% may occur among various parts of the shell (Goodfriend et al., 1997). Thus, some variability in the A/I values may occur independently of age. Samples were cleaned mechanically, using a Dremel motorized tool under a stereo microscope, then given a brief (5–10 sec) dip in approx. 1 N HCl, washed 3 times in distilled water, and dried under vacuum. Hydrolysis in 6 N HCl was carried out at 100°C for 20 hr. The samples were desalted with HF (1.25 μl/mg shell), centrifuged, and the supernatant frozen and dried under vacuum. Analyses were carried out using high-performance liquid chromatography (HPLC) using post-column derivization with *o*-phthaldialdehyde (OPA) (Hare et al., 1985). Peaks areas were integrated using Waters Millennium software and A/I values were calibrated against a standard with A/I = 0.30, prepared from D-alloisoleucine and L-isoleucine reagents. The calibration procedure generally increases the A/I values by 4–5%. Errors in A/I values (based on separate preparations and analyses of a standard shell powder) average 4%.

Selected shell samples were radiocarbon dated by accelerator mass spectrometry (AMS) at the Woods Hole Oceanographic Institution NOSAMS facility. Conventional radiocarbon analysis of bulk samples of rhodoliths was also carried out, by Beta Analytic Inc. Calibration of Holocene radiocarbon ages was carried out using the CALIB 3.0 program (Stuiver and Reimer, 1993). For calibration of marine radiocarbon dates, the ΔR value (deviation of the reservoir age from the worldwide average) was taken as 400 ± 110 yr, based on analysis of live-collected pre-bomb shells from the central Gulf of California (Goodfriend and Flessa, 1997). Pleistocene radio-

TABLE 22-1 List of cores analyzed, their lengths, locations, and water depths

Core no.	Decompacted length (cm)	Latitude (N)	Longitude (W)	Water depth (m)
CSG 3	5.71	24°25.867″	110°20.931″	1
CSG 4	4.87	24°25.699″	110°20.994″	2
CSG 5	5.79	24°26.091″	110°21.643″	1.5
CBal 4	5.51	24°19.153″	110°19.414″	0.5
CBal 8	3.10	24°19.277″	110°19.769″	2

carbon ages are presented uncorrected for reservoir age and uncalibrated and are thus comparable to other Pleistocene radiocarbon ages that have been reported for the region. Uranium–thorium (U-Th) analyses were carried out by thermal ionization mass spectometry (TIMS) by P. Holden at the University of California, Santa Cruz.

CORE SEDIMENTOLOGY AND STRATIGRAPHY

Core CSG 3

Core CSG 3 can be subdivided into two parts: the upper 4.12 m consists of carbonate sand and gravel, whereas the lower 1.59 m is composed of organic-rich mud (figure 22-3). From 0 to 0.90 m and from 3.10 to 4.12 m, a fine to coarse carbonate sand with only few larger mollusk or coral fragments is present. Between these two carbonate sand layers, from 0.90–3.10 m, the sediments consist of gravel-sized mollusk, coral, and algal fragments. One prominent 9-cm-thick shell layer occurs at a depth of 0.90 m. The lower 1.59 m shows a gradual transition from a carbonate sand facies to a mangrove mud facies below. Individual thin carbonate sand layers are present within the muddy, organic-rich sediment. The amount of carbonate within the mud decreases downcore. At 5.51 m, a complete pulmonate gastropod shell (*Melampus* sp.) was recovered. *Melampus* lives in high-intertidal muddy salt marsh or mangrove environments. Grain mount point-counting analysis revealed a steady downcore decrease in coral fragments, acccompanied by an increase in foraminifera. However, the most important biogenic component of CSG 3 is coral fragments, which comprise up to 60% of the sand-sized components of the core.

Core CSG 4

The upper (0–0.43 m) and lower (1.90–4.87 m) units of this core consist of fine to coarse carbonate sand; the lower unit contains abundant mollusk and coral fragments. The central part (0.43–1.90 m) of the core consists of coarse carbonate gravel (0.43–1.05 m) and sand to medium gravel (1.05–1.90 m), with complete mollusk shells and coral clasts. Biogenic grain counts were not carried out on this core.

Core CSG 5

The upper 0.60 m of this core consist of carbonate gravel with abundant shell fragments. This part also contains two prominent shell layers 10–12 cm thick. Below 0.60 m, the gravel grades into a medium carbonate sand. There are two more shell layers at depths of 1.80 and 2.00 m. The abundant mollusk fragments in the upper 2.00 m of the core give way to increasingly abundant coral pieces below. Below a depth of 3.00 m, the amount of coralline algae increases in the sand-sized fraction, whereas both foraminifera and coral fragments decrease. The decreased abundance of foraminifera could be the result of poor preservation of their fragile tests in this older portion of the core.

Core CBal 4

The sediments in this core alternate between units of medium sand and fine gravel, with abundant algal, coral, and mollusk debris. No trend is observable in the distribution of the biogenic constituents in this core. All three major components—coral, rhodolith, and mollusk shell fragments—are present in roughly equal amounts averaging approx. 20–25% throughout the core. Foraminifera are scarcer than in the San Gabriel Bay cores.

Core CBal 8

CBal 8 consists entirely of fine to coarse carbonate sand, with shell layers at 0.90 m and 1.40 m. As in CBal 4, the sediments consists of equal amounts of coral, rhodolith, and mollusk fragments, without any depth trend.

INITIAL RADIOCARBON-DATING RESULTS AND PURPOSE OF THE PRESENT STUDY

Under the assumption that the cores would probably penetrate only Holocene sediments, Halfar initially attempted to establish the chronology of core CSG 5 by conventional radiocarbon analysis of bulk sediment carbonate from depths of 4.2 and 5.7 m. Each sample consisted of 10 g of sand-sized rhodolith fragments. The ages obtained were far older than anticipated: 18.980 (±390) yr BP at 4.2 m and 28,190 (±1000) yr BP at 5.7 m (ages not calibrated) (table 22-2). These ages suggest that Pleistocene sediments had been penetrated. However, the radiocarbon ages are clearly problematic, as the eustatic sea level at those times should have been much lower (Bard et al., 1990), so the coring locations should then have been emergent. Three possible explanations were considered: (1) that the radiocarbon ages are accurate and a complicated uplift history for this tectonically active region has to be invoked; (2) that the ages represent reworking of sediments, with Pleistocene rhodolith fragments from adjacent uplifted marine terraces being mixed with Holocene fragments, resulting in an intermediate apparent age; or (3) that the core had penetrated into a submarine Pleistocene terrace (similar adjacent uplifted terraces have been dated at 125,000 yr BP; Sirkin et al., 1990) but exchange of carbonate carbon or recrystallization of carbonate has resulted in apparent radiocarbon ages that are too young.

At this point, it was decided to turn to AAR analysis in order to help solve this dating problem

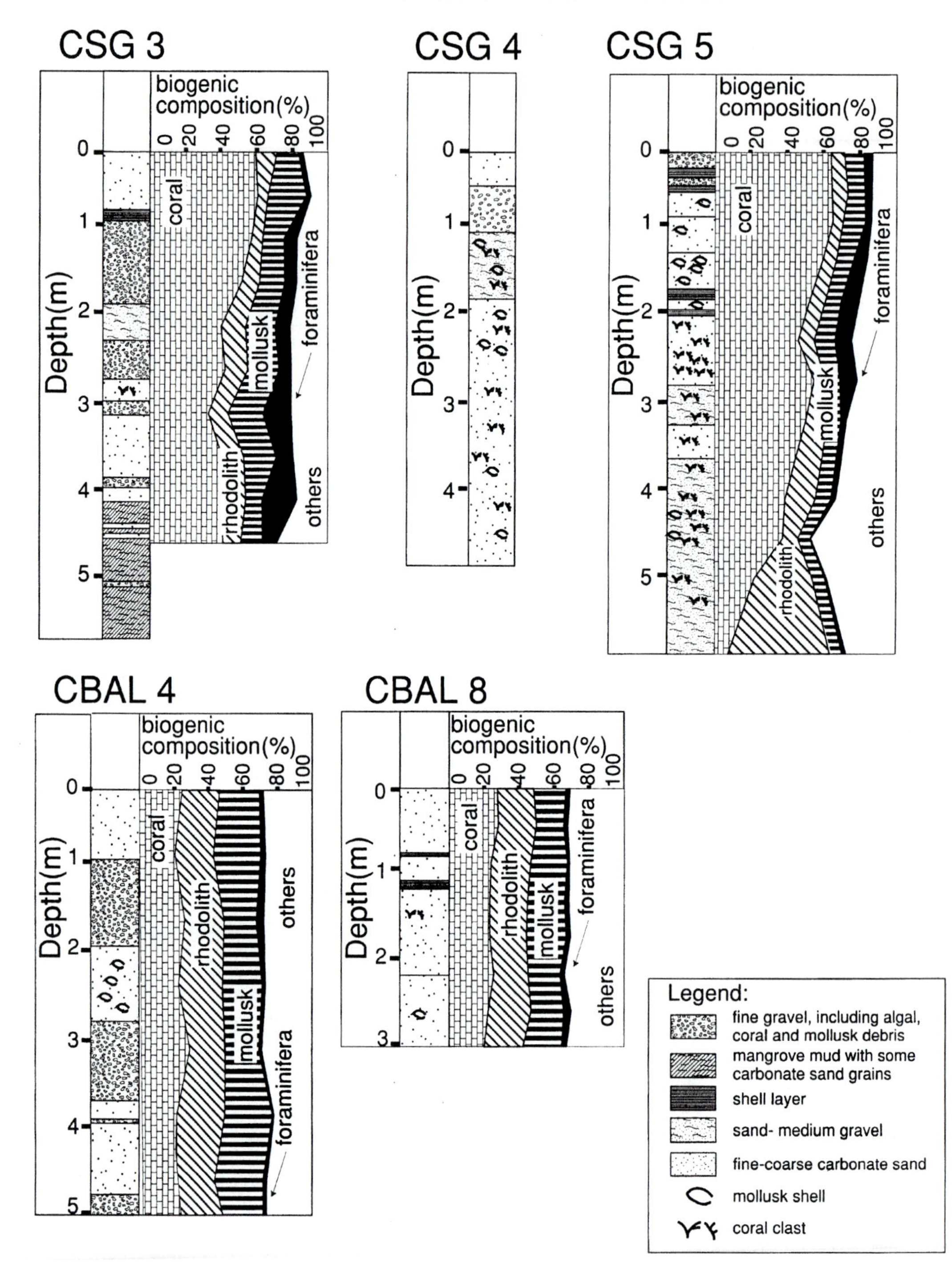

Fig. 22-3. Lithostratigraphic diagrams of cores and relative amounts of biogenic constituents.

and to elucidate the chronology of the cores more thoroughly. Because AAR analyses are inexpensive to carry out relative to AMS radiocarbon analyses, they enable the core chronologies to be examined in more detail. Multiple samples per level can be analyzed to check whether age mixtures are present, and denser vertical sampling of cores can be carried out to provide a more detailed picture of the age–depth relationships of the cores. Frequent age control is especially important in shallow-water settings, where bioturbations and wave action commonly cause stratigraphic disorder (Flessa et

TABLE 22-2 Radiocarbon dates for core samples, based on analysis of carbonates. Samples with OS-laboratory numbers were analyzed by AMS, and those with Beta- numbers were analyzed conventionally

Core	Depth (m)	Sample type	Lab. no.	$\delta^{13}C$ (‰)	^{14}C age (yr BP)	Calibrated ^{14}C age (yr BP)[a]
CSG 3	1.76	mollusk (*Chione*)	OS-14685	+1.9	4170 ± 35	3690 ± 145
CSG 3	2.38	mollusk (*Chione*)	OS-14687	+2.3	5660 ± 35	5605 ± 140
CSG 3	2.66	mollusk (*Chione*)	OS-14686	+1.5	6250 ± 65	6235 ± 140
CSG 3	3.94	mollusk (*Melampus*)	OS-13483	−9.1	7120 ± 30	7885 ± 40
CSG 5	4.17	alga (*Lithophyllum*)	Beta-10008	−1.5	$18,980 \pm 390$	–
CSG 5	5.74	alga (*Lithophyllum*)	Beta-10009	−2.2	$28,190 \pm 1000$	–

[a]*Chione* samples calibrated according to the marine calibration (see *Materials and Methods* for details), and a *Melampus* sample calibrated according to the terrestrial calibration.

al., 1993) of the sediment columm. In addition, racemization analyses provide age resolution for samples too old for radiocarbon dating. By radiocarbon-dating a few selected samples that were also analyzed for racemization, a racemization-rate calibration could be developed to provide estimates of ages for all samples analyzed by racemization. More specifically, our goals were as follows.

1. To establish whether mixed-age assemblages were present in the cores.
2. To evaluate the quality of the stratigraphy of the cores.
3. To establish the location of the Pleistocene–Holocene boundary in cores containing Pleistocene sediments.
4. To date selected levels by AAR calibrated against radiocarbon ages.
5. To evaluate the suitability of rhodoliths for AAR dating (apparently these have not been studied previously).

CHRONOSTRATIGRAPHY OF THE CORES

Calibration of AAR Rate

In order to estimate ages from AAR values, the racemization rate must be known. For *Chione*, the racemization rate for Holocene samples was estimated using three radiocarbon-dated samples from core CSG 3 (table 22-2), plus the A/I value measured in live-collected *Chione* specimens (representing the racemization induced during sample preparation; the value is 0.008 in *Chione fluctifraga* from the northern Gulf of California). A plot of the A/I values in relation to calibrated ages yields a slope of 4.54×10^{-5} (in A/I per yr), or 0.01 per

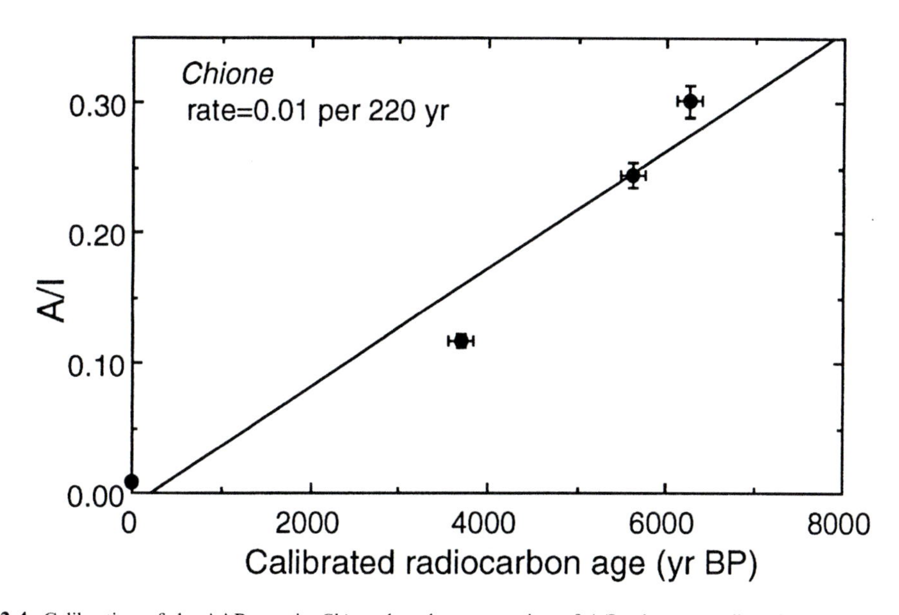

Fig. 22-4. Calibration of the AAR rate in *Chione*, based on regression of A/I values on radiocarbon ages corrected for the marine reservoir age and calibrated. Error bars for A/I values assume a 4% error.

220 yr (figure 22-4). To estimate ages from A/I values for Holocene samples, the net racemization (measured value minus the induced racemization of 0.008) is divided by this rate (or multiplied by the inverse);

$$\text{Age} = (\text{A/I} - 0.008)/(4.54 \times 10^{-5}) \quad (22\text{-}1)$$

It should be noted that this formula is *not* applicable to Pleistocene samples, because the temperature history of such samples would be very different (a cooler average temperature producing a lower racemization rate) and because the kinetics are non-linear at higher A/I values.

Core Chronologies

Core CSG 5

Our initial racemization work concentrated on core CSG 5, for which the Pleistocene radiocarbon ages (see above) had been obtained for the lower section. Racemization analyses of both a *Chione* and a rho-

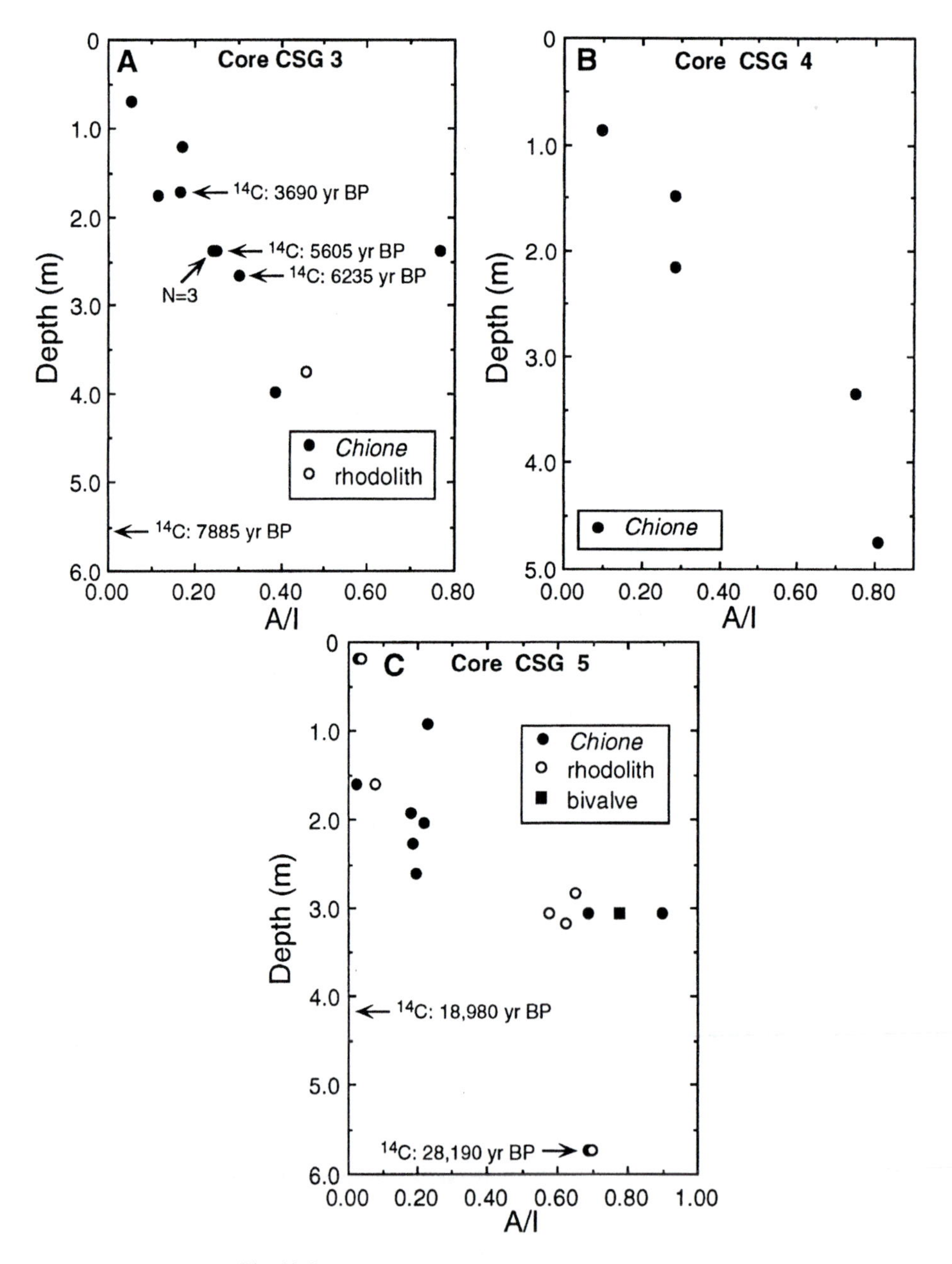

Fig. 22-5. AAR and radiocarbon results for CSG cores.

dolith sample from 5.77 m (the level at which the 28,190 yr BP date was obtained) gave identical A/I values of 0.69 (figure 22-5C). Similar values were also obtained for most rhodolith and mollusk samples from a depth of 2.83–3.14 m, except for one *Chione* sample with a higher A/I value (0.90). From a depth of 2.60 m up to the top of the core, dramatically lower A/I values (0.01–0.22) occur in both *Chione* and rhodoliths. Thus, there is a large jump in the racemization values for specimens located at depths of 2.60 and 2.83 m. In the upper 2.60 m of the core, the A/I values do not show any trends in relation to depth. Modern to near-modern shells are found at the top (0.008 at 0.20 m; 0 age) as well as lower down (0.023 at 1.60 m; estimated age, 300 yr BP), whereas values of approx. 0.19 (estimated age, 4000 yr BP) occur at intermediate depth (0.92 m) and at the interval from 1.93–2.60 m. Rhodolith and *Chione* A/I values are similar at a depth of 0.20 m.

The age of the lower part of the core (below 2.60 m) is problematic. U-Th analysis of a coral from 5.73 m gives an apparent age of 180,000 yr BP (table 22-3). However, x-ray diffraction analysis of the sample indicates that a significant amount of calcite is present, indicating recrystallization. Although the age can be taken only as a general indication, it suggests that the bulk radiocarbon ages of rhodolith samples from this level (28,000 yr BP), as well as the 19,000 yr age of the sample above it (table 22-2, figure 22-5C) are probably inaccurate; exchange of carbonate carbon may have made the radiocarbons ages too young. Radiocarbon ages in the range of 20,000–30,000 yr BP have been measured in numerous shell samples from emerent shorelines in the northern Gulf of California (Ortlieb and Malpica Cruz, 1978). As these presumably represent the last interglacial shoreline, the radiocarbon ages are clearly too young. Two coral samples from terraces surrounding the bay give ages of 143,000 and 103,000 yr BP (table 22-3). These may represent slightly altered Substage 5e samples (cf. Sirkin et al., 1990) and the core sample may represent a presently-submerged portion of this same Substage 5e fossil reef. Racemization results are also consistent with a last interglacial age for the lower portion of the cores, based on comparison with racemization values for *Chione* specimens from last interglacial terraces on the Pacific side. At Magdelena Bay, on the Pacific side at approximately the same latitude as La Paz Bay, Wehmiller and Emerson (1980) found A/I values of approx. 1.00 for *Chione* by using gas chromatography (GC). GC results for A/I values (but not for other amino acid D/L values) are biased toward higher values. If the GC calibration of Goodfriend and Mitterer (1993) is applied to these A/I results, a corrected value of 0.87 is obtained. This is within the range measured in the Gulf core samples. Ortlieb (1987) measured the racemization of leucine and valine in *Chione* specimens from three localities in the central region of the Baja Peninsula Pacific coast. Leucine racemizes slightly slower than isoleucine in such samples, according to the results of Wehmiller and Emerson (1980) for Magdalena Bay: the corrected A/I value averages 0.87, the leucine value averages 0.70. Ortlieb's (1987) leucine racemization results (his table 7.14) show slightly higher values than this, ranging from within-locality averages of 0.79 to 0.89. Thus it might be expected that the corresponding A/I values would be slightly higher than those measured in the Gulf core. Although Pacific temperatures are cooler than Gulf temperatures, the terrestrial temperatures to which these terrace samples had been exposed would be significantly warmer than the wholly submerged samples from the La Paz cores. Because of these differences between the samples, a precise comparison of racemization values would have little meaning. But it can be said that the racemization results for the lower part of the core are broadly consistent with the interpretation that it represents last interglacial sediments.

Core CSG3

A/I values for *Chione* show a reasonably regular trend of increases with increasing depth, ranging from 0.05 at 0.70 m to 0.39 at 4.34 m (figure 22-5). Three shells from 2.38 m give analytically identical A/I values (0.24, 0.24, 0.25) that fit well with the trend of A/I values with depth. However, a fourth shell from the same level gives a much higher A/I value (0.77), similar to the values seen in the lower part of CSG 5. A rhodolith sample from 3.78 m gives an A/I value similar to a *Chione* shell just below it in the core. The four radiocarbon dates for core CSG 3 (table 22-2, figure 22-5A) indicate

TABLE 22-3 TIMS U-Th dates on corals from core CSG 5 and coastal terraces along San Gabriel Bay, Espiritu Santo Island. Errors are $\pm 1\sigma$

Sample	[U] (ppm)	$^{230}Th/^{232}Th$	$^{234}U/^{238}U$	$^{230}Th/^{234}U$	Age (yr BP)
CSG 5, 5.73 m	4.37	124	1.141	0.832	$180,500 \pm 2900$
LSG 3, terrace at 1 m	3.90	315	1.151	0.748	$142,600 \pm 2000$
LSG 3 (rerun)	3.70	293	1.161	0.751	$143,300 \pm 2100$
LSG 15, terrace at 9 m	3.87	809	1.154	0.624	$103,200 \pm 1700$

that the sediments are entirely of Holocene age, starting at 7900 yr BP at 5.51 m near the base of the core, based on radiocarbon analysis of a *Melampus* shell, whose presence indicates an intertidal salt marsh or mangrove environment at that time. Calibration of the *Melampus* radiocarbon age was carried out using the database for terrestrial samples, as the relatively light $\delta^{13}C$ value (−9.1‰) indicates that its carbon is of terrestrial rather than marine origin. Sedimentation was rapid to 6200 yr BP, during which time 2.8 m of sediments accumulated. Over the rest of the Holocene, 2.7 m of sediments were deposited. The youngest shell analyzed for racemization, at a depth of 0.70 m, yields an estimated age of 1000 yr BP (based on equation 22-1).

Core CSG 4

Analyses of several *Chione* specimens from this core show a pattern similar to that seen in CSG 5: high A/I values are seen in the lower portion, similar to those in CSG 5, and low values are seen in the upper portion (figure 22-5B). The transition from lower to higher A/I values has not been located precisely, but lies somewhere between 2.16 and 3.36 m. *Chione* shells at 1.48 and 2.16 m both have identical A/I values (0.29), from which an age of 6100 yr BP is estimated.

Core CBal 4

Chione samples from this core all show very low A/I values—none great than 0.08 (estimated age, 1500 yr BP), and as low as 0.026 (estimated age, 400 yr BP) (figure 22-6A). There is no trend for A/I values in relation to depth, and in fact the shell with the lowest A/I value occurs at the greatest depth from which shells were analyzed (3.48 m). The uppermost shell at 1.15 m is of similar age (A/I = 0.032; estimated age of 500 yr).

Core CBal 8

As for CBal 4, there is no trend for A/I values with depth and all values are relatively low (figure 22-6B). The highest value (0.12) is found near the top of the core, at 0.42 m, and the lowest value (0.015; estimated age, 140 yr) is below that, at 0.81 m.

DISCUSSION OF CORE CHRONOLOGIES AND SEDIMENTARY HISTORY

One of the major patterns emerging from these analyses is that in no case is there an entirely regular pattern of increasing age of mollusks or calcareous algae with depth. Core CSG 3 comes closest to having such a regular pattern, but age reversals are seen at depths between 1 and 2 m, and a shell of apparently Pleistocene age is present at 2.38 m. However, other shells analyzed from the same level show uniform ages. The Pleistocene shell must have been reworked from an adjacent uplifted Pleistocene terrace or from eroding submerged Pleistocene sediments. The upper Holocene section of CSG 5 contains shells of the last few hundred years and shells approx. 4000 yr old, in no regular order—the younger shells are found as far as 1.60 m

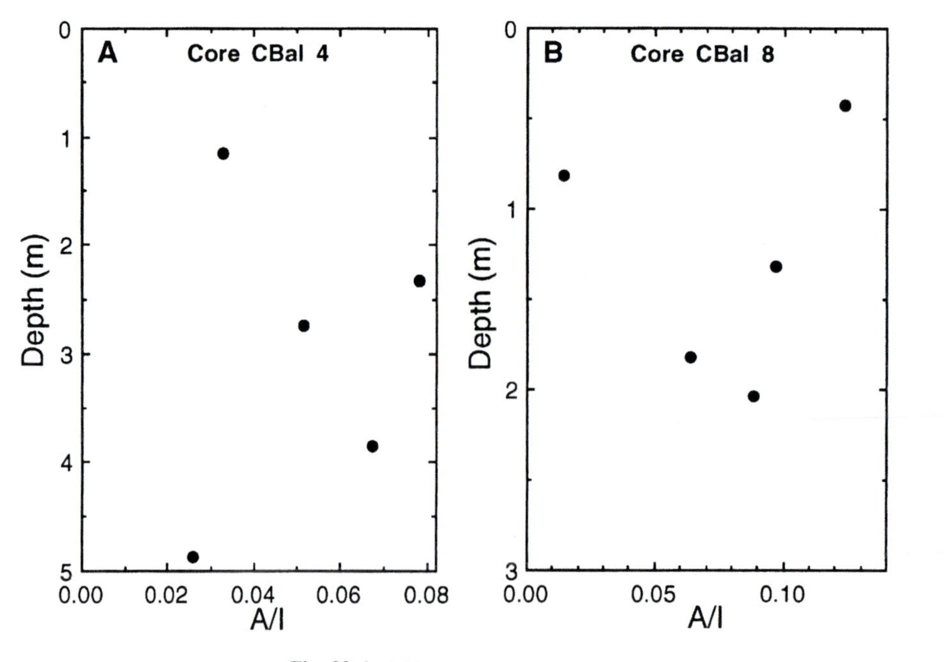

Fig. 22-6. AAR results for CBal cores.

below the surface. Shells dating to the last few hundred years are also found at 0.81 m in CBal 8 and down to 4.87 m in CBal 4.

These occurrences of relatively recent shells at depth in three of the five cores indicates that reworking or deposition of the sediments, at least to the depth of those shells, occurred just within the last few hundred years. Stratigraphic disorder might have been brought about by strong wave activity associated with hurricanes or other storm activity causing reworking of the sediments to some depth. But the young shell at a depth of 5 m (in CBal 4) seems to be too deep to be attributable to *in situ* reworking. Rather, it seems likely that sediments at this site represent the filling of some topographic feature within the last few hundred years. It should be noted here that AAR analysis provides good age resolution within this time period of the last few hundred years, whereas radiocarbon dates for these samples would be rather less useful. The poor precision of radiocarbon dating for the last several hundred years results mainly from two factors. First, the variability of the reservoir age (estimated at 110 yr for the central Gulf of California; see Goodfriend and Flessa, 1997) adds significantly to the overall uncertainty of the corrected age. Second, the calibration of radiocarbon ages adds even more uncertainty, as the calibration curve for this time period (after approx. 1600 AD) is relatively flat (see Goodfriend and Rollins, 1998).

The central and northwestern cores from San Gabriel Bay (CSG 4 and CSG 5) penetrate Pleistocene sediments of probable last interglacial age. Curiously, the Holocene–Pleistocene boundaries in CSG 4 and -5, which must represent a time gap of more than 100,000 yr, is not visible in the cores. Neither was it recognized originally when sampling of the core for dating was carried out, nor could it been seen later, even after its location had been determined from AAR analysis (with good precision in the case of CSG 5). If soil features formed during the intervening period of exposure, they must have eroded away during the subsequent Holocene inundation.

In the core with the best stratigraphy—CSG 3—mud representing mangrove or salt-marsh environments is present at the current depth of 5.5 m at 7900 yr BP. The rapidly rising sea level of the early Holocene (Lighty et al., 1982) inundated this intertidal mud and buried it under fine sands. Sedimentation is rapid until approx. 6200 yr BP, during the period of rapid sea-level rise and then slows as the sea-level rise slows thereafter. A similar history might be expected at the other core sites, but reworking appears to have removed or obscured this record.

Racemization in rhodoliths in the cores, though not analyzed as extensively as that in samples of *Chione*, show patterns very similar to those in *Chione* specimens. In some cases, analytically identical A/I values are found for *Chione* and rhodoliths from the same level, and in other cases the values are quite close. A rather narrow range of A/I values is seen in three rhodoliths from the top of the Pleistocene section of CSG 5 (at 3 m). All of the observations point to rhodoliths as being potentially good material for racemization dating. The similarity of the racemization kinetics to those of *Chione* would seem to be coincidental, given the wide range of racemization rates found among mollusk taxa and among different types of materials (Goodfriend, 1997). The coralline algae show promise for further racemization studies of chronostratigraphy.

Acknowledgments: We thank Alexander Van Geen (Lahmont–Doherty Earth Observatory) for sponsoring the dating of two radiocarbon samples, the Centro Interdisciplinario de Ciencias Marinas (CICIMAR) for providing office space and field equipment, Don Gorsline (University of Southern California) for lending the vibracore, Peter Holden (University of California, Santa Cruz) for providing U/Th dates, and P. E. Hare (Carnegie Institution of Washington) for providing analytical facilities for the amino acid racemization work, and J. C. Ingle, Jr (Stanford University) for his support for J. H. throughout this research. This project was supported by NSF grant No. 9722507 to Stanford University.

NOTES

1. Technically, the process involves epimerization rather than racemization, since D-alloisoleucine and L-isoleucine are epimers, not mirror-image enantiomers. However, for convenience, the process will be referred to as racemization.

REFERENCES

Alvarez-Borrego S. and Lara-Lara J. R. (1991) The physical environment and primary productivity of the Gulf of California: the Gulf and Peninsular Province of the Californias. *Amer. Assoc. Petrol. Geol. Mem.* **47**, 555–567.

Bard E., Hamelin B., and Fairbanks R. (1990) U-Th ages obtained by mass spectrometry in corals from Barbados: sea level during the past 130,000 years. *Nature* **346**, 456–458.

Bosellini A. and Ginsburg R. N. (1971) Form and internal structure of recent algal nodules (rhodolites) from Bermuda. *J. Geol.* **79**, 669–682.

Bosence D. W. J. (1983) Description and classification of rhodoliths (rhodoids, rhodolites). In *Coated Grains* (ed. T. M. Peryt), pp. 217–224. Springer-Verlag.

Bryan J. and Huddlestun P. F. (1991) Correlation and age of the Bridgeboro Limestone, a coralgal limestone from southwestern Georgia. *J. Paleont.* **65**, 864–868.

Carannante G. and Simone L. (1995) Chlorozoan versus foramol lithofacies in Upper Cretaceous rudist limestones. *Palaeogeog. Palaeoclimat. Palaeoecol.* **119**, 137–154.

Carannante G., Esteban M., Milliman J. D., and Simone L. (1988) Carbonate lithofacies as paleolatitude indicators: problems and limitation. *Sed. Geol.*, **60**, 333–346.

Flessa K. W., Cutler A. H., and Meldahl K. H. (1993) Time and taphonomy: quantitative estimates of time-averaging and stratigraphic disorder in a shallow marine habitat. *Paleobiology*.**19**, 266–286.

Fravega P., Piazza M., and Vannucci G. (1993) Importance and significance of the rhodolithic bodies in the Miocenic sequences of Tertiary Piedmont Basin. In *Studies on Fossil Benthic Algae* (eds F. Barattolo et al.), *Boll. Soc. Paleont. Ital.* Special vol. **1**, 197–210.

Goodfriend G. A. (1987) Chronostratigraphic studies of sediments in the Negev Desert, using amino acid epimerization analysis of land snail shells. *Quat. Res.* **28**, 374–392.

Goodfriend G. A. (1989) Complementary use of amino acid epimerization and radiocarbon analysis for dating of mixed-age fossil assemblages. *Radiocarbon* **31**, 1041–1047.

Goodfriend G. A. (1997) Aspartic acid racemization and amino acid composition of the organic endoskeleton of the deep-water colonial anemone *Gerardia*: determination of longevity from kinetic experiments. *Geochim. Cosmochim. Acta* **61**, 1931–1939.

Goodfriend G. A. and Flessa K. W. (1997) Radiocarbon reservoir ages in the Gulf of California: roles of upwelling and flow from the Colorado River. *Radiocarbon* **39**, 139–148.

Goodfriend G. A. and Mitterer R. M. (1993) A 45,000-yr record of a tropical lowland biota: the land snail fauna from cave sediments at Coco Ree, Jamaica. *Geol. Soc. Amer. Bull.* **105**, 18–29.

Goodfriend G. A. and Rollins H. B. (1998) Recent barrier beach retreat in Georgia: dating exhumed salt marshes by aspartic acid racemization and post-bomb radiocarbon. *J. Coastal Res.* **14**, 960–969.

Goodfriend G. A. and Stanley, D. J. (1996) Reworking and discontinuities in Holocene sedimentation in the Nile Delta: documentation from amino acid racemization and stable isotopes in mollusk shells. *Marine Geol.* **129**, 271–283.

Goodfriend G. A., Flessa K. W., and Hare P. E. (1997) Variation in amino acid epimerization rates and amino acid composition among shell layers in the bivalve *Chione* from the Gulf of California. *Geochim. Cosmochim. Acta* **61**, 1487–1493.

Hare P. E., St. John P. A., and Engel M. H. (1985) Ion-exchange separation of amino acids, In *Chemistry and Biochemistry of the Amino Acids* (ed. G. C. Barrett), pp. 415–425. Chapman and Hall.

Kowalewski M., Goodfriend G. A., and Flessa K. W. (1998) High-resolution estimates of temporal mixing within shell beds: the evils and virtues of time-averaging. *Paleobiology* **24**, 287–304.

Lighty R. G., Macintyre J. G., and Stuckenrath R. (1982) *Acropora palmata* reef framework: a reliable indicator of sea level in the Western Atlantic for the past 10,000 years. *Coral Reefs* **1**, 125–130.

Murray-Wallace C. V., Ferland M. A., Roy P. S., and Sollar A. (1996) Unravelling patterns of reworking in lowstand shelf deposits using amino acid racemisation and radiocarbon dating. *Quat. Sci. Rev.* **15**, 685–697.

Ortlieb L. (1987) *Neotectonique et Variations du Niveau Marin au Quaternaire dans la Région du Golfe de Californie, Mexique*. Collection Éditions et Thèses, ORSTOM, Institut Français de Recherche Scientifique pour le Développement en Coopération, Paris.

Ortlieb L. and Malpica Cruz V. (1978) Reconnaissance des dépôts Pléistocènes marins autour du Golfe de Californie, Mexique. *Cah. O.R.S.T.O.M., Ser. Geol.* **10**, 177–190.

Riosmena-Rodriguez R. (1997) A taxonomic reassessment of rhodolith forming species of *Lithophyllum*, in the Gulf of California, Mexico. *Phycologia* **36**, 93.

Sirkin L., Szabo B. J., Padilla A. G., Pedrin. A. S., and Diaz R. E. (1990) Uranium-series ages of marine terraces, La Paz Peninsula, Baja California Sur, Mexico. *Coral Reefs* **9**, 25–30.

Stuiver M. and Reimer P. J. (1993) Extended ^{14}C data base and revised CALIB 3.0 ^{14}C age calibration program. *Radiocarbon* **35**, 215–230.

Wehmiller J. F. and Emerson W. K. (1980) Calibration of amino acid racemization in late Pleistocene mollusks: results from Magdalena Bay, Baja California Sur, Mexico, with dating applications and paleoclimatic implications. *Nautilus* **94**, 31–36.

23. Aminostratigraphic correlation of loess–paleosol sequences across Europe

Eric A. Oches, William D. McCoy, and Dagmar Gnieser

Loess, a wind-blown silt deposit, blankets terraces and uplands adjacent to major river valleys throughout western, central, and eastern Europe and preserves a record of the climatic oscillations associated with repeated major glacial cycles. Details of the Pleistocene continental paleoclimate are revealed through the lithological, pedological and paleontological evaluation of alternating loess–paleosol cycles (e.g., Kukla, 1975; Pécsi, 1987a; Rousseau, 1987; Bronger and Heinkele, 1989, 1990; Ložek, 1990; Rousseau et al., 1990). Some of the most extensive, nearly continuous terrestrial sedimentary records of glacial to interglacial climatic and environmental changes come from the European loess sequences.

Amino acid epimerization studies using fossil gastropods have provided significant chronostratigraphic information for loess–paleosol sequences in central Europe (e.g., Oches and McCoy, 1995a, b), the midwestern U.S. (Clark et al., 1989), and China (W. D. McCoy, R. S. Bradley, and Z.-S. An, in preparation). The subdivision of Quaternary loess–paleosol sequences in central and eastern Europe has traditionally been based on national stratigraphic schemes, and correlation from one country to another has been problematic (c.f. Wintle and Packman, 1988; Frechen et al., 1997; Pécsi, 1992; Zöller et al., 1994). Success in the application of aminostratigraphic techniques to stratigraphic problems elsewhere in central Europe (Oches and McCoy, 1995a, b, c) led us to extend the region of stratigraphic correlations into the Dnieper region of central Ukraine (McCoy et al., 1996) and the Rhine valley and Alpine foreland of southern Germany (Gnieser, 1997) (sites shown in figure 23-1). Building upon our original data set, we further demonstrate the utility of amino acid geochronology in correlating loess–paleosol units across Europe, and we offer clarification of the chronostratigraphic interpretation of loess deposits in this expanded region.

ESTABLISHMENT OF THE CENTRAL EUROPEAN AMINOSTRATIGRAPHY

The application of amino acid geochronological techniques to questions of age and correlation for central European loess deposits represents an independent means of evaluating the stratigraphy and chronology of these long terrestrial records of climatic and environmental change. Ratios of amino acid diastereoisomers D-alloisoleucine and L-isoleucine (A/I ratio) have been measured in fossil mollusks and are able to resolve differences in age between samples from loess of the last four glacial cycles, i.e., samples from the last approximately 450 kyr. Beyond cycle "E" (oxygen isotope stages 10 and 11), the epimerization reaction proceeds much more slowly, and shells tend to be poorly preserved and are often absent altogether (Oches and McCoy, 1995a).

In the Czech Republic, Slovakia, and Austria, A/I values are significantly higher in samples from the penultimate glacial loess (cycle C) relative to samples from loess of the last glacial cycle (cycle B). This increase in A/I is observed in nearly 200 samples from numerous sites throughout Bohemia, Moravia, and Slovakia (Oches and McCoy, 1995a, c). The increase in the ratio reflects both significant age differences between samples and the relatively high effective temperatures of the last interglaciation that affected cycle C samples. This strong distinction between loess of cycles B and C allowed for the aminostratigraphic correlation of the last interglacial paleosol throughout the region (Oches and McCoy, 1995a, b). Amino acid data, therefore, support previous correlations between last interglacial paleosol complexes ("pedocomplex," PK; Kukla, 1975) PK II+III in the Czech Republic and Slovakia and the "Stillfried A" pedocomplex in Austria. Correlations were extended into Hungary, and it was determined that the MF paleosol, as described at Mende and Basaharc (e.g., Pécsi, 1987b; Pécsi and Hahn, 1987), is the last interglacial paleosol in that part of the region (Oches and McCoy, 1995b).

Older loess units were also correlated on the basis of alloisoleucine/isoleucine (A/I) values, and it was determined that Czech and Slovak PKIV (corresponding to oxygen isotope stage 7) correlates with Hungarian BD 1+2; PKV (δ^{18}O-stage 9) = BA; and PKVI (δ^{18}O-stage 11) = MB. This correlation offered a significant reinterpretation of previously held Hungarian chronostratigraphic models (cf. Pécsi, 1987a, 1992). On the basis

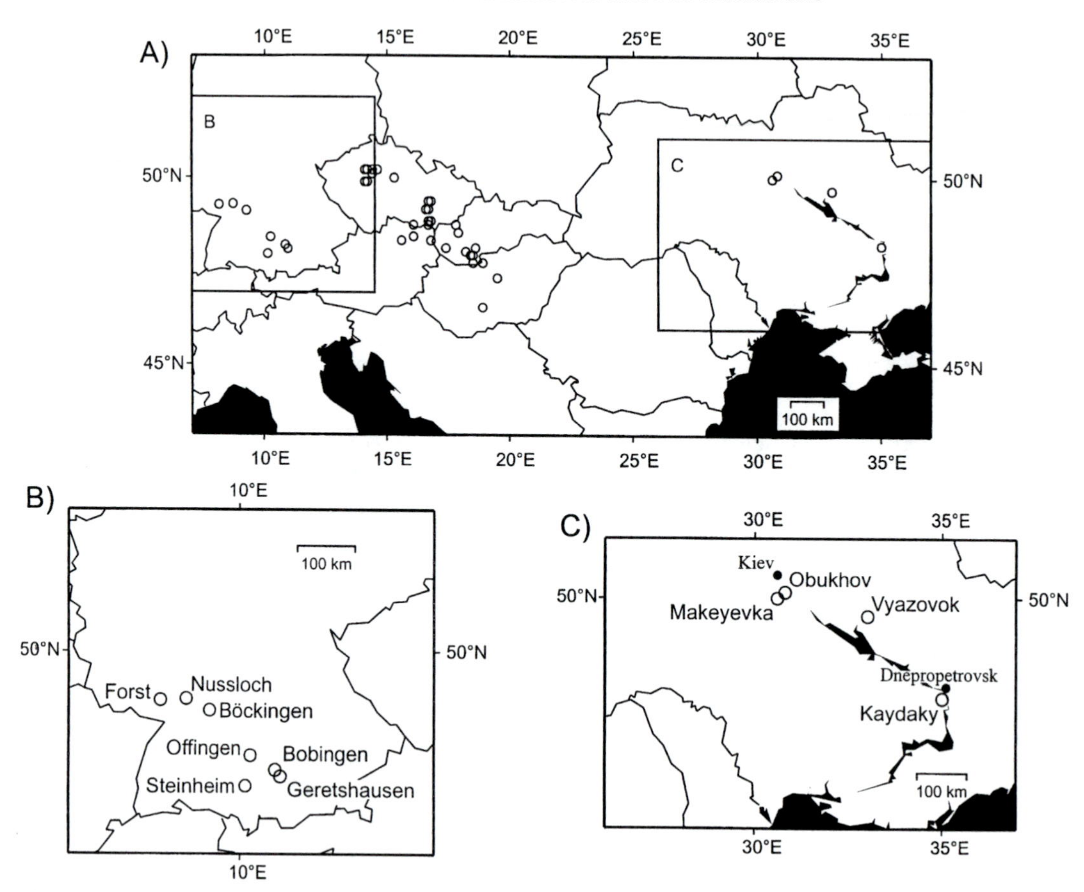

Fig. 23-1. Map of sampled loess–paleosol localities in Europe. Each dot represents one or more sites where the aminostratigraphic studies of loess have been carried out in the last dozen years, including the sites on which we report here. A. All studied sites in western, central and eastern Europe. B. Sampled localities in southern Germany. C. Sampled localities in the middle Dnieper region of Ukraine.

of aminostratigraphic evaluation of the "Young Loess" of Hungary, it was proposed that only the MF and the youngest loess are Late Pleistocene in age, and that the BD, BA, and MB paleosols, plus intervening loess units, are of Middle Pleistocene age (Oches and McCoy, 1995b). For details of the stratigraphic profiles sampled, the statistical evaluations, and the aminostratigraphic interpretations, refer to Oches (1994) and Oches and McCoy (1995a, b, c).

In previous investigations, we have attempted to document the ability of amino acid data to distinguish loess units of the last four glacial cycles. Total hydrolysate (HYD) and free amino acid (FREE) A/I results were evaluated both together and separately, and genera of widely occurring gastropods showing both high and low relative rates of isoleucine epimerization were studied. Data are most conclusive in distinguishing and correlating loess–paleosol units if multiple A/I analyses of HYD and FREE subsamples of different genera are available. However, in most cases, when sample size is limited, the analysis of either the free fraction or the total hydrolysate alone can provide useful information. The aminostratigraphic framework established for Czech, Slovak, Austrian, and Hungarian loess of glacial cycles B, C, D, and E using results from *Succinea* and *Pupilla* is summarized in figure 23-2. A/I data were measured in samples collected from stratigraphic units whose ages had been independently determined. This provides a valuable means by which samples of unknown age and stratigraphic position can be compared and correlated. In this investigation, we expand the existing aminostratigraphic framework for European loess–paleosol sequences eastward into the loess region of Ukraine and westward into southern Germany. We note, however, that the Ukrainian study has only been in the form of reconnaissance work and only tentative correlations can be made at this time. A much larger, systematic program of stratigraphic study and sampling must be carried out to confirm and extend the correlations presented here.

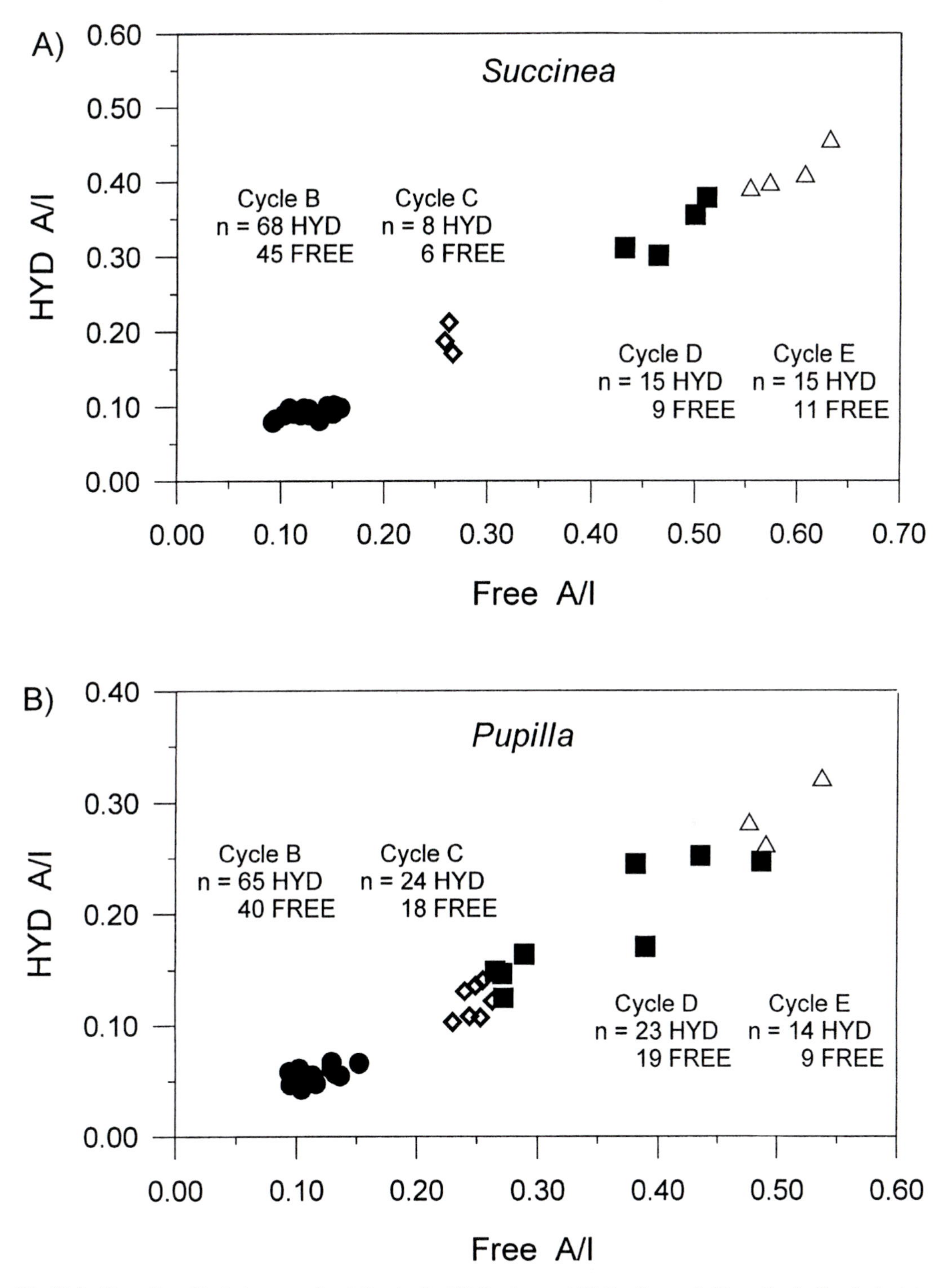

Fig. 23-2. Plots of total hydrolysate vs. free A/I ratios for (A) *Succinea* and (B) *Pupilla* sampled from Czech, Slovak, Austrian, and Hungarian loess units (Oches and McCoy, 1995a, b). Each point represents the mean of several subsamples from different localities. A/I data from *Succinea* specimens, which exhibit relatively rapid epimerization, show more distinct separation between glacial cycles, whereas ratios measured in *Pupilla*, a relatively slowly epimerizing genus, show greater overlap between cycles. The scatter of ratios within a single cycle mainly reflects the range of effective diagenetic temperatures across the region in which samples were collected. Samples from loess units of unknown age can be compared with these data in order to establish aminostratigraphic correlations between discontinuous outcrops.

METHODS

Fossil mollusk shells preserve indigenous proteins and their constituent amino acids, which were initially created as the L-stereoisomer by the living organism. After biomineralization, the protein within the carbonate shell begins to degrade through a series of complex diagenetic chemical reactions (e.g., Mitterer, 1993). Reactions of primary importance to amino acid geochronology include hydrolysis of peptide bonds and racemization or epimerization of constituent amino acids.

Measuring the natural extent of isoleucine epimerization in fossil gastropod shells provides a means by which the age of the enclosing sediments can be evaluated. Isoleucine, plus other protein amino acids, is synthesized by gastropods as part of the biomineralization process of carbonate shell formation. Over time, L-isoleucine epimerizes, or undergoes stereochemical conversion, resulting in the formation of its non-protein diastereoisomer D-alloisoleucine. The ratio of D-alloisoleucine/L-isoleucine (A/I) is therefore a measure of the extent of isoleucine epimerization in a sample and is measured using high-performance liquid chromatography (HPLC) (Hare, 1975).

The degree to which isoleucine has epimerized is primarily a function of the age and diagenetic temperature history of the sample. Within a limited geographic area, where it can reasonably be assumed that temperatures and temperature changes have been uniform across the region, samples with similar A/I values are interpreted as being approximately the same age. This approach to relative-age determination is termed aminostratigraphy (Miller and Hare, 1980). Numerical age estimation requires a knowledge of the post-depositional temperature history experienced by a sample, which is a complex variable for most Pleistocene samples. Alternatively, if independent numerical age-estimates are available, an estimate of the effective diagenetic temperature experienced by a sample since deposition can be calculated (McCoy, 1987).

At the University of Massachusetts Amino Acid Geochronology Laboratory (AGL), A/I ratios are measured in both the free fraction and the total acid hydrolysate in monogeneric gastropod shell samples. Thorough reviews of the principles and applications of amino acid geochronology are provided by Mitterer (1993), Wehmiller (1993), Miller and Brigham-Grette (1989) and McCoy (1987). Details of the methodology and sample preparation procedures of the University of Massachusetts AGL are available elsewhere (Oches, 1994; Oches and McCoy, 1995a, c). For comparison of AGL A/I data with that of other laboratories, the Interlaboratory Comparison sample standards of Wehmiller (1984) are presented in table 23-1.

Because taxonomic differences influence the rate of isoleucine epimerization, shells collected for amino acid analysis must be identified to the genus level (Lajoie et al., 1980). A/I data from three representative genera of fossil terrestrial gastropods have been selected for aminostratigraphic comparison of loess–paleosol units throughout the region. *Succinea* (a genus in which isoleucine epimerizes at a relatively high rate), *Trichia* (in which epimerization proceeds at a moderate rate), and *Pupilla* (in which epimerization is relatively slow) are found in a wide geographic range of Pleistocene loess sections. Though each genus is not present in abundance at each sampled locality, comparisons among these and other genera afford the opportunity to evaluate the relative ages of loess units based on a regional aminostratigraphic framework for the last several glacial cycles.

At each locality, weathered or slumped material was cleared from the profile to provide a fresh, uncontaminated surface for stratigraphic description and sampling. Bulk samples of loess ranging from 5 to 100 kg in size were taken from discrete horizons approximately 10–30 cm in thickness and washed through sieves in the field to recover mollusk shells. Samples were collected at least 1 m below the present surface and the tops of any paleosurfaces (e.g., paleosols) to minimize the effect of the higher effective diagenetic temperatures (EDTs) on shallowly buried shells. In the laboratory, shells from each sample were sorted by genus and prepared for amino acid analysis according to methods described by Oches and McCoy (1995c). In general, two or three replicate subsamples for each type of analysis (free fraction or total acid hydrolysate) for each genus from each sampled horizon were separately prepared and analyzed. However, in some cases only enough shell material for one subsample was recovered. The number of subsamples for each free fraction and for each total acid hydrolysate is given in the data tables. Each subsample contained approximately 10–20 mg of shell material. Therefore, each subsample would contain the entire shell, or fragments of shells, from one to several individuals, depending on their size.

AMINOSTRATIGRAPHIC CORRELATIONS

Aminostratigraphy of the Dnieper Region, Ukraine

Several loess exposures were studied and sampled along the middle Dnieper River Valley from Kiev to

TABLE 23-1 Results of University of Massachusetts AGL analyses of Wehmiller (1984) ILC sample standards

Standard no.	A/I (HYD)	No. of Runs	CV (%)[a]
81-ILC-A	0.149 ±0.006	26	4.0
81-ILC-B	0.489 ±0.015	27	3.0
81-ILC-C	1.070 ±0.023	28	2.2

[a] CV, Coefficient of variation = (S.D./mean) × 100.

Dnepropetrovsk (figure 23-1C). The stratigraphy of Quaternary loesses and paleosols in Ukraine has been summarized by Veklich (1979), and his stratigraphic nomenclature for the Middle and Late Quaternary succession is shown in table 23-2.

Although we washed many samples from sites throughout the region, we failed to recover any mollusk shells from loess deposits lying stratigraphically above the Dnieper loess (table 23-3). Mollusk fossils have previously been reported from the Prichernomorye, Bug, Uday, and Tiasmin loesses at various sites (Veklich, 1979), and perhaps more intensive sampling at the sites we visited would be worthwhile. We were able to collect mollusk shells from Dnieper and Tiligul loesses at various sites.

At Vyazovok near Lubny (figure 23-1C), mollusk shells were recovered from two stratigraphically distinct loess units (figure 23-3A). The upper mollusk-bearing unit is loess that directly underlies till of the Dnieper glaciation. Below that loess is the interglacial Zavadovka paleosol, which, in turn, overlies the Tiligul loess. Analyses of the fossil mollusk shells show that the loess units above and below the Zavadovka soil have distinctly different A/I ratios (table 23-3). The total acid hydrolysate of *Pupilla* shells from the Dnieper-age loess above the paleosol yields an A/I of 0.12 (there was only enough shell material for one subsample). The total acid hydrolysates of *Pupilla* shells taken from the Tiligul loess 6 m below the top of the paleosol give an average value of 0.21. Hydrolysates of *Vallonia* shells, which generally exhibit isoleucine-epimerization rates similar to those of *Pupilla* specimens (Oches, 1994), yield similar A/I ratios in the Tiligul loess. Hydrolysates of *Succinea*, in which isoleucine is known to epimerize at a higher rate (Oches, 1994), show higher A/I ratios from the Tiligul loess, as expected.

The A/I ratio in the free fraction of a mollusk shell is generally significantly higher than the ratio in the total acid hydrolysate of the same shell. At

TABLE 23-2 Stratigraphic nomenclature of Middle- andLate Quaternary loess in Ukraine (Veklich, 1979)

Holocene[a]	hl	Holocene
Prichernomorye[b]	pc	Upper Pleistocene
Dofinovka[c]	df	
Bug[b]	bg	
Vitachev[c]	vt	
Uday[b]	ud	
Priluky[c]	pl	
Tyasmin[b]	ts	
Kaydak[c]	kd	
Dnieper[d]	dn	Middle Pleistocene
Zavadovka[c]	zv	
Tiligul[b]	tl	
Lubny[c]	lb	
Sula[b]	sl	

[a]Soil. [b]Loess. [c]Paleosol. [d]Glacial deposits and loess.

TABLE 23-3 Summary of A/I data from Ukrainian sites

Field sample number	Laboratory number	Genus	A/I (free fraction) Mean	S.D.[a]	N[b]	A/I (hydrolysate) Mean	S.D.[a]	N[b]	Stratigraphic Unit
Vyazovok									
950816-1	AGL2670	*Pupilla*	–[c]			0.12		1	Dnieper
950817-2	AGL2689	*Pupilla*	0.36	0.01	3	0.21	0.01	3	Tiligul
950817-2	AGL2675	*Succinea*	0.48	0.03	3	0.31	0.03	3	Tiligul
950817-2	AGL2690	*Vallonia*	0.41		1	0.22	0.01	2	Tiligul
Obukhov									
950815-7	AGL2673	*Pupilla*	0.18		1	0.10		1	Dnieper
950815-7	AGL2672	*Succinea*	0.21	0.01	2	0.17	0.01	2	Dnieper
Makeyevka									
950815-4	AGL2687	*Pupilla*	0.19	0.02	3	0.11	0.01	3	Dnieper
950815-4	AGL2674	*Succinea*	0.22	0.02	3	0.16	0.03	3	Dnieper
950815-4	AGL2676	*Vallonia*	0.20	0.01	3	0.11		1	Dnieper
Kaydaky									
950819-1	AGL2668	*Pupilla*	–[c]			0.20		1	"Dnieper"[d]
950819-1	AGL2669	*Succinea*	0.46		1	0.31		1	"Dnieper"[d]

[a] S.D., Standard deviation.
[b] N, Number of subsamples prepared and analyzed.
[c] Fossils were not recovered from the unit.
[d] We interpret samples from "Dnieper" loess at Kaydaky to be aminostratigraphically equivalent to samples from Tiligul loess elsewhere.

Vyazovok, the A/I ratio in the free fraction of each subsample is consistent with the pattern established for those genera in central Europe (Oches and McCoy, 1995a). However, there was insufficient *Pupilla* shell recovered from the Dnieper loess at that locality to allow analysis of the free fraction.

At Obukhov and Makeyevka (figure 23-1c), shells were recovered only from what had previously been interpreted as Dnieper loess (Veklich, 1979; A. Ivchenko, personal communication, 1995). These deposits were clearly loesslike sediment associated with glacial till below a strongly developed soil at each site (figure 23-3B, C). The A/I values of the hydrolysates of *Pupilla* at both of these sites were very close to the A/I ratio determined in the same genus at Vyazovok (table 23-3). The A/I ratios of the free fraction of *Pupilla* shells from the Dnieper deposits were similar at Makeyevka and Obukhov. *Vallonia* samples were also collected from Dnieper loess at Makeyevka, and yielded A/I values comparable to those measured in *Pupilla* samples from all three sites. The ratios ascertained from the free fractions and hydrolysates, respectively, of *Succinea* shells from the two sites were also very similar.

Thus, the results of the aminostratigraphic studies at Vyazovok, Makeyevka, and Obukhov have supported the previous correlations of Dnieper deposits between those sites. Furthermore, the significantly higher A/I ratios in shells from loess below the strong soil that underlies the Dnieper loess at Vyazovok supports the interpretation of that soil (Zavadovka) as an interglacial soil.

At Kaydaky (figure 23-1C), south of the Dnieper glacial limit, mollusk shells were recovered from loess that was previously considered to be of Dnieper age (A. Ivchenko, personal communication, 1995) (figure 23-3D). However, the A/I ratios from *Pupilla* and *Succinea* correspond to the ratios found in shells from Tiligul loess at Vyazovok (table 23-3; figure 23-4). The A/I ratio in the total acid hydrolysate of *Succinea* shells taken at Kaydaky is 0.31, which is comparable to the 0.31 ± 0.03 for the three *Succinea* subsamples from Tiligul loess at Vyazovok, yet significantly higher than the means of 0.16 ± 0.03 and 0.17 ± 0.01 for the *Succinea* samples taken from Dnieper loess at Makeyevka and Obukhov, respectively (table 23-3, figure 23-4A). In the free fraction, the mean A/I ratio for *Succinea* from the sampled deposits at

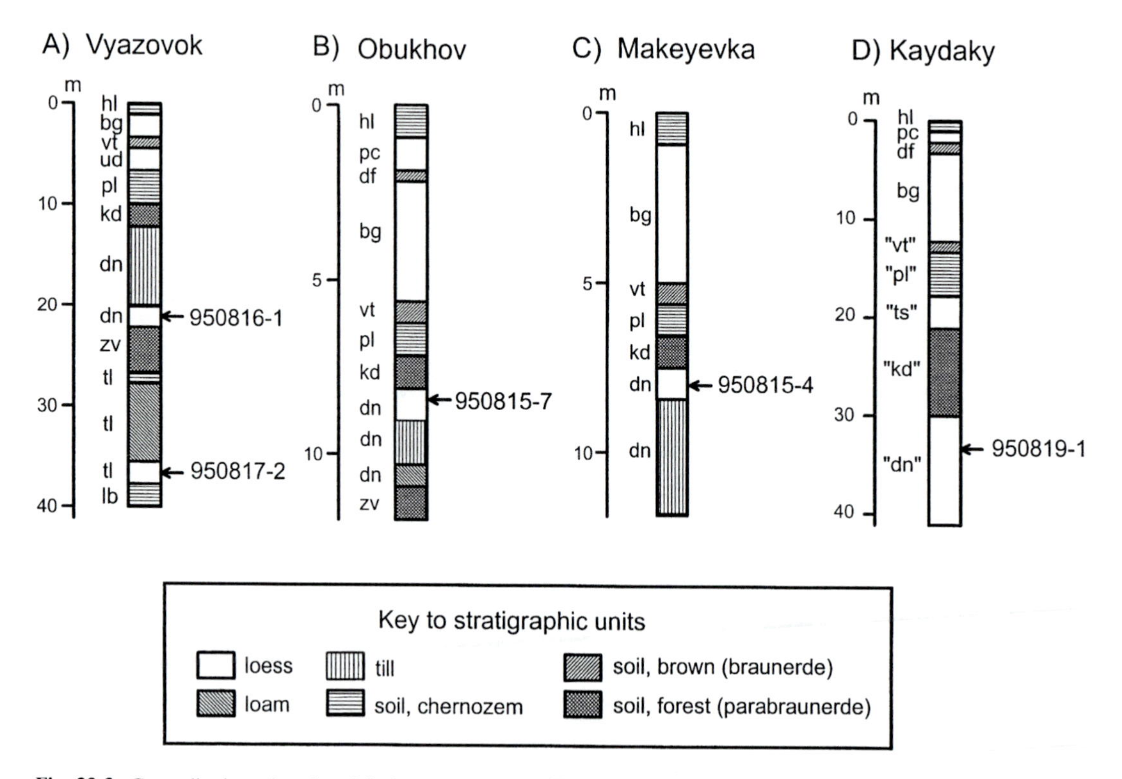

Fig. 23-3. Generalized stratigraphy of the loess exposures and locations of samples at (A) Vyazovok, based on our field notes and the description by Veklich (1979), (B) Obukhov, (C) Makeyevka, and (D) Kaydaky. Only one of many bulk samples from Kaydaky yielded any mollusk shells, and the location of that sample is indicated in the figure. The stratigraphic nomenclature follows that currently in use in Ukraine, which is described in English by Veklich (1979) and is summarized in table 23-2. A summary of the analytical results for all three sites is presented in table 23-3.

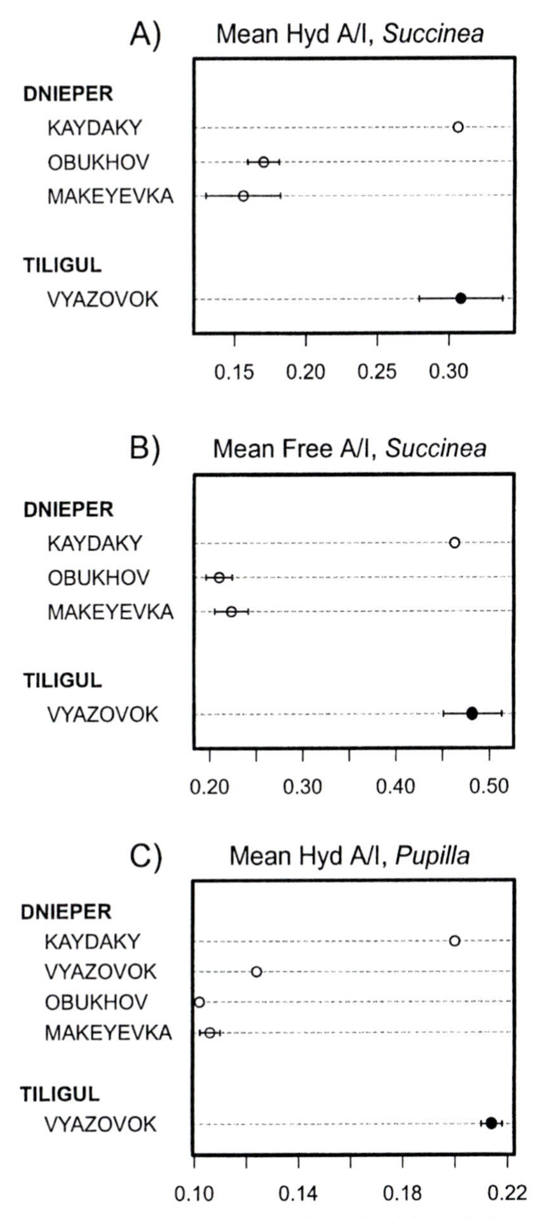

Fig. 23-4. Comparison of A/I ratios at Kaydaky with those at sites to the north of the Dnieper glacial limit. The A/I ratios in (A) the total acid hydrolysate of *Succinea* shells, (B) the free fraction of *Succinea* shells, and (C) the total acid hydrolysate of *Pupilla* shells from the sampled loess at Kaydaky are clearly much greater than those from Dnieper loess (open circles) at other sites and are virtually indistinguishable from those from the Tiligul loess (solid circles) at Vyazovok. Error bars reflect the means and standard deviations of measurements from multiple samples. Points without error bars indicate that only a single sample was available for analysis.

Kaydaky is 0.46, again very similar to the free A/I ratio of 0.48 ±0.03 from *Succinea* taken from the Tiligul loess at Vyazovok (table 23-3, figure 23-4B). The single A/I ratio from *Pupilla* shell samples at Kaydaky supports the *Succinea* evidence (table 23-3, figure 23-4C). Clearly, the sampled horizon at Kaydaky is older than the Dnieper loess sampled at all the other localities, but correlates well with the Tiligul loess at Vyazovok.

Throughout central Europe, the youngest major loess (cycle B) can be clearly distinguished from loess of the penultimate glacial cycle (cycle C) on the basis of A/I in either the free fraction or the total acid hydrolysate of mollusk shells (Oches, 1994; Oches and McCoy, 1995a, b). Because the region of the middle Dnieper is at the same latitude and experiences present-day mean annual temperatures that are roughly similar to those of the northernmost sites previously studied in central Europe, it is reasonable to examine the possible correlations of loess deposits of the major glacial cycles between the two areas, on the basis of isoleucine epimerization in fossil mollusks. Here, we propose correlations of all of the named Middle and Late Pleistocene loesses and paleosols of the middle Dnieper region with the loess stratigraphy of the Czech Republic (table 23-4).

The amino acid data from *Succinea* and *Pupilla* shells from the study sites in Ukraine reported here are compared with the range of A/I ratios found in shells from the Pleistocene loesses of central Europe (figure 23-5, table 23-5). The data points for the Ukrainian Dnieper loess fall within, or very near to, the range of values for cycle C loess in central Europe. Cycle C loess is correlated with marine oxygen isotope stage 6 by Kukla and Cílek (1996). The data points for the Tiligul loess samples from Vyazovok and Kaydaky fall within or near the range of values for cycle D loess from central Europe. Cycle D loess is correlated with marine oxygen isotope stage 8 by Kukla and Cílek (1996).

The interglacial soil complex that separates the Dnieper and Tiligul loesses, the Zavadovka soil complex, is correlated with the PKIV soil complex of marine isotope stage 7. The Zavadovka soil complex is a double soil (Veklich, 1979), as are PKIV and other soils correlated with stage 7, such as the Basaharc Double (BD1 + 2) of Hungary (Oches and McCoy, 1995b) and L2 of the Chinese Loess Plateau (Kukla, 1987). The interglacial soil overlying the Dnieper deposits, the Kaydaky soil, is correlated with PKIII of the Czech Republic.

The above correlations differ significantly from those of Pécsi (1992), who suggests that the Vitachev soil is the Eemian interglacial paleosol, and that the Zavadovka soil correlates with oxygen isotope stage 11. However, we feel that correlations proposed here are consistent with the general lithostratigraphic evidence as well as the aminostratigraphy. The Bug is generally a relatively thick loess and probably represents the major part of cycle B, equivalent to marine isotope stage 2. The Vitachev and Priluky soils are relatively weak and probably do not represent full interglacial conditions. Additional amino acid geochronological

TABLE 23-4 Suggested correlations of Ukrainian loess and paleosol units with the Czech stratigraphy

Ukraine		Czech Republic
Holocene (hl)	Cycle A	PK 0
Prichernomorye (pc)		
Dofinovka (df)		
Bug (bg)	Cycle B	loess
Vitachev (vt)		PK I
Ukay (ud)		loess
Priluky(pl)		PK II
Tyasmin (ts)		loess
Kaydaky (kd)		PK III
Dnieper (dn)	Cycle C	loess
Zavadovka (zv)		PK IV
Tiligul (tl)	Cycle D	loess
Lubny (lb)		PK V
Sula (sl)	Cycle E	loess

data presented by Ivchenko and Mirecki (1997) are in agreement with our chronostratigraphic interpretations.

In summary, although mollusk shells appear to be relatively sparse in the middle Dnieper exposures compared to most of the central European sites that we have visited, the shells that have been recovered have been very useful as an aid to stratigraphic correlation both within the region and between the middle Dnieper and central Europe.

Aminostratigraphy of Loess in southern Germany

Uncertainties exist in the correlation, dating, and climatostratigraphic interpretation of loess sequences in Germany (e.g., Tillmanns et al., 1986; Rögner et al., 1988; Schreiber and Müller, 1991; Frechen, 1999). In an attempt to address the stratigraphic and geochronologic problems of the region, several representative loess–paleosol sequences have been evaluated using amino acid

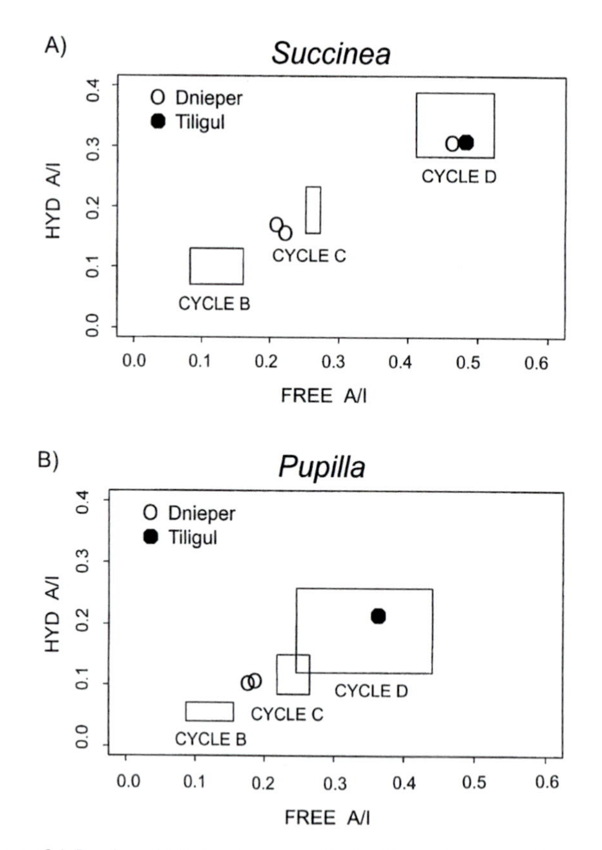

Fig. 23-5. Comparison of A/I ratios of (A) *Succinea* and (B) *Pupilla* shells recovered from middle-Dnieper sites (circles) with the range of ratios of shells recovered from loess of cycles B, C, and D in Austria, the Czech Republic, and Slovakia (boxes) (Oches and McCoy, 1995a). The mean A/I ratio of the total acid hydrolysate is plotted against the mean A/I ratio of the free fraction for each sample. The samples from the Dnieper loess at Obukhov and Makeyevka fall within or near the range of free and hydrolysate values for cycle C loess from central Europe. The Tiligul samples from Vyazovok and the "Dnieper" sample from Kaydaky (which we interpret to be equivalent to Tiligul elsewhere) fall within or near the range of A/I values for cycle D loess from central Europe.

TABLE 23-5A Summary of total hydrolysate D-alloisoleucine/L-isoleucine data

	Mean ± S.D. (no. of preps)					
Genus	Austria[a]	Czech Republic[a]	Slovakia[a]	Hungary[b]	Ukraine[c]	Germany[c]
Cycle B Loess						
Succinea	0.10 ±0.01 (11)	0.09 ±0.01 (30)	0.09 ±0.01 (20)	0.10 ±0.01 (10)	–	0.08 ±0.01 (27)
Trichia	0.07 ±0.01 (13)	0.08 ±0.01 (16)	0.06 ±0.01 (4)	0.10 ±0.02 (22)	–	0.07 ±0.01 (15)
Pupilla	0.06 ±0.01 (6)	0.06 ±0.01 (38)	0.05 ±0.01 (23)	–	–	0.06 ±0.01 (7)
Cycle C Loess						
Succinea	0.20 ±0.02 (5)	0.17 ±0.02 (2)	0.21 (1)	–	0.16 ±0.02 (5)	0.18 ±0.03 (13)
Trichia	0.16 ±0.02 (3)	–	–	0.20 ±0.03 (10)	–	0.15 ±0.02 (7)
Pupilla	0.13 ±0.01 (7)	0.11 ±0.01 (18)	0.12 ±0.01 (7)	–	0.11 ±0.01 (5)	0.12 ±0.02 (14)
Cycle D Loess						
Succinea	–	–	0.35 ±0.01 (7)	0.34 ±0.02 (2)	0.31 ±0.02 (4)	–
Trichia	–	–	–	0.31 ±0.02 (6)	–	–
Pupilla	–	0.16 ±0.01 (2)	0.23 ±0.03 (6)	0.23 ±0.02 (3)	0.21 ±0.01 (4)	–
Cycle E Loess						
Succinea	–	–	0.41 ±0.02 (3)	0.44 ±0.03 (8)	–	–
Trichia	–	–	0.30 ±0.02 (3)	0.37 ±0.02 (6)	–	0.37 ±0.01 (2)
Pupilla	–	–	0.26 ±0.02 (4)	0.28 ±0.06 (8)	–	0.29 ±0.01 (1x2)

[a] Oches and McCoy (1995a).
[b] Oches and McCoy (1995b).
[c] This investigation.

geochronology, and correlations are proposed with the established central European aminostratigraphy. Localities presented here include several sites within the Iller-Lech-Platte, located in the foothills of the German Alps, where Penck and Brückner (1909) developed their classical fourfold subdivision of the Alpine Pleistocene, based on the differentiation of glaciofluvial terraces. Since the fundamental work of Penck and Brückner, the Quaternary history of the German Alpine foreland has continued to be subject to debate and frequent revision (e.g., Sinn, 1972; Schaefer, 1973, 1979; Schreiner and Ebel, 1981). In this region, sediments often can be linked directly to glacial deposits in the Alps. In areas such as the upper Rhine and Neckar basins, the glacial reference is absent, and the chronostratigraphy of the fluvial terraces remains contentious. In those cases, interregional correlation of loess–paleosol sequences associated with the river terraces can provide key chronological data.

As part of our expanding aminostratigraphic investigation of the Pleistocene stratigraphy of Europe, numerous loess–paleosol sequences have been studied and sampled for fossil gastropods in southern Germany (Gnieser, 1997). Seven exposures yielded adequate data for aminostratigraphic comparison with sites elsewhere in the region (figure 23-1B). Data presented here include results from sections at Offingen, Steinheim, Bobingen, and Geretshausen in the Iller-Lech-Platte foothills of the German Alps, Nussloch and Forst in the upper Rhine graben, and Böckingen, in the Neckar Basin (Gnieser, 1997; figure 23-1B). Representative profiles showing typical stratigraphic subdivision and results of amino acid analyses are described for Nussloch, Böckingen, and Offingen (figures 23-6–23-8). Data from all seven sections are included in comprehensive regional aminostratigraphy (table 23-5, figures 23-9 and 23-10). Stratigraphic and geochronologic details of all sampled profiles are provided by Gnieser (1997).

Nussloch is located approximately 10 km south of Heidelberg on the eastern part of the upper Rhine graben (figure 23-1B). This site serves as an important reference locality for the subdivision of Weichselian deposits, and contains several important stratigraphic marker horizons (figure 23-6) including the Late Weichselian "Eltville Tephra", seven Middle to Late Weichselian fossil tundra–gley soils (Nassboden, N1-7), the Middle Weichselian "Lohner Boden" paleosol, an Early Weichselian humic steppe soil, and a Parabraunerde paleosol complex formed during the last interglacial period (Eem) (Bente and Löscher, 1987; Hatté et al., 1998). Luminescence dating by Zöller et al. (1988) provides numerical ages ranging from 22 ± 3 ka (at the level of the Eltville Tephra) to 66.9 ± 5.1 ka (from Lower Weichselian pleniglacial loess). Correlations with other northwest European loess sections, established on the basis of paleopedological, geochemical, and sedimentological analyses plus tephrostratigraphy, also support the chronology of the profile (Hatté et al., 1998).

TABLE 23-5B Summary of free amino acid D-alloisoleucine/L-isoleucine data

Genus	Mean ± Standard Deviation (no. of preps)					
	Austria[a]	Czech Republic[a]	Slovakia[a]	Hungary[b]	Ukraine[c]	Germany[c]
Cycle B Loess						
Succinea	0.12 ±0.03 (6)	0.12 ±0.03 (20)	0.12 ±0.02 (14)	0.13 ±0.02 (8)	–	0.10 ±0.02 (23)
Trichia	0.11 ±0.02 (9)	0.13 ±0.03 (11)	0.11 ±0.04 (4)	0.15 ±0.02 (16)	–	0.12 ±0.02 (13)
Pupilla	0.14 ±0.02 (3)	0.12 ±0.01 (22)	0.10 ±0.01 (15)	–	–	0.10 ±0.01 (5)
Cycle C Loess						
Succinea	0.26 ±0.01 (3)	0.27 (1)	0.26 (1)	–	0.22 ±0.02 (5)	0.23 ±0.03 (10)
Trichia	0.31 (1)	–	–	0.37 ±0.05 (10)	–	0.28 ±0.02 (5)
Pupilla	0.25 ±0.01 (7)	0.24 ±0.02 (12)	0.25 ±0.01 (6)	–	0.19 ±0.01 (4)	0.25 ±0.02 (13)
Cycle D Loess						
Succinea	–	–	0.50 ±0.02 (3)	–	0.48 ±0.02 (4)	–
Trichia	–	–	–	0.56 ±0.05 (4)	–	–
Pupilla	–	0.29 ±0.01 (2)	0.38 ±0.02 (5)	0.49 (1)	0.36 ±0.01 (3)	–
Cycle E Loess						
Succinea	–	–	0.61 ±0.02 (3)	0.62 ±0.04 (5)	–	–
Trichia	–	–	0.57 ±0.02 (3)	0.62 ±0.02 (4)	–	0.57 (1)
Pupilla	–	–	0.49 ±0.01 (3)	0.53 ±0.01 (6)	–	0.48 (1)

[a] Oches and McCoy (1995a).
[b] Oches and McCoy (1995b).
[c] This investigation.

Gastropod shells were sampled from the N5 Nassboden, from loess below the Lohner Boden, and from loess and sand bracketing the last interglacial Parabraunerde (figure 23-6). Total hydrolysate samples from above and below the *Lohner Boden* give comparable A/I ratios within each genus. Mean values for *Succinea* increase very little from an A/I ratio of 0.08 ±0.01 ($n = 4$) in tundra-gley N5 to 0.09 ±0.01 ($n = 3$) in loess directly below the Lohner Boden and 0.10 ($n = 1$) in the sand layer above the last interglacial paleosol complex. *Trichia* and *Pupilla* give results that are statistically indistinguishable across the same interval. When comparing samples bracketing the last interglacial paleosol, significant increases in A/I ratios are evident. *Succinea* HYD values increase to 0.19 ±0.01 ($n = 3$) in the sandy loess of the penultimate glaciation (Saal). *Trichia* HYD A/I values more than double across that same stratigraphic level (figure 23-6). A doubling in FREE A/I ratios is seen in both *Succinea* and *Trichia* collected from above and below the Eemian interglacial paleosol. This doubling in A/I ratio across the last interglacial paleosol is due, in part, to the difference in age between the two samples but is largely attributed to the accelerated epimerization within the older sample driven by the intervening warm interglacial climate. This phenomenon is observed in samples bracketing the last interglacial paleosol throughout the central European loess region and is a useful characteristic for distinguishing a full interglacial period in a stratigraphic profile.

The Böckingen loess exposure is located in an abandoned brickyard near Heilbronn, on top of an early Pleistocene terrace of the Neckar River (figure 23-1B). The well-subdivided exposure is an important section for Weichselian and Saalian stratigraphy in the Neckar basin (Semmel, 1968; Bibus, 1974). Significant horizons include a well-developed tundra–gley paleosol (E2), the Lohner Boden, the Neidereschbacher Zone (redeposited steppe soil), and the Mosbacher Humuszone, a strongly developed, Early Weichselian Chernozem-like forest-steppe paleosol (Bibus, 1989) (figure 23-7). At Böckingen the Lohner Boden is separated into two distinct horizons by a thin sandy layer. Intercalated in 6 m of Saalian loess are six tundra–gley paleosols (N1–N6) and a reworked humic paleosol (Bibus, 1989).

The stratigraphic position of the Eemian interglacial paleosol has recently been challenged (cf. Zöller and Wagner, 1990; Frechen, 1999). Two fossil B_t horizons (Parabraunerde) bracket what had been interpreted as the Saalian loess sequence (Bibus, 1989; Zöller and Wagner, 1990), and two additional interglacial paleosols of similar morphology are identified deeper in the profile. Thermoluminescence age estimates on sediment above the uppermost Parabraunerde, ranging from 22 ±2 ka to 68 ±6 ka (Zöller and Wagner, 1990) provide age control for the Weichselian sediments. These age estimates are generally similar to results obtained from Weichselian sediments at Nussloch (Zöller et al., 1988). A detailed optical-

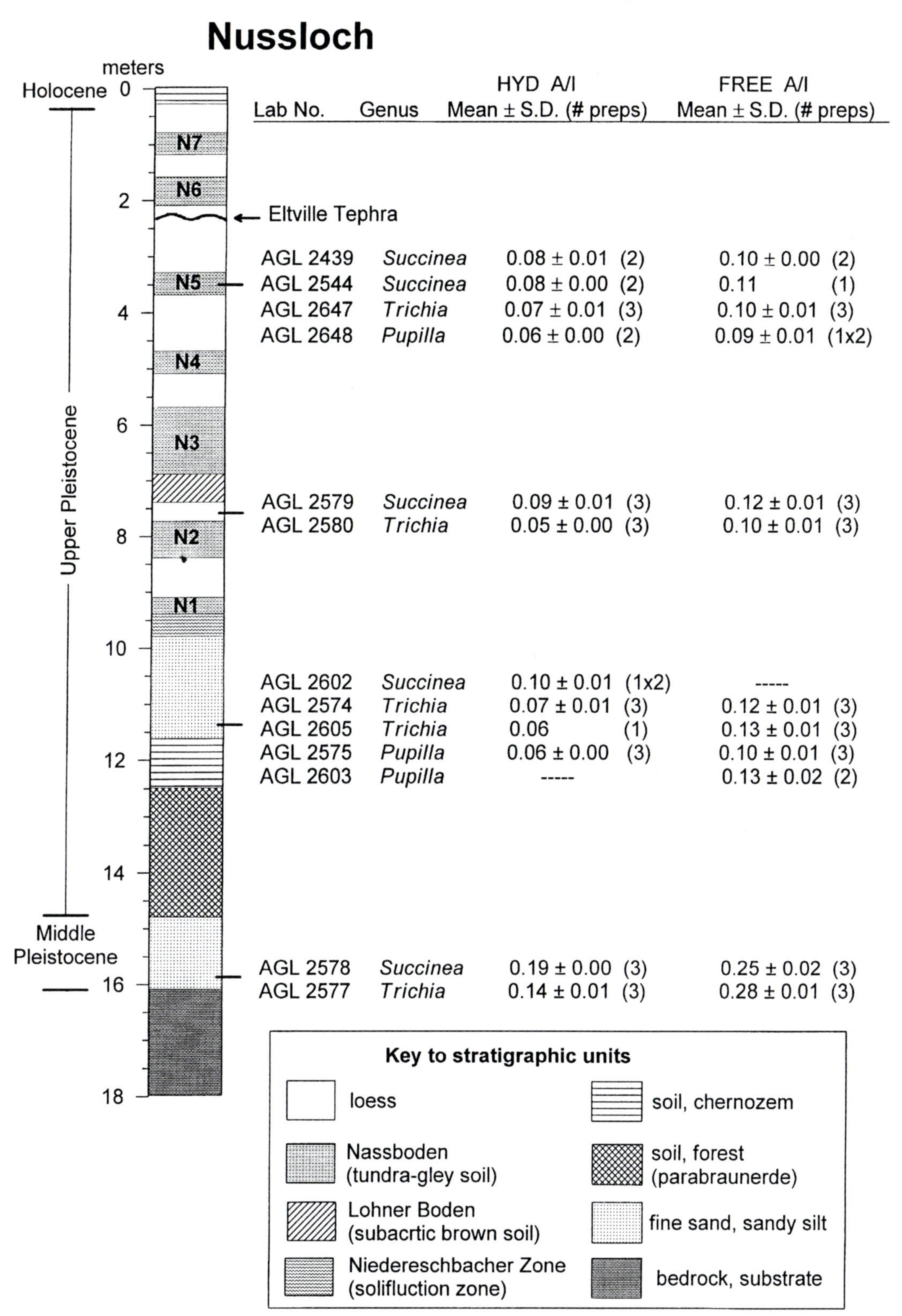

Fig. 23-6. Generalized composite stratigraphy (after Bente and Löscher, 1987; Hatté et al., 1998) and analytical results at Nussloch. Positions of samples are indicated, and A/I ratios are shown for *Succinea*, *Trichia*, and *Pupilla*.

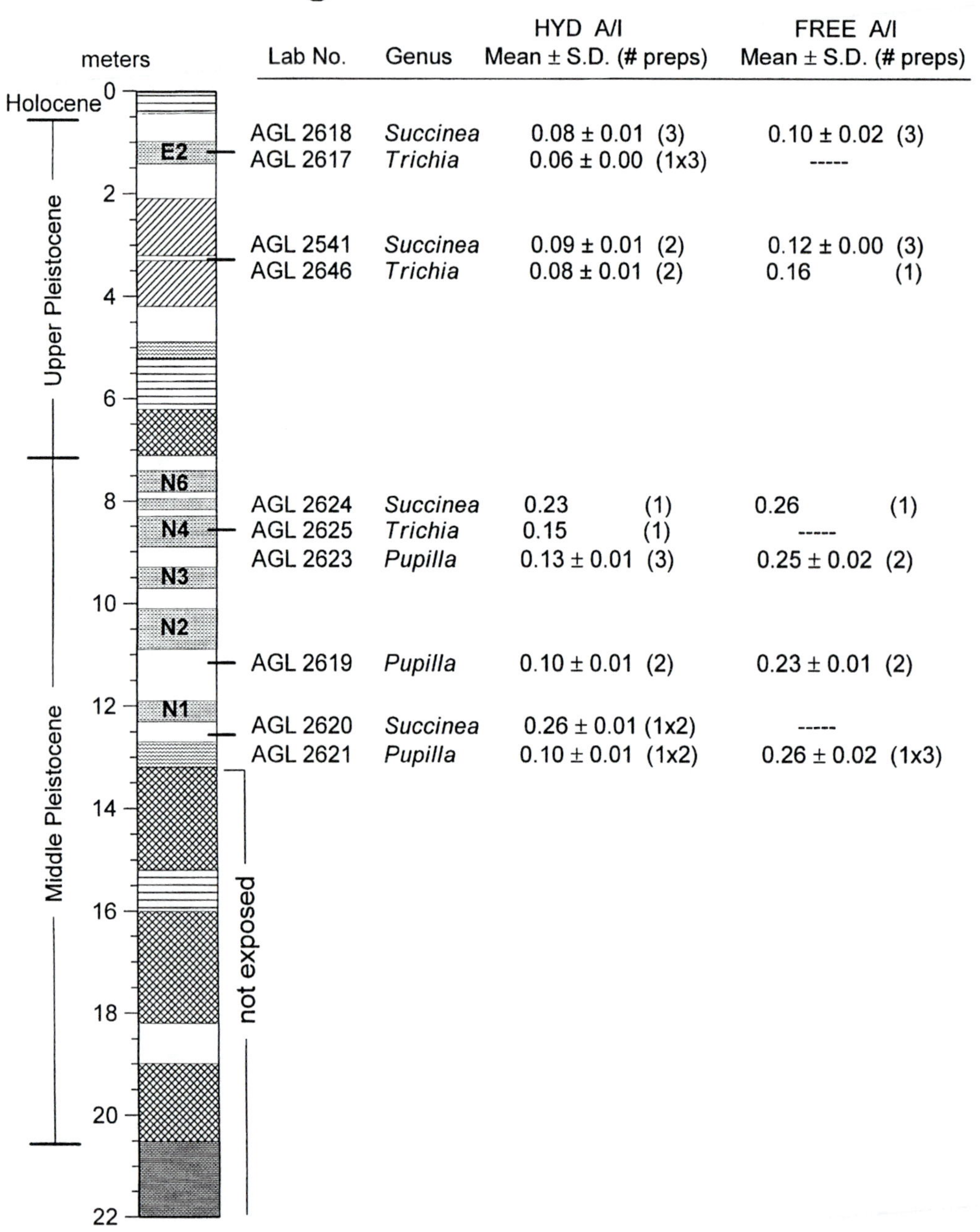

Fig. 23-7. Generalized composite stratigraphy (after Bibus, 1989) and analytical results at Böckingen. Positions of samples are indicated, and A/I ratios are shown for *Succinea*, *Trichia*, and *Pupilla*. A key to the stratigraphic symbols is shown in figure 23-6.

and thermoluminescence dating study by Frechen (1999) suggests that the second B_t horizon could be the Eemian interglacial paleosol, although results are inconclusive because of potential problems with age underestimation.

Shells were collected for amino acid analysis from three layers between the first and second Parabraunerde, and from several horizons within the Weichselian loess series (figure 23-7). Above the upper Parabraunerde, *Succinea* specimens

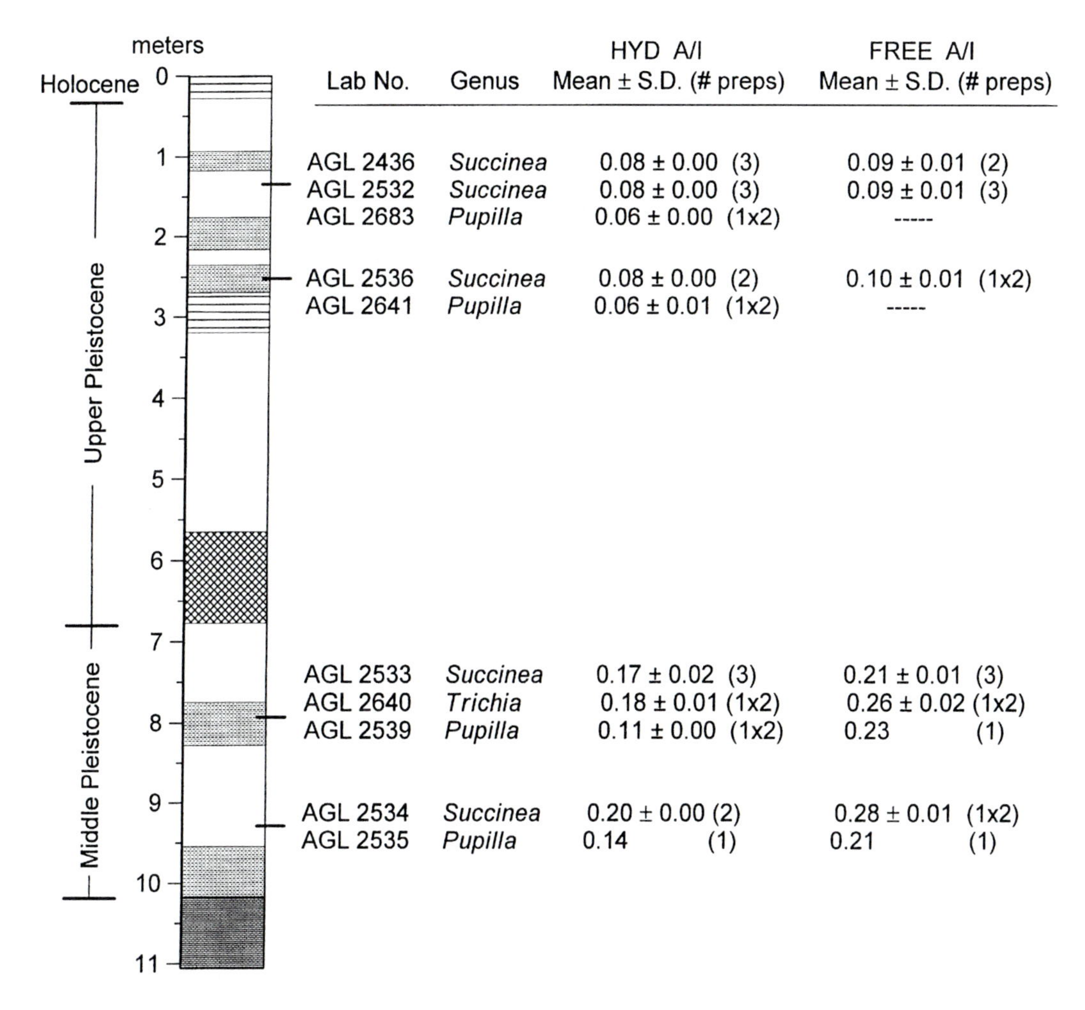

Fig. 23-8. Generalized composite stratigraphy and analytical results at Offingen. Sampled horizons are indicated, and A/I ratios are shown for *Succinea*, *Trichia*, and *Pupilla*. A key to the stratigraphic symbols is shown in figure 23-6.

were recovered from the well-developed tundra–gley paleosol (E2) and from the thin sandy layer subdividing the Lohner Boden. HYD A/I values of 0.08 ±0.01 ($n = 3$) and 0.09 ±0.01 ($n = 2$) were measured in the upper and lower samples, respectively. Free-fraction A/I ratios for *Succinea* and HYD A/I values for *Trichia* increase slightly over this same interval (figure 23-7).

Lower in the section, hydrolysate A/I values in *Succinea* from the upper Parabraunerde are nearly three times higher, with values of 0.23 ($n = 1$) in tundra–gley paleosol N4 and 0.26 ($n = 1$) in loess below N1. Higher A/I values were also measured in *Pupilla* and *Trichia* (figure 23-7). Taken together, the increase in A/I values observed in both HYD and FREE data suggest that the upper B_t horizon is the Eemian interglacial paleosol at this locality.

Offingen is located close to the junction of the Mindel and Danube rivers in the Alpine foreland (figure 23-1B). The loess–paleosol sequence at Offingen reveals loess of the last two glacial cycles (Würm and Riss), separated by a fossil Parabraunerde (Rögner et al., 1988; Stremme et al., 1991) (figure 23-8). Gastropod shells were obtained from several horizons above and below the last interglacial paleosol (figure 23-8). *Succinea* specimens were collected from two horizons above the Middle Würm Lohner Boden and show indistinguishable total hydrolysate A/I values of 0.08 ±0.01 ($n = 3$) and 0.08 ±0.00 ($n = 2$) in the

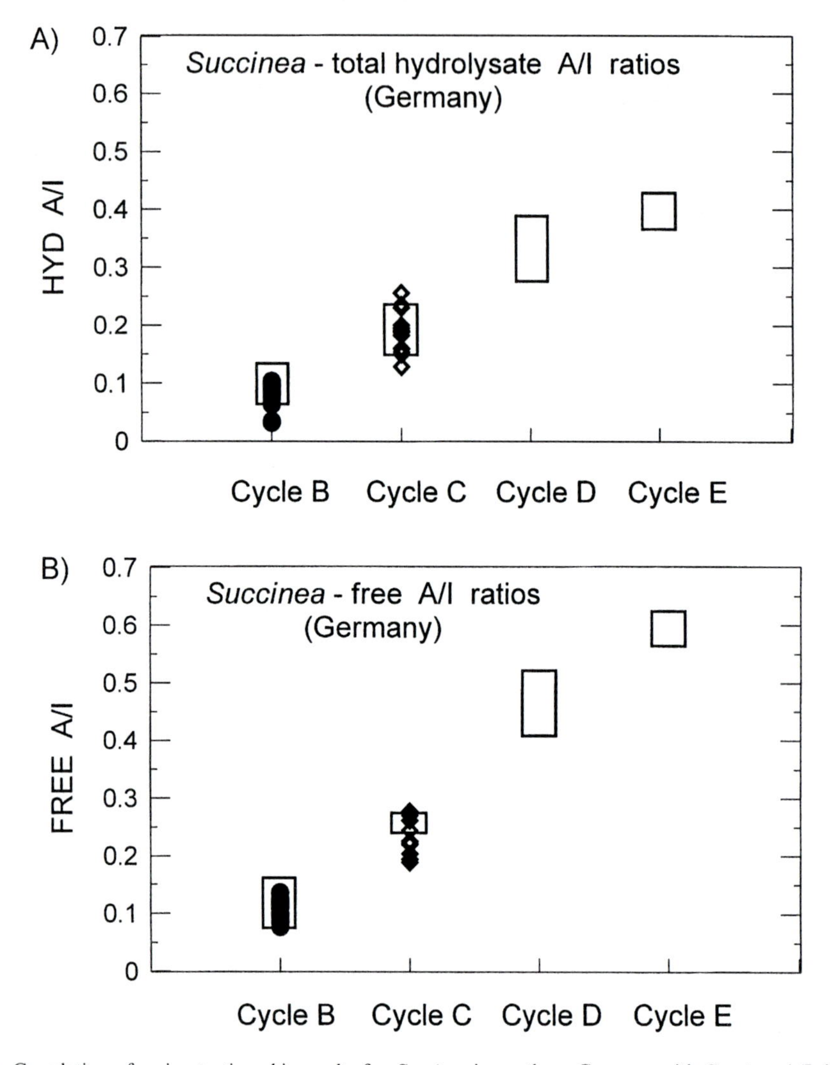

Fig. 23-9. Correlation of aminostratigraphic results for *Succinea* in southern Germany with *Succinea* A/I data from glacial cycles B, C, D, and E in Austria, Slovakia, and the Czech Republic. (A) Total hydrolysate A/I data. (B) Free A/I data. The mean A/I ratio measured in individual samples from southern Germany are plotted as discrete points for each glacial cycle. Data from Oches and McCoy (1995a) are shown as boxes, indicating the range of A/I ratios determined for each glacial cycle.

upper and lower samples, respectively. Below the last interglacial paleosol, *Succinea* specimens were sampled from a tundra–gley paleosol and the loess below, giving HYD A/I ratios of 0.17 ±0.02 ($n = 3$) and 0.20 ±0.01 ($n = 2$), respectively, providing aminostratigraphic confirmation for the last interglacial age of the Parabraunerde paleosol. Similar increases in isoleucine epimerization values are observed in *Pupilla* sampled from the same levels (figure 23-8).

Results from 26 sampled horizons at seven localities in southern Germany demonstrate consistent increases in A/I values in samples of increasing age (separated by tens of thousands of years) or separated by interglacial periods when buried shells would have been exposed to relatively high EDTs. Within a single glacial cycle, A/I ratios show only minor variations within any mollusk genus, suggesting that relatively little isoleucine epimerization occurred during the cold glacial-period climate. In order to demonstrate aminostratigraphic correlations with previously published results from central Europe (Oches and McCoy, 1995a), *Succinea* and *Pupilla* A/I data from German loess–paleosol sequences are plotted together with results from Austria, the Czech Republic, and Slovakia (figure 23-9 and 23-10; table 23-5). Because the study sites have comparable modern mean annual temperature ranges (approx. 8.1–10.1°C, central Europe; 7.8–10.7°C, southern Germany), direct comparison of data is possible across the larger region.

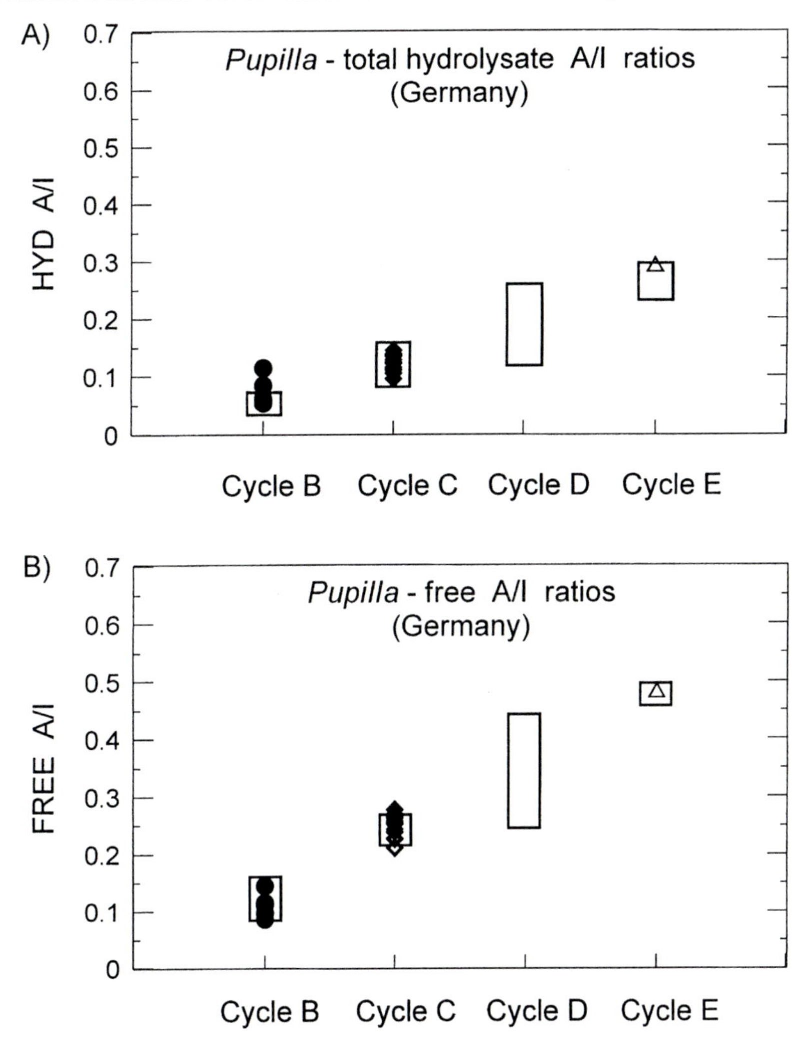

Fig. 23-10. Correlation of aminostratigraphic results for *Pupilla* in southern Germany with *Pupilla* A/I data from glacial cycles B, C, D, and E in Austria, Slovakia, and the Czech Republic. (A) Total hydrolysate A/I data. (B) Free A/I data. The mean A/I ratio measured in individual samples from southern Germany are plotted as discrete points for each glacial cycle. Data from Oches and McCoy (1995a) are shown as boxes, indicating the range of A/I ratios determined for each glacial cycle.

Amino acid data from southern Germany cluster into three distinct groups, correlating with glacial cycles B, C, and E in central Europe (figure 23-9 and 23-10, table 23-5). Comparisons with A/I ratios from central Europe support the German Weichselian and Saalian classification for sediments correlating with cycles B and C (oxygen isotope stages 2–4 and 6, respectively). Saalian-aged samples consistently yield significantly higher A/I values than those from Weichselian sediments, suggesting that there is a considerable age difference between them, and that the aminostratigraphic technique clearly distinguishes between these two glacial cycles. The third group of A/I ratios come from the basal loess at Forst (figure 23-1B) and correlates best with samples from cycle E (oxygen isotope stage 10) loess elsewhere in central Europe (figure 23-10, table 23-5). According to the correlations, none of the samples recovered from the seven study sites in southern Germany was collected from sediments deposited during cycle D.

CONCLUSIONS

The aminostratigraphic framework established for loess–paleosol sequences in Austria, the Czech Republic, Slovakia, and Hungary provides a valuable independent comparison by extending regional lithostratigraphic correlations into eastern and western Europe (figure 23-11). Data summarized for the entire region show comparable total hydro-

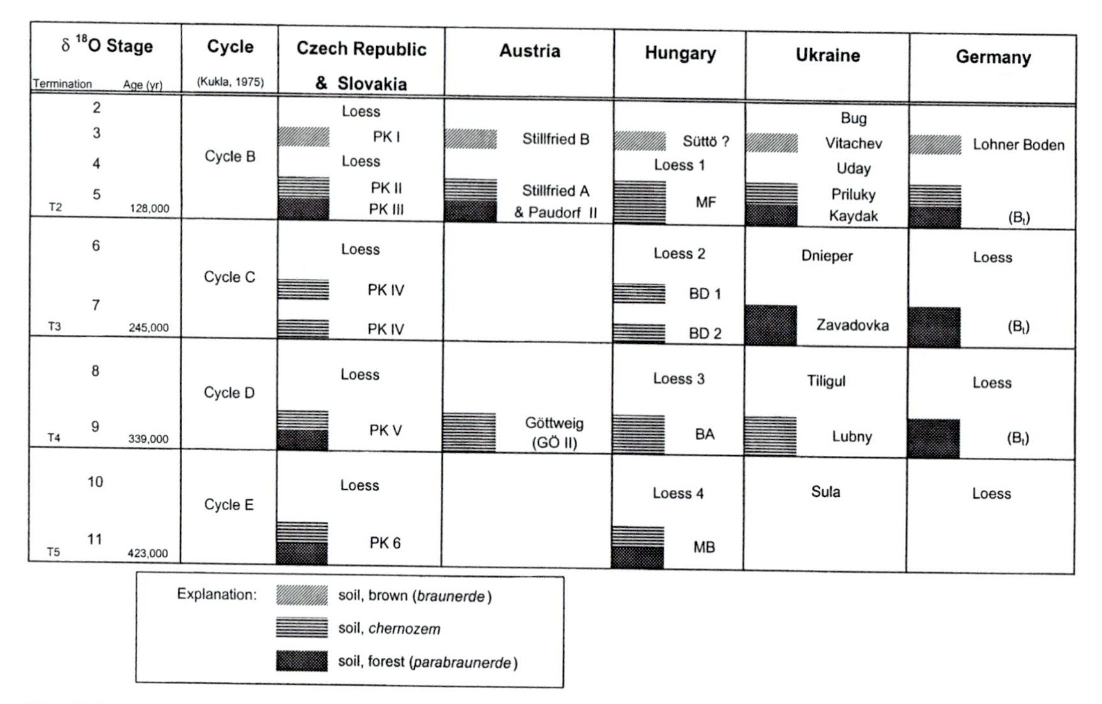

Fig. 23-11. Summary of proposed correlations of loess and paleosol units throughout Europe, based on previously published aminostratigraphic data (Oches and McCoy, 1995a, b) and new results presented here. Regional stratigraphic unit names are shown where they have been designated.

lysate and free-fraction A/I values in the representative mollusk genera *Succinea*, *Trichia*, and *Pupilla* collected from sediments deposited during individual glacial cycles (table 23-5). Consistent increases in A/I values are observed between samples collected from loess of glacial cycles B and C, reflecting the considerable difference in age between samples, as well as the effect of the relatively high interglacial temperatures on the epimerization reaction in buried cycle C shells. A/I measurements can be used confidently to distinguish among the last four glacial cycles. In samples older than those of cycle E (deposited prior to oxygen isotope stage 11), the epimerization reaction proceeds much more slowly, accompanied by a corresponding decrease in the temporal resolution of the method. Overall, these data allow the aminostratigraphic correlation of Middle to Late Pleistocene loess and paleosol units across the region (figure 23-11) and offer critical insights into the chronostratigraphic evaluation of sedimentary records of Quaternary climate and environmental change throughout Europe.

Acknowledgments: Much of this research was supported by NSF Grant INT-9012993, a University of Massachusetts Research Program grant to W.D.M., and an NSF–NATO fellowship to E.A.O. D.G. received support from a Sigma Xi Grant-in-Aid of Research and from the Geological Society of America. Numerous individuals representing geological programs in the Czech Republic, Slovakia, Hungary, Germany, and Ukraine deserve our thanks for their support in the field research. Special thanks go to Vaclav Cílek, Vojen Ložek, Marton Pécsi, Andreji Ivchenko, Ludwig Zöller, and George Kukla for their guidance in the field. We also thank Denis-Didier Rousseau and June Mirecki for their thorough and helpful reviews of the manuscript, and John Wehmiller for editorial support.

REFERENCES

Bente B. and Löscher M. (1987) Sedimentologische, pedologische und stratigraphische Untersuchungen an Lössen südlich Heidelberg. *Göttinger Geograph. Abhandl.* **84**, 9–17.

Bibus E. (1974) Abtragungs- und Bodenbildungsphasen im Risslöss. *Eiszeitalt. Gegenwart* **25**, 166–182.

Bibus E. (1989) 8. Tagung des Arbeitekreises "Paläobaöden" der Deutschen Bodenkundlichen Gessellschaft. *Geographisches Institut der Universität Tübingen, Programm und Exkursionsführer.*

Bronger A. and Heinkele T. (1989) Paleosol sequences as witnesses of Pleistocene climatic history. *Catena* Suppl. **16**, 163–186.

Bronger A. and Heinkele T. (1990) Mineralogical and clay mineralogical aspects of loess research. *Quat. Intl.* **7/8**, 37–51.

Clark P. U., Nelson A. R., McCoy W. D., Miller B. B., and Barnes D. K. (1989) Quaternary aminostratigraphy of Mississippi Valley loess. *Geol. Soc. Amer. Bull.*, **101**, 918–926.

Frechen M. A. (1999) Upper Pleistocene loess stratigraphy in southern Germany. *Quat. Geochronol.* **18**, 243–269.

Frechen M. A., Horváth E., and Gábris G. (1997) Geochronology of middle and upper Pleistocene loess sections in Hungary. *Quat. Res.*, **48**, 291–312.

Frechen M. A., Zander A., Cílek V., and Ložek V. (1999) Loess chronology of the last interglacial/glacial cycle in Bohemia and Moravia, Czech Republic. *Quat. Sci. Rev.* **18**, 1467–1493.

Gnieser D. (1997) Aminostratigraphy of loess–paleosol sequences in southern Germany. M.S. Thesis, University of Massachusetts, Amherst, USA.

Hare P. E. (1975) Amino acid composition by column chromatography. In *Molecular Biology, Biochemistry, and Biophysics*, vol. 8 (ed. S. B. Needleman), pp. 204–231. Springer-Verlag.

Hatté C., Fontugne M., Rousseau D. D., Antoine P., Zöller L., Tisnérat-Laborde N., and Bentaleb I. (1998) $\delta^{13}C$ variations of loess organic matter as a record of the vegetation response to climatic changes during the Weichselian. *Geology* **26**, 583–586.

Ivchenko A. and Mirecki J. (1997) Amino acid geochronology of the Middle–Upper Quaternary deposits associated with glacial deposits of Ukraine. In *Quaternary Deposits and Neotectonics in the Area of Pleistocene Glaciations*, Abstracts of the Field Symposium, Institute of Geological Sciences, p. 17. Academy of Sciences of Belarus, Minsk.

Kukla G. J. (1975) Loess stratigraphy of central Europe. In *After the Australopithecines* (eds K. W. Butzer and G. L. Isaac), pp. 99–188. Mouton.

Kukla G. (1987) Loess stratigraphy in central China. *Quat. Sci. Rev.* **6**, 191–219.

Kukla G. and Cílek V. (1996) Plio-Pleistocene megacycles: record of climate and tectonics. *Palaeogeog. Palaeoclim. Palaeoecol.* **120**, 171–194.

Lajoie K., Wehmiller J. F., and Kennedy G. J. (1980) Inter- and intrageneric trends in apparent racemization kinetics in Quaternary mollusks. In *Biogeochemistry of Amino Acids* (eds P. E. Hare, T. C. Hoering, and K. King, Jr), pp. 405–440. Wiley.

Ložek V. (1990) Mollusks in loess, their paleoecological significance and role in geochronology—principles and methods. *Quat. Intl.* **7/8**, 71–79.

McCoy W. D. (1987) The precision of amino acid geochronology and paleothermometry. *Quat. Sci. Rev.* **6**, 43–54.

McCoy W. D., Oches E. A., and Ivchenko A. (1996) Correlation of central Ukrainian loess–paleosol sequences with those of central Europe. *Geol. Soc. Amer. Abstr. Programs* **28**, A305.

Miller G. H. and Brigham-Grette J. (1989) Amino acid geochronology: resolution and precision in carbonate fossils. *Quat. Intl.* **1**, 111–128.

Miller G. H. and Hare P. E. (1980) Amino acid geochronology: Integrity of the carbonate matrix and potential of molluskan fossils. In: *Biogeochemistry of Amino Acids* (eds P. E. Hare, T. C. Hoering, and K. King, Jr), pp. 401–443. (Wiley).

Mitterer R. M. (1993) The diagenesis of proteins and amino acids in fossil shells. In: *Organic Geochemistry* (eds M. H. Engel and S. A. Macko), pp. 739–753. Plenum Press.

Oches E. A. (1994) Aminostratigraphy, geochronology, and paleoclimatology of Central European loess deposits. Ph.D. Thesis, University of Massachusetts, Amherst, USA.

Oches E. A. and McCoy W. D. (1995a) Amino acid geochronology applied to the correlation and dating of central European loess deposits. *Quat. Sci. Rev.* **14**, 767–782.

Oches E. A. and McCoy W. D. (1995b) Aminostratigraphic evaluation of conflicting age estimates for the "Young Loess" of Hungary. *Quat. Res.* **43**, 160–170.

Oches E. A. and McCoy W. D. (1995c) Aminostratigraphy of Central European loess cycles: introduction and data. *Geolines* **2**, 34–86.

Pécsi M. (1987a) The loess–paleosol and related subaereal sequence in Hungary. *GeoJournal* **15**, 151–162.

Pécsi M. (1987b) Type-locality of Young Loess in Hungary at Mende. In *Paleogeography and Loess* (eds M. Pécsi and A. A. Velichko), pp. 35–53. Akadémiai Kiadó.

Pécsi M. (1992) Loess of the last glaciation. In *Atlas of Paleoclimates and Paleoenvironments of the Northern Hemisphere* (eds B. Frenzel, M. Pécsi, and A. Velichko), pp. 111–119. Geographical Research Institute, Hungarian Academy of Sciences.

Pécsi M. and Hahn G. (1987) Paleosol stratotypes in the upper Pleistocene loess at Basaharc, Hungary. *Catena*, Suppl. **9**, 95–102.

Penck A. and Brückner E. (1909) *Die Alpen im Eiszeitalter*. Tauchnitz.

Rögner K., Löscher M., and Zöller L. (1988) Stratigraphie, Paläogeographie und erste Thermolumineszenzdatierungen in der westlichen Iller-Lech-Platte (Nördliches Alpenvorland, Deutschland). *Zeits. Geomorph, N.F.*, Suppl.-Bd. **70**, 51–73.

Rousseau D. D. (1987) Paleoclimatology of the Achenheim series (middle and upper Pleistocene, Alsace, France). A malacological analysis. *Palaeogeog., Palaeoclim., Palaeoecol.* **59**, 293–314.

Rousseau D. D., Puisségur J. J., and Lautridou J. P. (1990) Biogeography of the Pleistocene pleniglacial malacofaunas in Europe: stratigraphic and climatic implications. *Palaeogeog. Palaeoclim. Palaeoecol.* **80**, 7–23.

Schaefer I. (1973) Das Grönenbacher Feld—ein Beispiel für Wandel und Fortschritt der Eiszeitforschung seit Albrecht Penck. *Eiszeitalt. Gegenwart* **23/24**, 168–200.

Schaefer I. (1979) Das Warmisrieder Feld. *Quartär* **29/30**, 15–47.

Schreiber U. and Müller D. (1991) Mittel- und jungpleistozäne Ablagerungen zwischen Landsberg und Augsburg (LECH). *Sonderveröff. Geolog. Inst. Univ. Köln* **82**, 265–282.

Schreiner A. and Ebel R. (1981) Quartärgeologische Untersuchungen in der Umgebung von Interglazialvorkommen im östlichen Rheingletschergebiet (Baden-Württemberg). *Geol. Jahrbuch* **A59**, 3–64.

Semmel A. (1968) Studien über den Verlauf jungpleistozäner Formung in Hessen. *Frankfurter geographische Hefte* **45**, 133 pp.

Sinn P. (1972) Zur Stratigraphie und Paläogeographie des Präwürm im mittleren und südlichen Illergletscher-Vorland. *Heidelberger Geograph. Arbeiten* **37**, 159 pp.

Stremme H. E., Zöller L., and Krause W. (1991) Bodenstratigraphie und Thermolumineszenz-Datierungen für das Mittle- und Jungpleistozän desl Alpenvorlandes. *Sonderveröff. Geolog. Inst. Univ. Köln* **82**, 301–315.

Tillmanns W., Koci A., and Brunnacker K. (1986) Die Brunhes-Matuyam-Grenze in Rosshaupten (Bayerisch Schwaben). *Jahresber. Mitteil. oberrhein. geol. Vereins N.F.* **68**, 241–247.

Veklich M. F. (1979) Pleistocene loesses and fossil of the Ukraine. *Acta Geol. Acad. Sci. Hung.* **22**, 35–62.

Wehmiller J. F. (1984) Interlaboratory comparison of amino acid enantiomeric ratios in fossil Pleistocene mollusks. *Quat. Res.* **22**, 109–120.

Wehmiller J. F. (1993) Applications of organic geochemistry for Quaternary research: Aminostratigraphy and aminochronology. In: Organic Geochemistry (eds M. H. Engel and S. A. Macko), pp. 755–783. Plenum Press.

Wintle A. G. and Packman S. C. (1988) Thermoluminescence ages for three sections in Hungary. *Quat. Sci. Rev.* **7**, 315–320.

Zöller L. and Wagner G. (1990) Thermoluminescence dating of loess—recent developments. *Quat. Intl.* **7/8**, 119–128.

Zöller L., Stremme H., and Wagner G. (1988) Thermolumineszenz-Datierung an Löss-Paläoboden-Sequenzen von Nieder-, Mittel-, und Oberrhein/Bundesrepublik Deutschland. *Chem. Geology (Isotope Geosci. Sec.)* **73**, 39–62.

Zöller L., Oches E. A., and McCoy W. D. (1994) Towards a revised chronostratigraphy of loess in Austria with respect to key sections in the Czech Republic and in Hungary. *Quat. Sci. Rev.* **13**, 465–472.

24. Aspartic acid racemization and protein preservation in the dentine of Pleistocene European bear teeth

Trinidad Torres, Juan F. Llamas, Laureano Canoira, and Pilar García-Alonso

The dating of palaeontological materials from Pleistocene mammals has been considered one of the fundamental objectives for palaeontologists, archaeologists, and stratigraphers dedicated to analysis of the Quaternary. This is particularly true in Spain, which has an extraordinary abundance of fossil Pleistocene mammals, albeit without any stratigraphical context: the fossils are dispersed in fluvial deposits, in cave fillings, or on marine terraces. A curious feedback takes place in dating: the fossil mammals are used to date deposits whose topographical situation serves, in turn, to date the processes. On many occasions, the stratigraphic positioning of some faunal remains is no more than palaeontological conjecture. In fact, erroneous use is made of the Quaternary biostratigraphy and chronostratigraphy. The lack of volcanic materials suitable for radiometric dating leads to the employment of U-series analysis of teeth, bones, caliches, and nodules; however, such open geochemical systems, are subject to intense diagenesis, so the reliability of the data is questionable.

A large number of Spanish caves contain abundant remains of fossil mammals. In many cases the fossil remains are associated with lithic artifacts from the Upper Paleolithic, or ceramics, which allow precise dating. However, with increasing sample age, dating becomes less precise as lithic industries are less temporally resolved, while frequent monospecific faunal assemblages (e.g., of bears, hyenas, lions) preclude dating of the faunal assemblage.

Amino acid racemization (AAR) analysis offers one possible approach to this problem. There are previous reports of AAR analysis in bones: Bada (1972, 1973, 1985), Dungworth et al. (1976), King and Bada (1979), Belluomini (1981), McMenamin et al. (1982), Belluomini (1985), Belluomini and Bada (1985), Elster et al. (1991), Kimber and Hare (1992), and Meyer (1992). Belluomini and Bada (1985), Blackwell et al. (1990) and Bada et al. (1991) worked on AAR analysis of tooth enamel. Sinibaldi et al. (1998) worked on AAR in dentine.

We have chosen to work on the racemization of aspartic acid in the dentine of bears of the Pleistocene and Holocene because the genus is omnipresent throughout the European and Iberian Quaternary, and because large amounts of faunal material are often present. The Madrid School of Mines Museum conserves one of the most complete collections of fossil bears of the Quaternary. Dentine has been selected for the analysis of the racemization of the amino acids because enamel sampling is very destructive, and bones are open systems which we believe to be more prone to contamination than dentine.

In this study, we aim to clarify systematic and stratigraphical problems related with cave bears in Europe, and have therefore analyzed the aspartic acid ratios of teeth from almost all of the main Spanish localities.

PHYLOGENY AND GEOGRAPHICAL DISTRIBUTION OF CAVE BEARS

Although the evolutionary history of European Pleistocene bears began during the Pliocene, it was during the Middle Pleistocene that bears, mostly cave bears, became dominant in the European palaeontological record (figure 24-1). *Ursus deningeri* (Von Reichenau) was the Middle Pleistocene representative of this evolutionary line-

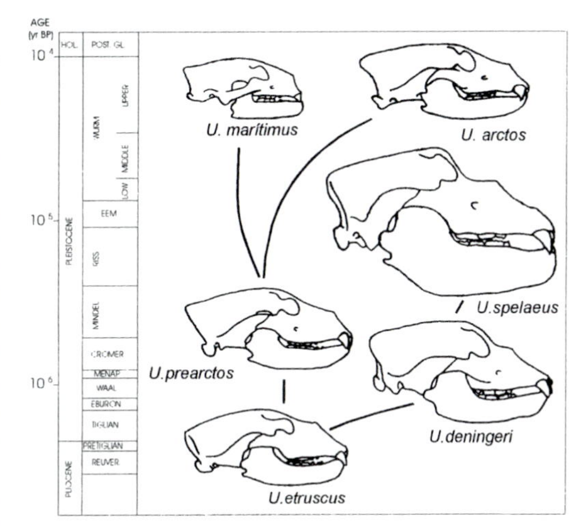

Fig. 24-1. Phylogeny of Pliocene and Pleistocene European bears.

age. The true cave bear, *Ursus spelaeus* (Rosenmüller-Heinroth), appeared during the Upper Pleistocene, vanishing before the Holocene. Speloid bears were coeval with the representatives of the arctoid lineage.

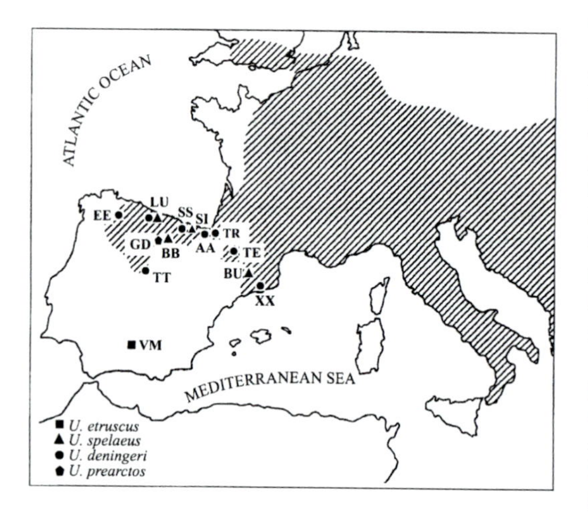

Fig. 24-2. Distribution of *U. deningeri* (von Reichenau) and *U. spelaeus* (Rosenmüller-Heinroth) in Europe and Spain. Locations of sample sites are indicated. (See table 24-1 for abbreviations.)

Although the arctoid bears (*Ursus etruscus*, *Ursus prearctos* and *Ursus arctos*) covered the whole of Europe, the distribution of speloid bears (*U. deningeri* and *U. spelaeus*) was restricted by the extent of the polar ice sheet during the glacial maximum. On the Iberian Peninsula, four areas of occupation have been defined (figure 24-2), sampling localities are given in table 24-1.

Distinguishing between arctoid and speloid representatives is straightforward, because of the marked morphological differences in their dentition. *U. etruscus* and the modern *U. arctos* can be distinguished by multivariate analysis of their dental morphology. *U. etruscus* and *U. arctos* appeared in very different geological/palaeontological contexts, which simplifies their separation. The task is more difficult when one tries to separate *U. prearctos* from *U. etruscus* and *U. arctos*, since intermediate forms are difficult to locate.

According to our data, there are striking morphological and dimensional differences between *U. deningeri* and *U. spelaeus* (Torres, 1989, 1992). Although the strong sexual dimorphism found in bears is recognized, dental and skeletal analyses reveal that *U. deningeri* was more primitive and slender. *U. deningeri* possesses strikingly long first metacarpal and metatarsal bones, differing radically from those of *U. spelaeus*; this facilitates identification. Numerous dimensional and morpho-

TABLE 24-1 Geographical situation of the localities sampled, current mean annual temperature (CMAT) range, and aspartic acid reacemization-ratio descriptive statistics

Localities	Coordinates		CMAT		D/L-Asp[a]				
	Latitude	Longitude	Altitude interval (m)	(C°)	*n*	m	S.D.	min.	max.
Ursus deningeri									
BB Sima Huesos	42° 20′ 55.4″ N	3° 0′ 4.3″ W	1022	10–12.5	15	0.32	0.059	0.23	0.43
SI Santa Isabel	43° 15′ 27.1″ N	3° 0′ 4.4″ W	300	10–12.5	6	0.37	0.071	0.24	0.42
BU Bunica	42° 23′ 45.3″ N	1° 48′ 59.8″ E	1200	7–10	2	0.27	0.035	0.25	0.30
LU La Lucia	44° 54′ 2.6″ N	4° 33′ 37.3″ W	510	12.5–15	3	0.33	0.030	0.30	0.35
Ursus spelaeus									
LU La Lucia	44° 54′ 2.6″ N	4° 33′ 37.3″ W	510	12.5–15	5	0.13	0.032	0.08	0.17
EE Eirós	42° 15′ 58.5″ N	7° 11′ 42.8″ W	780	10–12.5	5	0.07	0.008	0.05	0.08
SS La Pasada	44° 50′ 22.7″ N	1° 19′ 15.8″ W	460	12.5–15	4	0.10	0.038	0.06	0.15
AA Arrikrutz	43° 4′ 11.6″ N	3° 31′ 50.1″ W	453	10–12.5	6	0.19	0.048	0.14	0.26
TR Troskaeta	43° 0′ 7.3″ N	2° 15′ 1.4″ W	580	12.5–15	4	0.09	0.012	0.07	0.10
TE Coro Tracito	42° 37′ 10.2″ N	5° 49′ 8.2″ W	1600	7–10	5	0.08	0.023	0.05	0.11
XX El Toll	42° 11′ 5.7″ N	2° 12′ 31.4″ E	740	12.5–15	4	0.10	0.010	0.09	0.11
IT El Reguerillo	40° 53′ 13.6″ N	3° 36′ 33.8″ W	840	10–12.5	8	0.19	0.068	0.12	0.28
Ursus prearctos									
GD Gran Dolina	42° 20′ 34″ N	3° 0′ 4.3″ W	1000	10–12.5	7	0.23	0.079	0.14	0.36
Ursus etruscus									
VM Venta Micena	37° 44′ 32″ N	2° 24′ 55.3″ W	940	10–12.5	2		absent		

[a] *n*, Number of samples; m, mean; S.D., standard deviation.

logical characters also appear in the elements of their dentition (Torres and Cervera, 1995).

Comparison of the dental morphologies of *U. spelaeus* samples from different localities has also revealed the persistence of minor morphological features, allowing a certain "tribal" component to be identified, as a result of frequent isolation processes. In very craggy, isolated areas, the cave bears showed a very peculiar paw morphology (broader and shorter than "normal" animals). On the basis of this characteristic, a subspecies *Ursus spelaeus parvilatipedis* (Torres), was defined (Torres et al., 1991). Some authors (Grandal d'Anglade and Vidal Romaní, 1997) concluded, on the basis of tooth morphology, that *U. s. parvilatipedis*, *U. spelaeus* and *U. deningeri* were simply an expression of polytypism, belonging in fact to the same species. This assertion seems to be doubtful. Nevertheless, the Iberian Peninsula was the species border and the paleoenvironmental stresses were usually high, the caves were occupied for relatively short time periods, and population isolation phenomena (endogamy) were probably not uncommon. With a view to resolving the question of the status of this cave-bear species, we have attempted, using AAR analysis, to determine if these species were separated in time.

EXPERIMENTAL PROCEDURES

Sampling

Samples of the dentine or enamel of bear teeth from 13 localities on the Iberian Peninsula (mostly lying in the northern third of the Peninsula) were analyzed. Table 24-1 shows the geographical coordinates, the species sampled, and the current mean annual temperatures (CMATs) of the sites.

We believe that cave-bear caverns are ideal sites for racemization analysis because they are often monospecific, have a relatively low probability of secondary contamination, and have relatively stable and physical conditions. Local cave hydrogeological reactivation deposited a mud layer which buried the bear remains, and the saturation and the high $CaCO_3$ content of the enclosing sediments have helped to preserve the organic matter.

The sampled caves may be grouped into two different thermal zones, the CMATs ranging between 10 and 12.5°C (Arrikrutz, Sima de los Huesos, G. Dolina, Reguerillo, S. Isabel and Eirós) and between 12.5 and 15.0°C (La Lucia, La Pasada, Toll, Troskaeta, and Venta Micena). Thermal damping within the caves suppresses seasonal extremes of temperature. Hoyos et al. (1998) reported that in the Candamo cave, located in an area where the CMAT fluctuates between a winter low of 9.5°C and a summer high of 31.5°C, the temperature in the cave fluctuated by less than 0.5°C (14.5°C in the spring and 14.75°C in the fall). The Coro Tracito cave lies at a higher altitude and therefore in a cooler zone than the other sites (CMAT = 7.0–10.0°C) with an alpine-climate influence. As a result of the annual melting of ice/snow, very cold water infiltrates into the cavity, saturating sediments (and fossils).

Where possible, we have selected five canines or third upper incisors from each locality because their crown conic shape efficiently protects the dentine against contamination. We have also chosen, wherever possible, middle-aged specimens, neither very old (with strongly worn enamel and open pulp cavities) nor very young (with open apex roots and thin dentine layers). In all cases we have attempted to sample "sealed" teeth.

Powdered dentine samples (50 mg) were obtained from the innermost part of the crown by drilling the tooth with a diamond dental drill. Powder from the outer part of the root, to a depth of 1 mm, was rejected, so cement layers have not been sampled at all. This process produces an unavoidable slight heating of the samples.

Sample Preparation and Analysis

The glassware (except the Pasteur pipettes) used for the analyses was cleaned by baking in an oven at 500°C for about 2 hr. Eppendorf plastic micro test tubes, plastic micropipette tips, and Pasteur pipettes were all new from the factory. Teflon liners and septa were thoroughly washed with petroleum ether and acetone, and rinsed three times with ultra-clean water. All the water used in the analysis was Milli-Q quality from Millipore. Concentrated HF and HCl and trifluoroacetic acid anhydride were Merck analytical grade. Thionyl chloride was purchased from Fluka AG. Isopropyl alcohol and *n*-hexane were Merck high-performance liquid chromatography (HPLC) grade, and dichloromethane was Merck spectroscopy grade.

Before hydrolysis and amino acid derivatization were performed, the samples were treated to eliminate free amino acids. With this process, we aimed to remove foreign amino acids and to obtain a homogeneous "collagen extract". For this purpose, we have used a modification of the method proposed by Marzin (1990). The powder sample, 50 mg of dentine, was dissolved in 1 ml 2N HCl and then sonicated. Following the addition of 5 ml phosphate-buffered saline, the sample was dialyzed at 3500 Da (Spectra/Por mnco 3500 membrane) for a period of 20 hr in a buffered solution with magnetic stirring at room temperature.

Following dialysis, the samples were prepared and analyzed in accordance with the method of Goodfriend and Meyer (1991). Briefly, 20 hr of hydrolysis at 100°C was performed under N_2 in a mixture of 12 N HCl (2.9 μl/mg) and 6 N HCl (100 μl). Samples were desalted in HF and the resulting supernatant frozen and dried under vacuum, resuspended in 80 μl distilled water, and dried under vacuum.

Amino acids were derivatized in a two-step process. First, sterification was carried out with 3 M thionyl chloride in isopropanol for 1 hr at 100°C under N_2. Second, the samples were dried and acetylated by reaction with trifluoroacetic acid anhydride (25% in dichloromethane) for 5 min at 100°C. Excess solvent and derivative was evaporated under a gentle flow of nitrogen. The sample was taken up in 125 μl *n*-hexane; this was vortexed and the solvent evaporated in a stream of N_2 to a final volume of 15–25 μl.

A 0.2 μl sample of more recent bear species (*U. arctos* and *U. spelaeus*) or 2–4 μl from older species (*U. deningeri, U. prearctos, U. etruscus*) were injected into a Hewlett-Packard 5890 gas chromatograph. The injection port was kept at 215°C and set for splitless mode for the first 75 sec, at the beginning of which the sample was injected, and later set to split mode. We used helium as the carrier gas, at a column head pressure of 40 kPa, and a Chirasil-Val fused silica column (25 mm × 0.25 mm × 0.12 μm) from Chrompack. The gradients used were as follows: 50°C (1 min), heating at 40°C/min to 115°C, 12 min at 115°C, heating at 3°C/min to 190°C, 10 min at 190°C, cooling to 50°C, and maintenance at this temperature between runs (or at 80°C if the time between runs was longer, typically overnight). The detector was a nitrogen-phosphorus detector set at 300°C. Integration of the peak areas was carried out using the HP PEAK96 integration program from Hewlett-Packard, which runs on a PC computer. The sensitivity limits of the method could be fixed according to the induced racemization method (0.00–0.03, depending on the amino acid considered) and the minimum amino acid concentration detectable in the samples. As a laboratory routine, D/L-valine, D/L-alanine, D-alloisoleucine/L-isoleucine, D/L-proline, D/L-aspartic acid, L-hydroxyproline, D/L-phenylalanine and D/L-glutamic acid peaks were identified.

RESULTS AND CONCLUSIONS

Aminostratigraphy of the Pleistocene Bear

We believe that preliminary dialysis (which removed a low-molecular-weight component) improved the accuracy of our analytical results. This was not the case in all the previous analyses carried out on non-dialyzed samples, for which the racemization ratios for both free and bound amino acids were too variable to be reliable. An alternative problem with earlier studies is the possible interference of an L-hydroxyproline peak, which we observed in some analyses (typically those carried out with an old Chirasil L-Val column) to coelute with the L-Asp peak, giving rise to racemization-ratio distributions that were inadmissible for statistical calculation (cf. Waite et al., 1999). In our opinion, the use of dialysis coupled with the use of a newly purchased Chirasil L-Val column and an NPD detector were key factors in the success of the analyses.

The analysis of aspartic acid ratios in human dentine has been widely used in forensic analysis as an indicator of age at death (e.g., Gillard et al., 1990; Ohtani and Yamamoto, 1991; Carolan et al., 1997; Ohtani et al., 1998). This is the first study to report systematic amino acid racemization-ratio analysis of the dentine of a single fossil animal genus, and the results appear to be very encouraging.

Figure 24-3A shows the histograms of the racemization ratios of the analyzed samples of different species. The *U. deningeri* aspartic acid racemization ratios are (with only two exceptions) above 0.30, whereas those for *U. spelaeus* do not usually reach 0.30; overlap between the two species is almost negligible.

U. deningeri racemization ratios are quite similar in both the Sima de Los Huesos (BB) and La Lucia (LU) caves (table 24-1, figure 24-4). The average aspartic acid racemization ratio of the Santa Isabel (SI) cave samples is higher than both of these, and the lowest values are from the Bunica cave (BU). One possible explanation of the lower values for BU is that this site, located at the foot of the Pyrenees, has a cooler thermal history than the other sites. The Sima de los Huesos bear remains have been dated using the U-series and electron-spin resonance (ESR) methods at 320 kyr (cf. Bischoff et al., 1997).

We conclude that there is an identifiable *U. deningeri* aminozone on the Iberian Peninsula, with aspartic acid racemization ratios above 0.30 and average values close to 0.32. This fact is especially evident in comparisons of the mean values of racemization of the aspartic acid of *U. spelaeus* and *U. deningeri* from the LU Cave, where the mean values of racemization for aspartic acid are 0.13 and 0.33, respectively. Differences in the ratios cannot be attributed to thermal history and therefore probably reflect different periods of occupation. We conclude that materials classified purely on palaeontological grounds as *U. deningeri* also have a chronostratigraphic value and therefore cannot constitute a case of polytypism within a single species.

In *U. spelaeus* localities, most of the racemization ratios of aspartic acid fall below 0.10 (table 24-1, figure 24-4). Consequently, it is possible to establish a sub-aminozone of aspartic acid racemization, which would include most of the Iberian locations of *U. spelaeus*. The material from the Coro Tracito site (TE) has a low mean value of racemization of aspartic acid (0.08). We consider that this lower value may be due to the high altitude (1600 m) of this site, which would lower the rate of racemization. This site might therefore be older than the others in this group.

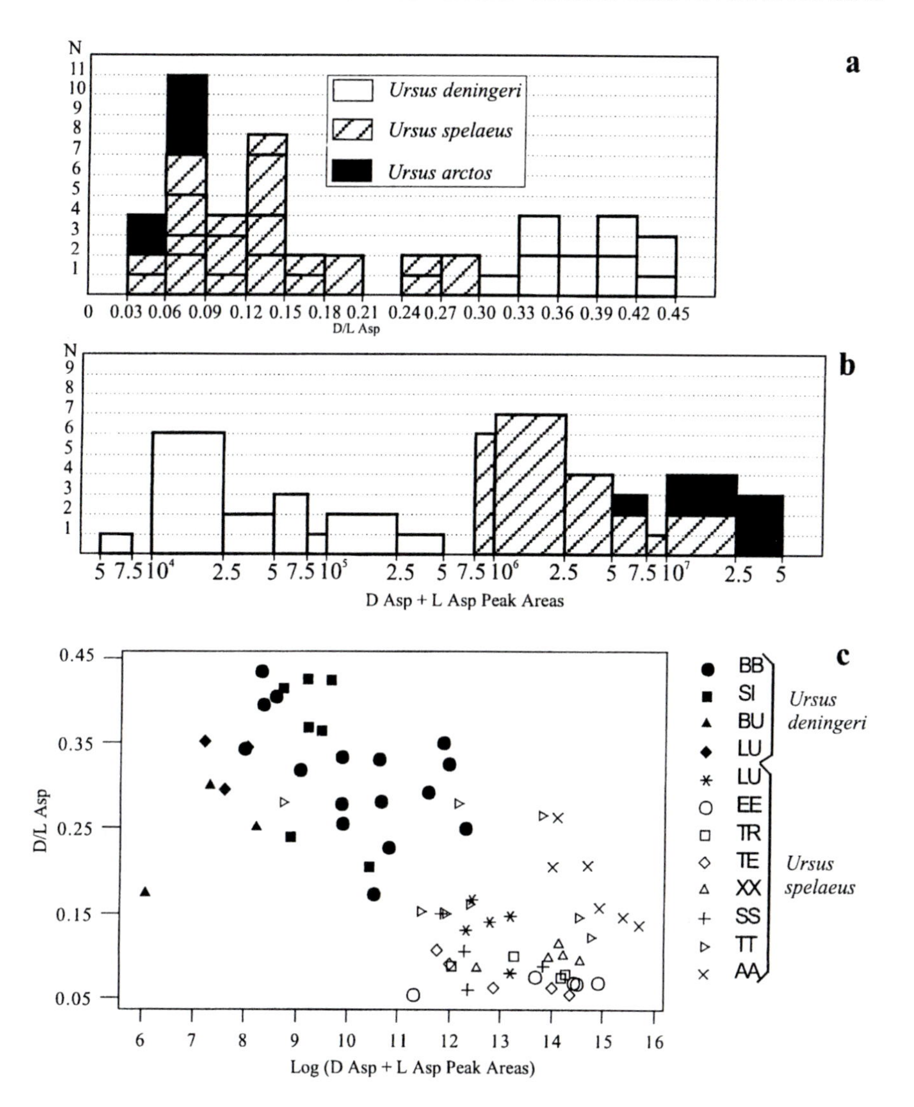

Fig. 24-3. A. Histogram of aspartic acid racemization ratios in samples of *U. deningeri*, *U. spelaeus* and *U. arctos*. B. Histogram of D+L aspartic acid peak areas (corrected for sample weight and injected sample volume) of *U. deningeri* and *U. spelaeus* samples. C. Plot of the aspartic acid racemization ratio against D+L aspartic acid chromatogram peak areas (corrected for sample weight and injected sample volume) of *U. deningeri* and *U. spelaeus* specimens from the different localities sampled).

The Arrikrutz (AA) and El Reguerillo (TT) caves have higher average aspartic acid racemization values (0.19). These lie outside the main *U. spelaeus* aspartic acid racemization "group", and in some cases may be considered along with the lowest aspartic acid racemization values of the *U. deningeri* group.

The *U. spelaeus* aspartic acid racemization ratios define a broad aminozone in which two sub-aminozones may be distinguished: one has two higher racemization ratio populations (El Reguerillo and Arrikrutz) and the other (Troskaeta, Coro Tracito, and La Pasada) has very low aspartic acid racemization ratios not very different from those of the Holocene brown bear (*U. arctos*) sample from Saldarrañao.

Skeletal material from smaller cave bears from the El Reguerillo and Arrikrutz caves have low aspartic acid racemization ratios. The dentine recovered from cave bears with short and broad paws have the lowest aspartic acid racemization ratios of all [La Pasada (SS), El Toll (XX) and Troskaeta (TR) caves]. We interpret these ratios as belonging to a relict Pleistocene subspecies (*U. spelaeus parvilatipedis* T.) existing alongside the extant Iberian brown bear but confined to less favorable craggy areas. The Eirós cave population (having the lowest aspartic acid racemization

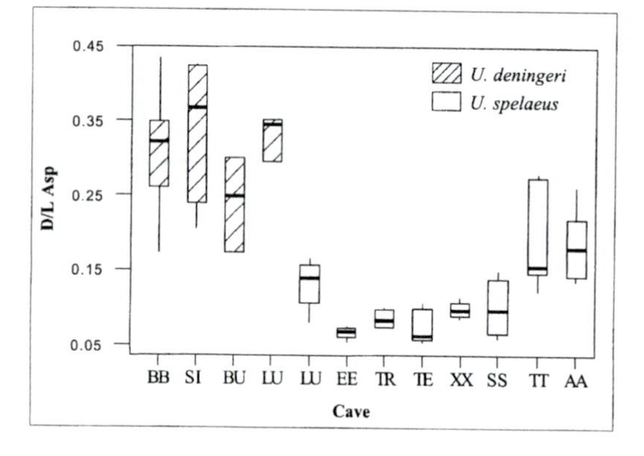

Fig. 24-4. Box-and-whisker plot of aspartic acid racemization ratios of specimens of *U. deningeri* and *U. spelaeus* from the different localities sampled (mean, Q_1, Q_3, and minimum and maximum values are shown).

ratios) would, in this interpretation, represent the last attempt to find refuge in an area without appropriate caves for hibernation and with soils and plants associated with siliceous rocks.

The La Lucia *U. spelaeus* samples show aspartic acid average racemization values of 0.13, almost half way between the two *U. spelaeus* sub-aminozones. In this particular case, we have a U-series-dated stalactite that was present in the muddy sediments containing the bear bones and teeth, and a spelothem that sealed the deposit. A dating of approx. 80 kyr was obtained for both.

Preservation of Protein and Amino Acids

Because of the high accuracy of our method, it has been possible to quantify the total amount of bound aspartic acid (>3500 Da fraction) that remained in fossil bear teeth (figure 24-3B). It may be appreciated that there is excellent differentiation between *U. deningeri* and *U. spelaeus* samples, because there is a dramatic decrease in the bound aspartic acid content. *U. deningeri* areas are between 5×10^3 and 2.5×10^5, whereas *U. spelaeus* samples range between 7.5×10^5 and 1×10^7. In fact, some *U. spelaeus* values are similar to those for the modern *U. arctos*. In our opinion there is no direct cause–effect relationship between aspartic acid racemization and loss of bound amino acids from collagen, although both were linked to diagenesis. In figure 24-3C we have plotted D/L aspartic acid ratios and D+L aspartic acid peak areas (corrected for weight and injected volume); a clear "biostratigraphic control" of the analytical results may be observed: points corresponding to dentine samples from older bear species (Middle Pleistocene *U. deningeri*) group together as well as those for more recent cave bears (Upper Pleistocene *U. spelaeus*).

We have also sampled dentine from a Lower Pleistocene bear, namely, *U. etruscus* from Venta Micena. Even without dialysis, amino acid concentrations are below detection limits, probably having been leached out during diagenesis from their unprotected situation in the collagen fibrils. However, analysis of specimens of Mollusca and the Ostracoda from Venta Micena revealed the persistence of considerable amounts of amino acids, mainly on ostracoda, gastropoda shells and gastropoda opercula, which were preserved in their intracrystalline position.

Acnowledgements: We are especially indebted to Dr Veronika Meyer of the University of Bern, now at the EMPA St. Gallen. She provided great help in the setting up of our laboratory. Dr Glenn Goodfriend, now at the Department of Earth and Environmental Sciences of the George Washington University of Washington D.C., sent us the analysis protocol and the gas chromatography program. The Biomolecular Stratigraphy Laboratory was partially funded by ENRESA (the Spanish National Radioactive Waste Management Company). The samples from el Toll and Bunica caves came from the Palaeontological Museum of the City of Barcelona. We are also deeply indebted to a large number of caring-society members who first discovered bear remains and provided very important help during the excavation campaigns. We would also like to thank the editor (Dr M. Collins) and the referees (Dr G. Belluomini and two anonymous referees) for their critical reading of the manuscript and their useful suggestions.

REFERENCES

Bada J. L. (1972) The dating of fossils using racemization of isoleucine. *Ann. Rev. Earth. Planet. Sci.* **15**, 223–231.

Bada J. L. (1973) Racemization of Amino Acids in Bones. *Nature*, **245**, 308–310.

Bada J. L. (1985) Amino acid racemization dating of fossil bones. *Ann. Rev. Earth. Planet. Sci.* **13**, 241–268.

Bada J. L., Belluomini G., Bonfiglio L., Branca M., Burgio E., and Delitalia L. (1991) Isoleucine ages of Quaternary mammals from Sicily. *Il Quaternario* **4**, 49–54.

Belluomini G. (1981) Direct aspartic acid racemization dating of human bones from archaeological sites of Central Southern Italy. *Archaeometry* **23**, 125–137.

Belluomini G. (1985) Risultate e prospettive di un nuovo metodo di datazione basato sulla racemizźacione degli aminoacidi. *Archeometria* **69**, 135–171.

Belluomini G. and Bada J. L. (1985) Isoleucine epimerization age of the dwarf elephants of Sicily. *Geology* **13**, 451–452.

Bischoff J. L., Fitzpatrick J. A., Falgueres C., Bahain J. J., and Bullen, T. T. (1997) Geology and preliminary dating of the hominid-bearing sedimentary fill of the Sima de los Huesos Chamber, Cueva Mayor of the Sierra de Atapuerca, Burgos, Spain. *J. Human Evol.* **33**, 129–154.

Blackwell B., Rutter N. W., and Debénath A. (1990) Amino acid racemization in mammalian bones and teeth from La Chaise-de-Vouton (Charente), France. *Geoarcheology* **5**, 121–147.

Carolan V. A., Gardner M. L.G., Lucy D., and Pollard A. M. (1997) Some considerations regarding the use of amino acid racemization in human dentine as an indicator of age at death. *J. Forensic Sci.* **42**, 10–16.

Dungworth G., Schwartz, A. W., and van de Leemput L. (1976) Composition and racemization of amino acids in mammoth collagen determined by gas and liquid chromatography. *Comp. Biochem. Physiol.* **53B**, 473–480.

Elster H., Gil-Av E., and Weiner, S. (1991) Amino acid racemization of fossil bones. *J. Arch. Sci.* **18**, 605–617.

Gillard R. D., Pollard A. M., Sutton P. A. and Whittaker D. K. (1990) An improved method for age at death determination from the measurement of D-aspartic acid in dental collage. *Archaeometry* **32**, 61–70.

Goodfriend G. A. and Meyer V. (1991) A comparative study of the kinetics of amino acid racemization/epimerization in fossil and modern mollusc shells. *Geochim. Cosmochim. Acta* **55**, 293–302.

Grandal d'Anglade A. and Vidal Romaní J. R. (1997) A population study on the cave bear (*Ursus spelaeus* Ros. Hein.) from Cova Eirós (Triacastela, Galicia, Spain). *Geobios* **30**, 723–731.

Hoyos M., Soler V,. Cañaveras J. C., Sánchez-Moral S., and Sanz-Rubio E. (1998) Microclimatic characterization of a karstic cave: human impact on microenvironmental parameters of a prehistoric rock art cave (Candamo Cave, Northern Spain). *Env. Geol.* **33**, 231–242.

Kimber R. W. L. and Hare P. E. (1992) Wide range of amino acids in peptides from human fossil bone and its implications for amino acid racemization dating. *Geochim. Cosmochim. Acta* **56**, 739–753.

King K. and Bada J. L. (1979) Effect of in situ leaching on amino acid racemization rates in fossil bone. *Nature* **281**, 135–137.

McMenamin M. A. S., Blunt D. J., Kvenvolden K. A., Miller S. E., Marcus L. F., and Pardi R. R. (1982) Amino acid geochemistry of fossil bones from the Rancho La Brea Asphalt Deposit, California. *Quat. Res.* **18**, 174–183.

Marzin E. (1990) Essai de normalisation du protocole d'analyse des taux de racémisation des acides aminés: applications a la datation d'ossements fossiles. *Trav. du LAPMO*, 167–178.

Meyer V. (1992) Amino acid racemization: a tool for dating. *Chemtech.* **22**, 412–417.

Ohtani S. and Yamamoto K. (1991) Age estimation using the racemization of amino acid in human dentine. *J. Forensic Sci.* **36**, 792–800.

Ohtani S., Matsushima Y., Kobayashi Y., and Kishi K. (1998) Evaluation of aspartic acid racemization ratios in the human femur for age estimation. *J. Forensic Sci.* **43**, 949–953.

Sinibaldi M., Ferrantelli P., Goodfriend G. A., and Barkay G. (1998) Aspartic acid racemization in teeth from late Holocene burial caves in Jerusalem. *Perspectives in Amino Acid and Protein Geochemistry Conference*, 64 (abstr.). American Geophysics Union.

Torres T. (1989) *Estudio de la Filogenia, Distribución Estratigráfica y Geográfica, Análisis Morfológico y Métrico del Esqueleto y Dentición de los Osos (Mammalia, Carnivora, Ursidae) del Pleistoceno de la Península Ibérica.* Instituto Geológico y Minero de España.

Torres T. (1992) Excavación en la Cueva de Santa Isabel. *Arkeoikuska* **92**, 303–310.

Torres T. and Cervera J. (1995) Multivariate analysis of the dental morphology of Plio-Pleistocene Ursids, with some observations concerning the phylogenetic position of *Ursus deningeri* (Von Reichenau) from Cueva Mayor (Sima de los Huesos, Atapuerca). In *Human Evolution in Europe and the Atapuerca evidence* (eds J. M. Bermúdez, J. L. Arsuaga, and E. Carbondell), pp. 123–136. Consejería de Cultura de la Junta de Castilla y León.

Torres T., Cobo R., and Salazar, A. (1991) La población del oso de las cavernas *(U. s. parvilatipedis* n. ssp.) de la Cueva de Troskaeta (Ataun, Guipuzcoa). *Munibe* **43**, 3–85

Waite E. R., Collins M. J., and van Duin A. T. C. (1999) Hydroxyproline interference during the gas chromatographic analysis of D/L aspartic acid in human dentine. *Intl. J. Legal Med.* **112**, 124–131.

Index